INTRODUCTION TO

Quantum Electrodynamics
AND
Particle Physics

INTRODUCTION TO Quantum Electrodynamics AND Particle Physics

Second Edition

Deep Chandra Joshi
Formerly
Professor and Head
Department of Physics
HNB Garhwal University, Srinagar, Garhwal
Uttarakhand

TECHSAR™
NEW DELHI

Published and Printed by

TechSar Pvt. Ltd.
4435–36/7, nsari Road, Daryaganj
New Delhi-110 002 (India)
E-mail: tech_info@techsarworld.com
Website: www.techsarworld.com

ISBN 978-93-90620-76-0

Reprint 2025

Preface to the Second Edition

The first edition of the book has been extended by including two new chapters. One is on Mandelstam Variables and Symmetries of Scattering Amplitude and the other is on Regge Poles. The former, that constitutes the seventh chapter of the book, introduces Mandelstam variables and describes at length the s-channel, t-channel and the u-channel processes for both the equal and unequal masses of participating particles. The conditions for the occurrence of these channel processes have been made explicit through the Mandelstam plot. Introducing scattering amplitude as the matrix element of S-matrix, the crossing and Bose symmetries of scattering amplitudes for s-, t-, u-channel scatterings have been explained and the analyticity of scattering amplitude has been elucidated through examples.

The topic Regge poles, that constitutes the eighth chapter, describes the study of resonances and Regge poles that can be undertaken through the scattering process. Due to the significant role of partial waves in the chapter, the scattering process has been explained through the partial wave analysis and the scattering cross section has been expressed in terms of scattering amplitude and by the optical theorem. Since, Regge poles have their conection with partial waves having complex angular momenta, the Watson-Sommerfeld transformation, that transforms partial waves with real angular momenta into partial waves having complex angular momenta, has been explaind pedagogically to make the appearance of Regge poles just natural. The characteristics of Chew-Frautschi diagrams of Regge trajectories have been explained with examples. As two different mechanisms, viz., the resonance formation and Regge pole mechanism guide the scattering amplitude at low and high energies respectively, the unified amplitude for the entire range from low to high energy is identified by Veneziano amplitude, the latter has been explained and described in detail. Exhibiting the Regge behaviour of Veneziano amplitude, its interpretation to describe the one-dimensional objects, the strings, has been analyzed and explained through bosonic and five superstring theories. The relative merits of these superstring theories have been analyzed.

After the publication of the first edition, the LIGO experiments, in 2015, confirmed the observation of gravitational waves.With this observation all the predictions of Einstein's General Theory of Relativity (EGTR) have been proved true. This fact has been incorporated in this edition by suitably revising the content on EGTR.

In seventh and eighth chapters too, at the end, assignments and objective questions have been given in large numbers. The contents of these chapters would be beneficial to a regular student, researcher and those preparing for competetive examinations.

I would like to remember my grandchildren Daksh and Mansi for their natural-innocent smiles.

Deep Chandra Joshi

Preface to the First Edition

Non-gravitational particle physics can be divided into three parts. There are, however, two ways of doing so—one is to divide it according to interactions, namely, the electromagnetic, weak and strong, and the other is its division according to particles, i.e., the photons, leptons and hadrons. The realm of particle physics, therefore, encompasses, both, particles and their interactions. Among the three interactions, a lot of significance is attached to the electromagnetic interaction, which incidentally is best understood too. The relativistic quantum theory of the electromagnetic interaction of charged particles, called quantum electrodynamics, therefore, acquires an identity of its own in particle physics. Keeping this point in view, the present book attempts to provide the essentials of particle physics.

Over the past many years, while teaching particle physics to postgraduate students, I have felt a serious need for a book on particle physics, which would not only inform but encourage students to pursue particle physics as their specialization and research subject. This book is an attempt towards achieving that objective. Though the book targets particle physics students, the contents would be quite useful for general postgraduate students in physics. Students preparing for various competitive examinations, particularly to higher education including research would also find the contents substantially meeting their requirements. Most of the topics forming the Model Syllabus of University Grants Commission for Postgraduate students of Physics in the papers of Nuclear and Particle Physics (III Semester), Nuclear and Particle Physics (IV Semester, Strong, Weak and Electromagnetic Interactions) and Quantum Electrodynamics have been covered in the book. The topics have been developed in a pedagogical manner by providing all possible algebraic details.

The book has been divided into six chapters. It begins with a chapter on Elementary Particles in which a fairly detailed account of these particles has been given. The descriptions range from the elementary particles of classical era to the present-day fundamental particles. The second chapter is on Quantum Electrodynamics. This chapter takes up the electromagnetic interactions of charged particles which include a detailed discussion on Compton scattering, electron-positron annihilation, and Moller and Bhabha scatterings. The positronium annihilation has been dealt with qualitatively. In spite of the fact that the calculations involved long expressions, maximum possible mathematical steps have been given to make the assimilation smooth.

In the third chapter, I have dealt with symmetries and groups. Lie groups that are used in particle physics have been defined and their representations have been explained. The Gell-Mann-Okubo mass formula and SU(6) quark model and its applications have also been explained in detail. The fourth chapter is devoted to a discussion of weak interactions. Beginning with the Fermi's theory of beta decay, the chapter goes through the *V-A* interactions, Cabibbo theory, quark-lepton universality and the GIM mechanism.

Since a complete and correct description of non-gravitational physics is possible through a theory that is based on the principle of gauge symmetry; the same has been taken up in the fifth chapter. The chapter describes global and local U(1), SU(2) and SU(3) gauge symmetries and their implications. Spontaneous symmetry breaking has been described in detail and the Higgs mechanism has been explained through the breaking of U(1) and SU(2) local gauge symmetries.

The last chapter has been devoted to the Standard Model in which the subject matter of GWS model and Quantum Chromodynamics has been developed step by step. Experimental evidences for the Standard Model have been explained. GUTs have been introduced as theories beyond the Standard Model, and the early universe has been discussed as a high energy experiment. The chapter closes by presenting a brief account of the theories beyond SM and GUTs.

Appended to the last chapter are the Appendices A, B, C, D, E whose contents find their use at various places in the main text .The Appendix A describes the conventions that have been adopted in the book, Appendix B discusses the Green's functions of the field theory, Appendix C explains the Schwinger's action principle, Appendix D describes the electromagnetic field and undertakes its Coulomb gauge quantization and Appendix E is on the continuous and discrete symmetries. These appendices acquaint the reader with the field theory methods that constitute the traditional tools for a particle physicist.

At the end of each chapter, assignments have been given. The assignments contain descriptive questions as well as problems. Most of the questions are based on the material of the chapter itself. However, for the questions asked beyond the chapter, brief notes have been provided. Since the derivations of equations in the main body are generally asked as problems, the same has not been done in large measure. This is because almost all the derivations have been done in the main body, where they have been worked out in the simplest possible manner. The assignments, apart from being the summary of the chapter, provide the student with an idea of the types of questions that can be set. A new feature of the book is that it has substantial number of objective questions that follow the assignments of each chapter. It is hoped that these objective questions would help prepare the student for GATE, NET and related examinations.

I owe a great deal to the texts mentioned in the bibliography, from where most of the inputs were drawn.

Needless to say, any task undertaken cannot be completed without the support and encouragement of the person's dear ones. My heartfelt thanks go to my wife Mrs. Manju Joshi, who constantly supported and reminded me of my task, and to my sons, Piyush and Saurabh, for their concern and encouragement. I also acknowledge the support and encouragement of the faculty members and research scholars of the Physics Department of our university throughout the writing and typing phases of this book.

Though every care has been taken that the mathematical expressions, which are many, are free from errors, I would be grateful to the readers if the errors that may have remained are brought to my notice. Any suggestions for the improvement and augmentation of the book would also be gratefully acknowledged.

Deep Chandra Joshi

CONTENTS

Preface to the Second Edition *v*
Preface to the First Edition *vii*

1. ELEMENTARY PARTICLES **1**

1.1 Introduction 1
1.2 The Classical Era (1897–1932) 1
1.3 The Years of Photon (1900–1924) 3
1.4 The Years of Mesons (1934-1947) 4
1.5 The Years of Antiparticles (1930–1956) 6
1.5.1 Antiparticles and Crossing Symmetry 9
1.6 The Years of Neutrinos (1930-1962) 10
1.6.1 Sources of Neutrinos 14
1.7 The Years of Strange Particles (1947-1960) 15
1.8 The Years of Resonances (1961–) 20
1.9 The Fundamental Interactions 29
1.9.1 Gravitational Interaction 29
1.9.2 Electromagnetic Interactions 31
1.9.3 The Feynman Diagrams 32
1.9.4 The Range and Strength of Electromagnetic Interaction 39
1.9.5 The Strong Interactions 41
1.9.6 The Weak Interactions 48
1.10 Conservation Laws 51
1.11 Classification of Elementary Particles 74
1.12 The Modern Era 77
1.12.1 The Eightfold Way 78
1.12.2 The Quark Structure of Hadrons 82
1.12.3 Intermediate Vector Bosons 96
1.13 The Ultimate Fundamental Particles 98
1.13.1 Basic Fermions 98
1.13.2 Leptons 98

1.13.3 Gauge Bosons 98
1.13.4 Higgs Boson 99
1.13.5 Antiparticles 99
1.14 Assignments 99
1.15 Objective Questions 101

2. QUANTUM ELECTRODYNAMICS **117**
2.1 Introduction 117
2.2 Preliminaries 117
2.2.1 Time Ordered Product and Propagators 117
2.2.2 Some Definitions 123
2.3 Quantum Electrodynamic Processes 129
2.3.1 Compton Scattering 129
2.3.2 Electron–Positron Annihilation 145
2.3.3 Moller (electron–electron) Scattering 156
2.3.4 Bhabha (positron–electron) Scattering 171
2.4 Assignments 178
2.5 Objective Questions 181

3. SYMMETRIES AND GROUPS **187**
3.1 Introduction 187
3.2 Groups and Their Properties 187
3.3 Lie Groups 190
3.3.1 Generators of Lie Groups and the Lie Algebra 191
3.3.2 Representations of Lie Groups 197
3.4 Group SU(2) and Its Irreducible Representations 201
3.4.1 SU(2) Product Representations 204
3.5 SU(3) Group and Its Irreducible Representations 210
3.5.1 SU(3) Shift Operators 211
3.5.2 Irreducible Representations of SU(3) 213
3.5.3 Product of SU(3) Representations 220
3.5.4 The Gell—Mann—Okubo Mass Formula 228
3.5.5 Applications of SU(3) Symmetry Group 231
3.5.6 Limitations of SU(3) Group 232
3.5.7 Extension of SU(3) Group: SU(4) 233
3.5.8 SU(6) Quark Model 239
3.6 Assignments 264
3.7 Objective Questions 266

4. WEAK INTERACTIONS **273**
4.1 Introduction 273
4.1.1 Purely Leptonic Interactions 273
4.1.2 Semileptonic Interactions 273

4.1.3 Non-Leptonic Weak Processes 274
4.2 Fermi's Theory of Beta Decay (Non-Relativistic) 275
4.2.1 The Allowed and Forbidden Transitions 280
4.3 Helicity and Handedness 283
4.4 The Two-Component Theory of Neutrinos 286
4.5 Fermi's Relativistic Theory of Beta Decay 290
4.5.1 The V–A Interaction 294
4.5.2 Universality of Weak Interactions 296
4.5.3 The Intermediate Vector Boson 299
4.5.4 Cabibbo Theory 303
4.5.5 Charged Weak Quark Current and Suppressed Weak Decays 320
4.5.6 GIM Mechanism and Charged Quarks 321
4.5.7 The Cabibbo—Kobayashi—Maskawa (CKM) Matrix 325
4.6 Assignments 327
4.7 Objective Questions 329

5. GAUGE SYMMETRIES 339

5.1 Introduction 339
5.2 Global Gauge Symmetry and Noether's Theorem 340
5.3 U(1) Local Gauge Symmetry and QED 343
5.4 SU(2) Global Gauge Symmetry 347
5.5 SU(2) Local Gauge Symmetry and Yang–Mills Fields 350
5.6 Properties of Yang–Mills Fields 355
5.6.1 Adjoint Representation 355
5.6.2 Electric and Magnetic Fields 357
5.7 SU(3) Global Gauge Invariance 359
5.8 SU(3) Local Gauge Symmetry and QCD 361
5.9 Spontaneous Symmetry Breaking 363
5.9.1 The Mass Term 364
5.9.2 Mass Generation through Spontaneous Symmetry Breaking 364
5.9.3 Spontaneous Breaking of Global Gauge Symmetry 369
5.9.4 The Spontaneous Breaking of U(1) Local Gauge Symmetry 374
5.9.5 Spontaneous Breaking of Local SU(2) Gauge Symmetry 380
5.10 Assignments 388
5.11 Objective Questions 391

6. THE STANDARD MODEL 401

6.1 Introduction 401
6.2 The GWS Model 401
6.2.1 Electromagnetic and Weak Embedding 402
6.2.2 The Combined Covariant Derivative 405
6.2.3 Lagrangian Density and Mass Term 406

6.2.4 Calculation of Interactions 408
6.2.5 Neutral Gauge Fields Z_μ and A_μ 410
6.2.6 Electromagnetic Coupling Constant e 412
6.2.7 Structure of Neutral Current 414
6.2.8 Spontaneous Breaking of Local $[SU_L(2) \otimes U_Y(1)]$ Gauge Symmetry 415
6.3 Renormalizability of Electroweak Gauge Theory 428
6.4 Quantum Chromodynamics 431
6.5 The Standard Model Lagrangian 433
6.6 Experimental Evidences for the Standard Model 434
6.6.1 Experimental Tests of QED 434
6.6.2 Evidences for Weak Interactions 435
6.7 Evidences for Quantum Chromodynamics 442
6.7.1 Evidence for Colour 442
6.7.2 Variation of Strong Coupling Constant (α_s) with Energy 445
6.8 Grand Unification 447
6.8.1 Implications of Grand Unification 451
6.8.2 Gravitational Interaction and the Planck Scales 453
6.9 Grand Unification and the Big Bang 454
6.9.1 $t = 0$ 455
6.9.2 From $t = 0$ to $t = 10^{-43}$ sec. 455
6.9.3 From 10^{-43} to 10^{-35} sec. 455
6.9.4 From 10^{-35} to 10^{-10} sec. 456
6.9.5 From 10^{-10} to 10^{-3} sec. 456
6.9.6 From 10^{-3} to 35 min. 457
6.9.7 From 35 min. to 10^5 years 457
6.9.8 From 10^5 years to 10^7 years 458
6.9.9 From 10^7 years to Present 458
6.10 Limitations of SM and GUTs 458
6.10.1 Limitations of the Standard Model 458
6.10.2 Limitations of GUTs 460
6.11 Beyond the SM and GUTs 460
6.11.1 Supersymmetry and Supergravity 461
6.11.2 Technicolour Theories 462
6.11.3 Superstring Theories 462
6.12 Assignments 463
6.13 Objective Questions 466

7. MANDELSTAM VARIABLES AND SYMMETRIES OF SCATTERING AMPLITUDE 481
7.1 Introduction 481
7.2 Mandelstam's s, t, u Variables 482

7.2.1 s, t, u for Equal Mass Case 483
7.2.2 s, t, u for Unequal Mass Case 492
7.3 Visualization of Mandelstam Variables 495
7.3.1 Equal Mass Case 495
7.3.2 Unequal Mass Case 498
7.4 Symmetries of Scattering Amplitude 499
7.4.1 Crossing Symmetry 499
7.4.2 Bose Symmetry 501
7.5 Analyticity 502
7.6 Assignments 504
7.7 Objective Questions 507

8. REGGE POLES **511**
8.1 Resonances and Regge Poles 511
8.2 Scattering Processes 512
8.2.1 Partial Waves and Scattering Amplitude 513
8.2.2 Scattering Cross Section 521
8.2.3 Optical Theorem 524
8.3 Breit-Wigner Resonances 525
8.4 Resonances as Poles of Scattering Matrix 530
8.5 Regge Theory 531
8.5.1 Concept of Complex Angular Momentum 531
8.5.2 One Particle Exchange Model 532
8.5.3 The Regge Pole Model 533
8.5.4 Duality 553
8.5.5 The Veneziano Amplitude 554
8.5.6 The Veneziano Amplitude and String Theory 562
8.5.7 The Spin and the Square of Mass Relation of a Particle 566
8.6 Assignments 568
8.7 Objective Questions 571

Appendix A: Conventions **577**
Appendix B: Green's Functions of Field Theory **595**
Appendix C: Schwinger's Action Principle **613**
Appendix D: Electromagnetic Field **627**
Appendix E: Continuous and Discrete Symmetries **643**

Bibliography **645**
Index **649**

1

ELEMENTARY PARTICLES

1.1 INTRODUCTION

The universe is filled with matter that exists in different states, namely, the solid, liquid, gas and the plasma. The question now is what the matter is made of? The answer to this question lies primarily in the realm of a particle physicist, who keeps on trying to know the ultimate constituent (or the constituents) of the matter. This ultimate or the fundamental constituent, which may also be called the constituent of matter at its smallest scale of size, is also known as the elementary particle or the fundamental particle. We mean by a particle an object that in an experimental layout behaves like a point. The point is supposed to carry a well-defined mass and a charge.

What must be considered an elementary particle depends on our knowledge on the subject in a particular period of time. If we go back to the period when only metaphysical thoughts without much scientific analyses dominated, we would say that air, water, fire and earth were the fundamental constituents of every matter. About 25 centuries later Mendeleev's periodic table listed the fundamental constituents of matter of his time. With the ever increasing scientific developments, our knowledge on the subject of fundamental particles kept on changing and what was an elementary particle during a particular period of time turned out to be composite, in the next. By composite, we mean a particle or a structure, which can be further splitted into smaller constituents. This all has been due to fast scientific developments in the field of particle physics both on theoretical and experimental fronts. In view of such a scenario, it is pertinent to go through the history of elementary particles according to the periods and see what was or were considered the elementary particles. In view of the above we describe below the important developments related to elementary particles during various periods.

1.2 THE CLASSICAL ERA (1897–1932)

In the development of the modern science, the beginning of classical era of the elementary particles can be attributed to the discovery of electron by J.J. Thomson in 1897. In his cathode ray experiments he observed that cathode rays emitted by a hot filament were deflected by a magnet. Whereas the deflection showed that the cathode rays carried electric charge, the direction of deflection indicated that the charge must be negative. The deflection of cathode rays by the magnetic field suggested that they consist of particles that were called corpuscles and their charge was called the electron. Later on the particles themselves were called electrons.

John Dalton had already put the atomic model of matter forward in 1808. Thomson conjectured that his electrons were the constituents of electrically neutral atoms of Dalton and suggested that in the atom there is a heavy positive charge paste on which are suspended the negatively charged electrons. This is known as the Thomson's model of atom in which the magnitudes of positive charge of paste and the negative charge of electrons were equal so as to make the atom electrically neutral and the heaviness imparted to the paste was to provide the mass of the atom as it had been guessed by Thomson that electrons were lighter in weight. However, Thomson's model of atom was not accepted because the famous alpha particle scattering experiments of Rutherford demonstrated that the heavy positive charge in the atom was not spread everywhere within the atom but confined to a small region around the center of the atom. This portion of atom which was very small, very hard, very heavy and had a positive charge equal to the magnitude of total negative charge of electrons in the atom, was known as nucleus and the nucleus of the lightest atom, i.e., hydrogen, was called proton by Rutherford. Thus was born the proton with a rest mass of 938.3 MeV/C^{2}.

However, proton alone could not explain the mass of heavier atoms. The next higher atom, the helium had two electrons hence two protons but its mass was four times the mass of hydrogen and lithium that has three electrons is seven times heavier than hydrogen. Rutherford in 1920 hypothesized the existence of a neutral particle inside the nucleus which in 1932 was discovered by Chadwick, when he repeated the Curie and Joliot experiment of bombarding the beryllium nucleus with alpha particles and successfully reasoned that the recoil of proton in the above experiment was due to its hitting a neutral particle of approximately protonic mass. Chadwick named the neutral particle neutron.

$$ {}_4\mathrm{Be}^9 + {}_2\mathrm{He}^4 \rightarrow {}_6\mathrm{C}^{12} + {}_0\mathrm{n}^1 \tag{1.1} $$

The discovery of neutron successfully explained the masses of higher atoms. The nucleus of an atom with Z number of protons and mass number A had $(A–Z)$ number of neutrons. The Helium nucleus had 2 protons and 2 neutrons, the Li nucleus, 3p and 4n and so on.

Thus till 1932, we had electrons, protons and neutrons as the elementary particles which constituted the matter. To this simple list of three we could add a particle of zero rest mass with energy given by Planck's equation $E = h\upsilon$. This particle that binds the electrons and the protons (nucleus) was given the name photon (γ) by the chemist Gilbert Lewis in 1926. We are however, talking about photon in the next subsection.

1.3 THE YEARS OF PHOTON (1900–1924)

In Rutherford's model of an atom, the electrons revolve round the nucleus in orbits but no restriction is imposed on the orbit itself and electrons were permitted to revolve in any orbit. Even if the magnitude of the velocity of electron is constant, the change of directions in velocity at different points in an orbit is equivalent to an acceleration that the electron gets subjected to. According to the rules of electrodynamics an accelerating charge radiates (i.e., loses) energy, the continuous loss of energy by electron would ultimately make it fall into the nucleus and the atom would not remain stable. To counter this problem, Bohr imposed restrictions on the orbits and postulated that the electrons do not revolve in every orbit. They revolve in those orbits in which they do not radiate energy, hence no question of loss of energy. The orbits, which are permitted to electron, are those for which the angular momentum of electron is integral multiple of $(h/2\pi)$.

The restrictions imposed on the orbits by Bohr were in fact the rules of quantization on the orbits and such orbits were the quantized orbits. The idea of this quantization came from the Planck's law for black body radiation. Whereas the Bohr's approach was like an improvement over the Rutherford's approach, the Planck's law was an improvement over Rayleigh–Jeans law. The Rayleigh–Jeans law predicted that the total power radiated by a black body must be infinite. This conclusion, that is also known as the *ultraviolet catastrophe* does not agree at higher frequencies (ultraviolet) with the experimental curve between the powers radiated versus the frequency of radiation.

The conditions imposed on the radiation by Planck were that the electromagnetic radiation does not come continuously from the black body; rather it comes in little packages of energy

$$E = h\upsilon \tag{1.2}$$

where υ is the frequency of the radiation and h is the Planck's constant. The coming of radiation from the black body in packages implies that the electromagnetic radiation is quantized. Adopting the Planck's idea of quantized radiation Einstein explained the photoelectric effect that further supported that the electromagnetic radiation has the quantum (particle) form. Finally, the results of Compton's experiments on the scattering of light from a material particle (electron) at rest agreed with the formula

$$\lambda' = \lambda + \lambda_c(1 - \cos\theta) \tag{1.3}$$

where λ' is the wavelength of light after scattering, λ the incident wavelength, θ is the scattering angle and

$$\lambda_c = \frac{h}{mc} \tag{1.4}$$

is the Compton wavelength of the target particle whose rest mass is m, and c in mc is the speed of light. The significant point in the agreement of the experimental observation with the formula (1.3) is that the eqn. (1.3) is derived by assuming light as a particle of zero rest mass and energy $E = h\upsilon$. This particle of zero rest mass and energy $h\upsilon$ was called photon.

In quantum field theory the electric field is quantized and obviously the quanta of electromagnetic field are photons. The attraction or the repulsion (interaction) between charges then means that a stream of photons is continually passing back and forth between them. Each charge continually emits virtual photons and also continually absorbs them. Classically we say that the action at a distance is mediated by a field, in quantum field theory we say that the action at a distance is mediated by the exchange of the quanta of the field. If the field in question is electromagnetic, the mediating particles would be photons and if we are talking of gravitational force experienced by a mass due to another at a distance, we say the mediating particle is graviton. Leaving gravity aside, we say that till 1932 photon had established itself, as an elementary particle that though did not contribute to the formation of matter, was responsible for the electromagnetic interactions involving the material particles.

1.4 THE YEARS OF MESONS (1934–1947)

As seen above, in an atom the positively charged nucleus is bound to the orbiting electrons due to the electromagnetic interaction (attractive Coulombian force) taking place between the orbiting electrons and protons in the nucleus. The messengers or the mediators that carry the message between electrons and protons to come close (attraction) are the photons. The message, which the photons carry between two protons or two electrons, would be to go away (repulsion). That is why we say that the photons mediate in electromagnetic interactions.

However, within the nucleus, we have protons and neutrons, which too are bound together. Whereas the atomic size is of the order of 10^{-50} m (sphere of 10^{-50} m radius), the size of nucleus is 10^{-15} meter. Thus protons and neutrons remain within a sphere of $\sim 10^{-15}$ m and naturally then, they must have attractive forces between the positively charged protons and neutral neutrons. If there is only the Coulombian repulsive force between two protons within the nucleus, they can't remain there, therefore there has to be an attractive force between protons as well. Since this force has also to counter the repulsive Coulombian force between the protons, the force has to be strong. This attractive force though strong in nature has a short range of the order of 10^{-15} m, which is also the size of the nucleus. Before the strong force, we only knew the electromagnetic and gravitational forces and had the concepts of corresponding fields as well. The quantum of electromagnetic field, the photon, had been identified and the quantum of the gravitational field was called graviton. Now there was the strong force, so it was natural to think of another field that is responsible for the attraction of proton and neutron. Also, after the quantization of this field we must have its quantum as well. The electromagnetic field has infinite range and its quantum the photon, has zero rest mass. As we shall see in the next subsections, the range of any field is inversely proportional to the rest mass of its quantum. Considering that the range of strong force is small (~10^{-15} m), Yukawa in 1934 calculated the mass of the quantum of this new field which caused proton and neutron attract each other. He found that the quantum should weigh nearly 300 times the electron. A proton weighs about 1840 times the electron. So the mass 300 m_e (m_e, the rest mass of electron) laid between the rest masses of electron (m_e) and that of proton (~1840m_e). So, the new quantum was named meson that means *of middleweight* (here the middleweight of electron and proton). Thus a new particle called meson, with a mass lying in between the masses of electron and proton was predicted which, however, awaited its experimental or the physical verification.

The earth is constantly bombarded with high-energy particles coming from outer space. These streams of high-energy particles are called cosmic rays. Around the year 1937 two groups of scientists, first was of C.D. Anderson and S.H. Neddermeyar and other group was of J.C. Street and E.C. Stevenson, were engaged in the studies of cosmic rays. During their studies, they identified particles, which resembled the Yukawa's description. From their observations it was concluded that the middleweight particle, the meson, has been found, though its source then was the cosmic rays. Later mesons were also produced in the laboratory by smashing the accelerated particle beams into a target.

A significant aspect of the mesons in cosmic rays is that when detailed studies of cosmic rays were made, it was found by Powell and his coworkers that two kinds of middleweight particles were present in the cosmic rays. One was called the π-meson and other was muon. The particle muon (μ particle) behaves as if it is a heavier electron (mass is 207 m_e) and hence should belong to the Lepton family to which electron also belongs. Since the particle muon was found in cosmic rays when mesons were being searched, it was erroneously called μ- meson. In fact, the so-called μ-meson is nothing but the muon particle, which is not a meson. The π-meson is in fact, the meson of Yukawa and this particle is produced in large amounts in upper atmosphere.

Thus, when mesons were discovered, π-mesons were considered the mediators of strong force between nucleons (protons and neutrons). Now a days, however, when we have the knowledge that protons and neutrons consist of three quarks and mesons are made of quark antiquark pair, the π-meson ceases to be the mediator of strong interactions. This is because, if π-meson is comparable to photon, it (π-meson) like photon should not have a substructure. By substructure we mean that the meson (or a particle) should not be composite or made up of other constituents. The mesons, however, have a substructure, the quark antiquark pair bound together. At present therefore, gluons, which bind quarks are the mediators of the strong interactions. If we compare the fo interactions, we would say that, what photon is to electromagnetic interactions, the gluon is to strong interactions, $W^{\pm}$ and Z^0 particles are to weak interactions and graviton is to the gravitational interactions. All mediators are bosons, i.e., they have integral spin. Photons, gluons and $W^{\pm}$, Z^0 particles all have spin 1, that is, they are vector particles and graviton is a spin 2 particle. Graviton, the quanta of the gravitational field has not been detected.

π-meson is the lightest of all mesons. When particle accelerators producing beams of higher and higher energies were invented, many more mesons were discovered. The list will be presented under the section *classification of elementary particles.* The meson family can itself be divided into subfamilies. One may be called the scalar (pseudo) meson family, other the vector meson family, and the third the tensor meson family. All the mesons of the scalar family have spin zero. Any spin zero particle is also called a scalar particle. To this family belong three kinds of π-mesons, the neutral and the charged, i.e., Π^0, Π^+, Π^-; K^+, K^-; K^0, $\bar{K}^0$ and η^0, $\eta^{0\prime}$. The vector meson family is one in which each meson has spin 1. The members of vector meson family are rho mesons (ρ^+, ρ^-, ρ^0), the omega meson (ω^0), phi meson (ϕ^0) and the psi meson (ψ^0). There are spin 2 mesons, which are also called the tensor mesons. The tensor meson family has members K^* mesons $[K^{*+}, K^{*-}, (K^{*0}, \bar{K}^{*0})]$, A_2 mesons (A_2^+, A_2^-, A_2^0), f^0 mesons, f^0 prime mesons $(f^{0\prime})$, K_N mesons $[K_N^+, K_N^-, (K_N^0, \bar{K}_N^0)]$. Theoretically, we can have all the integer spin mesons. That is, we can have in addition to spin 0, spin 1 and spin 2 mesons, the spin 3, spin 4, etc. mesons as well.

1.5 THE YEARS OF ANTIPARTICLES (1930–1956)

Quantum mechanics is basically a theory of a single particle. When the speed of the particle is small compared to the speed of light in vacuum, the quantum mechanics is called the non-relativistic (*NR*) quantum mechanics and when the speed of that single particle is in the relativistic limits, the quantum mechanics so developed is called the relativistic quantum mechanics. The transition from the non-relativistic to relativistic quantum mechanics involves a significant shift in the energy equation for a free particle. Whereas, the *NR* quantum mechanics admits the equation

$$E_{NR} = \frac{p^2}{2m} \tag{1.5}$$

it is replaced by

$$E_R^2 = p^2c^2 + m^2c^4 \tag{1.6}$$

in relativistic quantum mechanics. Obviously, the relativistic energy E_R has both positive and negative roots. The negative energy posed serious problems. If negative energy states are allowed to exist, then a positive energy particle may after coming to ground state (zero energy state) may further lose energy and may go to one negative energy state and then to other and so on. This would mean that even a particle of initially zero energy might emit radiations and become a source of energy. Such a thing does not happen in nature. So this possibility has to be checked. Schrödinger and Klein–Gordon's single particle relativistic wave equations could not provide an explanation to this problem. Dirac, while developing a relativistic wave equation for electron tackled the negative energy problem by providing an explanation to it. He said, the negative energy states are very much there, in fact there is a sea of negative energy states. However, none of the negative energy states is empty, all are occupied by negative energy electrons according to the Pauli's exclusion principle, which says that at one one electron can occupy an energy state. Dirac proposed that the sea of negative energy states in fact was the vacuum. Therefore, in a situation when all the negative energy states are fully occupied there is no question that a positive energy electron would ever jump into a negative energy state and create problems talked above. However, when a high-energy ($\geq 2mc^2$, m the rest mass of electron) photon excites a negative energy electron; it can go to a positive energy state. In such a situation two things happen, one the electron which had a negative energy say ($-mc^2$) absorbs $+ 2mc^2$ energy and becomes an electron of energy $-mc^2 + 2mc^2 = + mc^2$, and we observe an electron, where we note that only positive energy particles (electrons, etc.) can be observed. Other thing that happens is that a vacancy is created in that negative energy state from where the electron had left. This vacancy in the negative energy state is called the hole. Let us see how the charge of the hole is observed. It has however, to be noted that the proposed sea has to be of electrons, i.e., in general the fermions. This is because only the fermions obey the Pauli's exclusion principle that restricts the arrival of any other fermion. Had the negative energy state been occupied by bosons we could never say that a boson from positive energy or zero energy state cannot go to negative energy state as it was already occupied. The reason is that any number of bosons can occupy a state and occupancy is restricted only to fermions. The Dirac's vacuum therefore contains only the negative energy fermions, the electrons.

All the negative energy electrons (the sea) constitute the vacuum. Each electron whether positive energy or negative energy, has charge ($-|e|$). Let the total charge of all the negative energy electrons when all the negative energy states are occupied is Q. In other words, the charge of vacuum is Q. Till now, no electron in the negative energy states has been excited to leave the negative energy state, and we can call Q as the pre-excitation charge of vacuum. Now, an electron in a negative energy state absorbs energy from the photon, gets excited and leaves the negative energy state. The pre-excitation vacuum now becomes the post-excitation vacuum. The charge of the post-excitation vacuum is obviously $Q - (-|e|) = Q + |e|$, where $(-|e|)$ is the charge of electron.

The observed charge of hole is then

$$= \text{Post-excitation charge of vacuum minus the pre-excitation charge of vacuum}$$

$$= Q + |e| - Q = + |e| \tag{1.7}$$

Similarly

The observed energy of hole is

$$= \text{Post-excitation energy of vacuum minus the pre-excitation energy of vacuum}$$

$$= E_V - (-mc^2) - E_V = + mc^2 \tag{1.8}$$

where E_V is the energy of vacuum. If the three momentum of negative energy electron before excitation was $(+\boldsymbol{p})$, then the observed three momentum of hole

$$= \text{Post-excitation three momentum of vacuum minus the pre-excitation three momentum of vacuum}$$

$$= \boldsymbol{p}_v - (+\boldsymbol{p}) - \boldsymbol{p}_v = -\boldsymbol{p} \tag{1.9}$$

where $\boldsymbol{p}_v$ is the total three momentum of vacuum. From the above, we find that when a hole is created by the vacancy of an electron of charge $(-|e|)$, energy $-mc^2$ and three momentum $+\boldsymbol{p}$, we must observe a particle of charge $(+|e|)$, energy $+mc^2$ and three momentum $-\boldsymbol{p}$. Dirac made the prediction of such a particle called positron or the antiparticle of electron in 1929. Two years later in late 1931, C.D. Anderson analysed the tracks of a cosmic ray charged particle and confirmed that the charged particle was positron, the antiparticle of electron. Thus we say that when a high-energy photon knocks out a negative energy electron, we get an electron of energy $+mc^2$ and a positron of energy $+mc^2$. This is known as pair creation. However, a photon cannot produce an $e^- e^+$ pair in free space without violating the energy and momentum conservation. The process $\gamma \rightarrow e^- + e^+$, where both have positive energies, takes place in the Coulomb field of the nucleus.

Anderson's experiments therefore led to the discovery of antiparticles. The antiparticle of proton is antiproton, which was experimentally observed at the Berkeley Bevatron in 1951. It is easy to identify the antiparticles of charged particles, for example, π^+ meson has π^- as its antiparticle, μ^- has μ^+ and so on. However, in the neutral particles, there are some, which are their own anti-particles. For example, photon (γ) is its own antiparticle (γ) and the antiparticle of π^0 meson is π^0 itself. Other neutral particles, e.g., neutron has its antiparticle $\overline{n}$, which is different from n. The antineutron $\overline{n}$ was also experimentally observed at the Berkeley Bevatron in 1956. The reason that the neutron and antineutron are different is that the neutron has a charged structure, it is not uncharged, and it has positive charge at its centre and edges and equal negative charge in between. The antineutron would be made of the reverse combination.

Another distinguishing feature between n and $\overline{n}$ is that both differ in their

Baryon number quantum number. The Baryons have a quantum number called the Baryon number as +1 and antibaryons have –1, as their Baryon number. No such number is associated with photons or mesons. For Leptons $(e^{\mp}, \mu^{\mp}, \nu_e.\bar{\nu}_e, \nu_\mu.\bar{\nu}_\mu)$ we associate a Lepton number, which is again +1, for Leptons and –1, for antileptons. The particles e^- and μ^- and their antiparticles e^+ and μ^+ are easy to be distinguished by their charges. However, the distinguishing feature between neutrinos and antineutrinos, both spin (1/2) and massless, is the helicity. All neutrinos have –1 helicity and all antineutrinos +1. In other words, for neutrinos the component of their spin along the direction of motion is opposite to their direction of motion and for all antineutrinos it is along the direction of their motion. The +1 helicity corresponds to right-handed and –1 helicity to the left-handed. There are no right-handed neutrinos and no left-handed antineutrinos, hence neutrinos and their antiparticles, the antineutrinos, are distinguishable.

Another important particle–antiparticle pair is of K^0 and $\bar{K}^0$ mesons. It is only the strangeness quantum number that distinguishes K^0 and $\bar{K}^0$. The strangeness quantum number is +1 for K^0 and it is –1 for $\bar{K}^0$ mesons.

With the discovery of antiparticles, starting from positron, all the antiparticles were also included in the list of elementary particles. The hole theory of positron started with the single particle Dirac equation and then the sea of negative energy electrons was contemplated. The presence of so many electrons violates the one particle requirement of the Dirac equation. This was remedied by the development of quantum field theory in which both particles and antiparticles are treated on equal footing and the theory of positrons can be developed without using the hole concept.

1.5.1 Antiparticles and Crossing Symmetry

Let us suppose that a nuclear reaction

$$a + b \rightarrow c + d \tag{1.10}$$

is known to occur. The reaction of course would obey the energy and other conservation laws. The principle of crossing symmetry says that in the above process any of the particles *a*, *b*, *c* and *d* can crossover the arrow and go to other side provided it is replaced by its antiparticle. Suppose the particle *b* crossesover to the right of arrow than the reaction (1.10), would become

$$a \rightarrow \bar{b} + c + d \tag{1.11}$$

where $\bar{b}$ (*b* bar) is the antiparticle of *b*. If the mass of *a* is greater than the sum of the masses of *b*, *c* and *d*, then above reaction would also occur. In case it weighsless, then *a* must have sufficient kinetic energy for $a \rightarrow \bar{b} + c + d$ to take place. The important point is, if energy conservation law holds then following reactions obtained by using the principle of crossing symmetry would also occur

$$a+b \rightarrow c+d \tag{1.12a}$$

$$a+\bar{c} \rightarrow \bar{b}+d \tag{1.12b}$$

$$a+\bar{d} \rightarrow c+\bar{b} \tag{1.12c}$$

$$\bar{c}+\bar{d} \rightarrow \bar{a}+\bar{b} \tag{1.12d}$$

etc. The replacement of particle by its antiparticle also implies that the four momentum of antiparticle is opposite to that of the particle.

The crossing symmetry finds its use in writing Feynman diagrams to study the various interaction processes. The reverse reaction

$$c+d \rightarrow a+b \tag{1.13}$$

will also occur. A physical example of crossing symmetry is

$$\gamma+e^{-} \rightarrow \gamma+e^{-} \tag{1.14}$$

$$e^{-}+e^{+} \rightarrow \gamma+\gamma \tag{1.15}$$

The first one is Compton scattering in which a photon is scattered by the electron and second one is annihilation of electron-positron pair. Though above two processes appear to be different, they are same because using the principle of crossing symmetry $e^{-}+e^{+} \rightarrow \gamma+\gamma$ can be obtained from $\gamma+e^{-} \rightarrow \gamma+e^{-}$ where we note that antiparticle of γ is γ itself.

1.6 THE YEARS OF NEUTRINOS (1930–1962)

In the early twentieth century following kinds of nuclear decays were observed

$${}_{5}\mathrm{B}^{12} \rightarrow {}_{6}\mathrm{C}^{12}+\beta^{-} \tag{1.16}$$

$${}_{7}\mathrm{N}^{12} \rightarrow {}_{6}\mathrm{C}^{12}+\beta^{+} \tag{1.17}$$

These decays were called beta (β) decays and the particles, which were emitted, called the β particles. The identity of these particles was then not known. Though these reactions were experimentally observed, it was found then that energy, momentum and angular momentum were not conserved in these reactions. The uncertainty about the validity of these conservation laws continued till Pauli proposed that a neutral massless particle must have been emitted along with the β particle in the above reactions. Pauli called the particle neutrino, which in the Italian language means the little neutral one. The recent researches on the mass of neutrino find that neutrino should not be massless but expect the electron-neutrino mass must be less than 8 eV/C^2. However, for our purpose we will continue to consider it massless. The emission of neutrino easily accounted for the conservation of energy and momentum but the conservation of angular momentum required that the neutrino must be a spin (1/2) particle. The β^- particle in the above reactions turned out to be the electron. Thus the emission of electron from a nucleus is called the β decay. Since there are no electrons inside the nucleus, the emission of electron from the nucleus is when a neutron converts into a proton, i.e.

$$n \rightarrow p + e^{-} + \overline{\nu} \tag{1.18}$$

where $\overline{\nu}$ is the antineutrino. There is another decay in which π mesons decay as

$$\Pi \rightarrow \mu + \upsilon \tag{1.19}$$

Though neutrino had been theoretically proposed to validate the energy momentum and angular momentum conservation laws, their experimental confirmation came through the experiments of F. Reines and C.L. Cowan in 1953 in which they observed the inverse beta decay reaction

$$\overline{\nu} + p^{+} \rightarrow n + e^{+} \tag{1.20}$$

and were able to identify the outgoing positron which confirmed that neutrino did exist. R. Davis and D.S. Harmer established the distinction between neutrino and antineutrino in 1959. They thought that if the reaction (1.20) is experimentally observed, it's crossed reaction, i.e., the reaction obtained by using the principle of crossing symmetry, namely

$$\nu + n \rightarrow p^{+} + e^{-} \tag{1.21}$$

must also be observed at the same rate at which $\overline{\nu}p^{+} \rightarrow ne^{+}$ is observed. This reaction was observed. They further noted that if ν and $\overline{\nu}$ are same then

$$\overline{\nu} + n \rightarrow p^{+} + e^{-} \tag{1.22}$$

must also be observed. However, this equation was not observed. This showed that υ and $\overline{\nu}$ are two distinct particles.

In a reaction where neutrinos or antineutrinos are involved, the Lepton number conservation law easily finds out whether the reaction should have a neutrino or an antineutrino. The neutrinos are Leptons with Lepton number $L = +1$ and antineutrinos are Leptons with Lepton number $L = -1$. Protons and neutrons are Baryons hence have no Lepton number. Accordingly, total Lepton number on the LHS of $\overline{\nu}n \rightarrow p^{+}e^{-}$ is -1, whereas it is $+1$, on the right hence Lepton number conservation is violated in

$$\overline{\nu} + n \rightarrow p^{+} + e^{-}$$

and the reaction is forbidden. In the neutron decay

$$n \rightarrow p^{+} + e^{-} + \overline{\nu} \tag{1.23}$$

it has to be antineutrino so that the Lepton number in the RHS is zero as in the LHS.

The Lepton number conservation needs to be further specified. This is because in the nuclear decay $\pi \rightarrow \mu\nu$, the muon, an unstable particle with a mean life of 2×10^{-6} s, decays as

$$\mu^{-} \rightarrow e^{-} + \nu + \overline{\nu} \tag{1.24a}$$

$$\mu^+ \rightarrow e^+ + \nu + \overline{\nu} \tag{1.24b}$$

We will just see that the neutrinos and antineutrinos produced in the above muon decays are of different kinds. Another possible decay for muon (say μ^-) is

$$\mu^- \rightarrow e^- + \gamma \tag{1.25}$$

which, however, is never observed. All the conservation laws including the Lepton number conservation law defined as above is obeyed in the above equation but still it is not observed. In order to explain its non-occurrence, we have to introduce a conservation law, which must be violated in the above decay $\mu^- \rightarrow e^- \gamma$. Therefore, an electron, instead of the Lepton number L, is assigned an electron number L_e that is +1 for electron and −1 for positron. Similarly, μ^- is assigned a muon number $L_\mu = +1$ and μ^+ is assigned $L_\mu = -1$. Now demand is made that in a nuclear decay involving Leptons both the electron number (L_e) conservation law and the muon number (L_μ) conservation law must be separately obeyed. It can be checked that in $\mu^- \rightarrow e^- \gamma$, both electron and muon number conservation laws are violated, the LHS has $L_\mu = +1$, RHS $L_\mu = 0$, L_e for LHS = 0 whereas L_e for RHS is +1. Thus two separate conservation laws of electron number and the muon number replace the Lepton number conservation law. Therefore, to have the same electron and muon number on both the sides of a nuclear decay we must replace.

$$n \rightarrow p^+ + e^- + \overline{\nu}$$

by

$$n \rightarrow p^+ + e^- + \overline{\nu}_e \tag{1.26}$$

where L_e for $e^- = +1$ and L_e for $\overline{\nu}_e = -1$. Similarly

$$\pi \rightarrow \mu + \nu$$

must be replaced by

$$\pi^- \rightarrow \mu^- + \overline{\nu}_\mu \tag{1.27a}$$

and

$$\pi^+ \rightarrow \mu^+ + \nu_\mu \tag{1.27b}$$

where $L_\mu = +1$ for μ^- and ν_μ and $L_\mu = -1$ for μ^+ and $\overline{\nu}_\mu$. The muon decay

$$\mu \rightarrow e + \nu + \overline{\nu}$$

in fact would mean

$$\mu^- \rightarrow e^- + \overline{\nu}_e + \nu_\mu \tag{1.28a}$$

and

$$\mu^+ \rightarrow e^+ + \nu_e + \overline{\nu}_\mu \tag{1.28b}$$

Therefore, the demand of obeying the electron and muon number conservation laws has shown us that there must be two kinds of neutrinos ν_e and ν_μ.

For the experimental verification of these two kinds of neutrinos, ν_e and ν_μ, we go back to the inverse beta decay reaction (1.20)

$$\overline{\nu} + p^+ \rightarrow e^+ + n$$

which was used to experimentally verify the neutrino's existence. We now rewrite this reaction as

$$\overline{\nu}_\mu + p^+ \rightarrow \mu^+ + n \tag{1.29a}$$

$$\overline{\nu}_\mu + p^+ \rightarrow e^+ + n \tag{1.29b}$$

$$\overline{\nu}_e + p^+ \rightarrow \mu^+ + n \tag{1.30a}$$

$$\overline{\nu}_e + p^+ \rightarrow e^+ + n \tag{1.30b}$$

From the electron and muon number conservation laws, only the first and the last of the above four possibilities are allowed, the remaining two, the second and the third, must be forbidden. Using the antineutrino beam from the decay of $\pi^- \rightarrow \mu^- \overline{\nu}_\mu$, which implied that the beam was of $\overline{\nu}_\mu$, the first two of the above four reactions were tested. It was found by Lederman, Schwartz and Steinberger et al, in 1962, that the decay $\overline{\nu}_\mu p^+ \rightarrow e^+ n$, was not observed. Had $\overline{\nu}_\mu$ and $\overline{\nu}_e$ been the same, this decay must have been experimentally observed. In that case, $\overline{\nu}_\mu$ could be replaced by $\overline{\nu}_e$ and the violation of L_e, L_μ would not have taken place. Similarly, using the beams of $\overline{\nu}_e$, say from $\mu^- \rightarrow e^- \overline{\nu}_e \nu_\mu$, the reaction $\overline{\nu}_e p^+ \rightarrow \mu^+ n$ is not observed, whereas $\overline{\nu}_e p^+ \rightarrow e^+ n$ is easily observed. These observations again confirm that ν_e and ν_μ are different kinds of neutrinos; ν_e is called the electron neutrino and ν_μ the muon neutrino. The experimental observation of two neutrinos also verifies the existence of electron number conservation law and the muon number conservation law.

In the family of leptons, so far we have the fundamental particles (e^-, ν_e), (μ^-, ν_μ) and their antiparticles. It may be recalled that muon is about 207 times heavier than the electron. In 1975, M. Perl discovered a new lepton called the tau (τ) particle, which is about 3500 times heavier than electron ($m_\tau \approx$ 1784 MeV) and corresponding to τ lepton there is a neutrino ν_τ. Like the electron number and muon number conservation there would now be tau number conservation law which has to be obeyed in the decays like

$$\tau^- \rightarrow e^- + \overline{\nu}_e + \nu_\tau \tag{1.31a}$$

$$\tau^+ \rightarrow \mu^+ + \nu_\mu + \overline{\nu}_\tau \tag{1.31b}$$

The electron and its neutrino (e, ν_e) are referred to as the first generation of leptons. The muon and its neutrino (μ, ν_μ) are called the second generation of leptons and similarly, tau and its neutrino (τ, ν_τ) are referred to as the third generation of leptons. In each generation the inclusion of corresponding antiparticle is implied. The name lepton was given to the family when τ lepton was not discovered and other members being lighter in weight compared to mesons and baryons were called leptons. However, with the discovery of tau particle the lepton family now has heavier members as well.

1.6.1 Sources of Neutrinos

Billions of neutrinos pass through every square centimetre of our body per second. These neutrinos come from the sun. In the night when sun has gone to other side of earth, the neutrinos hit us from below as they have the capability to pass through the earth to do so. The sun is a natural source of neutrinos, which reach us. The energy the sun releases is due to the thermonuclear reactions taking place inside it. The reactions in which neutrinos are produced are part of the carbon cycle that converts four protons into a He^4 nucleus giving the same energy release as in proton-proton cycle.

$$\begin{aligned}
&{}_6C^{12} + {}_1H^1 \to {}_7N^{13} + \gamma \\
&{}_7N^{13} \to {}_6C^{13} + {}_{-1}e^+ + \nu_e \\
&{}_6C^{13} + {}_1H^1 \to {}_7N^{14} + \gamma \\
&{}_7N^{14} + {}_1H^1 \to {}_8O^{15} + \gamma \\
&{}_8O^{15} \to {}_7N^{15} + {}_{-1}e^+ + \nu_e \\
&{}_7N^{15} + {}_1H^1 \to {}_6C^{12} + {}_2He^4
\end{aligned} \tag{1.32}$$

The neutrinos interact very weakly with matter; a neutrino with moderate energy can easily penetrate the lead of thickness of some thousand light years.

Artificial production of neutrinos is through the decay processes. The muon neutrino and antineutrino beams can be obtained by pion and kaon decays

$$\left.\begin{aligned} &\pi^+ \to \mu^+ + \nu_\mu \\ &\pi^- \to \mu^- + \bar{\nu}_\mu \end{aligned}\right\} \text{Branching ratio} \approx 100\% \tag{1.33}$$

$$\left.\begin{aligned} &K^+ \to \mu^+ + \nu_\mu \\ &K^- \to \mu^- + \bar{\nu}_\mu \end{aligned}\right\} \text{Branching ratio} \approx 64\% \tag{1.34}$$

The electron neutrinos and antineutrinos are obtained from the decays

$$\left.\begin{aligned} &K^+ \to e^+ + \pi^0 + \nu_e \\ &K^- \to e^- + \pi^0 + \bar{\nu}_e \end{aligned}\right\} \text{Branching ratio} \approx 5\% \tag{1.35}$$

$$\left.\begin{aligned} &\mu^+ \to e^+ + \nu_e + \bar{\nu}_\mu \\ &\mu^- \to e^- + \bar{\nu}_e + \nu_\mu \end{aligned}\right\} \text{Branching ratio} \approx 100\% \tag{1.36}$$

Branching ratio implies that, for example, almost all ($\approx$ 100%) π-mesons decay through $\mu\nu$ mode, whereas only about 5% of K-mesons decay through $e\pi^0\nu$ mode. In a neutrino beam, where ν_e, ν_μ are present, the ratio of electron neutrinos (ν_e) to muon neutrinos (ν_μ) is about 1% to 2%. This means the number of ν_μ or $\bar{\nu}_\mu$ is larger than ν_e or $\bar{\nu}_e$.

Thus, till 1962, the lepton family of the elementary particles had eight members, $e^-, e^+, \nu_e, \bar{\nu}_e$ and $\mu^-, \mu^+, \nu_\mu, \bar{\nu}_\mu$. Four new members, namely, the τ lepton and its neutrino; $\tau^-, \tau^+, \nu_\tau, \bar{\nu}_\tau$ joined the family in 1975.

1.7 THE YEARS OF STRANGE PARTICLES (1947–1960)

With the discovery of π-meson in 1947, it was assumed that the pion would provide the nuclear binding force and a relatively simple picture of elementary particles would come out. However, this comfortable state did not last long. Towards the end of 1947 itself, G.D. Rochester and C.C. Butler obtained a cloud chamber picture of cosmic rays that indicated the existence of new particle. In the pictures (Fig. 1.1) a *V-track* was seen which showed that a neutral particle had decayed into two charged particles. A detailed analysis of these charged particles showed that these were π^+ and π^-. This implied that the neutral particle, which decayed into these two pions, must have a mass at least twice the mass of the pion. Both π^+ and π^- will have the same mass. When (Dec. 1947) this event was observed, no such neutral particle was known, hence the name given to this new particle was kaon and being neutral it was denoted by K^0

$$K^0 \to \pi^+ + \pi^- \tag{1.37a}$$

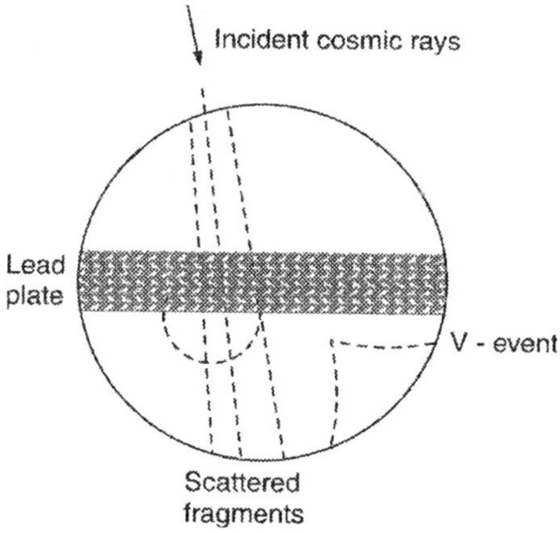

Fig. 1.1: The first strange particle. The cosmic ray shower strikes the 3 cm of lead plate and produces K^0 which subsequently decays as $K^0 = \pi^+ + \pi^-$. This decay is as the V-event in the figure

After two years Powell analysed another cloud chamber photograph and concluded that a charged particle called K^+ decayed into three pions, two positively charged and one negatively charged.

$$K^+ \to \pi^+ + \pi^+ + \pi^- \tag{1.37b}$$

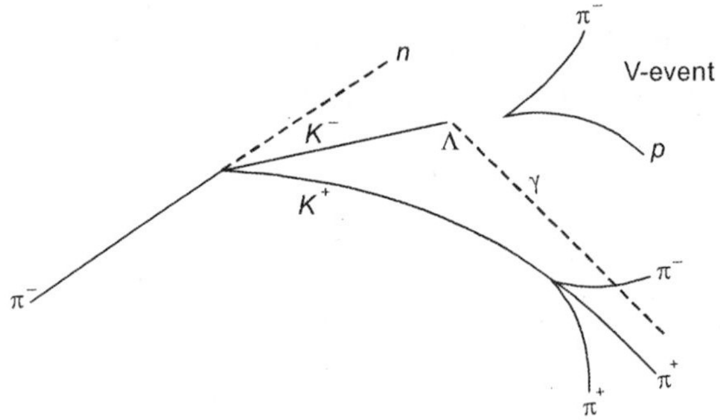

Fig. 1.2: Another V- event, a negative pion collides with a proton (nucleus of hydrogen atom which is already present in the bubble chamber). π^- disappears by decaying through $\pi^- \to n + K^- + K^+$ (the neutron *n*, leaves no tracks). K^- again collides with proton $K^- + p \to \Lambda + \gamma$ (both Λ and γ also leave no tracks). Λ decays into π^- and *p* as the V-event. [The magnetic field applied in the bubble chamber bends positively charged particles clockwise and negatively charged particles anticlockwise. Only charged particles produce tracks in the bubble chamber]

These kaons had some properties similar to pions hence they became the members of the meson family. Later, η, ϕ, ω, ρ etc., mesons joined them.

Another important hydrogen bubble chamber discovery was (Fig. 1.2) made when a negative pion of momentum 4 GeV/C interacted with proton (hydrogen in the bubble chamber), another *V-event* was observed. This *V-event*, which occurred in the bubble chamber at a point that was displaced from the interaction point of π with *p*, showed that a neutral particle had decayed and the branches of *V* represented the tracks of proton and the negative pion. Obviously, the neutral particle that decayed into *p* and π^- must be heavier than the proton. This neutral particle was called the *Lambda* (Λ) particle. The reactions involved were

$$\pi^- + p \to \Lambda + K^0 \tag{1.38a}$$

$$\Lambda \to p + \pi^- \tag{1.38b}$$

$$K^0 \to \pi^+ + \pi^- \tag{1.38c}$$

We have already mentioned that the elementary particles electron, muon, etc., are called the leptons because they have *lightweight* and mesons have weights between those of proton and electron. Then as such proton is a heavy particle. For a heavy particle, we have the word *Baryon* in Greek. Proton thus became the lightest member of the *Baryon family*. Neutron, a little heavier than proton was the other. The *Lambda* (Λ) particle also found its place in the Baryon family. Further cloud chamber observations on penetrating cosmic ray showers revealed the existence of *Sigma* (Σ) and *Xi* (Ξ) particles. The *Xi* (Ξ) particle was also called the cascade particle. The Σ particle has three states, $\Sigma^+, \Sigma^0, \Sigma^-$, the Ξ particle has been observed in Ξ^0 and Ξ^- states. The artificial production of these particles was also observed in a hydrogen diffusion cloud chamber by using a 1.4 $GeV/C\pi^-$ beam from the Brookhaven Cosmotron (proton synchrotron) in the year 1953. These particles having their mass higher than the proton, in fact higher than the neutron, were also included in the Baryon family. The reason for the inclusion of Σ, Ξ, etc., into the baryon family can be understood from the following.

Proton is considered a stable particle, so it must not decay into any of the other particles. However, the following decay equation,

$$p^+ \rightarrow e^+ + \gamma \tag{1.39}$$

which does not violate any of the conservation laws of energy momentum, angular momentum, charge, etc.,(the conservation laws known in late thirties), does not occur. For the stability of proton such decays indeed should not occur. Therefore to stop the decay $p^+ \rightarrow e^+\gamma$ from occurring, the baryon number conservation law was introduced. All baryons were allocated (+1) *baryon number* (B = +1) and their antibaryons (–1) (B = –1). In the decay (1.39), the LHS has +1 baryon number and in the RHS, there being no baryon, the baryon number is zero. Demanding that the baryon number conservation law should be observed, the decay (1.39) becomes forbidden. It can be observed that the lepton number conservation law that kept intact the stability of electron is also violated in the above proposed decay. However, there is no meson number required for mesons hence no meson number conservation law, then the reaction

$$\pi^- + p \rightarrow \Lambda + K^0 \tag{1.40}$$

which is observed, must obey the baryon number conservation law. Since, in the LHS we have proton (B = +1), the Λ-particle must also have (B = +1), which is possible only when the Λ-particle is a baryon. π^- and K^0 are mesons, hence cannot provide the baryon number. The charge conservation law makes the Λ-particle a neutral one. Similarly, the occurrence of the following decays:

$$\begin{aligned}
&\Sigma^+ \rightarrow p^+ + \pi^0 \quad \text{or} \quad \Sigma^+ \rightarrow n + \pi^0 \\
&\Sigma^0 \rightarrow \gamma + \Lambda \\
&\Xi^- \rightarrow \pi^- + \Lambda \quad \text{with} \quad \Lambda \rightarrow p^+ + \pi^- \\
&\Xi^0 \rightarrow \pi^0 + \Lambda
\end{aligned} \tag{1.41}$$

which have to respect the baryon number conservation law, demands that the Σ- and Ξ-particles must be the baryons. The baryons heavier than neutron are called hyperons. The protons and neutrons, the light baryons, are also given a common name, the nucleons.

Thus, in the above we have talked about the discovery of new particles the K-mesons and Λ, Σ and Ξ hyperons. These particles carried with them some special features. First, they appeared when no body had expected them. Second, these particles were produced in large amounts that too quite fast and third, they decayed relatively slowly. At nuclear level, the time scales vary from 10^{-23} second to 15 minutes. If any process (production or decay) takes place in a time interval (lifetime) of the order of 10^{-23} s, it is said to be fast. Such fast processes are through the strong interactions. Electromagnetic interactions take about 10^{-16} s, and the times of weak interactions range from 10^{-13} s, to 15 minutes ($\approx 10^3$ s, neutron decay). It must be clearly felt that 10^{-13} second is 10^{10} times the 10^{-23} second. This implies that if we imagine, a process through strong interaction is completed in say 1 second (highly exaggerated), then the same process if it can go through the weak interaction will be completed in 10^{10} s $\approx$ 3000 years.

Thus the production mechanism of these particles $(\Lambda,\ \Sigma,\ \Xi)$ is quite different from the one, which governs their disintegration. The production is through the strong interaction and decay is through the weak interaction. As already mentioned these particles decay as in eqn. (1.41), where despite the decay process-taking place through weak interaction, none of the product particle is neutrino or antineutrino. At the time of the discovery of these particles it was expected that ν or $\bar{\nu}$ would be emitted in weak decays, whereas here was none. Therefore, this feature of weak decays without neutrino or antineutrino emission was considered strange and hence the particles, $(K, \Lambda,\ \Sigma,\ \Xi)$, which decayed through such weak interactions were called strange particles.

Another important observation, which A. Pais made, was that these strange particles are produced in pairs. For example, in a pion-proton collision, two strange particles as given below may be produced.

$$\pi^- + p^+ \to K^0 + \Lambda \tag{1.42}$$

$$\pi^- + p^+ \to K^+ + \Sigma^- \tag{1.43}$$

$$\pi^- + p^+ \to K^0 + \Sigma^0 \tag{1.44}$$

and we do not find a production reaction in which a single strange particle is produced. For example following reactions do not take place.

$$\begin{aligned} \pi^- + p^+ &\not\to \pi^0 + \Lambda \\ &\not\to \pi^+ + \Sigma^- \\ &\not\to n + K^0 \end{aligned} \tag{1.45}$$

though they obey all the conservation laws including the baryon number. Again their occurrence has to be denied by introducing a new conservation law. Gell-Mann and Nishijima assigned to each strange particle a strangeness number S that could take values +1 and –1. However, unlike baryon or charge number the assignment of S is arbitrary. The strangeness values of the strange particles have been given in the following table:

Particle	Strangeness S
K^0	+ 1
$\bar{K}^0$	– 1
K^+	+ 1
K^-	– 1
Λ^0	– 1
Σ^+	– 1
Σ^0	– 1
Σ^-	– 1
Ξ^0	– 2
Ξ^-	– 2

For non-strange particles the S value is zero. Thus in view of above strangeness values, reactions (1.45) violate the strangeness conservation law, π^-, p^+, π^0, π^+ and n each has S = 0, hence LHS of eqn. (1.45) has total S = 0 and RHS have total S = –1, –1 and +1 respectively. Hence reactions (1.45) are disallowed. Reactions (1.42) – (1.44) are allowed, they obey S conservation. S conservation law also forbids the reaction

$$\pi^- + p^+ \rightarrow K^- + \Sigma^+ \qquad (1.46)$$

We have seen in the above that production reactions of strange particles, which are through the strong interactions, obey strangeness conservation law. Let us see their decay reactions (1.41). In eqn. (1.41) only the decay

$$\Sigma^0 \rightarrow \Lambda + \gamma \qquad (1.47)$$

which is in fact the decay through electromagnetic (γ) interaction, conserves the strangeness. None of the other decays of eqn. (1.41) obeys the strangeness conservation law.

Thus we conclude that strong and electromagnetic interactions conserve strangeness but weak interactions do not. The strangeness is assigned only to mesons and baryons because they are the particles that interact strongly. The leptons do not interact strongly; therefore the concept of strangeness is not applicable to them.

Thus by 1960, the strange particles had made their way into the list of fundamental particles. With the introduction of more and more energetic beams, available from accelerators, the resonances that were interpreted as particles began to pour in, and the list of elementary particles particularly the meson and baryon families grew by leaps and bounds.

1.8 THE YEARS OF RESONANCES (1961–)

The availability of high-energy accelerators initiated the systematic searches for new particles. With the production of pion beams, it was possible to perform pion nucleon scattering experiments. In particular, $\pi^+ p$ and $\pi^- p$ scattering experiments were performed in which beams of charged pions were scattered by the targets of hydrogen nuclei. Liquid hydrogen was the very commonly used target. The pion nucleon scattering experiments were expected to lead us to an understanding of the strong interactions to which all hadrons are subjected. The pion nucleon (πN) scattering is a particular example of more general hadron-hadron reactions. Thus the reaction of one hadron with other becomes significant from the point of view of understanding the strong interactions and during the study of these hadron-hadron reactions; the discovery was made of what we call the resonances.

In a hadron-hadron reaction, the probability that the two participating hadrons will really interact depends on the energy involved in the interaction process. The energy involved means the rest energies plus the kinetic energies of hadrons participating in the reactions. If the amount of energy involved changes, the probability of interaction of two hadrons also changes; it may increase or decrease with increasing energy. However, at certain values of energy, the reaction (interaction) probability is observed to increase sharply and the reaction is much more likely to occur at these values than at other energies. Let us assume that in a hadron-hadron interaction the energy involved was such that the interaction did take place. If the interaction process is scattering, the incoming and outgoing hadrons would be same and in the case of other reactions outgoing particles are different from the incoming. The mechanism is that when incoming hadrons have interacted they first join together to form a new hadron, which lives for a very short time ($\approx 10^{-23}$ s) and then decays. The decaying of this intermediate hadron may again produce the incoming pair of hadrons or new particles may come out.

These short-lived hadron states are called *resonances*. They are called resonances because the probability of their formation is maximum (i.e., it peaks) at certain values of energy and if we observe, for example, the $\pi^+ p$ scattering experiments, the graph between the total cross section and energy of $\pi^+ p$ system shows peaks at different values of energy. The curves with peaks resemble a resonant curve with which we encounter in mechanical or electrical oscillations where at a particular frequency the amplitude becomes maximum. Here at particular energy, the scattering cross section has become maximum. The rough sketch for $\pi^{\pm} p$ scattering has been shown in Fig. (1.3).

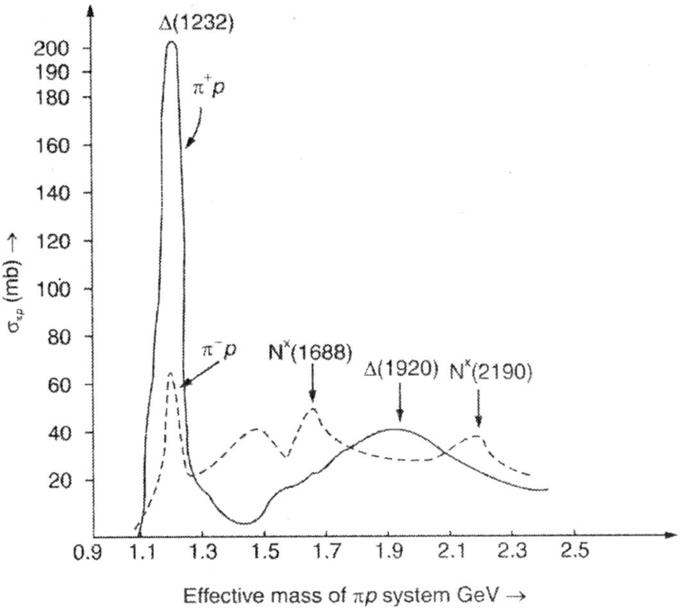

Fig. 1.3: Variation of total cross section for π^+ and π^- mesons on protons with incident pion energy. Δ are $I = \frac{3}{2}$ and $N^*, I = \frac{1}{2}$ resonances

From the figure we observe that there is a large peak in π^+p scattering curve at energy of 1232 MeV. The shape of the curve with a maximum at 1232 MeV is like the Breit–Wigner type of resonance with which we come across in nuclear reactions. This resonance curve is expressed by the equation

$$\sigma(E) = \sigma_{max} \frac{\frac{\Gamma^2}{4}}{(E-M)^2 + \frac{\Gamma^2}{4}} \tag{1.48}$$

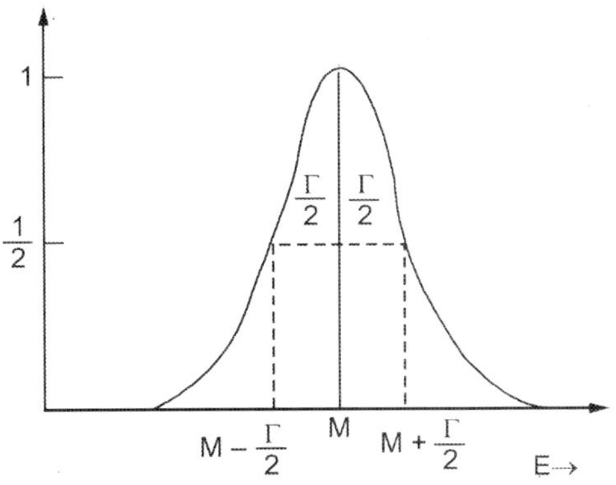

Fig. 1.4: Breit–Wigner resonance

and is drawn in Fig. (1.4). In eqn. (1.48) $\sigma(E)$ is the cross section of the target for the incident particle of energy E. At energy $E = M$ of the incident particle, the cross section is maximum, i.e., for $E = M$, $\sigma(E) = \sigma_{max}$. M is called the mass of the resonance and Γ its width. Γ is the difference in energies of two points on the curve corresponding to $\sigma(E) = \frac{\sigma_{max}}{2}$. In the reaction

$$\pi^+ + p \to \Delta(1232) \to \pi^+ + p \tag{1.49}$$

the resonance is called $\Delta(1232)$, where 1232 MeV/C^2 indicates the mass of the resonance. The width of the resonance is about 115 MeV. The lifetime of this resonance

$$\tau = \frac{\hbar}{\Gamma} \approx \frac{6.58\times 10^{-22}\,\text{MeV S}}{115\ \text{MeV}} \approx 5\times 10^{-24}\ \text{s} \tag{1.50}$$

The scattering reaction (1.49) can than be read as – when π^+ and p^+ interacted and their centre of mass energy was 1232 MeV an excited state denoted by Δ was formed which lived for about 5×10^{-24} s, and then decayed back into π^+ and p^+ implying that for about 10^{-24} s, we had a new particle called $\Delta(1232)$. The mass of the particle, i.e. the resonance, is the total energy of the products into which the particle decays and this energy has to be measured in the rest frame of the particle. The spin of this new particle can also be calculated. For that we use the formula

$$\sigma_{max} = \frac{4\pi}{k^2}\frac{(2J+1)}{(2s_\pi+1)(2s_p+1)} \tag{1.51}$$

where s_π and s_p are the spins of π^+ and p^+ respectively which we know are 0 and 1/2 respectively. J is the spin of the particle, i.e. the resonance. Since, $\sigma_{max} \approx \frac{8\pi}{k^2}$ eqn. (1.51) gives

$$J = \frac{3}{2} \tag{1.52}$$

Thus we have the description of $\Delta(1232)$ as the particle of mass 1232 MeV, and spin 3/2 which lasts only for about 10^{-24} s. Since the reacting particles π^+ and p both are positively charged, we expect a double charge (++) in the resonance Δ. We explain in the following that $\Delta(1232)$ particle exists in four states and Δ^{++} is one of them.

The common name for both proton and neutron is nucleon. Whether the nucleon is a proton or a neutron is decided by a degree of freedom called isotopic spin or just isospin. This is an internal degree of freedom as compared to external. When we say that a nucleon is at say (x, y, z, t) and has spin 1/2, we are referring to external degrees of freedom. The ordinary spin refers to rotation in external four-dimensional space, whereas isospin refers to rotation in internal space whose dimensions may be any number. Here of course in the case of nucleon the internal space which is also called the charge space or isospace has three dimensions. The isospin may be denoted by I and the ordinary spin by J. In later chapters we have denoted isospin by T as well.

The isospin I for the nucleon is 1/2 and when the third component I_z has value +1/2, the nucleon becomes a proton and when the I_z of nucleon is –1/2, it becomes a neutron. The I_z values range from $-I$ to $+I$ differing by one. Similarly, the pion having isospin equal to 1 has I_z values as + 1, 0 and – 1. Corresponding to these three I_z values the π-meson has three states namely, $\pi^+ (I_z = +1)$, $\pi^0 (I_z = 0)$, and $\pi^- (I_z = -1)$. Thus I tells the number of states, $(2I + 1)$, the particle will possess and I_z identifies them. The isospin states of a particle differ in their masses by only small amounts.

The isospin of Δ(1232) may then be guessed from the isospins of π^+ and p which are 1 and 1/2 respectively. Thus

$$I\Delta(1232) = 1 \pm \frac{1}{2} = \frac{3}{2}, \frac{1}{2}$$

and

$$I_z\Delta(1232) = +\frac{3}{2},\ \frac{1}{2}, -\frac{1}{2}, -\frac{3}{2} \tag{1.53}$$

correspondingly, we have four $(2 \times \frac{3}{2} + 1)$ states of Δ particles denoted by $\Delta^{++}(I_z = +\frac{3}{2}), \Delta^{+}(I_z = +\frac{1}{2}), \Delta^{0}(I_z = -\frac{1}{2}), \Delta^{-}(I_z = -\frac{3}{2})$. I for each of these four would be 3/2. It can be noted that just as nucleon with $I_z = +1/2$ is proton (p^+) and one with $I_z = -1/2$ is neutron (n^0), here we have Δ^+ and Δ^0. It may be pointed out that Δ^{++} particle is quite a noteworthy particle because, as we shall see later on, its existence was responsible to demand that quarks must have *colours*.

Let us look at $\pi^- p^+$ curve. π^- has $I = 1$, $I_z = -1$ and proton has $I = 1/2$, $I_z = +1/2$. Hence I of $\pi^- p^+$ resonance is $\frac{3}{2}$ or $\frac{1}{2}$. $I = \frac{3}{2}$, $I_z = -\frac{1}{2}$ state in $\pi^+ p^+$ curve is the Δ^0 state. Thus Δ^0 state is found in both $\pi^+ p^+$ scattering curve as well as in $\pi^- p^+$ curve at the same energy 1232 MeV. However, in $\pi^- p^+$ curve there are peaks at energies for which there are no peaks in $\pi^+ p^+$ curve. They are the N^* states, where N implies the nucleon. They are called N^* states, * denotes instability, because they correspond to $I = 1/2$ resonances. The mass of one of such important resonance is 1688 MeV/C². In fact, how the I's and I_z's will combine is determined by using Clebsch–Gordan coefficients. It may also be added that there exists a fairly large uncertainty in the observed mass of the resonance. This is because the lifetime of resonance is short and the uncertainty in its mass

$$\Delta M \approx \frac{\hbar}{\Delta t} \tag{1.54}$$

is obviously large.

$\Delta(1232)$ and $N^*(1688)$ are fermions and because they are heavy enough, they are baryons. Another baryon resonance $\Sigma(1520)$ is produced in the reaction

$$\begin{aligned} &K^- + p \to \Sigma \ (1520) \\ &\qquad\qquad \downarrow \longrightarrow \Lambda + \pi^- + \pi^+ \end{aligned} \tag{1.55}$$

In reactions that we use to produce particles (strange), a resonance is formed in the final state, and at least one, other particle is emitted. For example

$$\begin{aligned} &K^+ + p \to p + K^{*+} \ (890) \\ &\qquad\qquad\quad \downarrow \longrightarrow K^0 + \pi^+ \end{aligned} \tag{1.56}$$

$$\begin{aligned} &K^- + p \to K^+ + \pi^0 + \Xi^{*-} \\ &\qquad\qquad\qquad\qquad \downarrow \longrightarrow \Xi^0 + \pi^- \end{aligned} \tag{1.57}$$

these resonances differ from the Δ and N^* type of resonances in that $\Delta(1232)$ and $N^*(1688)$ type of resonances are formed by the colliding particles hence called resonances in *formation* experiments whereas, resonances in reactions (1.56) and (1.57) are formed in the final state and are called resonances in *production* experiments.

Above are the few examples of baryon resonances. Since, baryon is a fermion, it may be said that fermion resonances represent baryons. When fermion resonances have been observed, it becomes natural that we look for boson resonances as well. An important discovery in this regard was made in 1961 when following reactions were studied in a hydrogen bubble chamber

$$\pi^+ + p \to \pi^+ + \pi^0 + p \tag{1.58a}$$

$$\to \pi^+ + \pi^+ + \pi^- + p \tag{1.58b}$$

$$\to \pi^+ + \pi^+ + \pi^- + \pi^0 + p \tag{1.58c}$$

It was thought, e.g., in eqn. (1.58a) that the pions π^+ and π^0 may have been produced in two ways. One, they were produced directly and second the pions π^+ and π^0 were not produced straightaway, instead, there took place an emission of a resonance state and a proton. The resonance state lived for a very short time ($\approx 10^{-24}$ s) and then decayed into π^+ and π^0. Since both π^+ and π^0 are bosons (spin $J = 0$), the resonant state must also be a boson. This was the manner the existence of boson resonances was considered. The experimental discovery of such resonances is based on calculating the Lorentz invariant mass of π^+ and π^0 (when we are observing 1.58a) for different energies of π^+ and π^0. Now note the number of $\pi^+ \pi^0$ pairs for each Lorentz invariant mass. Plotting a graph between number of pairs and the Lorentz invariant mass, it was found that a broad peak was centred at about 765 MeV value of the invariant mass. The width of the peak was 150 MeV. For comparison we recall that width of $\Delta(1232)$ resonance was $\approx$115MeV. The existence of the peak at 765 MeV suggested that a large number of $\pi^+ \pi^0$ pairs were formed due to the decay of resonance of mass about 765 MeV (Fig.1.5).

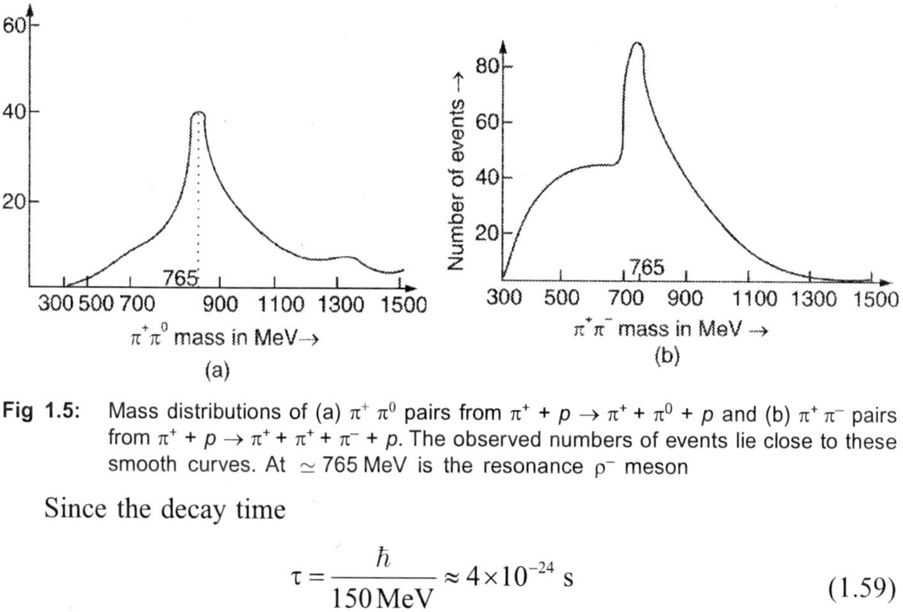

Fig 1.5: Mass distributions of (a) $\pi^+ \pi^0$ pairs from $\pi^+ + p \to \pi^+ + \pi^0 + p$ and (b) $\pi^+ \pi^-$ pairs from $\pi^+ + p \to \pi^+ + \pi^+ + \pi^- + p$. The observed numbers of events lie close to these smooth curves. At $\simeq$ 765 MeV is the resonance ρ^- meson

Since the decay time

$$\tau = \frac{\hbar}{150\,\text{MeV}} \approx 4\times10^{-24}\ \text{s} \tag{1.59}$$

belonged to the lifetime of strong interaction, implying that the bosonic resonance ($\pi^+ \pi^0$) decayed through strong interactions. Thus the resonance ($\pi^+ \pi^0$) must be a hadron and only the mesons are the bosonic hadrons. The resonance 765 must therefore be a meson and was called ρ-meson. Its mass is estimated to be 765 ± 10 MeV, and width 125 ± 20 MeV. A similar curve between the invariant mass of $\pi^+ \pi^0$ pairs and the number of ($\pi^+ \pi^0$) pairs also showed a peak at similar energy showing that ρ-meson was also produced as resonance in the final state of the reaction (1.58b). No peak in ($\pi^+ \pi^+$) mode however, was observed. In the reaction (1.58c) we can pick up ($\pi^+ \pi^0$) pair. The ρ-meson resonance is also called the pion-pion or two pion ($\pi^+ \pi^0$, $\pi^+ \pi^-$) resonance.

In view of the discovery of the ρ-meson, the reactions (1.58a) and (1.58b) can be expressed as

$$\begin{aligned} &\pi^+ + p \to \rho^+ + p \\ &\qquad\qquad \downarrow \longrightarrow \pi^+ + \pi^0 \end{aligned} \tag{1.59a}$$

$$\begin{aligned}\pi^- + p \rightarrow \rho^- &+ p \\ &\downarrow \longrightarrow \pi^- + \pi^0\end{aligned} \tag{1.59b}$$

we can also have

$$\begin{aligned}\pi^- + p \rightarrow \rho^0 &+ n \\ &\downarrow \longrightarrow \pi^+ + \pi^-\end{aligned} \tag{1.59c}$$

Since, in strong interactions isospin is conserved, reactions (1.59) suggest that the isospin I of ρ meson must be 1, implying that ρ meson must appear in three states corresponding to $I_z = +1, 0, -1$ denoted by ρ^+, ρ^0 and ρ^- respectively. Again ρ meson is a pion-pion resonance, where I of $\rho = 1$. The I of each of the pions that make ρ is also 1. This means that the two *isospin one* objects are forming another *isospin one* object. This implies that if we interchange the isospin coordinates of two pions (which make the ρ) the isospin wavefunction must become antisymmetric. For example if π^+ and ρ^0 form ρ^+ then isospin wavefunction must be

$$|\psi_I\rangle = \frac{1}{\sqrt{2}}\{|\pi^+, \pi^0\rangle - |\pi^0, \pi^+\rangle\} \tag{1.60}$$

which obviously changes sign when isospin coordinates of π^+ are replaced by those of π^0. Replacing π^0 by π^- in eqn. (1.60) we will have similar antisymmetric isospin wavefunction for $\pi^+ \pi^-$, forming ρ^0. What we get from above is the isospin wavefunction of two pions which form the ρ-meson is antisymmetric under the interchange of their isospin coordinates. In contrast, for $\pi^+ \pi^+$, $\pi^0 \pi^0$ and $\pi^0 \pi^0 + \pi^+ \pi^+$ the isospin wavefunction is symmetric.

Now consider that, whether $\pi^+ \pi^0$ or $\pi^+ \pi^-$, these pions are distinguishable only because of their charges that arise due to isospin. The ordinary spin for every pion is zero. Thus these pions are identical except for the isospin state. Since pions are bosons and we know that for bosons the total wavefunction must be symmetric under the interchange of all (space and ordinary spin) coordinates. This symmetry principle can be extended to include isospin coordinates as well, and accordingly in the above symmetry principle, the phrase *all coordinates* would mean the space, the ordinary spin and the isospin coordinates. Since the ordinary spin is zero, the symmetric behaviour of total wavefunction would be guided by the interchange of space and the interchange of isospin coordinates of two pions.

Therefore, in view of the antisymmetry of isospin wavefunction, the space wavefunction of two pions must also be antisymmetric under the interchange of their spatial coordinates. This means the ρ meson formed by these two pions must have odd orbital momentum i.e., J value of ρ would be 1, 3, 5.... The simplest is $J = 1$. Due to $J = 1$, the ρ meson is called the vector meson. It may be noted that for a composite system of two particles, J is the vector sum of spin S of the composite and the relative orbital angular momentum L of the two particles.

In the year 1961 itself, another bosonic resonance, called the omega (ω) meson was discovered. The reaction studied was

$$p+\bar{p}\to\omega+2\pi$$
$$\downarrow\longrightarrow\pi^{+}+\pi^{-}+\pi^{0} \qquad (1.61)$$

where $\bar{p}$ is antiproton. Similar to the case of ρ meson the invariant mass of three pions namely,

$$M^{*}\equiv(E_1+E_2+E_3)^2-(\boldsymbol{p}_1+\boldsymbol{p}_2+\boldsymbol{p}_3)^2 \qquad (1.62)$$

was calculated and for each value of M^*, the number of three pion events was observed. The graph between M^* values and the number of events showed a peak at about 780 MeV with a width of about 9 MeV. Obviously the life of this resonance is

$$\tau=\frac{6.58\times10^{-22}\,\text{MeV s}}{9\,\text{MeV}}\approx7\times10^{-23}\,\text{s} \qquad (1.63)$$

Such a three-pion resonance was called the ω meson. As compared to 10^{-24} s of the two-pion resonance (ρ meson) the time 10^{-23} s is long enough to allow ω meson decay into three pions. Again the ($\pi^+ + \pi^- + \pi^0$) state is a neutral state so, ω meson must be a neutral particle and hence its isospin $I = 0$. The spin (J) of ω meson is also 1, i.e., it is a neutral vector meson.

Another important bosonic resonance is the η meson. The reaction studied was

$$\pi^{+}+d\to\eta+2p$$
$$\downarrow\longrightarrow\pi^{+}+\pi^{-}+\pi^{0} \qquad (1.64)$$

where d is deuteron. The distribution in invariant mass of three pions showed a peak at about 550 MeV with a width less than 10 MeV, which gives a lifetime of $\approx10^{-21}$ s. This bosonic resonance was called the η meson.

Obviously η, (a three pion ($\pi^+\pi^-\pi^0$) resonance) is a neutral particle. It is important to see its decay modes. The prominent ones are

39% η mesons decay as

$$\eta\to2\gamma \qquad (1.65)$$

32% η mesons decay as

$$\eta\to3\pi^{0} \qquad (1.66)$$

24% η mesons decay as

$$\eta\to\pi^{+}+\pi^{-}+\pi^{0} \qquad (1.67)$$

In addition to these observed decays there are also the following decay reactions of η mesons.

$$\eta \rightarrow \pi^+ + \pi^- + \gamma$$
$$1.5\% \tag{1.68}$$

and

$$\eta \rightarrow \pi^0 + 2\gamma$$
$$2.5\% \tag{1.69}$$

Above decay reactions suggest that η is seen only in the neutral mode, hence its isospin $I = 0$. If we assume that the isospin must be conserved in η decay, then, involving three pions, only the combination $\pi^+\pi^-\pi^0$ has $I = 0$. Therefore, to preserve isospin conservation η meson should decay only into $\pi^+\pi^-\pi^0$ mode. But we find from decays (1.65) to (1.69) that only 24% of η mesons decay into this three pion mode, for other modes we do not have $I = 0$ in the right hand side. Moreover, photon (γ) is emitted in all other decays except $3\pi^0$ mode. The emission of photon is an indication that η meson has decayed through electromagnetic interactions and it is known about electromagnetic interactions that they do not preserve isospin conservation law. Thus the conclusion emerges is that most of the η-decay modes are electromagnetic hence do not obey isospin conservation. Only thing that comes in the way is the lifetime. The lifetime of η meson is about 10^{-21} s which is close to 10^{-23} s the lifetime of strong interactions. This implies that η meson should have decayed through isospin conserving strong interactions but the percentage of such decay mode is only 24%. Hence we look for the width of the η resonance and conclude that the width of η resonance must be much less than 10 MeV in fact of KeV order, so that the lifetime is around 10^{-16} s, the lifetime of electromagnetic interactions. The uniformity of experimental events leads us to believe that spin (J) of η meson is zero because any non-zero orbital angular momentum of product pions should produce non-uniformity in the experimental events.

Thus η resonance is a scalar meson of mass 550 MeV and width about 2.6 KeV.

It has to be added that, τ-lepton, W^+, W^- and Z^0 bosons, all are resonances. The number of resonances and hence the number of corresponding short-lived particles is large enough and in fact depends on the energy of the projectile. It may also be noted that whereas the particles (if charged) leave tracks in the detecting device, e.g., the bubble chamber, the resonances due to their very short lifetime do not leave tracks and are guessed on the basis of their decay products.

With the inclusion of resonances, the number of elementary particles became too large to be counted and in fact the question arose whether such a large number of particles should be the basic building blocks of matter! The answer to this question can be found in the sections beginning with the modern era. However, in between we incorporate three more sections namely the interaction strengths, the conservation laws and the classification of so far talked about elementary particles.

1.9 THE FUNDAMENTAL INTERACTIONS

There are four basic interactions in nature, in terms of which we can explain all observations of natural processes. They are:

1. Gravitational interaction
2. Electromagnetic interaction
3. Strong interaction and
4. Weak interaction

These four interactions are also called the *four fundamental forces in nature*. In the following, we talk about each.

1.9.1 Gravitational Interaction

Every particle feels the force of gravity; the gravitational interaction therefore is universal. The classical theory of gravity is the well-known Newton's law of gravitation, which says that the force of attraction between two particles of masses m_1 and m_2 at a distance r apart is given by

$$F = G\frac{m_1 m_2}{r^2} \tag{1.70}$$

where $G = 6.69 \times 10^{-11}$ Nm^{2}kg^{-2} is the gravitational constant. Newton's theory assumes that the interaction (here gravitational) takes place instantly, that is, the effect of attraction reaches from one mass to other in no time, implying that the effect travelled with an infinite speed. The postulate of special theory of relativity, which says that the maximum speed, which anything can have, is equal to the speed of light in vacuum, opposes the concept of infinite speed or the concept of action at a distance. Thus special theory of relativity contradicted the concept involved in the Newton's theory of gravity.

Special theory of relativity assumes that the space–time is flat. As an example we can say that, the surface of this page is flat whereas the surface of a ball is curved. When special theory of relativity could not bring about a generalization of Newton's law of gravitation, Einstein made a revolutionary suggestion that gravity, is not a force like other forces, but it is a consequence of the fact that the space–time is curved, not flat. The bodies like earth, which appear to move in curved paths in fact, move in straight paths in a curved space. These straight paths on the curved surface are called *geodesics*. A theory of relativity, which is based on the curved space–time, is called the general theory of relativity and hence Einstein's general theory of relativity is the relativistic generalization of Newton's law of gravitation. The basic equations of the theory of general relativity are Einstein's field equations

$$R_{\mu\nu} = k(T_{\mu\nu} - \frac{1}{2}g_{\mu\nu}T) \tag{1.71}$$

where $R_{\mu\nu}$ is called the Ricci tensor, $T_{\mu\nu}$ is the energy momentum tensor and T is the trace of the energy momentum tensor. $g_{\mu\nu}$ in eqn. (1.71) is the symmetric

metric tensor of the Riemannian curved space–time geometry and k is called the Einstein's gravitational constant. Since $T_{\mu\nu}$ describes the distribution of the matter, the Einstein field equations (1.71) contain the equations of motion of matter that produces the gravitational field. The important predictions of Einstein's general theory of relativity (EGTR) are:

1. The orbits of planets predicted by EGTR are slightly deviated from those predicted by Newtonian theory of gravity. In case of mercury planet the EGTR predicted that because of mercury's proximity to sun, it would experience strongest gravitational effects compared to the other planets and hence its elliptical orbit must be elongated and the long axis of this elongated ellipse should rotate about the sun at the rate of one degree in ten thousand years. This rotation has been observed and is the first confirmation of the validity of EGTR.
2. EGTR predicts that the gravitational fields should bend light. Black holes provide an example of this prediction.
3. EGTR's another prediction is that time should appear to run slower near a massive body like earth. Agreement with this prediction was found in year 1962.
4. EGTR also predicts that heavy objects that are moving will cause the emission of ripples in curvature of space. These ripples are called the gravitational waves. Einstein showed that massive accelerating objects, such as neutron stars or black holes orbiting each other, would disrupt space-time in such a way that space-time would continuously rise and fall and thus create gravitational waves that would travel with the speed of light. These waves carry with them the information about their origin and provide information about the nature of gravity. The strongest gravitational waves are produced by colliding black holes, supernovae and colliding neutron stars. Till the middle of 2015, the gravitational waves remained undetected leaving the last prediction of EGTR unconfirmed. Fortunately, on ***14 September 2015, the universe's gravitational waves were observed for the very first time.*** The waves, which were predicted by Albert Einstein a hundred years ago, came from a collision between two black holes. It took 1.3 billion years for the waves to arrive at the LIGO detector in the USA. **The detection of gravitational waves has conclusively proved the Einstein's Theory of General Relativity.** LIGO, the Laser Interferometer Gravitational Wave Observatory is a collaborative project with over one thousand researchers from more than twenty countries including India. The experimental data of LIGO experiment provides a new method of studying the distant astronomical objects by merely analyzing the gravitational waves they produce. In 2017, the Nobel Prize in Physics was awarded to Rainer Weiss, Kip Throne and Barry C. Barish for their decisive contributions to the LIGO detector and the observation of gravitational waves. With this, every prediction of Einstein's General Theory of Relativity stands confirmed.

Both Newton's law of gravitation and Einstein's theory of relativity are classical theories of gravitation. The gravitational interaction is felt only when masses are large enough. The effects of gravity can be neglected when we deal with elementary particles and atoms. For elementary particles its strength is about 10^{40} times weaker than the strength of electromagnetic force between the same elementary particls. Of course, we expect the elementary particles chosen for comparison as charged. Contrary to the nature of Coulombian force between the charges which could be both attractive and repulsive, the gravitational force is only attractive, and again in contrast to electromagnetic forces, whose algebraic sum may be zero, gravitational forces between particles of two larger bodies would add up and provide strength to the attractive force which would make one larger body attract other. For sufficiently large number of matter particles, the gravitational forces may dominate over all other forces.

For the quantum mechanical theory of gravity, we have to think of the gravitational field. In this description, the force between two matter particles is carried by the quantum of gravitational field, which is called graviton, an uncharged massless particle having spin-2. The gravational force between say earth and moon is described as the exchange of gravitons between the particles that make the earth and the moon. Though the exchanged gravitons are virtual, they produce a measurable effect that the moon orbits the earth. The real gravitons make up the gravitational waves, whose physical existence, as described above, has been verified by the recent LIGO experiments. Because of the massless quantum of gravitational field, its range is infinite. The expression for range will be discussed under EM interactions. A complete satisfactory quantum theory of gravity has not yet been developed and it is assumed that at the elementary particle level, the gravitational interaction is too weak to play a role, hence just ignored.

1.9.2 Electromagnetic Interactions

The electromagnetic force interacts with electrically charged particles like electrons and quarks but it does not interact with uncharged particles such as graviton. Its strength is far greater than the gravitational interaction at the elementary particle level. The electromagnetic force between two electrons is about 10^{40} times stronger than the gravitational force between them. Contrary to gravitational force, electromagnetic force is both attractive and repulsive. If we consider the large bodies like sun, moon or earth, they contain almost equal number of positively and negatively charged particles. Hence attractive forces between unlike charges almost cancel the repulsive forces between like charges and the large body does not experience the electromagnetic force. However, at the molecular, atomic or subatomic levels, electromagnetic forces dominate. This is opposite to gravitational force, which dominates at the level of very large masses. It is the electromagnetic force at the atomic level that makes electrons revolve round the nucleus.

The physical theory that describes electromagnetic forces is called *electrodynamics*. Maxwell formulated it classically some hundred years ago. Maxwell's theory was consistent with the special theory of relativity.

The quantum of electromagnetic field is photon. Thus quantum mechanically electromagnetic interaction is caused by the exchange of photons, which like gravitons are uncharged massless particles. Only difference is that photons have spin 1. The electromagnetic attraction, e.g., is seen as the exchange of large number of photons. The photons that are exchanged are virtual. However, when an electron goes from one allowed orbit to another orbit, which is nearer to nucleus, energy is released and real photon is emitted which can be observed as electromagnetic radiation. Conversely, if a real photon collides with an atom, it may make an electron move, from the near to the nucleus orbit, to a farther orbit. In this process the energy of photon is used up and the photon gets absorbed. **The quantum theory of electromagnetic interactions is Quantum Electrodynamics (QED)**, which was developed, by Tomonaga, Feynman and Schwinger in 1940s. The strength of electromagnetic interaction is determined by α, the fine structure constant and the range of electromagnetic interaction like the gravitational interaction is infinite. This is because the quantum of electromagnetic interaction, which is photon, has zero rest mass. We derive in the following the relationship between the range of electromagnetic interaction and the energy of its quantum. The relationship can be applied to other interactions as well. Since the derivation would require Feynman diagrams we describe them first.

1.9.3 The Feynman Diagrams

The interaction between elementary particles can be pictured in the form of space–time diagrams called the Feynman diagrams. Let us see what is a space–time diagram.

Space has three coordinates (x, y, z,) and time has one (t). Let us imagine all the three space coordinates are represented as one that is shown along horizontal direction and time along vertical direction as shown in Fig. (1.6).

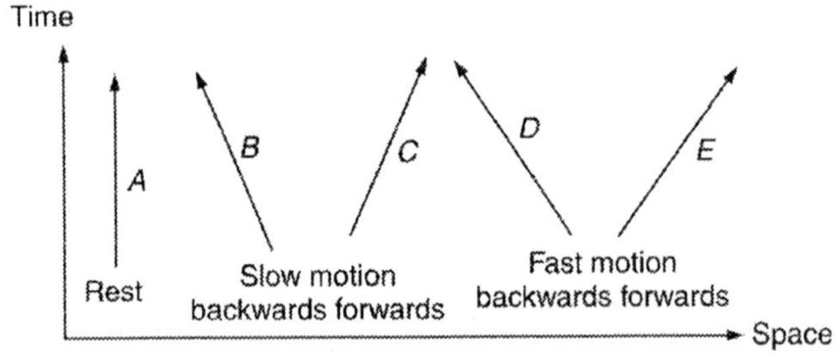

Fig. 1.6: World lines of particles

The particles can only move upwards in time but can move in forwards or backwards in space. A line on the space–time coordinate system shows the path of a particle. This line is called the world line of the particle. An arrowhead is put on the line. Now we look at the arrows A, B, C, D and E drawn in the space-time coordinate system (Fig.1.6).

Arrow A: It is parallel to time axis implying that this arrow has no component along space axis. Such a vertical arrow indicates that the particle moves through time only but not through space. Since, the motion through the space is the actual motion, the arrow A describes that the particle is at rest.

Arrow B: It is inclined; the inclination is leftwards or towards origin. This arrow indicates that the particle moves through space as well as through time. However, the motion through space is backwards.

Arrow C: The inclination of arrow is rightwards or away from origin. This indicates the forward motion of the particle.

Arrow D: It is similar to B but is more inclined towards space axis, implying that the particle now moves larger space in the same time as B. This means more inclined arrows like D indicate a fast backward motion of the particle.

Arrow E: It indicates fast forward motion of the particle.

We may now take up a physical example to draw and explain the space–time diagram. The example is electron–photon scattering and its space–time diagram is drawn ahead (Fig.1.7).

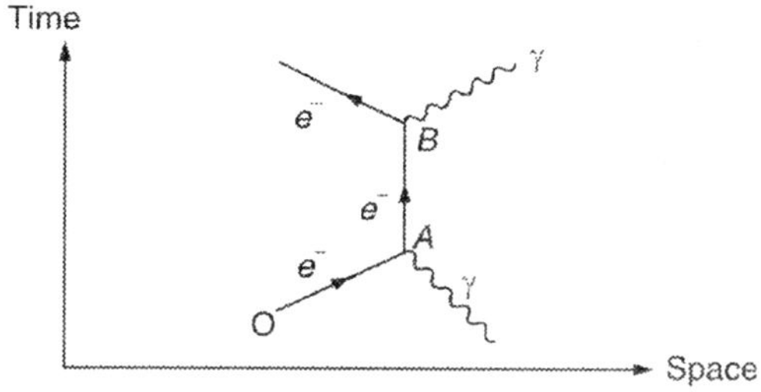

Fig. 1.7: Electron–photon scattering

In the diagram (e^-) denotes the electron, the thick line with arrowhead is the world line of electron and dashed line that of photon (γ). The diagram has to be read as: there takes place a collision of an electron and a photon at the space–time point A, the photon is absorbed by the electron. Now the velocity of electron changes, (AB has different inclination than OA) and this electron continues its path from A to B and at the space-time point B, the electron emits a photon and itself goes in reverse direction. This whole description is contained in the space–time diagram.

The quantum field theory describes all interactions through the exchange of bosons. In quantum field theory, all particle interactions can be pictured in space–time diagrams and a mathematical expression is associated with each diagram. From the mathematical expression one can calculate the probability of occurrence of the process corresponding to the diagram. Richard Feynman established in 1949, the exact correspondence between the diagrams and the mathematical expression, since then, the space–time diagrams are known as Feynman diagrams. In the Feynman diagrams the simplified version of space-time diagrams is used. The simplified space–time diagrams differ from the earlier ones in the following points:

In new space–time (Feynman) diagrams, time and space axes are generally not drawn on the diagram. It is implied that time flows forwards from the bottom to the top of the diagram and backwards from top to bottom. Arrowheads on the world lines do not indicate the direction of motion of the particle. Just as forward motion and backward motion in space are possible, it is taken that motion backwards through time is as much possible as the motion forwards through time. Generally, there is no arrowhead over the world line of a particle that is its own antiparticle. New interpretations are associated with the direction of arrowheads. If the arrowhead on the world line is upwards, it indicates that the particle is moving forwards in time. On the other hand, if the arrowhead is downwards, it can be interpreted either as the motion of the particle backwards in time or the motion of antiparticle forwards in time. Inclined world lines indicate motion through both space and time.

As an example, we draw here the *Feynman diagram* of a scattering process involving photon, electron and a positron (Fig. 1.8).

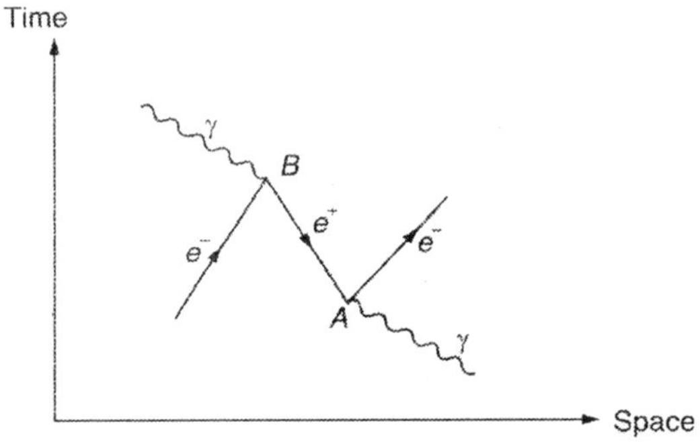

Fig. 1.8: Scattering process involving γ, e^-, e^+

Above diagram can be read as: we have an initial electron and an initial photon. The photon at the space–time 'A' creates an electron and positron pair. The electron created at 'A' flies to the right and positron to the left. The positron at the space–time point 'B' then gets annihilated by the initial electron and a photon is produced which flies off to left. The point 'A' is earlier with respect to time and point 'B' later. Another example is the interaction of two electrons (Fig. 1.9).

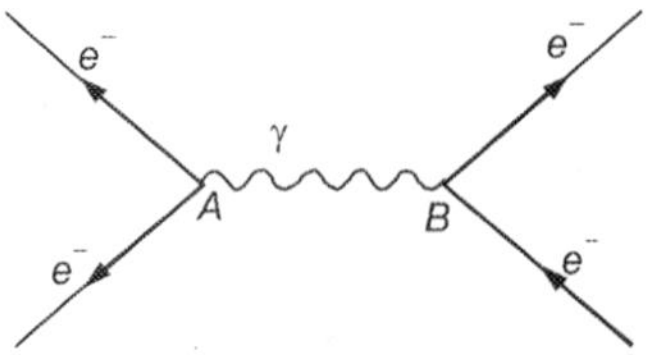

Fig. 1.9: Interaction of two electrons

In Fig. 1.9, the first electron is at 'A' and other is at 'B'. We may read the diagram as: the two electrons enter (the space–time) and photon is emitted by the electron at A and with the emission of photon, the electron at A is recoiled to the left. The emitted photon is absorbed by other electron at B, which then goes to right. It is equally probable that the photon was emitted at B, the second electron was recoiled to right, the photon was absorbed at A and the other (first) electron went to the left. Above diagram includes both the possibilities. It has to be noted that the photon involved is virtual and it will require energy from outside to become real, otherwise it will remain trapped between the electrons. At the space–time points A and B we have the following three world lines (Fig.1.10).

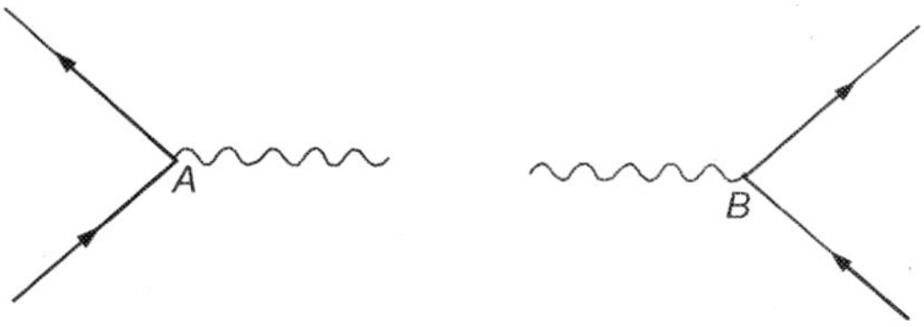

Fig. 1.10: The vertices A and B

Such points in the diagrams are called *vertices*. In the above diagram, we have two vertices. The first vertex is at A and other is at B.

The diagram (Fig. 1.9), in the classical theory describes the Coulomb repulsion and in quantum theory, i.e., quantum electrodynamics, the diagram represents the interaction mediated by the exchange of a photon. This is called $e^-e^- \rightarrow e^-e^-$ scattering or the Moller scattering. Yet another important example is $e^+e^- \rightarrow e^+e^-$ scattering, which is also called Bhabha scattering. The Feynman diagrams for Bhabha scattering are (Fig.1.11)

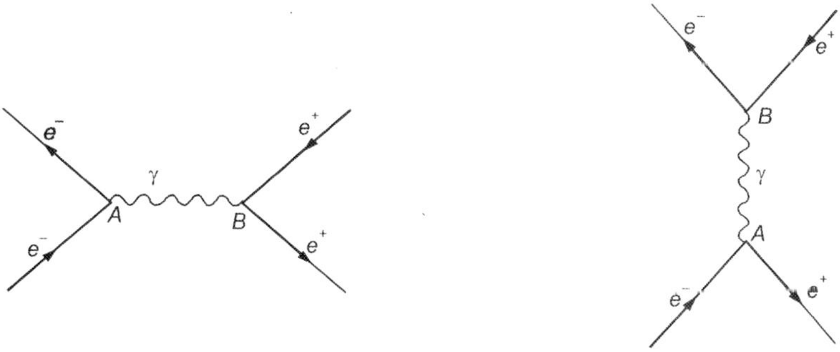

Fig. 1.11: Electron–positron scattering. (Bhabha scattering)

Both the diagrams drawn above are included in obtaining the cross section for Bhabha scattering. In the diagram (*a*) the photon exchanged between the vertices A and B is said to be spacelike and in diagram (*b*) the timelike photon gets exchanged. We may recall that the space axis is horizontal and time axis is vertical.

In the above, we have shown the Feynman diagrams with two vertices. Such diagrams are called the lowest order diagrams or the tree level diagrams. In a particular process, e.g., the Moller scattering ($e^-e^- \rightarrow e^-e^-$) the Fig. 1.9, is the tree level diagram. However, there are a number of diagrams with four vertices that contribute to a process. The following (Fig. 1.12) are the four–vertex Feynman diagrams for the Moller scattering.

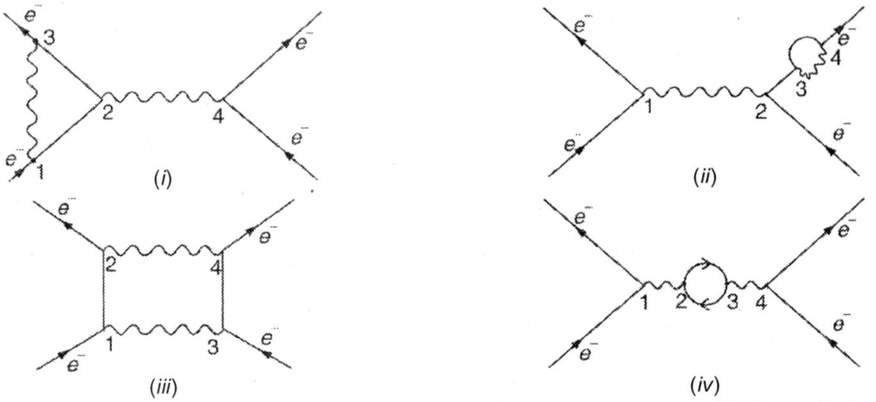

Figs. 1.12: [(*i*), (*ii*), (*iii*), (*iv*)] Four–vertex Feynmann diagrams $e^-e^- \rightarrow e^-e^-$ (Moller scattering)

In contrast to tree level Feynman diagram (Fig. 1.9), these are the *one loop Feynman diagrams* for the Moller scattering. If we want to describe, e.g., the Fig. (*iii*) above, we would say that first two electrons entered the reaction. At vertex 1, the electron emitted a virtual photon, the electron with lesser energy continued to vertex 2, there it again emitted a virtual photon and itself went to leftwards. The emitted virtual photons were absorbed at vertices 3 and 4 respectively by other electrons. Of course, the roles of emission and absorption by electrons are interchangeable. In figure (iv) the emitted virtual photon splits into an electron positron pair at vertex 2 and combines back into the virtual photon at vertex 3, which is absorbed at vertex 4. In making calculations using Feynman diagrams, we consider generally the tree level diagram and the four-vertex (one loop) diagrams and using Feynman rules calculate the contribution of each diagram. Each vertex within a diagram introduces a factor $e^2/4\pi\hbar c$ (the fine structure constant ($\approx 1/137$), which is small. Because of this small factor, the Feynman diagrams with more and more vertices make smaller and smaller contributions to the final result and thus higher vertices (vertices more than four) can be ignored. We generally do not require diagrams with more than four vertices in making calculations in quantum electrodynamics; the results though called approximate or quite accurate.

We have in the above talked briefly about the Feynman diagrams. The important point in all Feynman diagrams is the vertex. It is the point at which emission (of say photon, etc.) and absorbtion takes place. In other words, vertex is the point where the interaction takes place. For three interactions namely, weak, electromagnetic and strong, we denote vertices as below:

(A) Weak Interaction Vertices

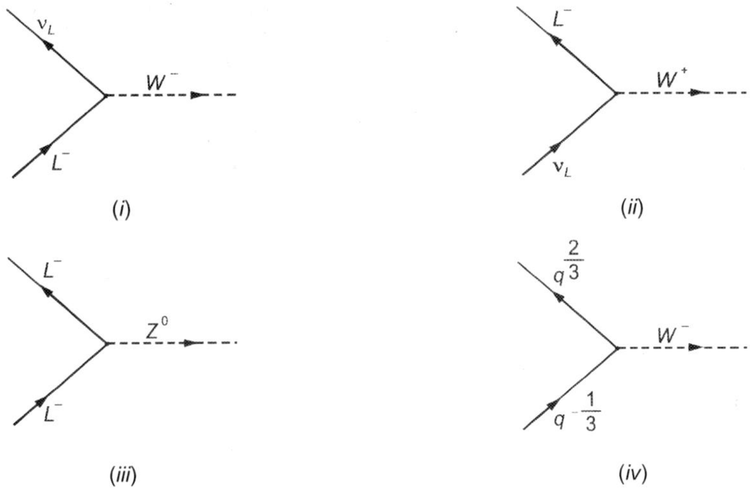

Figs. 1.13: Weak interaction vertices
L ⇒ charged lepton, q ⇒ quark.

Examples:

$(a)\ e^- + \nu_\mu \rightarrow \nu_e + \mu^-$

$(b)\ \nu_e + e^- \rightarrow e^- + \nu_e$

$(c)\ \nu_e + e^- \rightarrow \nu_e + e^-$

$(d)\ d + \nu_e \rightarrow u + e^-$

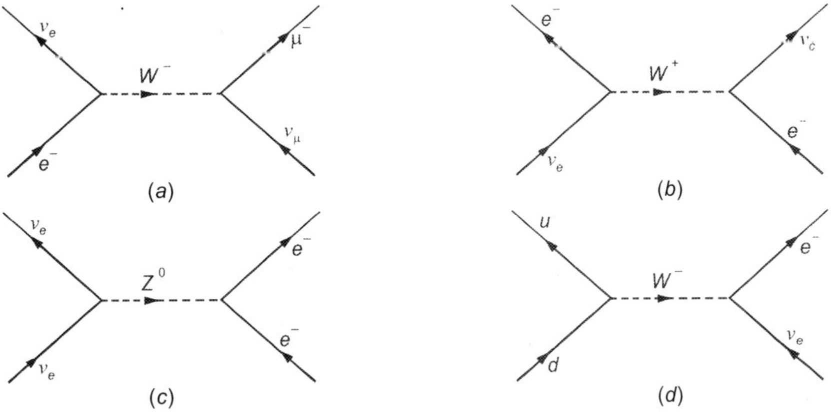

Figs. 1.14 [a, b, c, d]: Feynman diagrams showing weak interactions

In the above L stands for leptons, v_L for corresponding neutrino, $q^{\frac{2}{3}}$ is the quark with $+\frac{2}{3}e$ charge (up, charm or top quark); $q^{-\frac{1}{3}}$ is the quark with $-\frac{1}{3}e$ charge (down, strange or bottom quark). We may also have vertices involving quarks and neutral Z particle.

(B) Electromagnetic Interaction Vertices

The vertex for the electromagnetic interaction can have the following generalized form (Fig.1.15) in which $L^{\pm}$ represent the charged leptons. Examples of this type of vertex are found in Bhabha and Moller scatterings whose diagrams have already been drawn.

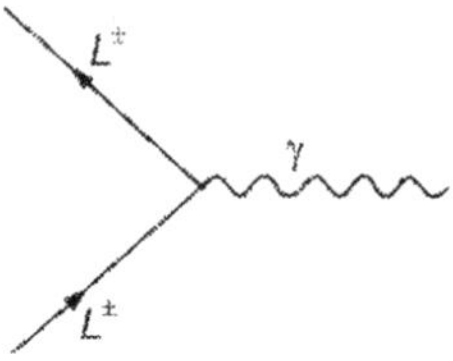

Fig. 1.15: Electromagnetic interaction vertex

Some of the other electromagnetic interaction vertices can be seen in the following interactions.

(a) $e^- + e^+ \rightarrow \gamma + \gamma$ (Pair annihilation)

(b) $\gamma + \gamma \rightarrow e^- + e^+$ (Pair creation)

(c) $\gamma + e^- \rightarrow e^- + \gamma$ (Compton scattering)

(d) $e + \mu \rightarrow e + \mu$ (Mott scattering)

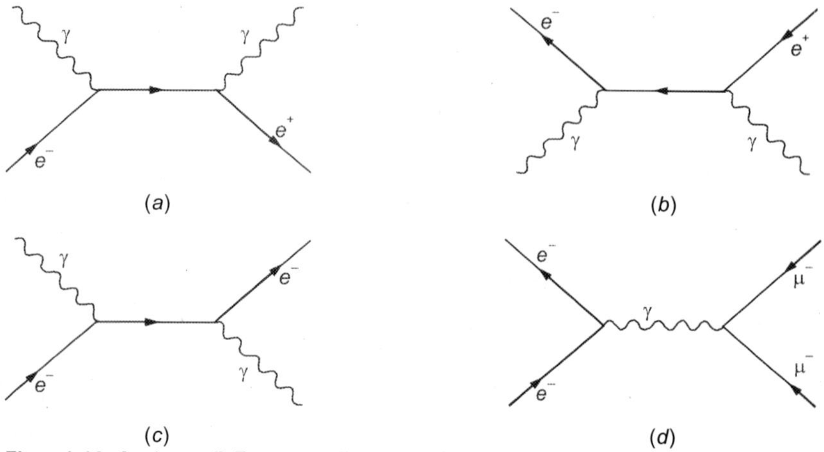

Figs. 1.16: [a, b, c, d] Feynman diagrams showing electromagnetic interactions

(C) Strong Interaction Vertices

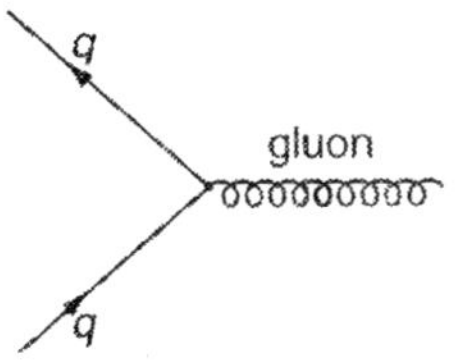

Fig. 1.17: Strong interaction vertex

Examples:

$$(a)\ u + d \rightarrow u + d$$
$$(b)\ u + \bar{d} \rightarrow u + \bar{d}$$

(a) (b)

Figs. 1.18: [a, b] Feynman diagrams showing strong interactions

where $\bar{d}$ is the antidown quark.

It may also be noted that, if each line of the Feynman diagram carries a definite momentum, the diagram is called the *momentum space Feynman diagram*. The momentum space diagrams offer a better description of a physical process that takes place between initially free particles. As an example, following figures (Fig. 1.19 a, b) show the electron–electron scattering diagrams for momentum space in lowest order.

(a) (b)

Figs. 1.19: [a, b,] Momentum space Feynman diagrams

1.9.4 The Range and Strength of Electromagnetic Interaction

Let us consider the following diagram (Fig. 1.20) as an example of electromagnetic interaction, which shows the binding of an electron and a proton via the virtual photon.

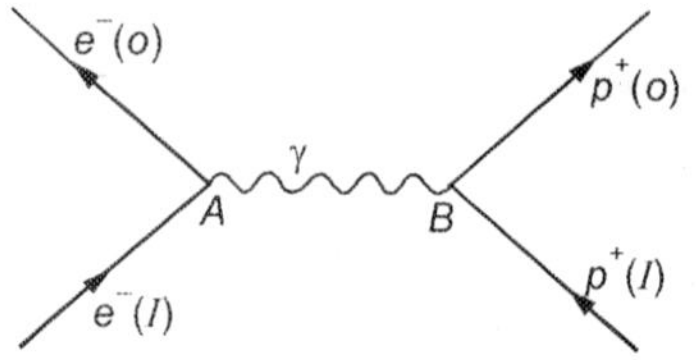

Fig. 1.20: Feynman diagram for electron–proton binding (I ⇒ incoming, O ⇒ outgoing)

Consider the vertex A. At vertex A, e^- i.e., the electron enters (indicated by I) the space–time and emits a virtual photon. The energy of the electron is diminished by the amount carried by the photon. This means the energy of electron before and after the emission of virtual photon is not equal. This loss of energy equal to say ΔE, is recovered when a virtual photon emitted by proton at vertex B reaches the electron, but then the energy of proton before and after the emission of virtual photon is not same. Thus, there is energy non-conservation at vertices A and B. Let this situation last up to Δt time. This means during the time Δt, there is energy non-conservation by an amount ΔE.

Now, ΔE and Δt are connected by the Heisenberg's uncertainty relation

$$\Delta E \Delta t \approx \hbar \tag{1.72}$$

The range of (electromagnetic) interaction is approximately

$$R \approx c\,\Delta t \tag{1.73}$$

where c is the photon speed. From equations (1.72) and (1.73), we get

$$R \cong \frac{c\hbar}{\Delta E} \tag{1.74}$$

From eqn. (1.74), we observe that when $\Delta E \to 0$; the range of the electromagnetic interaction tends to infinity. Now, ΔE is the energy taken away by the virtual photon, of course only for the time duration Δt, and since, virtual photons of arbitrary low amount of energy (implying $\Delta E \to 0$) can be created, we have the range of electromagnetic interactions as infinity or unlimited.

The strength of the electromagnetic interaction is characterized by the coupling constant

$$\alpha = \frac{e^2}{4\pi\hbar c} \tag{1.75}$$

where e is the magnitude of the charge of an electron. It is important to note that in fact; the charge e is the measure of the strength of electromagnetic interaction, i.e., the interaction of a charged particle with the electromagnetic field. However, the charge e, which is measured in Coulomb in SI units, is not dimensionless, hence not suitable for becoming a dimensionless coupling constant of electromagnetic interaction.

To obtain the dimensionless measure of the strength of electromagnetic interaction, the electrostatic potential energy (EPE) of repulsion between two electrons separated by a distance of $(\hbar/mc)$ is compared with the rest mass energy (RE) of electron. m is the rest mass of electron and c the speed of light in vacuum and $(\hbar/mc)$ is the Compton wavelength of the electron ($\approx 2.43\times10^{-12}$ m).

$$EPE = \frac{1}{4\pi\varepsilon_0}\frac{e^2}{(\hbar/mc)} \tag{1.76a}$$

$$RE = mc^2 \tag{1.76b}$$

$$\alpha = \frac{EPE}{RE} = \frac{1}{4\pi\varepsilon_0}\frac{e^2}{\hbar c} \tag{1.77}$$

Putting the numerical values in eqn. (1.77) as $\frac{1}{4\pi\varepsilon_0} = 9\times10^9\,\text{C}^2\text{N}^{-1}\text{m}^{-2}$, $e = 1.6\times10^{-19}$ C, $\hbar = 1.055\times10^{-34}$ Js and $c = 3\times10^8$ ms^{-1}, we get

$$\alpha = \frac{768}{1055\times100}\left(\frac{\text{C}^2}{\text{C}^2\text{N}^{-1}\text{m}^{-1}\text{J}}\right) \cong \frac{1}{137} \tag{1.78}$$

It has also to be noted that eqn. (1.77) is the expression for α in SI units. However, in general Heaviside–Lorentz system of units is employed for α in which the charge of electron e_{HL}, is related to e_{SI} as

$$e_{HL} = \frac{e_{SI}}{\sqrt{\varepsilon_0}} \tag{1.79}$$

and the expression for α is eqn. (1.75). The charge e in eqn. (1.75) is e_{HL}.

The coupling constant α is a measure of the probability of emission or absorption of photon from a charge. It is same as the fine structure constant, which determines the relative scale of fine structure in hydrogen. In Gaussian units of α, 4π disappears from eqn. (1.75).

1.9.5 The Strong Interactions

The strong interaction or the strong nuclear force is responsible for keeping protons and neutrons together in a nucleus. Since protons are charged and they experience Coulomb repulsive force, the strong force must be strong enough to keep protons together in a nucleus. That is the reason the nuclear force is called *the strong force*. Not only nucleons, but also the whole lot of baryons and mesons can be subjected to strong interactions and because of this mesons and baryons are called *hadrons*.

In any interaction a boson is exchanged between interacting particles. Photon, a boson, is the particle exchanged in electromagnetic interaction. In strong interactions, the exchanged boson is π-meson, K-meson or multimeson resonances. Just as hydrogen atom is an example of electromagnetic interactions, the similar example for strong interactions is provided by deuteron. Deuteron is the nucleus of deuterium atom and consists of one proton and one neutron.

The strong interaction between proton and neutron that keeps them inside the deuteron is shown below (Fig. 1.21).

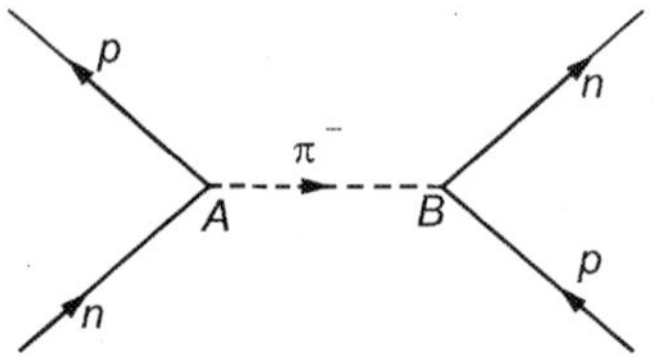

Fig. 1.21: Strong interaction between proton and neutron

The strength of the electromagnetic interaction, which is characterized by α, comes from the calculation of hydrogen atom ground state binding energy. In strong interactions, its strength would come from the calculation of ground state binding energy of deuteron. We compare the hydrogen atom and the deuteron and arrive at the strength of strong interaction.

The binding energy of the hydrogen atom ground state is

$$E_H = \frac{1}{2} mc^2 \alpha^2 \tag{1.80}$$

where mc^2 is the rest energy of electron in H-atom, α is the already introduced fine structure constant. E_H is the maximum energy absorbed at the electron vertex (Fig. 1.22).

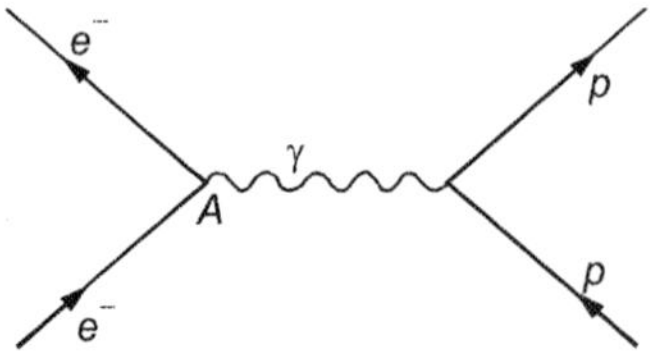

Fig. 1.22: Electron vertex (A) in electromagnetic interaction

Now consider the Fig. 1.21 for deuteron. The binding energy of the deuteron ground state can be written from eqn. (1.80), however, notice that the binding energy E_H is between two charged particles namely, the electron and proton. Then, if we consider that in Fig. 1.21 the neutron carried the negative charge and there were no strong interactions, Fig. 1.21 would be similar to Fig. (1.22), except that electron is replaced by neutron. When neutron is supposed to carry the negative charge and strong interactions turned off, electromagnetic interactions would take over and α would appear. Then the binding energy of deuteron would be

$$E_D = \frac{1}{2}\left(\frac{m_n c^2}{2}\right)\alpha^2 \tag{1.81}$$

where not m_n, the mass of neutron but the reduced mass of deuteron which is approximately equal to $\left(\frac{m_n}{2}\right)$ has been used. Putting the numerical values in eqn. (1.81) as $m_n c^2 \cong 10^3 \text{MeV}$, $\alpha^2 \cong 0.5 \times 10^{-4}$, we get

$$E_D = \frac{1}{8} \times 10^5 \text{eV} \cong 10^4 \text{eV} \tag{1.82}$$

However, the binding energy calculated from above which comes $\approx \frac{1}{100}$ MeV, does not match with the experimentally observed binding energy of about 2 MeV for deuteron ground state. Now we find out the cause of this discrepancy. The binding energy of the deuteron $\approx 10^4$ eV $= 10^{-2}$ MeV has been calculated by assuming that the coupling constant for *n-p* binding is also α, the same as for the electromagnetic interactions. The neutron was supposed to carry the charge hence use of α became justified. Now the value of $\alpha \approx 1/137$ does not give the observed value of binding energy, the coupling constant must have a value different from that of α. Recalling that in Heaviside–Lorentz units

$$\alpha = \frac{e^2}{4\pi\hbar c} \approx \frac{1}{137} \tag{1.83}$$

we may write the coupling constant for strong interactions as

$$\alpha_s = \frac{g^2}{4\pi\hbar c} \tag{1.84a}$$

whose value has to be determined. In eqn. (1.84a), g is the strong charge. In SI units

$$\alpha_s = \frac{1}{4\pi\varepsilon_{0g}} \frac{g^2}{\hbar c} \tag{1.84b}$$

where ε_{0g} is the dielectric constant of the vacuum with respect to strong interactions. Substituting equations (1.84) in eqn. (1.81) with α replaced by α_s, we get

$$E_D = \frac{1}{2}\left(\frac{m_n c^2}{2}\right)\alpha_s^2 \tag{1.85}$$

Putting E_D = 2MeV, $m_n c^2 \approx 10^3$MeV, we get

$$\alpha_s = \frac{1}{5}\sqrt{\frac{3}{10}} \approx 0.1 \tag{1.86}$$

implying that the coupling constant for strong interactions must be 0.1 instead of 1/137.

This value of strong coupling constant $\alpha_s = 0.1$ still requires improvement. In the diagram (Fig.1.21), π-meson is emitted at vertex A. π-meson is a massive particle and it absorbs at the vertex an energy equal to its rest energy, i.e., $m_\pi c^2 = 139$ MeV. This means at the vertex A, there is an energy non-conservation of $\Delta E = 139$ MeV. From the deuteron binding energy calculations, it is known that

$$\alpha_s \left(\frac{m_\pi}{2m_n} \right)^2 = 0.08 \tag{1.87}$$

from which using the values of m_π and m_n, we get

$$\alpha_s = \frac{g^2}{4\pi\hbar c} \approx 15 \tag{1.88}$$

Thus, when pion is the mediator between proton and neutron, the coupling constant for strong interactions is about 15 compared to 1/137 of electromagnetic interactions.

The range of strong interactions is calculated similar to that of electromagnetic. Using,

$$\Delta E\ \Delta t \approx \hbar \tag{1.89}$$

where $\Delta E = m_\pi c^2 \approx 140$ MeV, we get

$$\Delta t \approx \frac{\hbar}{140\ \text{MeV}} \tag{1.90}$$

Hence the range of strong interactions

$$R \approx v\Delta t \approx \frac{v\hbar}{140\ \text{MeV}} \tag{1.91}$$

Using $\hbar = 6.8 \times 10^{-22}\,\text{MeVs}$, $\text{v} \approx c = 3 \times 10^{10}\,\text{cm}$, we get

$$R \approx 1.4 \times 10^{-13}\ \text{cm} = 1.4\ \text{Fermi} \tag{1.92}$$

which is approximately equal to the radius of the nucleon.

Above description of strong interactions, which estimates the coupling constant for strong interactions approximately equal to 15, is based on the traditional description of interaction of nucleons via pions.

Modern description of strong interactions is based on the quark-gluon mechanism. The modern view of strong interaction is that, since the mediator (π-meson) as well as the interacting particles (n, p) have composite structures, the strong interaction must be defined in terms of more fundamental constituents of these particles. The more fundamental constituents, of which the hadrons are made of, are quarks. Among the hadrons, the baryons are made of three quarks and mesons are of a quark and an antiquark. The quarks are known to have six flavours. They are (up, down), (charm, strange) and (top, bottom). The up (u), charm (c) and top (t) quarks, each has an electric charge $+2/3e$ and each of down (d), strange (s) and bottom (b) quarks has electric charge $(-1/3e)$. The quarks also have to have a hypothetical quantum number called colour. Each quark exists in three colours, namely, Red (R), Blue (B) and Green (G). The three quarks that describe the composite structure of proton are $uud \equiv \frac{2}{3}e + \frac{2}{3}e - \frac{1}{3}e = +e$. The combination $udd \equiv \frac{2}{3}e - \frac{1}{3}e - \frac{1}{3}e = 0$, describes a neutron. The traditional mediator of strong interactions namely, the π-meson (π), which exists in three charge states, π^+, π^0 and π^- has quark and antiquark structures as $\pi^+ \rightarrow u\bar{d}$, $\pi^- \rightarrow \bar{u}d$ and $\pi^0 \rightarrow (u\bar{u} - d\bar{d}/\sqrt{2})$, where $\bar{u}$ is anti upquark (charge $-\frac{2}{3}e$) and $\bar{d}$ is antidown quark (charge $+\frac{1}{3}e$).

In view of the above, the force between neutron and proton in the deuterium nucleus, which is said to have been mediated by π-meson, is in fact the force between quarks mediated by massless particles called *gluons*. Thus, in this view, gluons not the π-mesons are the mediators of strong interaction. The strong interaction vertex is therefore like the one shown below (Fig. 1.23):

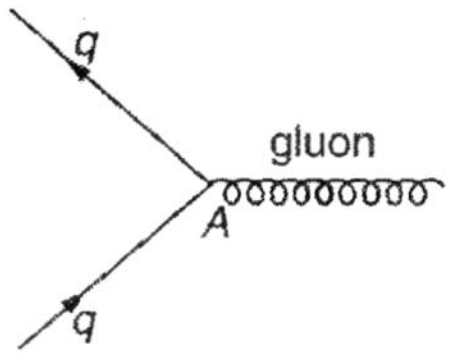

Fig. 1.23: Strong interaction vertex (A) of Fig. 1.21 with gluon emission

Unlike electromagnetic interactions in which the mediating particle, the photon, is uncharged, the gluon carries the charge. The charge carried by gluon is not electrical; it is the strong charge or the colour charge. In electromagnetic interactions, the electrical charge is conserved in the process, as well as at each interaction vertex of the corresponding Feynman diagram. It is the colour charge, which remains conserved at each of the strong interaction vertices. Since, the gluon emitted at the strong interaction vertex 'A' (Fig. 1.24) also carries colour, the out going quark will have a different colour than from the incoming. For example, in the Fig. (1.24), if the incoming quark is red and outgoing quark is blue, the gluon emitted at the vertex A would be red-antiblue. In fact each gluon carries one unit of colour and one unit of anticolour. The reverse of above interaction is that, when incoming red quark emits a red-antiblue gluon, the outgoing quark has blue colour. The reason that the gluon is bi-coloured (e.g., red plus antiblue) can be understood from the following. The interaction at the vertex A (Fig. 1.24) can be written as:

$$\textit{Incoming quark} \rightarrow \textit{Outgoing quark} + \textit{Gluon} \tag{1.93}$$

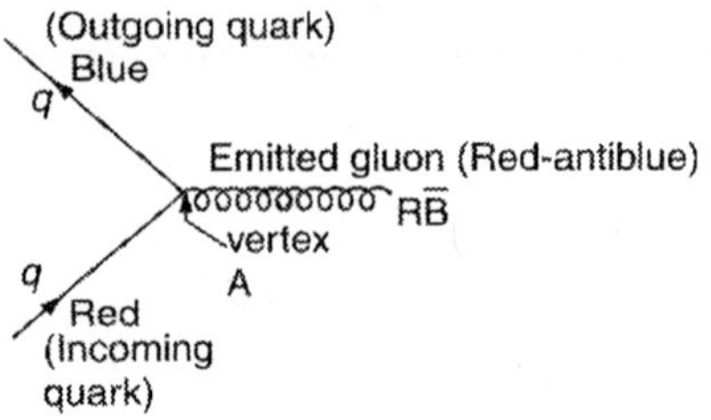

Fig. 1.24: Conservation of colour charge at strong interaction vertex A

Applying, colour conservation in the above equation with respect to the Fig. (1.24), we may write

$$\begin{array}{ccccc} Red & \rightarrow & Blue & + & (Red + Antiblue) \\ \text{(Incoming quark)} & \rightarrow & \text{(Outgoing quark)} & + & \text{(Gluon)} \end{array} \tag{1.94}$$

implying that

$$Red \rightarrow Blue \quad + Antiblue + Red \rightarrow Red \tag{1.95}$$

where blue and antiblue make a colourless combination. Eqn. (1.95) implies that to preserve the colour charge (*red* → *red*) the gluon had to have $R\overline{B}$ colour. Thus when two quarks interact, there is a flow of colour between them. The following figures (Fig.1.25) explain the colour flow.

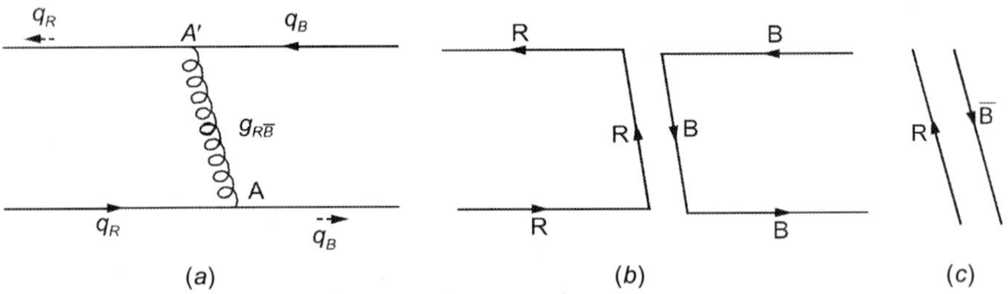

Fig. 1.25 (a, b, c): Colour flow during quark interaction

In Fig. (1.25a), a quark q_R of red colour moves from left to right and a blue quark q_B moves from right to left q_R emits at the vertex A the gluon $R\overline{B}$. As a result of $R\overline{B}$ gluon emission, the red quark has exchanged its colour with blue and has become blue quark beyond A. Similarly; the absorption of $R\overline{B}$ at A′ has converted q_B into q_R Fig. 1.25(b) is the colour flow diagram. The colour conservation equations at vertices A and A′ are

$$at\ \text{A} \rightarrow R \rightarrow (R + \overline{B}) + B \Rightarrow q_R \rightarrow g_{R\overline{B}} + q_B \tag{1.96}$$

$$at\ \text{A}' \rightarrow (R + \overline{B}) + B \rightarrow R \Rightarrow g_{R\overline{B}} + q_B \rightarrow q_R \tag{1.97}$$

In the above example, we chose a red quark and a blue quark. In fact, each quark can have three colours; *R*, *B* and *G*. Correspondingly, there are three anticolours $\overline{R}$, $\overline{B}$ and $\overline{G}$. The gluons can then have following nine colour-anticolour, i.e., bicolour, combinations

$$R\overline{R}, B\overline{B}, G\overline{G}, R\overline{B}, R\overline{G}, B\overline{R}, G\overline{R}, B\overline{G}, G\overline{B} \tag{1.98}$$

The colour wavefunctions of gluons are in general the linear combinations of these states. Accordingly, we can have in total nine gluon wavefunctions or gluon states. They are:

$$|g_1\rangle = \frac{1}{\sqrt{2}}(R\overline{B} + B\overline{R}),\ |g_2\rangle = \frac{-i}{\sqrt{2}}(R\overline{B} - B\overline{R})$$

$$|g_3\rangle = \frac{1}{\sqrt{2}}(R\overline{R} - B\overline{B}),\ |g_4\rangle = \frac{1}{\sqrt{2}}(R\overline{G} + G\overline{R})$$

$$|g_5\rangle = \frac{-i}{\sqrt{2}}(R\overline{G} - G\overline{R}),\ |g_6\rangle = \frac{1}{\sqrt{2}}(B\overline{G} + G\overline{B})$$

$$|g_7\rangle = \frac{-i}{\sqrt{2}}(B\overline{G} - G\overline{B}),\ |g_8\rangle = \frac{1}{\sqrt{6}}(R\overline{R} + B\overline{B} - 2G\overline{G}) \qquad (1.99)$$

and

$$|g_9\rangle = \frac{1}{\sqrt{3}}(R\overline{R} + B\overline{B} + G\overline{G})$$

where $|g_1\rangle, ..., |g_9\rangle$ are the gluon states. Equation (1.99) suggests that there must be nine gluons. However, there are only eight; one state in eqn. (1.99) must not be allowed. It is the ninth state namely, $|g_9\rangle$ because it is colour singlet. The gluons that are colour singlets do not play any role in strong interactions.

By colour singlet, it is meant that all the three colours red, blue, and green combine to form a colourless combination. Since, gluon is bicoloured; the totally symmetric combination of bicolours to form a colour singlet is the gluon state $|g_9\rangle$. Though, gluons $|g_3\rangle$ and $|g_8\rangle$ in eqn. (1.99) are colourless, they are not colour singlets.

All the particles, which are observed in nature, are colour singlets. The proton is a combination of three quarks namely *uud*. Because the proton is observed, net colour of *uud* must be so called *white*. The quarks must possess *RGB* (like $u_R u_G d_B$) combination to make proton colour singlet. Similar colour combinations can be applied to neutron, etc. According to above arguments then, the gluon $|g_9\rangle$ must be observed in nature. Since an individual quark is coloured i.e., it is not colour singlet, it cannot be observed, only proton and neutron (e.g. in deuteron) can be observed. Again, the proton and neutron are colour singlets and when they interact they will emit and absorb colour singlets only and $|g_9\rangle$ is that colour singlet. This colour singlet may be the pion, the mediator of conventional description of strong interactions.

Therefore, we can say that the strong interactions take place through the quark–gluon mechanism and nucleon–nucleon interaction via the exchange of virtual mesons is the observable form of this mechanism. Because of the involvement of colour in quark–gluon mechanism the quark theory of strong interactions is called *Quantum Chromodynamics* (QCD).

1.9.6 The Weak Interactions

The weak interactions or the weak nuclear force is responsible for the slow (lifetime $\geq 10^{-10}$ s) decays of unstable particles. A typical example of weak interaction is the nuclear β decay.

$$n \to p^+ + e^- + \bar{\nu}_e \tag{1.100}$$

Another example is

$$\pi^- \to \mu^- + \bar{\nu}_\mu \tag{1.101}$$

In the above examples only the particles $e^-, \bar{\nu}_e$; $\mu^-, \bar{\nu}_\mu$ are leptons and others are not, hence such weak decays are also called *Semileptonic decays*. The weak decays like

$$\mu^- \to e^- + \bar{\nu}_e + \nu_\mu \tag{1.102}$$

are called *leptonic decays*. There are weak decays, which do not involve leptons. Such weak decays are called *non-leptonic decays*. The examples are:

$$\begin{aligned} \Sigma^- &\to n + \pi^- \\ \Sigma^+ &\to p + \pi^0 \\ &\to n + \pi^+ \end{aligned} \tag{1.103}$$

The lifetime of weak decays $\tau_{\text{weak}} \approx 10^{-10}$ s 90° is long enough compared to $\tau_{\text{em}} \approx 10^{-16}$ s and $\tau_{\text{strong}} \approx 10^{-23}$ s. The reason is that in weak decays the difference between the masses of the decaying particle and the decay products is quite small. Larger the difference in masses quicker the decay takes place.

The strength of weak interaction can also be expressed in terms of its coupling constant. However, unlike electromagnetic and strong interactions, obtaining the dimensionless constant for weak interactions requires some manipulation. This is because in studying the weak interactions Fermi's constant G is used. It is the coupling constant for β decay, which Fermi used while formulating the theory of β decay. G is equal to 1.4×10^{-49} erg cm^3 = 1.4×10^{-62} Jm^3 $\approx 9 \times 10^{-50}$ MeV m^3. Since β decay process involves the interaction of four fermion fields, it differs from *EM* and strong interactions where in general two fermion and one-boson fields are involved. This demands manipulation in arriving at the dimensionless constant, which would characterize the weak interaction.

In deriving the dimensionless coupling constant for electromagnetic interaction the electrostatic potential energy of repulsion of two electrons separated by a distance equal to Compton length of electron was compared with the rest energy of electron. In weak interaction too, we need a length. Keeping in view the β decay process and the Fermi constant *G,* the length, which is utilized, is the Compton wavelength of pion which is

$$\frac{\hbar}{m_\pi c} = 1.4 \times 10^{-15}\,\text{m} \tag{1.104a}$$

Using the value of Fermi constant

$$G = 9 \times 10^{-50}\ \text{MeV m}^3$$
$$\hbar = 6.58 \times 10^{-22}\ \text{MeV s}$$

and
$$c = 3 \times 10^8\ \text{m s}^{-1} \tag{1.104b}$$

it can be calculated that

$$2.3 \times 10^{-7}\hbar c\left(\frac{\hbar}{m_\pi c}\right)^2 = 9 \times 10^{-50}\,\text{MeV m}^3 \tag{1.105}$$

which on using eqn. (1.100b) can be rewritten as

$$G = 2.3 \times 10^{-7}\hbar c\left(\frac{\hbar}{m_\pi c}\right)^2 \tag{1.106}$$

from which we may write

$$G^2(\hbar c)^{-2}\left(\frac{\hbar}{m_\pi c}\right)^{-4} = 5.29 \times 10^{-14} \tag{1.107}$$

Since RHS of eqn. (1.103) is dimensionless, the LHS is obviously dimensionless. Thus LHS of eqn. (1.103) provides the dimensionless constant characterizing the weak interaction. Its value is of the order of 10^{-14} as compared to 1/137 for the dimensionless constant for *EM* interaction and about 15 for strong interactions. Obviously, a too small value $\approx$ (10^{-14}) of the coupling constant implies that the interaction is weak. It may be noted that even if we use the Compton wavelength of some heavier particle in eqn. (1.103) the value of dimensionless constant for weak interaction would still be quite small compared to the constants for *EM* and strong interactions.

Just as electric charge e is associated with electromagnetic interaction and a strong charge g with strong interaction, the weak charge associated with weak interactions can be seen from eqn. (1.102). It is $\sqrt{G}$, because eqn. (1.102) can be written as

$$\frac{\left(\sqrt{G}\right)^2}{4\pi\hbar c} = 2.3 \times 10^{-7}\frac{1}{4\pi}\left(\frac{\hbar}{m_\pi c}\right)^2 \tag{1.108}$$

of which the LHS resembles with $\frac{e^2}{4\pi\hbar c}$ and $\frac{g^2}{4\pi\hbar c}$ of *EM* and strong interactions respectively.

The range of weak interaction can be estimated by considering that the intermediate vector bosons namely W^+, W^- and Z^0 are the mediators of weak interactions. The mass of W bosons is about 80 GeVC^{-2}. We may take for approximate calculations that the mass of the mediators of the weak interactions is about $100\text{GeVC}^{-2} = 10^5\text{GeVC}^{-2}$.

For the beta decay process namely, $n \to p^+ + e^- + \overline{\nu}_e$ we may draw the following Feynman diagram in which neutron and proton have been drawn in terms of their quark compositions $p \to uud$, $n \to udd$

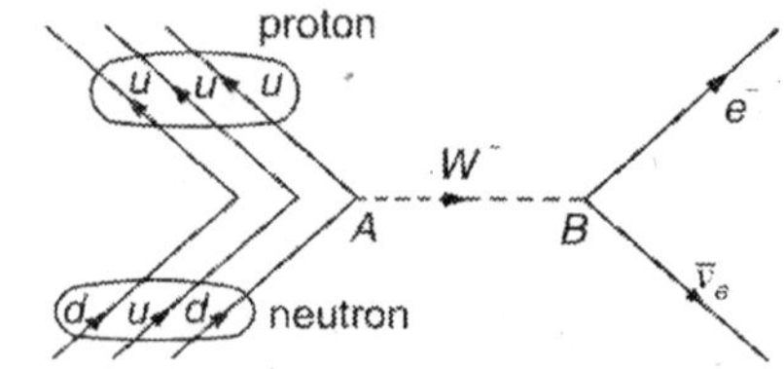

Fig. 1.26: Beta decay process at quark level

It has to be noted that when the W boson is emitted the flavour of the quark changes. In Fig. 1.26 at vertex A, a W^- boson (charge $-e$) is emitted and conservation of electrical charge at A gives $d \to u + W^- \Rightarrow -1/3e - e$ i.e., at vertex A, the outgoing quark is u whereas incoming was d quark. The colour of both incoming and outgoing quarks would however remain same. Because of the change of flavour the weak interaction theory is called *flavourdynamics*. At vertex A flavour is not conserved. If W^- particle carried the flavour, the flavour could be conserved but W^- cannot carry it because it is coupling at vertex B with leptons, which have no flavour. Thus in essence the β decay is

$$d \to u + e^- + \overline{\nu}_e \tag{1.109}$$

and W^- is the mediator.

The energy non-conservation of $\Delta E = m_W c^2$ $\Delta E = m_W c^2$ would take place at vertex A and last for time Δt with

$$\Delta E\ \Delta t \approx \hbar \tag{1.110}$$

The range of weak interaction

$$R \approx v\Delta t \approx \frac{v\hbar}{\Delta E} \approx \frac{v\hbar}{m_W\, c^2} \tag{1.111}$$

Taking $v \approx c$, $m_W c^2 \approx 10^5$ MeV, $\hbar = 6.58 \times 10^{-22}$MeV s, the range of weak interactions becomes

$$R_{\text{weak}} \approx 10^{-19} \text{ m} \tag{1.112}$$

This is too small, indicating that weak interactions have a range shorter than the strong interactions.

The decay (1.105) by applying the rules of crossing symmetry is

$$d + \nu_e \to u + e^- \tag{1.113}$$

The Feynman diagram for eqn. (1.109) is (Fig.1.27)

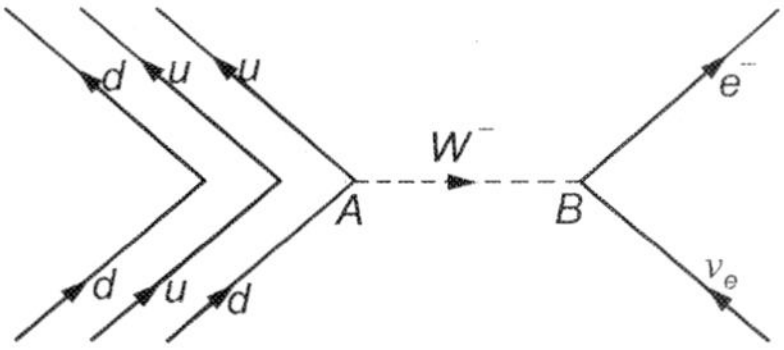

Fig. 1.27: Crossing symmetry equivalent of Fig. 1.26

We can consider either Fig. (1.26) or Fig. (1.27) to understand the change of flavour in β decay process.

1.10 CONSERVATION LAWS

The decay of elementary particles is determined by various conservation laws. We summarise here the various conservation laws, which determine the decay of elementary particles and also point out how the various conservation laws apply to three interactions.

(1) ***Energy and Momentum:*** In special theory of relativity, energy and momentum form a four-vector p^μ, the $\mu = 0$ component is energy and p^i ($i = 1, 2, 3$) form the momentum. The four-momentum p^μ remains the same before and after the decay process. This is called the energy and momentum conservation law. Due to this conservation law, a particle decays into lighter particles.

(2) ***Angular Momentum:*** Invariance under rotational symmetry leads to this conservation law. The total angular momentum of an isolated system remains conserved. The total angular momentum implies the vector sum of orbital and intrinsic angular momenta of the particles.

Above conservation laws are also called *kinematic conservation laws*. These apply to all the interactions, strong, *EM* and weak. However, there are dynamical conservation laws who do not uniformly apply to all the interactions. Following conservation laws may be attributed to this category.

(3) ***Electric charge Q:*** Every elementary particle or the short-lived state obeys this conservation law. The electric charge, before and after decay, (i.e. an interaction) must remain same. All the three interactions conserve electric charge.

(4) ***Electron number L_e:*** The leptons like electron, i.e., e^-, ν_e and their antiparticles $e^+, \overline{\nu}_e$ are associated with a number called electron number or lepton number L_e. e^-, ν_e have each $L_e = +1$ and $e^+, \overline{\nu}_e$ have $L_e = -1$. In any reaction the electron number L_e must be same before and after the reaction. Other leptons, i.e. $\mu^\mp$, ν_μ, $\overline{\nu}_\mu$ will have $L_e = 0$.

(5) ***Muon number L_μ:*** The leptons μ^-, ν_μ have muon number +1 and their antiparticles $\mu^+, \overline{\nu}_\mu$ have muon number −1. Obviously L_e for $\mu^-, \mu^+, \nu_\mu, \overline{\nu}_\mu$ will be zero and L_μ for e^-, e^+, ν_e and $\overline{\nu}_e$ will be zero. The muon number like the electron number must be conserved.

(6) *Tau number L_τ:* The tau (τ) lepton and its neutrino carry the tau number. τ^-, ν^-_τ have $L_\tau = +1$, and τ^+, $\bar{\nu}_\tau$ have $L_\tau = -1$. For τ lepton and its neutrino L_e and $L_\mu = 0$. The tau number must also be conserved in any decay.

The strong interactions do not touch the leptons. In an electromagnetic interaction a photon is emitted, the outgoing particle is same as the incoming and in the weak interactions only the leptons of same generations get mixed, i.e., $e^- \rightarrow \nu_e + W^-$, $\mu^- \rightarrow \nu_\mu + W^-$, $\tau^- \rightarrow \nu_\tau + W^-$ and there are no relations like $e^- \rightarrow \nu_\mu + W^-$ etc. Hence L_e, L_μ and L_τ all are conserved in all the three interactions. It may be noted that the conservations of numbers L_e, L_μ and L_τ are related to three generations of leptons namely, first generation leptons (e, ν_e), second-generation leptons (μ, ν_μ) and the third generation leptons (τ, ν_τ). Likewise, we have three generations of quarks first (u, d), second (c, s) and third (t, b), but there is no analogue of L_e, L_μ and L_τ conservation laws with respect to quarks. This is because cross-generational relations like $s \rightarrow u + W^-$ take place among quarks. Here s belongs to second generation of quarks and u to first. This is in contrast to a cross-generational relation between leptons, e.g., $e^- \rightarrow \nu_\mu + W^-$ which does not take place.

(7) *Baryon number B:* In any reaction or decay involving baryons, the baryon number B must be conserved. $p \rightarrow e^+ + \gamma$ is forbidden because of baryon number non-conservation. All baryons carry $B = +1$ and all antibaryons have $B = -1$. In terms of quarks we can associate a quark number of +1 with each quark and quark number –1 to each antiquark. A baryon, which consists of three quarks, will then have quark number +3 and for antibaryons the quark number will be –3. While describing the interactions through quarks, we may use the quark number conservation. Since quarks are not seen, the baryon number conservation replaces quark number conservation.

In contrast to baryons, mesons carry zero quark number, as they are made of a quark and an antiquark. There is no meson number conservation law and as such any number of mesons can be produced in a decay provided the energy conservation takes place.

(8) *Strangeness S:* The strangeness quantum number was introduced to account for the unusually long lifetime of the hyperons. We have talked at length under section (1.7) on strangeness. Many decays which, despite obeying all other conservation laws do not take place are explained on the basis of non-conservation of strangeness quantum number. For example,

$$\pi^- + p^+ \rightarrow \pi^0 + \Lambda \tag{1.114}$$

does not take place because it violates the conservation of strangeness quantum number. Λ is the strange particle with $S = -1$ whereas other particles in the above reaction are non-strange particles with $S = 0$. Strong and electromagnetic interactions conserve the strangeness while weak interactions do not.

The baryon number and strangeness combine to form

$$Y = B + S \quad (1.115)$$

This is called the hypercharge.

(9) *Isospin I and I_3:* About I and I_3 we have talked under the subsection *The Years of Resonances*. The isospin, which is also called isotopic spin, makes out that proton and neutron are the two charge states of a single particle called nucleon. I for nucleon is 1/2 and when its third component I_3 has the value +1/2, the nucleon is called proton and when $I_3 = -1/2$, the nucleon becomes a neutron. The value of I_3 therefore provides charge to the particles. For example, the relation for the charge of a particle was

$$Q = \left(I_3 + \frac{B}{2} \right) e \quad (1.116)$$

where B is the baryon number of the particle and e is the magnitude of electric charge. Above relation failed when applied to Λ particle which is a baryon ($B = +1$) with $I = 0$ and $I_3 = 0$ and charge Q also zero. This was another reason to introduce strangeness quantum number and also to define hypercharge $Y = B + S$. S for $\Lambda = -1$. Using eqn. (1.116) the charge of a particle is given by

$$Q = \left(I_3 + \frac{Y}{2} \right) e \equiv \left(I_3 + \frac{B+S}{2} \right) e \quad (1.117a)$$

Among the mesons, π, ρ, and A_2 particles have $I = 1$ and I_3 for each is +1, 0, –1. Mesons $K^+ \left(I_3 = +\frac{1}{2} \right)$, $K^- \left(I_3 = -\frac{1}{2} \right)$ have $S = +1$ and –1 respectively. Baryon number B for mesons is obviously zero. It can be noted that except for baryons, hypercharge is essentially the same as strangeness and hence for mesons eqn. (1.117a) is

$$Q = \left(I_3 + \frac{S}{2} \right) e \quad (1.117b)$$

Let us look at the relation (1.117a) from the interaction point of view. Since Q, B and S are all conserved by electromagnetic interactions, (1.117a) implies that I_3 must also be conserved for electromagnetic interactions. Since S is not conserved for weak interactions, B and Q have to be conserved and thus I_3 will not be conserved for weak interactions. The conservation of I_3, in for example electromagnetic interactions does not imply that I, i.e., total isospin is also conserved. I_1 and I_2, the two other components of I are not conserved. An example is $\pi^0 \to \gamma + \gamma$, this decay takes place, LHS has $I = 1$ and RHS has $I = 0$. I_3 of LHS is zero ($I_3 = +1(\pi^+)$, $I_3 = 0(\pi^0)$, $I_3 = -1(\pi^-)$) and so is of RHS. I_3 is conserved but not the I.

Isospin is not a relevant quantity for leptons. We can say that isospin of leptons is zero. The strong interactions, which are nearly independent of charge, obey the I-spin conservation law and as seen above electromagnetic and weak interactions do not obey isospin (I) conservation law, the electromagnetic interactions obey I_3, which provides charge (Q) conservation law. In later Chapters, the symbol (T) has also been used to denote isospin.

(10) *Charge Conjugation:* Charge conjugation operation is a discrete symmetry operation. In this operation a particle is changed into its antiparticle and the dynamical variables of the particle namely, the momentum, spin remain unchanged. If the original particle is positively charged, the charge conjugation operation or C-operation would change it into a negatively charged particle having momentum, spin, etc. that of original positively charged particle. If a reaction and its charge conjugate satisfy the same physical laws or behave in a similar manner, the reaction is said to be invariant under charge conjugation. Invariance under charge conjugation is also called *the conservation of C-parity*. Thus when a reaction involving particles is replaced by one involving corresponding antiparticles and both the reactions represent physical processes, we call C-parity of the reaction is conserved.

If a particle state is denoted by $|Q, B, L, S, ..., p, \sigma, ...\rangle$, where Q, denotes the charge on the particle, B its baryon number, L the lepton number and S the strangeness quantum number. Dots indicate other quantum numbers. p and σ denote the momentum and spin of the particle. Thus, C-operation will do the following:

$$C|Q, B, L, S, ..., p, \sigma, ...\rangle = |\bar{Q}, \bar{B}, \bar{L}, \bar{S}, ..., p, \sigma, ...\rangle \tag{1.118}$$

where bar denote the quantities for antiparticles. Notice that C-operation does not change p, σ... Thus under C-operation following changes take place ($\bar{Q} = -Q$, etc.).

$$\begin{aligned} Q &\rightarrow -Q \\ B &\rightarrow -B \\ L &\rightarrow -L \\ S &\rightarrow -S \end{aligned} \tag{1.119}$$

Denoting

$$|Q, B, L, S, ..., p, \sigma, ...\rangle = |particle\rangle \tag{1.120a}$$

and

$$|\bar{Q}, \bar{B}, \bar{L}, \bar{S}, ..., p, \sigma, ...\rangle = |antiparticle\rangle \tag{1.120b}$$

we may write

$$C|particle\rangle = |antiparticle\rangle \tag{1.121a}$$

and

$$C^2|particle\rangle = C|antiparticle\rangle = |particle\rangle \tag{1.121b}$$

implying that

$$C^2 | particle\rangle = | particle\rangle \tag{1.122}$$

which suggests

$$C^2 = I \tag{1.123}$$

where I is the unit operator. Eqn. (1.119) tells that the C-operator has eigenvalues (+1) and (–1) only. Thus eqn. (1.117a) reads

$$C| particle\rangle = \pm| antiparticle\rangle \tag{1.124}$$

It is easy to see that in eqn. (1.120), if the particle state and the antiparticle state are same, then it can be written as

$$C| particle\rangle = \pm| particle\rangle \tag{1.125}$$

which in turn is an eigenvalue equation. Thus eqn. (1.120) becomes an eigenvalue equation only when there is no difference between particle state and the antiparticle state. When this condition is obeyed the particle state under C-operation may remain the same state without changing the sign (+1 eigenvalue) or remain the same state with the change of sign (–1 eigenvalue). This change of sign justifies the use of the C-parity and also underlines the point that the C-parity is associated only with those particle states, which are their own antiparticles. The particles, which are their own antiparticles, are photon, π^0, η, η', ϕ, ω, ψ etc. mesons. The particle-antiparticle combination like e^+e^-, $\bar{p}p$ can also be the eigenstates of C. In contrast for other particles, say π^+ we have

$$C|\pi^+\rangle = |\pi^-\rangle \tag{1.126}$$

which is not an eigenvalue equation and C-parity making no sense for such particles. Only those particles would be the eigenstates of C for which $Q = B = L = S = 0$.

In the eigenvalue equation (1.121), plus sign denotes even C-parity and minus sign the odd C-parity. Since photons are the quanta of electromagnetic field whose sources are electrons, which change the sign of charge under C-operation, the C-parity of photon is therefore odd, i.e.

$$C|\text{photon}\rangle = -|\text{photon}\rangle \tag{1.127}$$

C is a multiplicative quantum number. The successive operation of one C after another implies $C \times C = C^2$ operation, not $C + C = 2C$ operation. Since

$$\pi^0 \rightarrow \gamma + \gamma = 2\gamma \tag{1.128}$$

$$C\text{-parity of } \pi^0 = C\text{-parity of } 2\gamma \text{ state} \tag{1.129a}$$

$$C\text{-parity of } 2\gamma \text{ state} = (C\text{-parity of one } \gamma\text{-state}) \times C\text{-parity of other } \gamma \text{ state})$$

$$= (1) \times (-1) = +1 \tag{1.129b}$$

implying

$$C|\pi^0\rangle = +|\pi^0\rangle \tag{1.130}$$

Similarly from

$$\omega^0 \rightarrow \pi^0 + \gamma \tag{1.131a}$$

$$C\text{-parity of } \omega^0 = (+1)\times(-1) = \text{odd} \tag{1.131b}$$

implying

$$C\left|\omega^0\right\rangle = -\left|\omega^0\right\rangle \tag{1.132}$$

The decay products help us to determine the C-parity of decaying particle.

Both the electromagnetic and strong interactions are invariant under charge conjugation but weak interactions are not. The violation of C-invariance in weak interactions can be understood by applying C-operation on a neutrino state. Remembering that a neutrino is a left-handed particle, the C-operation changes a neutrino into a left-handed antineutrino and since there is no left-handed antineutrino in nature, all antineutrinos are right-handed, the process

$$C\left|neutrino\right\rangle = \left|left\text{-}handed\text{-}antineutrino,\ which\ does\ not\ exist\right\rangle \tag{1.133}$$

is unphysical, hence not invariant under C. Thus leptonic and semileptonic weak decays where neutrinos or antineutrinos are present violate the C-invariance. Non-leptonic weak interactions too show violations of C.

(11) *Parity:* The operator, which describes the behaviour of a state vector $\left|\psi(x,y,z,t)\right\rangle$ under inversion of all space coordinates, is called *the parity operator* P and the eigenvalue of P is called *the parity of the state*. The state $\left|\psi(\boldsymbol{r},t)\right\rangle$ represents a system.

The inversion of all space coordinates namely, $x \to -x, y \to -y$ and $z \to -z$ is called the space inversion. The parity operator changes $\left|\psi(\boldsymbol{r},t)\right\rangle$ with $\boldsymbol{r} = (x, y, z)$ into $\left|\psi(-\boldsymbol{r},t)\right\rangle$, i.e.

$$P\left|\psi(\boldsymbol{r},t)\right\rangle \to \left|\psi(-\boldsymbol{r},t)\right\rangle = \alpha\left|\psi(\boldsymbol{r},t)\right\rangle \tag{1.134}$$

where α is a constant. A further operation of P would bring the system back to the original one, implying

$$P\left|P\left|\psi(\boldsymbol{r},t)\right\rangle\right\rangle \to P^2\left|\psi(\boldsymbol{r},t)\right\rangle = \alpha P\left|\psi(\boldsymbol{r},t)\right\rangle = \alpha^2\left|\psi(\boldsymbol{r},t)\right\rangle \tag{1.135}$$

Since, $\alpha^2\left|\psi(\boldsymbol{r},t)\right\rangle$ is the original state $\left|\psi(\boldsymbol{r},t)\right\rangle$, the constant

$$\alpha^2 = 1 \Rightarrow \alpha = \pm 1 \tag{1.136}$$

α is called the parity of the state $\left|\psi(\boldsymbol{r},t)\right\rangle$. If $\alpha = +1$, the state $\left|\psi(\boldsymbol{r},t)\right\rangle$ is said to have even parity, and $\alpha = -1$ denotes the state of odd parity.

The space inversion is equivalent to reflection in a plane followed by a rotation of 180° about an axis perpendicular to the plane. As a result, a right handed coordinate system is transformed into a left-handed system and since, no continuous transformation can convert a left-handed frame of reference to right-handed one, space inversion is a discrete transformation and parity is a multiplicative quantum number.

We may talk of two kinds of parities. One, associated with the spatial configuration and the other intrinsic. The parity associated with spatial configuration is described by the angular momentum (*l*) value and the intrinsic is one that we inherently associate with a particle. The relative intrinsic parity of a fermion-antifermion pair is negative. For examples, the particles (fermions)

$$p,\ n,\ \Lambda,\ \Sigma^+,\ \Sigma^0,\ \Sigma^-,\ \text{etc.}$$

have positive intrinsic parity, whereas their antiparticles

$$\bar{p},\ \bar{n},\ \bar{\Lambda},\ \bar{\Sigma}^+,\ \bar{\Sigma}^0,\ \bar{\Sigma}^-,\ \text{etc.}$$

have negative intrinsic parities. The quarks, which are also fermions, have positive intrinsic parity and antiquarks have negative.

The hadrons (mesons and baryons) are composite structures, for example, mesons consist of quarks and antiquarks. The net parity of the hadron, say the meson, would be the product of spatial parity and the intrinsic parity of the constituents. The pions are considered to consist of quark-antiquark pair in a state of relative angular momentum $l = 0$. Thus the parity of pion would be $(-1)^l(+1)(-1)$ with $l = 0$ which is (-1) or odd. The meson in its ground state would have odd or negative parity. This (–1) parity of mesons can be called their intrinsic parity. For excited states of mesons, $l \neq 0$ and the parity would be $(-1)^l(-1) = (-1)^{l+1}$. The baryons are made of three quarks, each having positive intrinsic parity, the parity of baryon in ground state would be $(+1)^3 = +1$ and in an excited state $(-1)^l(+1) = (-1)^l$. Contrary to a fermion, the particle and anti-particle of a boson have same parity.

The strong and electromagnetic interactions are found to be invariant under the parity operation. The weak interaction is not invariant under *P*. In a theory where invariance of parity is observed, we can never add a vector to a pseudovector. Vectors, which are also called *polar vectors*, like ***r***, ***p*** (linear momentum) ***E*** (electric field) change sign, i.e., $\boldsymbol{r} \to -\boldsymbol{r}$, $\boldsymbol{p} \to -\boldsymbol{p}$ under space inversion, whereas, pseudo vectors like $\boldsymbol{L} = \boldsymbol{r} \times \boldsymbol{p}$, ***B*** (magnetic field) do not change sign, i.e., $\boldsymbol{L} \to \boldsymbol{L}$, $\boldsymbol{B} \to \boldsymbol{B}$ under space inversion. The pseudo vectors are also called *axial vectors*. Similarly, the scalar product of two polar vectors would be a scalar that will not change sign under space inversion and the dot product of a polar vector and an axial vector would change sign under space inversion. The sign changing scalars are called *pseudoscalars*.

As told above, weak interactions violate parity conservation law. When we develop the theory of weak interactions, called *vector current minus axial vector current* ($V-A$), an axial vector is added to a polar vector that leads to the violation of parity. In the experimental detection of parity violation, it is however, a pseudoscalar quantity that is observed. If the expectation value of the pseudoscalars is different from zero, it implies parity non-conservation. C.S. Wu and other scientists in 1956 set up an experiment to detect a pseudoscalar in beta decay. The pseudoscalar was ***J.p***, where ***J*** an axial vector was the nuclear spin of parent

(Co-60) nucleus and ***p***, a polar vector, was the linear momentum of the emitted electron. The lack of symmetry in the angular distribution of electrons with respect to the nuclear spin direction gave a net (non-zero) value to pseudoscalars ***J.p***, which meant the non-conservation of parity.

In Wu's experiment the isotope of Cobalt namely, the Co^{60}, was used as the source. This isotope is radioactive and decays by emitting β-particles. It also possesses nuclear magnetic moment due to which the nuclear spin of Co^{60} nucleus can be aligned by an external magnetic field. At very low temperatures (0.01K), the external magnetic field aligns the nuclear spins of Co^{60} nuclei in one direction and we say that Co^{60} nuclei have magnetically polarized. Since Co^{60} nuclei decay as

$$\mathrm{Co}^{60} \rightarrow \mathrm{Ni}^{*60} + e^- + \overline{\nu} \tag{1.137a}$$

$$\mathrm{Ni}^{*60} \rightarrow \mathrm{Ni}^{60} + \gamma \tag{1.137b}$$

the behaviour of emitted γ-rays is the test of the polarization of Co^{60} nuclei. When emitted gamma rays show anisotropic behaviour, it is implied that Co^{60} nuclei have polarized.

When the nuclear spin of Co^{60} nuclei was aligned along the direction of ***J***, it was observed that more electrons (β-particles) were emitted *opposite to the direction of magnetic field*. If we look at the space-inverted image of this picture, we find as shown in Fig. (1.28) that in the space inverted image more electrons are emitted *along the direction of magnetic field*, i.e., in the space inverted image the direction of emitted electrons is opposite to that in the original. Had there been symmetry in the distribution of emitted electrons, i.e., equal number of electrons would have emitted in both the directions; we could not distinguish between the two situations. Since we are able to distinguish the two situations, it gets demonstrated that the parity is violated when the β-decay process takes place. The parity violation is not only limited to β-decay in Cobalt, it has also been observed in

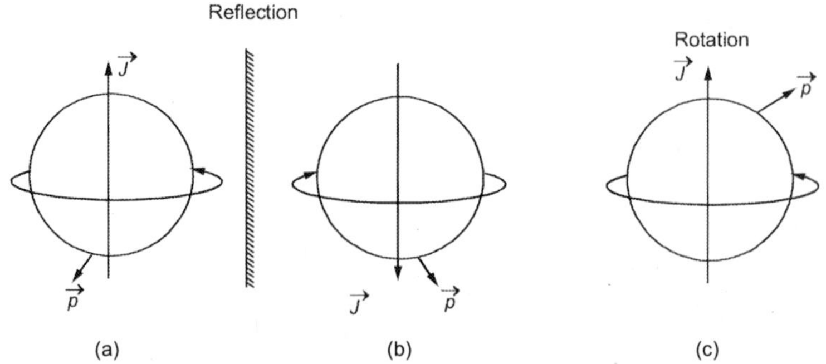

Figs. 1.28: Parity violation in beta decay of nuclei aligned by an external magnetic field. Parity operation is taken as a reflection in a plane (Fig. b) followed by a 180° rotation about the axis perpendicular to that plane (Fig. c). The direction of electron emission (p) with respect to the direction of ***J*** (nuclear spin of parent nucleus) changes when we go from Fig. (a) to Fig. (c). Fig. (a) tells electrons are emitted (preferentially) opposite to J, while (c) tells they are emitted along *J*

$$K^+ \rightarrow \mu^+ + \nu$$
$$\mu^+ \rightarrow e^+ + \nu + \bar{\nu} \quad (1.137c)$$

and

$$\Lambda \rightarrow p + \pi^-$$
$$\Lambda \rightarrow n + \pi^0 \quad (1.137d)$$

Thus, there exists ample evidence for the non-conservation of parity in weak interactions and the evidences lead us to conclude that the non-conservation of parity is the signature of weak interactions.

(11.1) ***Helicity of Neutrino:*** Parity violation also implies the violation of left-right symmetry. The violation of left-right symmetry is revealed in the behaviour of neutrino and antineutrino. All neutrinos are left-handed and all antineutrinos are right-handed. This also means that there is no right-handed neutrino and no left-handed antineutrino.

A particle whose spin vector is parallel to its direction of motion is called the right-handed particle and if the particle's spin vector is opposite to its direction of motion the particle is left-handed. If ***s*** denotes the spin vector and ***p*** the linear momentum vector (indicating direction of its motion) then

$$H = \frac{\boldsymbol{p}.\boldsymbol{s}}{|\boldsymbol{p}||\boldsymbol{s}|} \quad (1.138)$$

is called the helicity (H) of the particle. For right-handed $\boldsymbol{p}.\boldsymbol{s} = |\boldsymbol{p}||\boldsymbol{s}|$, but for left-handed $\boldsymbol{p}.\boldsymbol{s} = -|\boldsymbol{p}||\boldsymbol{s}|$, so

$$H_{\text{right-handed}} = +1 \quad (1.139a)$$

and

$$H_{\text{left-handed}} = -1 \quad (1.139b)$$

The helicity of neutrino is therefore (–1) and that of antineutrino (+1). For the measurement of helicities of neutrino, it is easier to adopt an indirect method like the decay of π^+ meson.

$$\pi^+ \rightarrow \mu^+ + \nu_\mu \quad (1.140)$$

If π^+ is at rest (Fig.1.29) the μ^+ and muon neutrino ν_μ go into opposite directions

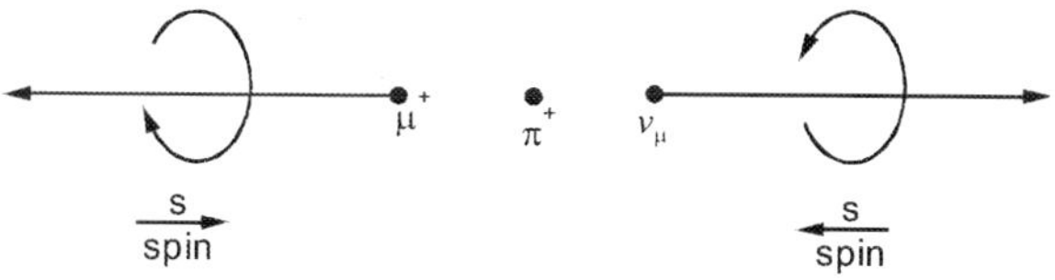

Fig. 1.29: Decay of π^+ at rest

and spins of μ^+ and ν_μ should also be opposite to each other so that the net spin is zero, which is the spin of parent particle π^+. If ν_μ is left-handed, in the rest frame of π^+, the μ^+ must also be left-handed and experiments do tell that in π^+

decay μ^+ is always left-handed implying that ν_μ is left-handed and its helicity is -1. Using the similar arguments for the decay

$$\pi^- \rightarrow \mu^- + \bar{\nu}_\mu \tag{1.141}$$

the μ^- are found to be right-handed and so are the $\bar{\nu}_\mu$ and hence $\bar{\nu}_\mu$ has $+1$ helicity.

(11.2) ***Helicity and Lorentz Invariance:*** If the helicity of a particle on observing from every inertial frame of reference remains same, we would then call, the helicity of the particle is Lorentz invariant. This means that observing from a particular frame of reference if a particle appears left-handed/right-handed then from any other inertial frame of reference too, it will look left-handed/right-handed. However, handedness changes for massive particles when we go from one Lorentz frame to other. This is because massive particles travel with a speed less than the speed of light and suppose observed from a frame moving less than the speed of light a massive particle (say electron) has its spin vector in the direction of its motion (i.e. electron appears right-handed), then for this case ($m_e \neq 0$) we can have a frame of reference moving with the speed of light (implying speed greater than particle's speed) and observing from this frame the electron would appear moving backwards while spin vector would be in the same direction. This would imply that the momentum and the spin vectors of the electron would appear opposite to each other and the electron, which looked right-handed, now appears left-handed. So, for massive particles the helicity is not a Lorentz invariant quantity. Due to this reason, to massive particles we cannot assign a definite helicity. However, for a massless particle like neutrino or antineutrino, which travels with speed of light, the upper limit of speeds, we cannot have a Lorentz frame moving faster than their speed and handedness of such particles would not change. Thus the helicities of neutrino and antineutrino are Lorentz invariant. If the neutrinos are found to have a non-zero mass, we can have both left- and right-handed neutrinos and antineutrinos. However, at present the neutrinos are considered massless and we have only left-handed neutrinos and right-handed antineutrinos.

(12) ***G-Parity:*** We have already noticed that the charge conjugation (C) operation reverses the sign of charge. In fact it reverses the sign of all additive quantum numbers, viz., charge, baryon number, lepton number, strangeness, charm, topness, bottomness etc. and also the I_3, which indicates upness and downness. Let us consider the states for which we have zero for all additive quantum numbers except for I_3. $I_3 \neq 0$ implies the charged states, e.g., in pion triplet $I_3 = +1$ and $I_3 = -1$ imply π^+ and π^- particles respectively. These are the charged states. Since C reverses the sign of I_3, only $I_3 = 0$ (implying $Q = I_3 + \dfrac{B+S}{2} = 0$ with $B, S = 0$) states are eigenstates of C. For non-zero I_3, the property of C can be expressed as

$$CI_3 = -I_3C \tag{1.142}$$

which implies that the eigenstates of I_3, viz., the charged particles are not the eigenstates of C. Thus because of only the neutral particles or the particle-antiparticle systems (e^+e^-, $\overline{p}p$) being the eigenstates of C, its use becomes limited. The strength of C-operator can however, be enhanced if we construct a product operator of isospin rotation and the C-operator. This product operator is called the G-parity operator. The charged particle states, which were not the eigenstates of C-operator, become the eigenstates of G-operator. For example, if we consider a rotation of 180° in pion isospin space about either I_2 (or I_1) axis, we find that as a result of such rotation the I_3, of pion gets reversed, i.e., π^+ becomes π^+ and vice versa. If we now apply the charge conjugation operation we get back the original charged state. Expressing rotation about I_2 axis by R_2 where

$$R_2 = e^{i\pi I_2} \tag{1.143}$$

and defining G-parity operator

$$G = CR_2 = Ce^{i\pi I_2} \tag{1.144}$$

we shall have

$$G\left|\pi^+\right\rangle \equiv Ce^{i\pi I_2}\left|\pi^+\right\rangle \equiv C\left|\pi^-\right\rangle = (\boldsymbol{e.v})\left|\pi^+\right\rangle \tag{1.145}$$

where $\boldsymbol{e.v}$ denotes the eigenvalue of G-operator. The eigenvalue of G-operator is expressed in terms of the C-parity (eigenvalue of C-operator) of the neutral member of a multiplet. The member of an isospin multiplet is denoted by ket $\left|I, I_3\right\rangle$ and the neutral member by $\left|I,0\right\rangle$.

$$G\left|I,0\right\rangle = c(-1)^I\left|I,0\right\rangle \tag{1.146}$$

and

$$G\left|I,I_3\right\rangle = c(-1)^I\left|I,I_3\right\rangle \tag{1.147}$$

where c is the C-parity of $\left|I,0\right\rangle$ state, i.e., the neutral state. Equation (1.147) tells that we must know c, i.e., the C-parity of neutral member of the multiplet to know the eigenvalue $c(-1)^I$ of the G-parity operator where I is the isospin of the multiplet.

(12.1) *G-parity of pion state:* The pion state is a triplet of isospin, expressed by

$$\left|\pi\right\rangle = \left|\pi_1, \pi_2, \pi_3\right\rangle \tag{1.148}$$

π_1, π_2 and π_3 are related to observed pions by

$$\pi^+ = \frac{\pi_1 + i\pi_2}{\sqrt{2}} \tag{1.149a}$$

$$\pi^- = \frac{\pi_1 - i\pi_2}{\sqrt{2}} \tag{1.149b}$$

$$\pi^0 = \pi_3 \tag{1.149c}$$

The isorotation by 180° about I_2-axis would lead to

$$|\pi_1\rangle \rightarrow -|\pi_1\rangle \Rightarrow R_2|\pi_1\rangle = -|\pi_1\rangle$$
$$|\pi_2\rangle \rightarrow +|\pi_2\rangle \Rightarrow R_2|\pi_2\rangle = +|\pi_2\rangle$$

and (1.150)

$$|\pi_3\rangle \rightarrow -|\pi_3\rangle \Rightarrow R_2|\pi_3\rangle = -|\pi_3\rangle$$

Since the charge conjugation operator C reverses the charges of π^+ and π^- i.e.

$$C|\pi^{\pm}\rangle = |\pi^{\mp}\rangle \tag{1.151}$$

we must have

$$C\left|\frac{\pi_1 \pm i\pi_2}{\sqrt{2}}\right\rangle = \left|\frac{\pi_1 \mp i\pi_2}{\sqrt{2}}\right\rangle \tag{1.152}$$

which demands

$$C|\pi_1\rangle = +|\pi_1\rangle \tag{1.153a}$$

$$C|\pi_2\rangle = -|\pi_2\rangle \tag{1.153b}$$

and obviously for neutral component

$$C|\pi_3\rangle = +|\pi_3\rangle \tag{1.153c}$$

Thus

$$\begin{aligned} G|\pi\rangle &= G|\pi_1, \pi_2, \pi_3\rangle \\ &= CR_2|\pi_1, \pi_2, \pi_3\rangle \\ &= C|-\pi_1, \ \pi_2, \ -\pi_3\rangle \\ &= |-\pi_1, \ -\pi_2, \ -\pi_3\rangle \\ &= (-1)|\pi\rangle \end{aligned} \tag{1.154}$$

where equations (1.150) and (1.153) have been used and R_2 implies $R_2\,(180°)$ i.e., a rotation of 180° about I_2-axis.

Thus

$$G|\pi\rangle = (-1)|\pi\rangle \tag{1.155}$$

tells that the G-parity (eigenvalue of G-operator) of a single pion state $(\pi \rightarrow \pi_1, \pi_2, \pi_3)$ is -1, i.e. it is negative or odd.

If the state has n pions eqn. (1.155) would become (G is a multiplicative quantum number)

$$G|n\pi\rangle = (-1)^n|n\pi\rangle \tag{1.156}$$

Thus if we have a 2π state (even pion state) the *G*-parity would be positive or even and for odd pion state the *G*-parity would be odd.

The fact, that the *G*-parity of *n* pion state is $(-1)^n$, can be used to tell the number of pions which may be emitted in a particular decay. Let us consider the ρ meson decay. This particle is a spin-one, isospin (*I*)-one, meson with mass ≈ 770 MeV. The *G*-parity of this ρ meson state is calculated from eqn. (1.147) where *c* is the *C*-parity (eigenvalue of *C*-operator) of ρ-meson state. Due to $I = 1$ and *c* for ρ-meson $= -1$, we have

$$G|\rho\rangle = c(-1)^I|\rho\rangle = (+1)|\rho\rangle \tag{1.157}$$

telling that *G*-parity of ρ-meson is even. Now the ρ-meson would decay (it decays through strong interactions) into pions, whose number must be such that the product state too has even *G*-parity according to $(-1)^n$ with *n* as number of pions. Obviously ρ-meson must decay into two pions.

The *G*-parity of *n* pions $= (-1)^n$ can also be used to determine the *G*-parity of ω and ϕ-mesons. ω-meson (spin 1, $I = 0$, $m \approx 783$ MeV), ϕ-meson (spin 1, $I = 0$, $m \approx 1020$ MeV) decay strongly.

$$\omega \rightarrow 3\pi(\pi^+ + \pi^- + \pi^0) \tag{1.158}$$

$$\phi \rightarrow \rho\pi \tag{1.159}$$

Obviously $|3\pi\rangle$ will have $(-1)^3 =$ odd *G*-parity and $|\rho\pi\rangle$ too will have $(+1)(-1)$ $=$ odd *G*-parity implying both ω and ϕ vector mesons shall have negative *G*-parities.

(12.2) *G-Parity in Nucleon-antinucleon systems:* For ($N\bar{N}$) nucleon-antinucleon systems ($p\bar{p}, n\bar{n}, p\bar{n}, n\bar{p}$) the *G*-parity is

$$G(N\bar{N}) = (-1)^{l+s+I} \tag{1.160}$$

where *l* denotes the relative angular momentum and *s* the total spin of nucleon-antinucleon state. *I* is the isospin of $N\bar{N}$ system which can be zero or one.

The relation (1.160) can provide selection rules, that is, it can tell into how many pions the nucleon-antinucleon system can annihilate. Suppose ($N\bar{N}$) system annihilates into *n* pions then the product state should have a *G*-parity of $(-1)^n$. From eqn. (1.160) we should then have, (*G*-parity is conserved)

$$(-1)^{l+s+I} = (-1)^n \tag{1.161}$$

which tells that if $l + s + I$ of ($N\bar{N}$) system is even (odd), it will annihilate into even (odd) number of pions.

Since *G*-operator is a product of isospin rotation operator and charge conjugation operator, both being the symmetries of strong interactions, the *G* parity remains conserved in strong interactions. Because isospin rotations are not invariant in *QED*; the *G*-parity is not a symmetry of electromagnetic interactions.

(13) *CP Invariance:* We have observed that the weak interactions are not invariant under parity transformations (*P*) and the charge conjugation transformation (*C*). For example in the pion decays

$$\pi^+(\text{spin } 0) \to \mu^+(RH) + \nu_\mu(LH) \tag{1.162}$$

$$\pi^-(\text{spin } 0) \to \mu^-(LH) + \bar{\nu}_\mu(RH) \tag{1.163}$$

where $RH \Rightarrow$ right-handed and *LH* the left-handed. Now applying *P*-operation to eqn. (1.162) and *C* to eqn. (1.163) we get, noting that *P*-operator does not affect scalar particles.

$$\begin{aligned} &P|\pi^+\rangle \to P|\mu^+(RH)\rangle + P|\nu_\mu(LH)\rangle \\ \Rightarrow \\ &|\pi^+\rangle \to |\mu^+(LH)\rangle + |\nu_\mu(RH)\rangle \end{aligned} \tag{1.164}$$

where each state on the RHS is unphysical indicating parity violation. Similarly

$$\begin{aligned} &C|\pi^-\rangle \to C|\mu^-(LH)\rangle + C|\bar{\nu}_\mu(RH)\rangle \\ \Rightarrow \\ &|\pi^+\rangle \to |\mu^+(LH)\rangle + |\nu_\mu(RH)\rangle \end{aligned} \tag{1.165}$$

in which too the RHS has unphysical states, indicating *C*-violation. It may be recalled that eqns. (1.162) and (1.163) are weak decays.

We now apply the combined operator *CP* to eqns. (1.162) and (1.163) and find that

$$\begin{aligned} CP|\pi^+\rangle &\to CP|\mu^+(RH)\rangle + CP|\nu_\mu(LH)\rangle \\ &\to C|\mu^+(LH)\rangle + C|\nu_\mu(RH)\rangle \\ \text{or} \\ |\pi^-\rangle &\to |\mu^-(LH)\rangle + |\bar{\nu}_\mu(RH)\rangle \end{aligned} \tag{1.166}$$

Since left-handed μ^- and right-handed antineutrino represent physical states, eqn. (1.166) implies that the combined operation *CP* is a symmetry of weak interactions. Similarly $CP|\pi^-\rangle$ would yield eqn. (1.162), a decay realized in nature.

(13.1) *CP violation:* Most of the weak interactions are invariant under the combined operation of *CP*. However, in the case of the decay of neutral *K*-mesons i.e., K^0 mesons which too decay through weak interactions, the *CP* invariance gets violated.

K^0 are the spin zero mesons with mass ($\approx 498\,\text{MeV/C}^2$). $\overline{K}^0$ is the antiparticle of K^0. Isospin of K^0 is 1/2. K^0 and $\overline{K}^0$ form an isospin doublet. The Gell-Mann and Nishijima relation

$$Q = I_3 + \frac{S+B}{2} \tag{1.167}$$

with $B = 0$ for mesons, provides $S = +1$ to K^0 ($I_3 = -1/2$, $Q = 0$, $B = 0$) and $S = -1$ to $\overline{K}^0$(I_3 + 1/2, $Q = 0$, $B = 0$). K^0 and $\overline{K}^0$ are lightest strange particles. Strong interaction processes produce them

$$\pi^- + p \rightarrow K^0 + \Lambda^0 \tag{1.168a}$$

$$\pi^+ + p \rightarrow K^+ + \overline{K}^0 + p \tag{1.168b}$$

A detailed analysis of angular momentum states involved in the K^0 production process tells that K^0 has odd intrinsic parity. Since K^0 is a boson and for bosons the particle and antiparticle have same parity, the intrinsic parity of $\overline{K}^0$ is also (–1). This implies that

$$P\left|K^0\right\rangle = -\left|K^0\right\rangle \tag{1.169a}$$

also

$$P\left|\overline{K}^0\right\rangle = -\left|\overline{K}^0\right\rangle \tag{1.169b}$$

Due to the change of sign upon space inversion, K^0 and $\overline{K}^0$ are therefore pseudoscalar particles. Since, they are the lightest strange particles; they cannot decay through strangeness conserving strong interactions, instead K^0 and $\overline{K}^0$ decay only via weak interactions, which do not conserve strangeness. The neutral kaons decay into two or three pions. The two pions could be $\pi^+ \pi^-$ or $\pi^0 \pi^-$ and the three pions could be $\pi^+ \pi^- \pi^0$ or $3\pi^0$.

We have noticed in the previous section that weak interactions observe *CP* conservation law. This implies that the operator *CP* commutes with the weak Hamiltonian (*H*), hence both *CP* and *H* can be diagonalized simultaneously and states of definite *CP* value and definite energy or mass can be physically realized. In the laboratory, we can observe neutral kaons as the physical particles, then if K^0 and $\overline{K}^0$ are these physical particles they must be the eigen states of operator *CP*. Let us check that the *C* operation onto $\left|K^0\right\rangle$ and $\left|\overline{K}^0\right\rangle$ states yields

$$C\left|K^0\right\rangle = \left|\overline{K}^0\right\rangle \tag{1.170a}$$

$$C\left|\overline{K}^0\right\rangle = \left|K^0\right\rangle \tag{1.170b}$$

Using equations (1.169) and (1.170) we can have

$$CP\left|K^0\right\rangle = -C\left|K^0\right\rangle = -\left|\overline{K}^0\right\rangle \tag{1.171a}$$

and

$$CP\left|\overline{K}^0\right\rangle = -C\left|\overline{K}^0\right\rangle = -\left|K^0\right\rangle \tag{1.171b}$$

Equations (1.171) clearly tell that they are not the eigenvalue equations implying that $\left|K^0\right\rangle$ and $\left|\overline{K}^0\right\rangle$ are not the eigenstates of the operator *CP*, implying that what we observe in the laboratory as neutral kaons are not K^0 and $\overline{K}^0$.

To find the observable neutral kaon states we construct two linear combinations of K^0 and $\overline{K}^0$ and check if these linear combinations are the eigenstates of *CP*. Let us denote these linear combinations by K_1^0 and K_2^0 where

$$\left|K_1^0\right\rangle = \frac{1}{\sqrt{2}}\left(\left|K^0\right\rangle - \left|\overline{K}^0\right\rangle\right) \tag{1.172a}$$

and

$$\left|K_2^0\right\rangle = \frac{1}{\sqrt{2}}\left(\left|K^0\right\rangle + \left|\overline{K}^0\right\rangle\right) \tag{1.172b}$$

in which $\frac{1}{\sqrt{2}}$ is the normalization constant. Using relations (1.171), we can easily find that

$$CP\left|K_1^0\right\rangle = \left|K_1^0\right\rangle \tag{1.173a}$$

and

$$CP\left|K_2^0\right\rangle = -\left|K_2^0\right\rangle \tag{1.173b}$$

implying that $\left|K_1^0\right\rangle$ and $\left|K_2^0\right\rangle$ are the eigenstates of *CP* operator. Thus what we observe in laboratory as neutral kaons are not K^0 and $\overline{K}^0$ but K_1^0 and K_2^0. The statement, that neutral kaons decay into 2π and 3π modes, would now mean that the particles K_1 and K_2 decay into these modes. Let us analyse these decay modes.

The parity of $2\pi \Rightarrow \pi^0\pi^0$ or $\pi^+\pi^-$ state is even and the parity of $3\pi \Rightarrow 3\pi^0$ or $\pi^+\pi^-\pi^0$ is odd. Eqn. (1.173a) tells that for $\left|K_1^0\right\rangle$ the eigenvalue of *CP* is +1 and since *C*-operation only converts particle into its antiparticle, (+1) eigenvalue comes due to *P*-operation. This implies that $\left|K_1^0\right\rangle$ is the even parity state and likewise $\left|K_2^0\right\rangle$ is the odd parity state. Thus even parity state $\left|K_1^0\right\rangle$ would decay into even parity state $\left|2\pi\right\rangle$ and odd parity state $\left|K_2^0\right\rangle$ into odd parity state $\left|3\pi\right\rangle$. That is

$$K_1^0 \rightarrow 2\pi \tag{1.174a}$$

$$K_2^0 \rightarrow 3\pi \tag{1.174b}$$

The 2π decay mode due to greater energy release is faster than the 3π mode. The lifetimes (τ) of these particles are experimentally found to be

$$\tau(K_1^0) = 0.9 \times 10^{-10} \text{ sec} \tag{1.175}$$

and

$$\tau(K_2^0) = 0.5 \times 10^{-7} \text{ sec} \tag{1.176}$$

i.e., if we have a beam containing both K_1^0 and K_2^0, both K_1^0 and K_2^0 would decay but decay of K_2^0 would be so slow that they will live about 10^3 second ($\approx$ 17 minutes) more than K_1^0 particles. Thus if we have a beam of K^0's

$$|K^0\rangle = \frac{1}{\sqrt{2}}(|K_1^0\rangle + |K_2^0\rangle) \tag{1.177}$$

K_1^0's have mostly decayed after few centimetres ($\approx$ 2.7 cm at high energies) whereas K_2^0's can travel many meters ($\approx$ 15 m). Near the source of K^0 we shall have many 2π events but away from the source we shall notice only 3π events.

If *CP* is conserved in weak interactions, the K_2^0 state, which has *CP* eigenvalue $= -1$, cannot decay into 2π; it must decay only into three pions. If we observe that K_2^0's have decayed into two pions also, then that would mean *CP* conservation in weak interactions has been violated.

Christenson, Cronin, Fitch and Turley in 1964, performed an experiment in which they found that at the end of about 17 meter beam, where only 3π events must have been observed, 2π events are also detected. They analysed about 23000 decays and found that out of these, 45 decays produced two pions. Though the number of 2π events was small, it was there. The presence of 2π events, where we should have observed only 3π events indicated that at 17 meters the beam was not pure K_2^0, but it contained a small admixture of K_1^0.

Following the discovery of *CP* violation, the names of neutral kaons were changed. Before this we had K_1^0, (*CP* eigenvalue $= +1$) and K_2^0 (*CP* eigenvalue $= -1$). Now, the neutral kaons were called short-lived kaons (K_S^0) and long lived kaons (K_L^0). However, K_S^0 and K_L^0 were not K_1^0 and K_2^0 respectively, instead they were mixed states

$$|K_S^0\rangle = \frac{|K_1^0\rangle + \varepsilon |K_2^0\rangle}{\sqrt{1+\varepsilon^2}} \tag{1.178a}$$

$$|K_L^0\rangle = \frac{|K_2^0\rangle + \varepsilon |K_1^0\rangle}{\sqrt{1+\varepsilon^2}} \tag{1.178b}$$

where, ε is a small complex number which measures the degree of *CP* violation due to kaon state mixing. Experimentally its magnitude is about 2.3×10^{-3}. $(1+\varepsilon^2)^{-\frac{1}{2}}$ is the normalization constant. *CP* violation in the 2π decay modes is usually expressed as the amplitude ratio

$$|\eta_{+-}| = \frac{Ampl.\ \ K_L^0 \to \pi^+\pi^-}{Ampl.\ \ K_S^0 \to \pi^+\pi^-} = 2.268\times10^{-3} \tag{1.179a}$$

$$|\eta_{00}| = \frac{Ampl.\ \ K_L^0 \to 2\pi^0}{Ampl.\ \ K_S^0 \to 2\pi^0} = 2.253\times10^{-3} \tag{1.179b}$$

One of the important consequences of *CP* violation in neutral kaon decays is that, we have to consider a mixing of states while dealing with weak interactions. An example of this mixing is the Cabibbo-Kobayashi-Maskawa (*CKM*) matrix, which mixes the generations of quarks and the mixed quark states participate in weak interactions. Another conclusion that can be derived from *CP* violation relates to *CPT* invariance. All interactions are assumed to be invariant under *CPT* operation taken in any order. If *CP* violation has taken place, the violation under *T* (time reversal) operator is natural. So *CP* violation in weak interactions implies their *T* violation as well.

(14) Time Reversal (T): Classically, time reversal transformation is expressed by

$$x \to x'$$

with

$$x_0' = -x_0$$

and

$$\boldsymbol{x}' = \boldsymbol{x} \tag{1.180}$$

where $x = (x_0, \boldsymbol{x})$ and $x = (x_0', \boldsymbol{x}')$; $x_0 \Rightarrow t,\ x_0' \Rightarrow t'$. However, time reversal ($t' \to -t$) does not mean that the time is running backwards into the past. Below we list the transformation of common quantities under Time Reversal (*T*) as well as under space inversion (*P*)

Quantity	*Behaviour under Time Reversal (T)*	*Behaviour under Space Inversion (P)*
Position $\boldsymbol{r}$	$\boldsymbol{r}$	$-\boldsymbol{r}$
Linear Momentum $\boldsymbol{p}$	$-\boldsymbol{p}$	$-\boldsymbol{p}$ (polar vector)
Angular Momentum $\boldsymbol{L}$	$-\boldsymbol{L}$	$\boldsymbol{L}$ (axial vector)
Spin (angular momentum) σ	$-\sigma$	σ (axial vector)
Electric Field $\boldsymbol{E}$	$\boldsymbol{E}$	$-\boldsymbol{E}\left(\boldsymbol{E} = -\frac{\partial \boldsymbol{V}}{\partial \boldsymbol{r}}\right)$

Magnetic Field ***B***	–***B***	***B***
σ.***E***	– σ.***E***	– σ.***E*** (Electric dipole moment)
σ.***B***	σ.***B***	σ.***B*** (magnetic dipole moment)

(14.1) Microscopic Reversibility: At microscopic level, time reversal invariance appears natural. For example, we may consider classically the collision between two molecules p and q whose velocities before collision are u_p and u_q and u_q and after collision v_p and v_q respectively. We may show the collision by Fig. (1.30a). The time inverted diagram of the above process can be drawn by making

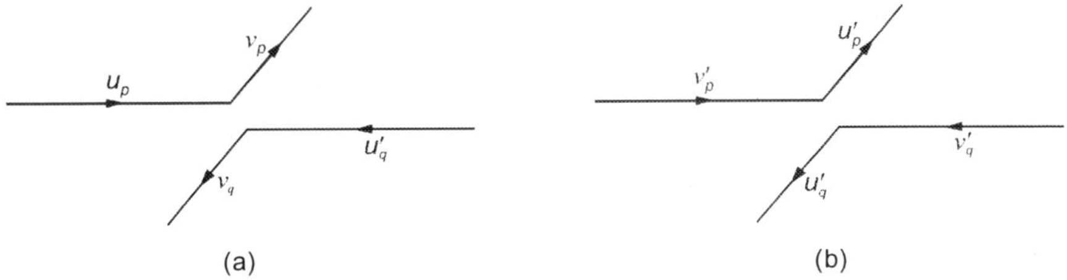

Figs. 1.30 (a, b): Diagram (**a**) shows collision of two molecules p and q to explain microscopic reversibility. Diagram (**b**) is the time inverted diagram of (**a**). Collision processes indicated by both the diagrams are possible at microscopic (molecular) level

the replacements (Fig. 1.30) $v_p \to -u'_p$, $v_q \to -u'_q$ and $u_p \to -v'_p$, $u_q \to -v'_q$. It can be easily understood that in a gas above two collision processes are equally probable and we say the collision of gas molecules (microscopic) is time invariant. This implies that at microscopic level the backward motion is just as possible as the original. This is known as the ***principle of microscopic reversibility*** and *states that the individual collisions between atoms or molecules are indistinguishable under the operation of time reversal.*

For macroscopic systems we can usually tell which is the right way. For example the fall of a glass from a table top and its breaking into pieces is the forward motion which can always occur but the probability of occurring its reverse motion, in which the broken pieces gather into a glass and then glass rises onto the table top, is extremely small. Thus at macroscopic level, we encounter with the non-invariance under time reversal. At this level, the statistical laws or the laws of increase of entropy, *a system goes from ordered to disordered state*, guides the systems.

***(14.2) Time reversal transformation in Quantum Mechanics*:** The time development of a state $|\psi(\boldsymbol{r},t)\rangle$ is given by Schrödinger's Equation (SE)

$$i\frac{\partial}{\partial t}|\psi(\boldsymbol{r},t)\rangle = H|\psi(\boldsymbol{r},t)\rangle \tag{1.181}$$

Now consider above equation under $t \rightarrow -t$, we get

$$-i\frac{\partial}{\partial t}|\psi(\boldsymbol{r},-t)\rangle = H|\psi(\boldsymbol{r},-t)\rangle \tag{1.182}$$

Obviously because of minus sign ($-i$), eqn. (1.182) is not the Schrödinger's Equation. For the state $|\psi(\boldsymbol{r},-t)\rangle$, eqn. (1.182) may look like (1.181), the SE, provided it's $-i$ gets replaced by $+i$, implying complex conjugation, but we can't complex conjugate i alone, every quantity in (1.182) will have to be complex conjugated.

Thus the time reversal operator T would first change t to $-t$ in the SE (1.181) and then complex conjugate (*CC*) it (eqn. 1.182). Accordingly we get the CC of (1.182) as

$$i\frac{\partial}{\partial t}|\psi^*(\boldsymbol{r},-t)\rangle = H^*|\psi^*(\boldsymbol{r},-t)\rangle \tag{1.183}$$

which is of course the SE. Due to $t \rightarrow -t$ and complex conjugation done by T, we may write

$$T|\psi(\boldsymbol{r},t)\rangle = |\psi^*(\boldsymbol{r},-t)\rangle \tag{1.184}$$

Substituting eqn. (1.184) into eqn. (1.183), we get

$$i\frac{\partial}{\partial t}[|T\psi(\boldsymbol{r},t)\rangle] = H^*[|T\psi(\boldsymbol{r},t)\rangle] \tag{1.185}$$

where $T\psi \Rightarrow T|\psi\rangle$. Thus if pure $t \rightarrow -t$ transformation is carried by the unitary operator U, and complex conjugation is done by the antiunitary operator K, then above discussion tells that the time reversal operator T would be

$$T = UK \tag{1.186}$$

Obviously T is an antiunitary operator.

If the state $|\psi(\boldsymbol{r},t)\rangle$ is a linear combination

$$|\psi(\boldsymbol{r},t)\rangle = C_1|\psi_1(\boldsymbol{r},t)\rangle + C_2|\psi_2(\boldsymbol{r},t)\rangle \tag{1.187}$$

Then using eqn. (1.184) in RHS

$$T|\psi(\boldsymbol{r},t)\rangle = C_1^*|\psi_1^*(\boldsymbol{r},-t)\rangle + C_2^*|\psi_2^*(\boldsymbol{r},-t)\rangle$$

which is also expressed as

$$T|\psi(\boldsymbol{r},t)\rangle = C_1^*T|\psi_1(\boldsymbol{r},t)\rangle + C_2^*T|\psi_2(\boldsymbol{r},t)\rangle \tag{1.188}$$

Comparing equations (1.187) and (1.188), whereas eqn. (1.187) is a linear relation, eqn.(1.188) is an antilinear relation. So, we observe that the operator T is an antilinear operator. The antiunitarity of T can also be expressed by

$$\langle\psi'|\phi'\rangle = \langle\psi T^\dagger|T\phi\rangle = \langle\psi|\phi\rangle^* = \langle\phi|\psi\rangle$$

However

$$|\langle\psi'|\phi'\rangle| = |\langle\phi|\psi\rangle| = |\langle\psi|\phi\rangle| \tag{1.189}$$

implying that the operation by T does not affect the physical content of quantum mechanics.

Thus classically time reversal implies $t \to -t$ but quantum mechanically time reversal transformation would be

$$t \to -t \tag{1.190a}$$

and complex conjugation

$$i \to -i \tag{1.190b}$$

Under the transformations (1.190), the quantum mechanical principle of microscopic reversibility holds and the basic relations

$$\begin{aligned}[x_i, p_j] &= i\hbar\delta_{ij} \\ [L_i, L_j] &= i\hbar\varepsilon_{ijk}L_k\end{aligned} \tag{1.191}$$

are preserved. The table of *subsection 14* (Time Reversal) can be used for verification.

(14.3) Experimental Verification of T-invariance: The experimental verification of T-invariance or T-non-invariance is more difficult than testing the P-invariance or C-invariance. This is because there is no particle, which is the eigenstate of the time reversal operator T. All particles are eigenstates of P some are of C but none are of T. So by observing the pseudoscalars we cannot verify the conservation of T.

For testing the time reversal invariance we check the reaction rates of reactions, for example, of

$$n + p \to d + \gamma \tag{1.192}$$

and its reverse

$$d = \gamma \to n + p \tag{1.193}$$

The time reversal invariance demands that for corresponding conditions of momentum, energy and spin, the reaction rate of direct reaction (1.192) and its reverse (1.193) must be same. This is also called the *principle of detailed balance*. However, such tests work well for strong and electromagnetic interactions but in case of weak interactions, the inverse reactions are difficult to be carried and hence the principle of detailed balance cannot be easily applied to them.

A different method to test T-invariance is therefore employed. It is based on the measurement of a quantity which is zero if the system is T-invariant and 7 non-zero when T-invariance is violated. The electric dipole moment is the quantity, which is measured to test the validity of time reversal invariance in electromagnetic interactions.

Let us consider a particle, which possesses the electric dipole moment μ_e and also the magnetic dipole moment μ_m. The particle is placed in an electromagnetic field ($\boldsymbol{E}$, $\boldsymbol{B}$). The interaction Hamiltonian then is

$$H_{\text{int}} = \mu_e\boldsymbol{\sigma}.\boldsymbol{E} + \mu_m\boldsymbol{\sigma}.\boldsymbol{B} \tag{1.194}$$

where σ is the spin of the particle. From the table (see section Time Reversal), we notice that $\sigma.E$ changes sign under time reversal as well as under P, and $\sigma.B$ does not. So after T-inversion

$$(H_{\text{int}})_T = -\mu_e(\boldsymbol{\sigma}.\boldsymbol{E}) + \mu_m \boldsymbol{\sigma}.\boldsymbol{B} \tag{1.195}$$

Equation (1.195), tells that the T-violation is due to μ_e term and not due to μ_m term. In other words, if we have $\mu_e \neq 0$, we will have a T-violating term. Of course $\sigma.E$ changes sign under P operation as well. Thus a measurement of non-zero value of μ_e would mean T as well as P-violation in electromagnetic interactions. To check that, the electric dipole moment of neutron was measured, this gave a value

$$\mu_e < 1.2\times10^{-25} \text{ e.cm} \tag{1.196}$$

which is quite small implying almost no T and P violation in electromagnetic interactions.

Strong interactions have been found to conserve T-invariance. Electromagnetic interactions preserve T-invariance fairly well. The violations in EM interactions are below one per cent. The CP violation in neutral kaons, which also suggests T-violation, is an example of T-violation in weak decays. In general weak interactions too are invariant under time reversal upto a few per cent. This observation is in contrast to P and C violations in weak interactions, which are maximal.

(15) CPT Theorem: The CPT theorem states that the combined operation of charge conjugation, space inversion and time reversal carried out in any order is an exact symmetry of any interaction. The CPT theorem is based on very general assumption of quantum field theory and relativity namely, the Lorentz invariance, quantum mechanics and the idea that fields carry the interactions. The CPT theorem also states that the local Lagrangian of a system of quantum fields is invariant under the simultaneous application of CPT.

To illustrate this let us take the example of complex scalar field whose Lagrangian density is

$$£(x) = -m^2\phi^\dagger(x)\phi(x) + \partial_\mu\phi^\dagger(x)\partial^\mu\phi(x) \tag{1.197a}$$

where

$$x = (x^0, x) \tag{1 .197b}$$

It is obvious that CPT operation on $x = (x^0, x)\, x = (x^0, \boldsymbol{x})$ would change it into $(-x^0, -x)$, C being ineffective. Let us then write the Lagrangian density (1.197a) at

$$x' = -x = (-x^0, -x) \tag{1.198}$$

and consider

$$\begin{aligned} £_{CPT}(x') &= [-m^2\phi^\dagger(x')\phi(x') + \partial'_\mu\phi^\dagger(x')\partial'^\mu\phi(x')]_{CPT} \\ &= CPT[-m^2\phi^\dagger(x')\phi(x') + \partial'_\mu\phi^\dagger(x')\partial'^\mu\phi(x')]T^{-1}P^{-1}C^{-1} \end{aligned} \tag{1.199}$$

Now

$$
\begin{aligned}
&CPT\phi^{\dagger}(x')\phi(x')T^{-1}P^{-1}C^{-1} \\
&\quad = CPT\phi^{\dagger}(-x^0,-\boldsymbol{x})T^{-1}P^{-1}C^{-1}.CPT\phi(-x^0,-\boldsymbol{x})T^{-1}P^{-1}C^{-1} \\
&\quad = CP\phi^{\dagger}(x^0,-\boldsymbol{x})P^{-1}C^{-1}.CP\phi(x^0,-\boldsymbol{x})P^{-1}C^{-1} \\
&\quad = C\phi^{\dagger}(x^0,+\boldsymbol{x})C^{-1}.C\phi(x^0,+\boldsymbol{x})C^{-1} \\
&\quad = \phi(x)\phi^{\dagger}(x)
\end{aligned}
\tag{1.200}
$$

Similarly using

$$\partial'_{\mu} = \frac{\partial}{\partial x'^{\mu}} = \frac{\partial}{\partial(-x^0,-\boldsymbol{x})} \tag{1.201a}$$

and

$$\partial'^{\mu} = \frac{\partial}{\partial x'_{\mu}} = \frac{\partial}{\partial(-x^0,+\boldsymbol{x})} \tag{1.201b}$$

we have

$$
\begin{aligned}
CPT\partial'_{\mu}T^{-1}P^{-1}C^{-1} &= CPT\frac{\partial}{\partial(-x^0,-\boldsymbol{x})}T^{-1}P^{-1}C^{-1} \\
&= CP\frac{\partial}{\partial(x^0,-\boldsymbol{x})}P^{-1}C^{-1} \\
&= C\frac{\partial}{\partial(x^0,+\boldsymbol{x})}C^{-1} \\
&= \frac{\partial}{\partial x^{\mu}} = \partial_{\mu}
\end{aligned}
\tag{1.202a}
$$

and

$$
\begin{aligned}
CPT\partial'^{\mu}T^{-1}P^{-1}C^{-1} &= CPT\frac{\partial}{\partial(-x^0,+\boldsymbol{x})}T^{-1}P^{-1}C^{-1} \\
&= CP\frac{\partial}{\partial(x^0,+\boldsymbol{x})}P^{-1}C^{-1} \\
&= C\frac{\partial}{\partial(x^0,-\boldsymbol{x})}C^{-1} \\
&= \frac{\partial}{\partial x_{\mu}} = \partial^{\mu}
\end{aligned}
\tag{1.202b}
$$

Hence

$$
\begin{aligned}
&CPT\partial'_{\mu}\phi^{\dagger}(x')\partial'^{\mu}\phi(x')T^{-1}P^{-1}C^{-1} \\
&\quad = CPT\partial'_{\mu}T^{-1}P^{-1}C^{-1}.CPT\phi^{\dagger}(x')T^{-1}P^{-1}C^{-1}. \\
&\quad\quad CPT\partial'^{\mu}T^{-1}P^{-1}C^{-1}.CPT\phi(x')T^{-1}P^{-1}C^{-1} \\
&\quad = \partial_{\mu}\phi(x).\partial^{\mu}\phi^{\dagger}(x)
\end{aligned}
\tag{1.203}
$$

Using equations (1.197) and (1.200) in eqn. (1.199) we find that for $x' = -x$

$$\pounds_{CPT}(x') = \pounds(x) \tag{1.204}$$

implying invariance of the field Lagrangian density under combined *CPT* operation.

(15.1) Physical Consequences of CPT Theorem: The *CPT* theorem involves three operations. The *C*-operation changes particles into antiparticles. The *P*- operation replaces the three vectors into their opposites and *T*-operation inverts time, i.e., *T* makes us start from the *end* in reverse direction to reach at the *beginning*. This implies that *T* interchanges initial and final states and vice versa and also changes the direction of momenta and spins.

CPT invariance implies that if we have a process Ω and another process *CPT* Ω that is obtained from Ω by replacing particles by antiparticles, by inverting the spins and by replacing final states by initial states, then the square of the matrix elements of the processes Ω and *CPT*Ω would be equal. From this equality it is concluded that: (1) The masses and lifetimes of particles and their antiparticles must be exactly same. In the list of elementary particles an exception to this exactness of mass is K^0 (particle), K^0 (antiparticle) mass difference. But $|m_{K^0} - m_{\bar{K}^0})/m_{K^0}| < 6\times10^{-19}$ which is too small to show a difference in the masses of particle and its antiparticle. It may be recalled that $|m_{e^-} - m_{e^+})/m_{e^-}| < 4\times10^{-8}$, which too is negligibly small. (2) The magnetic moments of particle and antiparticle must be equal. They should differ in sign only. (3) The violation of any one symmetry or a pair of symmetries in *C*, *P*, and *T* would be followed by a compensatory violation of remaining symmetry such that *CPT* remains inviolate. For example, the *CP* violation in neutral kaon decay implies the *T* violation as well.

1.11 CLASSIFICATION OF ELEMENTARY PARTICLES

We now connect ourselves to section (1.8) and those preceding it, to classify the elementary particles described in those sections. The particles may be grouped into four classes, viz., Leptons, Mesons, Baryons and Gauge Bosons. Mesons and baryons can be further separated into their strange and nonstrange members. We list them in the form of tables.

(A) Broad Classification

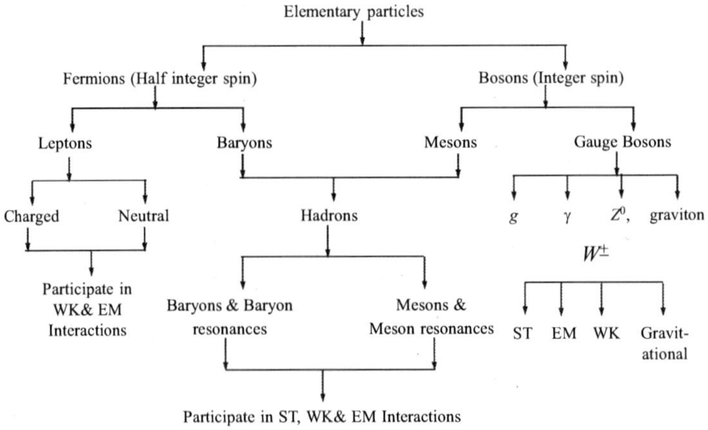

(B) Leptons

Particle	Spin J	Mass MeV/C²	Principal Decaymodes	Mean Lifetime/sec	Remarks
$\nu_e, \bar{\nu}_e$	$\frac{1}{2}$	$< 7.3\times10^{-3}$	—	stable	Mass of ν_e, $\nu_\mu \approx 0$
e^-, e^+	$\frac{1}{2}$	0.511	—	stable	Applying electron, muon
$\nu_\mu, \bar{\nu}_\mu$	$\frac{1}{2}$	< 0.27	—	stable	and tauon number
μ^-, μ^+	$\frac{1}{2}$	105.66	$\mu^- \to e^- \bar{\nu}_e \nu_\mu$	2.20×10^{-6}	conservation laws, the
			$\mu^+ \to e^+ \nu_e \bar{\nu}_\mu$		decay modes can be
$\nu_\tau, \bar{\nu}_\tau$	$\frac{1}{2}$	< 35	—	stable	written. Antiparticles
τ, τ^+	$\frac{1}{2}$	1784.1	$\tau^- \to \mu^- \bar{\nu}_\mu \nu_\tau$	3.1×10^{-13}	have opposite charge
			$\tau^- \to e^- \bar{\nu}_e \nu_\tau$ $\tau \to$ *hadrons*		and lepton numbers

(C) Mesons

Particle	Spin Parity J^P	Isospin G-parity I^G	Mass MeV/C²	Principal Decay modes	Mean lifetime Sec	Strangeness S	Remarks
π^0	0^-	—	135	$\gamma\gamma$	0.8×10^{-16}	0	Nonstrange ps mesons
π^+, π^-	0^-	1^-	139.6	$\mu^+\nu_\mu$, $e^+\nu_e$	2.6×10^{-8}	0	
K^+, K^-	0^-	$\frac{1}{2}$	493.7	$\mu^+\nu_\mu$, $\pi^\pm\pi^0$, 3π	1.24×10^{-8}	+1, −1	Strange ps mesons
$K^0, \bar{K}^0$	0^-	$\frac{1}{2}$	497.7	$50\% K_S^0$, $50\% K_l^0$	—	+1, −1	
K_S^0	0^-	—	—	$\pi^+\pi^-$, $\pi^0\pi^0$	0.89×10^{-10}	—	
K_l^0	0^-	—	—	$\pi^+\pi^-\pi^0$, $3\pi^0$, $\pi\mu^\mp\nu_\mu$, $\pi^\pm e^\mp\nu_e$	5.2×10^{-8}	—	
η^0	0^-	0^+	549	$\gamma\gamma$, $\pi^\pm\pi^0$, $3\pi^0$	7.7×10^{-19}	0	Nonstrange vector mesonres
ρ^+, ρ^-, ρ^0	1^-	1^+	765	$\pi\pi$	4×10^{-24}	0	Nonstrange vector mesonres
ω^0	1^-	0^-	784	$\pi\pi\pi$,	7×10^{-23}	0	

K^{*+}, K^{*-} $(K^{*0}, \bar{K}^{*0})$	1^-	$\frac{1}{2}$	893	$K\pi$,	1×10^{-23}	+1, −1	Strange vector mesonres
ϕ^0	1^-	0^-	1019	$KK, \pi\pi\pi$,	2×10^{-22}	0	Nonstrange vector mesonres
f^0	2^+	0^+	1270	$\pi\pi$,	4×10^{-24}	0	Nonstrange tensor mesonres
A_2^+, A_2^-, A_2^0	2^+	1^-	1310	$\rho\pi$	7×10^{-24}	0	
K_N^+, K_N^-, $(K_N^0, \bar{K}_N^0)$	2^+	$\frac{1}{2}$	1408	$K\pi\pi$	6×10^{-24}	+1, −1	Strange tensor mesonres
J/ψ	1^-	0^-	3100	e^+e^-, $\mu^+\mu^-$ 5π, 7π	1×10^{-20}	0	Charm anticharm state
Y	1	—	9460	e^+e^-, $\mu^+\mu^-, \tau^+\tau^-$	2×10^{-20}	0	Bottom antibottom state

(D) Baryons

Particle		Spin Parity J^P	Isospin I	Mass MeV/C²	Principal Decay modes	Mean life-time Sec	Strangeness S	Remarks
N	p	$(\frac{1}{2})^+$	$\frac{1}{2}$	938.3	–	stable ($>10^{32}$years)	0	Non strangeBaryons
	n	$(\frac{1}{2})^+$	$\frac{1}{2}$	939.6	$pe^-\bar{\nu}_e$	900(15minutes)	0	
Λ	Λ	$(\frac{1}{2})^+$	0	1115.6	$p\pi^-, n\pi^0$	2.6×10^{-10}	-1	Strangeness(−1)Baryons
Σ	Σ^+	$(\frac{1}{2})^+$	1	1189.4	$p\pi^0, n\pi^+$	0.8×10^{-10}	-1	
	Σ^0	$(\frac{1}{2})^+$	1	1192.6	$\Lambda\gamma$	7.4×10^{-10}	-1	
	Σ^-	$(\frac{1}{2})^+$	1	1197.4	$n\pi^-$	1.5×10^{-10}	-1	
Ξ	Ξ^0	$(\frac{1}{2})^+$	$\frac{1}{2}$	1314.9	$\Lambda\pi^0$	2.9×10^{-10}	−2	Strangeness(−2)Baryons
	Ξ^-	$(\frac{1}{2})^+$	$\frac{1}{2}$	1321.3	$\Lambda\pi^-$	1.6×10^{-10}	−2	
Ω	Ω^-	$(\frac{3}{2})^+$	0	1672.4	$\pi k^-, \Xi^0\pi^-$, $\Xi^-\pi^0$	0.82×10^{-10}	-3	Strangeness(−3)Baryon

(E) Gauge Bosons

Particle	Spin Parity J^P	Mass MeV/C^2	Principal Decay modes	Mean Life Time (τ)
γ (photon)	1^-	0	—	*Stable*
$W^{\pm}$	1	80.22	$e\nu, \mu\nu, \tau\nu$	*Width of Resonance* $\Gamma = 2.12\text{GeV}, \tau = \frac{\hbar}{\Gamma}$
Z^0	1	91.17	$e^+e^-, \mu^+\mu^-, \tau^+\tau^-$ $\nu\bar{\nu}$, *hadrons*	$\Gamma = 2.487\text{GeV}$
g (gluon)	1^-	0	—	*Stable*
Graviton	2	0	—	*Stable*

(F) Some Baryon Resonances

Resonances	Mass MeV/C^2	Spin J	Isospin I	Lifetime sec	Decay Modes
N	1450–3030	$\frac{1}{2}-\frac{9}{2}$	$\frac{1}{2}$	$\approx 10^{-23}$	$N\pi$, $N\pi\pi$, $N = n$ or p
Δ	1230–3230	$\frac{1}{2}-\frac{11}{2}$	$\frac{3}{2}$	$\approx 10^{-23}$	$N\pi$, $N\pi\pi$
Λ	1405–2600	$\frac{1}{2}-\frac{7}{2}$	0	$\approx 10^{-23}$	$\Sigma\pi$, $N\bar{K}$, $\Lambda\pi\pi$
Σ	1385–2600	$\frac{1}{2}-\frac{7}{2}$	1	$\approx 10^{-23}$	$\Sigma\pi$, $N\bar{K}$, $\Lambda\pi\pi$
Ξ	1530–1940	$\frac{3}{2}$	$\frac{1}{2}$	$\approx 10^{-23}$	$\Xi\pi$

1.12 THE MODERN ERA

We have noticed in the previous sections that the modern particle accelerators with their energetic beams produced a large number of baryons and mesons or called collectively the hadrons. But the hadrons, which were being observed, could not be arranged in an ordered way. This situation was similar to the situation, which existed before Mendeleev introduced the Periodic table of elements. During that period, i.e., some hundred years ago, many elements had been identified but there was no order to arrange them. They were finally arranged in the form of a Periodic table in which the properties of elements became the basis of their location in the Periodic table.

Here, in elementary particle physics too, an ordered arrangement was required to place the already discovered baryons and mesons and make predictions about new ones. Murray Gell–Mann in 1961 introduced a way to arrange the baryons and mesons and the way is called the *Eightfold Way.* The Eightfold Way arranges the baryons and mesons in geometrical patterns according to the charge and strangeness (hypercharge) of the particles. It is the Eightfold Way that initiated the Modern Era in particle physics.

1.12.1 The Eightfold Way

Let us look at the following table of eight baryons, which are lightest and have the spin-parity configuration $J^P = \left(\frac{1}{2}\right)^+$

Particle	Mass	Isospin I	I_3	Baryon no. B	Charge no. Q	Strangeness S	Hyper charge Y = B + S
p	938.3	$\frac{1}{2}$	$+\frac{1}{2}$	+ 1	+ 1	0	1
n	939.6	$\frac{1}{2}$	$-\frac{1}{2}$	+ 1	0	0	1
Λ	1115.6	0	0	+ 1	0	− 1	0
Σ^+	1189.4	1	+ 1	+ 1	+ 1	− 1	0
Σ^0	1192.6	1	0	+ 1	0	− 1	0
Σ^-	1197.4	1	− 1	+ 1	− 1	− 1	0
Ξ^0	1314.9	$\frac{1}{2}$	$+\frac{1}{2}$	+ 1	0	− 2	− 1
Ξ^-	1321.3	$\frac{1}{2}$	$-\frac{1}{2}$	+ 1	− 1	− 2	− 1

Now plot I_3 versus Y graph. Why we have chosen I_3 and Y-axes will become clear when we take SU (3) representations. The pattern, which we get, is drawn below [Figs. 1.31 (a, b)].

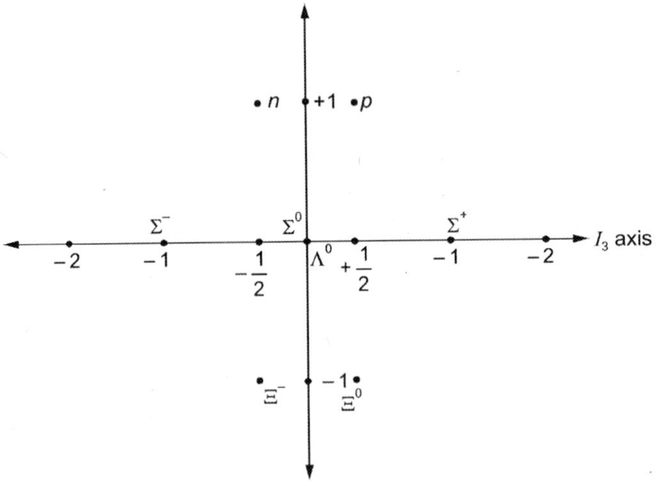

Fig. 1.31(a): I_3 –Y plot of $J^P = \frac{1^+}{2}$ baryons

Joining the points, we get a hexagonal pattern as in Fig. (1.31b) from which we can notice that the $J^P = \left(\frac{1}{2}\right)^+$ eight (lightest) baryons fit into a hexagonal

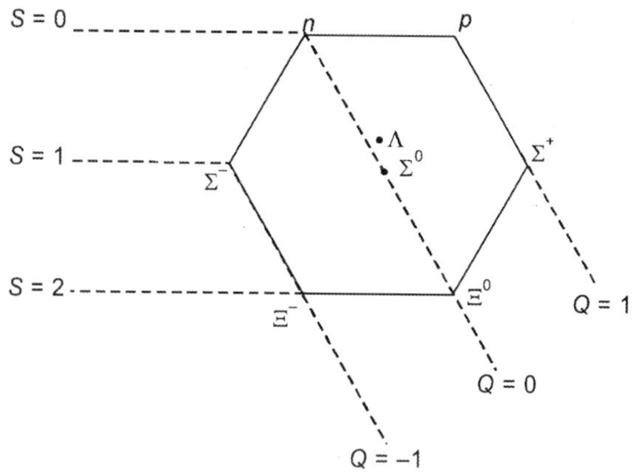

Fig. 1.31(b): Hexagonal pattern of baryon octet of Fig. 1. 31(a)

array with two baryons at the center. It is also observed that the particles of like charges lie along the diagonals shown by $Q = +1$, $Q = 0$ and $Q = -1$ axes.

The hexagonal pattern suggests that the eight baryons are different states of a single entity just as proton and neutron are two states of N (nucleon), $\Sigma^{\pm 0}$ are three states of a single entity Σ and so on. In N and Σ, etc.; however, a single parameter I_3 separates the members, whereas, two parameters I_3 and Y differentiate the members of the octet. If this analogy is correct the members of the octet must have nearly the same mass. For instance, we know that the neutron-proton mass difference is only about 1.3 MeV/C^2 and that between Σ^0 and $\Sigma^+ \approx 3.2$ MeV/C^2, $m(\Sigma^-) - m(\Sigma^0) \approx 4.8$ MeV/C^2 and even $m(\Sigma^-) - m(\Sigma^+) \approx 8$ MeV/C^2. Similar conclusions may be drawn for Ξ^0 and Ξ^- as well. The average mass of nucleon (lightest) is ≈ 938.9 MeV/C^2 and the average mass of Ξ particle (heaviest among the octet) ≈ 1318 MeV/C^2. Compared to the difference of ≈ 379 MeV/C^2 between the lightest and heaviest baryons of the octet, the mass differences of 1.3, 4.8 and 8 MeV/C^2 are small. These mass differences between two charge states of an isospin doublet or triplet are attributed to electromagnetic self-energy effects produced by virtual interactions with the electromagnetic field. However, in the case of baryon octet, where mass difference between the lightest and heaviest member is as large as 379 MeV/C^2 the breaking of flavour SU (3) symmetry is regarded as the cause of such large mass difference. We shall read about SU (3) (flavour) and SU (3) (colour) symmetries in next Chapters. The presence of large mass difference also implies that the flavor SU (3) symmetry is approximate. For exact symmetry the masses must have been equal. Therefore, the approximate nature of flavour SU (3) symmetry allows us to regard the baryons of the octet as different states of one single entity. The eight particle states of this octet are also called $J^P = \left(\frac{1}{2}\right)^+$ SU (3) multiplets.

The heavier baryons, which have been given in the following table are represented by ten states, hence are called *decuplet*. Each baryon state has spin $\frac{3}{2}$ and parity even (+) and therefore the decuplet states can also be called $J^P = \left(\frac{3}{2}\right)^+$ SU (3) multiplets.

Name of Multiplet	Particles	Mass MeV/C²	I	I_3	B	S	Y
Δ (Delta)	Δ^{++}	1230	$\frac{3}{2}$	$+\frac{3}{2}$	1	0	1
	Δ^{+}	1231	$\frac{3}{2}$	$+\frac{1}{2}$	1	0	1
	Δ^{0}	1233	$\frac{3}{2}$	$-\frac{1}{2}$	1	0	1
	Δ^{-}	1234	$\frac{3}{2}$	$-\frac{3}{2}$	1	0	1
Σ* (Sigma Star)	Σ^{*+}	1383	1	+ 1	1	– 1	0
	Σ^{*0}	1385	1	0	1	– 1	0
	Σ^{*-}	1387	1	– 1	1	– 1	0
Ξ* (Xi Star)	Ξ^{*0}	1532	$\frac{1}{2}$	$+\frac{1}{2}$	1	– 2	– 1
	Ξ^{*-}	1535	$\frac{1}{2}$	$-\frac{1}{2}$	1	– 2	– 1
	Ω^{-}	1672	0	0	1	– 3	– 2

Above ten heavy baryon states are however, not get arranged in a hexagonal

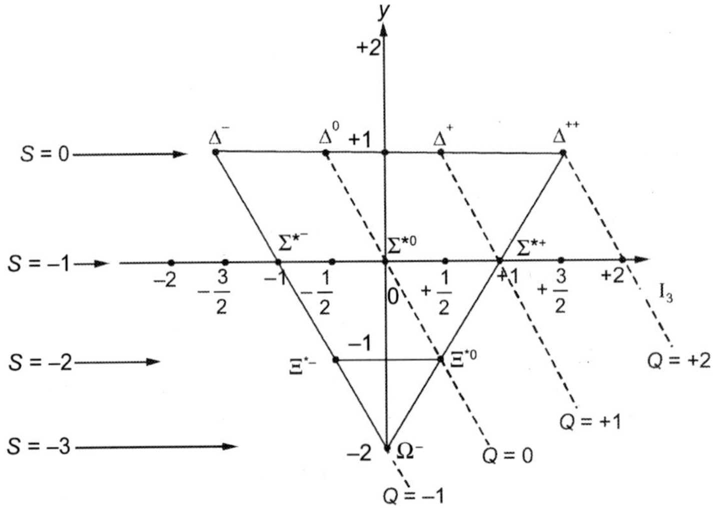

Fig. 1.32: The $J^P = \frac{3^+}{2}$ baryon decuplet. Isospin multiplets lie along the horizontal lines Members along dashed lines have constant electric charges

array but in a triangular array as can be seen by drawing the $I_3 - Y$ plot (Fig. 1.32).

The tenth particle in the decuplet is Ω^- particle. Before the discovery of Ω^- particle, only nine were experimentally known and $I_3 - Y$ arrangement of these nine particles left a vacancy at $I_3 = 0$, $Y = -2$. Gell–Mann predicted that a negatively charged particle with $(I_3, Y) = (0, -2)$ must be discovered and certainly, in 1964 such a particle was discovered. This imparted validity to geometrical arrangements of particles, which were provided by the Eightfold Way.

Not only baryons but also mesons can be arranged in hexagonal patterns. The only difference is, whereas, both mesons and their antiparticles find a place in a particular geometrical pattern (hexagon), the baryons and their antiparticles get arranged in separate geometrical patterns. It may be noted that in baryon octet (hexagon) (and decuplet also) all particles are baryons and no antibaryons. Σ^- is not the antiparticle of Σ^+. The antiparticles of Σ^+, Σ^0 Σ^- are $\overline{\Sigma}^+, \overline{\Sigma}^0$ and $\overline{\Sigma}^-$ respectively. Similar is the case of baryon *decuplet*.

Similar to the baryon octet, the Eightfold Way provides a hexagon for the lightest pseudoscalar (spin $J = 0$, parity P = negative) mesons, but here at the center we have three mesons. Hence because of nine particles, the meson hexagon is called a *nonet*. The mesons, which constitute the lightest nonet, are given below. The pattern is drawn in Fig. (1.33).

Particle	Mass MeV/C²	I	I_3	Y = S
π { π^+	140	1	+ 1	0
π^0	135	1	0	0
π^-	140	1	– 1	0
K { K^+	494	$\frac{1}{2}$	$+\frac{1}{2}$	1
K^0	498	$\frac{1}{2}$	$-\frac{1}{2}$	1
$\bar{K}$ { $\bar{K}^0$	498	$\frac{1}{2}$	$+\frac{1}{2}$	– 1
K^-	494	$\frac{1}{2}$	$-\frac{1}{2}$	– 1
η	549	0	0	0
η'	958	0	0	0

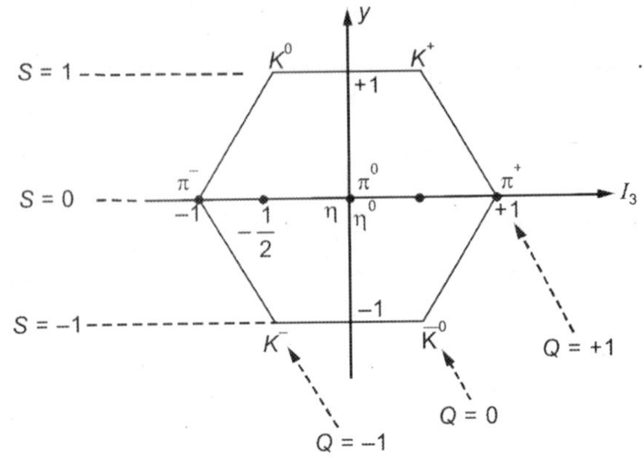

Fig. 1.33: $J^P = 0^-$ Meson nonet

Likewise, the vector mesons $J^P = 1^-$, also form a nonet (Fig. 1.34)

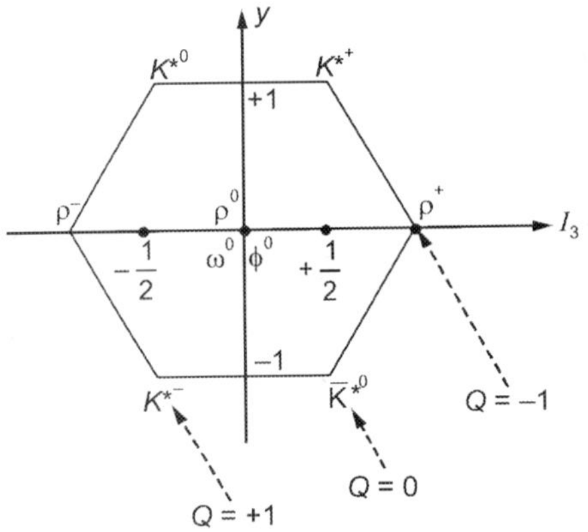

Fig. 1.34: Nine vector mesons (1^- meson nonet). Horizontal lines connect the same isospin multiplets

At least six sets of meson nonets are known. In baryons, however, beyond octet and decuplet, we have only partially filled diagrams but no complete baryon multiplets.

From the above we observe that both baryons and mesons can be fitted into geometrical patterns from which the properties of hadrons can be read. The Eightfold Way, therefore, assumes significance in that it provided organizational structure to hadrons.

1.12.2 The Quark Structure of Hadrons

The Eightfold Way successfully classified hadrons into hexagonal and triangular patterns. A question arose naturally—why the particles in a multiplet which otherwise look different, are seen as different states of one entity? This meant that there are more fundamental particles of which each of the particles in SU (3) multiplet is made of. Gell–Mann called these fundamental particles quarks. The model of hadrons, which showed hadrons are made of quarks, is called the quark model of hadrons.

It was in fact not that the quarks were proposed straightaway. Before this Fermi and Yang had proposed that all the elementary particles are made of nucleon-antinucleon ($N\overline{N}$) pairs. However, the Fermi–Yang model suffered a problem that strange particles could not be made from ($N\overline{N}$) (N & $\overline{N}$ are nonstrange particles) combination. To make for the deficiency, Sakata in 1956 proposed a model called Sakata model in which proton (p), neutron (n) and lambda (Λ) particle made up the fundamental triplet. All the elementary particles were supposed to be made up of p, n and Λ. It did incorporate strange particles, as one of the fundamental constituents Λ was strange. The mesons were described fairly well by Sakata model but when it came to baryons, the combination like $pn\Lambda$ produced the baryon number $B = 3$ and no such state was observed. The combinations which could give $B = 1$ were $p\bar{n}\Lambda$ and $pn\overline{\Lambda}$. Since $\overline{\Lambda}$ has $S = +1$, so if $pn\overline{\Lambda}$ exists we should have a baryon with $S = +1$. Such state too does not exist in nature. So, Sakata model too did not work. Despite this inadequacy, however, the Sakata model paved way for the quark model.

In quark model, the base triplet is not p, n, Λ but particles with fractional baryon number. Gell–Mann and Zweig called each object with a baryon number 1/3 as quark. They showed that the octet of the Eightfold Way could be thought in terms of combinations of a basic triplet of quarks. The triplet consisted of three types of quarks, u, d and s. The three types were called the three flavours of quarks. Thus upness, downness and strangeness were called the initial three flavours of quarks. Later three more flavours namely, charmness, topness and bottomness were discovered. The properties of the fundamental triplet (u, d, s) and their antiparticles ($\bar{u}, \bar{d}, \bar{s}$) can be listed as:

Flavour	**Spin** J	**Charge** Q **in units of** (+e)	I	I_3	B	S	$Y = B + S$
u	$\frac{1}{2}$	$+\frac{2}{3}$	$\frac{1}{2}$	$+\frac{1}{2}$	$\frac{1}{3}$	0	$\frac{1}{3}$
d	$\frac{1}{2}$	$-\frac{1}{3}$	$\frac{1}{2}$	$-\frac{1}{2}$	$\frac{1}{3}$	0	$\frac{1}{3}$
s	$\frac{1}{2}$	$-\frac{1}{3}$	0	0	$\frac{1}{3}$	-1	$-\frac{2}{3}$
$\bar{u}$	$\frac{1}{2}$	$-\frac{2}{3}$	$\frac{1}{2}$	$-\frac{1}{2}$	$-\frac{1}{3}$	0	$-\frac{1}{3}$
$\bar{d}$	$\frac{1}{2}$	$+\frac{1}{3}$	$\frac{1}{2}$	$+\frac{1}{2}$	$-\frac{1}{3}$	0	$-\frac{1}{3}$
$\bar{s}$	$\frac{1}{2}$	$+\frac{1}{3}$	0	0	$-\frac{1}{3}$	$+1$	$+\frac{2}{3}$

The $I_3 - Y$ plots of u, d, s and $\bar{u}, \bar{d}, \bar{s}$ form triangular Eightfold Way patterns (Fig. 1.35).

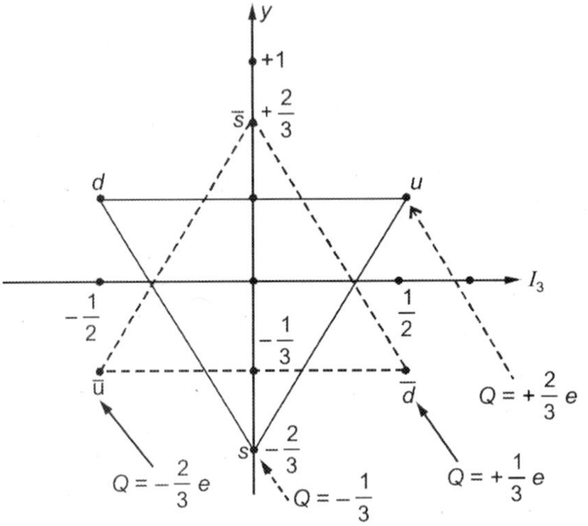

Fig. 1.35: Quark, antiquark triangles. Quark triangle is with continuous lines and antiquark triangle has broken sides. The triangles are inverted

The quark model of hadrons tells that

1. Every baryon is made of three quarks and every antibaryon is composed of three antiquarks.
2. Every meson is made of a quark and an antiquark.

Above prescription can be used to assign quark composition to baryons and mesons. The quark contents of some mesons and baryons and their resonances have been listed below:

(A) Quark composition of mesons and meson resonances

Meson	J^P	Quark composition	Meson reasonance	J^P	Quark composition
π^+	0^-	$u\bar{d}$	η	0^-	$(u\bar{u} + d\bar{d} - 2s\bar{s})/\sqrt{6}$
π^0	0^-	$(u\bar{u} - d\bar{d})/\sqrt{2}$	η'	0^-	$(u\bar{u} + d\bar{d} + s\bar{s})/\sqrt{3}$
π^-	0^-	$\bar{u}d$	ρ^+	1^-	$u\bar{d}$
K^+	0^-	$u\bar{s}$	ρ^0	1^-	$(u\bar{u} - d\bar{d})/\sqrt{2}$
K^-	0^-	$\bar{u}s$	ρ^-	1^-	$\bar{u}d$
K^0	0^-	$d\bar{s}$	ω	1^-	$(u\bar{u} + d\bar{d})/\sqrt{2}$
$\bar{K}^0$	0^-	$\bar{d}s$	ϕ	1^-	$s\bar{s}$
			K^{*+}	1^-	$u\bar{s}$
			K^{*-}	1^-	$\bar{u}s$
			K^{*0}	1^-	$d\bar{s}$
			$\bar{K}^{*0}$	1^-	$\bar{d}s$

(*B*) Quark content of baryons and baryon resonances

Baryon	Quark content	Baryon reasonance	Quark content
p	uud	Δ^{++}	uuu
n	udd	Δ^{+}	uud
Λ	uds	Δ^{0}	udd
Σ^{+}	uus	Δ^{-}	ddd
Σ^{0}	uds	Σ^{*+}	uus
Σ^{-}	dds	Σ^{*0}	uds
Ξ^{0}	uss	Σ^{*-}	dds
Ξ^{-}	dss	Ξ^{*0}	uss
		Ξ^{*-}	dss
		Ω^{-}	sss

1.12.2.1 Quark Colours

In the above table ***B***, we focus our attention on baryon resonance Δ^{++}. Δ^{++} has charge (+ 2*e*), spin $J = \frac{3}{2}$ and is composed of three spin (J) $\frac{1}{2}$, up (u) quarks.

This means Δ^{++} state contains three identical spin 1/2 particles (i.e. u quarks).

Having three identical fermions in one state is against the Pauli's exclusion principle. In order to justify the reality Δ^{++}, the quarks are assigned a hypothetical quantum number, the colour, and it is supposed that quarks come in three primary colours; Red (R), Green. (G) and Blue (B). With the introduction of colour quantum number, the identical character of three u quarks in Δ^{++} can be broken and we write Δ^{++} is composed of $\mu_R G \mu_B$. The colour quantum numbers save the violation of Pauli's exclusion principle. However, with the introduction of colour quantum number proton $\rightarrow$*uud* can have many possible combinations of *uud*. They could be $u_R u_G d_{G'}$ $u_R u_B d_{B'}$ $u_G u_G d_G$ etc., which would have implied that so many proton states exist. Since only one proton state exists, we impose an additional condition on colours that for a particle to become observable it must be colourless or white or better a colour singlet (sec. 1.9.5). Thus in *uud* each of the three quarks must have a different colour. Keeping flavours as such (*uud*) and assigning permutations of colours *RGB* to quarks, the possibilities may be found. They are $u_R u_G d_{B'} u_R$ $u_B d_{G'}$ $u_G u_B d_{R'}$ $u_G u_R d_{B'}$ $u_B u_R d_{G'} u_B u_G d_R$ which are obtained by interchanging the colours of two quarks from the preceding combination. But quarks are spin 1/2 objects; the interchange of any quantum number of two spin 1/2 objects must make the state (wavefunction) antisymmetric. Then if we say that ψ (colour) is the colour wavefunction of proton, it must be antisymmetric under the interchange of any two colours and will be expressed by

$$\psi\ (\text{colour}) = (u_R u_G d_B - u_R u_B d_G + u_G u_B d_R - u_G u_R d_B + u_B u_R d_G - u_B u_G d_R)/\sqrt{6} \quad (1.205)$$

Since the flavour in each of the six terms is same (uud), we may retain only the colour indices and write

$$\psi(\text{colour}) = (RGB - RBG + GBR - GRB + BRG - BGR)/\sqrt{6} \quad (1.206)$$

where $\frac{1}{\sqrt{6}}$ is the normalization constant. Eqn. (1.205) is for proton but eqn. (1.206) is for any baryon (three quark state). If we interchange any two colours in each term of eqn. (1.206), what we get shall be the negative of eqn. (1.206) telling that eqn. (1.206) is antisymmetric. In fact the proton state or any baryon state would be

$$\psi = \psi(\text{space})\psi(\text{spin})\psi(\text{flavour})\psi(\text{colour}) \quad (1.207)$$

where ψ(space is the wave function related to spatial location of three quarks of the baryons, ψ(spin) represents the wavefunction related to spins of quarks, i.e., whether they are up (↑) or down (↓). ψ (flavour) is the wavefunction, which tells which combination of u, d, s is involved. Out of these four wavefunctions ψ(colour) is antisymmetric. Since, ψ belongs to a baryon, which is a fermion, it (the ψ) must be antisymmetric. This implies that for a baryon, the product ψ(space)ψ(spin)ψ(flavour) must be symmetric. In the ground state, ψ(space) is symmetric, hence for the ground state of the baryon, the product ψ(spin) ψ(flavour) must also be symmetric, which can be obtained by having both ψ(spin)ψ(flavour)ψ(spin) and ψ(flavour) symmetric or by having both antisymmetric. In the case of spin-3/2 baryon decuplet, we have [ψ(spin)

$$\psi(\text{flavour})]_{\text{baryon decuplet}} = \psi_S(\text{spin})\psi_S\ \psi(\text{flavour}) \quad (1.208)$$

where S stands for symmetric. In the case of baryon octet however, we have

$$[\psi(\text{spin})\psi(\text{flavour})]_{\text{baryon octet}} = \frac{\sqrt{2}}{3}[\psi_{12}\ (\text{spin})\psi_{12}\ (\text{flavour}) + \psi_{23}\ (\text{spin})\psi_{23}\ (\text{flavour}) + \psi_{13}\ (\text{spin})\psi_{13}\ (\text{flavour})] \quad (1.209)$$

where both ψ_{12}(spin) and ψ_{12} (flavour) are antisymmetric under the interchange of quark 1 and quark 2. The product (first term) is therefore symmetric. Similarly second and third terms become symmetric which make the wavefunction (1.209) symmetric. When eqs. (1.208) and (1.209) are multiplied by the antisymmetric ψ(colour) the total wavefunction becomes antisymmetric, which should be the case for baryon (a fermion).

In the case of mesons, which are bosons and have quark-antiquark structure, the colour wavefunction can be $\bar{R}R, \bar{G}G$ *or* $\bar{B}B$ each of which is colourless. But it is only the combination

$$\psi(\text{colour}) = \frac{1}{\sqrt{3}}(\bar{R}R + \bar{G}G + \bar{B}B) \quad (1.210)$$

a colour singlet that remains invariant by rotation in R, G, B colour space, which is the colour wavefunction for bosons. Obviously eqn. (1.210) is symmetric under the interchange of colours.

1.12.2.2 Experimental Support to Quark Model

According to quark model, hadrons are composed of quarks. A serious limitation of quark model is that no isolated quarks have ever been observed experimentally. The explanation of this limitation is however, given in terms of quark confinement which tells that quarks are totally confined within baryons and mesons and no free quarks can come out of them. The validity of this picture demands that the charge and mass of hadrons should not be uniformly distributed within their size instead lumps must be felt inside the hadrons. To detect the possibility of these lumps, experiments analogous to Rutherford's alpha scattering experiments were carried out.

In the Rutherford's alpha scattering experiment, alpha particles were targeted at a thin gold foil. It was observed that most of the alpha particles passed straight through the gold foil or suffered small deflection. A few alpha particles, however, deflected through large angles. The large cross section observed at large angles was due to alpha particles deflected by the nuclei deep inside the gold atoms. The experimental results of this experiment have been shown in the following figure (Fig. 1.36).

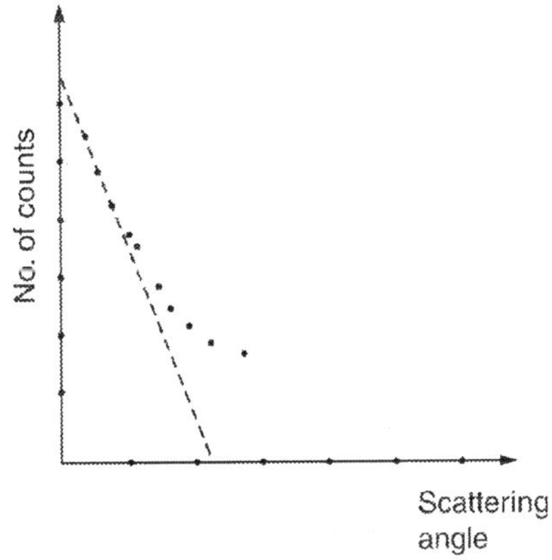

Fig. 1.36: Inelastic scattering of charged particles by atoms. The graph indicating experimental results shows that atom has substructure

Similar to Rutherford scattering experiment, high-energy scattering experiments have been conducted. In these experiments, proton is the target; the high-energy beams of electrons, neutrinos and protons have been used as different projectiles. As an example, the following diagram (Fig. 1.37) shows the electron

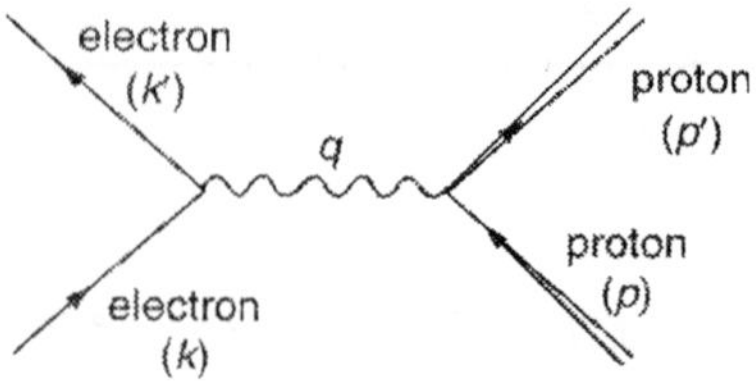

Fig.1.37: Electron-proton scattering diagram

proton scattering in which a virtual photon of four momentum q is being exchanged between the electron and proton. As a result of this exchange the final four momenta of electron and proton become $k' = k \mp q$ and $p' = p \pm q$. Instead of q, however, the scattering process is discussed in terms of q^2, a Lorentz invariant quantity. When $q^2 \rightarrow 0$, the virtual photon has a very long wavelength and does not respond to the structure of proton. In this limit of q^2, the virtual photon sees proton just as a particle of charge $+e$ and tells nothing about the constituents of proton. When q^2 increases but is still relatively small, the proton may be only excited into a resonant state, which may then decay. For instance the resonant state may decay into a nucleon and a pion. When q^2 values are high enough, the wavelength of virtual photon becomes small and thus has the capacity to probe the inside of proton. This $q^2 \rightarrow \infty$ scattering is called deep *inelastic scattering*. Its results are drawn in Fig. (1.38). This plot is similar to the α-particle scattering

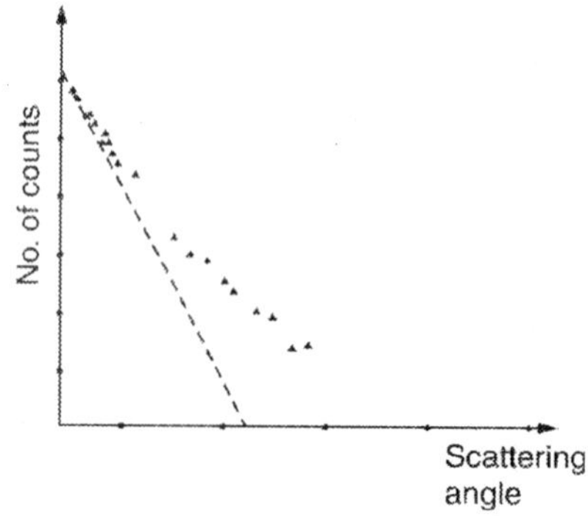

Fig. 1.38: A graph similar to Fig. 1.36: but obtained by observing the inelastic scattering of proton beam by proton targets. The experimental results indicated in the graph show that proton has substructure

plot suggesting that proton too has a structure and deep inelastic scattering of electrons is due to charge centre or centres present inside the proton. A detailed analysis showed that there are three charge centres inside the proton. These scattering centres were called *partons* and it was found that they have the properties similar to quarks. The deep inelastic experiment by establishing structure of proton came strongly in support of the quark model. At such higher values of q^2 the energy transferred to proton is so great that it breaks up into many hadrons and the proton completely loses its identity (Fig. 1.39).

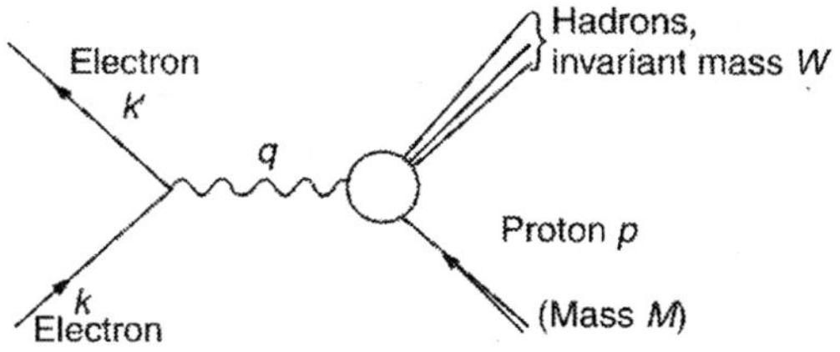

Fig. 1.39: Feynman diagram for electron-proton inelastic scattering

In addition to deep inelastic scattering experiments, there are other ways to probe quarks. If very high-energy electron and positron beams were made to collide, they would produce photons. These high-energy photons can decay into $q\overline{q}$ pair (shown in Fig. 1.40). When this pair separates by a distance of about 1 Fermi, the colour interactions between q and $\overline{q}$ become strong enough to retard them. Similar to the deceleration of electric charge emitting photon, these retarded quarks emit hadrons, which are mostly pions. Though, in this process no free quark is seen but the detection of π-mesons implies that they were made by $q\overline{q}$ pairs which themselves were produced by the high-energy photons, generated by e^+e^- collider (Fig. 1.40).

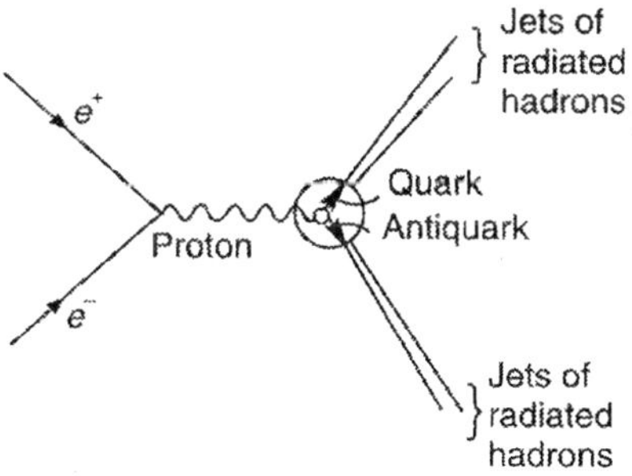

Fig. 1.40: Colliding electron-positron beams. The photon produced by colliding e^- and e^+ beams decay into $q\overline{q}$ pair. When the pair separates by a distance $\simeq$ 1 Fermi, the colour interactions between q and $\overline{q}$ become strong, resulting in the deceleration of quarks. The decelerated quarks radiate jets of hadrons, which are mostly π-mesons. Using this technique, quarks can be probed.

1.12.2.3 Quark Confinement

No free quarks are observed in nature. The reason is that quarks remain confined within the hadrons and no free quarks can come out of them. This is called *quark confinement*. A qualitative description of the quark confinement mechanism can be given by observing the contrasts between the electric charge screening in *QED* and the colour charge (anti) screening in *QCD*.

In Quantum Electrodynamics for electrons virtual electron-positron pairs are produced around any given electron. Such virtual pairs are produced in even a pure vacuum state. The presence of a given charge polarizes the vacuum and such effects are called *vacuum polarization effects*. When virtual $e^- e^+$ pairs have been produced around an electron, they will align themselves (Fig. 1.41) such that positrons surround the electron at the centre. Coming of positrons closer to bare electron is in fact the polarization of vacuum. Now we try to measure the charge of the given electron, which has been shown at the centre of the circle

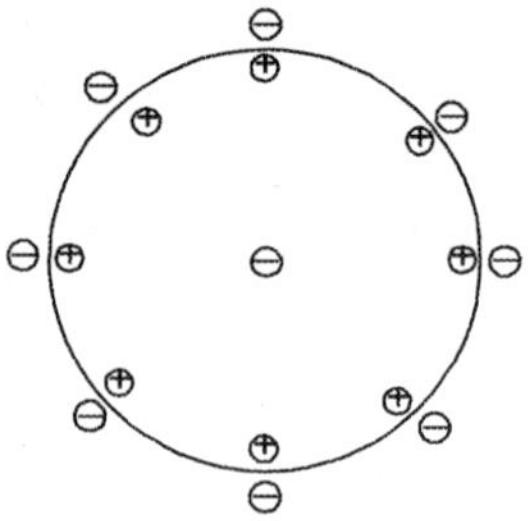

Fig. 1.41: Virtual e^-e^+ pairs around an electron

in Fig. (1.41). To measure the charge of electron we would take a test charge ⊕ and the measurement of the charge of electron would depend on the location of the test charge. We may have two kinds of situations as shown in figures (1.42 a, b).

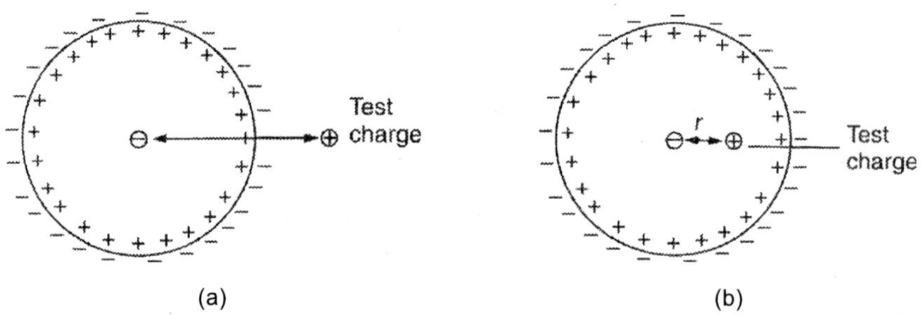

Fig. 1.42: Measurement of charge of electron. (a) The probe, i.e., the test charge is at a long distance. From such long distances the virtual electron-positron pairs screen the electron charge and the probe's measurements give a less amount of electron's charge. In figure (b) the probe has crossed over the e^-e^+ pairs and hence it measures the charge of electron larger than measured in (a). The measurement of electron's charge depends on the place from where the charge is being measured

When the test charge is away from the electron, the electron as viewed by the test charge gets screened by positrons, which have gathered around the electron. So in this position of test charge (Fig. 1.42a), a lesser amount of electron charge is seen by the test charge and we say that if the measurements were made from Fig. (1.42a) type position, i.e., long distances, the effective charge of electron would be less than its bare charge. On the other hand, if the test charge is brought close to the electron (Fig. 1.42b), the test charge would penetrate the cloud of positrons that screen the charge of electron. Obviously from closer distances, the test charge will see a larger amount of electron charge.

What we conclude from above is that, in *QED*, the electron charge appears less from larger distances and large from closer distances. If we plot this observation, we would get the Figure like (1.43). The measured charge is also called *the effective charge*. The measure of distance is with respect to the Compton wavelength $\lambdabar = \frac{\hbar}{mc}$ of electron.

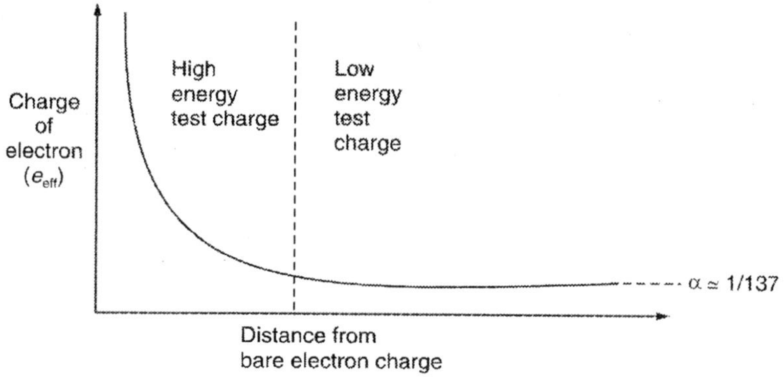

Fig. 1.43: Variation in the electron's charge with respect to the place of probe. High energy ⇒ the probe has penetrated the virtual electron-positron pairs and has come close to bare electron.

Short distances are those, which are comparable to $\lambdabar$. At distances large compared with the Compton wavelength the effective charge of electron is the standard charge e, corresponding to a fine structure constant $\alpha = \frac{e^2}{\hbar c} = \frac{1}{137}$. Since charge is referred to as the coupling constant, we can say that effective coupling constant (e_{eff}) runs with p^2 where p is the four momentum of the virtual photon exchanged between the electron and the test charge. Thus we have a concept of running coupling constant $e_{effective}$ or α_{eff} which in the case of electromagnetic interaction (discussed above) increases when interacting particles are brought close.

In the case of Quantum Chromodynamics, the colour charge of quarks is the analogue of electric charge in *QED*. Two quarks interact (strong interaction) with the exchange of a virtual gluon which itself carries colour (colour charge). The photons of *QED* do not carry electric charge, so there is no photon-photon interaction. That is why the field theories, in which no interaction between the quanta of the field takes place, are called *the Abelian field theories*. In *QCD*, the colour field binds quarks in nucleons and also the nucleons in nuclei. The gluons are the quanta of colour field and due to carrying of colour charge we have gluon—gluon interactions. Therefore, when there is interaction (as in *QCD*) between the quanta of a field, the field theory is called *non-Abelian field theory*.

Now, imagine a quark (free quarks are not a reality) in vacuum. The bare colour charge of the quark is modified by vacuum polarization and we have quark—antiquark pairs around a given quark. The quark-antiquark pairs interact via virtual gluons. The gluons, being coloured, interact among themselves. In *QED* we had $e^+ e^-$ pairs, which were bound via virtual photons, but photons did not interact. D.J. Gross, F. Wilczek and H.D. Politzer have worked it out on the basis of interaction diagrams that the effect of gluon-gluon interaction is to enhance the colour charge of the quark. Then, if the colour of a given quark is blue, it will be surrounded preferentially by blue colour as shown in Fig. (1.44). When the test colour charge is at *P* like points, i.e., the points away from the quark, the test colour charge would measure larger amount of colour (blue) charge.

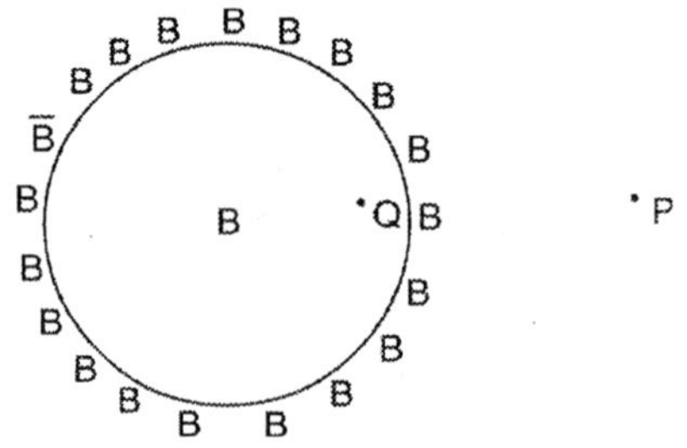

Fig. 1.44: Measurement of colour charge. Due to gluon-gluon interaction a colour charge (blue) is surrounded by similar colour charges. A long distance probe (at *P*) measures larger colour charge than a close distance (at Q) probe. The result of the 'measurements' of colour charge are opposite to those of electron charge

On the other hand, if the test colour charge is at location like *Q*, it penetrates the sphere of colour charge and some of it is left behind and thus the test charge measures a lesser amount of colour charge. Thus in *QCD*, from larger distances larger amount of colour charge is measured and from distances close to the given quark lesser amount of colour charge is measured. The result has been drawn in Fig. (1.45).

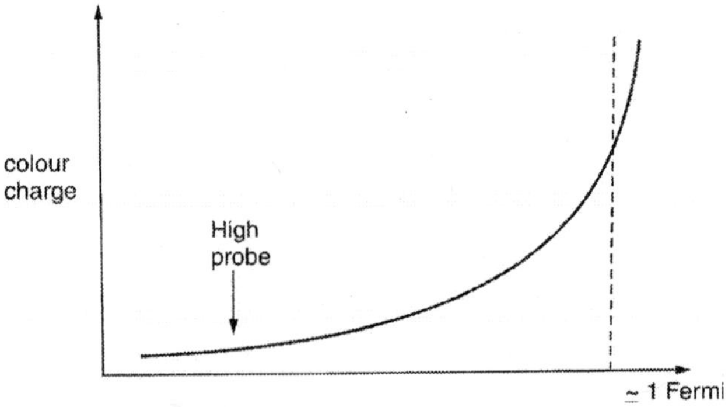

Fig. 1.45: Variation in colour charge with respect to the energy of probe. The high energy probe penetrates the colour cloud (Fig. 1.44) of quarks and sees less colour charge. From distances (Q of Fig. 1.44) the coupling between the probe and bare colour charge becomes feeble. This is asymptotic freedom. The large distances from where the colour charge appears large are about 1 Fermi only

This observation is just opposite to that in *QED*, (Fig. 1.43) where screening of the electron charge led to a lesser measurement at long distances. In *QCD*, we have therefore antiscreening effect, which enhances the colour charge for long distances and diminishes for short distances. The colour charge would be a measure of strong coupling constant. Therefore, when quarks are brought together, each quark sees a lesser amount of colour charge on other and hence the strong coupling constant will have smaller and smaller values, closer we bring the quarks. Small coupling constants would bring a feeble binding and hence at short distances the quarks behave as almost free particles. This is called *asymptotic freedom*. On the other hand, when the separation between quarks increases the strong coupling constant would increase and attraction between quarks would increase. Normally we would have expected that when separation between quarks increases, they would become free but here opposite happens. When separation between quarks increases the strong attractive force comes into play and brings them closer. In this way due to antiscreening of colour charge, quarks remain confined within the hadron.

The origin, of total confinement of quarks to colourless hadrons, is believed to be the so-called *infrared slavery*. To understand it's meaning, let us consider a quark-antiquark pair and compare it with an electron-positron pair. In the latter case, the potential energy is inversely proportional to the separation r, between the electron and positron. The lines of force between e^- and e^+ bulge out because photons that are exchanged between e^- and e^+ do not have self-interaction. In contrast, the colour field lines of force, between quark and antiquark squeeze into a tube like region. This is due to the gluons, which are exchanged between q and and $\bar{q}$, interact among them. Due to this tube like shape of colour field lines of force, the potential energy between q and $\bar{q}$ increases with separation. Whereas, with r, the separation, $\rightarrow \infty$, the (Coulomb field) potential energy of $e^-e^+ \rightarrow 0$, for colour field ($q\bar{q}$) for $r \rightarrow \infty$, $V(r) \rightarrow \infty$. This is called infrared (large r) slavery due to which the quarks (and gluons) never escape.

1.12.2.4 Quark Masses

We have talked about the three flavours u, d and s. These are the flavours of lighter quarks. Charmness (c), bottomness (b) and topness (t), are other three flavours, which belong to heavy quarks. The quarks have two types of masses. One bare other effective. The intrinsic mass of quark is its bare mass. The mass of a quark flavour in a hadron is different from its bare mass and is called the effective mass of a particular quark flavour. Since free quarks do not exist, we cannot make direct experimental measurement of their masses. Only their masses can be estimated. Following table gives an idea about the quark masses.

Quark flavours	Bare mass MeV/C²	Effective mass MeV/C²			
		In mesons		In baryons	
u d	4.2 7.5	310			363
s	150	483			538
c	1100		1500		
b	4200		4700		
t			174000		

The light quarks are intrinsically very light but in hadrons they have a large mass. To estimate the masses of quarks we look for the masses of particles and their quark structure, e.g. within $J^P = 1^-$ meson multiplet we have (1GeV = 10^3 MeV).

$$\begin{aligned} m_\omega \approx m_\rho(u\bar{u}) &= 0.78 \text{ GeV} \Rightarrow m_u = 0.39 \text{ GeV} \\ m_\phi(s\bar{s}) &= 1.02 \text{ GeV} \Rightarrow m_s = 0.51 \text{ GeV} \\ m_u &\approx m_d \end{aligned} \tag{1.211}$$

1.12.2.5 The Discovery of J or ψ Meson

The proton, neutron and pion are the bound states of u and d quarks. This means that from the existence of p, n and π particles we infer the existence of u and d flavours. The discovery of $\eta[(u\bar{u} + d\bar{d} - 2s\bar{s})/\sqrt{6}], \phi(s\bar{s}), K^{*+}(u\bar{s})$, etc. mesons provides evidence for the s flavour of quarks. Till 1964, only three flavours (u, d & s) were proposed. Since quarks had no substructure and all the hadrons could be seen as made of qqq or $q\bar{q}$; the quarks emerged as the fundamental particles. In addition leptons, which too were structureless, were other fundamental particles. Obviously quarks and leptons made up the whole matter. However, till 1964, the situation was

$$\begin{aligned} &\textit{Quarks} \quad - \quad u, d, s \\ &\textit{Leptons} \quad - \quad e, \nu_e; \mu, \nu_\mu \end{aligned} \tag{1.212}$$

implying that there appears to be a symmetry between quarks and leptons but this is lacking because to four leptons we have only three quark flavours. To maintain the symmetry with respect to number of quark flavours and the number of leptons, Bjorken and Glashow proposed a new flavour for quarks.

$$\begin{aligned} &\textit{Quarks} \quad - \quad u, d, s, c \\ &\textit{Leptons} \quad - \quad e, \nu_e; \mu, \nu_\mu \end{aligned} \tag{1.213}$$

It was called charm flavour (c) and with this proposal we had a perfect symmetry between Quarks and Leptons. If the charm flavour existed, we must be able to discover a, $c\bar{c}$, meson. A discovery of $c\bar{c}$ meson would not only confirm the fourth flavour, it would validate the quark model as well.

It was in the November of 1974 that a new meson (resonance) was discovered by two independent groups of scientists. One led by C.C.Ting called this electrically neutral heavy meson as *J* meson and other led by Burton Richter, announced it as ψ meson. It is referred in literature as *J*/ψ particle.

The *J*/ψ particle is electrically neutral, its spin is one and parity odd. It was called a meson though it did not conform to the masses of mesons, which lay, between leptons and baryon masses. The mass of 3100 MeV/C^2 of *J*/ψ particles, which was more than three times the mass of proton, was quite large. Its lifetime was 1000 times greater than the lifetime (10^{-23} S) of the hadrons in this mass range. These properties made the *J*/ψ an unusual particle whose mass (3.1 GeV) could not be explained by existing *u*, *d*, *s* flavours and new flavour which was heavier than s was invented and *J*/ψ particle was universally accepted as the bound state of charm quark (c) and its antiquark ($\bar{c}$), i.e. $J/\psi = c\bar{c}$. With this fourth flavour, a series of new hadrons called charmed hadrons were predicted and indeed charmed baryons like $\Lambda_c^+ = udc, \Sigma_c^{++} = uuc$ were observed in 1975 itself. $D^0 = c\bar{u}$ and $D^+ = c\bar{d}$ were the first charmed mesons.

Before the discovery of *J*/ψ particle the quark model was not drawing the serious attention of scientists, but following the discovery of *J*/ψ particles and its implications, the quark model came not only into limelight but seemed to have found a firm ground to stand. Since, the discovery of *J*/ψ particle was announced in November (10), 1974, this discovery which virtually revolutionized the particle physics, was known as *November Revolution*.

1.12.2.6 The Upsilon (ϒ) Discovery

After discovering the *J*/ψ resonance at 3.1 GeV in e^+e^- annihilation, the total energy of the colliding e^+ and e^- beams was increased. At 9.46 GeV a narrow resonance was observed. This was called *Upsilon resonance*.

The mass of Upsilon particle was thus about ten times the mass of proton. Such heavy mass predicted another flavour, the bottom (b) flavour and upsilon was the $b\bar{b}$ meson. Thus just as *J*/ψ particle gave fourth flavour (charm) of quarks the upsilon (ϒ) particle gave the fifth flavour (bottom) of quarks. The discovery of heavy leptons the τ lepton and its neutrino increased the number of leptons to six and here we had only five quark flavours. To maintain the symmetry between the quarks and leptons a sixth flavour called topness was predicted. Direct observation of top events have put the top quark mass as 174.3 ± 5.1 GeV.

Thus we have three generations of quarks-first (*u*, *d*), second (*c*, *s*) and third (*t*, *b*). No evidence of higher generations of quarks has been found in any experiment. These generations of quarks are in perfect correspondence with the three-lepton generation, i.e.,

$$\begin{pmatrix} u \\ d \end{pmatrix}\begin{pmatrix} c \\ s \end{pmatrix}\begin{pmatrix} t \\ b \end{pmatrix} : \begin{pmatrix} \nu_e \\ e \end{pmatrix}\begin{pmatrix} \nu_\mu \\ \mu \end{pmatrix}\begin{pmatrix} \nu_\tau \\ \tau \end{pmatrix} \tag{1.214}$$

1.12.3 Intermediate Vector Bosons

The slow decay of unstable particles is caused by weak interactions. Weak interactions were first studied in beta decay of unstable nuclei. In his original theory of beta decay, Fermi treated the process as a contact interaction or point interaction in which all the particles interact at a single point in space-time.

$$n \to p + e^- + \bar{\nu}_e \tag{1.215}$$

For the beta decay process (1.215), the point interaction is drawn as in Fig. (1.46).

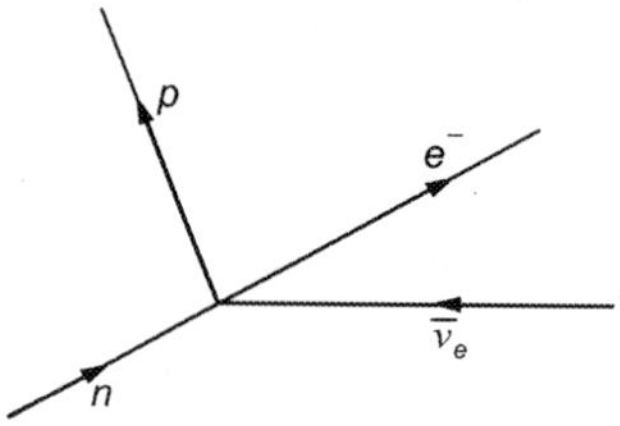

Fig. 1.46: The single point interaction shown through the example of neutron decay $n \to p + e^- + \bar{\nu}_e$

If weak interactions are explained by point interactions, they give acceptable results at low energies but at higher energies the point interaction mechanism leads to differential cross sections, which rise with energy, without any limit. In order to avoid such a catastrophe, the particles are considered to interact at different (two) space-time points and the interactions at these two space-time points are mediated by the exchange of some (vector) particle, which is known as intermediate vector boson. The beta decay process can now be drawn as Fig. (1.47),

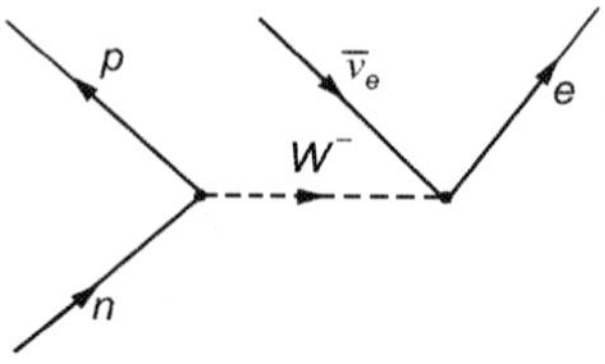

Fig. 1.47: The neutron decay shown taking place via the intermediate vector boson (W^- boson). Compared to (Fig. 1.46), this is a two-point interaction. The intermediate vector boson propagates between the two (space-time) points.

Thus the mediator of weak interaction is called the Intermediate Vector Boson (IMB). In the above example of β-decay, the conservation of electric charge at each vertex demands that the IMB be charged and here it is W^-. In the following weak interaction (scattering) processes, the intermediate vector bosons are W^+ and Z^0. The process

$$\nu_e + e^- \to e^- + \nu_e \tag{1.216}$$

drawn as (Fig. 1.48), is mediated by IMB, W^+, whereas, in the process

$$\nu_e + e^- \to \nu_e + e^- \tag{1.217}$$

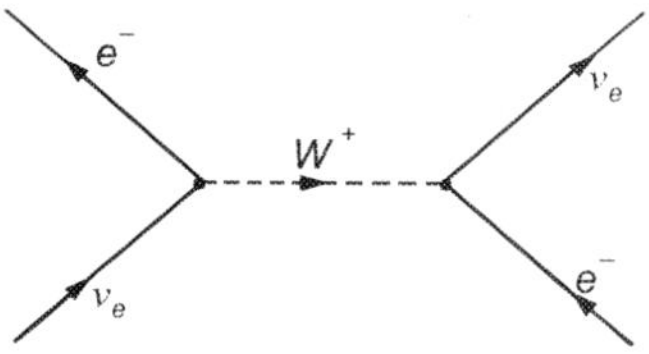

Fig. 1.48: The process $\nu_e + e^- \rightarrow e^- + \nu_e$ shown via the intermediate vector boson W^+

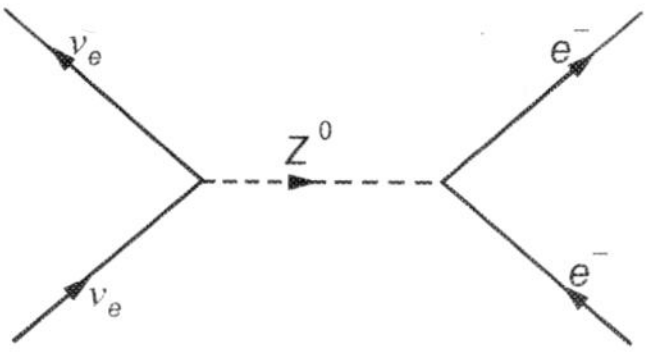

Fig. 1.49: The process $\nu_e + e^- \rightarrow \nu_e + e^-$ shown taking place via the Z^0 boson.

drawn as (Fig. 1.49), the Z^0 (the neutral Z boson) is the intermediate vector boson. The Glashow, Weinberg and Salam model, which unifies electromagnetic and weak interactions, predicted the mass of these IMB's and these particles were observed in 1983 in the proton-antiproton $(p\bar{p})$ colliders. The masses of these particles were observed as

$$M_{W^\pm} = 81 \pm 5 \text{ GeV}/\text{C}^2$$

and

$$M_{Z^0} - 95 \pm 3 \text{ GeV}/\text{C}^2 \tag{1.218}$$

which were in agreement of the predicted masses of $M_{W^\pm} = 81 \pm 2$ GeV/C^2 and $M_{Z^0} = 92 \pm 2$ GeV/C^2.

Not only the weak interactions, but every interaction has its mediator. Photon is the mediator of electromagnetic interaction and we have seen earlier that gluon is the mediator of the strong interactions. Graviton is the mediator of gravitational interactions. The number of gluons is eight. If we keep away the gravitational interaction, the number of mediators in E.M, weak and strong interactions is

$$1\ Photon + 3\ IMBs + 8\ Gluons = 12 \tag{1.219}$$

It has to be noted that (we shall later see it in gauge theories) the mass of any mediating particle must be zero. The photon is massless and so are gluons (refer section 1.9.5). However, we have told that $W^\pm$ and Z^0 are quite massive, though they too are mediators. As we shall see in the later Chapters that, it is the mechanism of spontaneous symmetry breaking that provides mass to massless vector mesons and because of such symmetry breaking we find the mediators of weak interactions as massive.

The spontaneous symmetry breaking is induced by a spin-0 (scalar) field, called the *Higgs field* and the quantum of Higgs field (a scalar particle) is called the *Higgs particle*. We can therefore say that the Higgs scalar boson is responsible for the masses of $W^\pm$ and Z^0 particles.

1.13 THE ULTIMATE FUNDAMENTAL PARTICLES

In the beginning of this Chapter, we started with electron, proton and neutron as elementary particles and then added to it photon, mesons, neutrinos, strange particles and antiparticles. Then came the resonances. If the above particles are called elementary, or fundamental, we have a very long list and following the discovery of quark structure, the mesons and baryons can't be called *fundamental*. Therefore, we now talk of those fundamental particles, which have not shown composite structure or are structureless. Such particles may be called the *Ultimate Fundamental Particles*, which constitute all the so-called elementary particles we know.

The ultimate fundamental particles can be categorized as

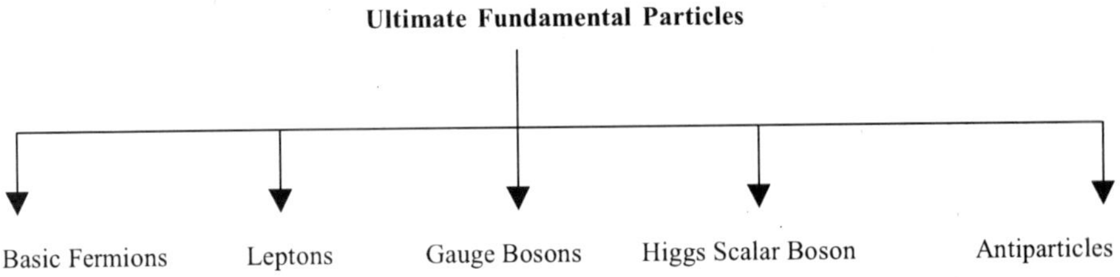

1.13.1 Basic Fermions

The three generations of quarks constitute the basic fermions.

Generation	**First**	**Second**	**Third**
Electric Charge $+\frac{2}{3}e$	*up* u	*charm* c	*top* t
Electric Charge $-\frac{1}{3}e$	*down* d	*strange* s	*bottom* b

$$up\begin{pmatrix} u \\ d \end{pmatrix}down \qquad charm\begin{pmatrix} c \\ s \end{pmatrix}strange \qquad top\begin{pmatrix} t \\ b \end{pmatrix}bottom$$

1.13.2. Leptons

Leptons too have three generations.

Generation	**First**	**Second**	**Third**
Electric Charge $= 0$	ν_e	ν_μ	ν_τ
Electric Charge $= -e$	e^-	μ^-	τ^-

$$\begin{pmatrix} \nu_e \\ e^- \end{pmatrix} \qquad \begin{pmatrix} \nu_\mu \\ \mu^- \end{pmatrix} \qquad \begin{pmatrix} \nu_\tau \\ \tau^- \end{pmatrix}$$

1.13.3. Gauge Bosons

Photon (γ), $W^{\pm}$, Z^0 and eight gluons are the Gauge Bosons. The graviton has not been included as we are restricting to three interactions, the EM, weak and strong only.

1.13.4. Higgs Boson

Higgs scalar (spin-0) field is required to initiate spontaneous symmetry breaking, which provides mass to $W^{\pm}$, Z^0 mediators. The quantum of this field is called *Higgs scalar boson*. This particle has yet to be found. It may be composite as well. Till this particle's structure is known, the Higgs boson as such will comprise the ultimate fundamental particles.

1.13.5. Antiparticles

To every particle talked above there is an antiparticle. Some particles like π^0 and γ are their own antiparticles. W^+ is the antiparticle of W^-.

The total number of ultimate fundamental particles is therefore 61, as given below:

Quarks + antiquarks	=	(6 + 6) 3 = 36 (each $q, \bar{q}$ has 3 colours)
Leptons + antileptons	=	6 + 6 = 12
Gauge bosons	=	1 + 3 + 8 =12
Higgs boson	=	1

The number sixty-one is the number of ultimate fundamental particles that describes the Standard Model, a model that provides us a successful theory of strong, weak and electromagnetic interactions. We shall talk more about Standard Model in later Chapters.

1.14 ASSIGNMENTS

1. What is an elementary particle? Can we say that the word elementary is absolute? Support your answer by quoting developments in the field of elementary particles.
2. Give an account of the developments, which led to the discovery of proton, neutron and photon. How photons explain the attraction and repulsion between electric charges?
3. Give a summary of the scientific developments that led to the discovery of mesons. Comment on the mu-meson, muon controversy. Name mesons of various spins.
4. Explain Dirac's hole theory, which predicts antiparticles. How the various antiparticles are distinguished from the particles? Describe briefly the experiments that made the discovery of some antiparticles.
5. Give an account of the theoretical and experimental developments that led to positron discovery. Explain the connection between antiparticles and crossing symmetry.
6. Write a note on neutrinos and its sources.
7. What are strange particles? List them. Why they are called strange? Describe the V-events and discuss the formation and decay processes of these particles.

8. What are resonances? Comment on their being called particles. Describe the formations of baryon and meson resonances. Mention some important resonances and their decay modes.
9. Describe in brief the various kinds of interactions. Workout their strengths and ranges. The strong interaction is mediated by gluons, which are massless, hence the range of strong interactions should be infinite, but they are of short range. Explain this paradox.
10. Give short qualitative accounts of
    (a) Predictions of Einstein's general theory of relativity.
    (b) Nine gluon states and quark gluon mechanism.
11. Applying conservation laws show whether the following decays and particle reactions are possible or not. State the type of interaction involved in each process. Mention the quark content of the hadron involved.

$$\Sigma^0 \to \Lambda^0 + \gamma$$

$$p + \nu_e \to e^+ + K^0 + \Lambda^0$$

$$p + \pi^- \to \Lambda^0 + \overline{\Sigma}^0$$

$$p + \overline{p} \to \pi^0 + \pi^+ + \pi^- + \pi^+ + \pi^-$$

$$p + \overline{\nu}_\mu \to \mu^+ + n$$

$$p + K^- \to \Sigma^+ + \pi^- + \pi^+ + \pi^- + \pi^0$$

[**Hint:** Neutrinos participate in weak interactions, photons in electromagnetic. Baryon number violates in third reaction. $p + \overline{p}$ and $p + K^-$ are strong interactions]

12. Give a comprehensive account of conservation laws, which particle interactions obey.
13. What are discrete symmetry operations? Describe charge conjugation operation. Explain that $|\gamma\rangle, |\pi^0\rangle$ are the eigenstates of C-operator and $|\pi^+\rangle, |\pi^-\rangle$ are not the eigenstates of the charge conjugation operator. Is the state $\{|\pi^+\rangle - |\pi^-\rangle\}$, the eigenstate of C. Find out the eigenvalues of C-eigenstates.
14. Define and explain parity. Show that parity is not conserved by weak interactions. Describe an experiment to prove that.
15. Define helicity of a particle. Explain that the helicity is generally assigned to massless particles, but not to massive. What is meant by the chiral states?
16. Define G-parity and provide details to prove that the G-parity greatly enhances the strength of C-operator. Show that G-parity of $|n\pi\rangle$ is $(-1)^n$ and apply it to determine the G-parity of ω and ϕ mesons.

17. Show that CP is a symmetry of most of the weak interactions. Cite an example of CP violation in weak interactions and provide theoretical and experimental support to it.
18. Explain the concept of time reversal. Obtain time reversal transformation and explain, how time reversal invariance is verified experimentally. What is meant by microscopic reversibility?
19. State and prove CPT theorem and discuss its consequences.
20. Classify elementary particles in different families with increasing mass. Mention one decay mode each of the unstable particles.
21. In reference to the elementary particles, what is eight fold way? What light does it throw on the nature of SU (3) symmetry?
22. Explain that the baryons and mesons can be arranged in certain geometrical patterns. Discuss the conclusions that can be derived from such patterns.
23. Mention the names and properties of ten baryon states of the decuplet. How the tenth particle of this pattern was discovered?
24. Discussing the implications of the classification of hadrons into geometrical patterns, develop the quark model of hadrons. Mention the quark contents of $\frac{1}{2}^+$ baryons and 0^- and 1^- mesons.
25. Explain the necessity of introducing colour to quarks. Obtain colour wavefunction of a baryon state and discuss its implication on the total wavefunction of the baryon.
26. Discuss the experiments that support the quark structure of hadrons.
27. Describe the variations of the coupling constants of quantum electrodynamics and quantum chromodynamics with energy; hence explain asymptotic freedom and infrared slavery. The infrared slavery is believed to be the origin of total confinement of quarks to colourless hadrons. Discuss.
28. How the estimation of quark masses can be made? Discuss the experimental techniques.
29. Discuss the significance of J/ψ and upsilon discoveries. How did these discoveries provide a firm footing to quark model?
30. Write notes on
    (a) Intermediate vector boson
    (b) Ultimate fundamental particles
    (c) Top quark mass
    (d) Colliders, and their impact on high energy physics.

1.15 OBJECTIVE QUESTIONS

1. The electron is an elementary particle because
   (a) it is the constituent of matter and has electric charge.

(b) it is the constituent of matter and has a small mass.
(c) it is the constituent of matter and has no composite structure.
(d) it binds with proton to form an atom.

2. The statement that the mediating particle of electromagnetic field is photon is
(a) a quantum field theory concept.
(b) a classical field theory concept.
(c) a classical particle theory concept.
(d) a concept of gauge theory.

3. The particles π^0, $\pi^{\pm}$ are
(a) spin-$\frac{1}{2}$ leptons.
(b) spin-$\frac{1}{2}$ baryons.
(c) spin-0 mesons.
(d) spin-1 mesons.

4. The K^* mesons are
(a) scalar mesons.
(b) vector mesons.
(c) tensor mesons.
(d) pseudoscalar mesons.

5. The vector mesons are
(*a*) $\rho,\ \omega,\ \phi$.
(*b*) $\rho,\ \omega,\ \pi^+$.
(*c*) $\rho,\ \omega,\ K^0$.
(*d*) $\pi^0,\ K^0,\ \phi$.

6. If the spin of electron is taken away, the electron would become
(a) scalar electron.
(b) vector electron.
(c) pseudovector electron.
(d) photon

7. K^0 and $\overline{K}^0$ are distinguished by the
(a) parities of these particles.
(b) strangeness of these particles.
(c) spin of these particles.
(d) charges of these particles.

8. Neutrinos and antineutrinos are distinguished by
(a) their spin.
(b) their charges.
(c) their strangeness.
(d) their helicities.

9. What among the following is true about neutrinos and antineutrinos?
   (a) All neutrinos are left handed and all antineutrinos right handed.
   (b) All neutrinos are right handed and all antineutrinos left handed.
   (c) There are both left and right handed neutrinos and antineutrinos.
   (d) No handedness is associated with neutrinos and antineutrinos.
10. If neutrinos have mass, we will have
   (a) no change in the handednesses of neutrinos and antineutrinos.
   (b) there will be both left and right handed neutrinos and antineutrinos.
   (c) the handednesses of neutrinos and antineutrinos will be interchanged.
   (d) handednesses will remain same, only the spin would change.
11. Match the following Lists I and II and select the correct code.

	List I		**List II**
A	Anderson and Neddermeyer	1	Neutrino
B	Perl	2	Strange particles
C	Reines and Cowan	3	Tau lepton
D	Rochester and Butler	4	Yukawa's meson

Codes:

(a)	A	B	C	D
	1	3	2	4
(b)	A	B	C	D
	2	1	4	3
(c)	A	B	C	D
	4	3	1	2
(d)	A	B	C	D
	3	4	1	2

12. The crossing symmetry equivalent of Compton effect is
   (a) Compton effect itself.
   (b) electron-positron annihilation.
   (c) electron-electron scattering.
   (d) positron-positron scattering.
13. The experimental status of the reactions $\bar{\nu}_\mu p \rightarrow e^+ p$ and $\bar{\nu}_e p \rightarrow \mu^+ p$ is that they
   (a) are observed because ν_e and ν_μ are same.
   (b) are not observed because ν_e and ν_μ are different.
   (c) are observed if e^+ and μ^+ are interchanged.
   (d) obey both (b) and (c) above.
14. The ratio of the sum of a particular subset of cross sections to the total cross section is called the
   (a) branching ratio for that subset of processes.
   (b) mixing ratio for that subset of processes.
   (c) inverse ratio for that subset of processes.
   (d) composite ratio for that subset of processes.

15. In literature, the strange particles are those particles, which are
    (a) produced through weak interaction and decay through electromagnetic interaction.
    (b) produced through weak interaction and decay through strong interaction.
    (c) produced through strong interaction and decay through electromagnetic interaction.
    (d) produced through strong interaction and decay through weak interaction.
16. The typical time scales for weak, electromagnetic and strong interactions are
    (a) 10^{-10} sec, 10^{-29} sec and 10^{-36} sec respectively.
    (b) 10^{-29} sec, 10^{-36} sec and 10^{-10} sec respectively.
    (c) 10^{3} sec, 10^{-16} sec and 10^{-23} sec respectively.
    (d) 10^{-16} sec, 10^{-23} sec and 10^{3} sec respectively.
17. Δ (1232) has
    (a) spin $J = \frac{1}{2}$ hence has two states.
    (b) spin $J = \frac{3}{2}$ hence has four states.
    (c) spin $J = \frac{5}{2}$ hence has six states.
    (d) spin $J = \frac{7}{2}$ hence has eight states.
18. The particles heavier than neutron are called
    (a) tachyons
    (b) mesons
    (c) hyperons
    (d) baryons
19. The elementary spin-$\frac{1}{2}$ particles, which feel the weak interaction, if charged feel electromagnetic interaction, but not the strong, are called
    (a) photons
    (b) leptons
    (c) mesons
    (d) baryons
20. The tau particle was discovered at
    (a) SPEAR e^+e^- storage ring at SLAC.
    (b) HERA e^-p^+ collider at DESY, Hamburg.
    (c) PEP e^+e^- collider at Stanford.
    (d) LEP e^+e^- collider at CERN.

21. The energy dependence of the cross section of a reaction between two particles, close to resonance energy E_0 is described by
(a) Bethe–Bloch formula.
(b) Breit–Wigner formula.
(c) Gamow–Teller formula.
(d) Weizsaker formula.

22. In a storage ring, electrons and positrons orbit with the same energy E but in opposite directions and collide head on. The square of the centre of mass energy s is
(a) $s = E^2$ (b) $s = 2E^2$
(c) $s = 3E^2$ (d) $s = 4E^2$

23. Choose the correct answer.
(a) ρ meson is a three pion resonance and ω two pion resonance. η is a vector meson.
(b) ρ meson is a two pion resonance and ω three pion resonance. η is a vector meson.
(c) ρ meson is a two pion resonance, ω three pion resonance and η is a scalar meson.
(d) ρ meson is a three pion resonance, ω two pion resonance and η is a vector meson.

24. Ripples caused in the curvature of space by heavy moving objects are called
(a) gravitational waves.
(b) matter waves.
(c) tachyonic waves.
(d) pressure waves.

25. Below are given the ranges and strengths of coupling constants for three interactions. Pick up the correct combination from the codes.

Ranges $R_1 = 10^{-15}$ m, $R_2 = 10^{-19}$ m, R_3 = unlimited

Coupling constants $C_1 = 10^{-14}$, $C_2 = 15$, $C_3 = 10^{-2}$

Codes:	*Electromagnetic*	*Weak*	*Strong*
(a)	R_3C_3	R_2C_2	R_1C_1
(b)	R_1C_2	R_3C_1	R_2C_3
(c)	R_3C_1	R_2C_3	R_1C_2
(d)	R_3C_3	R_2C_1	R_1C_2

26. The nucleus of $_5B^{11}$ consists of
(a) 16 up quarks and 16 down quarks.
(b) 16 up quarks and 17 down quarks.
(c) 17 up quarks and 16 down quarks.
(d) 17 up quarks and 17 down quarks.

27. Which one of the following nuclear reactions is possible?
    (a) ${}_7N^{14} \rightarrow {}_6C^{13} + \beta^+ + \nu_e$
    (b) ${}_7N^{13} \rightarrow {}_6C^{13} + \beta^+ + \nu_e$
    (c) ${}_7N^{13} \rightarrow {}_6C^{13} + \beta^+$
    (d) ${}_7N^{13} \rightarrow {}_7N^{13} + \beta^+ + \nu_e$
28. If a neutron at rest in free space decays into a proton and an electron, which of the following conservation laws is violated
    (a) conservation of charge.
    (b) conservation of energy.
    (c) conservation of linear momentum.
    (d) conservation of angular momentum.
29. At an interaction vertex a down quark converts into an up quark. The interaction and the particle absorbed at the vertex are
    (a) electromagnetic interaction and photon.
    (b) weak interaction and W^+ particle.
    (c) weak interaction and W^- particle.
    (d) strong interaction and gluon.
30. At an interaction vertex a gluon of colour $R\overline{B}$ is emitted. The interaction and the participating particles are
    (a) weak interaction. Up quark enters and down quark leaves the vertex.
    (b) electromagnetic interaction. The particles entering and leaving the vertex are up quarks.
    (c) strong interaction. The particle entering is up quark of red colour and leaving is up quark of blue colour.
    (d) strong interaction. The entering particle is up (red) quark and leaving is down (blue) quark.
31. At an interaction vertex electron neutrino converts into an electron. The interaction and the particle absorbed at the vertex are
    (a) electromagnetic, photon.
    (b) strong, gluon.
    (c) weak, W^+.
    (d) weak, W^-.
32. Select the correct statement about the mediators of three interactions.
    (a) Photon has electric charge, $W^{\pm}$ have colour charges and gluon has no charge.
    (b) Photon and gluon have colour charges and $W^{\pm}$ have electric charges.
    (c) Photon has no charge, $W^{\pm}$ have electric charges and gluon has colour charge.
    (d) Photon has no charge, $W^{\pm}$ have colour charges and gluon has electric charge.

33. The gluons which play no role in strong interactions are
   (a) monocoloured.
   (b) bicoloured.
   (c) multicoloured.
   (d) colour singlets.
34. The proton state is
   (a) $u_R u_G d_B$
   (b) $u_R u_G d_G$
   (c) $u_G u_B d_G$
   (d) $u_R u_B d_B$
35. In weak interactions, for the incoming and outgoing quarks, the
   (a) colour changes but the flavour does not.
   (b) flavour changes, the colour does not.
   (c) both colour and flavour change.
   (d) neither the colour nor the flavour changes.
36. The reaction $e^- \rightarrow \nu_\mu + W^-$
   (a) can take place as it does not violate any conservation law.
   (b) cannot take place as it violates the muon number conservation law.
   (c) cannot take place as it violates the electron number conservation law.
   (d) cannot take place as it violates both electron and muon number conservation law.
37. The isospin conservation law is obeyed by
   (a) all the four interactions.
   (b) the EM, weak and strong only.
   (c) weak and strong only.
   (d) strong interactions only.
38. The quarks and antiquarks have
   (a) opposite intrinsic parities.
   (b) similar intrinsic parities.
   (c) intrinsic parities depending on their relative motion.
   (d) no intrinsic parities.
39. Pseudoscalar mesons have
   (a) spin 0, parity even.
   (b) spin 0, parity odd.
   (c) spin 1, parity even.
   (d) spin 1, parity odd.
40. The interactions that remain invariant under parity operation are
   (a) all the four interactions.
   (b) the EM, weak and strong only.
   (c) EM and strong.
   (d) weak and strong only.

41. The operation of parity operator P on B (magnetic field) would lead to
    (a) $P B = B$, because P has no effect on B.
    (b) $P B = B$, because both the position vector and the direction of motion of moving charges that produce the magnetic field are inverted.
    (c) $P B = E$, because both electric and magnetic fields are produced by charges.
    (d) $P B = -E$, because P reverses the positions of negative and positive charges.
42. The lightest hadrons are
    (a) sigma particles.
    (b) resonances.
    (c) pions.
    (d) Xi-particles.
43. Baryon number conservation law means
    (a) baryons can only be created.
    (b) baryons cannot be created, only annihilated.
    (c) baryons can be created as well as annihilated.
    (d) baryons can neither be created nor annihilated, only transformed into each other.
44. Conventionally, the nucleon is ascribed a positive parity. The ground state deuteron's parity would be
    (a) positive.
    (b) negative.
    (c) both positive and negative.
    (d) cannot be obtained.
45. G-parity is a combination of
    (a) parity and isospin symmetry.
    (b) C-parity and isospin symmetry.
    (c) parity and C-parity.
    (d) C-parity and strangeness.
46. ρ meson decays into two pions but ω meson is not allowed to decay into two mesons. This is a consequence of the conservation of
    (a) parity.
    (b) strangeness.
    (c) G-parity.
    (d) isospin.
47. A pion at rest decays into a muon and its antineutrino. The spins of these two leptons would be
    (a) in the same directions.
    (b) opposite to each other.
    (c) in different directions.
    (d) C-parity and strangeness.

48. The helicity is always Lorentz invariant for
    (a) massless particles.
    (b) massive particles.
    (c) both massless and massive particles.
    (d) charged particles.
49. *CP* violation takes place in
    (a) π meson decay.
    (b) K^0 meson decay.
    (c) ω meson decay.
    (d) ρ meson decay.
50. While dealing with weak interactions, the states are required to be mixed. This requirement is a consequence of
    (a) *CP* invariance.
    (b) violation of parity conservation.
    (c) violation of charge conjugation invariance.
    (d) *CP* violation.
51. The time reversal invariance is natural at
    (a) microscopic level only.
    (b) macroscopic level only.
    (c) at both micro and macro levels.
    (d) none of the above.
52. The equality of masses and lifetimes of particles and their antiparticles comes from the physical consequences of
    (a) *CP* invariance.
    (b) *CP* violation.
    (c) *CPT* Theorem.
    (d) *T* invariance.
53. The eight baryons of the eightfold way are
    (a) $p, n, \Sigma^{\pm, 0}, \Xi^{-, 0}, \Lambda$.
    (b) $p, n, \Sigma^{\pm, 0}, \Delta^{++}, \Delta^{+}, \Lambda$.
    (c) $p, n, \Delta^{++}, \Delta^{+}, \Delta^{-}, \Delta^{0}, \Sigma^{0}, \Lambda$.
    (d) $p, n, \Delta^{++}, \Delta^{+}, \Delta^{-}, \Delta^{0}, \Lambda, \Omega$.
54. The hexagonal pattern of the baryon octet is a
    (a) two dimensional graph between I_3 and I.
    (b) two dimensional graph between I_3 and Y.
    (c) three dimensional graph between I, I_3 and Y.
    (d) three dimensional graph between I_1, I_2 and I_3.
55. Select the correct statement from the following
    (a) Octet and decuplet baryons are $(J)^P = (\frac{1}{2})^-$ and $(\frac{3}{2})^-$ baryons respectively.

(b) Octet and decuplet baryons are $(J)^P = (\frac{1}{2})^-$ and $(\frac{3}{2})^+$ baryons respectively.

(c) Octet and decuplet baryons are $(J)^P = (\frac{1}{2})^+$ and $(\frac{3}{2})^+$ baryons respectively.

(d) Octet and decuplet baryons are $(J)^P = (\frac{1}{2})^+$ and $(\frac{3}{2})^-$ baryons respectively.

56. Match the following two columns and select the correct code. Column II denotes (I_3, Y) values.

	Column I		*Column II*
A	Λ	1	$(+\frac{1}{2}, -1)$
B	Σ^+	2	$(+\frac{3}{2}, 1)$
C	Ξ^0	3	$(0, 0)$
D	Δ^{++}	4	$(0, -2)$
E	Δ^-	5	$(+1, 0)$
F	Σ^{*-}	6	$(-1, 0)$
G	Ω^-	7	$(-\frac{3}{2}, 1)$

Codes:

	A	B	C	D	E	F	G
(a)	1	3	4	2	7	5	6
(b)	5	3	1	2	6	7	4
(c)	3	5	2	1	6	7	4
(d)	3	5	1	2	7	6	4

57. Match the particles in column I below with their quark contents in column II.

	Column I		*Column II*
A	Σ^0	1	sss
B	Ξ^-	2	$d\bar{u}$
C	Δ^0	3	dss
D	Ω^-	4	$(u\bar{u} - d\bar{d})/\sqrt{2}$
E	π^0	5	uds
F	ρ^-	6	$s\bar{s}$
G	φ	7	udd

Codes:

(a)

A	B	C	D	E	F	G
4	3	2	1	7	6	5

(b)

A	B	C	D	E	F	G
5	3	7	1	4	2	6

(c)

A	B	C	D	E	F	G
5	7	3	1	4	6	2

(d)

A	B	C	D	E	F	G
4	5	2	1	3	7	6

58. The particles observed in nature are
    (a) monocoloured.
    (b) bicoloured.
    (c) colour singlets
    (d) multicoloured.
59. The colour wavefunction

$$\psi_{\text{colour}} = \frac{1}{\sqrt{6}}(RGB - RBG + GBR - GRB + BRG - BGR) \text{ belongs to}$$

    (a) proton.
    (b) neutron.
    (c) any meson.
    (d) any baryon.
60. The colours (R, G, B) form a fundamental triplet of SU (3) colour symmetry. This symmetry is
    (a) approximate because colours R, G, B belong to different flavours which have unequal masses.
    (b) approximate because R, G, B are different colours.
    (c) exact because colours R, G, B belong to the same flavour.
    (d) none of the above.
61. The π^+ wavefunction expressed in terms of spin, flavour and colour of the component quarks is

$$(a)\ \left|\pi^+\right\rangle = \frac{1}{\sqrt{6}} \sum_{c=R,G,B} [u_c(\uparrow)\bar{d}_c(\downarrow) + u_c(\downarrow)\ \bar{d}_c(\uparrow)]$$

$$(b)\ \left|\pi^+\right\rangle = \frac{1}{\sqrt{6}} \sum_{c=R,\,G,\,B} [u_c(\uparrow)\bar{d}_c(\downarrow) - u_c(\downarrow)\ \bar{d}_c(\uparrow)]$$

$$(c)\ \left|\pi^+\right\rangle = \frac{1}{\sqrt{6}} \sum_{c=R,G,B} [d_c(\uparrow)\bar{s}_c(\downarrow) + d_c(\downarrow)\ \bar{s}_c(\uparrow)]$$

$$(d)\ \left|\pi^+\right\rangle = \frac{1}{\sqrt{6}} \sum_{c=R,G,B} [d_c(\uparrow)\bar{s}_c(\downarrow) - d_c(\downarrow)\ \bar{s}_c(\uparrow)]$$

62. Up quark and down quark form an isospin doublet because
   (a) they have unequal fractional charges.
   (b) the difference of their charges is e.
   (c) the difference of their masses is small.
   (d) they are fermions.

63. For the particle isospin doublet $\begin{pmatrix} p \\ n \end{pmatrix}$, the corresponding antiparticle isospin doublet is

   (a) $\begin{pmatrix} \overline{p} \\ \overline{n} \end{pmatrix}$, (b) $\begin{pmatrix} -\overline{p} \\ \overline{n} \end{pmatrix}$, (c) $\begin{pmatrix} \overline{n} \\ \overline{p} \end{pmatrix}$, (d) $\begin{pmatrix} -\overline{n} \\ \overline{p} \end{pmatrix}$

64. The strong coupling constant α_s
   (a) is a constant, which does not vary.
   (b) is called a constant, but in fact it depends on Q^2 [Lorentz invariant four-momentum (transferred)].
   (c) has a fixed value for a given set of quarks.
   (d) is similar to electromagnetic coupling constant.

65. In the inter-quark coupling high Q^2 values imply
   (a) large separation between quarks.
   (b) small separation between quarks.
   (c) no change in separation with Q^2.
   (d) the separation depends on the type of quarks.

66. The strong coupling constant α_s obtained through the first order perturbative calculation in QCD has the form

$$\alpha_s(Q^2) = \frac{12\pi}{(33-n_f)\log\left(\dfrac{Q^2}{\Lambda^2}\right)},$$ where n_f is the number of quark types involved and Λ is the free parameter of QCD ($\approx 250 \dfrac{\text{MeV}}{C}$). The application of perturbative expansion procedures in QCD would be valid only when

   (a) $Q^2 >> \Lambda^2$
   (b) $Q^2 << \Lambda^2$
   (c) $Q^2 = \Lambda^2$
   (d) $Q^2 = i\Lambda^2$

67. Read the following statements and choose the correct answer from the codes.

Statement 1: The strong coupling constant $\alpha_s \to \infty$ for $Q^2 \to \infty$. This is called asymptotic freedom.

Statement 2: The strong coupling constant $\alpha_s \to 0$ when $Q^2 \to \infty$. This is called asymptotic freedom.

Statement 3: When the separation $r \to \infty$, the quark—antiquark potential $V \to \infty$. This is called quark confinement.

Statement 4: When $r \to 0$, the quark—antiquark potential $V \to \frac{1}{r}$. This is called asymptotic freedom.

Statement 5: When $r \to \infty$, the quark—antiquark potential $V \to 0$. This is called quark confinement.

Statement 6: When $r \to 0$, the quark—antiquark potential $V(r) \to kr$. This is called asymptotic freedom.

Codes:

(a) Statements 1,3,4,5 are correct.
(b) Statements 2, 3, 4 are correct.
(c) Statements 1, 5, 6 are correct.
(d) Statements 2,3,4,6 are correct.

68. The quark—antiquark potential that can display both asymptotic freedom andquark confinement (question 67) can be expressed as

(a) $V = -\frac{4}{3}\frac{\alpha_s(r)\hbar c}{r} + kr$

(b) $V = -\frac{4}{3}\frac{\alpha_s(r)\hbar c}{r^2} + kr^3$

(c) $V = -\frac{4}{3}\frac{\alpha_s(r)\hbar c}{r^3} + kr^3$

(d) $V = [-\frac{4}{3}\frac{\alpha_s(r)\hbar c}{r} + kr^2]r$

69. The electric lines of force between a quark and antiquark are
(a) similar to the electric lines of force between an electron and positron, spreading out in space.
(b) not like (a) above, but like a tube. The lines of force are stretched.
(c) spirals spreading out into space.
(d) emanate radially outward like the lines of force from isolated charges.

70. The difference between photons and gluons is that
(a) photons have self-interactions and gluons have not.
(b) both have self-interactions.
(c) gluons have self-interactions and photons do not have.
(d) photons are coloured and gluons are not.

71. The bare colour charge of a quark is modified
    (a) by gluon—gluon interaction such that the bare colour charge gets enhanced.
    (b) by gluon—gluon interaction such that the bare colour charge gets diminished.
    (c) by photon—photon interaction to get enhanced.
    (d) by photon—photon interaction to get diminished.
72. The top quark mass is
    (a) 310 MeV
    (b) 1100 MeV
    (c) 4700 MeV
    (d) 174 GeV
73. *J* and ψ particles
    (a) are the names of same meson.
    (b) are the names of same baryon.
    (c) are the names of different mesons.
    (d) are the names of different baryons.
74. The discovery of *J* particle is significant because it
    (a) predicted the up flavour of quarks.
    (b) predicted the down flavour of quarks.
    (c) predicted the strange flavour of quarks.
    (d) predicted the charm flavour of quarks.
75. The discovery of ψ particle is known as November revolution because it
    (a) predicted the new hadrons called strange hadrons.
    (b) predicted the new hadrons called charmed hadrons.
    (c) predicted the new leptons called tau and its neutrino.
    (d) predicted the new quark flavors called top and bottom.
76. Upsilon is a resonance at
    (a) 92 GeV
    (b) 82 GeV
    (c) 9.46 GeV
    (d) 3.1 GeV
77. The discovery of Upsilon particle is significant because it
    (a) predicted the strange flavour of quarks.
    (b) predicted the charm flavour of quarks.
    (c) predicted the bottom flavour of quarks.
    (d) predicted the top flavour of quarks.

78. Match the following columns and select the correct answer from the codes.

Column I		*Column II*	
A	ρ^0	1	$c\bar{c}$
B	ω	2	$b\bar{b}$
C	φ	3	$s\bar{s}$
D	J	4	$c\bar{c}$
E	Y′	5	$\frac{1}{\sqrt{2}}(u\bar{u}-d\bar{d})$
F	ψ	6	$\frac{1}{\sqrt{2}}(u\bar{u}+d\bar{d})$

Codes:

	A	B	C	D	E	F
(a)	5	6	2	1	3	4
(b)	3	2	6	4	5	1
(c)	6	3	2	1	5	4
(d)	5	6	3	4	2	1

79. In parton model, the constituents of proton (baryon) are called partons. What is true among the following?
   (a) The charged partons are identified with quarks and neutral ones with gluons.
   (b) The charged partons are identified with gluons and neutral ones with quarks.
   (c) The charged partons are identified with electric charges and neutral ones with photons.
   (d) The charged partons are identified with electric charges and neutral ones with magnetic charges.

80. A stationary particle in free space is observed to spontaneously decay into two photons. This implies that
   (a) the particle carries electric charge.
   (b) the spin of the particle must be greater than or equal to 2.
   (c) the particle is a boson.
   (d) the mass of the particle must be greater than or equal to the mass of the hydrogen atom.

81. According to the present knowledge, the number of ultimate fundamental particles is
(a) 18
(b) 81
(c) 61
(d) 101

ANSWERS

1. c, 2. a, 3. c, 4. c, 5. a, 6. a, 7. b, 8. d, 9. a, 10. b, 11. c,12. b, 13. d, 14. a, 15. d, 16. c, 17. b, 18. c, 19. b, 20. a, 21. b, 22. d, 23. c, 24. a, 25. d, 26. b, 27. b, 28. d, 29. b, 30. c, 31. d, 32. c, 33. d, 34. a, 35. b, 36. d, 37. d, 38. a, 39. b, 40. c, 41. b, 42. c, 43. d, 44. a, 45. b, 46. c, 47. b, 48. a, 49. b, 50. d, 51. a, 52. c, 53. a, 54. b, 55. c, 56. d, 57. b, 58. c, 59. d, 60. c, 61. b, 62. c, 63. d, 64. b, 65. b, 66. a, 67. b, 68. a, 69. b, 70. c, 71. a, 72. d, 73. a, 74. d, 75. b, 76. c, 77. c, 78. d, 79. a, 80. c, 81. c.

2

QUANTUM ELECTRODYNAMICS

2.1 INTRODUCTION

The relativistic quantum theory of the electromagnetic interaction of charged particles is called the *Quantum Electrodynamics* (*QED*). Among the three interactions, the electromagnetic interaction is the best understood. The theoretical calculations for the processes in which electromagnetic interaction is dominant agree very well with the experimental results. The success of *QED* depends on the point like behaviour of leptons and the small value of the electromagnetic coupling constant ($\alpha \approx \frac{1}{137}$). Due to the smallness of a, the perturbation approach becomes applicable. Using the perturbation approach, Feynman derived the rules for *QED*, which we will take up in this Chapter, before that; however, we shall deal with some preliminaries required in this Chapter.

2.2 PRELIMINARIES

2.2.1 Time Ordered Product and Propagators

The time ordered product of n field operators, $\psi_a(x^a), \psi_b(x^b), \psi_c(x^c), \psi_d(x^d)$, $\ldots\psi_i(x^i), \psi_j(x^j), \ldots\psi_n(x^n)$, where $x^a \Rightarrow (x_0^a, \boldsymbol{x}^a)$ etc. is defined as the product of these operators taken in order corresponding to the time arguments increasing from right to left. In such arrangement due regard has to be paid to the overall sign which may change if some of the operators are fermionic. Denoting the time ordered product by $T[\ ]$, it is then expressed by

$$T[\psi_a(x^a), \psi_b(x^b), \psi_c(x^c), \psi_d(x^d), \ldots, \psi_i(x^i), \psi_j(x^j), \ldots \psi_n(x^n)]$$
$$\equiv (-1)^P[\psi_j(x^j), \psi_d(x^d), \psi_b(x^b), \ldots, \psi_a(x^a), \psi_c(x^c), \psi_n(x^n), \ldots, \psi_i(x^i)] \qquad (2.1)$$

where

$$x_0^j \geq x_0^d \geq x_0^b, \ldots, \geq x_0^a \geq x_0^b \geq x_0^n, \ldots, \geq x_0^i \qquad (2.2)$$

and p is the number of transpositions among the fermion operators that are required to achieve time ordering. It is implied that among operators $\psi_a(x^a)$ to $\psi_n(x^n)$ some are bosonic and some are fermionic. When we rearrange bosonic operators in terms of their increasing time from right to left, there is no change of sign but if among the fermion operators, we make even number of changes to achieve time ordering, no change of sign will take place but for odd number of changes, we shall have a minus sign in eqn. (2.1). For example, if ψ_a and ψ_b be two bosonic field operators, then

$$T(\psi_a(x)\psi_b(y)) = \theta(x-y)\psi_a(x)\psi_b(y) + \theta(y-x)\psi_b(x)\psi_a(y) \tag{2.3}$$

where θ function has the property

$$\theta(x-y) \equiv \theta(x^0 - y^0) = \begin{cases} +1 & \text{for} \quad x^0 > y^0 \\ 0 & \text{for} \quad x^0 < y^0 \end{cases} \tag{2.4}$$

$$\theta(y-x) \equiv \theta(y^0 - x^0) = \begin{cases} +1 & \text{for} \quad y^0 > x^0 \\ 0 & \text{for} \quad y^0 < x^0 \end{cases} \tag{2.5}$$

In eqn. (2.3), $\psi_a(x)$ is the bosonic operator defined at time x^0 at the space point $\boldsymbol{x}$. $\psi_a(y)$ has been defined at time $y^0$and space point $\boldsymbol{y}$. If $x^0 > y^0$ then the time ordered product would be given by the first term of the RHS of eqn. (2.3) and if $y^0 > x^0$, it will be given by the second term. At equal times $x^0 = y^0$, defining $\theta(0) = \frac{1}{2}$, eqn. (2.3) becomes

$$T[\psi_a(x)\psi_b(y)]_{x^0=y^0} = \tfrac{1}{2}[\psi_a(x)\psi_b(y) + \psi_b(y)\psi_a(x)] \tag{2.6}$$

But at equal times operators $\psi_a(x)$ and $\psi_b(y)$ commute, therefore, the equal time *time ordered* product does not require rearrangement of operators.

$$T(\psi_a(x)\psi_b(y))^{\text{bosonic}}_{x^0=y^0} = \psi_a(x)\psi_b(y) = \psi_b(y)\psi_a(x) \tag{2.7}$$

For fermionic operators, the equation corresponding to eqn. (2.3) is (now $\psi_a(x)$ and $\psi_b(y)$ are fermionic operators)

$$T(\psi_a(x)\psi_b(y)) = \theta(x-y)\psi_a(x)\psi_b(y) - \theta(y-x)\psi_b(x)\psi_a(y) \tag{2.8}$$

implying that if $x^0 > y^0$, first term in the RHS of eqn. (2.8), using eqn. (2.4), will give the time ordered product. But, when $y^0 > x^0$, the second term of eqn. (2.8), which is obtained by one (odd) transposition of, $\psi_a(x)$ and $\psi_b(y)$, gives the time ordered product. The single (odd) transposition therefore brings a minus sign. For equal time ($y^0 > x^0$), the anticommuting $\psi_a(x)$ and $\psi_b(y)$, bring, corresponding to eqn. (2.7), the relation

$$T(\psi_a(x)\psi_b(y))^{\text{fermionic}}_{x^0=y^0} = \psi_a(x)\psi_b(y) = -\psi_b(y)\psi_a(x) \tag{2.9}$$

The T-product of two field operators differs from their ordinary product by a c-number. For example, if $\psi_a(x) = \phi(x)$, the real scalar field and $\psi_b(y) = \phi(y)$, then

$$T(\phi(x)\phi(y)) = \theta(x-y)\phi(x)\phi(y) + \theta(y-x)\phi(y)\phi(x) \tag{2.10}$$

Using the commutation relation

$$[\phi(x), \phi(y)] = -i\Delta(x-y) \tag{2.11}$$

where $\Delta(x-y)$ is the Schwinger's delta function (see Appendix B)

$$\Delta(x-y) = \Delta^+(x-y) + \Delta^-(x-y) \tag{2.12}$$

equation (2.10) becomes

$$\begin{aligned} T(\phi(x)\phi(y)) &= (\theta(x-y) + \theta(y-x))\phi(x)\phi(y) + i\theta(y-x)\Delta(x-y) \\ &= \phi(x)\phi(y) + i\theta(y-x)\Delta(x-y) \end{aligned} \tag{2.13}$$

where the inference from eqns. (2.4) and (2.5) that whether $x^0 > y^0$ or $y^0 > x^0$

$$\theta(x-y) + \theta(y-x) = 1 \tag{2.14}$$

has been used. In eqn. (2.13), the LHS is the T-product, $\phi(x)\phi(y)$ is the ordinary product and the second term in the RHS is a c-number. It is important to note that if two space—time points are separated by a spacelike interval, *i.e.*,

$$(x^a - x^b)^2 = (x_0^a - x_0^b)^2 - (\mathbf{x}^a - \mathbf{x}^b)^2 < 0 \tag{2.15}$$

their time ordering is not relativistically invariant. In such situations, the requirement of Lorentz invariance is that the field operators at these space—time points must commute or anticommute, depending on their bosonic or fermionic character. In other words, the time ordering say $x_0^a - x_0^b > 0$ would be Lorentz invariant if the interval $(x^a - x^b)$ is not spacelike.

The vacuum expectation value of normal ordered product vanishes but the vacuum expectation value of time ordered product of operators plays an important role in the theory. From eqn. (2.10), we write the vacuum $(|0\rangle)$ expectation value

$$\langle 0|T\phi(x)\phi(y)|0\rangle = \theta(x-y)\langle 0|\phi(x)\phi(y)|0\rangle + \theta(y-x)\langle 0|\phi(y)\phi(x)|0\rangle \tag{2.16}$$

To understand the normal product, we first explain the ordinary product of operators. For that, we consider the decomposition of the field operator $\phi(x)$ into its positive frequency part $\phi^+(x)$ and the negative frequency part $\phi^-(x)$ as

$$\phi(x) = \phi^+(x) + \phi^-(x) \tag{2.17}$$

and write in terms of the above decomposition a product

$$\begin{aligned} \phi(x)\phi(y) &= [\phi^+(x) + \phi^-(x)][\phi^+(y) + \phi^-(y)] \\ &= [\phi^+(x)\phi^+(y) + \phi^+(x)\phi^-(y) + \phi^-(x)\phi^+(y) + \phi^-(x)\phi^-(y)] \end{aligned} \tag{2.18}$$

in which care has been taken to maintain the left and right positions (i.e., the order) of operators ϕ^+ and ϕ^-. Such a product of operators, like eqn. (2.18), is called the *ordinary product*. If we look at the second and third terms of eqn. (2.18), we find that in the second term ϕ^+ is on the left of ϕ^- while in the third it is on the right of ϕ^-. Now, if we write the so-called ordinary product of operators $\phi(x)$ and $\phi(y)$ as given by eqn. (2.18), in such a way that all positive frequency parts of the operators (ϕ^-) are invariably on the left and the negative frequency parts(ϕ^-) are kept on the right, then the third term of eqn. (2.18) shall have to be written as $\phi^+(y)\phi^-(x)$ and then, the ordinary product (2.18) shall be called the normal ordered product or normal product or even just the ordered product of operators. We denote such normal ordered product by putting N on the ordinary product, i.e., writing it as $N(\phi(x)\phi(y))$. Sometimes a colon (:) is also used

$$N(\phi(x)\phi(y)) \equiv :\phi(x)\phi(y) \tag{2.19}$$

Thus

$$\begin{aligned} N(\phi(x)\phi(y)) = [&\phi^+(x)\phi^+(y) + \phi^+(x)\phi^-(y) \\ &+ \phi^+(y)\phi^-(x) + \phi^-(x)\phi^-(y)] \end{aligned} \tag{2.20}$$

is the normal product of operators $\phi(x)$ and $\phi(y)$. However, interchanging the position of two operators, here $\phi^-(x)$ and $\phi^+(y)$, is possible when these two operators either commute or their commutator is a c-number operator. This is because the c-number operator being the multiple of identity or unit operator when operates on a state, brings no change in it. Again, since, the Fourier expansion of ϕ^+ carries the creation operator $C_{\boldsymbol{k}}^\dagger$ and that of ϕ^- carries the annihilation operator C_k, the definition of normal product (2.20) may also be put as the product in which creation operators are on the left of annihilation operators. It can be seen that

$$[\phi^-(x), \phi^+(y)] = -i\left[\frac{+i}{(2\pi)^3}\int \frac{d^3\boldsymbol{k}}{2\omega_k} e^{-i[\omega_k(x_0-y_0)-\boldsymbol{k}\cdot(\boldsymbol{x}-\boldsymbol{y})]}\right] \tag{2.21}$$

or (refer equation Appendix B63)

$$[\phi^-(x), \phi^+(y)] = -i\,\Delta^-(x-y) \tag{2.22}$$

which implies that

$$\phi^+(y)\phi^-(x) = \phi^-(x)\phi^+(y) + i\,\Delta^-(x-y)\,I \tag{2.23}$$

Similarly, using the equation (B74), we can express

$$\phi^+(x)\phi^-(y) = \phi^-(y)\phi^+(x) - i\Delta^+(x-y)I \tag{2.24}$$

where I is the identity operator and $i\Delta^-(x-y)$ and $i\,\Delta^-(x-y)$ are c-numbers. Thus we can write

$$N(\phi(x)\phi(y)) = \phi(x)\phi(y) + i\Delta^-(x-y) \tag{2.25a}$$

i.e., $$\langle 0|N\phi(x)\phi(y)|0\rangle = \langle 0|\phi(x)\phi(y)|0\rangle + i\Delta^-(x-y)\langle 0|0\rangle \tag{2.25b}$$

Since, the vacuum expectation value of the normal product vanishes, i.e., $\langle 0|N\psi(x)\psi(y)|0\rangle = 0$ we infer from equation (2.25b)

$$\langle 0|\phi(x)\phi(y)|0\rangle = -i\Delta^{-}(x-y) \tag{2.26a}$$

Likewise

$$\langle 0|\phi(y)\phi(x)|0\rangle = i\Delta^{+}(x-y) \tag{2.26b}$$

Using equation (2.26) in equation (2.16), the vacuum expectation value of the time ordered product becomes

$$\langle 0|T\phi(x)\phi(y)|0\rangle = -i\theta(x-y)\Delta^{-}(x-y) + i\theta(y-x)\Delta^{+}(x-y) \tag{2.27}$$

Let us now recall the following two equations (B89 and B94) from Appendix B

$$\Delta^{1}(x) = i[\Delta_F(x) - \Delta_D(x)] \tag{2.28}$$

$$\Delta^{1}(x) = i[\Delta^{+}(x) - \Delta^{-}(x)] \tag{2.29}$$

A comparison of above equations gives us

$$\Delta_F(x) = \Delta^{+}(x) \tag{2.30}$$

or

$$\Delta_F(x) = -\Delta^{-}(x) \tag{2.31}$$

Now, from equations (B62) and (B63), Δ^{+} contains positive frequencies ($+\omega_k$) and Δ^{-} contains negative frequencies ($-\omega_k$). Also from the path of integration for Δ_F redrawn below, (Fig. 2.1),

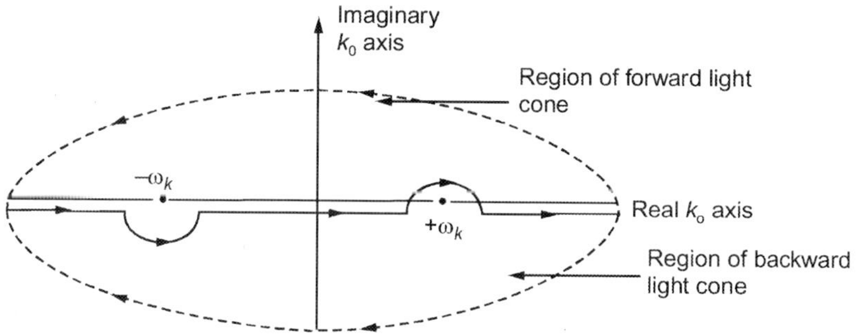

Fig. 2.1: Path of integration for Feynman function $\Delta_F(x)$ Dashed paths indicate the infinite semi-circle.

the ($-\omega_k$) lies in the future lightcone, i.e., $x^0 > y^0$ in $(x-y)$ and ($+\omega_k$) lies in the past light cone, i.e., $x^0 > y^0$ in $(x-y)$. So, we may write from eqns. (2.30) and (2.31) that

$$\Delta_F(x-y) = -\Delta^{-}(x-y) \text{ for } x^0 > y^0 \tag{2.32}$$

and

$$\Delta_F(x-y) = +\Delta^{+}(x-y) \text{ for } x^0 < y^0 \tag{2.33}$$

Using equations (2.32) and (2.33) in eqn. (2.26), we get

$$\langle 0|T\phi(x)\phi(y)|0\rangle = i[(\theta(x-y)+\theta(y-x)]\Delta_F(x-y) \tag{2.34}$$

which on using eqns. (2.14) becomes

$$\langle 0|T\phi(x)\phi(y)|0\rangle = i\Delta_F(x-y) \tag{2.35}$$

where the exponential form of Feynman function $\Delta_F(x-y)$ from equation (B39) is

$$\Delta_F(x-y) = \frac{1}{(2\pi)^4}\int \frac{e^{ik(x-y)}}{m^2-k^2-i\varepsilon}\,dk \tag{2.36}$$

It has to be recalled (Appendix B) that Δ_F is a Green's function of the scalar Klein–Gordon equation,

$$\phi(x) = \int \Delta_F(x-y)\rho(y)\,dy \tag{2.37}$$

To bring about the significance of $\Delta_F(x-y)$, we take an example of two-nucleon interaction, which takes place via the exchange of pions between the nucleons. One nucleon creates a pion, which is absorbed (destroyed) by the other. However, in the interaction process no free pions are available. The potential energy operator between two pion sources, namely the two nucleons, is E_{OP}, given by

$$E_{OP} \approx \int\int \rho(x)\Delta_F(x-y)\rho(y)dyd^3\mathbf{x} \tag{2.38}$$

A pion may be seen created by the source operator $\rho(x)$ at space-time x and the same is destroyed at the space—time y by the source operator $\rho(y)$. The reverse may also take place. Pions not being free are just virtual. The function $\Delta_F(x-y)$ describes the propagation of pions between x and y and therefore $\Delta_F(x-y)$ may be called the *propagator*.

Looking again at eqn. (2.35), we then find that since $i\Delta_F$ comes as the vacuum expectation value (*VEV*) of time ordered free scalar meson field operators, we call $i\Delta_F(x-y)$ as the free meson propagator. Because they describe a propagator, the vacuum expectation values of time ordered product of operators are very important. Since, the propagators are described by the vacuum expectation values of the time ordered product of operators (*TOPO*); the latter (*VEV* of *TOPO*) acquires a considerable significance. Below we list the propagators for various free fields.

Field	**T-ordered VEV**	**Propagator**
Scalar meson	$\langle 0\|T\phi(x)\phi(y)\|0\rangle$	$i\Delta_F(x-y)$
Charged Scalar meson	$\langle 0\|T\phi(x)\phi^*(y)\|0\rangle$	$i\Delta_F(x-y)$

Vector meson	$\langle 0 \vert T\phi_\mu(x)\phi_\nu(y) \vert 0\rangle$	$i(g_{\mu\nu} + \frac{\partial_\mu \partial_\nu}{m^2})\Delta_F(x-y)$
Massless Vector meson (EM)	$\langle 0 \vert TA_\mu(x)A_\nu(y) \vert 0\rangle$	$ig_{\mu\nu}\,[\Delta_F(x-y)]_{m=0}$
Dirac	$\langle 0 \vert T\psi(x)\bar{\psi}(y) \vert 0\rangle$	$i(i\gamma_\mu\,\partial^\mu + m)\,\Delta_F\,(x-y)$

For brevity, the propagators for the massive vector (vector meson) field and the Dirac field are denoted by $i\Delta_{\mu\nu F}\,(x-y)$ and $iS_F\,(x-y)$ respectively.

2.2.2 Some Definitions

2.2.2.1 *Differential Scattering Cross Section*

In classical scattering theory, we consider a uniformly dense beam of particles, which are travelling in the same direction with velocity v. If we consider a unit area perpendicular to the direction of beam, then the number of particles crossing this unit area per unit time is called the *flux* (F) of the beam. The flux (F) is also the number of particles in a volume of unit crosssection and length v. If ρ be the particle density

$$F = \rho v \tag{2.39}$$

This beam of particles approaches a target or the scatter and experiences a force that results in the beam's deflection (scattering). Some particles of the transmitted beam may remain unscattered. Let us picturise this and assume that the target is the origin of coordinates and the direction of beam is the z-axis or the polar axis, θ is the scattering angle and θ the azimuthal angle (Fig.2.2).

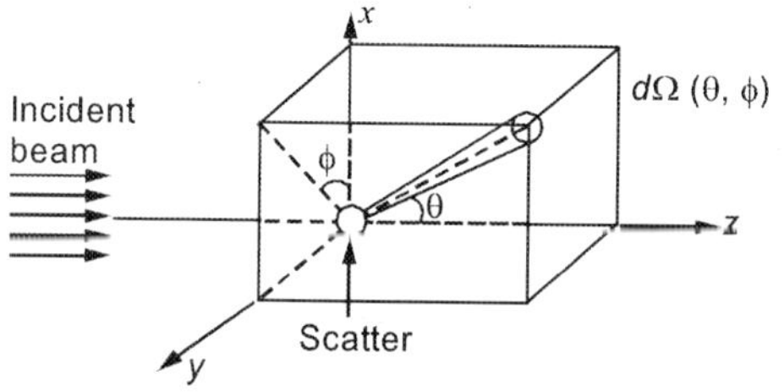

Fig. 2.2: Description of scattering in polar coordinates. Origin of coordinates is at the target and beam direction is z-axis

In Fig.(2.2) the incident particles have scattered into a solid angle $d\Omega$. When ρ is small, the mutual interactions of the incident particles can be neglected, hence each particle can be seen scattering independently.

Let us consider a classical system, which possesses azimuthal symmetry and hence the deflections produced by the target are independent of ϕ. Let the range of impact parameter extend from $b(\theta)$ to $b(\theta + d\theta)$ for the trajectories, which deflect through angles θ to $\theta + d\theta$. The effective area of the target, which produces such deflections, is $d\sigma(\theta, \phi)$ where from the Fig. (2.3)

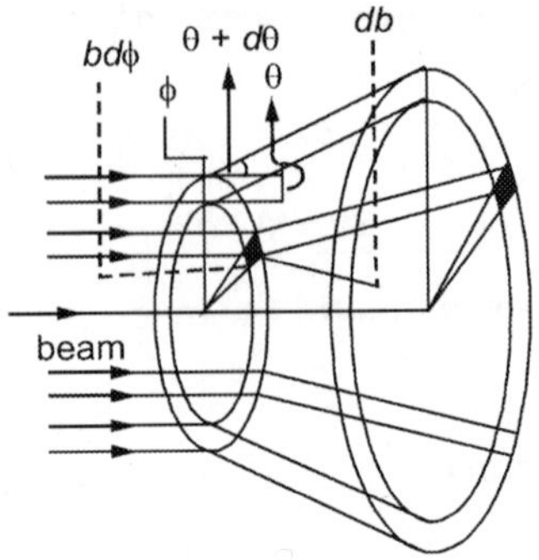

Fig. 2.3: Diagram showing that [$b\ db\ d\phi$] is the effective area of the target that deflects the incident particles into the solid angle $d\Omega$ b implies $b(\theta)$ and $d\Omega \Rightarrow d\Omega(\theta, \phi)$

$$d\sigma(\theta, \phi) = b(\theta)\, db d\phi \tag{2.40}$$

Obviously, larger we make $d\sigma$, larger will be $d\Omega$ into which the particles scatter. The proportionality factor is called the differential scattering cross section $\sigma(\theta, \phi)$, i.e,

$$d\sigma(\theta,\phi) = \sigma(\theta,\phi) d\Omega(\theta,\phi) \tag{2.41}$$

which gives the number of particles into the solid angle $d\Omega(\theta, \phi)$ per unit time per unit flux. From eqn. (2.41), the differential scattering cross section

$$\frac{d\sigma}{d\Omega} = \sigma(\theta,\phi) \tag{2.42}$$

and its dimensions are of an area.

$$\left[\frac{d\sigma(\theta,\phi)}{d\Omega(\theta,\phi)}\right] = [\sigma(\theta,\phi)] = T^{-1}(L^{-2}T^{-1})^{-1} = L^2 \tag{2.43}$$

The total cross section σ is

$$\sigma = \int \sigma(\theta,\phi)\, d\Omega \tag{2.44}$$

where the integration is over all solid angles.

When we have the azimuthal symmetry, the deflection produced is independent

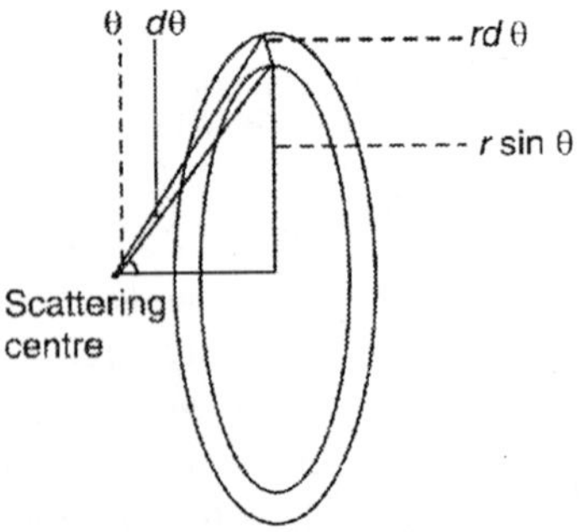

Fig. 2.4: Figure indicating $d\Omega = \frac{dS}{r^2} = \frac{(2\pi r \sin\theta)\, r\, d\theta}{r^2} = 2\pi \sin\theta\, d\theta$

of ϕ. For example, a spherically symmetric potential, which is mostly used, does not depend on ϕ. From the Fig. (2.3), the solid angle

$$d\Omega = \sin\theta d\theta d\phi \quad (2.45)$$

Integrating this equation over ϕ from 0 to 2π, the solid angle (2.45) becomes

$$d\Omega = 2\pi \sin\theta d\theta \quad (2.46)$$

which can be easily worked out from the following figure (Fig.2.4).

$$d\Omega = \frac{dS}{r^2} = \frac{(2\pi r \sin\theta) r\, d\theta}{r^2} = 2\pi \sin\theta\, d\theta \quad (2.47)$$

Using equation (2.40) with $\int d\phi = 2\pi$, and equations (2.41) and (2.46), we get for the differential scattering cross section

$$\sigma(\theta) = \frac{d\sigma(\theta)}{d\Omega} = -\frac{b(\theta)}{\sin\theta}\frac{db}{d\theta} \quad (2.48)$$

where due to ϕ independence $\sigma(\theta, \phi) \to \sigma(\theta)$ and minus sign put by hand indicates that, when scattering angle increases the impact parameter b decreases and vice versa.

2.2.2.2 Matrix Element

Let us consider the example of scattering. A particle described by the state $|i\rangle = |\boldsymbol{p}_i\rangle$ gets scattered by the potential between target and the projectile and acquires the final state $|f\rangle = |\boldsymbol{p}_f\rangle$. The states $|\boldsymbol{p}_i\rangle$ and $|\boldsymbol{p}_f\rangle$ respectively specify the energy and direction of the particle before and after the scattering. The amplitude for a transition from $|i\rangle$ to $|f\rangle$ depends upon the matrix element of the interaction term in the Lagrangian or the Hamiltonian H′ between the two states, i.e., the matrix element M_{fi} is

$$M_{fi} = \int \psi_f^* \mathrm{H}' \psi_i dV = \langle f | \mathrm{H}' | i \rangle \quad (2.49)$$

where the operator H′ comes when we split the total Hamiltonian operator H into H^0, the Hamiltonian operator for free system and H′, the interaction Hamiltonian operator.

$$\mathrm{H} = \mathrm{H}^0 + \mathrm{H}' \quad (2.50)$$

For scattering

$$\frac{d\sigma}{d\Omega} \approx |M_{fi}|^2 \quad (2.51)$$

Since, the differential scattering cross section is proportional to the square of the matrix element; the matrix element is also called the *amplitude* in analogy with the optical interactions, where the rate of reaction is proportional to the square of the amplitude.

2.2.2.3 Fermi's Golden Rule

The transition rate for a given process is determined by the matrix element and the phase space according to Fermi's Golden Rule

$$\text{Transition Rate} = \frac{2\pi}{\hbar}\left|M_{fi}\right|^2 \times \text{(phase space)} \tag{2.52}$$

The given process may be a decay process or scattering. In the β-decay process, e.g., ${}_5B^{12} \rightarrow {}_6C^{12} + e^- \bar{\nu}e$ the initial state is the state of Boron nucleus and the final state is the product of the states of Carbon nucleus, electron and antineutrino. The Golden Rule (2.52) would determine the rate of transition of nucleus from initial to final state.

Whereas, the transition matrix element M_{fi} contains the dynamical information, the phase space factor contains only the kinematical information. The phase space factor gives an idea of *likelihood of a particular process*—taking place. Larger the phase space factor, greater the likelihood and lesser the phase space factor, it is less likely that the process would take place. The masses, energies and momenta determine the phase space factor. The larger the mass difference between the decaying particle and its products, the phase space is said to be larger for the process. In the β-decay process $n \rightarrow p + e^- + \bar{\nu}e$, the mass difference being small, the phase space is small, indicating that the beta decay process is quite constrained. The matrix element on the other hand is calculated from the relevant Feynman diagrams by using the Feynman Rules.

2.2.2.4 Feynman Rules

In Chapter one, we introduced the Feynman diagrams. The evaluation of a Feynman diagram, which shows a particular decay or the scattering process, is done by using the Feynman rules which help calculate the matrix element without making the actual derivation. Let us take up the example of a scattering process to understand the Feynman rules. The tree level Feynman diagram in momentum space is drawn in Fig. (2.5). Two particles having momenta p_1 and p_2 enter the reaction and after interaction are scattered with respective momenta of p_3 and p_4. In the above diagram, the lines indicating p_1, p_2, p_3 and p_4 are external lines. Points 1 and 2 are vertices. The line joining the vertices indicated by momentum q is internal line. In the following figure (Fig. 2.6), there is only one vertex and no internal line; all the three are the external lines.

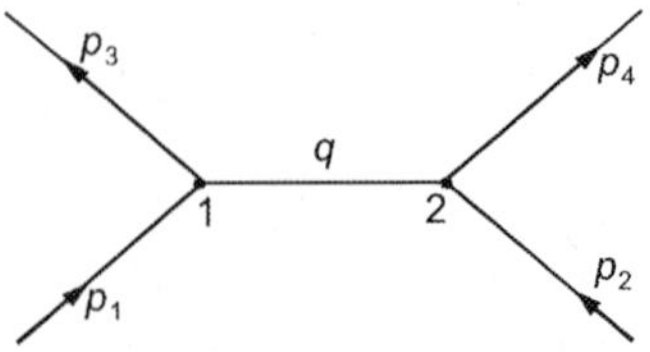

Fig. 2.5: Scattering process Feynman diagram in momentum space

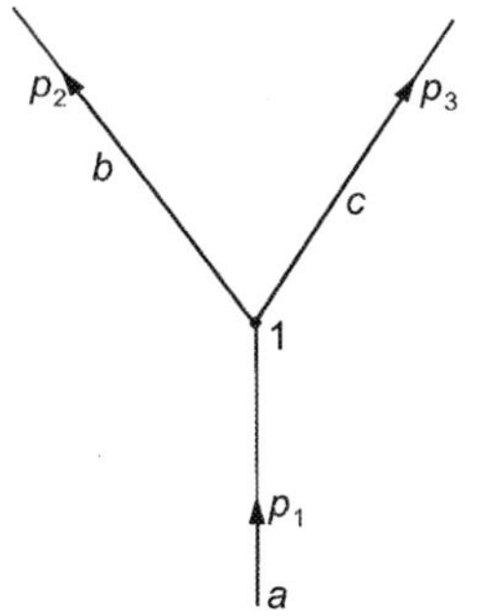

Fig. 2.6: A vertex without internal line. The process is decay $a \to b + c$

The diagrams (2.5) and (2.6) are lowest order diagrams. They are also called the *tree level diagrams*. The four, six, and more vertices' diagrams are called the *higher order diagrams* (see Chapter 1).

To calculate the matrix element for a process, we adopt the following Feynman Rules. For applying these rules, we have to notice in the Feynman diagram:

I. The kind of external line: Whether it is representing the spin-0 boson or antiboson; fermion (spin $-\frac{1}{2}$) or antifermion; spin-1 boson or antiboson (this includes the photon as well). Accordingly, from the table given below, note the multiplicative factor that is used in the construction of the matrix element.

II. The kind of internal line: Whether the internal line (i.e., the interaction is via) is representing spin-0 boson or fermion or massive boson or photon. Accordingly, the multiplicative factor has to be picked up. This factor comes from the corresponding propagator and

III. The kind of vertex: Whether the vertex is made by two spin-0 lines, or by two spin $-\frac{1}{2}$ lines, or by two spin-1 lines. The table gives the multiplicative factor for the vertices, which we may encounter in quantum electrodynamics. The *QED* vertex multiplicative factors then can be used in constructing the matrix element.

Following is the table that provides the multiplicative factors for the situations mentioned above. The table is named *Rules for Feynman diagrams*.

Rule No.	Spin	Part of Feynman diagram	Graphic Representation	Description	Multiplicative factor for constructing the matrix element $-i\, M_{fi}$
1. (a)	0	External line		Boson	
(b)	0	Do		Do	1
2. (a)	0	Do		Antiboson	
(b)	0	Do		Do	

3. (a)	$\frac{1}{2}$	Do		Fermion (initial)	$u\,(p)$
(b)	$\frac{1}{2}$	Do		Fermion (final)	$\bar{u}(p)$
4. (a)	$\frac{1}{2}$	Do		Antifermion (initial)	$\bar{v}(p)$
(b)	$\frac{1}{2}$	Do		Antifermion (final)	$v(p)$
5.	1	Do		Boson including Photon (initial)	ε^{μ}
6.	1	Do		Antiboson (initial)	
7.	1	Do		Boson (final)	$\varepsilon^{\mu *}$
8.	1	Do		Antiboson (final)	
9.	0	Internal line		Boson	$\frac{i}{m^2 - k^2}$
10.	$\frac{1}{2}$	Do		Fermion	$\frac{i(\gamma_{\mu}k^{\mu} + m)}{m^2 - k^2}$
11.	1	Do		Massive Vector Boson	$\frac{i\left(g_{\mu\nu} - \frac{k_{\mu}k_{\nu}}{m^2}\right)}{m^2 - k^2}$
12.	1	Do		Massless Vector Boson	$-\frac{ig_{\mu\nu}}{k^2}$
13.	0-0	QED Vertex		Point like particles	$-ie(p_1 + p_2)^{\mu}$
14.	$\frac{1}{2}$-$\frac{1}{2}$	QED Vertex		Do	$-ie\gamma^{\mu}$
15.	1-1	QED Vertex	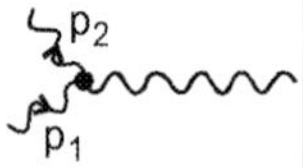	Do $\{-ie[(p_1 + p_2)^{\mu}g\alpha\beta - p_{2\alpha}g^{\mu}_{\beta} - p_{1\beta}g^{\alpha}_{\mu}]\}$	
16.		Each Vertex		Conservation of energy and momentum $(2\pi)^4\delta^4(k_1 + k_2 + k_3)$	
17.	Each internal momentum q			Integration over internal momenta $\frac{d^4q}{(2\pi)^4}$	
18.	Obtain delta function in the form $(2\pi)^4\delta^4(p_1 + p_2 + \cdots - p_n)$ and then cancel it				

Above rules would become clear when we take in the following the specific examples of quantum mechanical processes.

2.3 QUANTUM ELECTRODYNAMIC PROCESSES

The above-mentioned Feynman rules can be applied to various quantum electrodynamic processes. We shall consider the following processes:

(a) Compton scattering
(b) Electron–positron (e^-e^+) annihilation
(c) Moller (electron–electron) scattering
(d) Bhabha (electron–positron) scattering

For each process, we shall draw the all possible (there is more than one Feynman diagram for each process) Feynman diagram and using the Feynman rules, calculate the transition matrix element M_{fi} and then calculate the differential cross section for the process. We shall, however, restrict to tree level (two—vertex) Feynman diagrams.

2.3.1 Compton Scattering

The scattering of a photon by an electron is called *Compton scattering*. A.H. Compton made this experimental discovery in 1923. It can be expressed by

$$e+\gamma \rightarrow e+\gamma \tag{2.53}$$

where $e \Rightarrow e^-$. We can consider three kinds of Compton scatterings: (a) Scattering of unpolarized photon by unpolarized electron. (b) Scattering of polarized photon by unpolarized electron and (c) scattering of polarized photons by polarized electrons. We would consider the case (b). The tree level Feynman diagrams for the Compton scattering can be drawn in the following two ways (Figs.2.7, a and b).

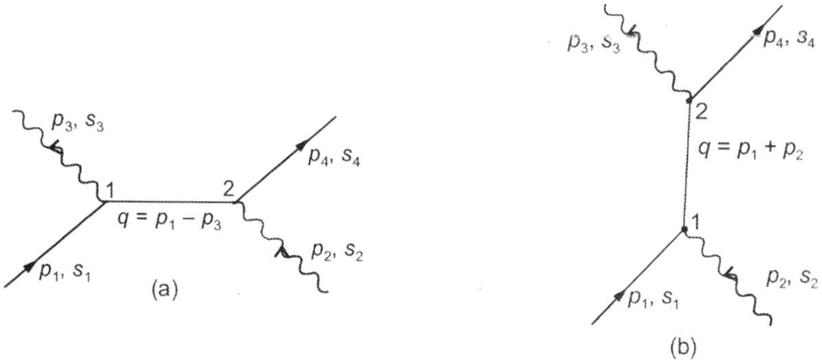

Fig. 2.7: Feynman diagrams for Compton scattering

Fig. (2.7a) reads – an electron with four-momentum p_1 and spin s_1 (up or down) and a photon with four momentum p_2 and polarization s_2 (photon has also two polarization states, so s_2, like s_1 has also only two values) enter the reaction. At the vertex 1, the electron emits a photon with momentum p_3 and polarization s_3 and itself propagates with a momentum $q = p_1 - p_3$, which in turn at vertex 2 absorbs the photon (p_2, s_2) and goes away as (p_4, s_4) electron. In Fig. (2.7b), however reverse happens. At vertex 1, first the absorption of photon (p_2, s_2) by the electron (p_1, s_1) takes place, which propagates with ($q = q_1 + q_2$) between vertices 1 and 2 to release photon (p_3, s_3) and itself recoiling as(p_4, s_4) electron.

In both the diagrams there is a virtual electron propagating between vertices 1 and 2. Thus the Compton Effect provides an example of electron (fermion) propagator and photon polarization.

Now we apply Feynman rules to calculate the matrix element. Diagram (2.7a) would give the matrix element M_1 and diagram (2.7b) would give the matrix element M_2. The total matrix element for the Compton scattering would be

$$M_{fi} = M_1 + M_2 \tag{2.54}$$

It has however, to be noted in advance that in case of e^-e^- or e^-e^+ scattering where both the particles are fermions we shall have minus sign in between M_1 and M_2. This is because of the antisymmetrization effect. Here, since photon is boson, due to the interchange of ingoing and outgoing photon $M_1 + M_2$ must be symmetric hence plus sign.

To calculate M_1, we look at the diagram (2.7a) and proceed in the backward direction, i.e., proceed opposite to the direction of arrows. Start from vertex 2 and apply Feynman rules given in the table and note the following multiplicative factors.

1. (p_4, s_4) is the outgoing (final) electron (fermion) line. It denotes an external line going away from the vertex 2. Feynman rule (3b) is applicable which provides a multiplicative factor $\bar{\mu}(p_4)$, which we may write as $\bar{\mu}(4)$.
2. The photon (p_2, s_2), implying the four momentum of photon is p_2 and its state of polarization is denoted by s_2, is entering the vertex 2. Rule (5) is applicable which provides a multiplicative factor $\varepsilon_\mu(s_2)$ written as $\varepsilon_\mu(2)$.
3. The vertex 2. Two fermion lines and one photon make this vertex. One fermion line (p_4, s_4) is external; other (q) is internal. This is spin $\left(\frac{1}{2}-\frac{1}{2}\right)$ vertex. Rule (14) is applicable and it provides a multiplicative factor of $(-ie\gamma^\mu)$.
4. Apply rule (16) at the vertex 2. Rule (16) demanding conservation of energy and momentum provides a multiplicative factor $(2\pi)^4\delta^4$ $(k_1 + k_2 + k_3)$ where k_1, k_2, k_3 are the three–four component momenta entering or leaving the vertex. The momenta, which enter the vertex, carry a positive sign and momenta, which leave the vertex, carry a negative sign. Hence rule (16) provides a multiplicative factor of $(2\pi)^4\delta^4$ $(q_2 + q - q_4)$.

5. Coming towards vertex 1, we now have the internal line denoted by four-momentum q. It denotes a virtual electron of four-momentum q. We therefore have a fermion internal line inviting Feynman rule (10) which provides a multiplicative factor of

$$\frac{i(\gamma_\mu k^\mu + m)}{m^2 - k^2} \equiv \frac{i(\gamma_\mu q^\mu + m)}{m^2 - q^2} \tag{2.55}$$

where q will be determined by δ^4 functions coming from vertices 2 and 1.

6. Now the vertex 1. Like vertex 2, it is also a spin $(\frac{1}{2} - \frac{1}{2})$ vertex and hence rule (14), similar to Para 3 above, provides a multiplicative factor of $(-ie\gamma^\mu)$.

7. At vertex 1, the four momentum p_1 enters the vertex hence is $+p_1$; p_3 and q leave the vertex hence each carries a minus sign. Rule (16) (see Para 4 above) provides a multiplicative factor
$$(2\pi)^4\delta^4(p_1 - p_3 - q)$$

8. At vertex 1, (p_3, s_3) denotes the emitted photon with polarization denoted by s_3. Rule 7 provides a multiplicative factor of

$$\varepsilon^*_\mu(s_3) \equiv \varepsilon^*_\mu(3) \tag{2.56}$$

9. At vertex 1, (p_1, s_1) denotes the external line of electron (fermion), which is entering the vertex. Rule (3a) provides a multiplicative factor of

$$u(p_1) \equiv u(1) \tag{2.57}$$

10. Multiplying all the above nine factors we apply rule 17. Integrate it over the internal momenta $\dfrac{d^4q}{(2\pi)^4}$.

11. After Para 10, we apply rule (18).

Above-mentioned 11 steps will give $-iM_1$ for the Feynman diagram (2.7a) from which canceling $(-i)$ we get the matrix element M_1.

Thus

$$-iM_1 = \int [\overline{u}(4)\varepsilon_\mu(2)(-ie\gamma^\mu)(2\pi)^4\delta^4(p_2 + q - p_4)$$
$$\frac{i(\gamma_\mu q^\mu + m)}{m^2 - q^2}(-ie\gamma^\nu)(2\pi)^4\delta^4(p_1 - p_3 - q)\varepsilon^*_\nu(3)u(1)]\frac{d^4q}{(2\pi)^4} \tag{2.58}$$

from which we get canceling $(-i)$

$$M_1 = (2\pi)^4 e^2 \int \overline{u}(4)\varepsilon_\mu(2)\gamma^\mu \frac{(\gamma_\mu q^\mu + m)}{m^2 - q^2}\delta^4(p_1 - p_3 - q)$$
$$\gamma^\nu\varepsilon^*_\nu(3)u(1)\delta^4(p_2 + q - p_4)d^4q \tag{2.59}$$

In the above equation we may first use the following property of δ function

$$f(x)\delta(x-a) = f(a)\delta(x-a) \tag{2.60}$$

to write

$$\frac{(\gamma_\mu q^\mu + m)}{m^2 - q^2}\delta^4(p_1 - p_3 - q) = \frac{[\gamma_\mu (p_1 - p_3)^\mu + m]}{m^2 - (p_1 - p_3)^2}\delta^4(p_1 - p_3 - q) \tag{2.61}$$

and then use the property

$$\int \delta(a-x)\delta(x-b) = \delta(a-b) \tag{2.62}$$

to write

$$\int \delta^4[(p_1 - p_3) - q]\delta^4[q - (p_4 - p_2)]d^4q = \delta^4(p_1 - p_3 - p_4 + p_2) \tag{2.63}$$

Using equations (2.61) and (2.63), eqn. (2.59) becomes

$$M_1 = e^2\overline{u}(4)\varepsilon^\mu(2)\gamma_\mu \frac{[\gamma_\mu p_1^\mu - \gamma_\mu p_3^\mu + m]}{m^2 - (p_1 - p_3)^2}\gamma^\nu \varepsilon_\nu^*(3)u(1)$$
$$(2\pi)^4\delta^4(p_1 - p_3 + p_2 - p_4) \tag{2.64}$$

After obtaining δ function in terms of all ps, we may erase it along with $(2\pi)^4$. So finally the diagram (2.7a) makes M_1 contribution to the matrix element, which is given by

$$M_1 = \frac{e^2}{m^2 - (p_1 - p_3)^2}[\overline{u}(4)\varepsilon_\mu(2)\gamma^\mu(\gamma_\mu p_1^\mu - \gamma_\mu p_3^\mu + m)\gamma^\nu\varepsilon_\nu^*(3)u(1)] \tag{2.65}$$

Likewise we may employ Feynman rules to calculate $-iM_2$ from the diagram (2.7b). We write few steps only. Proceed from vertex 2 towards vertex 1.

$$-iM_2 = \int [\overline{u}(4)\varepsilon_\mu^*(3)(-ie\gamma^\mu)(2\pi)^4\delta^4(q - p_3 - p_4)$$
$$\frac{i(\gamma_\mu q^\mu + m)}{m^2 - q^2}(-ie\gamma^\nu)\varepsilon_\nu(2)u(1)(2\pi)^4\delta^4(p_1 + p_2 - q)]\frac{d^4q}{(2\pi)^4} \tag{2.66}$$

which, on using

$$\frac{(\gamma_\mu q^\mu + m)}{m^2 - q^2}\delta^4(p_1 + p_2 - q) = \frac{(\gamma_\mu p_1^\mu + \gamma_\mu p_2^\mu + m)}{m^2 - (p_1 + p_2)^2}\delta^4(p_1 + p_2 - q) \tag{2.67}$$

and

$$\int \delta^4(p_1 + p_2 - q)\delta^4[q - (p_3 + p_4)]d^4q = \delta^4(p_1 + p_2 - p_3 - p_4) \tag{2.68}$$

and then canceling $(2\pi)^4\delta^4(p_1 + p_2 - p_3 - p_4)$, gives

$$M_2 = \frac{e^2}{m^2 - (p_1 + p_2)^2}[\overline{u}(4)\varepsilon_\mu^*(3)\gamma^\mu(\gamma_\mu p_1^\mu + \gamma_\mu p_2^\mu + m)\gamma^\nu\varepsilon_\nu(2)u(1)] \tag{2.69}$$

The total matrix element for the Compton scattering would be

$$M_{fi} = M_1 + M_2 \tag{2.70}$$

where M_1 and M_2 are given by eqns. (2.65) and (2.69) respectively.

2.3.1.1 Calculation of $|M_{fi}|^2$

We require the square of matrix element, i.e., $|M_{fi}|^2$ to obtain the differential scattering cross section. Suppose we have a reaction in which particles 1 and 2 with four momenta p_1 and p_2 collide to produce particles 3, 4,..., $n+2$ with four momenta $p_3, p_4, ...p_{n+2}$ respectively, implying (Fig. 2.8), where the particles in the final state, i.e., particles 3, 4...$n+2$ ($n = 1, 2, 3...$) are assumed distinct and spinless. The differential scattering cross section $d\sigma$ is given by the Golden rule

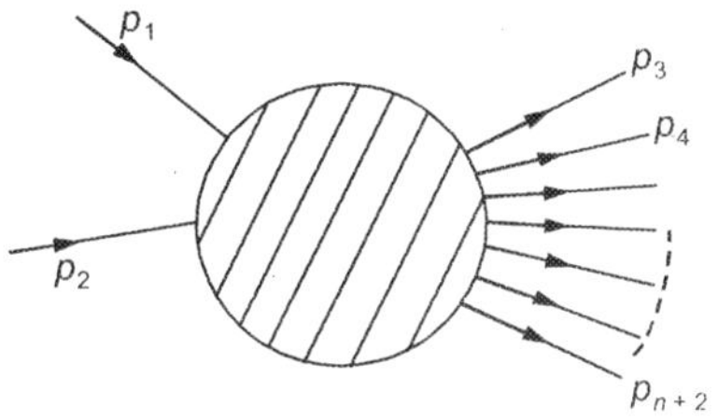

Fig. 2.8: The scattering process 1 + 2 → 3 + 4 + ... + (*n* + 2)

$$d\sigma = \frac{1}{4[(p_1 \cdot p_2)^2 - m_1^2 m_2^2]^{\frac{1}{2}}} \int_{\Delta p} \frac{d^3 \boldsymbol{p}_3}{2p_3^0(2\pi)^3} \frac{d^3 \boldsymbol{p}_4}{2p_4^0(2\pi)^3} \cdots \frac{d^3 \boldsymbol{p}_{n+2}}{2p_{n+2}^0(2\pi)^3} |M_{fi}|^2$$
$$\times (2\pi)^4 \delta^4(p_1 + p_2 - p_3 - p_4 \cdots - p_{n+2}) \tag{2.71}$$

where the one particle states are normalized in accordance with the relation

$$\langle p' | p \rangle = 2p^0 (2\pi)^3 \delta^3(p' - p) \tag{2.72}$$

The interval Δp implies that the energy momentum of a particle lies in the range d^3p about p. It has however, to be noted that if we are dealing with massive fermions the factor $2p^0$ is replaced by $\frac{p^0}{m}$. If the colliding particles 1 and 2 are fermions, the factor $\frac{1}{4}$ in front of eqn. (2.71) is replaced by $m_1 m_2$. If there are fermions in the final state (i.e. among particles 3, 4... $n+2$) the corresponding phase space element $\frac{d^3\boldsymbol{p}}{2p^0(2\pi)^3}$ is replaced by $\left(\frac{m}{p^0}\right)\frac{d^3\boldsymbol{p}}{(2\pi)^3}$. However, if there are massless fermions among the final state particles we retain the phase element as $\frac{d^3\boldsymbol{p}}{2p^0(2\pi)^3}$. If in the final state there are identical particles we put a factor

$s = \frac{1}{j!}$ in the numerator of eqn. (2.71) where j is the number of identical particles.

Eqn. (2.71) then tells that to obtain $d\sigma$, we need to first calculate $|M_{fi}|^2$. To calculate M_{fi} for Compton scattering, we recall equations (2.65) and (2.69). We rewrite them here as

$$M_1 = \frac{e^2}{2p_1p_3}\bar{u}(4)\Gamma_1 u(1) \tag{2.73}$$

and

$$M_2 = \frac{e^2}{-2p_1p_2}\bar{u}(4)\Gamma_2 u(1) \tag{2.74}$$

in which

$$\Gamma_1 = \varepsilon_\mu(2)\gamma^\mu(\gamma_\mu p_1^\mu - \gamma_\mu p_3^\mu + m)\gamma^\nu \varepsilon_\nu^*(3) \tag{2.75}$$

$$\Gamma_2 = \varepsilon_\mu^*(3)\gamma^\mu(\gamma_\mu p_1^\mu + \gamma_\mu p_2^\mu + m)\gamma^\nu \varepsilon_\nu(2) \tag{2.76}$$

and the denominator of eqn. (2.65) (M_1) has been written as

$$m^2 - (p_1 - p_3)^2 = m^2 - p_1^2 - p_3^2 + 2p_1 \cdot p_3 = 2p_1 \cdot p_3 \tag{2.77}$$

where

$$p_1^2 = E_{\text{electron}}^2 - \boldsymbol{p}_1^2 = \boldsymbol{p}_1^2 + m^2 - \boldsymbol{p}_1^2 = m^2 \tag{2.78}$$

and

$$p_3^2 = E_{\text{photon}}^2 - \boldsymbol{p}_3^2 = \boldsymbol{p}_3^2 + 0 - \boldsymbol{p}_3^2 = 0 \tag{2.79}$$

have been used. Similarly, the denominator of M_2 (eqn. (2.69)) has been rewritten in eqn. (2.74) as

$$m^2 - (p_1 - p_2)^2 = m^2 - p_1^2 - p_2^2 - 2p_1 \cdot p_2 = -2p_1 \cdot p_2 \tag{2.80}$$

in which like eqn. (2.78) and (2.79) we have used

$$p_1^2 = m^2 \tag{2.81a}$$

and

$$p_2^2 = 0 \tag{2.81b}$$

It is to be noted that p_1, p_2, p_3 have four components whereas $\boldsymbol{p}_1, \boldsymbol{p}_2, \boldsymbol{p}_3$ have three components. We may now begin with eqns. (2.73) and (2.74) to write

$$\begin{aligned} M_{fi} &= M_1 + M_2 \\ &= e^2\left[\bar{u}(4)\left(\frac{\Gamma_1}{2p_1 \cdot p_3} - \frac{\Gamma_2}{2p_1 \cdot p_2}\right)u(1)\right] \end{aligned} \tag{2.82}$$

and

$$|M_{fi}|^2 = M_{fi}\, M_{fi}^\dagger \tag{2.83}$$

Using $\bar{u} = u^\dagger \gamma^0, \gamma^{0^\dagger} = \gamma^0$ and $\gamma^{0^2} = 1$ in eqn. (2.83),

$$\begin{aligned}(M_{fi})^\dagger &= e^2 u(1)^\dagger \left(\frac{\Gamma_1^\dagger}{2p_1 \cdot p_3} - \frac{\Gamma_2^\dagger}{2p_1 \cdot p_2} \right) \gamma^0 u(4) \\ &= e^2 u(1)^\dagger \gamma^0 \left(\frac{\gamma^0 \Gamma_1^\dagger \gamma^0}{2p_1 \cdot p_3} - \frac{\gamma^0 \Gamma_2^\dagger \gamma^0}{2p_1 \cdot p_2} \right) u(4) \\ &= e^2 \bar{u}(1) \left(\frac{\bar{\Gamma}_1}{2p_1 \cdot p_3} - \frac{\bar{\Gamma}_2}{2p_1 \cdot p_2} \right) u(4)\end{aligned} \tag{2.84}$$

where

$$\bar{\Gamma}_1 = \gamma^0 \Gamma_1^\dagger \gamma^0 \tag{2.85a}$$

and

$$\bar{\Gamma}_2 = \gamma^0 \Gamma_2^\dagger \gamma^0 \tag{2.85b}$$

Using eqns. (2.84) and (2.82), we may write for $|M_{fi}|^2$

$$|M_{fi}|^2 = e^4 \left[\bar{u}(4) \left(\frac{\Gamma_1}{2p_1 \cdot p_3} - \frac{\Gamma_2}{2p_1 \cdot p_2} \right) u(1)\bar{u}(1) \left(\frac{\bar{\Gamma}_1}{2p_1 \cdot p_3} - \frac{\bar{\Gamma}_2}{2p_1 \cdot p_2} \right) u(4) \right] \tag{2.86}$$

Now $u(1)$ is the spin $-\frac{1}{2}$ (Dirac) wavefunction for the electron of mass (rest) m having the four-momentum p_1. The electron may have spin up or down which may be denoted by $s_1 = 1$ and $s_1 = 2$. Equation (2.86) may therefore be summed over the spin orientation $s_1 = 1, 2$ to write

$$\begin{aligned}\sum_{s_1} |M_{fi}|^2 = e^4 \Bigg[\bar{u}(4) &\left(\frac{\Gamma_1}{2p_1 \cdot p_3} - \frac{\Gamma_2}{2p_1 \cdot p_2} \right) \left\{ \sum_{s_1=1,2} u^{s_1}(1)\bar{u}^{s_1}(1) \right\} \\ &\left(\frac{\bar{\Gamma}_1}{2p_1 \cdot p_3} - \frac{\bar{\Gamma}_2}{2p_1 \cdot p_2} \right) u(4) \Bigg]\end{aligned} \tag{2.87}$$

These spinors $u^{s_1}(1)\bar{u}^{s_1}(1)$ obey the following matrix relation

$$\sum_{s_1=1,2} u^{s_1}(1)\bar{u}^{s_1}(1) = \frac{1}{2m}(\gamma^\mu p_{1\mu} + m) = \frac{\not{p}_1 + m}{2m} \tag{2.88}$$

where m is the rest mass of electron. Similar matrix relation can be written for spinors $u(4) \equiv u^{s_4}(4)$ and $\bar{u}(4) \equiv \bar{u}^{s_4}(4)$

$$\sum_{s_4=1,2} u^{s_4}(4)\bar{u}^{s_4}(4) = \frac{1}{2m}(\gamma^{\mu} p_{4\mu} + m) = \frac{\not{p}_4 + m}{2m} \tag{2.89}$$

Using eqn. (2.88) we may write from eqn. (2.87)

$$\sum_{s_4}\sum_{s_1} |M_{fi}|^2 = \frac{e^4}{2m} \sum_{s_4=1,2} \bar{u}^{s_4}(4) Q u^{s_4}(4) \tag{2.90}$$

where Q is the 4×4 matrix

$$Q = \left(\frac{\Gamma_1}{2p_1 \cdot p_3} - \frac{\Gamma_2}{2p_1 \cdot p_2} \right)(\not{p}_1 + m)\left(\frac{\bar{\Gamma}_1}{2p_1 \cdot p_3} - \frac{\bar{\Gamma}_2}{2p_1 \cdot p_2} \right) \tag{2.91}$$

Equation (2.90) may be expressed as

$$\begin{aligned}\sum_{s_4}\sum_{s_1} |M_{fi}|^2 &= \frac{e^4}{2m} \sum_{s_4=1,2} [\bar{u}^{s_4}(4)]_{\alpha} \; Q_{\alpha\beta} \; [u^{s_4}(4)]_{\beta} \\ &= \frac{e^4}{2m} Q_{\alpha\beta} \sum_{s_4=1,2} [u^{s_4}(4)\bar{u}^{s_4}(4)]_{\beta\alpha}\end{aligned} \tag{2.92}$$

where α and β have been summed over from 1 to 4. Using eqn. (2.89), eqn. (2.92) becomes

$$\sum_{s_4}\sum_{s_1} |M_{fi}|^2 = \frac{e^4}{4m^2} Q_{\alpha\beta}[\not{p}_4 + m]_{\beta\alpha} \tag{2.93}$$

where $\not{p}_4 + m$ with $m = m1$, where 1, is also a 4×4 matrix. The RHS of equation (2.93) can be expressed in terms of trace of the product of two matrices. For the two matrices P and R we have

$$\text{Trace}[PR] = (PR)_{\alpha\alpha} = P_{\alpha\beta} R_{\beta\alpha} \tag{2.94a}$$

or in general

$$\text{Trace}[AB \ldots YZ] = [AB \ldots YZ]_{\alpha\alpha} = A_{\alpha\beta} B_{\beta\gamma} \ldots Y_{\delta\varepsilon} Z_{\varepsilon\alpha} \tag{2.94b}$$

Therefore comparing eqn. (2.94a) with eqn. (2.93), we may write by dropping the summation signs

$$|M_{fi}|^2 = \frac{e^4}{4m^2} \text{Trace}\{Q(\not{p}_4 + m)\} \tag{2.95}$$

where the matrix Q is given by eqn. (2.91), Γ_1, Γ_2 by eqs. (2.75) and (2.76) respectively and $\bar{\Gamma}_1, \bar{\Gamma}_2$ are obtained by using eqn. (2.85). We calculate the trace (2.95) under the following subtitle.

2.3.1.2 Calculation of Trace {Q($\not{p}_4$ + m)}

We rewrite Γ_1 and Γ_2 from eqns. (2.75) and (2.76) as

$$\Gamma_1 = \not{\varepsilon}_2(\not{p}_1 - \not{p}_3 + m)\not{\varepsilon}_3^* \tag{2.96a}$$

$$\Gamma_2 = \not{\varepsilon}_3^*(\not{p}_1 + \not{p}_2 + m)\not{\varepsilon}_2 \tag{2.96b}$$

where slash notations imply $\not{\varepsilon}_2 = \varepsilon_\mu(2)\gamma^\mu = \gamma^\mu\varepsilon_\mu(2)$, also $\not{p} \Rightarrow p_\mu\gamma^\mu$ etc. Using the identity (to calculate $\not{p}_1\not{\varepsilon}_3^*$ and $\not{p}_3\not{\varepsilon}_3^*$)

$$\not{a}\not{b} = -\not{b}\not{a} + 2a\cdot b \tag{2.97}$$

and choosing p_1, p_2, ε_2 and ε_3^* such that

$$\not{p}_1\not{\varepsilon}_3^* = \not{p}_3\not{\varepsilon}_3^* = p_1\cdot\varepsilon_2 = p_2\cdot\varepsilon_2 \tag{2.98}$$

we may write

$$\begin{aligned}\Gamma_1 &= \not{\varepsilon}_2(-\not{\varepsilon}_3^*\not{p}_1 + \not{\varepsilon}_3^*\not{p}_3 + m\not{\varepsilon}_3^*)\\ &= -\not{\varepsilon}_2\not{\varepsilon}_3^*(\not{p}_1 - m) + \not{\varepsilon}_2\not{\varepsilon}_3^*\not{p}_3\\ &= \not{\varepsilon}_2\not{\varepsilon}_3^*\not{p}_3\end{aligned} \tag{2.99}$$

where Dirac equation [see eqn.(2.86)]

$$(\not{p}_1 - m)u(1) = 0 \tag{2.100}$$

has been used. Likewise, using the identity (2.97) to calculate $\not{p}_1\not{\varepsilon}_2$ and $\not{p}_2\not{\varepsilon}_2$ and employing the relevant conditions from (2.98) we get

$$\begin{aligned}\Gamma_2 &= -\not{\varepsilon}_3^*\not{\varepsilon}_2(\not{p}_1 - m) - \not{\varepsilon}_3^*\not{\varepsilon}_2\not{p}_2\\ &= -\not{\varepsilon}_3^*\not{\varepsilon}_2\not{p}_2\end{aligned} \tag{2.101}$$

where too we have used the Dirac equation (2.100).

The $\overline{\Gamma}_1$ and $\overline{\Gamma}_2$ can likewise be simplified by using eqns. (2.85), (2.99) and (2.101)

$$\begin{aligned}\overline{\Gamma}_1 &= \gamma^0(\not{\varepsilon}_2\not{\varepsilon}_3^*\not{p}_3)^\dagger\gamma^0\\ &= \gamma^0(\gamma^\mu\varepsilon_{2\mu}\gamma^\mu\varepsilon_{3\mu}^*\gamma^\mu p_{3\mu})^\dagger\gamma^0\\ &= \gamma^0(p_{3\mu}\gamma^{\mu^\dagger}\varepsilon_{3\mu}^*\gamma^{\mu\dagger}\varepsilon_{2\mu}\gamma^{\mu^\dagger})\gamma^0\\ &= \gamma^0[p_{3\mu}(\gamma^0\gamma^\mu\gamma^0)\varepsilon_{3\mu}^*(\gamma^0\gamma^\mu\gamma^0)\varepsilon_{2\mu}(\gamma^0\gamma^\mu\gamma^0)]\gamma^0\\ &= [p_{3\mu}\gamma^{0^2}\gamma^\mu\varepsilon_{3\mu}^*\gamma^{0^2}\gamma^\mu\varepsilon_{2\mu}\gamma^{0^2}\gamma^\mu\gamma^{0^2}]\\ &= [\not{p}_3\not{\varepsilon}_3^*\not{\varepsilon}_2]\end{aligned} \tag{2.102}$$

where $\gamma^{\mu^\dagger} = \gamma^0\gamma^\mu\gamma^0$ and $\gamma^{0^2} = 1$ have been used. Likewise we get

$$\begin{aligned}\bar{\Gamma}_2 &= -\gamma^0(\not{\epsilon}_3^*\not{\epsilon}_2\not{p}_3)^\dagger\gamma^0 \\ &= -\not{p}_2\not{\epsilon}_2\not{\epsilon}_3^*\end{aligned} \tag{2.103}$$

Thus using equations (2.99), (2.101), (2.102), (2.103) and (2.91), we may write from eqns. (2.95)

$$Trace\{Q(\not{p}_4+m)\}$$
$$= Trace\left[\left(\frac{\not{\epsilon}_2\not{\epsilon}_3^*\not{p}_3}{2p_1\cdot p_3}+\frac{\not{\epsilon}_3^*\not{\epsilon}_2\not{p}_2}{2p_1\cdot p_2}\right)(\not{p}_1+m)\left(\frac{\not{p}_3\not{\epsilon}_3^*\not{\epsilon}_2}{2p_1\cdot p_3}+\frac{\not{p}_2\not{\epsilon}_2\not{\epsilon}_3^*}{2p_1\cdot p_2}\right)(\not{p}_4+m)\right] \tag{2.104}$$

which on using

$$Trace\,(P+Q) = Trace\;P + Trace\;Q \tag{2.105}$$

can be separated into four traces

$$Trace\;\{Q(\not{p}_4+m)\} = Trace\;I + Trace\;II + Trace\;III + Trace\;IV \tag{2.106}$$

where

$$Trace\;I = \frac{1}{4(p_1\cdot p_3)^2}Tr[\not{\epsilon}_2\not{\epsilon}_3^*\not{p}_3(\not{p}_1+m)\not{p}_3\not{\epsilon}_3^*\not{\epsilon}_2(\not{p}_4+m)] \tag{2.107a}$$

$$Trace\;II = \frac{1}{4(p_1\cdot p_3)(p_1\cdot p_2)}Tr[\not{\epsilon}_2\not{\epsilon}_3^*\not{p}_3(\not{p}_1+m)\not{p}_2\not{\epsilon}_2\not{\epsilon}_3^*(\not{p}_4+m)] \tag{2.107b}$$

$$Trace\;III = \frac{1}{4(p_1\cdot p_2)(p_1\cdot p_3)}Tr[\not{\epsilon}_3^*\not{\epsilon}_2\not{p}_2(\not{p}_1+m)\not{p}_2\not{\epsilon}_3^*\not{\epsilon}_2(\not{p}_4+m)] \tag{2.107c}$$

$$Trace\;IV = \frac{1}{4(p_1\cdot p_2)^2}Tr[\not{\epsilon}_3^*\not{\epsilon}_2\not{p}_2(\not{p}_1+m)\not{p}_2\not{\epsilon}_2\not{\epsilon}_3^*(\not{p}_4+m)] \tag{2.107d}$$

We may now obtain the above traces. Omitting the multiplying factor for the moment we may write

$$\begin{aligned}Tr\;I\quad &= Tr\{\not{\epsilon}_2\not{\epsilon}_3^*\not{p}_3\not{p}_1\not{p}_3\not{\epsilon}_3^*\not{\epsilon}_2\not{p}_4\} + mTr\{\not{\epsilon}_2\not{\epsilon}_3^*\not{p}_3\not{p}_1\not{p}_3\not{\epsilon}_3^*\not{\epsilon}_2\} \\ &+mTr\{\not{\epsilon}_2\not{\epsilon}_3^*\not{p}_3\not{p}_3\not{\epsilon}_3^*\not{\epsilon}_2\not{p}_4\} + m^2Tr\{\not{\epsilon}_2\not{\epsilon}_3^*\not{p}_3\not{p}_3\not{\epsilon}_3^*\not{\epsilon}_2\}\end{aligned} \tag{2.108}$$

We now use one of the trace theorems that the trace of the product of an odd number of gamma matrices vanishes. Hence in eqn. (2.108), the traces which are multiplied by m vanish as they contain seven (odd) gamma matrices. Note that each slashed parameter has one γ^μ. We also find that the last, the m^2 trace of eqn. (2.108) also vanishes because it contains $\not{p}_3\not{p}_3$ which on using the identity (2.97) becomes

$$\not{p}_3\not{p}_3 = p_3\cdot p_3 = p_3^2 = 0 \tag{2.109}$$

where we have used eqn. (2.79). Thus from eqn. (2.108) we have

$$Tr\ I = Tr\{\not{\varepsilon}_2 \not{\varepsilon}_3^* \not{p}_3 \not{p}_1 \not{p}_3 \not{\varepsilon}_3^* \not{\varepsilon}_2 \not{p}_4\} \tag{2.110}$$

To calculate (2.110) we keep on using the identity (2.97) to shift any parameter to a desired place. For example in eqn. (2.110) we would like that two $\not{\varepsilon}_3^*$ come together to allow us use

$$\not{\varepsilon}_3^* \not{\varepsilon}_3^* = \varepsilon_3^* \cdot \varepsilon_3^* = -1 \tag{2.111a}$$

and similarly we could use

$$\not{\varepsilon}_2 \not{\varepsilon}_2 = \varepsilon_2 \cdot \varepsilon_2 = -1 \tag{2.111b}$$

The identity (2.97) is thus helpful in forming a useful togetherness of any two slashed parameters.

Thus

$$\begin{aligned}
Tr\ I &= Tr\{\not{\varepsilon}_2 \not{\varepsilon}_3^* \not{p}_3 (-\not{p}_3 \not{p}_1 + 2p_1 \cdot p_3) \not{\varepsilon}_3^* \not{\varepsilon}_2 \not{p}_4\} && (2.112a)\\
&= -Tr\{\not{\varepsilon}_2 \not{\varepsilon}_3^* \not{p}_3 \not{p}_3 \not{p}_1 \not{\varepsilon}_3^* \not{\varepsilon}_2 \not{p}_4\} + (2p_1 \cdot p_3) Tr\{\not{\varepsilon}_2 \not{\varepsilon}_3^* \not{p}_3 \not{\varepsilon}_3^* \not{\varepsilon}_2 \not{p}_4\} && (2.112b)\\
&= 0 - (2p_1 \cdot p_3) Tr\{\not{\varepsilon}_2 \not{\varepsilon}_3^* \not{\varepsilon}_3^* \not{p}_3 \not{\varepsilon}_2 \not{p}_4\} && (2.112c)\\
&= 2(p_1 \cdot p_3) Tr\{\not{\varepsilon}_2 \not{p}_3 \not{\varepsilon}_2 \not{p}_4\} && (2.112d)\\
&= 8(p_1 \cdot p_3)\{2(\varepsilon_2 \cdot p_3)(\varepsilon_2 \cdot p_4) + (p_3 \cdot p_4)\} && (2.112e)\\
&= -8(p_1 \cdot p_3)\{2(\varepsilon_2 \cdot p_3)^2 + p_1 \cdot p_2\} && (2.112f)
\end{aligned}$$

In the above calculations we have used the identity (2.97) for $\not{p}_1 \not{p}_3$ to obtain equation (2.112a), used $\not{p}_3 \not{p}_3 = 0$ and identity (2.97) for $\not{p}_3 \not{\varepsilon}_3^*$ with $\not{p}_3 \not{\varepsilon}_3^* = 0$ to obtain (2.112c), used $\not{\varepsilon}_3^* \not{\varepsilon}_3^* = -1$ to obtain (2.112d) and used the trace theorem

$$Trace\ (\not{a} \not{b} \not{c} \not{d}) = 4[(a \cdot b)(c \cdot d) - (a \cdot c)(b \cdot d) + (a \cdot d)(b \cdot c)] \tag{2.113}$$

to obtain the eqn. (2.112e). Finally we used $p_3 \cdot p_4 - p_1 \cdot p_2$ obtained from the momentum conservation $(p_1 + p_2)^2 = (p_3 + p_4)^2$ by using $p_1^2 = p_4^2 = m^2$, $p_2^2 = p_3^2 = 0$ and $\varepsilon_2 \cdot p_4 = \varepsilon_2 \cdot (p_1 + p_2 - p_3) = -\varepsilon_2 \cdot p_3$ to arrive at the value for *Trace I* given by equation (2.112f).

The *Tr II* (eqn. (2.107b)) can be worked out as follows.

$$\begin{aligned}
Tr\ II &= Tr\{\not{\varepsilon}_2 \not{\varepsilon}_3^* \not{p}_3 \not{p}_1 \not{p}_2 \not{\varepsilon}_2 \not{\varepsilon}_3^* \not{p}_4\} + mTr\{\not{\varepsilon}_2 \not{\varepsilon}_3^* \not{p}_3 \not{p}_2 \not{\varepsilon}_2 \not{\varepsilon}_3^* \not{p}_4\}\\
&\quad + mTr\{\not{\varepsilon}_2 \not{\varepsilon}_3^* \not{p}_3 \not{p}_1 \not{p}_2 \not{\varepsilon}_2 \not{\varepsilon}_3^*\} + m^2 Tr\{\not{\varepsilon}_2 \not{\varepsilon}_3^* \not{p}_3 \not{p}_2 \not{\varepsilon}_2 \not{\varepsilon}_3^*\}\\
&= Tr\{\not{\varepsilon}_2 \not{\varepsilon}_3^* \not{p}_3 \not{p}_1 \not{p}_2 \not{\varepsilon}_2 \not{\varepsilon}_3^* \not{p}_4\} + m^2 Tr\{\not{\varepsilon}_2 \not{\varepsilon}_3^* \not{p}_3 \not{p}_2 \not{\varepsilon}_2 \not{\varepsilon}_3^*\} && (2.114)
\end{aligned}$$

where traces with m have vanished because of odd number of gamma matrices. To obtain above traces of eqn. (2.114) we may use another trace theorem, which is obviously applicable to even number of slashed parameters. The theorem is

$$Tr(\not{a}_1, \not{a}_2, \not{a}_3 \dots \not{a}_{2n}) = (a_1 \cdot a_2) Tr(\not{a}_2, \not{a}_4 \dots \not{a}_{2n})$$
$$-(a_1 \cdot a_3) Tr(\not{a}_2, \not{a}_4, \not{a}_5 \dots \not{a}_{2n})$$
$$+(a_1 \cdot a_4) Tr(\not{a}_2, \not{a}_3, \not{a}_{5.} \dots \not{a}_{2n}) \quad -\dots. \tag{2.115}$$
$$+(a_1 \cdot a_{2n}) Tr(\not{a}_2, \not{a}_3, \not{a}_4 \dots \not{a}_{2n-1})$$

Using the theorem (2.115), the eight-parameter trace of eqn. (2.114) can be reduced to seven–six parameter traces, few of which will vanish because of eqn. (2.98). Using again theorem (2.115) to each of the remaining six-parameter traces, the six parameter traces can be reduced to four-parameter traces. To each four-parameter trace, we may apply the theorem (2.113). Similar method may be applied to solve the m^2 Trace of eqn. (2.114). The results are:

$$Tr\{\not{\varepsilon}_2 \not{\varepsilon}_3^* \not{p}_3 \not{p}_1 \not{p}_2 \not{\varepsilon}_2 \not{\varepsilon}_3^* \not{p}_4\}$$
$$= 8(\varepsilon_2 \cdot \varepsilon_3^*)^2 [(p_1 \cdot p_3)(p_2 \cdot p_4) - (p_2 \cdot p_3)(p_1 \cdot p_4) + (p_1 \cdot p_2)(p_3 \cdot p_4)]$$
$$+8(p_1 \cdot p_1)(\varepsilon_2 \cdot \varepsilon_3^*)(\varepsilon_2 \cdot p_3)(\varepsilon_3^* \cdot p_2)$$
$$+8(\varepsilon_2 \cdot p_3)^2 (p_1 \cdot p_2) - 8(\varepsilon_3^* \cdot p_2)^2 (p_1 \cdot p_3)$$
$$-4[(p_1 \cdot p_3)(p_2 \cdot p_4) - (p_2 \cdot p_3)(p_1 \cdot p_4) + (p_1 \cdot p_2)(p_3 \cdot p_4)]$$
$$= 8 \ (\varepsilon_2 \cdot \varepsilon_3^*)^2 \cdot 2(p_1 \cdot p_2)(p_1 \cdot p_3) - 8m^2 (\varepsilon_2 \cdot \varepsilon_3^*)^2 (p_2 \cdot p_3)$$
$$+8m^2 (\varepsilon_2 \cdot \varepsilon_3^*)(\varepsilon_2 \cdot p_3)(\varepsilon_3^* \cdot p_2) + 8(\varepsilon_2 \cdot p_3)^2 (p_1 \cdot p_2)$$
$$-8(\varepsilon_3^* \cdot p_2)^2 (p_1 \cdot p_3) - 8(p_1 \cdot p_2)(p_1 \cdot p_3) + 4m^2 (p_2 \cdot p_3) \tag{2.116}$$

where $p_4 = p_1 + p_2 - p_3, p_1 \cdot p_1 = m^2, \varepsilon_2 \cdot p_4 = -\varepsilon_2 \cdot p_3$ and $\varepsilon_3^* \cdot p_4 = \varepsilon_3^* \cdot p_2$ have been used. The other trace

$$m^2 Tr\{\not{\varepsilon}_2 \not{\varepsilon}_3^* \not{p}_3 \not{p}_2 \not{\varepsilon}_2 \not{\varepsilon}_3^*\} = 8m^2 (\varepsilon_2 \cdot \varepsilon_3^*)^2 (p_2 \cdot p_3) - 8m^2 (\varepsilon_2 \cdot \varepsilon_3^*)(\varepsilon_2 \cdot p_3)(\varepsilon_3^* \cdot p_2)$$
$$-4m^2 (p_3 \cdot p_2) \tag{2.117}$$

Adding equations (2.116) and (2.117) we get *Tr II* as

$$Tr\ II = 8(p_1 \cdot p_2)(p_1 \cdot p_3)[2(\varepsilon_2 \cdot \varepsilon_3^*)^2 - 1] + 8(\varepsilon_2 \cdot p_3)^2 (p_1 \cdot p_2)$$
$$-8(\varepsilon_3^* \cdot p_2)^2 (p_1 \cdot p_3) \tag{2.218}$$

The *Tr III* of eqn. (2.107c) is equal to *Tr II* because eqn. (2.107c) can be obtained from (2.107b) by replacing $\varepsilon_2 \leftrightarrow \varepsilon_3^*$ and $p_2 \leftrightarrow -p_3$. This interchange (↔) comes from the fact that the total contribution to the matrix element from the two Feynman diagrams (Fig 2.7) is invariant under $\varepsilon_2 \leftrightarrow \varepsilon_3^*$ and $p_2 \leftrightarrow -p_3$. Thus

$$Tr\ III = Tr\ II \tag{2.119}$$

Likewise the *Tr IV* of (2.107d) can be obtained from *Tr I* derived in eqn. (2.112) by interchanging $\varepsilon_2 \leftrightarrow \varepsilon_3^*$ and $p_2 \leftrightarrow -p_3$. The result is

$$Tr\ IV = 8(p_1 \cdot p_2)\{2(\varepsilon_3^* \cdot p_2)^2 + p_1 \cdot p_3)\} \tag{2.220}$$

Substituting *Tr I*, *Tr II*, *Tr III* and *Tr IV* from eqns. (2.112f), (2.118), (2.119) and (2.120) in equations (2.107a) to (2.107d) respectively, we get from eqn. (2.106)

$$Trace\,\{Q(\not{p}_4 + m)\} = Trace\ I + Trace\ II + Trace\ III + Trace\ IV$$

$$= 2\left[\frac{p_1 \cdot p_2}{p_1 \cdot p_3} + \frac{p_1 \cdot p_3}{p_1 \cdot p_2} + 4(\varepsilon_2 \cdot \varepsilon_3^*)^2 - 2\right] \tag{2.121}$$

where p_1, p_2 and p_3 are the four momenta. In the laboratory coordinates, we take that the electron is initially at rest which means that it's four-momentum $p_1 = (m,0,0,0)$. From eqns. (2.79) and (2.81) $p_2^2 = 0 = p_3^2$ from which

$$p_2^2 = p_2^{0\,2} - \boldsymbol{p}_2^2 = 0 \Rightarrow p_2^0 = \omega_2 = |\boldsymbol{p}_2| \tag{2.122a}$$

and

$$p_3^0 = \omega_3 = |\boldsymbol{p}_3| \tag{2.122b}$$

Thus in eqn. (2.121) we have due to $\boldsymbol{p}_1 = 0$ and $p_1^0 = m$

$$p_1 \cdot p_2 = p_1^0 p_2^0 - \boldsymbol{p}_1 \cdot \boldsymbol{p}_2 = m\omega_2 \tag{2.123a}$$

and

$$p_1 \cdot p_3 = p_1^0 p_3^0 - \boldsymbol{p}_1 \cdot \boldsymbol{p}_3 = m\omega_3 \tag{2.123b}$$

and consequently eqn. (2.121) in laboratory coordinates becomes

$$Trace\{Q(\not{p}_4 + m)\} = 2\left[\frac{\omega_2}{\omega_3} + \frac{\omega_3}{\omega_2} + 4(\varepsilon_2 \cdot \varepsilon_3^*)^2 - 2\right] \tag{2.124}$$

where ω_2 (read $\hbar\omega_2$) and ω_3 (read $\hbar\omega_3$) are the energies of photons (incoming and outgoing).

The $|M_{fi}|^2$ can now be calculated by substituting eqn. (2.124) into eqn. (2.95) to give

$$|M_{fi}|^2 = \frac{e^4}{2m^2}\left[\frac{\omega_2}{\omega_3} + \frac{\omega_3}{\omega_2} + 4(\varepsilon_2 \cdot \varepsilon_3^*)^2 - 2\right] \tag{2.125}$$

2.3.1.3 Calculation of Differential Scattering Section for Compton Scattering. The Klein–Nishina Relation

The square of the matrix element $|M_{fi}|^2$ obtained in eqn. (2.125) may now be used to calculate the scattering of photons on electrons. The target electron is considered unpolarized and the polarization of final (recoiling) electron too is not taken into account.

Using eqn. (2.71) for the scattering $1 + 2 \rightarrow 4$, where particles 1 and 4 are electrons with four momenta p_1 and p_4 and the particles 2 and 3 are photons with four momenta p_2 and p_3 respectively. For this case in laboratory frame where $p_1 = 0$, i.e., the initial electron is at rest, we have, by taking into the account the facts $m_2(photon) = 0$ and p_1^0 to be replaced by $\frac{p_1^0}{2m}$ (see below eqn (2.72) for the initial electron (fermion))

$$4[(p_1 \cdot p_1)^2 - m_1^2 m_2^2]^{\frac{1}{2}} = 4[(p_1^0 p_2^0 - \mathbf{p}_1 \cdot \mathbf{p}_2) - m_1^2 m_2^2]^{\frac{1}{2}}$$

$$= 4\frac{p_1^0 p_2^0}{2m} = \frac{2}{m} p_1^0 p_2^0 \tag{2.126}$$

and hence eqn. (2.71) gives for Compton scattering

$$d\sigma = \frac{m}{2p_1^0 p_2^0} \int \frac{d^3 \mathbf{p}_3}{2p_3^0 (2\pi)^3} \left(\frac{m}{p_4^0} \right) \frac{d^3 \mathbf{p}_4}{(2\pi)^3} \times \frac{1}{2} |M_{fi}|^2$$

$$\times (2\pi)^4 \delta^4 (p_1 + p_2 - p_3 - p_4) \tag{2.127}$$

where the factor $\frac{1}{2}$ before $|M_{fi}|^2$ comes due to the fact that the initial electron has two spin states. Since, we have taken that the initial electron is unpolarized and being unpolarized is equivalent to taking average over the initial electron polarization directions, i.e., two spin states, a factor of $\frac{1}{2}$ is included.

The δ^4 function from eqn. (2.127) can be eliminated by the following process

$$\delta^4 (p_1 + p_2 - p_3 - p_4) = \delta(p_1^0 + p_2^0 - p_3^0 - p_4^0) \delta^3 (\mathbf{p}_1 + \mathbf{p}_2 - \mathbf{p}_3 - \mathbf{p}_4) \tag{2.128a}$$

In the laboratory frame, the electron is at rest, i.e., $\mathbf{p}_1 = 0$ and carrying out the $\mathbf{p}_4$ integration we may write

$$\int d^3 \mathbf{p}_4 \delta^3 (\mathbf{p}_2 - \mathbf{p}_3 - \mathbf{p}_4) = 1 \tag{2.128b}$$

provided, $p_4 = p_2 - p_3$. Thus p_4 integration sets this value for p_4. The eqn. (2.127), then becomes

$$d\sigma = \frac{m^2}{8(2\pi)^2 p_1^0 p_2^0} \int \delta(p_1^0 + p_2^0 - p_3^0 - p_4^0) \frac{d^3 \mathbf{p}_3}{p_3^0 p_4^0} |M_{fi}|^2$$

$$= \frac{m^2}{8(2\pi)^2 p_1^0 p_2^0} \int \delta(p_1^0 + p_2^0 - W) \frac{dW}{p_3^0 p_4^0} \frac{d^3 \mathbf{p}_3}{dW} |M_{fi}|^2 \tag{2.129}$$

where $W = p_3^0 + p_4^0$. Here dW has been introduced because we want to integrate $\delta(p_1^0 + p_2^0 - W)$ like eqn. (2.128b). Writing

$$d^3\boldsymbol{p}_3 = |\boldsymbol{p}_3|^2 d^3|\boldsymbol{p}_3| \sin\theta d\theta d\phi = p_3^{0^2} dp_3^0 d\Omega \tag{2.130}$$

where for the outgoing photon $|\boldsymbol{p}_3| = p_3^0$ has been used. Substituting eqn. (2.130) into eqn. (2.129), we get.

$$\frac{d\sigma}{d\Omega} = \frac{m^2}{8(2\pi)^2 p_1^0 p_2^0} \int \delta(p_1^0 + p_2^0 - W)dW \cdot \frac{p_3^{0^2}}{p_3^0 p_4^0} \frac{d^3 p_3^0}{dW} \cdot |M_{fi}|^2$$

$$= \frac{m^2}{32\pi^2 p_1^0 p_2^0} \cdot \frac{p_3^{0^2}}{p_3^0 p_4^0} \cdot \frac{dp_3^0}{dp_3^0 + dp_4^0} \cdot |M_{fi}|^2 \tag{2.131a}$$

where $\int \delta(p_1^0 + p_2^0 - W)dW = 1$ and $W = p_3^0 + p_4^0$ have been used. Now

$$\frac{dp_3^0 + dp_4^0}{dp_3^0} = 1 + \frac{1}{2p_4^0} \frac{d}{dp_3^0}\left(p_4^0\right)^2$$

$$= 1 + \frac{\boldsymbol{p}_3}{2p_3^0 \; p_4^0} \frac{d}{dp_3^0}(\boldsymbol{p}_2 - \boldsymbol{p}_3)^2$$

$$= 1 + \frac{\boldsymbol{p}_3 \cdot (\boldsymbol{p}_2 - \boldsymbol{p}_3)}{p_3^0 \; p_4^0} \frac{d}{dp_3^0}(\boldsymbol{p}_2 - \boldsymbol{p}_3)$$

$$= 1 - \frac{\boldsymbol{p}_3 \cdot \boldsymbol{p}_4}{p_3^0 \; p_4^0}$$

$$= \frac{p_3 \cdot p_4}{p_3^0 \; p_4^0} \tag{2.131b}$$

where in the second step we used $p_4^0 = |\boldsymbol{p}_4| = |\boldsymbol{p}_2 - \boldsymbol{p}_3|$. It may also be noted that $|\boldsymbol{p}_3| = p_3^0$. Since $p_3 \cdot p_4 = p_1 \cdot p_2 = p_1^0 p_2^0 - \boldsymbol{p}_1 \cdot \boldsymbol{p}_2 = p_1^0 p_2^0$ where $\boldsymbol{p}_1 = 0$(lab) has been used, we may write from eqn. (2.131b) that

$$\frac{dp_3^0}{dp_3^0 + dp_4^0} = \frac{p_3^0 p_4^0}{p_1^0 p_2^0} \tag{2.131c}$$

Substituting eqn. (2.131c) into eqn. (2.131a), we get

$$\left(\frac{d\sigma}{d\Omega}\right)_{\text{lab}}^{e\gamma} = \frac{1}{32\pi^2} \frac{m^2}{p_1^{0^2}} \left(\frac{p_3^0}{p_2^0}\right)^2 |M_{fi}|^2$$

$$= \frac{1}{32\pi^2} \left(\frac{\omega_3}{\omega_2}\right)^2 |M_{fi}|^2 \tag{2.132}$$

where $p_1^{0^2} = m^2$(electron mass), $p_3^0 = \omega_3$ and $p_2^0 = \omega_2$, have been used.

Substituting eqn. (2.125) for $\left|M_{fi}\right|^2$, the differential scattering cross section for the plane polarized photons on unpolarized electrons in the laboratory system, i.e., equation (2.132) , becomes

$$\frac{d\sigma}{d\Omega}=\frac{\alpha^2}{4m^2}\left(\frac{\omega_3}{\omega_2}\right)^2\left[\frac{\omega_2}{\omega_3}+\frac{\omega_3}{\omega_2}+4(\varepsilon_2\cdot\varepsilon_3^*)^2-2\right] \tag{2.133}$$

where $\alpha=\frac{e^2}{4\pi}\simeq\frac{1}{137}$ is the fine structure constant. The relation (2.133) is known as the **Klein–Nishina formula** for scattering of polarized photons by unpolarized electrons.

Equation (2.133) gives the relativistic differential scattering cross section for Compton scattering and from this equation the non-relativistic (*nr*) limit of $\frac{d\sigma}{d\Omega}$ can be obtained. In the non-relativistic limit or the low energy limit, the energies of incident (ω_2) and scattered (ω_3) photons are quite small compared to the energy of electron *m*, i.e., ω_2, $\omega_3 << m$ suggesting that $\omega_3 \simeq \omega_2$. As a result eqn. (2.133) gives

$$\left(\frac{d\sigma}{d\Omega}\right)_{nr}=\frac{\alpha^2}{m^2}(\varepsilon_2\cdot\varepsilon_3^*)^2=r_0^2(\varepsilon_2\cdot\varepsilon_3^*)^2 \tag{2.134}$$

where

$$\frac{\alpha}{m}=\frac{e^2}{4\pi m}=r_0=2.8\times10^{-13}\quad\text{cm} \tag{2.135}$$

is the classical electron radius. The result (2.134) is similar to classical Rayleigh –Thomson result. We may therefore conclude that the relativistic (high energy) Klein—Nishina relation in its non—relativistic (low energy) limit reduces to classical Rayleigh—Thomson formula.

In the above-considered Compton scattering the initial electron was unpolarized. A further extension could be to consider the scattering of photons by polarized electrons. As already noted the calculation of traces even in unpolarized electron case involves lot of labour, the trace calculation becomes quite cumbersome in case of polarized electrons. Still we may point out that:

1. The polarization of the electron does not affect the total cross section for the scattering of unpolarized photons.
2. In the scattering of unpolarized photons, the recoil electron is unpolarized.
3. The scattering cross—section for the case of polarized electron contains terms bilinear in the polarizations of electron and photon. But only such terms are present which correspond to circular polarization of photon.
4. The cross section for scattering of a polarized photon by a polarized electron differs from the scattering of polarized photon by unpolarized

electron only in that the photon is circularly polarized and the electrons have a non-zero projection of the mean spin on the scattering plane.

5. The recoil electron is polarized only if the photon is circularly polarized.
6. The requirements of symmetry determine the qualitative properties of the polarization effects in the scattering of photons by electrons.

2.3.2 Electron–Positron Annihilation

As another example of quantum electrodynamic process we consider the annihilation of a pair of electron and positron into two photons. The tree level Feynman diagrams for the process

$$e^+ + e^- \rightarrow \gamma + \gamma \tag{2.136}$$

are given below (Fig. 2.9).

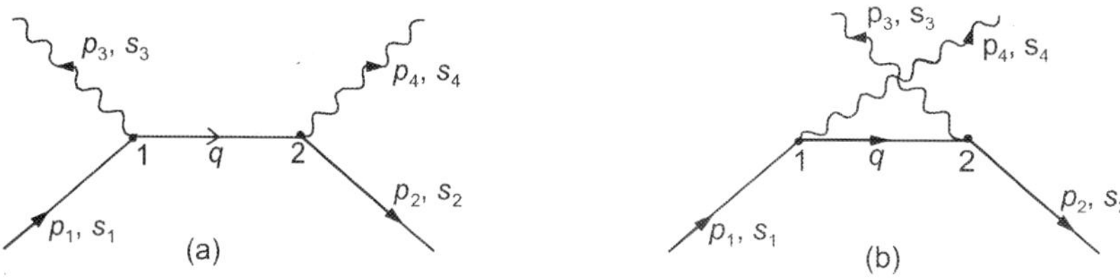

Fig. 2.9(a, b): Feynman diagrams for electron–positron annihilation

(p_1, s_1) denotes the electron, (p_2, s_2), the positron. Both the diagrams contribute to the matrix element between initial $|i\rangle$ and final $|f\rangle$ state. The initial state has e^+ and e^- and final state has two photons. Since photons are bosons the total matrix element

$$M = M_1 + M_2 \tag{2.137}$$

where $M_1 \equiv (M_1)_{fi}$ is the matrix element corresponding to Fig (2.9a) and $M_2 \equiv (M_2)_{fi}$ is the matrix element for e^+, e^- annihilation process shown by Fig. (2.9b).

Similar to Compton scattering, we assume that the electron and positron are unpolarized and photons are polarized. The polarization vectors of photons are ε_3 and ε_4. p_1 is the four-momentum of electron, p_2 that of positron, p_3 is the four-momentum of emitted photon whose polarization vector is ε_3 and p_4 that of photon with polarization vector ε_4. The electron is assumed at rest. It is also assumed that electron has up spin and positron has down spin such that the pair has net spin 0 (singlet) state. It is the singlet state that can decay into even (usually two) number of photons. The net spin 1 (triplet) state decays into odd (generally three) photons. The process $e^+ + e^- \rightarrow \gamma + \gamma$ then automatically selects the singlet state.

Applying Feynman rules of article 2.2.2.4 to the diagram (2.9a), starting from vertex 2 we shall have following multiplicative factors to calculate M_1.

1. The outgoing photon (p_4, s_4), a boson, due to rule 7 provides a multiplicative factor $\varepsilon^*_\mu(s_4)$, which we write as $\varepsilon_\mu(4)$. Since both the photons are outgoing we may omit the star.
2. The positron (p_2, s_2), i.e., an antiparticle, is entering at vertex 2. Rule (4a) is applicable and provides a multiplicative factor $\bar{v}(p_2) = \bar{v}(2)$.
3. The vertex 2, similar to the case of Compton scattering is a spin $(\frac{1}{2}-\frac{1}{2})$ vertex. Rule (14) provides a multiplicative factor of $(-ie\gamma_\mu)$.
4. Applying rule (16) at vertex 2, the conservation of energy and momentum provides a multiplicative factor of $(2\pi)^4\delta^4$ $(q_2 + q - q_4)$, where q and p_2 are the entering momenta and p_4 the leaving momentum. Being the momentum of positron p_2 has opposite arrow.
5. Coming towards vertex 1 from 2, we have the internal line denoting four-momentum q. Rule (10) provides a multiplicative factor $\dfrac{i(\gamma_\mu q^\mu + m)}{m^2 - q^2}$
6. The vertex 1 is also a spin $(\frac{1}{2}-\frac{1}{2})$ vertex. Rule (14) here too provides a multiplicative factor of $(-ie\gamma^\mu)$.
7. Applying rule (16) at vertex 1, we have the multiplicative factor $(2\pi)^4\delta^4$ $(p_1 + q - p_3)$.
8. At vertex 1, the emitted photon (p_3, s_3) has polarization denoted by s_3. Rule (7) provides a multiplicative factor of $\varepsilon^*_\mu(s_3) \equiv \varepsilon^*_\mu(3) \rightarrow \varepsilon_\mu(3)$ where too, like para (1), we have omitted the star as both the photons in Fig. (6.8) are outgoing.
9. At vertex 1, due to (p_1, s_1), rule (3a) provides a multiplicative factor $u(p_1) \equiv u(1)$.
10. Applying rule (17), we have the multiplicative factor $\dfrac{d^4q}{(2\pi)4}$
11. Apply rule (18)

From the above, similar to Compton scattering case we get

$$-iM_1 = \int [\varepsilon_\mu(4)\bar{v}(2)(-ie\gamma^\mu)(2\pi)^4\delta^4(p_2 + q - p_4)$$

$$\times \frac{i(\gamma_\mu q^\mu + m)}{m^2 - q^2}(-ie\gamma^\nu)(2\pi)^4\delta^4(p_1 - p_3 - q)\varepsilon_\nu(3)u(1)]\frac{d^4q}{(2\pi)^4} \tag{2.138}$$

from which we get

$$M_1 = (2\pi)^4 e^2 \int \bar{v}(2)\varepsilon_\mu(4)\gamma^\mu \frac{(\gamma_\mu q^\mu + m)}{m^2 - q^2}\delta^4(p_1 - p_3 - q)$$

$$\gamma^\nu\varepsilon_\nu(3)u(1)\delta^4(p_2 + q - p_4)d^4q \tag{2.139}$$

We may now use equations (2.61) and (2.63) to write eqn. (2.139) as

$$M_1 = e^2\overline{v}(2)\varepsilon_\mu(4)\gamma^\mu \frac{[\gamma_\mu p_1^\mu - \gamma_\mu p_3^\mu + m]}{m^2 - (p_1 - p_3)^2}\gamma^\nu\varepsilon_\nu(3)u(1)$$
$$(2\pi)^4\delta^4(p_1 - p_3 + p_2 - p_4) \qquad (2.140)$$

from which erasing $(2\pi)^4\delta^4$ function we write

$$M_1 = \frac{e^2}{m^2 - (p_1 - p_3)^2}\overline{v}(2)\not{\varepsilon}_4(\not{p}_1 - \not{p}_3 + m)\not{\varepsilon}_3 u(1) \qquad (2.141)$$

where $\not{\varepsilon}_\mu = \varepsilon_\mu(4)\gamma^\mu, \not{p} = \gamma_\mu p_1^\mu$ etc.

Likewise, adopting the similar procedure for calculating M_2 from the diagram (2.9b) we may write proceeding from vertex 2 towards vertex 1.

$$-iM_2 = \int[\overline{v}(2)\varepsilon_\mu(3)(-ie\gamma^\mu)(2\pi)^4\delta^4(p_2 + q - p_3)$$
$$\frac{i(\gamma_\mu q^\mu + m)}{m^2 - q^2}(-ie\gamma^\nu)\varepsilon_\nu(4)u(1)(2\pi)^4\delta^4(p_1 - p_4 - q)]\frac{d^4q}{(2\pi)^4} \qquad (2.142)$$

from which

$$M_2 = (2\pi)^4 e^2\int\overline{v}(2)\varepsilon_\mu(3)\gamma^\mu \frac{(\gamma_\mu q^\mu + m)}{m^2 - q^2}\delta^4(p_1 - p_4 - q)$$
$$\gamma^\nu\varepsilon_\nu(4)u(1)\delta^4(p_2 - p_3 + q)d^4q \qquad (2.143a)$$

$$= e^2\overline{v}(2)\varepsilon_\mu(3)\gamma^\mu \frac{(\gamma_\mu(p_1 - p_4)^\mu + m)}{m^2 - (p_1 - p_4)^2}$$
$$\gamma^\nu\varepsilon(4)u(1)(2\pi)^4\delta^4(p_1 - p_4 + p_2 - p_3) \qquad (2.143b)$$

where equations similar to (2.61) and (2.62) have been used. Eliminating $(2\pi)^4\delta^4$ we finally get from eqn. (2.143b)

$$M_2 = \frac{e^2}{m^2 - (p_1 - p_4)^2}\overline{v}(2)\not{\varepsilon}_3(\not{p}_1 - \not{p}_4 + m)\not{\varepsilon}_4 u(1) \qquad (2.144)$$

2.3.2.1 $|M_{fi}|^2$ *for* $e^+e^- \to 2\gamma^0$

Substituting equations (2.141) and (2.144) in (2.137) we may write for the total matrix element M_{fi} for the electron–positron pair annihilation

$$M = e^2\overline{v}(2)\left[\frac{\not{\varepsilon}_4(\not{p}_1 - \not{p}_3 + m)\not{\varepsilon}_3}{m^2 - (p_1 - p_3)^2} + \frac{\not{\varepsilon}_3(\not{p}_1 - \not{p}_4 + m)\not{\varepsilon}_4}{m^2 - (p_1 - p_4)^2}\right]u(1) \qquad (2.145)$$

where p_1 is the four-momentum of ingoing electron; p_3, p_4 are the momenta of outgoing photons and m is the rest mass of electron (positron). The four-momentum of positron is p_2. The numerators and denominators of eqn. (2.145) can be separately solved. For the denominators we use the relativistic energy momentum relations similar to equation (2.78) and those following it. For this case they are

$$\begin{aligned} p_1^2 &= m^2 = p_2^2 \\ p_3^2 &= 0 = p_4^2 \end{aligned} \tag{2.146}$$

and hence

$$m^2 - (p_1 - p_3)^2 = 2p_1 \cdot p_3$$

and

$$m^2 - (p_1 - p_4)^2 = 2p_1 \cdot p_4 \tag{2.147}$$

For simplifying the numerators, we use the conditions

$$p_1 \cdot \varepsilon_3 = 0 = p_1 \cdot \varepsilon_4 \tag{2.148}$$

$$p_3 \cdot \varepsilon_3 = 0 = p_4 \cdot \varepsilon_4 \tag{2.149}$$

The first of these conditions is a consequence of the assumption that the electron (p_1) is at rest, implying that temporal component of p_1 is m and space components are zero. The polarization vectors ε_3 and ε_4 have also four components. In the Coulomb gauge the temporal component of the polarization vector is zero and then ε_3 and ε_4 shall have only space components. Both of the above considerations lead to condition (2.148). Equation (2.149) is Lorentz condition, which requires that

$$p_3 \cdot \varepsilon_3 \equiv p_3^\mu \varepsilon_\mu(3) = 0 \tag{2.150}$$

and

$$p_4 \cdot \varepsilon_4 \equiv p_4^\mu \varepsilon_\mu(4) = 0 \tag{2.151}$$

where p_3 and p_4 are the four-momenta of two outgoing photons.

The first numerator [eqn. (2.145)]

$$\begin{aligned} (\not{p}_1 - \not{p}_3 + m)\not{\varepsilon}_3 u(1) &= (-\not{\varepsilon}_3\not{p}_1 + \not{\varepsilon}_3\not{p}_3 + \not{\varepsilon}_3 m)u(1) \\ &= -\not{\varepsilon}_3(\not{p}_1 - m)u(1) + \not{\varepsilon}_3\not{p}_3 u(1) \\ &= \not{\varepsilon}_3\not{p}_3 u(1) \end{aligned} \tag{2.152}$$

where in the first step we made use of the identity (2.97) and (2.148)& (2.149) to write $\not{p}_1\not{\varepsilon}_3 = -\not{\varepsilon}_3\not{p}_1$ and $-\not{p}_3\not{\varepsilon}_3 = \not{\varepsilon}_3\not{p}_3$. In the second step we made use of the Dirac equation

$$(\not{p}_1 - m)u(1) = 0 \tag{2.153}$$

Likewise the second numerator

$$(\not{p}_1 - \not{p}_4 + m)\not{\epsilon}_4 u(1) = (-\not{\epsilon}_4\not{p}_1 + \not{\epsilon}_4\not{p}_4 + \not{\epsilon}_4 m)u(1) \\ = \not{\epsilon}_4\not{p}_4 u(1) \tag{2.154}$$

Thus, using equations (2.147), (2.152) and(2.154), the M_{fi} from eqn. (2.145) become.

$$M_{fi} = e^2\overline{v}(2)\left[\frac{\not{\epsilon}_4\not{\epsilon}_3\not{p}_3}{2p_1 \cdot p_3} + \frac{\not{\epsilon}_3\not{\epsilon}_4\not{p}_4}{2p_1 \cdot p_4}\right]u(1) \tag{2.155}$$

which we may write as

$$M_{fi} = e^2\overline{v}(2)\left[\frac{\Gamma_1}{2p_1 \cdot p_3} + \frac{\Gamma_2}{2p_1 \cdot p_4}\right]u(1) \tag{2.156}$$

with

$$\Gamma_1 = \not{\epsilon}_4\not{\epsilon}_3\not{p}_3$$

and

$$\Gamma_2 = \not{\epsilon}_3\not{\epsilon}_4\not{p}_4 \tag{2.157}$$

The square of the matrix element

$$\left|M_{fi}\right|^2 = M_{fi}M_{fi}^\dagger \tag{2.158}$$

where $M_{fi}^\dagger$ similar to eqn. (2.84) can be calculated as

$$(M_{fi}) = e^2\overline{u}(1)\left[\frac{\overline{\Gamma}_1}{2p_1 \cdot p_3} + \frac{\overline{\Gamma}_2}{2p_1 \cdot p_4}\right]v(2) \tag{2.159}$$

with

$$\overline{\Gamma} = \gamma^0\Gamma^\dagger\gamma^0 \tag{2.160}$$

The $\left|M_{fi}\right|^2$ then becomes

$$\left|M_{fi}\right|^2 = e^4\left[\overline{v}(2)\left(\frac{\Gamma_1}{2p_1 \cdot p_3} + \frac{\Gamma_2}{2p_1 \cdot p_4}\right)u(1)\overline{u}(1)\left(\frac{\overline{\Gamma}_1}{2p_1 \cdot p_3} + \frac{\overline{\Gamma}_2}{2p_1 \cdot p_4}\right)v(2)\right] \tag{2.161}$$

Summing over the spins $s_1 = 1, 2$ and using

$$\sum_{s_1=1,2} u^{s_1}(1)\overline{u}^{s_1}(1) = \frac{\not{p}_1 + m}{2m} \tag{2.162}$$

we may write eqn. (2.161) as

$$\sum_{s_1}\left|M_{fi}\right|^2 = \frac{e^4}{2m}\overline{v}(2)Qv(2) \tag{2.163}$$

where

$$Q = \left(\frac{\Gamma_1}{2p_1 \cdot p_3} + \frac{\Gamma_2}{2p_1 \cdot p_4} \right) (\not{p}_1 + m) \left(\frac{\overline{\Gamma}_1}{2p_1 \cdot p_3} + \frac{\overline{\Gamma}_2}{2p_1 \cdot p_4} \right) \tag{2.164}$$

Now we sum over the spins $s_2 = 1, 2$ and write eqn. (2.163) as

$$\begin{aligned} \sum_{s_2} \sum_{s_1} \left| M_{fi} \right|^2 &= \frac{e^4}{2m} \sum_{s_2=1,2} [\overline{v}^{s_2}(2)]_\alpha Q_{\alpha\beta} [v^{s_2}(2)]_\beta \\ &= \frac{e^4}{2m} Q_{\alpha\beta} \sum_{s_2=1,2} [v^{s_2}(2) \overline{v}^{s_2}(2)]_{\beta\alpha} \\ &= \frac{e^4}{4m^2} Q_{\alpha\beta} [\not{p}_2 - m]_{\beta\alpha} \end{aligned} \tag{2.165}$$

where the matrix relation for positron spinors

$$\sum_{s_2} v(p_2, s_2) \overline{v}(p_2, s_2) = \frac{\gamma_\mu p_2^\mu - m}{2m} \equiv \frac{\not{p}_2 - m}{2m} \tag{2.166}$$

has been used. Using equation (2.94a), namely

$$Trace[AB] = [AB]_{\alpha\alpha} = A_{\alpha\beta} B_{\beta\alpha} \tag{2.167}$$

we may write eqn. (2.165) as

$$\left| M_{fi} \right|^2 = \frac{e^4}{4m^2} Trace\{Q(\not{p}_2 - m)\} \tag{2.168}$$

where Q is given by eqn. (2.164) with Γ_1 and Γ_2 by eqs. (2.157) and

$$\overline{\Gamma}_1 = \not{p}_3 \not{\epsilon}_3 \not{\epsilon}_4$$

and

$$\overline{\Gamma}_2 = \not{p}_4 \not{\epsilon}_4 \not{\epsilon}_3 \tag{2.169}$$

Above relations for $\overline{\Gamma}_1$ and $\overline{\Gamma}_2$ can easily be obtained by carrying out the derivations similar to eqns. (2.102) and (2.103).

2.3.2.2 Calculation of Trace $\{Q(\not{p}_2 - m)\}$

From equations (2.164), (2.157) and (2.169) we have

$$Trace\{Q(\not{p}_2 - m)\} = -Tr\{Q(-\not{p}_2 + m)\} \tag{2.170}$$

$$= -Tr\left\{ \left(\frac{\not{\epsilon}_4 \not{\epsilon}_3 \not{p}_3}{2p_1 \cdot p_3} + \frac{\not{\epsilon}_3 \not{\epsilon}_4 \not{p}_4}{2p_1 \cdot p_4} \right) (\not{p}_1 + m) \left(\frac{\not{p}_3 \not{\epsilon}_3 \not{\epsilon}_4}{2p_1 \cdot p_3} + \frac{\not{p}_4 \not{\epsilon}_4 \not{\epsilon}_3}{2p_1 \cdot p_4} \right) (-\not{p}_2 + m) \right\}$$

Let us compare it with the corresponding trace arrived at in the Compton scattering in eqn. (2.104). We rewrite that

$$Trace\{Q(\not{p}_4 + m)\}_{\text{Compton}} \tag{2.171}$$

$$= \left\{ \left(\frac{\not{\varepsilon}_2 \not{\varepsilon}_3^* \not{p}_3}{2p_1 \cdot p_3} + \frac{\not{\varepsilon}_3^* \not{\varepsilon}_2 \not{p}_2}{2p_1 \cdot p_2} \right) (\not{p}_1 + m) \left(\frac{\not{p}_3 \not{\varepsilon}_3^* \not{\varepsilon}_2}{2p_1 \cdot p_3} + \frac{\not{p}_2 \not{\varepsilon}_2 \not{\varepsilon}_3^*}{2p_1 \cdot p_2} \right) (\not{p}_4 + m) \right\}$$

A comparison of equations (2.170) and (2.171) leads to the observation that

Compton Scattering parameter	is replaced by parameter	Pair annihilation
$\not{\varepsilon}_2(\varepsilon_2)$	$\rightarrow$	$\not{\varepsilon}_4(\varepsilon_4)$
$\not{\varepsilon}_3^*(\varepsilon_3^*)$	$\rightarrow$	$\not{\varepsilon}_3(\varepsilon_3)$
p_2	$\rightarrow$	$-p_4$
$\not{p}_2$	$\rightarrow$	$-\not{p}_4$

(2.172)

The trace calculation of eqn. (2.170) may also be done similar to one we did in Compton scattering. But instead of making the detailed calculations all over again, we make the replacements (2.172) in the result (2.121) obtained for Compton scattering, which too we rewrite

$$Trace\{Q(\not{p}_4 + m)\}_{\text{Compton}} = 2 \left[\frac{p_1 \cdot p_2}{p_1 \cdot p_3} + \frac{p_1 \cdot p_3}{p_1 \cdot p_2} + 4(\varepsilon_2 \cdot \varepsilon_3^*)^2 - 2 \right] \tag{2.173}$$

and make the replacements (2.172) to get

$$Trace\{Q(\not{p}_2 - m)\} = -2 \left[-\frac{p_1 \cdot p_4}{p_1 \cdot p_3} - \frac{p_1 \cdot p_3}{p_1 \cdot p_4} + 4(\varepsilon_3 \cdot \varepsilon_4)^2 - 2 \right] \tag{2.174}$$

which on using the laboratory coordinates, i.e., taking $p_1 = (m, 0, 0, 0)$, $p_3^0 = \omega_3$, $p_4^0 = \omega_4$, becomes

$$Trace\{Q(\not{p}_2 - m)\} = 2 \left[\frac{\omega_4}{\omega_3} + \frac{\omega_3}{\omega_4} - 4(\varepsilon_3 \cdot \varepsilon_4)^2 + 2 \right] \tag{2.175}$$

and consequently using eqn. (2.168) the square of the matrix element for $e^+ + e^- \rightarrow \gamma + \gamma$ is

$$|M_{fi}|^2 = \frac{e^4}{2m^2} \left[\frac{\omega_4}{\omega_3} + \frac{\omega_3}{\omega_4} - 4(\varepsilon_3 \cdot \varepsilon_4)^2 + 2 \right] \tag{2.176}$$

2.3.2.3 The Differential Scattering Cross Section for Pair Annihilation Process

We consider the pair annihilation as a process in which the electron with four-momentum p_1 is at rest and the positron with four-momentum p_2 impinges on it.

The rest state of electron obviously implies that $p_1^0 = m$ and $\mathbf{p}_1 = 0$. It is also assumed that both electron and the positron are unpolarized. We may write $p2 = (p_2^0, \mathbf{p}_2) \equiv (E_2, \mathbf{p}_2)$.

To calculate $d\sigma$ we shall use eqn. (2.71) along with the prescriptions given below this equation. In the present case, $1 + 2 \rightarrow 3 + 4$, the particles 1 and 2 are fermions (e^- and e^+ each of rest mass m) hence 1/4 factor of eqn. (2.71) would be replaced by $m_1 m_2 = m^2$. However, another factor of 1/4 would come because of spin considerations of electron and positron. Since both are spin 1/2 particles, the pair can exist in singlet (net spin zero) and triplet (net spin 1) states, the singlet state would carry a statistical weight of 1/4 and triplet 3/4. Again, since the singlet state of pair can only decay into even (usually two) number of photons and triplet into odd (usually three) number of photons, the process $e^+ + e^- \rightarrow \gamma + \gamma$ provides a factor of only 1/4 as the statistical weight.

Thus for $e^+e^- \rightarrow 2\gamma$

$$d\sigma = \frac{m^2}{[(p_1 \cdot p_2)^2 - m^4]^{\frac{1}{2}}} \int \frac{1}{4} \frac{d^3\mathbf{p}_3}{2p_3^0(2\pi)^3} \frac{d^3\mathbf{p}_4}{2p_4^0(2\pi)^3} |M_{fi}|^2 (2\pi)^4 \delta^4(p_1 + p_2 - p_3 - p_4)$$

$$= \frac{m^2}{64\pi^2[(p_1 \cdot p_2)^2 - m^4]^{\frac{1}{2}}} \int \frac{d^3\mathbf{p}_3}{p_3^0} \frac{d^3\mathbf{p}_4}{p_4^0} |M_{fi}|^2 \delta^4(p_1 + p_2 - p_3 - p_4) \tag{2.177}$$

where eqn. (2.176) gives $|M_{fi}|^2$. We obtain the $d\sigma$ in laboratory coordinates in which either of the incident particles (we have chosen electron) is at rest implying $p_1^0 = m$ and $\mathbf{p}_1 = 0$ and we take $p_2 = (E_2, \mathbf{p}_2)$. Therefore when electron is at rest

$$\begin{aligned}\left[(p_1.p_2)^2 - m^4\right]^{\frac{1}{2}} &= \left[(p_1^0 p_2^0 - \mathbf{p}_1.\mathbf{p}_2)^2 - m^4\right]^{\frac{1}{2}} \\ &= \left[m^2 E_2^2 - m^4\right]^{\frac{1}{2}} = m|\mathbf{p}_2|\end{aligned} \tag{2.178}$$

where

$$p_2^2 = E_2^2 - \mathbf{p}_2^2 = m^2 \tag{2.179}$$

has been used . Then in laboratory coordinates eqn. (2.177) becomes

$$d\sigma = \frac{m}{64\pi^2} \cdot \frac{1}{|\mathbf{p}_2|} \frac{|M_{fi}|^2}{p_3^0 p_4^0} d\Omega \cdot \frac{p_3^{0\,2} dp_3^0}{dp_3^0 + dp_4^0} \tag{2.180}$$

where eqns. (2.128) have been used . Now, similar to eqn. (2.131b)

$$\begin{aligned}\frac{dp_3^0 + dp_4^0}{dp_3^0} &= 1 + \frac{1}{2p_4^0} \frac{d}{dp_3^0} \left(p_4^0\right)^2 \\ &= 1 + \frac{\mathbf{p}_4}{p_4^0} \frac{d}{dp_3^0} (\mathbf{p}_2 - \mathbf{p}_3)\end{aligned} \tag{2.181}$$

$$= 1 - \frac{\boldsymbol{p}_3 \cdot \boldsymbol{p}_4}{p_3^0 p_4^0}$$

$$= \frac{p_3 \cdot p_4}{p_3^0 p_4^0}$$

where in the second step we used $p_4^{0^2} = \boldsymbol{p}_4^2 \; (p_4^2 = 0)$ and the momentum conservation in laboratory coordinates $\boldsymbol{p}_1 = 0, \boldsymbol{p}_4 = \boldsymbol{p}_2 - \boldsymbol{p}_3$. In the third step we have used $p_3^0 = |\boldsymbol{p}_3| \, (p_3^2 = 0)$. The quantity $p_3 . p_4$ from the energy momentum conservation

$$(p_3 + p_4)^2 = (p_1 + p_2)^2$$

is

$$p_3 \cdot p_4 = p_1 \cdot p_2 + m^2$$

$$= m(m + E_2) \tag{2.182}$$

where $p_1^2 = p_2^2 = m^2$ and $p_1^0 = m, p_2^0 = E_2$ and $\boldsymbol{p}_1 = 0$ have been used. Thus

$$\frac{dp_3^0}{dp_3^0 + dp_4^0} = \frac{p_3^0 \; p_4^0}{m(m + E_2)} \tag{2.183}$$

Substituting eqn. (2.183) into eqn. (2.180), we get

$$\frac{d\sigma}{d\Omega} = \frac{1}{64\pi^2} \cdot \frac{1}{|\boldsymbol{p}_2|} \frac{\omega_3^2}{(m + E_2)} |M_{fi}|^2 \tag{2.184}$$

where $\omega_3 = p_3^0$ is the energy of one of the outgoing photons. Putting the value of $|M_{fi}|^2$ from eqn. (2.176) we finally obtain $\left(\frac{d\sigma}{d\Omega}\right)_{\text{lab}}$, i.e., the differential cross section in the laboratory coordinates for the emission of one of the photons in the direction expressed by the solid angle Ω when a positron $(E_2, \boldsymbol{p}_2)$ impinges on an electron (at rest), as

$$\left(\frac{d\sigma}{d\Omega}\right)_{\text{lab}} = \frac{1}{8m^2} \left(\frac{e^2}{4\pi}\right)^2 \frac{\omega_3^2}{|\boldsymbol{p}_2|(m + E_2)} \left[\frac{\omega_4}{\omega_3} + \frac{\omega_3}{\omega_4} - 4(\varepsilon_3 \cdot \varepsilon_4)^2 + 2\right] \tag{2.185}$$

which can also be expressed in terms of r_0, the classical electron radius $r_0 = \frac{\alpha}{m}$, with $\alpha = e^2/4\pi$, the fine structure constant.

$$\left(\frac{d\sigma}{d\Omega}\right)_{\text{lab}} = \frac{\omega_3^2 r_0^2}{8|\boldsymbol{p}_2|(m + E_2)} \left[\frac{\omega_4}{\omega_3} + \frac{\omega_3}{\omega_4} - 4(\varepsilon_3 \cdot \varepsilon_4)^2 + 2\right] \tag{2.186}$$

Above is a relativistic differential cross section for electron—positron pair annihilation. In the nonrelativistic (*nr*) or the low energy limit, $\omega_4 \simeq \omega_3 = 2$ and the incident positron is very slow, as a result of which electron—positron pair is effectively at rest and photons come out back to back. Hence we have $|\boldsymbol{p}_2| << m = \beta_2 m$ $(\beta_2 << 1)$ and $\boldsymbol{p}_3 = -\boldsymbol{p}_4 = \boldsymbol{p}$. Also we shall have from the energy conservation $p_1^0 + p_2^0 = p_3^0 + p_4^0$ implying $m + E_2 = \omega_3 + \omega_4 \simeq 2\omega, m \to \omega$. Incorporating these limiting values in eqn. (2.186), we get the nonrelativistic limit of eqn. (2.186) as

$$\left(\frac{d\sigma}{d\Omega}\right)^{nr}_{\text{lab}} = \frac{r_0^2}{4\beta_2}\left[1 - (\varepsilon_3 \cdot \varepsilon_4)^2\right] \tag{2.187}$$

in which ε_3 and ε_4 are the polarization vectors of emitted photons. It is easy to observe from eqn. (2.187) that when ε_3 and ε_4 are parallel, the differential scattering cross section becomes zero implying that the transition probability vanishes if the polarization vectors of emitted photons are parallel. Thus for a nonzero transition probability ε_3 and ε_4 should not have parallel orientations.

In the pair annihilation process, if we are interested say only in the number of emitted photons not on the orientations of their polarization vectors, we are free to choose the directions of ε_3 and ε_4 and then sum over these photon polarization states. If we choose that $\boldsymbol{p} = \boldsymbol{p}_3$ is along the z-axis, the ε_3 and ε_4 can be along x or y-axis. Thus the value of $[1 - (\varepsilon_3 \cdot \varepsilon_4)^2]$ is zero when both ε_3 and ε_4 are along x-axis or both are along y-axis. Its value would be one when ε_3 is along x-axis but ε_4 along y-axis and vice versa. The sum over all polarization states (four possibilities) gives the value 2, to the factor $[1 - (\varepsilon_3 \cdot \varepsilon_4)^2]$. Since in the final state we have two indistinguishable photons, we shall have a factor of 1/2. Thus, in view of the above, the integration of eqn. (2.187) gives

$$\begin{aligned}\sigma^{nr}_{\text{lab}} &= \frac{1}{2}\int\left(\frac{d\sigma}{d\Omega}\right)^{nr}_{\text{lab}} d\Omega \\ &= \frac{1}{2}\cdot\frac{r_0^2}{4\beta_2}\cdot 2\cdot 4\pi \\ &= \frac{\pi r_0^2}{\beta_2} \qquad (\beta_2 << 1)\end{aligned} \tag{2.188}$$

Since $|p_2| = \beta_2 m$, β_2 provides a measure to the velocity of positron. From eqn. (2.188), we observe that σ^{nr}_{lab} increases with the decrease in the positron velocity and becomes infinite for $\beta_2 \to 0$. However, the number of electron—positron annihilation per unit time remains finite. This is due to the flux factor, which is proportional to β_2 and determines the number of incident (positron) particles incident upon the target (electron) per unit area per unit time. As $\beta_2 \to 0$, he positron beam also slows down (in the matter), and we have finite annihilations.

2.3.2.4 Qualitative Discussion of Positronium Annihilation

In the above articles we have discussed the annihilations of electron and positron pair into two photons. Essentially, there are three different kinds of pair annihilations. The first kind, which has been discussed above, is the annihilation of free electron and free positron and may be named as annihilation of free $e^+ e^-$ pair.

If the pair $e^+ e^-$ forms a bound state, a hydrogenlike atom with positive charge of positron and negative charge of an electron, such bound state is called the *positronum*. The difference between hydrogen atom and the positronium is that in the former a heavy proton makes the nucleus around which the electron revolves but in positronium both the charges have equal masses and both e^+ and e^- orbit around the common centre. This bound state of e^+ and e^- can also decay into photons and the process is called *positronium annihilation*. The third kind of pair annihilation takes place in the presence of an external field, for example, the Coulomb field of a nucleus. In this process called the pair annihilation in external field, the incident positron annihilates with an atomic electron that is obviously bound (tightly) with the nucleus and the nucleus gets the recoiled momentum. In this kind only one photon is emitted.

In positronium annihilation, the positron can annihilate the electron by emitting two or more photons. The positronium annihilation is governed by selection rules and therefore it cannot decay into single photon. The charge conjugation number for positronium is $(-1)^{l+s}$, while for n photons it is $(-1)^n$. Thus if the positronium decays into n photons, the charge conjugation invariance must be obeyed which prescribes the selection rule

$$(-1)^{l+s} = (-1)^n \tag{2.189}$$

Since for annihilation positron and electron must come close, they (e^+ and e^-) overlap only when $l = 0$. This implies that the positronium annihilation takes place only from s ($l = 0$) states. For $l = 0$, the selection rule (2.189) would read.

$$(-1)^s = (-1)^n \tag{2.190}$$

Since for positronium $s = 0$ or 1, the annihilation would take place with the emission of even (usually two, $n = 2$) photons when $s = 0$, and when $s = 1$ odd (usually 3) number of photons. For $s = 1$, the $n = 1$ condition is denied by energy momentum conservation. Thus we may conclude that for $l = 0$ the singlet ($s = 0$) states must go to two photons and triplet ($s = 1$) to go to three photons. The positronium can also decay directly from a state $l > 0$ but, in reality it does not annihilate from $l \neq 0$ state, it undergoes a radiative transition to lower states and finally to s state from where it annihilates. The lifetime of the two-photon annihilation of ${}^1s_0$ state is 1.25×10^{-10} sec and for three-photon decay of ${}^3s_1$ state is 1.48×10^{-7} sec. The state ${}^1s_0$ of positronium is called *para-positronium* and the state ${}^3s_1$ the *ortho-positronium*. The above-mentioned lifetimes of para and ortho-positronium have been further corrected and have shown agreement with experimental values.

2.3.3 Moller (electron–electron) Scattering

The scattering of an electron by an electron

$$e^- + e^- \rightarrow e^- + e^-$$

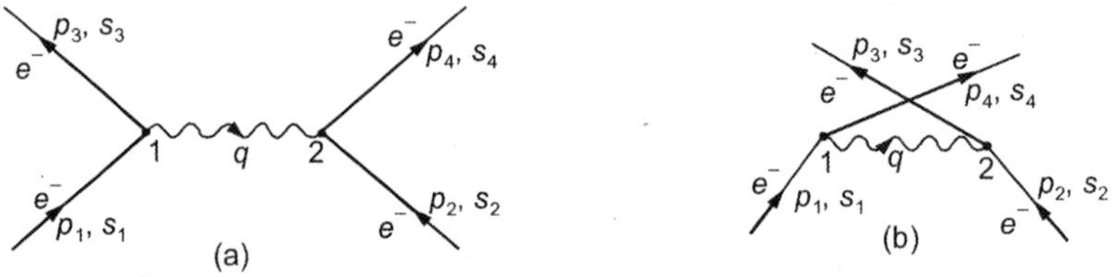

Fig. 2.10 (a, b): Feynman diagrams for Moller scattering

is also known as *Moller scattering*. Let us consider that an electron with four-momentum p_1 and spin s_1 collides with another electron with four-momentum p_2 and spin s_2. After scattering they emerge as the electrons with four-momentum p_3 and spin s_3 and four-momentum p_4 and spin s_4. The Feynman diagrams are drawn as Fig. (2.10).

The electrons with four-momenta p_1 and p_2 enter the reaction. In diagram (a) the electron recoiled after the emission of photon, goes away with four-momentum p_3 whereas in (b) it goes away with p_4. The photon with four momentum q is absorbed at vertex 2 to allow other electron to go away with four momentum p_4 in Fig. (2.10a) and with p_3 in Fig. (2.10b).

The electron—electron scattering ($e^- e^-$) is the quantum electrodynamic expression given to the Coulomb repulsion of like charges (at rest) of the classical theory. The Coulomb repulsion is instantaneous without the exchange of any mediating particle, the *QED* process, $e^- e^-$ scattering, is mediated by a photon. This scattering is different from Compton scattering or pair annihilation process in that there we had an electron propagator; here we have a photon propagator. Another difference comes due to the fact that both the particles are fermions. We shall have a minus sign between M_1 and M_2 or in other words diagrams (2.10a) and (2.10b) will have to be subtracted. In Compton scattering and pair annihilation the diagrams had been added.

We may now proceed to apply Feynman rules to calculate the matrix elements $(M_1)_{fi}$ and $(M_2)_{fi}$ from figures (2.10a) and (2.10b). In diagram (2.10a), in the initial state, we have four-momenta p_1, p_2 and in the final state they are p_3, p_4. The spins are s_1, s_2 and s_3, s_4 in initial and final states respectively. We first apply Feynman rules to calculate M_1 (diagram a). Start from vertex 2. Following table gives the multiplicative factors (mf) for calculating M_{fi}.

1	(p_4, s_4)	Rule (3b)	mf	$\bar{u}(p_4) \equiv \bar{u}(4)$
2	Vertex 2	Rule (14)	mf	$-ie\gamma^\mu$
3	(p_2, s_2)	Rule (3a)	mf	$u(p_2) \equiv u(2)$

4	Vertex 2	Rule (16)	mf	$(2\pi)^4\delta^4(p_2+q-p_4)$
5	Internal line Photon	Rule (12)	mf	$-ig_{\mu\nu}/q^2$
6	(p_3, s_3)	Rule (3b)	mf	$\bar{u}(p_3)\equiv\bar{u}(3)$
7	Vertex 1	Rule (14)	mf	$-ie\gamma^{\nu}$
8	(p_1, s_1)	Rule (3a)	mf	$u(p_1)\equiv u(1)$
9	Vertex 1	Rule (16)	mf	$(2\pi)^4\delta^4(p_1-p_3-q)$
10		Rule (17)	mf	$(d^4q)/(2\pi)^4$
11		Rule (18)		

Thus

$$-iM_1 = \int \bar{u}(4)(-ie\gamma^{\mu})u(2)(2\pi)^4\delta^4(p_2+q-p_4)\frac{(-ig_{\mu\nu})}{q^2}$$
$$\bar{u}(3)(-ie\gamma^{\nu})u(1)(2\pi)^4\delta^4(p_1-p_3-q)\frac{d^4q}{(2\pi)^4} \tag{2.191}$$

Using

$$\left(\frac{g_{\mu\nu}}{q^2}\right)\delta^4(p_1-p_3-q)=\frac{g_{\mu\nu}}{(p_1-p_3)^2}\delta^4(p_1-p_3-q)$$

and then using the property

$$\int\delta^4[(p_1-p_3)-q]\delta^4[q-(p_4-p_2)]d^4q=\delta^4(p_1-p_3-p_4+p_2) \tag{2.192}$$

we can write eqn. (2.191), after deleting $(2\pi)^4\delta^4(p_1-p_3-p_4+p_2)$, as

$$M_1=-e^2\bar{u}(4)\gamma^{\mu}u(2)\frac{g_{\mu\nu}}{(p_1-p_3)^2}\;\bar{u}(3)\gamma^{\nu}u(1) \tag{2.193}$$

Similarly, we can apply Feynman rules to get first

$$-iM_2=\int\bar{u}(3)(-ie\gamma^{\mu})u(2)(2\pi)^4\delta^4(p_2+q-p_3)\frac{(-ig_{\mu\nu})}{q^2}$$
$$\bar{u}(4)(-ie\gamma^{\nu})u(1)(2\pi)^4\delta^4(p_1-p_4-q)\frac{d^4q}{(2\pi)^4} \tag{2.194}$$

and then using

$$\frac{g_{\mu\nu}}{q^2}\delta^4(p_1-p_4-q)=\frac{g_{\mu\nu}}{(p_1-p_4)^2}\delta^4(p_1-p_4-q) \tag{2.195}$$

and

$$\int\delta^4[q-(p_3-p_2)]\delta^4[p_1-p_4-q]d^4q=\delta^4(p_1-p_4-p_3+p_2) \tag{2.196}$$

and also deleting $(2\pi)^4\delta^4(p_1 - p_4 - p_3 + p_2)$, we obtain

$$M_2 = -e^2\bar{u}(3)\gamma^\mu u(2)\frac{g_{\mu\nu}}{(p_1 - p_4)^2}\ \bar{u}(4)\gamma^\nu u(1) \tag{2.197}$$

Since both the particles are fermions, the total matrix element is

$$M = M_1 - M_2 \tag{2.198}$$

where, M implies M_{fi} and so do M_1 and M_2. Thus for electron—electron scattering

$$M_{fi} = -e^2\bar{u}(4)\gamma^\mu u(2)\frac{g_{\mu\nu}}{(p_1 - p_3)^2}\ \bar{u}(3)\gamma^\nu u(1)$$

$$+e^2\bar{u}(3)\gamma^\alpha u(2)\frac{g_{\alpha\beta}}{(p_1 - p_4)^2}\ \bar{u}(4)\gamma^\beta u(1) \tag{2.199}$$

in which $g_{\mu\nu}$ and $g_{\alpha\beta}$ set $\mu = \nu$ and $\alpha = \beta$ respectively. We may therefore write

$$M_{fi} = -\frac{e^2}{(p_1 - p_3)^2}[\bar{u}(4)\gamma^\mu u(2)][\bar{u}(3)\gamma_\mu u(1)]$$

$$+\frac{e^2}{(p_1 - p_4)^2}[\bar{u}(3)\gamma^\alpha u(2)][\bar{u}(4)\gamma_\alpha u(1)] \tag{2.200}$$

2.3.3.1 Calculation of $|M_{fi}|^2$

$$|M_{fi}|^2 \equiv M_{fi}M_{fi}^\dagger$$

Let us write M_{fi} [eqn. (2.200)] as

$$M_{fi} = -\frac{e^2}{(p_1 - p_3)^2}AB + \frac{e^2}{(p_1 - p_4)^2}CD \tag{2.201}$$

where

$$\begin{aligned} A &= \bar{u}(4)\gamma^\mu u(2) \\ B &= \bar{u}(3)\gamma_\mu u(1) \\ C &= \bar{u}(3)\gamma^\alpha u(2) \\ D &= \bar{u}(4)\gamma_\alpha u(1) \end{aligned} \tag{2.202}$$

From eqn. (2.201)

$$(M_{fi})^\dagger = -\frac{e^2}{(p_1 - p_3)^2}B^\dagger A^\dagger + \frac{e^2}{(p_1 - p_4)^2}D^\dagger C^\dagger \tag{2.203}$$

where

$$\begin{aligned} A^\dagger &= [\bar{u}(4)\gamma^\mu u(2)]^\dagger \\ &= u^\dagger(2)(\gamma^\mu)^\dagger(u^\dagger(4)\gamma^0)^\dagger \\ &= u^\dagger(2)\gamma^0\gamma^\mu\gamma^0\gamma^0 u(4) \\ &= \bar{u}(2)\gamma^\mu u(4) \end{aligned} \tag{2.204a}$$

where $\bar{u} = u^\dagger\gamma^0$, $(\gamma^\mu)^\dagger = \gamma^0\gamma^\mu\gamma^0$ and $(\gamma^0)^2 = 1$ have been used. Similarly,

$$B^\dagger = \bar{u}(1)\gamma_\mu u(3) \tag{2.204b}$$

$$C^\dagger = \bar{u}(2)\gamma^\alpha u(3) \tag{2.204c}$$

$$D^\dagger = \bar{u}(1)\gamma_\alpha u(4) \tag{2.204d}$$

Equations (2.201) and (2.203) give

$$\begin{aligned} M_{fi}M_{fi}^\dagger = &\frac{e^4}{(p_1-p_3)^4}ABB^\dagger A^\dagger + \frac{e^4}{(p_1-p_4)^4}CDD^\dagger C^\dagger \\ &-\frac{e^4}{(p_1-p_3)^2(p_1-p_4)^2}ABD^\dagger C^\dagger \\ &-\frac{e^4}{(p_1-p_3)^2(p_1-p_4)^2}CDB^\dagger A^\dagger \end{aligned} \tag{2.205}$$

We calculate one (e.g., $ABD^\dagger C^\dagger$) explicitly and shall write for others accordingly.

$$\begin{aligned} ABD^\dagger C^\dagger &= \bar{u}(4)\gamma^\mu u(2)\cdot\bar{u}(3)\gamma_\mu u(1)\cdot\bar{u}(1)\gamma_\alpha u(4)\cdot\bar{u}(2)\gamma^\alpha u(3) \\ &= \bar{u}(4)\gamma^\mu u(2)\cdot\bar{u}(2)\gamma^\alpha u(3)\cdot\bar{u}(3)\gamma_\mu u(1)\cdot\bar{u}(1)\gamma_\alpha u(4) \end{aligned} \tag{2.206}$$

Now we sum it over the spins s_1 and s_2 and use the relations

$$\sum_{s_i=1,2} \bar{u}^{s_i}(i)\, u^{s_i}(i) = \frac{\not{p}_i + m}{2m} \tag{2.207}$$

where $i = 1, 2, 3, 4$ to get

$$\begin{aligned} \sum_{s_1}\sum_{s_2}\sum_{s_3} ABD^\dagger C^\dagger &= \bar{u}(4)\gamma^\mu \frac{\not{p}_2 + m}{2m}\gamma^\alpha \frac{\not{p}_3 + m}{2m}\gamma_\mu \frac{\not{p}_1 + m}{2m}\gamma_\alpha u(4) \\ &= \bar{u}(4)Qu(4) \end{aligned} \tag{2.208}$$

where

$$Q = \frac{1}{8m^3}[\gamma^\mu(\not{p}_2 + m)\gamma^\alpha(\not{p}_3 + m)\gamma_\mu(\not{p}_1 + m)\gamma_\alpha] \tag{2.209}$$

Summing over s_4, eqn. (2.208) becomes

$$\begin{aligned} \sum_{s_1}\sum_{s_2}\sum_{s_3}\sum_{s_4} ABD^\dagger C^\dagger &= \sum_{s_4=1,2} [\bar{u}^{s_4}(4)]_a Q_{ab}[u^{s_4}(4)]_b \\ &= Q_{ab}\sum_{s_4}[u^{s_4}(4) \quad \bar{u}^{s_4}(4)]_{ba} \\ &= \frac{1}{2m}Q_{ab}(\not{p}_4 + m)_{ba} \end{aligned}$$

$$= \frac{1}{2m}[Q(\not{p}_4 + m)]_{aa}$$

$$= \frac{1}{2m} Trace\{Q(\not{p}_4 + m)\} \tag{2.210}$$

where eqn. (2.207) with $i = 4$ and the eqn. (2.94a) have been used. Using the similar technique, we can calculate

$$\sum_{s_1}\sum_{s_2} ABB^\dagger A^\dagger = \sum_{s_1 s_2}\{\overline{u}(4)\gamma^\mu u(2)\cdot\overline{u}(3)\gamma_\mu u(1)$$
$$\times\overline{u}(1)\gamma_\nu u(3)\cdot\overline{u}(2)\gamma^\nu u(4)\}$$
$$= \sum_{s_2}[\overline{u}(4)\gamma^\mu u^{s_2}(2)\overline{u}^{s_2}(2)\gamma^\nu u(4)]$$
$$\times\sum_{s_1}[\overline{u}(3)\gamma_\mu u^{s_1}(1)\overline{u}^{s_1}(1)\gamma_\nu u(3)]$$
$$= \frac{1}{2m}[\overline{u}(4)\gamma^\mu(\not{p}_2 + m)\gamma^\nu u(4)]$$
$$\times\frac{1}{2m}[\overline{u}(3)\gamma_\mu(\not{p}_1 + m)\gamma_\nu u(3)] \tag{2.211a}$$

from which on summing over s_3 and s_4 we get

$$\sum_{s_1}\sum_{s_2}\sum_{s_3}\sum_{s_4} ABB^\dagger A^\dagger = \frac{1}{4m^2}\{[\gamma^\mu(\not{p}_2 + m)\gamma^\nu]_{ab}$$
$$\times\sum_{s_4}[u^{s_4}(4)\overline{u}^{s_4}(4)]_{ba}\}$$
$$\times\frac{1}{4m^2}\{[\gamma_\mu(\not{p}_1 + m)\gamma_\nu]_{ab}$$
$$\times\sum_{s_3}[u^{s_3}(3)\overline{u}^{s_3}(3)]_{ba}\} \tag{2.211b}$$
$$= \frac{1}{16m^4}[Trace\{\gamma^\mu(\not{p}_2 + m)\gamma^\nu(\not{p}_4 + m)\}]$$
$$\times[Trace\{\gamma_\mu(\not{p}_1 + m)\gamma_\nu(\not{p}_3 + m)\}]$$

where matrices $\gamma^\mu(\not{p}_2 + m)\gamma^\nu$ and $\gamma_\mu(\not{p}_1 + m)\gamma_\nu$ are like matrix Q of say eqn. (2.210). Likewise we may obtain

$$\sum_{s_1}\sum_{s_2}\sum_{s_3}\sum_{s_4} CDD^\dagger C^\dagger = \frac{1}{16m^4}[Trace\{\gamma^\alpha(\not{p}_2 + m)\gamma^\rho(\not{p}_3 + m)\}]$$
$$\times[Trace\{\gamma_\alpha(\not{p}_1 + m)\gamma_\rho(\not{p}_4 + m)\}] \tag{2.212}$$

and similar to eqn. (2.210)

$$\sum_{s_1}\sum_{s_2}\sum_{s_3}\sum_{s_4} CDB^\dagger A^\dagger = \frac{1}{2m} Trace\{R(\not{p}_3 + m)\} \tag{2.213a}$$

with

$$R = \frac{1}{8m^3}[\gamma^\alpha(\not{p}_2 + m)\gamma^\mu(\not{p}_4 + m)\gamma_\alpha(\not{p}_1 + m)\gamma_\mu] \qquad (2.213b)$$

2.3.3.1.1 Evaluation of traces

We now calculate the traces from equations (2.210) to (2.213) to obtain $|M_{fi}|^2$. Equation (2.210) reads

$$\begin{aligned}\Sigma ABD^\dagger C^\dagger = \frac{1}{16m^4} &Trace\,[\gamma^\mu(\not{p}_2 + m)\gamma^\alpha(\not{p}_3 + m)\gamma_\mu \\ &\times(\not{p}_1 + m)\gamma_\alpha(\not{p}_4 + m)]\end{aligned} \qquad (2.214)$$

Now

$$\begin{aligned}\gamma^\mu(\not{p}_2 + m)\gamma^\alpha(\not{p}_3 + m)\gamma_\mu = &\gamma^\mu \not{p}_2\gamma^\alpha\not{p}_3\gamma_\mu \\ &+m(\gamma^\mu\not{p}_2\gamma^\alpha\gamma_\mu + \gamma^\mu\gamma^\alpha\not{p}_3\gamma_\mu) \\ &+m^2\gamma^\mu\gamma^\alpha\gamma_\mu\end{aligned} \qquad (2.215)$$

in which we may now use $\not{p}_2 = \gamma^\lambda p_{2\lambda}$, $\not{p}_3 = \gamma^\rho p_{3\rho}$ and then apply the following properties of gamma matrices

$$\gamma^\mu\gamma^\lambda\gamma^\alpha\gamma^\rho\gamma_\mu = -2\gamma^\rho\gamma^\alpha\gamma^\lambda \qquad (2.216a)$$

$$\gamma^\mu\gamma^\lambda\gamma^\alpha\gamma_\mu = 4g^{\lambda\alpha} \qquad (2.216b)$$

$$\gamma^\mu\gamma^\alpha\gamma^\rho\gamma_\mu = 4g^{\alpha\rho} \qquad (2.216c)$$

$$\gamma^\mu\gamma^\alpha\gamma_\mu = -2\gamma^\alpha \qquad (2.216d)$$

in respectively the first , second, third and fourth term of eqn. (2.215), to write the RHS of eqn. (2.215) as

$$\begin{aligned}\text{RHS} &= -2p_{3\rho}\gamma^\rho\gamma^\alpha\gamma^\lambda p_{2\lambda} + 4m(p_{2\lambda}g^{\lambda\alpha} + p_{3\rho}g^{\alpha\rho}) - 2m^2\gamma^\alpha \\ &= -2\not{p}_3\gamma^\alpha\not{p}_2 + 4m(p_2^\alpha + p_3^\alpha) - 2m^2\gamma^\alpha\end{aligned} \qquad (2.217)$$

Substituting eqn. (2.217), equation (2.214) reads

$$\begin{aligned}\Sigma ABD^\dagger C^\dagger = \frac{1}{16m^4} &Tr\Big[-2\not{p}_3\gamma^\alpha\not{p}_2\not{p}_1\gamma_\alpha\not{p}_4 \\ &-2m^2\big\{\not{p}_3\gamma^\alpha\not{p}_2\gamma_\alpha + \gamma^\alpha\not{p}_1\gamma_\alpha\not{p}_4 \\ &-2(p_2^\alpha + p_3^\alpha)(\not{p}_1\gamma_\alpha + \gamma_\alpha\not{p}_4)\big\} - 2m^4\gamma^\alpha\gamma_\alpha\Big]\end{aligned} \qquad (2.218)$$

in which the terms with coefficients m and m^3 have vanished because these terms contain odd number of gamma matrices whose traces vanish. Note that $\not{p}$ contains one gamma matrix (γ^μ). Eqn. (2.218) has five traces we evaluate them individually.

$$Trace[-2\not{p}_3\gamma^\alpha\not{p}_2\not{p}_1\gamma_\alpha\not{p}_4]$$
$$= -8(p_1 \cdot p_2)Tr(\not{p}_3\not{p}_4) \qquad [\because \gamma^\mu \not{a}\not{b}\gamma_\mu = 4a.b]$$
$$= -32(p_1 \cdot p_2)(p_3 \cdot p_4) \qquad [\because Tr(\not{a}\not{b}) = 4a.b] \qquad (2.219a)$$
$$= -32(p_1 \cdot p_2)^2$$

where we have used the energy momentum conservation equation

$$(p_1 + p_2) = (p_3 + p_4) \qquad (2.220)$$

which when squared and using $p_1^2 = p_2^2 = p_3^2 = p_4^2 = m^2$ gives $p_1 \cdot p_2 = p_3 \cdot p_4$. The next trace of eqn. (2.218)

$$Tr[\not{p}_3\gamma^\alpha\not{p}_3\gamma_\alpha] = -2Tr[\not{p}_3\not{p}_2][\because \gamma^\mu\not{a}\gamma_\mu = -2\not{a}]$$
$$= -8(p_2 \cdot p_3) \qquad (2.219b)$$

The third trace of eqn. (2.218)

$$Tr[\gamma^\alpha\not{p}_1\gamma_\alpha\not{p}_4] = -2Tr[\not{p}_1\not{p}_4] \quad [\because \gamma^\alpha\not{p}_1\gamma_\alpha = -2\not{p}_1]$$
$$= -8(p_1 \cdot p_4) \qquad (2.219c)$$

The fourth trace of eqn. (2.218) is

$$-2(p_2^\alpha + p_3^\alpha)Tr(\not{p}_1\gamma_\alpha + \gamma_\alpha\not{p}_4)$$
$$= -2(p_2^\alpha + p_3^\alpha)Tr(p_{1\lambda}\gamma^\lambda\gamma_\alpha + \gamma_\alpha\gamma^\rho p_{4\rho})$$
$$= -2(p_2^\alpha + p_3^\alpha)Tr(p_{1\lambda}g_\alpha^\lambda \cdot 1 + 1 \cdot g_\alpha^\rho p_{4\rho})$$
$$= -8(p_2^\alpha + p_3^\alpha)(p_{1\alpha} + p_{4\alpha}) \qquad [\because Tr[1] = 4] \qquad (2.219d)$$
$$= -8[(p_1 \cdot p_2) + (p_2 \cdot p_4) + (p_1 \cdot p_3) + (p_3 \cdot p_4)]$$

where $p_2^\alpha p_{1\alpha} = (p_1 \cdot p_2)$ etc. are implied. The last of the traces of eqn. (2.218) is simply

$$-2m^4 Tr \cdot (\gamma^\alpha\gamma_\alpha) = -32m^4 \qquad [Tr(\gamma^\alpha\gamma_\alpha) = 16] \qquad (2.219e)$$

Substituting the trace values from equations (2.219a) to (2.219e) in eqn. (2.218), we get

$$\sum ABD^\dagger C^\dagger$$
$$= \frac{1}{16m^4}[-32(p_1 \cdot p_2)^2 + 16m^2(p_2 \cdot p_3) + 16m^2(p_1 \cdot p_4)$$
$$+16m^2\{(p_1 \cdot p_2) + (p_2 \cdot p_4) + (p_1 \cdot p_3) + (p_3 \cdot p_4)\} - 32m^4]$$
$$= \frac{1}{16m^4}[-32(p_1 \cdot p_2)^2 + 32m^2(p_1 \cdot p_4) + 32m^2(p_1 \cdot p_2 + p_1 \cdot p_3) - 32m^4]$$
$$= \frac{2}{m^4}[-(p_1 \cdot p_2)^2 + m^2(p_1 \cdot p_4) + m^2\{p_1 \cdot p_2 + p_1 \cdot (p_1 + p_2 - p_4)\} - m^4]$$
$$= -\frac{2}{m^4}[(p_1 \cdot p_2)^2 - 2m^2(p_1 \cdot p_2)] \qquad (2.221)$$

where we have used

$$p_2 \cdot p_3 = p_1 \cdot p_4; \; p_2 \cdot p_4 = p_1 \cdot p_3; \; p_3 \cdot p_4 = p_1 \cdot p_2, \; p_3 = p_1 + p_2 - p_4$$

and $p_1^2 = m^2$.

These relations follow from eqn. (2.220). Likewise equations (2.213) tell that

$$\Sigma CDB^\dagger A^\dagger = \frac{1}{16m^4} Tr\Big[\gamma^\alpha (\not{p}_2 + m)\gamma^\mu (\not{p}_4 + m) \times \gamma_\alpha (\not{p}_1 + m)\gamma_\mu (\not{p}_3 + m)\Big] \tag{2.222}$$

Comparing this equation with eqn. (2.214), we find that the eqn. (2.222) differs from eqn. (2.214) only in the exchange $\not{p}_3 \leftrightarrow \not{p}_4$. A look at the first of equations . . . . p_3 by p_4 and vice versa. So the trace of eqn. (2.222) shall have the same value as has $\Sigma ABD^\dagger C^\dagger$. We write then

$$\Sigma CDB^\dagger A^\dagger = -\frac{2}{m^4}[(p_1 \cdot p_2)^2 - 2m^2 (p_1 \cdot p_2)] \tag{2.223}$$

We now calculate the traces in equation (2.211b)

$$\Sigma ABB^\dagger A^\dagger = \frac{1}{16m^4}[Tr\gamma^\mu (\not{p}_2 + m)\gamma^\nu (\not{p}_4 + m)] \times [Tr\gamma_\mu (\not{p}_1 + m)\gamma_\nu (\not{p}_3 + m)] \tag{2.211b}$$

in which

$$Tr[\gamma^\mu (\not{p}_2 + m)\gamma^\nu (\not{p}_4 + m)] = Tr(\gamma^\mu \not{p}_2 \gamma^\nu \not{p}_4) + m^2 Tr(\gamma^\mu \gamma^\nu) \tag{2.224}$$

where the terms with the coefficients m have vanished because of odd number of gamma matrices. RHS of eqn. (2.224) may be written as

$$\begin{aligned} \text{RHS} &= (p_2)_\lambda (p_4)_\rho Tr(\gamma^\mu \gamma^\lambda \gamma^\nu \gamma^\rho) + 4m^2 g^{\mu\nu} \\ &= (p_2)_\lambda (p_4)_\rho \cdot 4(g^{\mu\lambda} g^{\nu\rho} - g^{\mu\nu} g^{\lambda\rho} + g^{\mu\rho} g^{\lambda\nu}) + 4m^2 g^{\mu\nu} \\ &= 4 \;\; (p_2^\mu p_4^\nu - g^{\mu\nu} p_2 \cdot p_4 + p_4^\mu p_2^\nu) + 4m^2 g^{\mu\nu} \end{aligned} \tag{2.225}$$

where the following trace theorems have been used

$$Tr(\gamma^\mu \gamma^\nu \gamma^\lambda \gamma^\sigma) = 4(g^{\mu\nu} g^{\lambda\sigma} - g^{\mu\lambda} g^{\nu\sigma} + g^{\mu\sigma} g^{\nu\lambda})$$

and

$$Tr(\gamma^\mu \gamma^\nu) = 4g^{\mu\nu} \tag{2.226}$$

Thus from eqn. (2.224) we have

$$Tr[\gamma^\mu (\not{p}_2 + m)\gamma^\nu (\not{p}_4 + m)] = 4[p_2^\mu p_4^\nu + p_4^\mu p_2^\nu + g^{\mu\nu}(m^2 - p_2 \cdot p_4)] \tag{2.227}$$

Likewise the other trace of eqn. (2.211b)

$$Tr[\gamma_\mu(\not{p}_1+m)\gamma_\nu(\not{p}_3+m)] = 4[p_{1\mu}p_{3\nu}+p_{3\mu}p_{1\nu} + g_{\mu\nu}(m^2 - p_1\cdot p_3)] \quad (2.228)$$

Substituting above equations in eqn. (2.211b) we get

$$\begin{aligned}\Sigma ABB^\dagger A^\dagger &= \frac{2}{m^4}[(p_1\cdot p_2)(p_3\cdot p_4)+(p_1\cdot p_4)(p_2\cdot p_3)+(p_2\cdot p_4)(m^2-p_1\cdot p_3) \\ &\quad +(p_1\cdot p_3)(m^2-p_2\cdot p_4)+2(m^2-p_1\cdot p_3)(m^2-p_1\cdot p_4)] \\ &= \frac{2}{m^4}[(p_1\cdot p_2)^2+(p_1\cdot p_4)^2+2m^4-2m^2 p_1\cdot p_3] \\ &= \frac{2}{m^4}[(p_1\cdot p_2)^2+(p_1\cdot p_4)^2+2m^2(p_1\cdot p_4-p_1\cdot p_2)] \quad (2.229)\end{aligned}$$

where in the first step we have used

$$\begin{aligned} p_2^\mu p_{1\mu} p_4^\nu p_{3\nu} &= (p_2\cdot p_1)(p_4\cdot p_3) \\ g_{\mu\nu}p_2^\mu p_4^\nu &= p_{2\nu}p_4^\nu = p_2\cdot p_4 \\ g_{\mu\nu}g^{\mu\nu} &= \delta^\mu_\mu = 4 \end{aligned} \quad (2.230)$$

and similar other relations. In the second step we used $p_3\cdot p_4 = p_1\cdot p_2$ etc. and in the last step used $p_3 = p_1+p_2-p_4$ and $p_1^2 = m^2$.

Likewise, the equation (2.212), i.e,

$$\Sigma CDD^\dagger C^\dagger = \frac{1}{16m^4}[Trace\{\gamma^\alpha(\not{p}_2+m)\gamma^\rho(\not{p}_3+m)\}] \times[Trace\{\gamma_\alpha(\not{p}_1+m)\gamma_\rho(\not{p}_4+m)\}] \quad (2.212)$$

when compared with eqn. (2.211b) suggests that if $\not{p}_3$ and $\not{p}_4$ are interchanged, the eqn. (2.212) becomes eqn. (2.211b). Thus replacing p_4 by p_3 in eqn. (2.229), it would give the value of $\Sigma CDD^\dagger C^\dagger$. That is

$$\Sigma CDD^\dagger C^\dagger = \frac{2}{m^4}[(p_1\cdot p_2)^2+(p_1\cdot p_3)^2+2m^2(p_1\cdot p_3-p_1\cdot p_2)] \quad (2.231)$$

Thus, we have evaluated all the traces, which gave the values of $ABB^\dagger A^\dagger$, $CDD^\dagger C^\dagger$ $ABD^\dagger C^\dagger$ and $CDB^\dagger A^\dagger$ required in eqn. (2.205), which gives $|M_{fi}|^2$. The respective values are rewritten below:

$$ABB^\dagger A^\dagger = \frac{2}{m^4}[(p_1\cdot p_2)^2+(p_1\cdot p_4)^2+2m^2(p_1\cdot p_4-p_1\cdot p_2)] \quad (2.229)$$

$$CDD^{\dagger}C^{\dagger} = \frac{2}{m^4}[(p_1 \cdot p_2)^2 + (p_1 \cdot p_3)^2 + 2m^2(p_1 \cdot p_3 - p_1 \cdot p_2)] \quad (2.230a)$$

$$ABD^{\dagger}C^{\dagger} = -\frac{2}{m^4}[(p_1 \cdot p_2)^2 - 2m^2(p_1 \cdot p_2)] \quad (2.230b)$$

$$CDB^{\dagger}A^{\dagger} = -\frac{2}{m^4}[(p_1 \cdot p_2)^2 - 2m^2(p_1 \cdot p_2)] \quad (2.231)$$

where summation over all the spins is implied on the LHS of above four equations.

Substituting these values in eqn. (2.205), we get on using

$$\overline{|M_{fi}|^2} = \frac{1}{4} \sum_{s_1, s_2, s_3, s_4} |M_{fi}|^2 \quad (2.232)$$

where the factor of 1/4 is due to averaging over the two spin-1/2 electrons in the initial state, $(2s + 1)^2 = 4$

$$\overline{|M_{fi}|^2} = \frac{e^4}{2m^4}\left[\frac{(p_1 \cdot p_2)^2 + (p_1 \cdot p_4)^2 + 2m^2(p_1 \cdot p_4 - p_1 \cdot p_2)}{(p_1 - p_3)^4} + \frac{(p_1 \cdot p_2)^2 + (p_1 \cdot p_3)^2 + 2m^2(p_1 \cdot p_3 - p_1 \cdot p_2)}{(p_1 - p_4)^4} + 2\frac{(p_1 \cdot p_2)^2 - 2m^2(p_1 \cdot p_2)}{(p_1 - p_3)^2(p_1 - p_4)^2}\right] \quad (2.233)$$

Since the differential scattering cross section $\frac{d\sigma}{d\Omega}$ requires $|M_{fi}|^2$ and $\frac{d\sigma}{d\Omega}$ is calculated in centre of mass and laboratory coordinate systems, we choose the centre of mass system. In centre of mass system the $e^- + e^- \rightarrow e^- + e^-$ scattering could be depicted as in the following figure (Fig. 2.11). In the centre of mass

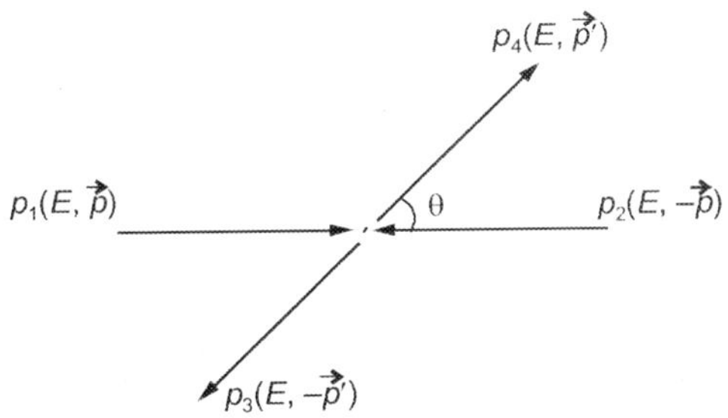

Fig. 2.11: Electron–electron scattering in centre of mass frame $|p| = |p'|$

frame $\boldsymbol{p}_1 + \boldsymbol{p}_2 = 0$ which is obeyed in above figure as $\boldsymbol{p} - \boldsymbol{p} = \boldsymbol{p}' - \boldsymbol{p}' = 0$. The angle θ is the scattering angle. In the centre of mass frame, the Lorentz invariants $(p_1 \cdot p_2)$, $(p_1 \cdot p_3)$ and $(p_1 \cdot p_4)$, of eqn. (2.233) become

$$p_1 \cdot p_2 = p_1^0 p_2^0 - \boldsymbol{p}_1 \cdot \boldsymbol{p}_2 = E^2 + \boldsymbol{p}^2 \tag{2.234}$$

where $p_1^0 = p_2^0 = E, \boldsymbol{p}_1 = \boldsymbol{p}$ and $\boldsymbol{p}_2 = -\boldsymbol{p}$ (see Fig. 2.11) have been used:

$$\begin{aligned} p_1 \cdot p_3 &= p_1^0 p_3^0 - \boldsymbol{p}_1 \cdot \boldsymbol{p}_3 = E^2 + \boldsymbol{p} \cdot \boldsymbol{p}' \\ &= E^2 - \boldsymbol{p}^2 \cos\theta \end{aligned} \tag{2.235}$$

where $\boldsymbol{p}_3 = -\boldsymbol{p}', |\boldsymbol{p}'| = |\boldsymbol{p}|$ and from Fig. (2.11) $\boldsymbol{p} \cdot \boldsymbol{p}' = -\boldsymbol{p}^2 \cos\theta$ have been used. Similarly,

$$\begin{aligned} p_1 \cdot p_4 &= p_1^0 p_4^0 - \boldsymbol{p}_1 \cdot \boldsymbol{p}_4 = E^2 - \boldsymbol{p} \cdot \boldsymbol{p}' \\ &= E^2 + \boldsymbol{p}^2 \cos\theta \end{aligned} \tag{2.236}$$

where $\boldsymbol{p}_4 = \boldsymbol{p}'$ and above-mentioned relations have been used. Likewise

$$\begin{aligned} (p_1 - p_3)^2 &= p_1^2 + p_3^2 - 2p_1 \cdot p_3 \\ &= -2\boldsymbol{p}^2(1 - \cos\theta) \end{aligned} \tag{2.237}$$

and

$$\begin{aligned} (p_1 - p_4)^2 &= p_1^2 + p_4^2 - 2p_1 \cdot p_4 \\ &= -2\boldsymbol{p}^2(1 + \cos\theta) \end{aligned} \tag{2.238}$$

where $p_1^2 = p_3^2 = p_4^2 = E^2 - \boldsymbol{p}^2$ and eqns. (2.235) and (2.236) have been used.

Thus substituting eqns. (2.234) to (2.238) in eqn. (2.233) we get $\overline{|M_{fi}|^2}$ in centre of mass frame as

$$\begin{aligned} \overline{|M_{fi}|^2_{CM}} = \frac{e^4}{2m^4}\frac{1}{4\boldsymbol{p}^4}\Bigg[&\frac{(E^2 + \boldsymbol{p}^2)^2 + (E^2 + \boldsymbol{p}^2\cos\theta)^2 + 2m^2\boldsymbol{p}^2(\cos\theta - 1)}{(1 - \cos\theta)^2} \\ &+ \frac{(E^2 + \boldsymbol{p}^2)^2 + (E^2 - \boldsymbol{p}^2\cos\theta)^2 - 2m^2\boldsymbol{p}^2(\cos\theta + 1)}{(1 + \cos\theta)^2} \\ &+ \frac{2(E^2 + \boldsymbol{p}^2)^2 - 4m^2(E^2 + \boldsymbol{p}^2)}{1 - \cos^2\theta}\Bigg] \\ = \frac{e^4}{4m^4\boldsymbol{p}^4}&\left[\frac{\begin{array}{r} 2(E^2 + \boldsymbol{p}^2)^2 + (E^2 + \boldsymbol{p}^2\cos^2\theta)^2 + (E^2 + \boldsymbol{p}^2)^2\cos^2\theta \\ -2m^2\boldsymbol{p}^2\sin^2\theta - 2m^2(E^2 + \boldsymbol{p}^2)\sin^2\theta \end{array}}{\sin^4\theta}\right] \\ = \frac{e^4}{4m^4\boldsymbol{p}^4}&\left[\frac{4(E^2 + \boldsymbol{p}^2)^2 - 3(E^2 + \boldsymbol{p}^2)^2\sin^2\theta + \boldsymbol{p}^4\sin^4\theta + 4\boldsymbol{p}^4\sin^2\theta}{\sin^4\theta}\right] \end{aligned}$$

$$= \frac{e^4}{4m^4}\frac{(E^2+\boldsymbol{p}^2)^2}{\boldsymbol{p}^4}\left[\frac{4}{\sin^4\theta}-\frac{3}{\sin^2\theta}+\left(\frac{\boldsymbol{p}^2}{E^2+\boldsymbol{p}^2}\right)^2\left(1+\frac{4}{\sin^2\theta}\right)\right]$$

(2.239a)

$$\equiv \frac{e^4}{4m^4}\left(\frac{2E^2-m^2}{E^2-m^2}\right)^2\left[\frac{4}{\sin^4\theta}-\frac{3}{\sin^2\theta}+\left(\frac{E^2-m^2}{2E^2-m^2}\right)^2\left(1+\frac{4}{\sin^2\theta}\right)\right]$$

(2.239b)

2.3.3.2 The Differential Scattering Cross Section in Centre of Mass Frame

We use the general equation (2.71) and incorporate in it the necessary changes for fermions as in this scattering 1 + 2 → 3 + 4 all the particles are electrons (fermions). Accordingly, the factor 1/4 in eqn. (2.71) is replaced by, $m_1 m_2 = m^2, m_1 = m_2, 2p_3^0$ is replaced by $\frac{p_3^0}{m}$ and $2p_4^0$ by $\frac{p_4^0}{m}$ and then eqn. (2.71) for $e^- e^- \to e^- e^-$ scattering reads

$$d\sigma = \frac{m^4}{[(p_1\cdot p_2)^2 - m^4]^{\frac{1}{2}}}\int\left[\frac{d^3\boldsymbol{p}_3}{(2\pi)^3 p_3^0}\frac{d^3\boldsymbol{p}_4}{(2\pi)^3 p_4^0}\right.$$
$$\left.\times(2\pi)^4\delta^4(p_1+p_2-p_3-p_4)\overline{|M_{fi}|^2}\right] \tag{2.240}$$

where $\overline{|M_{fi}|^2}$ is given by eqn. (2.239). In the centre of mass frame $p = p_2 = -p_1$, $E = E_1 = E_3 = E_2 = E_4$. In addition to these $p' = p_4 = -p_3$ with $|\boldsymbol{p}'| = |\boldsymbol{p}|$ in CM frame. We shall find the condition $p_4 = -p_3$ when we solve the three- dimensional delta function δ^3. The four dimensional delta function δ^4 of eqn. (2.240) can be separated as

$$\delta^4(p_1+p_2-p_3-p_4) = [\delta(E_1+E_2-E_3-E_4)$$
$$\times\delta^3(\boldsymbol{p}_1+\boldsymbol{p}_2-\boldsymbol{p}_3-\boldsymbol{p}_4)] \tag{2.241}$$

which on using $p_1 + p_2 = 0$ (CM frame) reads that, $\delta_3 \Rightarrow \delta^3(-p_3 - p_4)$. Note that we have d^3p_3 and d^3p_4 in the integral. We first carry out the p_4 integration i.e., use

$$\int d^3\boldsymbol{p}_4\delta^3(-\boldsymbol{p}_3-\boldsymbol{p}_4) = 1 \tag{2.242}$$

provided $p_4 = -p_3$. Thus the result of using up d^3p_4 (i.e. carrying out p_4 integration) is that we get $p_4 = -p_3$. Then what remains is d^3p_3 and $\delta(E_1 + E_2 - E_3 - E_4)$. Equation (2.240) on using eqn. (2.234) then becomes

$$d\sigma = \frac{m^4}{2E|\boldsymbol{p}|(2\pi)^2}\int \delta(W - E_1 - E_2)\frac{d^3\boldsymbol{p}_3}{E_3E_4}\overline{|M_{fi}|^2} \tag{2.243}$$

where $\delta(x) = \delta(-x)$ and $E_3 + E_4 = W$ have been employed. To eliminate the remaining delta function in eqn. (2.243), we need to have a dW so that we may apply

$$\int \delta(W - E_1 - E_2)dW = 1 \tag{2.244}$$

provided $W = E_3 + E_4 = E_1 + E_2$. Accordingly, we write eqn. (2.243) as

$$\begin{aligned} d\sigma &= \frac{m^4}{2E|\boldsymbol{p}|(2\pi)^2}\int \frac{\delta(W - E_1 - E_2)dW}{E_3E_4}\cdot\frac{d^3\boldsymbol{p}_3}{dW}\overline{|M_{fi}|^2} \\ &= \frac{m^4}{2(2\pi)^2E|\boldsymbol{p}|}\frac{1}{E_3E_4}\frac{d^3\boldsymbol{p}_3}{dW}\overline{|M_{fi}|^2} \end{aligned} \tag{2.245}$$

where in the second step we used eqn. (2.244). Now

$$d^3\boldsymbol{p}_3 = |\boldsymbol{p}_3|^2\, d|\boldsymbol{p}_3|\sin\theta\, d\theta\, d\phi = \boldsymbol{p}_3^2 d|\boldsymbol{p}_3|d\Omega \tag{2.246}$$

using which we may write

$$d\sigma = \frac{m^4}{2(2\pi)^2E|\boldsymbol{p}|}\frac{\boldsymbol{p}_3^2}{E_3E_4}\frac{d|\boldsymbol{p}_3|}{dW}\overline{|M_{fi}|^2}d\Omega \tag{2.247}$$

Now

$$\begin{aligned} \frac{d|\boldsymbol{p}_3|}{dW} &= \left[\frac{dW}{d|\boldsymbol{p}_3|}\right]^{-1} = \left[\frac{dE_3}{d|\boldsymbol{p}_3|} + \frac{dE_4}{d|\boldsymbol{p}_3|}\right]^{-1} \\ &= \left[\frac{dE_1}{d|\boldsymbol{p}_1|} + \frac{dE_2}{d|\boldsymbol{p}_2|}\right]^{-1} \\ &= \left[\frac{d}{d|\boldsymbol{p}_1|}(\boldsymbol{p}_1^2 + m^2)^{1/2} + \frac{d}{d|\boldsymbol{p}_2|}(\boldsymbol{p}_2^2 + m^2)^{1/2}\right]^{-1} \\ &= \left[\frac{|\boldsymbol{p}_1|}{E_1} + \frac{|\boldsymbol{p}_2|}{E_2}\right]^{-1} = \frac{E}{2|\boldsymbol{p}|} \end{aligned} \tag{2.248}$$

where in the second step we have used $W = E_3 + E_4 = E_1 + E_2$ and $|\boldsymbol{p}_3| = |\boldsymbol{p}_1|$ as well as $|\boldsymbol{p}_3| = |\boldsymbol{p}_2|$. In the last step we have used $E_1 = E_2 = E$ and $|\boldsymbol{p}_1| = |\boldsymbol{p}_2| = \boldsymbol{p}$. Substituting eqn. (2.248) in eqn. (2.247) and using $E_3 = E_4 = E$, $\boldsymbol{p}_3^2 = \boldsymbol{p}^2$ we get

$$\left(\frac{d\sigma}{d\Omega}\right)_{CM} = \frac{m^4}{4E^2(2\pi)^2}\overline{|M_{fi}|^2} \tag{2.249}$$

from which on substituting $\overline{|M_{fi}|^2}$ from eqn. (2.239b), we get the relativistic differential cross section for Moller scattering in the center of mass frame as

$$\left(\frac{d\sigma}{d\Omega}\right)_{CM} = \frac{\alpha^2}{4E^2}\left(\frac{2E^2-m^2}{E^2-m^2}\right)^2$$

$$\left[\frac{4}{\sin^4\theta} - \frac{3}{\sin^2\theta} + \left(\frac{E^2-m^2}{2E^2-m^2}\right)^2\left(1+\frac{4}{\sin^2\theta}\right)\right] \tag{2.250}$$

where $\alpha = (e^2/4\pi) \simeq$ is the fine structure constant.

The nonrelativistic and extreme relativistic limits of the cross section (2.250) may now be obtained.

In the nonrelativistic (NR) limit p and $p' \to 0$ and hence

$$E^2 = p^2 + m^2 \to m^2 \tag{2.251a}$$

and the velocity v of one of the electrons is

$$v^2 = \frac{p^2}{m^2} \to \frac{E^2-m^2}{E^2} \tag{2.251b}$$

Accordingly,

$$\left(\frac{2E^2-m^2}{E^2-m^2}\right)^2 = \frac{(1+v^2)^2}{v^4} \to \frac{1}{v^4} \tag{2.251c}$$

and

$$v^4 \to 0 \tag{2.251d}$$

Hence eqn. (2.250) becomes

$$\begin{aligned}\left(\frac{d\sigma}{d\Omega}\right)_{CM}^{NR} &= \frac{\alpha^2}{m^2}\frac{1}{4v^4}\left[\frac{4}{\sin^4\theta} - \frac{3}{\sin^2\theta}\right] \\ &= \frac{\alpha^2}{m^2}\frac{1}{v^4}\left[\frac{\cos^2\theta}{\sin^4\theta} + \frac{\sin^2\theta}{4\sin^4\theta}\right] \\ &= \frac{\alpha^2}{m^2}\frac{1}{16v^4}\left[\frac{(\cos^2\frac{\theta}{2} - \sin^2\frac{\theta}{2})^2}{\sin^4\frac{\theta}{2}\cos^4\frac{\theta}{2}} + \frac{\sin^2\frac{\theta}{2}\cos^2\frac{\theta}{2}}{\sin^4\frac{\theta}{2}\cos^4\frac{\theta}{2}}\right] \\ &= \frac{r_0^2}{(2v)^4}\left[\frac{1}{\sin^4\frac{\theta}{2}} + \frac{1}{\cos^4\frac{\theta}{2}} - \frac{1}{\sin^2\frac{\theta}{2}\cos^2\frac{\theta}{2}}\right]\end{aligned} \tag{2.252}$$

where

$$r_0 = \frac{\alpha}{m} = \frac{e^2}{4\pi m} \tag{2.253}$$

with e, m the charge and rest mass, is the classical radius of electron. It was Mott, who first obtained the result (2.252). The first term in the above Mott's formula is the Rutherford term. The second term is due to the requirement that the elastic scattering of two identical particles must be symmetric under $\theta \rightarrow \pi - \theta$. The third term is the interference term arising as a result of interference between the direct and the exchange scatterings. At scattering angles $\theta \simeq \frac{\pi}{2}$, the cross term tends to reduce the cross section and thus showing the effect of Pauli's principle.

In the ultra relativistic (UR) limit of energy of electron we have E large compared to its rest mass m

$$\frac{m}{E} \rightarrow 0 \tag{2.254a}$$

and hence

$$\frac{2E^2 - m^2}{E^2 - m^2} \rightarrow 2 \tag{2.254b}$$

and thus the cross section (2.250) for UR case becomes

$$\left(\frac{d\sigma}{d\Omega}\right)_{CM}^{UR} = \frac{\alpha^2}{4E^2}\frac{(3+\cos^2\theta)^2}{\sin^4\theta} \tag{2.255a}$$

$$\left(\frac{d\sigma}{d\Omega}\right)_{CM}^{UR} = \frac{\alpha^2}{E^2}\left[\frac{4}{\sin^4\theta} - \frac{3}{\sin^2\theta} + \frac{1}{4}\right]$$

$$= \frac{\alpha^2}{E^2}\left[\frac{(\cos^2\frac{\theta}{2} + \sin^2\frac{\theta}{2})^2}{8\sin^4\frac{\theta}{2}\cos^4\frac{\theta}{2}} + \frac{(\cos^2\frac{\theta}{2} - \sin^2\frac{\theta}{2})^2}{8\sin^4\frac{\theta}{2}\cos^4\frac{\theta}{2}} + \frac{1}{4}\right]$$

$$= \frac{\alpha^2}{4E^2}\left[\frac{1}{\sin^4\frac{\theta}{2}} + \frac{1}{\cos^4\frac{\theta}{2}} + 1\right] \tag{2.255b}$$

Expression (2.255a) can also be expressed as

$$\left(\frac{d\sigma}{d\Omega}\right)_{CM}^{UR} = \frac{\alpha^2}{4E^2}\frac{(3+\cos^2\theta)^2}{\sin^4\theta} \tag{2.255c}$$

The non-relativistic limit of Moller scattering in laboratory coordinates can be obtained as

$$\left(\frac{d\sigma}{d\Omega}\right)_{L}^{NR} = \frac{4r_0^2}{v^4}.\cos\beta\left[\frac{1}{\sin^4\beta} + \frac{1}{\cos^4\beta} - \frac{1}{\sin^2\beta.\cos^2\beta}\right] \tag{2.256}$$

where v and β refer to the laboratory system. In the NR limit, the scattering angle θ in the CM system is twice the angle in the laboratory system β.

2.3.4 Bhabha (positron–electron) Scattering

The elastic scattering of a positron (e^+) by an electron (e^-) is called the *Bhabha scattering*, after H.J. Bhabha who treated the problem in 1935.

$$e^+ + e^- \to e^+ + e^-$$

The Bhabha scattering is the quantum electrodynamic description of Coulomb attraction between unlike charges in the classical theory. If one of the particles of Moller scattering is replaced by its antiparticle, we have Bhabha scattering. The Feynman diagrams for $e^+e^- \to e^+e^-$ are drawn below (Fig. 2.12) .In the first diagram (2.12a), the electron and positron scatter via the exchange of a space like photon and in the second (diagram b), the electron and positron first annihilate into a time like photon of four-momentum q, which then reappears as electron—positron pair. The spacelike and timelike character of photons in these diagrams can be easily understood by considering the energy momentum conservation at vertices, say at vertex 2. In diagram (a), at vertex 2

$$q = p_2 - p_4$$

$\Rightarrow$

$$q^2 = (E_2 - E_4)^2 - (\boldsymbol{p}_2 - \boldsymbol{p}_4)^2 < 0 \tag{2.257a}$$

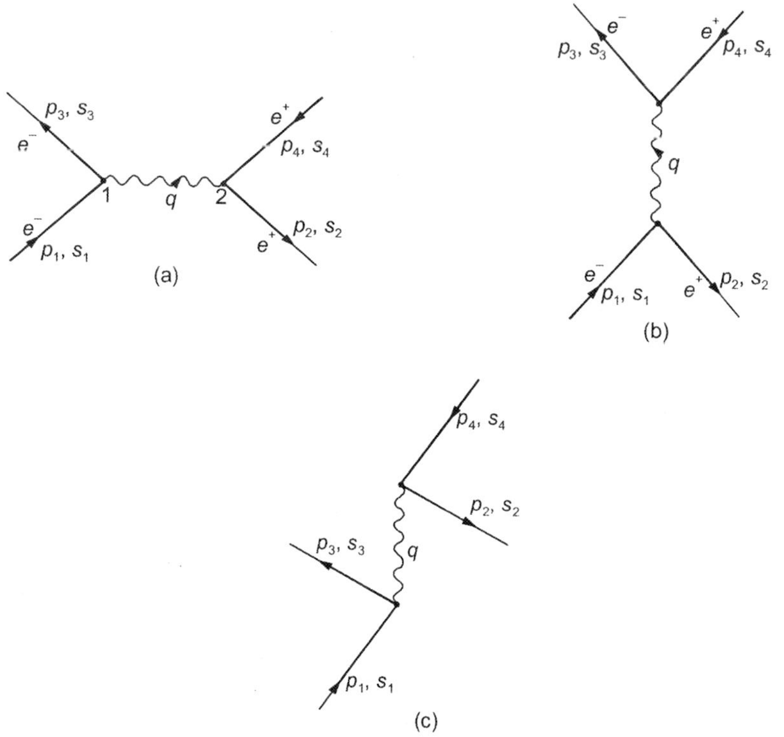

Fig. 2.12 (a, b, c): Feynman diagrams for Bhabha scattering

where $E_2 = E_4$ in CM frame. $q^2 < 0$ implies spacelike four-vector q. In diagram (b), however we have

$$q = p_1 + p_2$$

$\Rightarrow$

$$q^2 = (E_1 + E_2)^2 - (\boldsymbol{p}_1 + \boldsymbol{p}_2)^2 = 2E^2 > 0 \tag{2.257b}$$

where in CM frame $E_1 = E_2$ and $p_1 + p_2 = 0$ have been used. $q^2 > 0$ implies a timelike four-vector q. Like Moller scattering, here too we have both the particles as fermions and the rule of antisymmetrization tells that if two diagrams differ only in the interchange of two incoming or outgoing electrons or positrons or of an incoming electron (positron) with an outgoing positron (electron), then there shall be a minus sign between the two diagrams. Here, if we interchange in diagram (2.12b), the incoming positron (p_2, s_2) and outgoing electron (p_3, s_2), we get a diagram as Fig. (2.12c), which can be easily redrawn as Fig. (2.12a)

We may now proceed, as usual, to apply Feynman rules to calculate $(M_1)_{fi}$ and $(M_2)_{fi}$ from the Figs. (2.12a) and (2.12b) respectively. It has to be noted here that because of antiparticle (positron) we shall also have $v(p)$ and $\bar{v}(p)$. We apply Feynman rules to diagram (2.12a) and note down the multiplicative factors.

S. No	Diagram site	Rule	Multiplicative factor
1.	(p_2, s_2)	4 (a)	$\bar{v}(p_2) = \bar{v}(2)$
2.	Vertex 2	14	$-ie\gamma^{\mu}$
3.	(p_4, s_4)	4 (b)	$v(p_4) = v(4)$
4.	Vertex 2	16	$(2\pi)^4 \delta^4 (p_2 + q - p_4)$
5.	Internal line (Photon) q	12	$-ig_{\mu\nu}/q^2$
6.	(p_3, s_3)	3 (b)	$\bar{u}(p_3) \equiv \bar{u}(3)$
7.	Vertex 1	14	$-ie\gamma^{\nu}$
8.	(p_1, s_1)	3 (a)	$u(p_1) \equiv u(1)$
9.	Vertex 1	16	$(2\pi)^4 \delta^4 (p_1 - p_3 - q)$
10.		17	$d^4 q/(2\pi)^4$
11.		18	

Thus

$$-iM_1 = \int \bar{v}(2)(-ie\gamma^{\mu})v(4)(2\pi)^4 \delta^4 (p_2 + q - p_4).(-ig_{\mu\nu}/q^2).$$
$$\bar{u}(3)(-ie\gamma^{\nu})u(1)(2\pi)^4 \delta^4 (p_1 - p_3 - q).[d^4 q/(2\pi)^4] \tag{2.258}$$

Using the property mentioned below the eqn. (2.191) and then deleting the δ^4 function we get

$$M_1 = -e^2\, \overline{v}(2)\gamma^{\mu}v(4)\frac{g_{\mu\nu}}{(p_1 - p_3)^2}\overline{u}(3)\gamma^{\nu}u(1) \tag{2.259}$$

Likewise, applying the relevant Feynman rules to diagram (2.12b) we can get

$$-iM_2 = \int \overline{u}(3)(-ie\gamma^{\mu})v(4)(2\pi)^4\delta^4(q - p_3 - p_4).(-ig_{\mu\nu}/q^2).$$
$$\overline{v}(2)(-ie\gamma^{\nu})u(1)(2\pi)^4\delta^4(p_1 + p_2 - q).[d^4q/(2\pi)^4] \tag{2.260}$$

which on using

$$\frac{g_{\mu\nu}}{q^2}\delta^4(p_1 + p_2 - q) = \frac{g_{\mu\nu}}{(p_1 + p_2)^2}\delta^4(p_1 + p_2 - q) \tag{2.261a}$$

and

$$\int \delta^4[q - (p_3 + p_4)]\delta^4[p_1 + p_2 - q]d^4q = \delta^4(p_1 + p_2 - p_3 - p_4) \tag{2.261b}$$

and then deleting $(2\pi)^4\, \delta^4\, (p_1 + p_2 - p_3 - p_4)$ we obtain

$$M_2 = -e^2\overline{u}(3)\gamma^{\mu}v(4)\frac{g_{\mu\nu}}{(p_1 + p_2)^2}\overline{v}(2)\gamma^{\nu}u(1) \tag{2.262}$$

and the total matrix element

$$\begin{aligned} M_{fi} &= M_1 - M_2 \\ &= \frac{-e^2}{(p_1 - p_3)^2}\left[\overline{v}(2)\gamma^{\mu}v(4)\right]\left[\overline{u}(3)\gamma_{\mu}\, u(1)\right] \\ &\quad + \frac{e^2}{(p_1 + p_2)^2}\left[\overline{u}(3)\gamma^{\mu}v(4)\right]\left[\overline{v}(2)\gamma_{\mu}\, u(1)\right] \end{aligned} \tag{2.263}$$

where $g_{\mu\nu}\gamma^{\nu} = \gamma_{\mu}$ has been used.

As told earlier, Bhabha scattering differs from the Moller scattering in that one e^- of latter is replaced by e^+. Most of the calculations in Bhabha scattering would be similar to those of Moller scattering. The differential scattering cross section in centre of mass frame would here also, like eqn. (2.249), be given by

$$\left(\frac{d\sigma}{d\Omega}\right)_{CM} = \frac{m^4}{4E^2(2\pi)^2}\overline{|M_{fi}|^2_{CM}} \tag{2.264}$$

where $\overline{|M_{fi}|^2}$ has now to be calculated from eqn. (2.263). To evaluate it, we make a comparison between eqn. (2.200) and eqn. (2.263) and notice that if in the M_{fi} (Moller), we make the replacements

$$\bar{u}(4) \to \bar{v}(2) \Rightarrow p_4 \to -p_2$$
$$u(2) \to v(4) \Rightarrow p_2 \to -p_4 \qquad (2.265)$$

the M_{fi} for Moller scattering becomes the M_{fi} for Bhabha scattering. The minus sign in the above replacements of four-momenta is due to the fact that the electron (particle) spinor $u(p)$ obeys the momentum space Dirac equation

$$(\gamma^\mu p_\mu - m)m = 0 \qquad (2.266a)$$

while the positron (antiparticle) spinor $v(p)$ obeys

$$(\gamma^\mu p_\mu - m)v = 0 \qquad (2.266b)$$

the latter is obviously obtained from the former by replacing $m(p)$ by $v(p)$ and p_μ by $-p_\mu$.

Thus substituting the changes (2.265) in eqn. (2.233) we get

$$\overline{|M_{fi}|^2_{e^+e^-}} = \frac{e^4}{2m^4}\left[\frac{(p_1.p_2)^2+(p_1.p_4)^2-2m^2(p_1.p_2-p_1.p_4)}{(p_1-p_3)^4}\right.$$
$$+\frac{(p_1.p_3)^2+(p_1.p_4)^2+2m^2(p_1.p_3+p_1.p_4)}{(p_1+p_2)^4}$$
$$\left.+\frac{2(p_1.p_4)^2+4m^2p_1.p_4}{(p_1-p_3)^2(p_1+p_2)^2}\right] \qquad (2.267)$$

whose value in the center of the mass frame has now to be calculated. Figure (2.11) can be referred here also, and we shall have in the center of the mass frame in which

$$\boldsymbol{p}_1 = -\boldsymbol{p}_2 \equiv |\boldsymbol{p}| \equiv \boldsymbol{p}_3 = -\boldsymbol{p}_4,$$

$$p_1.p_2 = E^2 + \boldsymbol{p}^2$$
$$p_1.p_3 = E^2 - \boldsymbol{p}^2\cos\theta$$
$$p_1.p_4 = E^2 + \boldsymbol{p}^2\cos\theta$$
$$(p_1 - p_3)^2 = -2\boldsymbol{p}^2(1-\cos\theta) \qquad (2.268)$$
$$(p_1 + p_2)^2 = 4E^2$$

In Fig. (2.11), however, interchanges (2.265) are implied. The interchanges have already provided us the $\overline{|M_{fi}|^2}$ for e^+e^- scattering as eqn. (2.267), in which when we substitute the values (2.268), we get

$$\overline{|M_{fi}|^2_{e^+e^-}} = \frac{e^4}{2m^4}\left[\frac{(E^2+\boldsymbol{p}^2)^2+(E^2+\boldsymbol{p}^2\cos\theta)^2+2m^2\boldsymbol{p}^2(1+\cos\theta)}{4\boldsymbol{p}^4(1-\cos\theta)^2}\right.$$

$$+\frac{(E^2-\boldsymbol{p}^2\cos\theta)^2+(E^2+\boldsymbol{p}^2\cos\theta)^2+4m^2E^2}{16E^4}$$

$$\left.-\frac{2(E^2+\boldsymbol{p}^2\cos\theta)^2+4m^2\left(E^2+\boldsymbol{p}^2\cos\theta\right)}{8\boldsymbol{p}^2E^2\left(1-\cos\theta\right)}\right] \qquad (2.269)$$

To simplify the above equation we use

$$\begin{aligned}
(E^2+\boldsymbol{p}^2\cos\theta)^2 &= (E^2+\boldsymbol{p}^2)^2+4\boldsymbol{p}^4\sin^4\frac{\theta}{2}-4\boldsymbol{p}^2(E^2+\boldsymbol{p}^2)\sin^2\frac{\theta}{2}\\
(E^2-\boldsymbol{p}^2\cos\theta)^2 &= (E^2-\boldsymbol{p}^2)^2+4\boldsymbol{p}^4\sin^4\frac{\theta}{2}+4\boldsymbol{p}^2(E^2-\boldsymbol{p}^2)\sin^2\frac{\theta}{2}\\
\left(1-\cos\theta\right) &= 2\sin^2\frac{\theta}{2}\\
\boldsymbol{p}^2 &= E^2-m^2
\end{aligned} \qquad (2.270)$$

and write $\overline{\left|M_{fi}\right|^2_{e^+e^-}}$ as

$$\overline{\left|M_{fi}\right|^2_{e^+e^-}}=\frac{e^4}{2m^4}\left[\frac{A}{\sin^4\frac{\theta}{2}}-\frac{B}{\sin^2\frac{\theta}{2}}+C-D\sin^2\frac{\theta}{2}+G\sin^4\frac{\theta}{2}\right] \qquad (2.271)$$

where

$$A=\frac{2(E^2+\boldsymbol{p}^2)^2}{16\boldsymbol{p}^4}$$

$$B=\frac{4\boldsymbol{p}^2(E^2+\boldsymbol{p}^2)+4m^2\boldsymbol{p}^2}{16\boldsymbol{p}^4}+\frac{2(E^2+\boldsymbol{p}^2)^2+4m^2(E^2+\boldsymbol{p}^2)}{16\boldsymbol{p}^2E^2}$$

$$C=\frac{1}{4}+\frac{(E^2-\boldsymbol{p}^2)^2+(E^2+\boldsymbol{p}^2)^2+4m^2E^2}{16E^4}+\frac{8(E^2+\boldsymbol{p}^2)+8m^2}{16E^2}$$

$$D=\frac{8\boldsymbol{p}^4}{16E^4}+\frac{8\boldsymbol{p}^2}{16E^2}$$

and

$$G=\frac{8\boldsymbol{p}^4}{16E^4} \qquad (2.272)$$

which can be further solved by using $p^2 = E^2 - m^2$ or $E^2 = p^2 + m^2$ etc. to get the following forms.

$$B=\frac{8E^4-m^4}{8\boldsymbol{p}^2E^2}$$

$$C = \frac{12E^4 + m^4}{8E^4}$$

$$D = \frac{4\boldsymbol{p}^2\left(E^2 + \boldsymbol{p}^2\right)}{8E^4}$$

$$G = \frac{4\boldsymbol{p}^4}{8E^4} \tag{2.273}$$

Substituting these values along with the value of A, in equation (2.271), we get

$$\overline{\left|M_{fi}\right|^2_{e^+e^-}} = \frac{e^4}{16m^4}\left[\frac{\left(E^2+\boldsymbol{p}^2\right)^2}{\boldsymbol{p}^4}\cdot\frac{1}{\sin^4\dfrac{\theta}{2}} - \frac{8E^4 - m^4}{\boldsymbol{p}^2E^2}\cdot\frac{1}{\sin^2\dfrac{\theta}{2}} + \frac{12E^4+m^4}{E^4} - \frac{4\boldsymbol{p}^2\left(E^2+\boldsymbol{p}^2\right)}{E^4}\cdot\sin^2\frac{\theta}{2} + \frac{4\boldsymbol{p}^4}{E^4}\cdot\sin^4\frac{\theta}{2}\right] \tag{2.274a}$$

$\Rightarrow$

$$\overline{\left|M_{fi}\right|^2_{e^+e^-}} = \frac{e^4}{16m^4}W \tag{2.274b}$$

where

$$W = [\ \ ] \tag{2.274c}$$

in which [ ] contains the terms within the square bracket of equation (2.274a).

The differential cross section for the e^+e^- scattering (elastic), i.e. the Bhabha scattering, can now be obtained by substituting equations (2.274) into equation (2.264) to give

$$\left(\frac{d\sigma}{d\Omega}\right)^{e^+e^-}_{CM} = \frac{m^4}{4E^2(2\pi)^2}\cdot\frac{e^4}{16m^4}W = \frac{\alpha^2}{16E^2}W \tag{2.275}$$

where $\alpha = [e^2/4\pi] \approx \dfrac{1}{137}$ is the fine structure constant. The result (2.275) can also be expressed as

$$\left(\frac{d\sigma}{d\Omega}\right)^{e^+e^-}_{CM} = \frac{r_0^2m^2}{16E^2}\left[\frac{\left(E^2+\boldsymbol{p}^2\right)^2}{\boldsymbol{p}^4}\cdot\frac{1}{\sin^4\dfrac{\theta}{2}} - \frac{8E^4 - m^4}{\boldsymbol{p}^2E^2}\cdot\frac{1}{\sin^2\dfrac{\theta}{2}} + \frac{12E^4+m^4}{E^4} - \frac{4\boldsymbol{p}^2\left(E^2+\boldsymbol{p}^2\right)}{E^4}\cdot\sin^2\frac{\theta}{2} + \frac{4\boldsymbol{p}^4}{E^4}\cdot\sin^4\frac{\theta}{2}\right] \tag{2.276}$$

where $r_0 = \alpha/m$ is the classical radius of electron. As in the case of Moller scattering, the result (2.276) gives the relativistic differential Bhabha scattering cross section. The non-relativistic and ultra relativistic limits of above relativistic $(d\sigma/d\Omega)_{CM}$ may now be obtained.

In the non-relativistic limit, the three momentum $\boldsymbol{p}$ of the particles is small and hence $E^2 = p^2 + m^2$ becomes

$$E^2 \approx m^2$$

and

$$v^2 = \frac{\boldsymbol{p}^2}{E^2} = \frac{E^2 - m^2}{E^2} \tag{2.277}$$

where v is the velocity of say, positron, v is not the relative velocity of two particles e^+ and e^-. The non-relativistic approximations (2.277) imply that v^2 is small. These considerations lead to the vanishing of all terms except the first and it can be seen that the non-relativistic limit of eqn. (2.276) on using eqn. (2.277) is

$$\left(\frac{d\sigma}{d\Omega}\right)_{CM}^{e^+e^-}(NR) = \frac{r_0^2}{16v^4} \cdot \frac{1}{\sin^4 \frac{\theta}{2}} \tag{2.278}$$

A comparison of this non-relativistic result with the corresponding non-relativistic differential scattering cross section for the Moller scattering derived in eqn. (2.252) tells that in the case of Bhabha scattering there is no $\left[\cos^4 \frac{\theta}{2}\right]^{-1}$ term which in Moller scattering indicated the symmetry of the expression under $\theta \to \pi - \theta$, because of the indistinguishability of the particles (the electrons). In Bhabha scattering the positron and electron being distinguishable, the $\left(\cos^4 \frac{\theta}{2}\right)^{-1}$ term and hence the interference term as well, have to be absent. Therefore it is only the Rutherford term, which determines the non-relativistic differential Bhabha scattering cross section.

In the ultra relativistic limit of eqn. (2.276) we shall have

$$\frac{m}{E} \to 0 \tag{2.279a}$$

which implies

$$\boldsymbol{p}^2 \approx E^2, \tag{2.279b}$$

consequently, the relativistic expression (2.276) under above ultra relativistic (UR) limits becomes

$$\left(\frac{d\sigma}{d\Omega}\right)_{CM}^{e^+e^-}(UR) = \frac{\alpha^2}{16E^2}\left[\frac{4}{\sin^4\frac{\theta}{2}} - \frac{8}{\sin^2\frac{\theta}{2}} + 12 - 8\sin^2\frac{\theta}{2} + 4\sin^4\frac{\theta}{2}\right]$$

$$= \frac{\alpha^2}{4E^2}\left[\sin^2\frac{\theta}{2} + \frac{1}{\sin^2\frac{\theta}{2}} - 1\right]^2$$

$$= \frac{\alpha^2}{4E^2}\left[\frac{4-\sin^2\theta}{4\sin^2\frac{\theta}{2}}\right]^2$$

$$= \frac{\alpha^2}{64E^2}\frac{\left(3+\cos^2\theta\right)^2}{\sin^4\frac{\theta}{2}} \tag{2.280}$$

Comparing this result with the ultra relativistic differential Moller scattering cross section [eqn. (2.255c)] we find that in *UR* limit, they differ only by a factor of $\cos^4\frac{\theta}{2}$ that is

$$\left(\frac{d\sigma}{d\Omega}\right)_{UR}^{\text{Bhabha}} = \cos^4\frac{\theta}{2}\cdot\left(\frac{d\sigma}{d\Omega}\right)_{UR}^{\text{Moller}} \tag{2.281}$$

where both the cross sections are in center of mass frame.

The theoretical results obtained above for Moller and Bhabha scatterings are found to be in excellent agreement with experimental results.

2.4 ASSIGNMENTS

1. Explain the following
   (a) Time ordered operators and propagators
   (b) Differential scattering cross section
   (c) Fermi's Golden rule
2. Explain the significance of the matrix element in decay or a scattering process. Using a Feynman diagram and explaining the rules, work out the matrix element of a general scattering process. How do the matrix elements differ if the participants are (a) bosons (b) fermions?
3. Describe Compton scattering and obtain its relativistic scattering cross section. Show that in the non-relativistic limit, the relativistic scattering cross section reduces to classical Rayleigh—Thomson result.
4. Deduce Klein—Nishina relation for the scattering of polarized photons by unpolarized electrons. Discuss the effect of polarization of electrons on the above relation.

5. Obtain relativistic differential cross section for electron—positron pair annihilation and show that in the non-relativistic limit, the cross section increases with the decrease in positron velocity.
6. Show that for the process

$$e^+ + e^- \rightarrow \gamma + \gamma$$

the square of the matrix element is

$$|M_{fi}|^2 = \frac{e^2}{2m^2}\left[\frac{\omega_4}{\omega_3} + \frac{\omega_3}{\omega_4} - 4(\varepsilon_3.\varepsilon_4)^2 + 2\right]$$

7. Describe the mechanism involved in the formation of bound e^+e^- state. Provide a qualitative discussion of positronium annihilation. What are ortho and para positronium states?
8. An electron beam of energy E undergoes elastic scattering by a heavy nucleus. Calculate
(1) The maximum momentum transfer.
(2) The energy and momentum of the backwardly scattered nucleus.
[**Answer:** Let the four-momentum of projectile (electron) before interaction is $p = \left(\frac{E}{c}, \boldsymbol{p}\right)$ after interaction $p' = \left(\frac{E'}{c}, \boldsymbol{p}'\right)$. Similarly, for the target nucleus $P = (Mc, 0)$, $P' = \left(\frac{E'_p}{c}, \boldsymbol{p}'\right)$. The energy momentum conservation leads to $E.Mc^2 = E'E - \boldsymbol{pp}'c^2 + E'Mc^2 - m^2c^4$ (1) (m = mass of electron) and using the approximation $E \approx |\boldsymbol{p}|.c$ for $E >> mc^2$

$$E' = \frac{EMc^2}{E(1-\cos\theta) + Mc^2} \qquad (2)$$

The momentum transfer is $q^2 = -(p - p')^2$. Using equation (1)
$q^2 = 2M(E - E')$, q^2 is largest for smallest E', i.e., when $\theta = 180°$

$$(E')_{\text{smallest}} = \frac{EMc^2}{2E + Mc^2}$$

From equations (3) and (4) $(q)^2_{\max} = \dfrac{4ME^2}{2E + Mc^2}$

Ans. (2): From eqn. (2) for $\theta = 180°$, the energy transfer is

$$E_T = E - E' = \frac{2E^2}{2E + Mc^2}$$

The energy of backwardly scattered nucleus is $E_T + Mc^2$. The maximum momentum transfer is

$$|\boldsymbol{P}'| = \sqrt{q_{\max}^2 + \frac{E_T^2}{c^2}} = \sqrt{\frac{4ME^2}{2E + Mc^2} + \frac{4E^4}{c^2(2E + Mc^2)^2}}$$

9. Similar to problem (8), obtain the energy and momentum of backwardly scattered nucleus if photon of energy E is scattered by the nucleus (heavy). This is called nuclear Compton Effect.

   [**Hint:** Use $h\nu = \dfrac{h\nu_0 Mc^2}{Mc^2 + h\nu_0(1 - \cos\theta)}$ which for $h\nu \to E'$, $h\nu_0 \to E$ becomes equation (2) of problem (8). Hence the results will be same as in problem (8)]

10. Show that in Compton scattering, the energy of electron is maximum in the forward direction and is equal to

$$E = \frac{h\nu_0}{1 + (mc^2 / 2h\nu_0)}$$

   [**Hint:** Kinetic energy of electron $= \dfrac{h\nu_0 . 2\alpha \ \cos^2\theta}{(1+\alpha)^2 - \alpha^2 \cos^2\theta}$

   where $\alpha = (h\nu_0/mc^2)$, θ scattering angle]

11. Deduce the differential scattering cross section for Moller scattering in centre of mass frame. Show that the non-relativistic limit of this scattering cross section is Mott's formula. Analyse the ultra relativistic limit of the deduced differential scattering cross section.
12. Show that the matrix element of the Moller scattering in the centre of mass frame is expressed as

$$\overline{|M_{fi}|^2_{CM}}$$

$$= \frac{e^4}{4m^4}\left(\frac{2E^2 - m^2}{E^2 - m^2}\right)^2 \left[\frac{4}{\sin^4\theta} - \frac{3}{\sin^2\theta} + \left(\frac{E^2 - m^2}{2E^2 - m^2}\right)^2 \left(1 + \frac{4}{\sin^2\theta}\right)\right]$$

   and explain the meanings of symbols.
13. Evaluate the traces that appear in Moller scattering.
14. Differentiate between Moller and Bhabha scatterings. Mention their classical analogues. Compute the differential scattering cross section for Bhabha scattering and obtain its non-relativistic and ultra relativistic limits.
15. Show that the matrix element for the Bhabha scattering is expressed as equation (2.267).
16. Derive equations (2.270).
17. Show that the replacements (2.265) made in the matrix element of Moller scattering lead to the matrix element of Bhabha scattering. Analyse equation (2.265).

2.5 OBJECTIVE QUESTIONS

1. Fermi's Golden Rule expresses the
   (a) rate of transition of a system from the initial state to final state.
   (b) total transition probability that the system goes from its initial to final state.
   (c) total differential cross section for the system in going from one state to other.
   (d) total momentum transferred when a system goes from its initial to final state.
2. The uncertainty relation $\Delta t\,\Delta E \geq \hbar$, implies that
   (a) Δt and ΔE are not related to each other.
   (b) E is not the canonical conjugate of t.
   (c) for an accurate determination of energy, infinite time is required.
   (d) accurate measurement of E can be made during any time interval.
3. Fermi's Golden Rule is expressed as

   (a) Transition rate $= \dfrac{2\pi}{\hbar}\left|M_{fi}\right|^2 \times$ phase space.

   (b) $\dfrac{d\sigma}{d\Omega} = \left(\dfrac{\hbar}{8\pi Mc}\right)^2 \left\langle \left|M_{fi}\right|^2 \right\rangle$

   (c) $\Gamma = \dfrac{4\pi}{\cdot c}\left(\dfrac{\hbar\alpha}{m}\right)^2 \left|\psi(0)\right|^2$

   (d) $\left|M_{fi}\right|^2 = \left|M_{fi}^1\right|^2 + \left|M_{fi}^2\right|^2 + M_{fi}^1 M_{fi}^{2*} + M_{fi}^{1*} M_{fi}^2$
4. M_{fi} and phase space are known as
   (a) matrix element and density of initial states respectively.
   (b) matrix element and density of final states respectively.
   (c) matrix element and amplitude respectively.
   (d) amplitude and dynamical factor respectively.
5. Matrix element is also called
   (a) amplitude (b) kinematical factor
   (c) transition rate (d) impact parameter
6. The formula

$$d\sigma = \left|M_{fi}\right|^2 \frac{\hbar^2 S}{4\sqrt{(p_1 \cdot p_2)^2 - (m_1 m_2 c^2)^2}} \times$$

$$\left[\left(\frac{cd^3\boldsymbol{p}_3}{(2\pi)^3 2E_3}\right)\left(\frac{cd^3\boldsymbol{p}_4}{(2\pi)^3 2E_4}\right)\cdots\left(\frac{cd^3\boldsymbol{p}_n}{(2\pi)^3 2E_n}\right)\right]$$

$$\times (2\pi)^4 \delta^4 (p_1 + p_2 - p_3 - p_4 \cdots - p_n)$$

where, $p_i = (\frac{E_i}{c}, \boldsymbol{p}_i), S = \frac{1}{j!}$ with j, number of identical particles in the final state, is the Golden rule for

(a) decay $1 \to 2 + 3 + 4 + ... + n$.
(b) scattering $1 \to 2 + 3 + 4 + ... + n$.
(c) decay $1 \to 2 + 3$
(d) scattering $1 + 2 \to 3 + 4$

7. If p_1, p_2 and p_3 represent the four-momenta and $p_1 = 0$, $E_1 = mc^2$, then $\delta^4(p_1 - p_2 - p_3)$ is equal to

(a) $\delta\left(mc - \frac{E_2}{c} - \frac{E_3}{c}\right)\delta^3(\boldsymbol{p}_2, \boldsymbol{p}_3)$

(b) $\delta\left(mc - \frac{E_2}{c} - \frac{E_3}{c}\right)\delta^3(-\boldsymbol{p}_2, -\boldsymbol{p}_3)$

(c) $\delta\left(\frac{E_2}{c} - \frac{E_3}{c}\right)\delta^3(-\boldsymbol{p}_2, -\boldsymbol{p}_3)$

(d) $\delta\left(\frac{E_2}{c} + \frac{E_3}{c}\right)\delta^3(\boldsymbol{p}_2, \boldsymbol{p}_3)$

8. Which of the following statements is true?
   (a) If two particles collide with momenta equal in magnitude but opposite in direction, we call this a centre of mass frame and if one of the particles is at rest the collision is viewed in laboratory frame.
   (b) In centre of mass frame, one of the particles is at rest and in laboratory frame the particles have equal but opposite momenta.
   (c) In the centre of mass frame the momenta are equal but not opposite and in the laboratory frame one of the particles is at rest.
   (d) In the centre of mass frame the momenta are opposite but not equal and in the laboratory frame the momenta are equal and opposite.

9. The value of $\sqrt{(p_1 \cdot p_2)^2 - (m_1 m_2 c^2)^2}$ (question 6) in centre of mass and laboratory frames (particle 1 in motion) is

(a) $\frac{(E_1 + E_2)|\boldsymbol{p}_1|}{c}$ and $m_2 c|\boldsymbol{p}_1|$ respectively.

(b) $m_2 c|\boldsymbol{p}_2|$ and $\frac{(E_1 + E_2)|\boldsymbol{p}_2|}{c}$ respectively.

(c) $\frac{E_2|\boldsymbol{p}_1|}{c}$ and $m_2 c|\boldsymbol{p}_2|$ respectively.

(d) $\frac{E_2|\boldsymbol{p}_2|}{c}$ and $m_2 c|\boldsymbol{p}_1|$ respectively.

10. In an electromagnetic interaction the exchanged particle is photon. This photon is
    (a) real, because it can be observed.
    (b) real, because the energy conservation takes place at its emission and absorption points.
    (c) virtual, because the square of its four-momentum $q^2 = 0$.
    (d) virtual, because the square of its four-momentum $q^2 \neq 0$.
11. Which statement is true among the following?
    (a) For a real particle $E^2 - \mathbf{p}^2c^2 = m^2c^4$, but for a virtual particle $E^2 - p^2c^2$ = any value.
    (b) For a real particle $E^2 - p^2c^2$ = any value, but for a virtual particle $E^2 - p^2c^2 = m^2c^4$.
    (c) Real particles possess electric charge while the virtual particles are neutral.
    (d) Real particles are neutral, whereas all the virtual particles are charged.
12. The propagator
    (a) propagates real states in an interaction process forward in time.
    (b) propagates virtual states in an interaction process forward in time.
    (c) propagates both real and virtual states forward in time.
    (d) propagates all states backward in time.
13. The propagators for various fields are described by the
    (a) vacuum expectation value of field operators.
    (b) normal ordered vacuum expectation value of field operators.
    (c) time ordered vacuum expectation value of field operators.
    (d) time ordered expectation value of field operators for any state.
14. In Column I below, there are multiplicative factors for constructing matrix element and in Column II are parts of the Feynman diagram. Match them and select the correct code.

Column I	*Column II*
A $-ie\gamma^{\mu}$	1 *Bosonic external lines.*
B $-ig_{\mu\nu}/k^2$	2 *Final antifermion external line*
C $\dfrac{i\ (\gamma_{\mu}k^{\mu}+m)}{m^2-k^2}$	3 *0 – 0 QED vertex.*
D $-ie(p_1+p_2)^{\mu}$	4 *Final antiboson external line.*
E $v(p)$	5 $\frac{1}{2}-\frac{1}{2}$ *QED vertex.*
F ε^{μ}	6 *Fermion internal line.*
G 1	7 *Photon internal line.*

Codes:

(a)	A	B	C	D	E	F	G
	5	7	6	2	3	1	4
(b)	A	B	C	D	E	F	G
	5	6	7	1	2	3	4
(c)	A	B	C	D	E	F	G
	5	7	6	3	2	4	1
(d)	A	B	C	D	E	F	G
	7	6	5	3	2	1	4

15. Which of the following statements regarding Compton effect is not true?
   (a) The wavelength of the scattered photon is greater than or equal to the wavelength of incident photon.
   (b) The electron can acquire a kinetic energy equal to the energy of incident photon.
   (c) The energy of incident photon equals the kinetic energy of the electron plus the energy of the scattered photon.
   (d) The kinetic energy acquired by the electron is largest when the incident and scattered photons move in opposite directions.
16. If q represents the four momentum of a photon, then which of the following is true.

   (a) $q^2 = 0$ implying $m_\gamma = 0.$

   (b) $q^2 \neq 0$ implying $m_\gamma = 0.$

   (c) $q^2 = 0$ implying $m_\gamma \neq 0.$

   (d) $q^2 \neq 0$ implying $m_\gamma = 0.$

17. Choose the correct relationship.

   (a) $\not{a}\not{b} = \not{b}\not{a} + 2ab$ hence $\not{p}\not{p} = p^2$

   (b) $\not{a}\not{b} = -\not{b}\not{a} + 2ab$ hence $\not{p}\not{p} = p^2$

   (c) $\not{a}\not{b} = -\not{b}\not{a} + 2ab$ hence $\not{p}\not{p} \neq p^2$

   (d) $\not{a}\not{b} = 2ab$ hence $\not{p}\not{p} = p^2$

18. The trace of the product of n gamma matrices is zero if
   (a) n is greater than 4 (b) n is any even number
   (c) n is any odd number (d) n is any even or odd number
19 Given the relation, Trace $(\gamma^\mu\gamma^\nu\gamma^\lambda\gamma^\sigma) = 4\ (g^{\mu\nu}g^{\lambda\sigma} - g^{\mu\lambda}g^{\nu\sigma} + g^{\mu\sigma}g^{\nu\lambda})$, the value of Trace $(\gamma^\mu \not{p}_1 \gamma^\nu \not{p}_3)$ is equal to

   (a) $4[p_1^\mu p_3^\nu + g^{\mu\nu}(p_1 \cdot p_3) - p_3^\mu p_1^\nu]$

   (b) $4[p_1^\mu p_3^\nu - g^{\mu\nu}(p_1 \cdot p_3) + p_3^\mu p_1^\nu]$

   (c) $4[p_1^\mu p_3^\nu + p_3^\mu p_1^\nu]$

   (d) $4[p_3^\mu p_{3\mu} + p_1^\mu p_{1\mu}]$
20. The relation

$$\frac{d\sigma}{d\Omega} = \frac{\alpha^2}{4m^2}\left(\frac{\omega_3}{\omega_2}\right)^2\left[\frac{\omega_2}{\omega_3} + \frac{\omega_3}{\omega_2} + 4(\varepsilon_2.\varepsilon_3^*)^2 - 2\right]$$ is the

(a) Klein—Nishina formula for the scattering of unpolarized photon by unpolarized electron.
(b) Klein—Nishina formula for the scattering of polarized photon by unpolarized electron.
(c) Klein—Nishina formula for the scattering of polarized photon by polarized electron.
(d) Gell—Mann—Nishijima formula.

21. $|M_{fi}|^2 = \frac{e^4}{2m^2}\left[\frac{\omega_4}{\omega_3} + \frac{\omega_3}{\omega_4} - 4(\varepsilon_3.\varepsilon_4^*)^2 + 2\right]$ is for
(a) $e^- + \gamma \rightarrow e^- + \gamma$
(b) $e^+ + e^- \rightarrow \gamma + \gamma$
(c) $e^+ + e^- \rightarrow e^+ + e^-$
(d) $e^- + e^- \rightarrow e^- + e^-$

22. Positronium is a bound state of e^+ and e^- in which
(a) e^+ orbits around e^-.
(b) e^- orbits around e^+.
(c) both orbit around a common center.
(d) sometimes e^- orbits around e^+ and sometimes e^+ around e^-.

23. Which of the following is true?
(a) In Moller scattering there is electron propagator, whereas in Compton scattering we have photon propagator.
(b) In Moller scattering we have photon propagator, whereas in Compton scattering we have electron propagator.
(c) We have electron propagator in both.
(d) There is photon propagator in both the scatterings.

24. Positronium annihilation takes place from its
(a) s state (b) p state
(c) d state (d) f state

25. The total matrix element M for Moller scattering is obtained by M_1 and M_2 corresponding to its two types of Feynman diagrams. What is true?
(a) $M = M_1$ (b) $M = M_2$
(c) $M = M_1 + M_2$ (d) $M = M_1 - M_2$

26. Trace $[-2\not{p}_3\gamma^\alpha \not{p}_2\, \not{p}_1\gamma_\alpha \not{p}_4]$ is equal to
(a) $-8(p_1.p_2)$ (b) $16(p_1.p_2)^2$
(c) $-32(p_1.p_2)^2$ (d) $48(p_1.p_2)$

27. Bhabha scattering is
(a) $e^- + e^- \rightarrow e^- + e^-$
(b) $e^+ + \gamma \rightarrow e^+ + \gamma$
(c) $e^+ + e^- \rightarrow e^+ + e^-$
(d) $e^+ + e^- \rightarrow \gamma + \gamma$

28. Which is the correct expression?

(a) $$\left(\frac{d\sigma}{d\Omega}\right)_{\text{UR}}^{\text{Bhabha}} = \sin^4\frac{\theta}{2}\left(\frac{d\sigma}{d\Omega}\right)_{\text{UR}}^{\text{Moller}}$$

(b) $$\left(\frac{d\sigma}{d\Omega}\right)_{\text{NR}}^{\text{Bhabha}} = \left(\frac{d\sigma}{d\Omega}\right)_{\text{NR}}^{\text{Moller}}$$

(c) $$\left(\frac{d\sigma}{d\Omega}\right)_{\text{UR}}^{\text{Bhabha}} = \cos^4\frac{\theta}{2}\left(\frac{d\sigma}{d\Omega}\right)_{\text{NR}}^{\text{Moller}}$$

(d) $$\left(\frac{d\sigma}{d\Omega}\right)_{\text{UR}}^{\text{Bhabha}} = \cos^4\frac{\theta}{2}\left(\frac{d\sigma}{d\Omega}\right)_{\text{UR}}^{\text{Moller}}$$

ANSWERS

1a, 2c, 3a, 4b, 5a, 6b, 7b, 8a, 9a, 10d, 11a, 12b, 13c, 14c, 15b, 16a, 17b, 18c, 19b, 20b, 21b, 22c, 23a, 24b, 25d, 26c, 27c, 28d.

3

SYMMETRIES AND GROUPS

3.1 INTRODUCTION

The simplest definition of symmetry is that – it is the non-distinguishability of an object or structure across a dividing line or a point. This implies that by doing something (here by bringing in the dividing line or the point) we wish to observe the change in the object. If the operation (doing something) does not bring about any change in the object, the object is said to be symmetric under that particular operation. We have discussed in Appendix C, that if a system remains unchanged (invariant) under the operations of translations in time, translations in space and rotation about a point, the system is said to be symmetrical with respect to these operations and follow from these three symmetries, the conservation laws of energy, linear momentum and angular momentum respectively.

In this Chapter we focus our attention on the properties of operations or transformations which when applied on the system, do not change it. Such transformations are called *symmetry transformations* and we say in other words that a system remains invariant under the application of symmetry transformations. If the system remains invariant on applying the translations or rotations, the system is said to be having a translational or a rotational symmetry. By the phrase—the system remains invariant—we generally refer to the laws of nature or say laws of physics that remain invariant under symmetry transformations. As discussed in Appendix C, the Lagrangian (or Hamiltonian) and hence the equations of motion are invariant under the symmetry transformations. The continuous and discrete nature of symmetries too has been described in Appendix E.

Now consider that we have a set of symmetry transformations in which each transformation can be seen as an element of the set. Obviously, each transformation leaves the system invariant. It is found that the set of symmetry transformations obeys the properties of a group and we call such groups as symmetry groups. These groups are very important in physics and chemistry. In the following we will briefly describe the properties of various symmetry groups and also their representations.

3.2 GROUPS AND THEIR PROPERTIES

We take the general transformations and consider a set of transformations in which each transformation constitutes an element of the set. If the set possesses the properties of closure, identity, inverse and associativity, it is called a *group*. These properties are explained below. Let us denote the *i*th transformation by the element M_i.

1. Closure

If M_i and M_j are in the set, then the product M_iM_j is also in the set. This implies that there exists some $M_k = M_iM_j$. By M_iM_j we mean that we first perform the transformation M_j and then M_i. Thus the existence of the composite M_iM_j in the set describes its closure property.

2. Identity

If in the set there is an identity element I such that $IM_i = M_i$, $IM_i = M_iI = M_i$, for all elements M_i of the set, it is said to possess the property of having identity element.

3. Inverse

If for every element M_i of the set, there is an inverse, M_i^{-1} such that $M_iM_i^{-1}$ $M_i = I$, the set is said to possess the property of having inverses.

4. Associativity

The set would possess this property if the elements obey

$$M_i(M_jM_k) = (M_iM_j)M_k \tag{3.1}$$

In addition to the above defining properties, a group may possess properties defined by the following terms.

1. Abelian/ Non-Abelian Groups

If the elements of a group commute, i.e., if $M_iM_j = M_jM_i$, the group is called *Abelian*. If $M_iM_j \neq M_jM_i$, the group is called *non-Abelian*.

2. Finite/Infinite Groups

If the number of elements of a group is finite, the group is called *finite group* and a group with infinite number of elements is called an *infinite group*.

3. Discrete/Continuous Groups

If the group is infinite, it may be discrete or continuous. When the infinite number of elements in a group is denumerably infinite, the group is called *discrete*. If the elements are non-denumerably infinite, the group is called *continuous*. By denumerable, it is meant that the number of elements can be counted by correspondence with infinite set of integers. ... $-4, -3, -2, -1, 0, 1, 2, 3, 4$... is the infinite set of integers. The group of all integers, in which addition replaces the multiplication to define the closure property, is a discrete group and the group of all real numbers or a group of all rotations in a plane is continuous. Obviously, all finite groups are discrete.

A group is said to be continuous if the elements of the group observe the definition of nearness or continuity. The elements of a continuous group can be characterized by or labeled by a set of real parameters $a_1, a_2, a_3, ..., a_n$, in which at least one parameter varies continuously over a certain interval. For example the set of transformations

$$x' \; ax + b, \;\; a \neq 0 \tag{3.2}$$

in which a and b are two parameters, forms a group. The parameters a and b vary continuously from $-\infty$ to $+\infty$ and we say that the group of transformations (3.2) is a two parameter continuous group. The set of real parameters must be such that all the elements of the group can be characterized by it. If the number of continuous parameters is r where $1 \leq r \leq n$, we call that r is the order of continuous group. It may be recalled that if the infinitesimal change in one of the factors of the product produces only an infinitesimal change in the product, the definition of continuity is said to have been observed. Under this definition of continuity if the group manifold forms a topological space, the group is said to be a topological group. The simplest topological groups are those in which the elements can be put in one to one correspondence with the points of a subset of a $r-$dimensional real inner product space. This subset is called a *parameter space.*

If the two elements of a topological group denoted by e_1 and e_2 have one to one correspondence with points $I(e_1)$ and $I(e_2)$ in the parameter space and if these points in the parameter space can be connected by one or more paths lying within the parameter space only, the parameter space is said to be connected. If no such paths exist within the parameter space, it is disconnected. Accordingly then a group is said to be connected if its parameter space is connected otherwise the group is called *disconnected.* The range of variation of parameters may vary between $-\infty$ to $+\infty$ or may be confined to some finite domain. If the domain of the variation of parameters is finite, the group manifold is called *closed.* If the parameter space of a topological group is closed and bounded, the parameter space is said to be a compact space and topological group a **compact group**.

4. Simple and Semisimple Groups

The notions of simple and semisimple groups play an important role in group theory. To define them, we have to understand the invariant subgroups. It is easy to understand the meaning of a subgroup. If we select from the elements of a group G a subset H, i.e., H is contained in G and set H itself forms a group under the same law of combination that was used in G, then H is said to be a subgroup of G.

Starting from a subgroup H of G we may denote any element of H by h and write g for any element of G. Spanning over all the elements of H we can construct a set of elements $\{ghg^{-1}\} = gHg^{-1}$. This set of elements gHg^{-1} is again a subgroup of G and is called a conjugate subgroup of H in G. Different elements of G, i.e., different g_s would form different conjugate subgroups. It may happen that

$$gHg^{-1} = H \tag{3.3}$$

implying that all the conjugate subgroups of H in G are identical with H. When this happens, we say that the subgroup H is an *invariant subgroup* or *self-conjugate subgroup* in G.

A group G is said to be simple, if it has no invariant subgroups. A group is said to be semisimple, if it has no Abelian invariant subgroups. Important groups of elementary particle physics like the unitary symmetry group and the Lorentz group belong to these categories.

3.3 LIE GROUPS

Lie groups are those continuous groups whose parameter space is locally Euclidean. By locally Euclidean, we mean that the parameter space within the infinitesimal vicinity of any element of the continuous group is taken as finite dimensional Euclidean space. We may recall that the elements of an n-parameter continuous group are labeled by n continuous real parameters. That is, any element $R(a)$ of the group is $R(a) = R(a_1, a_2, ..., a_n)$ and the elements of the continuous groups are in one to one correspondence with the points of a subset of n-dimensional real inner product (scalar product) space, which is called the *parameter space*.

In other words, we may say that a continuous group G is called an n-parameter Lie group if a certain *nearness* of its elements can be mutually uniquely and mutually continuously mapped into a certain region of n-dimensional real Euclidean space. The n-parameter Lie group is also called a *Lie group of dimension n*. Practically, all the continuous groups appearing in particle physics are Lie groups and generally, the most important ones are the U(n), SU(n), O(n) and SO(n) groups. We briefly describe them.

1. Unitary group in *n* dimensions U(n)

A set of all $n \times n$ unitary matrices, obeying the defining properties of a group, forms the U(n) group. The unitary matrices, for which the inverse is equal to transpose conjugate; $U^{-1} = \tilde{U}^*$, constitute the elements of the U(n) group. For $n = 1$, the group is
U(1), which is Abelian. For $n > 1$, U(n) group is non-Abelian. The number of essential real parameters for this group is n^2, which vary over a finite range and hence the Lie group U(n) is closed. The U(1) group consists of 1×1 unitary matrices, i.e., they are phase transformations $e^{i\theta}$. The U(1) group appears as the Abelian group of gauge (phase) transformations connected with the conservation of additive charges. The groups SU(n) and O(n) are the subgroups of U(n).

2. Special Unitary group in *n* dimensions SU(n)

If the unitary matrices of U(n) group have their determinant equal to one, they are called *special (det U = 1) matrices*, and the group whose elements are these special (S) matrices is called *the SU(n) group*. In the SU(n) family, we frequently come across with SU(2), SU(3), SU(4), SU(5) and SU(6) groups which are the special unitary groups respectively in two, three, four, five and six dimensions.

SU(2) group is important as it can describe both spin and isospin. We may therefore have SU(2) spin and SU(2) isospin groups. SU(3) group is the generalization of SU(2) isospin group where we include hypercharge. There is SU(3) of flavour (u, d, s) and SU(3) of colour (R, G, B). When spin is included in the consideration of SU (2) isospin group, we have the SU (4) group of isospin and spin. The unification of strong, weak and electromagnetic interactions requires the SU(5) group. When SU(3) of flavour and SU (2) of spin are combined (direct product), we get the SU(6) group, which forms the basis of quark model of hadrons. We shall separately deal with SU(2) and SU(3) groups.

3. Orthogonal group in *n* dimension O(n)

If the group were formed by $n \times n$ real unitary matrices, we would have the O(n) group. An orthogonal (O) matrix obeys, *inverse of* O = *transpose of* O. If the determinant of an O(n) group is equal to one, such O(n) group too becomes special and called *SO(n) group*. O(n) group is a compact Lie group which is however, not connected.

SO(n) is obviously its subgroup. The group SO(3)—the special orthogonal group in three dimensions describes the rotational symmetry of our world. The group SU(2) is homomorphic with SO(3).

3.3.1 Generators of Lie Groups and the Lie Algebra

An element of a Lie group is continuous and continuously differentiable (analytic) function of its parameters and in the infinitesimal vicinity of an element of a Lie group the parameter space is Euclidean. In the Lie group, usually, we consider the infinitesimal vicinity of the identity element of the group. Any element of the Lie group is expressed in terms of the n-continuous real parameters as $R(a) = R(a_1, a_2, ..., a_n)$ and the identity element e is chosen as $e = R(0, 0 ..., 0) \equiv R(0)$. Now, we can write the element that is infinitesimally close to e and then we may write the next close element and carrying out this process successively, we can generate an element of the Lie group which is at a finite distance from the identity element e. Let the element generated at finite distance from e be denoted by $R(0, 0\ a_j, ..., 0)$. Starting from the identity element $R(0)$, the element $R(0, 0, a_j, ..., 0)$ would be generated, say in N steps, where N is a large positive integer and in each step we generate an element ε_j (infinitesimal) close to preceding element. The first element generated ε_j close to identity element $R(0)$ can be written as

$$R(0, 0, \varepsilon_j, ..., 0) = R(0) + i\varepsilon_j X_j(0, 0, ..., 0) \tag{3.4}$$

where

$$X_j = \lim_{\varepsilon_j \to 0}\left[\frac{1}{i\varepsilon_j}\{R(0,0,...,\varepsilon_j 0) - R(0)\}\right] \equiv -i\frac{\partial R(0,0,...,\varepsilon_j 0)}{\partial \varepsilon_j}\bigg|_{\varepsilon_j = 0} \tag{3.5}$$

From the LHS of eqn. (3.4), we get the element R(0, 0, ..., 0), in N steps, so we may write

$$a_j = N\varepsilon_j \tag{3.6}$$

and hence

$$\begin{aligned} R(0,0,...a_j,...,0) &= [R(0,0,..., \varepsilon_j,...,0)]^N \\ &= [R(0) + i\varepsilon_j X_j]^N \\ &= [e + i\frac{a_j}{N} X_j]^N \end{aligned} \tag{3.7}$$

where $R(0) = e$ and eqn. (3.6) have been used. Allowing $N \to \infty$ and using the identity

$$\lim_{N\to\infty}\left(1 + \frac{\theta}{N}\right)^N = \exp\theta \tag{3.8}$$

we may write

$$R(0, 0, ..., a_j, ..., 0) = \exp(ia_j X_j) \tag{3.9}$$

and hence for the general element of the Lie group we have

$$R(a_1, a_2, ..., a_n) = \exp\left[\sum_{j=1}^{n} ia_j X_j\right] \tag{3.10}$$

where X_j can be written as (like eqn 3.5)

$$X_j = -i\frac{\partial R(a_1, a_2, ..., a_n)}{\partial a_j}\bigg|_{a_1 = a_2 = ... = a_n = 0} \tag{3.11}$$

Thus, X_j (eqn.3.11) generate the elements of the Lie group. We, therefore call X_j the infinitesimal generators of the Lie group. In fact the elements generated above belong to the subset, which contains the identity. The (infinitesimal) generators X_j are also called simply the generators of the Lie group. If the Lie group has n continuous parameters, the number of its generators would also be n. It is desired that in physically interesting cases the generators X_j (eqn. 3.11) become the Hermitian operators of physical quantities like energy, momentum, spin, etc., an imaginary unit (i) is introduced in the expression (3.11). The generators of some important Lie groups have been given in the following.

(a) U(n) group: It is $n^2 n^2 = N$ parameter group, so we shall have N generators. Any unitary matrix of the order n can be written as

$$U = \exp\left[i\sum_{j=1}^{N} a_j K_j\right] \tag{3.12}$$

where $K_1, K_2, ..., K_j, ..., K_N$ is a set of n^2 independent Hermitian matrices of order n and a_j are N real parameters. By giving all possible values to a_j, all the elements of U(n) can be generated from the RHS of eqn. (3.12). Comparing eqn. (3.12) with eqn. (3.10) we can easily say that the generators of U(n) group are n^2 (we have put $n^2 = N$) independent Hermitian matrices K_j. For example, U(2) group will have $2^2 = 4$, (2×2) independent Hermitian matrices as its generators. These are:

$$I = \begin{bmatrix} 1 & 0 \\ 0 & 1 \end{bmatrix}, \quad \sigma_1 = \begin{bmatrix} 0 & 1 \\ 1 & 0 \end{bmatrix}, \sigma_2 = \begin{bmatrix} 0 & -i \\ i & 0 \end{bmatrix}, \sigma_3 = \begin{bmatrix} 1 & 0 \\ 0 & -1 \end{bmatrix} \tag{3.13}$$

where σ_1, σ_2, s_3 are the Pauli matrices.

(b) SU(n) group: The elements (matrices) of this group have their determinant = +1. Using eqn. (3.12), we may denote an element (matrix) of SU(n) by

$$U_0 = e^{iK_0} \tag{3.14}$$

Using det $U_0 = +1$ and for a square matrix K_0, det $e^{iK^0} = e^{trace\ K^0}$, we can see that K_0 is traceless. So the generators of SU(n) group must be traceless Hermitian matrices of order n. Now, the maximum number of traceless Hermitian matrices of order n can be $n^2 - 1$. Therefore, SU(n) group shall have $(n^2 - 1)$ traceless Hermitian matrices as its generators. SU(2) group shall have 3, traceless Hermitian matrices of order 2 (i.e. 2×2) as its generators. If the group is SU(2) spin, then the Pauli spin matrices $\frac{\sigma_1}{2}, \frac{\sigma_2}{2}, \frac{\sigma_3}{2}$ (eqn. 3.13) are the generators. However, for SU(2) isospin group $\frac{\tau_1}{2}, \frac{\tau_2}{2}, \frac{\tau_3}{2}$ matrices are the generators. τ matrices are similar to σ matrices; only to distinguish SU(2) isospin group from SU(2) spin group we use τ matrices in place of σ matrices. For SU(3) group we have 8, (3×3) traceless Hermitian matrices in terms of which its generators are expressed. They are called λ *matrices*. Following are the *eight* generators ($\frac{\lambda_a}{2}$ matrices) of SU(3) group.

$$\lambda_1 = \begin{bmatrix} 0 & 1 & 0 \\ 1 & 0 & 0 \\ 0 & 0 & 0 \end{bmatrix} \quad \lambda_2 = \begin{bmatrix} 0 & -i & 0 \\ i & 0 & 0 \\ 0 & 0 & 0 \end{bmatrix} \quad \lambda_3 = \begin{bmatrix} 1 & 0 & 0 \\ 0 & -1 & 0 \\ 0 & 0 & 0 \end{bmatrix} \quad \lambda_4 = \begin{bmatrix} 0 & 0 & 1 \\ 0 & 0 & 0 \\ 1 & 0 & 0 \end{bmatrix},$$

$$\lambda_5 = \begin{bmatrix} 0 & 0 & -i \\ 0 & 0 & 0 \\ i & 0 & 0 \end{bmatrix} \lambda_6 = \begin{bmatrix} 0 & 0 & 0 \\ 0 & 0 & 1 \\ 0 & 1 & 0 \end{bmatrix} \lambda_7 = \begin{bmatrix} 0 & 0 & 0 \\ 0 & 0 & -i \\ 0 & i & 0 \end{bmatrix} \lambda_8 = \frac{1}{\sqrt{3}}\begin{bmatrix} 1 & 0 & 0 \\ 0 & 1 & 0 \\ 0 & 0 & -2 \end{bmatrix} \tag{3.15}$$

It can be observed that (**a**) above matrices have been constructed from the Pauli matrices, which provide the generators of SU(2) and (**b**) only λ_3 and λ_3 are diagonal which commute with each other. This sets the rank of SU(3) as two.

(c) O(*n*) group: This is the group of those transformations, which leave

$$x_1^2 + x_2^2 + x_3^2 + ... + x_n^2 = \text{invariant} \tag{3.16}$$

Though there are n^2 parameters, the requirement (3.16) imposes $n + \frac{n(n-1)}{2} = \frac{n(n+1)}{2}$ conditions on them leaving $n^2 - \frac{n(n+1)}{2} = \frac{n(n-1)}{2}$ essential parameters of the group O(n). Thus there will be $\frac{n(n-1)}{2}$ generators of O(n) group.

The orthogonal group in three dimensions O(3) would leave $x^2 + y^2 + z^2$ invariant whereas O(2) would require $x^2 + y^2 = \text{const}$. Let a_1, a_2, a_3 and a_4 be the four (2^2) parameters of O(2) group. Then

$$x'^2 + y'^2 = (a_1x + a_2y)^2 + (a_3x + a_4y)^2 = x^2 + y^2 \tag{3.17}$$

Equating like terms on both sides we find that the parameters a_1, ..., a_4 obey the following three conditions

(1) $a_1^2 + a_3^2 = 1$

(2) $a_2^2 + a_4^2 = 1$

(3) $a_1a_2 + a_3a_4 = 0$ (3.18)

Thus the essential parameter of O(2) group is $4 - 3 = 1$ only. Similarly we can see that 9 parameters of O(3) group obey 6 conditions leaving only three essential parameters. The one-parameter O(2) group is the group of rotations about z-axis. The angle θ (the angle of rotation about z-axis) is that one parameter. $0 \le \theta \le 2\pi$. That is the transformations

$$x' = x \cos\theta - y \sin\theta$$
$$y' = x \sin\theta + y \cos\theta \tag{3.19}$$

form the O(2) group.

The generators of O(n) group would therefore be $\frac{n(n-1)}{2}$, that of O(2) only one and O(3) group will have three generators. O(3), is called the three dimensional rotation-inversion group. The pure three-dimensional rotational group is SO(3) whose three generators are the Hermitian matrices J_1, J_2 and J_3, which are the usual angular momentum operators.

Every Lie group has its Lie algebra. Let us suppose there is a Lie group with n continuous parameters a_4. The group will obviously have n generators X_4. Now if X_j and X_k are the two generators of the Lie group then their linear combination with real coefficients would also be a generator of the group. Recalling the properties of linear vector space, the generators X_k would also form a n-dimensional real linear vector space.

The Lie algebra of a Lie group is therefore a real n-dimensional linear vector space L with generators (X_i, X_j, X_k ...) as its elements where the law of composition for any two elements of L is denoted by the commutator $[X_i, X_j]$ such that

(1) $[X_i, X_j]$ belongs to L

(2) $[X_i, X_j] = -[X_j, X_i]$

(3) $[X_i, [X_j, X_k]] + [X_j, [X_k, X_i]] + [X_k, [X_i, X_j]] = 0$ (3.20)

for all X_i, X_j, X_k belonging to L. The third of the above is the Jacobi identity for the generators. It may be recalled that if the law of composition is—the product of two elements of Linear vector space instead of the commutator, we shall have just algebra and hence we may say that whereas in algebra taking the product of two elements is the law of composition, *in Lie algebra taking the commutator of two elements provides the law of composition*. For all practical purposes the commutation relation between the generators of a Lie group gives its Lie algebra.

The commutator of the generators of a Lie group vanishes, if the Lie group is Abelian and in that case the Lie algebra of the group (Abelian) is simplest. An example is the translation group, which is the group of translations along x, y, z, t axes. The group has four generators P_μ with $\mu = 0, 1, 2, 3$. The commutator

$$[P_\mu, P_\nu] \tag{3.21}$$

gives the Lie algebra of translation group. In general, however,

$$[X_i, X_j] = i\sum_{k=1}^{n} C_{ij}^k X_k \tag{3.22}$$

where C_{ij}^k are a set of numbers called the structure constants of the Lie group.

The hermiticity of the generators demands that C_{ij}^k be real. They are called structure constants because the set of structure constants determines completely the structure of Lie algebra, i.e., the local structure of the Lie group. The structure constants are a characteristic property of the Lie group; however, they are not unique because the generators themselves are not unique. The structure constants obey the properties

(*i*) $C_{ij}^k = -C_{ji}^k$ (3.23a)

(*ii*) $C_{ij}^m C_{mk}^p + C_{jk}^m C_{mi}^p + C_{ki}^m C_{mj}^p = 0$ (3.23b)

The first property describes that C_{ij}^{k} is antisymmetric in indices ijk and it can be written that

$$C_{ij}^{k} = C_{ijk} \tag{3.23c}$$

The second property (3.23b) follows from the Jacobi identity (3.20).

The Lie algebra of SO(3), the rotation group in three dimensions, is

$$\left[J_i, J_j\right] = i\varepsilon_{ijk} J_k \tag{3.24}$$

ε_{ijk} are the structure constants of the rotation group. ε_{ijk}, also called the Levi-Civita tensor, is totally antisymmetric in i, j, k each of which can take values 1, 2, 3. It has the property

$$\begin{aligned} \varepsilon_{ijk} &= +1 \quad \text{if } i, j, k \text{ take values in cyclic order} \\ &= -1 \quad \text{if } i, j, k \text{ are anticyclic} \\ &= 0 \quad \text{if any of the two indices take the same value} \end{aligned} \tag{3.25}$$

Hence we may write

$$\begin{aligned} \varepsilon_{123} &= \varepsilon_{312} = \varepsilon_{231} = +1 \\ \varepsilon_{132} &= \varepsilon_{213} = \varepsilon_{321} = -1 \\ \varepsilon_{112} &= \varepsilon_{212} = etc. = ... = 0 \end{aligned} \tag{3.26}$$

It has to be noted that the coefficient i in RHS of eqn. (3.24) is $\sqrt{-1}$.

The Lie algebra of SU(2) spin group is obtained by replacing the generators J_i by $\frac{\sigma_i}{2}$ and for the Lie algebra of SU(2) isospin J_i is replaced by $\frac{\tau_i}{2}$ and accordingly we have the Lie algebras

$$\left[\frac{\sigma_i}{2}, \frac{\sigma_j}{2}\right] = i\varepsilon_{ijk} \frac{\sigma_k}{2} \qquad \text{SU(2) spin} \tag{3.27a}$$

$$\left[\frac{\tau_i}{2}, \frac{\tau_j}{2}\right] = i\varepsilon_{ijk} \frac{\tau_k}{2} \qquad \text{SU(2) isospin} \tag{3.28a}$$

which can also be written as

$$[\sigma_i, \sigma_j] = 2i\, \varepsilon_{ijk}\, \sigma_k \tag{3.27b}$$

$$[\tau_i, \tau_j] = 2i\, \varepsilon_{ijk}\, \tau_k \tag{3.28b}$$

In the above, ε_{ijk} would be called the structure constants of SU(2) group.

Replacing J_i of eqn. (3.24) by $\frac{1}{2}\lambda_a$, where λ_a with a = 1, 2, ... 8, denote the eight Gell—Mann matrices (eqn. 3.15), we may write the Lie algebra of SU(3) group as

$$\left[\frac{\lambda_a}{2}, \frac{\lambda_b}{2}\right] = if_{abc} \frac{\lambda_c}{2} \tag{3.29a}$$

or

$$[\lambda_a, \lambda_b] = 2if_{abc}\lambda_c \tag{3.29b}$$

where f_{abc} are the structure constants of the SU(3) group. Each of the indices a, b, c can take values from 1 to 8. f_{abc}, are antisymmetric under the interchange of any two indices. The non-zero values of SU(3) structure constants are:

$$\begin{aligned} f_{123} &= 1 \\ f_{147} &= f_{246} = f_{257} = f_{345} = f_{516} = f_{637} = \tfrac{1}{2} \\ f_{458} &= f_{678} = \frac{\sqrt{3}}{2} \end{aligned} \tag{3.30}$$

3.3.2 Representations of Lie Groups

A representation of a group G is a mapping, i.e., the correspondence, which associates, with every element R of G, a linear operator $D(R)$ in a certain vector space V, such that group multiplication is preserved and the identity E of G is mapped into the identity 1 in V.

Above definition of representation means that if we have a group G in which R, S,... etc. are the elements and also we have another group $D(G)$ in which the linear operators $D(R)$, $D(S)$, etc. are the elements and then if it happens that the group G maps homomorphically on the group of operators $D(G)$ in the vector space V, we say that operator group $D(G)$ is a representation of group G in the representation space V. Group G may be called the group of elements R,S, etc. and group $D(G)$ may be called the *group of linear operators D(R), D(S)*, etc., and vector space V is the linear vector space of these operators. If the vector space V has n dimensions, the representation is said to be of n degree or n dimensions. By homomorphism we mean that there may be only one operator say $D(R)$ in the group $D(G)$ corresponding to more than one elements say R, S, etc., of the group G. The operators of group $D(G)$ obey following properties:

$$\begin{aligned} D(RS) &= D(R)D(S) \\ D(R^{-1}) &= [D(R)]^{-1} \\ D(E) &= 1 \end{aligned} \tag{3.31}$$

where $D(RS)$ means the operator corresponding to RS element of group G. E is the identity element of G. Focusing our attention on the group $D(G)$ of operators we can define

(1) Linear representation

If the elements of the group $D(G)$ are linear operators then we have the linear representation of group G.

(2) Matrix representation

If the operators (elements) of group $D(G)$ are $n \times n$ matrices, then we have the matrix representation of the group G.

(3) Faithful representation

If the homomorphic mapping of G on $D(G)$ becomes isomorphic, the group G is said to have a faithful representation. In this case the matrices (operators) of $D(G)$ are distinct. A faithful representation completely describes the structure of the Lie group.

(4) Equivalent representation

If the basis of n dimensional vector space V were changed, the matrices $D(R)$ would become $D'(R) = AD(R)A^{-1}$ where A is transforming matrix. The matrices $D'(R)$ also provide a representation of G, which is equivalent to the representation provided by $D(R)$ matrices.

(5) Unitary representation

If the operators are unitary (unitary matrices) the representation is said to be unitary. Such representations are important in particle physics because they conserve the normalization of state vectors.

(6) Defining representation

We have seen that to define a group representation we correspond the elements of group G with the elements of group $D(G)$ and the elements of group $D(G)$ are matrices. Now if the elements of G itself were matrices then their correspondence is with G itself, i.e., the group G itself acts as group $D(G)$. Then we say if G is a group of matrices such as U(n), SU(n), O(n), etc., and then this group is a faithful representation of itself. We call this representation the defining representation or the fundamental representation. The set of orthogonal 2×2 matrices is called the defining representation of O(2) group. The two dimensional representation of SU(2), i.e., the set of σ (or τ) matrices, is the defining representation of SU(2). The set of λ matrices, the unitary unimodular three-dimensional matrices, form the defining representation of SU(3) group. The unimodular 3×3 orthogonal matrices form the defining representation of SO(3). The fundamental representation is the lowest dimensional faithful representation and by taking the repeated direct products of this representation with itself, other (higher dimensional) representations can be obtained.

(7) Trivial representation

With every group G, we can associate a group D(G) whose elements are 1×1 matrices (a constant number). Then, the $D(G) = (1, 1, \ldots 1)$ forms the trivial or the identity representation of the group. Since, the order of matrices determines the dimensions of the representation; *the identity representation is the one-dimensional representation.*

(8) Adjoint representation

An irreducible representation, which is completely determined by the structure constants of the group, is adjoint representation. Its dimensionality is equal to the number of generators of the group. The irreducible representations have been explained in section (3.3.2.1). The group SU(2) has three generators hence its adjoint representation is three-dimensional and we say *the adjoint representation of* SU(2) *is* **3**. Likewise, the group SU(3) has eight generators; hence we say that the *adjoint representation of* SU(3) *is an* **8**. Obviously, the group SU(3) has eight dimensional adjoint representation. It may also be noted that the defining representations of SU(2) and SU(3) are 2 and 3 respectively, whereas, their adjoint representations are respectively **3** and **8**.

3.3.2.1 Reducible and Irreducible Representations

Before explaining the above representations let us define the direct sum of matrices.

The direct sum of square matrices

$$P = p,\ Q = \begin{bmatrix} q & r \\ s & t \end{bmatrix} \text{ and } R = \begin{bmatrix} u & v & w \\ x & y & z \\ a & b & c \end{bmatrix}$$

is defined as

$$T = P \oplus Q \oplus R = \begin{bmatrix} \boldsymbol{p} & 0 & 0 & 0 & 0 & 0 \\ 0 & \boldsymbol{q} & \boldsymbol{r} & 0 & 0 & 0 \\ 0 & \boldsymbol{s} & \boldsymbol{t} & 0 & 0 & 0 \\ 0 & 0 & 0 & \boldsymbol{u} & \boldsymbol{v} & \boldsymbol{w} \\ 0 & 0 & 0 & \boldsymbol{x} & \boldsymbol{y} & \boldsymbol{z} \\ 0 & 0 & 0 & \boldsymbol{a} & \boldsymbol{b} & \boldsymbol{c} \end{bmatrix} \tag{3.32}$$

where $\oplus$ is the symbol for the direct sum. We find that the matrix T, which is the direct sum of matrices P, Q and R, has non-zero elements in square blocks along the main diagonal and zeros elsewhere. Such a form, which the matrix T has – is said to be the *block-diagonal form.*

Now we apply the concept of block-diagonal form to group representation. For this purpose we look at the matrices (operators) of the group $D(G)$. We recall that $D(G)$ is a representation of group G. Let us again denote an element of $D(G)$, which is a matrix, by $D(R)$. Suppose the matrix $D(R)$ can be expressed in following two ways:

$$(i) \qquad D(R) = \begin{bmatrix} D^1(R) & A(R) \\ 0 & D^2(R) \end{bmatrix} \tag{3.33}$$

and

$$(ii) \qquad D(R) = \begin{bmatrix} D^1(R) & 0 \\ 0 & D^2(R) \end{bmatrix} \tag{3.34}$$

If the representation $D(G)$ is n dimensional, the matrix $D(R)$ is $n \times n$. In eqn. (3.33) the matrix $D^1(R)$ is say $m \times m$, $D^2(R)$ matrix is $(n - m) \times (n - m)$, $A(R)$ matrix has m rows and $(n - m)$ columns and the null matrix O has $(n - m)$ rows and m columns. Each element of O matrix is zero. When we decompose one matrix $D(R)$ of $D(G)$ as eqn. (3.33), it is implied that all the matrices of group $D(G)$ can be decomposed like equation (3.33). If matrices $D(R)$ can be written as eqn. (3.33) we say that matrices $D(R)$ are reducible. Initially $D(R)$ were n dimensional, after decomposition $D^1(R)$ is m, which is less than n, dimensional and $D^2(R)$ too has its dimensions less than n(i.e., $n - m$), so we call $D(R)$ as reducible. In eqn. (3.33) there are $A(R)$ matrices as well. If the decomposition of $D(R)$ is such that $A(R) = 0$, the matrices $D(R)$ are fully reducible as expressed in eqn. (3.34) and we call that the representation of group G is fully reducible.

It may happen that the matrices $D^1(R)$ and $D^2(R)$ can be further decomposed till further decomposition is impossible. When such a state is reached, the matrices $\boldsymbol{D}(R)$ can be expressed in a block diagonal form like eqn. (3.32), in which each matrix along the principle diagonal can no longer be reduced as eqn. (3.33) or eqn. (3.34). Each matrix along the principle diagonal is now irreducible. Assuming full reduction $A(R)$ matrices can be taken as zero. In this case we can express

$$D(R) = \begin{bmatrix} D^1(R) & 0 & & & & & & & \\ & D^2(R) & 0 & & & & & & \\ & & D^3(R) & 0 & & & & & \\ & & & .. & 0 & & & & \\ & & & & .. & 0 & & & \\ & & & & & .. & 0 & & \\ & & & & & & D^{n-1}(R) & 0 & \\ 0 & 0 & 0 & 0 & 0 & 0 & 0 & D^n(R) \end{bmatrix} \tag{3.35}$$

in which each of the matrices $D^1(R)$, $D^2(R)$, $D^3(R)$, ..., $D^n(R)$ is irreducible. In case of full reduction [eqn. (3.35)], we can write the representation $D(R)$ as the direct sum of irreducible representations $D^1(R)$, $D^2(R)$, etc.

i.e., $$D(R) = D^1(R) \oplus D^2(R) \oplus D^3(R) \oplus ... \oplus D^n(R) \tag{3.36}$$

It may happen that in carrying out successive decompositions an irreducible matrix may get repeated. Let $D^1(R)$ irreducible representation appears a_1 times, $D^2(R)$, a_2 times and so on, we may then express

$$D(R) = a_1 D^1(R) \oplus a_2 D^2(R) \oplus \ldots \oplus a_n D^n(R) = \sum_{k=1}^{n} a_k D^k(R) \tag{3.37}$$

where a_1 are positive integers and Σ implies direct sum.

It may be mentioned here that if the matrices $D(R)$ are unitary, then $A(R) = 0$, implying that the unitary $D(R)s$ in an appropriate basis can be fully decomposed into irreducible matrices and hence keeping in view the relations (3.32), (3.36) and (3.37), the unitary representations can be written as the direct sum of irreducible representations. Thus it is important to see that the matrices $D(R)$ are unitary.

3.4 GROUP SU(2) AND ITS IRREDUCIBLE REPRESENTATIONS

It has already been mentioned that SU(2) is the special unitary group in two dimensions and this group describes both the spin and isospin.

The spin operator J is an angular momentum operator and J^2 and J_z are two commuting operators having simultaneous eigenstate $|j,m\rangle$. For a spin $-\frac{1}{2}$ particle, $m = +\frac{1}{2}$ denotes spin up state and $m = -\frac{1}{2}$, the spin down state. If these eigenstates are written as $\begin{pmatrix} 1 \\ 0 \end{pmatrix}$ and $\begin{pmatrix} 0 \\ 1 \end{pmatrix}$, we may connect them by raising and lowering operators

$$J_+ = \begin{pmatrix} 0 & 1 \\ 0 & 0 \end{pmatrix} \text{ and } J_- = \begin{pmatrix} 0 & 0 \\ 1 & 0 \end{pmatrix} \tag{3.38}$$

from which using $J \pm J_x = \pm iJ_y$, we get $J_x = \frac{\sigma_x}{2}, J_y = \frac{\sigma_y}{2}$ and $J_z = \frac{\sigma_z}{2}$, where $\sigma's$ are 2×2 Pauli matrices (eqn.3.13, $\sigma_1 = \sigma_x$, etc.) A general spin state $|\psi\rangle = \begin{pmatrix} \alpha \\ \beta \end{pmatrix} = \alpha \begin{pmatrix} 1 \\ 0 \end{pmatrix} + \beta \begin{pmatrix} 0 \\ 1 \end{pmatrix}$ can be transformed into $|\psi'\rangle$ by a unitary matrix U such that

$$|\psi'\rangle = U|\psi\rangle \tag{3.39}$$

The unitarity $UU^\dagger = 1$ provides

$$\det(UU^\dagger) = (\det U)(\det U^\dagger) = |\det U|^2 = 1$$

implying that

$$\det U = e^{i\theta} \tag{3.40}$$

where θ = real. A set of all unitary 2×2 matrices is the $U(2)$ group but when det

$U = +1$, $U(2)$ becomes SU(2) and the set of unitary matrices

$$'U_i(\theta) = e^{+\frac{1}{2}i\sigma_i\theta} \tag{3.41}$$

forms a representation (fundamental) of SU(2) and the σ_i matrices ($i = 1, 2, 3$) obey the commutation relation.

$$\left[\frac{\sigma_i}{2}, \frac{\sigma_j}{2}\right] = i\varepsilon_{ijk}\frac{\sigma_k}{2} \tag{3.42}$$

Defining the generators of SU(2) spin group as

$$J_i = \frac{\sigma_i}{2} \tag{3.43}$$

the Lie algebra (3.42) of SU(2) spin group reads

$$[J_i, J_j] = i\varepsilon_{ijk}J_k \tag{3.44}$$

which is also the Lie algebra of rotation group SO(3) (eqn. 3.24). Thus SU(2) spin is also a rotation group which is found to be homomorphic on SO(3). The generators of SO(3), i.e., the angular momentum operators J_1, J_2, J_3 (matrices) can be obtained by employing the usual theory of angular momentum. Similar theory is also applicable to SU(2) group and we can obtain the relations

$$J_3|j,m\rangle = m|j,m\rangle$$
$$J_\pm|j,m\rangle = [(j \mp m)(j \pm m + 1)]^{\frac{1}{2}}|j,m \pm 1\rangle \tag{3.45}$$

which tell that the SU(2) irreducible representations are characterized by j, which is either an integer or half integer. The states, $|j,m\rangle$ with $m = j, j-1, \ldots -j$, form the basis of an SU(2) irreducible representation. The operator J_+ raises the value of m by one unit till the upper bound is reached and J_- lowers the m values upto the lower bound of m. The dimension of the SU(2) irreducible representations would be $2j + 1$. When $j = 0$, the representation is one-dimensional, for $j = \frac{1}{2}$ it is two dimensional and so on. The $j = \frac{1}{2}$ (2 dimensional) representation is the defining representation of SU(2). The representation matrices are obtained as (using eqn. 3.45) $j = \frac{1}{2}$, $m = \pm\frac{1}{2}$

$$J_1 = \frac{1}{2}\begin{pmatrix} 0 & 1 \\ 1 & 0 \end{pmatrix}, J_2 = \frac{1}{2}\begin{pmatrix} 0 & -i \\ i & 0 \end{pmatrix}, J_3 = \frac{1}{2}\begin{pmatrix} 1 & 0 \\ 0 & -1 \end{pmatrix}, J_\pm = J_1 \pm iJ_2 \tag{3.46}$$

Thus J_1, J_2, J_3 of eqn. (3.46) give the two-dimensional irreducible representation of SU(2) spin group. For,

$j = 1, m = 1, 0, -1$

$$J_1 = \frac{1}{\sqrt{2}}\begin{pmatrix} 0 & 1 & 0 \\ 1 & 0 & 1 \\ 0 & 1 & 0 \end{pmatrix}, J_2 = \frac{1}{\sqrt{2}}\begin{pmatrix} 0 & -i & 0 \\ i & 0 & -i \\ 0 & i & 0 \end{pmatrix}, J_3 = \begin{pmatrix} 1 & 0 & 0 \\ 0 & 0 & 0 \\ 0 & 0 & -1 \end{pmatrix} \tag{3.47}$$

implying that the matrices (3.47) provide the three dimensional $(2j+1)$ irreducible representation of SU(2) spin. It may be noted, e.g., for $j = 1$, that $j = 1$ $j = 1$ is an irreducible representation of SU(2) and $|j, m\rangle$ i.e. $|1,1\rangle, |1,0\rangle$ and $|1,-1\rangle$ are the three basis states which cover (span) the space of $j = 1$ irreducible representation and because of three basis states (coordinates) the $j = 1$ irreducible representation space is three-dimensional. The basis states of an irreducible representation are connected through the raising and lowering operators $J_{\pm}$. Similarly we can have $j = \frac{3}{2}$ (4-dimensional), $j = 2$ (5-dimensional) irreducible representations of SU(2) spin group.

In an exactly similar manner the irreducible representations of SU(2) isospin rotation group can be obtained. Whereas the spin refers to the rotation in spin space, the isospin refers to the rotation in a different space called *isospin space* or *the charge space* or *the internal space*. Since proton and neutron are two isospin states of nucleon where proton refers to isospin *up* ($I_3 = +\frac{1}{2}$) state and neutron to isospin *down* ($I_3 = -\frac{1}{2}$) state, the situation is exactly similar to spin *up* electron (here proton) and spin *down* electron (here neutron). Instead of angular momentum operator J we shall now have isospin operator T with components T_1, T_2, T_3. The commutating operators are T^2 and T_3 having a common eigenstate $|\lambda, t_3\rangle$ with eigenvalue equations

$$\begin{aligned} T^2|\lambda, t_3\rangle &= \lambda|\lambda, t_3\rangle \\ T_3|\lambda, t_3\rangle &= t_3|\lambda, t_3\rangle \end{aligned} \tag{3.48}$$

For a given λ, values of t_3 are bounded, i.e., they have an upper limit and a lower limit determined by

$$\lambda - t_3^2 \geq 0 \tag{3.49}$$

In terms of largest value of $t_3 = t$, $\lambda = t, (t + 1)$ and in terms of smallest value of $t_3 = t'$, $\lambda = t'(t' - 1)$ which give

$$t' = -t \tag{3.50}$$

implying that t_3 values range from its highest of t to its lowest of $-t$, i.e. $t_3 = t$, $t - 1$ .... $- t$. The operator $T_- = T_1 - iT_2$ lowers the t_3 value by one unit so $t - t' = 2t$ must be an integer implying that t can take on an integer or half integer values so

$$t = 0, \frac{1}{2}, 1, \frac{3}{2}, 2, \ldots \tag{3.51}$$

The eigenvectors $|\lambda, t_3\rangle$, for a given value of t, which are now denoted by $|t, t_3\rangle$, form a $(2t + 1)$ dimensional basis for the irreducible representations of SU(2) isospin rotation group. The defining or the fundamental representation of SU(2) isospin group belongs to $t = \frac{1}{2}$, i.e., it is two-dimensional and the matrices

$$U_i(\theta) = \exp\,(+\frac{1}{2} i\tau_i \theta)\ (\mathrm{i} = 1, 2, 3) \tag{3.52}$$

form this (fundamental) representation. The τ matrices obey the commutation relations

$$\left[\frac{\tau_i}{2}, \frac{\tau_j}{2}\right] = i\varepsilon_{ijk} \frac{\tau_k}{2} \tag{3.53}$$

from which the generators of the SU(2) isospin group defined by

$$T_i = \frac{\tau_i}{2} \tag{3.54}$$

provide the Lie algebra of SU(2) isospin group as

$$[T_i, T_j] = i\varepsilon_{ijk} T_k \tag{3.55}$$

where ε_{ijk} are the structure constants of the SU(2) isospin group. Various irreducible representations of SU(2) isospin group belong to different values of t (eqn. 3.51) among which the $t = \frac{1}{2}$ gives the defining representation.

3.4.1 SU(2) Product Representations

Before explaining the SU(2) product representations, let us define the direct product of matrices. Let us have two matrices A and B given by

$$A = \begin{bmatrix} a_1 & a_2 & a_3 \\ b_1 & b_2 & b_3 \end{bmatrix} \quad \text{and} \quad B = \begin{bmatrix} c_1 & d_1 \\ c_2 & d_2 \\ c_3 & d_3 \end{bmatrix} \tag{3.56}$$

then their direct product ($\otimes$) is

$$A \otimes B = \begin{bmatrix} a_1c_1 & a_1d_1 & a_2c_1 & a_2d_1 & a_3c_1 & a_3d_1 \\ a_1c_2 & a_1d_2 & a_2c_2 & a_2d_2 & a_3c_2 & a_3d_2 \\ a_1c_3 & a_1d_3 & a_2c_3 & a_2d_3 & a_3c_3 & a_3d_3 \\ b_1c_1 & b_1d_1 & b_2c_1 & b_2d_1 & b_3c_1 & b_3d_1 \\ b_1c_2 & b_1d_2 & b_2c_2 & b_2d_2 & b_3c_2 & b_3d_2 \\ b_1c_3 & b_1d_3 & b_2c_3 & b_2d_3 & b_3c_3 & b_3d_3 \end{bmatrix} \tag{3.57}$$

Regarding the direct product of representations we mention an important observation that—if matrices D^a are the matrices of one representation of a group G and matrices D^b are matrices of other representation of G then the direct product of matrices D^a and $D^b = D^a \otimes D^b$ would generate another representation of G.

Having defined the direct product in the above, we work out the irreducible representations of SU(2) product representation.

Let us recall from previous sections that for SU(2) (say isospin group) the generators T_i obey

$$[T_i, T_j] = i\varepsilon_{ijk}T_k \tag{3.58}$$

Let us here denote the set of matrices, which denote the fundamental representation of SU(2) by $D^2(\theta)$, where

$$D^2(\theta) = \exp(+i\theta_i \frac{\tau_i}{2}) \equiv \exp(+i\theta.\boldsymbol{T}) \tag{3.59}$$

Now, let us construct a four-component column vector ψ by taking the direct product of two, two-component column vectors (spinors)

$$\chi = \begin{pmatrix} \chi_1 \\ \chi_2 \end{pmatrix}, \quad \eta = \begin{pmatrix} \eta_1 \\ \eta_2 \end{pmatrix} \tag{3.60}$$

such that

$$\chi \otimes \eta = \begin{pmatrix} \chi_1\eta_1 \\ \chi_1\eta_2 \\ \chi_2\eta_1 \\ \chi_2\eta_2 \end{pmatrix} = \begin{pmatrix} \psi_1 \\ \psi_2 \\ \psi_3 \\ \psi_4 \end{pmatrix} = \psi \tag{3.61}$$

Now our aim is to know how ψ transforms, when we employ matrices $D^2(\theta)$. Since components of ψ are made of χ and η, we have to use transformation laws of χ and η under the operation of $D^2(\theta)$. We take the transformations as

$$\chi' = D^2(\theta)\chi$$

and

$$\eta' = D^2(\theta)\eta \tag{3.62}$$

so that

$$\begin{aligned} \chi' \otimes \eta' &= [D^2(\theta)\chi] \otimes [D^2(\theta)\eta] \\ &= [D^2(\theta) \otimes D^2(\theta)][\chi \otimes \eta] \end{aligned} \tag{3.63}$$

where we have used the property of the direct product

$$AB \otimes CD = (A \otimes C)(B \otimes D) \tag{3.64}$$

Now using the first property from eqn. (3.31) to write

$$D^2(\theta) \otimes D^2(\theta) = D^{2\otimes 2}(\theta) \tag{3.65}$$

and writing similar to eqn. (3.61)

$$\chi' \otimes \eta' = \psi' \tag{3.66}$$

equation (3.63) becomes

$$\psi' = D^{2\otimes 2}(\theta)\psi \tag{3.67}$$

thus giving the transformation of ψ into ψ' by four-dimensional matrix $D^{2\otimes 2}(\theta)$.

Now we make a change of basis. For this purpose, we would require a unitary matrix, which transforms ψ into $\tilde{\psi}$ i.e.

$$\tilde{\psi} = U\psi \tag{3.68}$$

Using $U^{\dagger}U = 1$, eqn. (3.68) may be written as

$$\psi = U^{\dagger}\tilde{\psi} \tag{3.69}$$

Similar to eqn. (3.68), we may write

$$U\psi' = \tilde{\psi}' \tag{3.70}$$

The equations (3.68) and (3.70) tell that, on making the change of basis ψ becomes $\tilde{\psi}$ and ψ' becomes $\tilde{\psi}'$. Now, we carryout the U-operation on eqn. (3.67) as

$$U\psi' = U[D^{2\otimes 2}(\theta)\psi]$$
$$= [UD^{2\otimes 2}(\theta)U^{\dagger}][U\psi]$$

or

$$\tilde{\psi}' = \tilde{D}^{2\otimes 2}(\theta)\tilde{\psi} \tag{3.71}$$

where as told before $U^{\dagger}U = 1$, $U\psi = \tilde{\psi}$ and $U\psi' = \tilde{\psi}'$ have been used. Also, we have defined

$$\tilde{D}^{2\otimes 2}(\theta) = UD^{2\otimes 2}(\theta)U^{\dagger} \tag{3.72}$$

which is the transformation law for $D^{2\otimes 2}$ into $\tilde{D}^{2\otimes 2}$. Now suppose, there exists a unitary matrix U such that $\tilde{D}^{2\otimes 2}(\theta)$ matrix for all θ can acquire a block diagonal form with elements (square matrices) only along the principal diagonal and zeros elsewhere, we would then say that the matrix $\tilde{D}^{2\otimes 2}(\theta)$ has been fully reduced and the elements in the principal diagonal represent irreducible representations. Keeping in mind the relation (3.36), for fully reduced representations, we may write

$$\tilde{D}^{2\otimes 2}(\theta) = \begin{pmatrix} [D^{3}(\theta)] & 0 \\ 0 & [D^{1}(\theta)] \end{pmatrix} \tag{3.73}$$

in which $D^3(\theta)$ is the three dimensional irreducible representation and $D^1(\theta)$, the one dimensional irreducible representation. Relation (3.36) helps us to write for eqn. (3.73) the reduction of the direct product as

$$\mathbf{2} \otimes \mathbf{2} = \mathbf{4} = \mathbf{1} \oplus \mathbf{3} \tag{3.74}$$

In terms of dimensions LHS of above equation must be read as the direct product of two, two-dimensional, implying the direct product of two fundamental representations of SU(2) group. Inference that we draw from eqn. (3.74) is that—from the product (direct) of two defining representations of SU(2) group; we get a three-dimensional representation and a one-dimensional representation. This means the higher dimensional (here three) representations can also be obtained by taking the successive product representations of the fundamental representation. Another example is the direct product of three two-dimensional representations

$$\mathbf{2} \otimes \mathbf{2} \otimes \mathbf{2} = \mathbf{4} \oplus \mathbf{2} \oplus \mathbf{2} \tag{3.75}$$

The physical meaning of eqn. (3.74) can be understood as follows. If we combine two spin-half objects (spin-1/2 ⇒ two dimensions), we can either get a spin zero or spin one. Spin zero corresponds to a singlet and spin one to a triplet. We may also read eqn. (3.74) as, when two doublets **combine,** we would get a singlet and a triplet. In other words, reductions like eqn. (3.74) provide a means to combine multiplets to form new multiplets. Similar result regarding combining the two spin 1/2 states is obtained in quantum mechanics, where the triplet states are $|1,1\rangle, |1,0\rangle$ and $|1,-1\rangle$ and the singlet is $|0,0\rangle$. The combination (3.75) has its physical content in the formation of three-quark state. Quark too is a spin-half object. The **4** belongs to four Δ particle states and two doublets belong to symmetric nucleon doublet and antisymmetric nucleon doublet.

The product representations can also be represented graphically. For SU(2) we note that among its three generators $J_i = \frac{\sigma_i}{2}$ only one namely, J_3 is diagonal. Hence SU(2) is a group of rank one. Due to this, its irreducible representations are one-dimensional straight lines. Corresponding to particular *j* value we have a line segment and *m* values are denoted by dots (.) on it. The dots denote the states or the dimension of the irreducible representation. An $J = \frac{3}{2}$ irreducible representation will be a straight line with 4 equally spaced dots on it, indicating it a four-dimensional irreducible representation. The diagonal elements of the diagonal generators are eigenvalues and when the two SU(2) representations (generators $J_3^{(1)}$ and $J_3^{(2)}$) are combined the eigenvalues of $J_3^{(1)}$ and $J_3^{(2)}$ simply add. This addition can be represented graphically by repeatedly placing the $m = 0$ point of one representation over on every point of the other representation. For example two representations are corresponding to $j_1 = \frac{1}{2}$ and $j_2 = \frac{1}{2}$, hence *m*

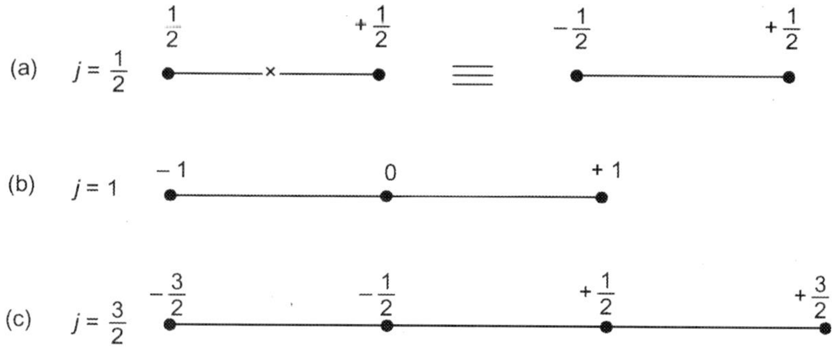

Fig. 3.1: Graphical representation of SU(2) multiplets. The dots denote the *m* values. The cross in Fig. (a) indicates the centre (×) . Fig. (a) is the graphical representation of $j = \frac{1}{2}$ and Figs. (b) and (c) show the graphical representations of *j* = 1 and $j = \frac{3}{2}$ respectively

values in each case are – 1/2 and + 1/2. The graphical representations of some j values have been shown in Fig. (3.1).
Thus

$$j_1 = \frac{1}{2} \text{ representation is denoted by Fig. 3.1(a)}$$

$$j_2 = \frac{1}{2} \text{ representation is also denoted by Fig. 3.1(a)}$$

where $\times$ denotes the centre of the lines. Keeping j_2 line fixed, place the $\times$ of j_1 line on each of the dots of j_2 line. The result is

$$\frac{1}{2} \otimes \frac{1}{2} = 0 \oplus 1 \tag{3.76}$$

which has been shown in Fig. (3.2).

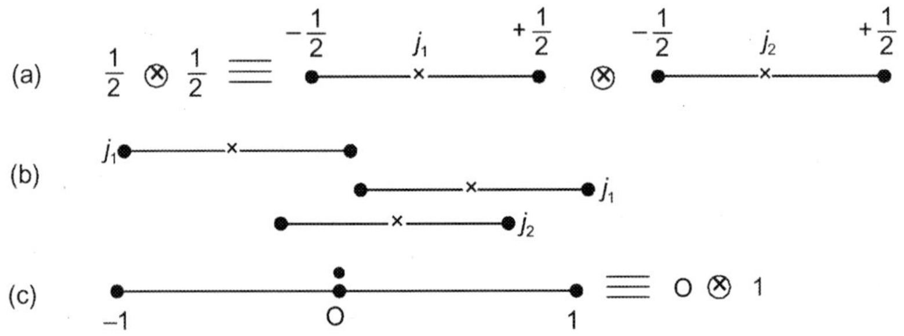

Fig. 3.2: Graphical representation (*GR*) of $\frac{1}{2}\otimes\frac{1}{2}=0\oplus 1$. Fig. (a) denotes GRs of $j_1 = \frac{1}{2}$ and $j_2 = \frac{1}{2}$. Fig. (b) shows that on each dot of j_2, the centre of j_1 has been placed resulting in Fig. (c), which is the GR of $0 \oplus 1$

Likewise

$$\frac{1}{2} \otimes 1 = \frac{1}{2} \oplus \frac{3}{2} \tag{3.77a}$$

or

$$\mathbf{2 \otimes 3 = 2 \oplus 4} \tag{3.77b}$$

i.e., when a spin $\frac{1}{2}$ and a spin 1 object combine we get a spin $\frac{1}{2}$ (doublet) and spin $\frac{3}{2}$ (quartet) object. The graphical representation of eqn. (3.77a) has been shown in Fig. (3.3).

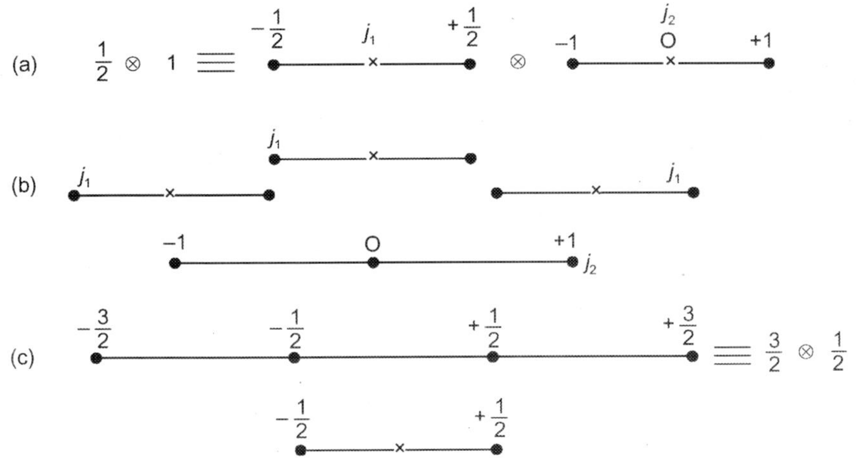

Fig. 3.3: Graphical representation of $\frac{1}{2} \otimes 1 = \frac{1}{2} \oplus \frac{3}{2}$. Fig. (a) is the GR of $j_1 = \frac{1}{2}$ and $j_2 = \frac{1}{2}$. In Fig. (b) on each dot of j_2 is placed the centre ($\times$) of j_1 resulting in $\frac{1}{2} \oplus \frac{3}{2}$ represented graphically in Fig. (c)

Another interesting method to know the irreducible representations of product representations of unitary groups is *Young's tableau* technique. In this technique, the fundamental representation is denoted by a box and the conjugate representation (for example, $\bar{2}$ is the conjugate of 2) is denoted by a column of boxes whose number is one less than the dimension of the group. In SU(2) we will denote

$$2 = \square \text{ (Row)}, \quad \bar{2} = \square \text{ (Column)} \tag{3.78}$$

In constructing $2 \otimes 2$ we write

$$\boxed{a} \otimes \boxed{b} = \boxed{a}\boxed{b} \oplus \begin{array}{|c|}\hline a \\ \hline b \\ \hline \end{array} \tag{3.79}$$

where RHS is obtained by seeing that in how many ways the box b can be attached to box a. Now we have to calculate the dimensions in the RHS, LHS is obviously $2 \otimes 2$. In the RHS, the dimensions of $\square\square$ and $\begin{array}{|c|}\hline \ \\ \hline \ \\ \hline \end{array}$ are calculated by finding the numerator (N) and the denominator (D) of each tableau. The dimension of each tableau would be equal to its $\frac{N}{D}$. Since we are dealing with SU(2) representations, we have to note 2 of SU(2). So for N, of first tableau write

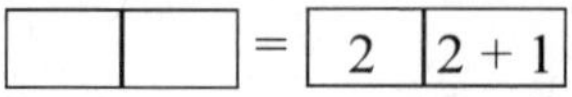

implying that $N = 2(2+1) = 6$. For D, draw lines 1 and 2 and notice

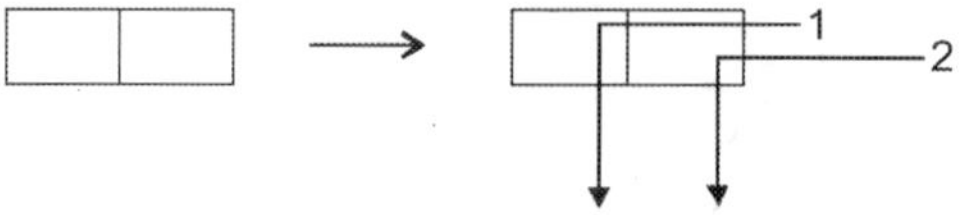

that the line 1 crosses through two (2) boxes and line 2 through 1 box. So, $D = 2 \times 1 = 2$. The difference between line 1 and line 2 is that they are passing through different columns. Therefore the dimensions of SU(2) 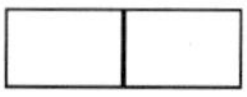 = 6/2 = 3. Likewise, N for

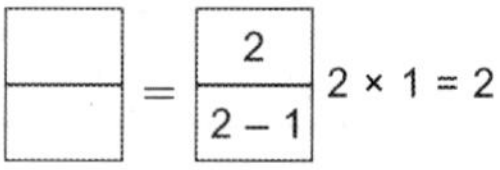

where it has to be noted that the number decreases while going down a column. The D for

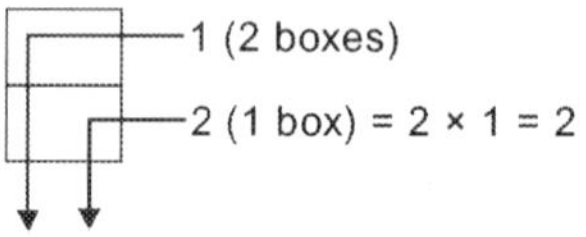

Therefore, the dimensions of SU(2)  = 2/2 = 1. Hence in SU(2) the representation (3.79) implies

$$\mathbf{2} \otimes \mathbf{2} = \mathbf{3} \oplus \mathbf{1}$$

It can be noted that in SU(2) 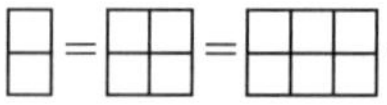 = ... all have dimension 1, because for example N for

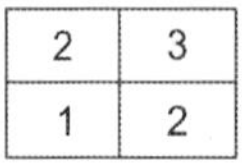

$= 2 \times 3 \times 1 \times 2 = 12$ and for the D for these four boxes we have

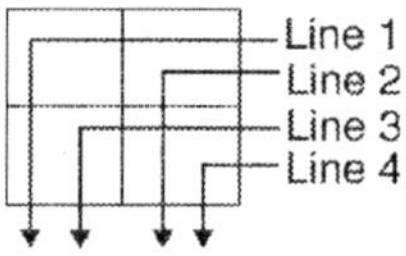

The lines 1, 2, 3 and 4 cross respectively, 3, 2, 2 and 1 box. Hence, $D = 3 \times 2 \times 2 \times 1 = 12$ and the dimension for these four boxes, $N/D = 12/12 = 1$. Young's Tableau (YT) is more useful to know the irreducible representations of higher dimensions. We will use it in SU(3), where a better understanding of YT would follow.

3.5 SU(3) GROUP AND ITS IRREDUCIBLE REPRESENTATIONS

Basically, the group SU(3) originated as an extension of SU(2) isospin group in which Λ particle was included to p and n to form the fundamental triplet. Since,

Λ is a strange particle, the SU(3) group includes, strangeness (or hypercharge $Y = B + S$) with the isospin of SU(2). In terms of quark composition $p = uud$, $n = udd$, $\Lambda = uds$, the fundamental triplet is (u, d, s) and the SU(3) group that uses (u, d, s) as the fundamental triplet is called SU(3) flavour. With the introduction of color, (R, G, B) also became a fundamental triplet and we have SU(3) colour group. Let us use a general fundamental triplet

$$\phi = \begin{pmatrix} \phi_1 \\ \phi_2 \\ \phi_3 \end{pmatrix} \tag{3.80}$$

and assume that it transforms as

$$\phi \to \phi' = U\phi \tag{3.81}$$

where U matrices are arbitrary, unitary 3×3 matrices with det $U = +1$. A set of matrices

$$U(\theta) = \exp(-\tfrac{i}{2}\theta_a \lambda_a) \tag{3.82}$$

where θ_a, with $a = 1, 2, ..., 8$ are eight parameters of SU(3) group and λ_a are eight 3×3 traceless matrices, form the fundamental representation of SU(3). The Gell—Mann matrices have been given in eqn. (3.15). The eight generators of this group are

$$F_a = \frac{\lambda_a}{2} \tag{3.83}$$

whose commutation relations

$$[F_a, F_b] = if_{abc}F_c \tag{3.84}$$

provide the Lie algebra of the SU(3) group. The constants f_{abc} are antisymmetric under the interchange of any two indices.

3.5.1 SU(3) Shift Operators

Among the eight Gell—Mann matrices (3.15), only two namely, λ_3 and λ_8 are diagonal. Therefore the rank of SU(3) group is two. The diagonal matrices λ_3 and λ_8 make the generators F_3 and F_8 commute, i.e.,

$$[F_3, F_8] = 0 \tag{3.85}$$

implying that the generators F_3 and F_8 would have simultaneous eigenvectors. If we look at the first three Gell—Mann matrices, we find that they can be written in terms of τ_i (Pauli matrices)

$$\lambda_i = \left[\begin{array}{cc|c} \multicolumn{2}{c|}{\tau_i} & 0 \\ & & 0 \\ \hline 0 & 0 & 0 \end{array}\right] \quad (i = 1, 2, 3) \tag{3.86}$$

where the only difference is that there is an extra column and extra row of zeros. We therefore expect that $\lambda_1, \lambda_2, \lambda_3$ behave similar to τ_1, τ_2, τ_3, the latter providing the generators $T_{\pm} = T_1 \pm iT_2$ and T_3, where in SU(2) isospin group $T_i = \frac{\tau_i}{2}$ $(i = 1, 2, 3)$. Therefore, in SU(3) the generators F_1, F_2 and $F_3 (F_i = \frac{\lambda_i}{2})$ would be related to isospin and we shall have

$$T_{\pm} = F_1 \pm iF_2$$
$$T_3 = F_3 \tag{3.87}$$

The matrix λ_8 is also diagonal and its eigenvalues are related to hypercharge Y. We define that

$$Y = \frac{2}{\sqrt{3}} F_8 \equiv \frac{1}{\sqrt{3}} \lambda_8 \tag{3.88}$$

From the remaining four operators F_4 to F_7, are defined U-spin operators $U_{\pm}$ and V spin operators $V_{\pm}$ as

$$U_{\pm} = F_6 + iF_7$$

and

$$V_{\pm} = F_4 + iF_5 \tag{3.89}$$

We know the raising and lowering property of $T_{\pm}$, likewise there are $U_{\pm}$ and $V_{\pm}$ operators. These six operators are called the *Shift operators of SU(3) group* and F_3 and F_8 are the diagonal operators. To obtain SU(3) representations we require the commutation relations between these operators, which can be worked out by using equations (3.84) and (3.30). For example,

$$\begin{aligned}
[U_+, U_-] &= [F_6 + iF_7, F_6 - iF_7] \\
&= -2i\,[F_6, F_7] \\
&= 2\sum_{c=1}^{8} f_{67c} F_c = 2 f_{673} F_3 + 2 f_{678} F_8 \\
&= \sqrt{3}\, F_8 - F_3 = \frac{3}{2} Y - T_3 \equiv 2U_3
\end{aligned} \tag{3.90}$$

where though summation over all the eight possible values of c has been taken but, from eqn. (3.30) we can see that for 6,7 only $c = 3$ and 8 have non-vanishing structure constants. Likewise, other commutation relations can also be obtained. Putting them together, we have

$$\begin{aligned}
&[T_3, T_{\pm}] = \pm T_{\pm}, \quad [Y, T_{\pm}] = 0 \\
&[T_3, U_{\pm}] = \mp \frac{1}{2} U_{\pm}, \quad [Y, U_{\pm}] = \pm U_{\pm} \\
&[T_3, V_{\pm}] = \pm \frac{1}{2} V_{\pm}, \quad [Y, V_{\pm}] = \pm V_{\pm}
\end{aligned} \tag{3.91}$$

$$\begin{aligned}[T_+, T_-] &= 2T_3 \\ [U_+, U_-] &= \frac{3}{2}Y - T_3 \equiv 2U_3 \\ [V_+, V_-] &= \frac{3}{2}Y + T_3 \equiv 2V_3\end{aligned} \tag{3.92}$$

$$\begin{aligned}[T_+, V_+] &= [T_+, U_-] = [U_+, V_+] = 0 \\ [T_+, V_-] &= -U_-, \qquad [T_+, U_+] = V_+ \\ [U_+, V_-] &= T_-, \qquad [T_3, Y] = 0\end{aligned} \tag{3.93}$$

where T_3, U_3, and V_3 are the third components of isospin, U-spin and V-spin respectively.

3.5.2 Irreducible Representations of SU(3)

The two diagonal operators T_3 and Y commute and therefore have a common eigenvector say $|t_3, y\rangle$ obeying

$$T_3|t_3, y\rangle = t_3|t_3, y\rangle$$

and

$$Y|t_3, y\rangle = y|t_3, y\rangle \tag{3.94}$$

where t_3 and y are respective eigenvalues. The SU(3) irreducible representation is therefore labeled by two eigenvalues t_3 and y. This fact differs from SU(2), where only one eigenvalue j or t characterized the SU(2) irreducible representations. The graphical representations therefore were one-dimensional straight lines. In SU(3), because of two characters t_3 and y, the SU(3) irreducible representations are two-dimensional geometrical figures like a triangle, or a hexagon on a t_3–y plane. As in SU(2), all states in an irreducible representation can be generated from a particular state by applying repeatedly the shift operators. Notice however, that here we have three raising (T_+, U_+, V_+) and three lowering (T_-, U_-, V_-) operators in SU(3). The properties of these operators are determined from their commutation relations, which we do in the following.

From the first commutation relation of eqn. (3.91)

$$T_3T_+ = T_+T_3 + T_+$$

$$\Rightarrow \qquad T_3T_+|t_3, y\rangle = T_+(T_3 + 1)|t_3, y\rangle$$

$$= (t_3 + 1)T_+|t_3, y\rangle$$

$$\Rightarrow \qquad T_3\big|T_+|t_3, y\rangle\big\rangle = (t_3 + 1)\big|T_+|t_3, y\rangle\big\rangle \tag{3.95}$$

which is an eigenvalue equation and tells that when T_+ operator operates on a state $(|t_3, y\rangle)$, the t_3 value of the new state $(|T_+|t_3, y\rangle\rangle)$ is raised by 1. Similarly from $[Y, T_\pm] = 0$, we get

$$YT_+|t_3, y\rangle = T_+Y|t_3, y\rangle = yT_+|t_3, y\rangle$$

$$\Rightarrow \quad Y|T_+|t_3, y\rangle\rangle = y|T_+|t_3, y\rangle\rangle \tag{3.96}$$

which tells that the y value of the new state does not change. Choosing the appropriate commutation relation from equations (3.91) to (3.93) and operating them on the state $|t_3, y\rangle$, the properties of all the raising and lowering operators can be derived. How t_3 values and y values change on the operation of shift operators can be compiled in the form of a table given below:

Operator	t_3		y	
	Raises t_3 by	**Lowers t_3 by**	**Raises y by**	**Lowers y by**
T_+	1 unit	—	0	0
U_+	—	$\frac{1}{2}$ unit	1 unit	—
V_+	$\frac{1}{2}$ unit	—	1 unit	—
T_-	—	1 unit	0	0
U_-	$\frac{1}{2}$ unit	—	—	1 unit
V_-	—	$\frac{1}{2}$ unit	—	1 unit

If we plot the t_3 and y values according to above table by using proper scales for $\frac{1}{2}$ and 1 unit, we would get the graph as in Fig. (3.4).

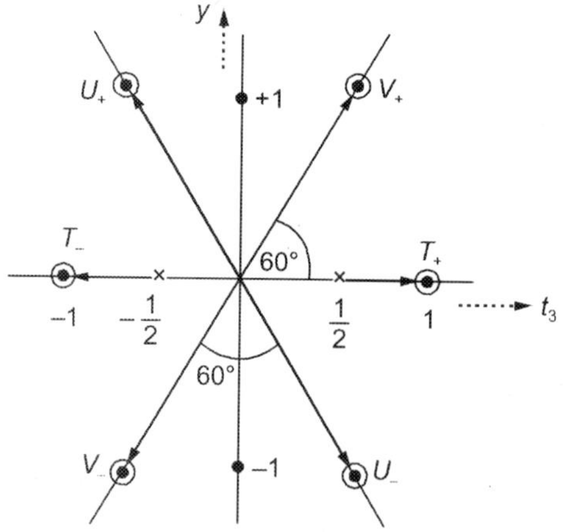

Fig. 3.4: SU (3) shift operators $T_\pm$, $U_\pm$ and $V_\pm$ drawn on a t_3–y plane. It is understood that values get lowered along negative t_3 and y axes

Above graph suggests a hexagon as the t_3—y plot. A hexagon in the t_3—y plane is an irreducible representation of SU(3) group. As an example, we consider the following hexagon (Fig.3.5), which represents the octet representation of

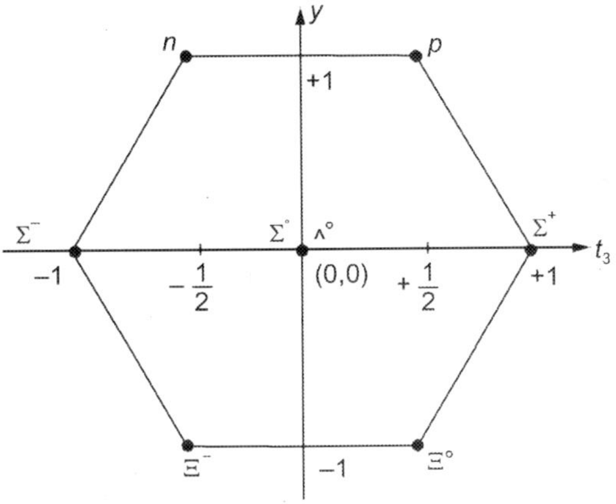

Fig. 3.5: Octet representation of SU(3). Joining the circled points in Fig. (3.4), we get Fig. (3.5) as an irreducible representation of SU(3) called octet (eight baryons occupy the sites) representation

SU(3) in which eight baryons occupy the sites of the hexagon. Sites are the points in the hexagon, which correspond to various admissible values of t_3 and y. We have already seen that t_3 can increase or decrease in units of $\frac{1}{2}$ and y in units of 1. Because eight particle states are occupying the sites of the hexagon, this irreducible representation is also called a *multiplet of SU(3)*. Another irreducible representation would likewise mean another multiplet. Depending on the dimensions of the irreducible representation, the sites may be singly, doubly or multiply occupied. By double occupation of a site, we mean that two particles have the same t_3 and y. The occupancy of the sites may also be called weights and the geometrical figure whose sites have been occupied by particle states (singly or multiply) is called the *weight diagram*. The SU(3) multiplets in general possess hexagonal weight diagrams. In some cases the weight diagram is triangular.

For a particular irreducible representation of SU(3), all the particle states of the representation can be constructed by making use of shift operators. Starting from any of the particle state of the representation, the shift operators would take us to any other state. We have seen in the hexagonal weight diagram, that there are three axes; U, V and T. if the particle states lie along the V-axis then $V_\pm$ operator will shift the particle states along that axis. For example, in the octet representation p, (Σ^0, Λ^0), Ξ^- lie along the V-axis ($\Lambda^0 \Rightarrow \Lambda$); so, $V_+|\Sigma^0\rangle = |p\rangle$ and $V_-|p\rangle = |\Sigma^0\rangle$, where $|p\rangle = \left|\frac{1}{2}, 1\right\rangle$ and $|\Sigma^0\rangle = |0, 0\rangle$ in which any state is denoted by $|t_3, y\rangle$. The particle state at the centre is a linear combination of Σ^0

and Λ^0 Similarly we can construct the particle states along the *U*-axis and *T*-axis. It is to be noted that to construct a state from any other state, a particular shift operator may have to be applied more than once, which includes the product of two different shift operators as well.

Considering the above situation and taking the hexagonal representation for SU(3) as in Fig. (3.6), we may characterize it by (p, q), where p and q are two

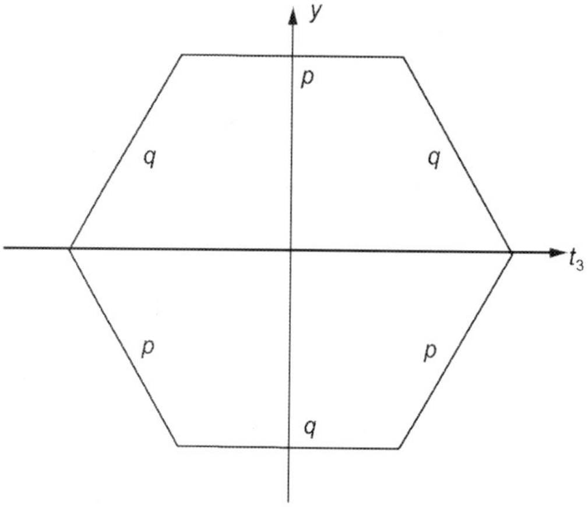

Fig. 3.6: The hexagonal representation of SU(3). For SU(3), the boundary of a representation is a hexagon specified by two lengths p and q. The number of substates of a multiplet is given by $n = \frac{1}{2}(p+1)(q+1)(p+q+2)$

integers. This means that if we start from a state of maximum t_3, the boundary of the representation will be reached in p applications of say, V_- operator and q applications of say T_- operator. The boundary of a SU(3) representation is a hexagon. All the sites on the boundary are occupied only by single particle states. The boundary of a representation implies that there are no particle states outside it. Considering the boundary as the layer no. 1, and moving towards interior, the $t_3 - y$ values would provide sites for layer 2 and layer 3, etc. The layers will be formed till the hexagon has become a triangle. The multiplicity (states with a given $t_3 - y$ value) of the layers increases by 1 from the boundary (the boundary has multiplicity 1). If the layer 3 is a triangle, each of its sites will be occupied by three particle states. The centre $(t_3 = 0, y = 0)$ will have the same multiplicity as of the triangle. The weight diagram of SU(3) for $p = 2$, and $q = 5$ (implying a (2, 5) representation) has been shown in the Fig. (3.7), where the number of dots indicates the multiplicity at each site. The maximum multiplicity is $p + 1$ if $q > p$ and it is $q + 1$ if $p > q$. In $p = 2$, $q = 5$, it is obviously 3.

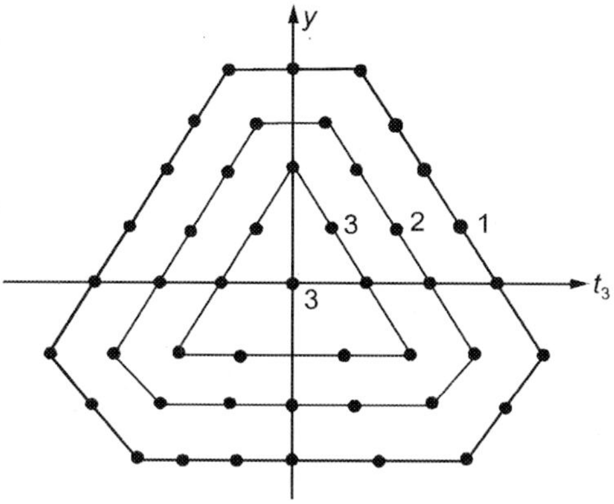

Fig. 3.7: Weight diagram for an SU(3) multiplet corresponding to $p = 2$, $q = 2$, (Fig. 3.6). The dots indicate the number of particles at that site. The numbers 1, 2, 3 indicate multiplicity. On the outer layer each dot means one particle. In layer two, each dot means two particles. On the triangle, each dot means three particles. The dot at the centre also means three particles. The total multiplicity corresponding to above weight diagram is 21 × 1 + 15 × 2 + 9 × 3 + 1 × 3 = 81. This figure shows the 81 dimensional irreducible representation of SU(3). dots indicates the multiplicity at each site. The maximum multiplicity is $p + 1$ if $q > p$ and it is $q + 1$ if $p > q$. In $p = 2$, $q = 5$ it is obviously 3

The sum of the particle states (multiplicities) at each site in all the layers and at the centre gives the dimension of the irreducible representation and is given by

$$n = \frac{1}{2}(p+1)(q+1)(p+q+2) \tag{3.97}$$

and accordingly Fig. (3.7) denotes the **81** dimensional SU(3) irreducible representation.

3.5.2.1 3 and 3* Representations of SU(3)

The **3** representation is the fundamental or the defining representation of SU(3) and is a triplet. This means the irreducible representation **3** has three particle states. When the irreducible representation has three antiparticle states, it is called an antitriplet and is denoted by **3*.3*** is a conjugate representation of **3**. We may recall that,

$$T_3 = \begin{bmatrix} \frac{1}{2} & 0 & 0 \\ 0 & -\frac{1}{2} & 0 \\ 0 & 0 & 0 \end{bmatrix} \quad \text{and} \quad Y = \begin{bmatrix} \frac{1}{3} & 0 & 0 \\ 0 & \frac{1}{3} & 0 \\ 0 & 0 & -\frac{2}{3} \end{bmatrix} \tag{3.98}$$

and the fundamental triplet

$$\phi = \begin{pmatrix} \phi_1 \\ \phi_2 \\ \phi_3 \end{pmatrix} = \phi_1 \begin{pmatrix} 1 \\ 0 \\ 0 \end{pmatrix} + \phi_2 \begin{pmatrix} 0 \\ 1 \\ 0 \end{pmatrix} + \phi_3 \begin{pmatrix} 0 \\ 0 \\ 1 \end{pmatrix} \tag{3.99}$$

is described in terms of three eigenvectors $\begin{pmatrix}1\\0\\0\end{pmatrix}, \begin{pmatrix}0\\1\\0\end{pmatrix}$ and $\begin{pmatrix}0\\0\\1\end{pmatrix}$. These eigenvectors are simultaneously the eigenstates of T_3 and Y with eigenvalues (t_3, y) as $(\frac{1}{2}, \frac{1}{3})$, $(-\frac{1}{2}, \frac{1}{3})$ and $(0, -\frac{2}{3})$ respectively. The plot $t_3 - y$ or the weight diagram for the fundamental triplet (**3**) is drawn in Fig. (3.8). We see that the triplet

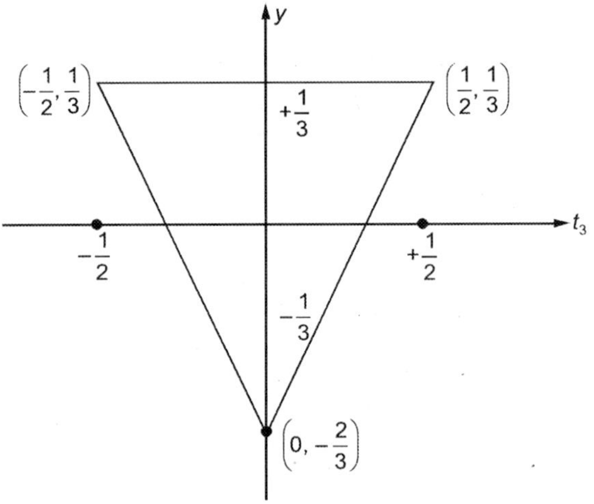

Fig. 3.8: The **3** representation or the fundamental representation of SU(3)

consists of three states; two are $T = \frac{1}{2}$, implying an isospin doublet with hypercharge $\frac{1}{3}$ each, and one $T = 0$, an isospin singlet with hypercharge $(-\frac{2}{3})$. The triangle with vertex downwards is therefore the irreducible representation of SU(3), called the *fundamental representation* and since three states are on the triangle, this representation is called a *triplet*. As in any multiplet of SU(3), here also, starting from any state $|t_3, y\rangle$ of the triplet, the shift operators $T_\pm$, $U_\pm$ and $V_\pm$ can construct other two states. The action of the shift or step operators on the fundamental triplet is shown in the following figure (Fig. 3.9).

The irreducible representation [Fig. (3.6)] was (p, q) representation and it can be seen that when ($p = 1$, $q = 0$), Fig. (3.6) reduces to Fig. (3.8), so the **3** representation is also called the (1, 0) representation. By the same token, the octet representation is $(p, q) = (1, 1)$ representation and it can be checked from the eqn. (3.97) that $n\,(1, 0) = 3$ and $n\,(1, 1) = 8$, as expected, where n denotes the dimension of the representation. Thus, when we call the irreducible representation as **3** representation, we indicate that the dimension of the representation is 3.

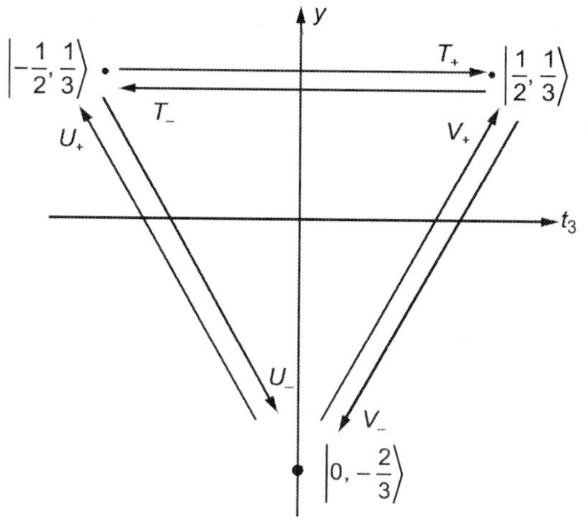

Fig. 3.9: Figure showing action of SU(3) shift operators on the fundamental triplet

For the irreducible representation **3*** of the SU(3) group, the particle states of **3** representation are replaced by their antiparticle states. These states are $\left|-\frac{1}{2},-\frac{1}{3}\right\rangle, \left|\frac{1}{2},-\frac{1}{3}\right\rangle$ and $\left|0,+\frac{2}{3}\right\rangle$ respectively. So the weight diagram for **3***would be as drawn in Fig. (3.10). The **3*** representation is also a triangle but its vertex is upwards.

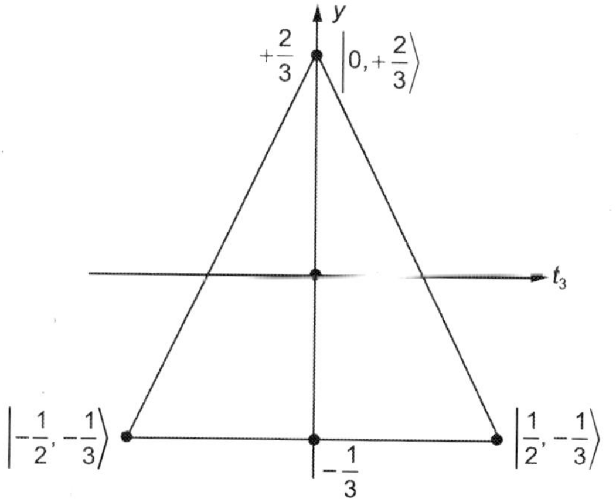

Fig. 3.10: Antitriplet or 3* representation of SU(3)

In (p, q) notation of the irreducible representations, it is (0, 1) representation and eqn. (3.97) gives its dimension too as three. Because of three antiparticle states **3*** representation is called antitriplet or just **3*** multiplet.

The physical nature of the particle and antiparticle states of respectively the **3** and **3*** representations can be examined through the Gell—Mann—Nishijima formula

$$q = t_3 + \frac{y}{2} \tag{3.100}$$

which all observed particles satisfy, qa being the charge (electric) of the particle. Applying this formula to **3** and **3***states, it can be found that the states refer to particles with electric charge $(\frac{2}{3}, -\frac{1}{3}, -\frac{1}{3})$ and antiparticles with electric charge $(-\frac{2}{3}, +\frac{1}{3}, +\frac{1}{3})$ of e, the basic unit of electric charge. Observing these fractional charges along with the corresponding t_3 and y values, we find that the particle states belong to u, d, s quarks and antiparticles to $\bar{u}, \bar{d}, \bar{s}$ antiquarks. However, no free quarks have been observed, we therefore say that no known elementary particles belong to the fundamental representation **3** or **3*** of SU(3). This observation differs from the one for the octet representation, where each of the particles is in the list of elementary particles. To find other elementary particles, we must therefore look for higher dimensional representations of SU(3).

3.5.3 Product of SU(3) Representations

The higher dimensional representations of SU(3) group too can be obtained by taking the direct products of its fundamental (**3**) representation with itself or with **3***. Since, in SU(3), the irreducible representation is a t_3–y plot; we have to construct a t_3–y plot for the product representation as well, from the t_3–y plots of given (fundamental) representations. For this purpose, we first derive the t_3 and y relations between the product and component representations.

Let us denote the column vector by χ. The number of components of χ, in case of a triplet would be three χ transforms as

$$\chi' = D^R(\theta)\chi \tag{3.101}$$

where $D^R(\theta)$ denote the set of unitary matrices

$$D^R(\theta) = \exp(i\theta_a F_a^R) \tag{3.102}$$

which form the R-dimensional representation of SU(3). θ_a are parameters, ($a = 1, \ldots, 8$. When $R = 3$, $F_a^3 = \frac{\lambda_a}{2}$, λ_a being the Gell—Mann 3×3 matrices.

Obviously, F_a^R are generators of R-dimensional SU(3) representation.

Now, let us take another column vector η, transforming as

$$\eta' = D^S(\theta)\eta \tag{3.103}$$

Taking the direct product of χ' and η', we get

$$\begin{aligned} \chi' \otimes \eta' &= [D^R(\theta)\chi] \otimes [D^S(\theta)\eta] \\ &= [D^R(\theta) \otimes D^S(\theta)][\chi \otimes \eta] \end{aligned} \tag{3.104}$$

which can also be written as (see eqn. 3.65)

$$\psi' = D^{R \otimes S}(\theta)\psi \tag{3.105}$$

where

$$D^{R\otimes S} = D^R \otimes D^S \tag{3.106a}$$

$$= \exp[i\theta_a F_a^R] \otimes \exp[i\theta_a F_a^S] \tag{3.106b}$$

$$= \exp[i\theta_a F_a^{R\otimes S}] \tag{3.106c}$$

If the parameters θ_a are small, we can see by equating the RHS of eqn. (3.106c) and RHS of eqn. (3.106b) that

$$1 + i\theta_a F_a^{R\otimes S} = (1 + i\theta_a F_a^R) \otimes (1 + \theta_a F_a^S)$$

$$\approx 1 + i\theta_a (F_a^R \oplus F_a^S) \qquad O(\theta) \tag{3.107}$$

from which we infer that

$$F_a^{R\otimes S} = F_a^R \oplus F_a^S \tag{3.108a}$$

where, LHS are the generators of $R \otimes S$ dimensional (product) representation of SU(3). The decomposition (3.108a) suggests that

$$F_a^{R\otimes S} = \left[\begin{array}{c|c} F_a^R & 0 \\ \hline 0 & F_a^S \end{array}\right] \tag{3.108b}$$

Since, F_a^R are the generators of SU(3) group and for the fundamental representation $R = 3$, only $a = 3$ and $a = 8$ generators in F_a^3 are diagonal. Again, $F_3 \to T_3$, $F_8 \to Y$ the decomposition (3.108a) suggests

$$T_3^{R\otimes S} = T_3^R \oplus T_3^S$$

and

$$Y^{R\otimes S} = Y^R \oplus Y^S \tag{3.109a}$$

and hence the eigenvalues

$$t_3^{R\otimes S} = t_3^R \oplus t_3^S$$

and

$$y^{R\otimes S} = y^R \oplus y^S \tag{3.109b}$$

which tell that the quantum numbers get added in the product representation. This additivity of t_3 and y can be used graphically to obtain the product of SU(3) multiplets.

3.5.3.1 3 ⊗ 3 Representation

We note that the representation **3** is an inverted triangle in t_3–y plane with $t_3 = 0$, $y = 0$ as its centre. Graphically, the **3 ⊗ 3** representation can be obtained by placing the centre of one triangle on each of the vertices of other triangle. The figure (Fig. 3.11) describes the process.

In the **3** ⊗ **3** representation, we can identify a **3*** representation and a **6** representation and we say **3** ⊗ **3** is reducible into **3*** and **6**, i.e.

$$\mathbf{3} \otimes \mathbf{3} = \mathbf{3}^* \oplus \mathbf{6} \tag{3.110a}$$

which in $D^n(p, q)$ notation ($D^n = n\ D^n = n$ dimensions) is

$$D^3(1, 0) \otimes D^3\ (1, 0) = D^3\ (0, 1) \oplus D^6\ (2, 0) \tag{3.110b}$$

Thus, when two triplets of SU(3) combine, we get an antitriplet and a hexad. The antitriplet is antisymmetric and **6** is symmetric. If the triplet is (d, u, s), then **3***is antisymmetric under the interchange of two quarks while the hexad is symmetric. The states in **3*** and **6** are diquark states, which do not belong to known elementary particles.

Young Tableau technique can provide the reduction of **3** ⊗ **3** in a simpler way. In SU(3) we denote

$$\mathbf{3} \text{ by } \square \text{ and } \mathbf{3}^* \text{ by } \begin{array}{c}\square\\[-4pt]\square\end{array} \tag{3.111}$$

$$\mathbf{3} \otimes \mathbf{3} = \square \otimes \square = \square\square \oplus \begin{array}{c}\square\\[-4pt]\square\end{array} \tag{3.112}$$

(a)

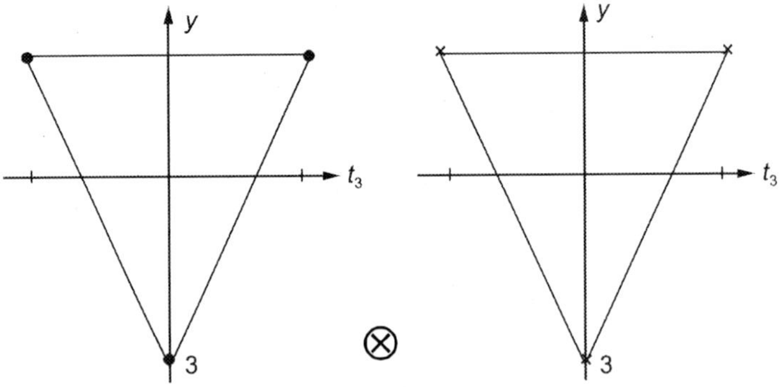

(b)

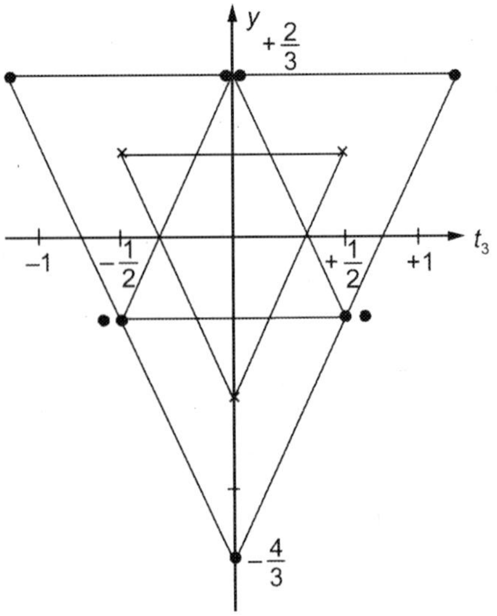

(c)

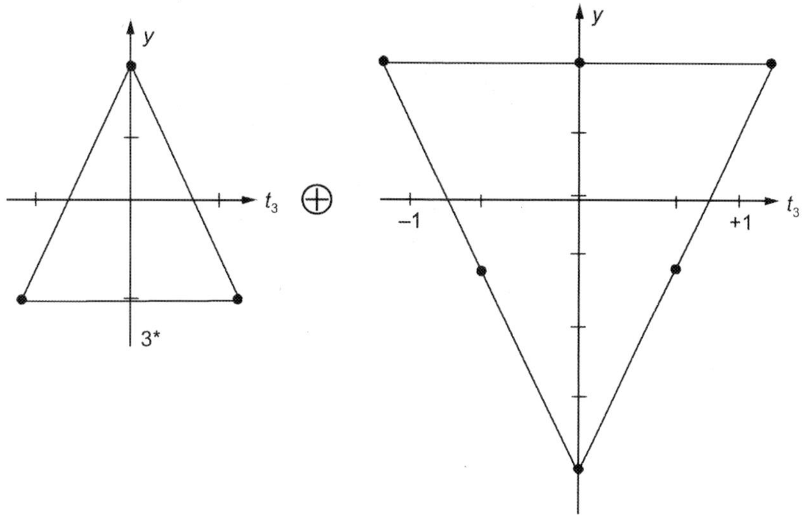

Fig. 3.11: The graphical representation of 3 ⊗ 3 = 3* ⊕ 6 Fig. (a) shows two, **3** representations. In Fig. (b) on every vertex of other **3** (× sites) the first triangle is placed. Fig. (b) can be separated into Fig. (c). 6 is a six dimensional multiplet

in which the dimensions of  are 6 and the dimensions of the column  are though 3, it is defined as **3***. We calculate these dimensions in the following. The dimensions of

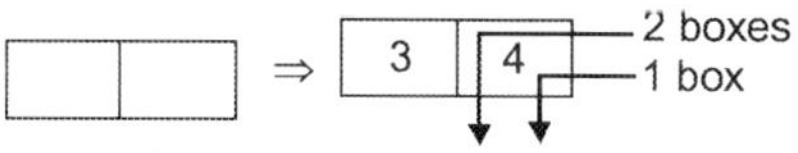

$$= \frac{3\times 4}{2\times 1} = \mathbf{6} \tag{3.113a}$$

and the dimensions of

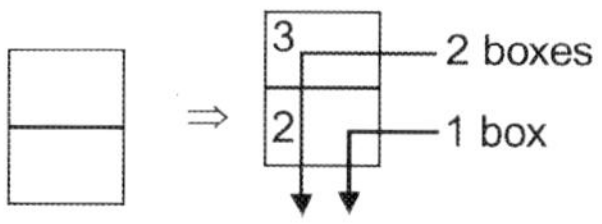

$$= \frac{3\times 2}{2\times 1} = 3 \Rightarrow \mathbf{3}^* \tag{3.113b}$$

Therefore, eqn. (3.112) is

$$\mathbf{3} \otimes \mathbf{3} = \mathbf{6} \oplus \mathbf{3}^* \tag{3.114}$$

It has to be noted that, while assigning numbers [(3, 4) and (3, 2)] to boxes the 3 of SU(3) was kept in mind. The first box was assigned the number 3. In SU(2), too, the number assigned to next box in the row increased from 2 by 1 while the number assigned to next box in the column decreased from 2 by 1.

3.5.3.2 3 ⊗ 3 Representation

Graphically (Fig. 3.12)

$$\mathbf{3} \otimes \mathbf{3}^* = \mathbf{9} = \mathbf{1} \oplus \mathbf{8}$$

or

$$D^3(1, 0) \otimes D^3(0,1) = D^1(0, 0) \oplus D^8(1, 1) \tag{3.115}$$

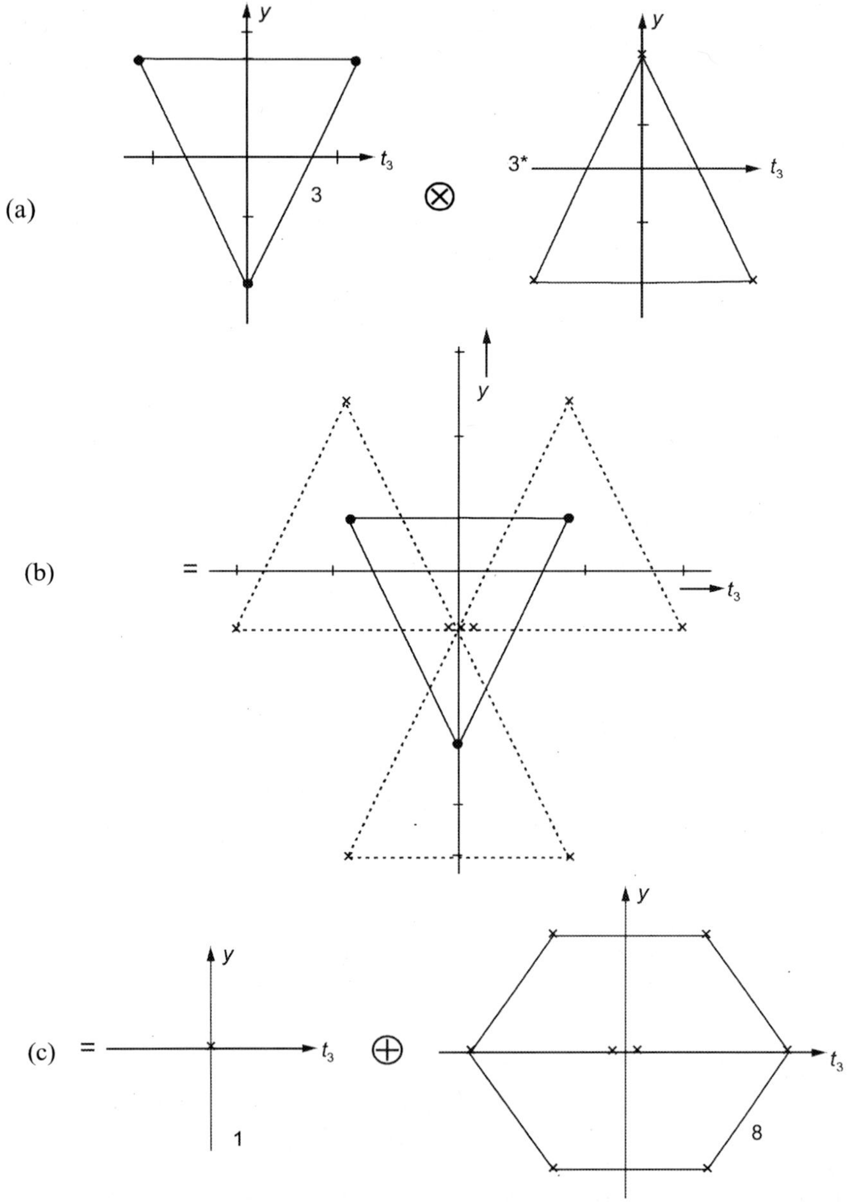

Fig. 3.12: Graphical representation of 3 ⊗ 3* = 1 ⊕ 8. Fig. (a) SU(3) representation 3 and 3* with dots on the sites of **3** and crosses on 3*. Fig. (b) 3* triangle is placed on each vertex of 3 triangle. The separation of superimposition (b) results in Fig. (c)

The physical meaning of **3** ⊗ **3*** may be given as the combination of an SU(3) triplet and antitriplet leading to a singlet and an octet. In terms of quark triplet and antiquark antitriplet, the graphical representation is as in Fig. (3.13).

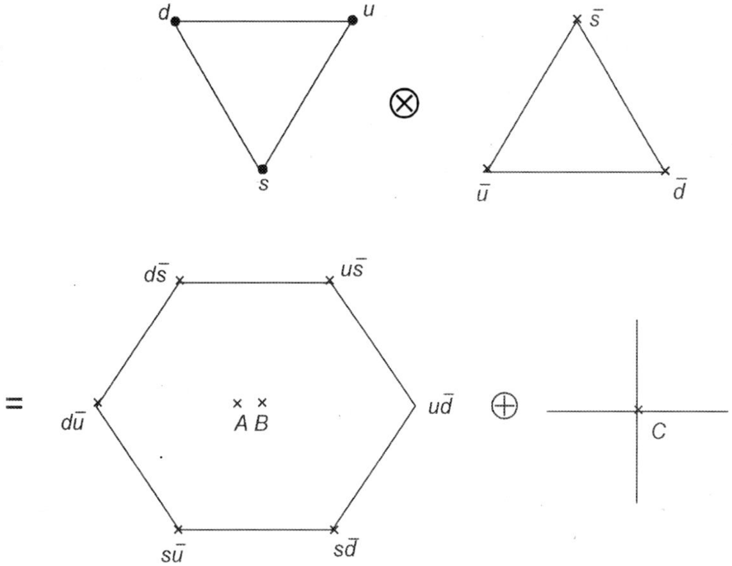

(3.116)

Fig. 3.13: The physical meaning of 3 ⊗ 3* = 8 ⊕ 1 of Fig. (3.12) is expressed by above figure equation no. (3.116). **8** represents eight mesons belonging to the octet, and meson **1** is singlet

Unlike **3** ⊗ **3** representation, in **3** ⊗ **3***, we have quark antiquark states, which constitute mesons. Thus the octet and the singlet of Fig. (3.13) refer to mesons. The eight states transform among themselves hence an octet, but do not mix with the ninth, hence the ninth (C) is a singlet. The mesons of the octet and the singlet are given in the following table.

Quark content	Meson	Mass in MeV/C^2
$d\bar{s}$	K^0	494
$u\bar{s}$	K^+	494
$d\bar{u}$	π^-	140
$\frac{1}{\sqrt{2}}(u\bar{u} - d\bar{d})$	$\pi^0(A)$	135
$\frac{1}{\sqrt{6}}(u\bar{u} + d\bar{d} - 2s\bar{s})$	$\eta\ (B)$	549
$u\bar{d}$	π^+	140
$s\bar{u}$	K^-	494
$s\bar{d}$	$\bar{K}^0$	494
$\frac{1}{\sqrt{3}}(u\bar{u} + d\bar{d} + s\bar{s})$	$\eta'\ (C)$	958

Figure (3.13) gives the spin zero, odd parity (0^-), SU(3) octet and singlet of measons. They are 1^- and 2^+ SU(3) octets and singlets as well.

Young Tableau gives **3** ⊗ **3*** reduction as below.

3 ⊗ **3*** = **8** ⊕ **1**

$$D^3\,(1, 0) \otimes D^3\,(0, 1) = D^8\,(1, 1) \oplus D^1\,(0, 0) \tag{3.117}$$

The dimensions **8** and **1** can be checked. The dimension of

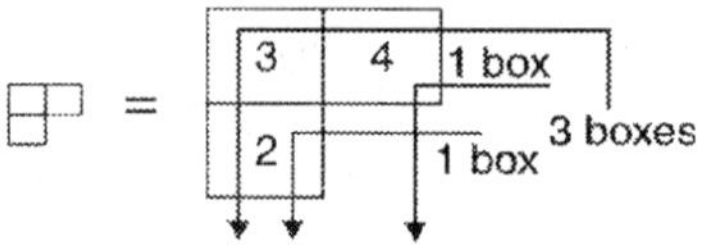

The dimension of

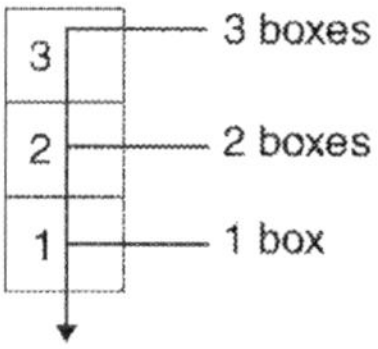

$$= \frac{3\times 2\times 1}{3\times 2\times 1} = 1$$

3.5.3.3 3 ⊗ 6 Representation

Graphically, the product representation **3** ⊗ **6** has been drawn in the following figure (Fig. 3.14).

Since each vertex of **3** is single quark state and each vertex of **6** is a two-quark state, the combination in **8** and **10** has three quark states at their sites. These three quark states are baryons, so **8** and **10** representations of SU(3) describe the baryons. The **8** of Fig. (3.14) is the baryon octet of Fig. (3.5) and representation **10** is the baryon decuplet of Chapter 1.

These hexagonal and triangular patterns for both mesons and baryons were described in Eightfold Way in Chapter one without advancing the reason as to why the hadrons are arranged in such geometrical patterns. Now, it can be said that these are the SU(3) product representations that provide such geometrical patterns to hadrons, thus confirming the SU(3)—classification scheme of hadrons.

The Young Tableau for **3** ⊗ **6** is given below:

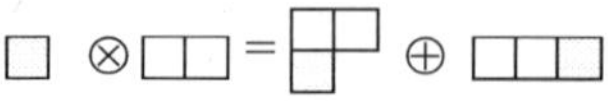

3 ⊗ **6** = **8** ⊕ **10**

$$D^3\,(1, 0) \otimes D^6\,(2, 0) = D^8\,(1, 1) \oplus D^{10}\,(0, 3) \tag{3.118}$$

The dimensions **6** and **8** have already been calculated in the previous sections. The dimension **10** can be calculated as

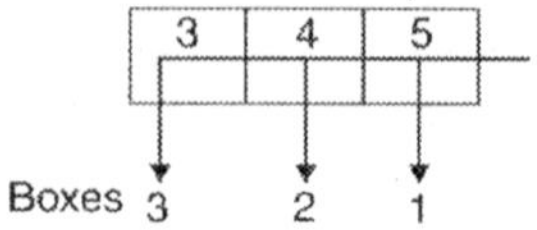

$$= \frac{3\times 4\times 5}{3\times 2\times 1} = \mathbf{10}$$

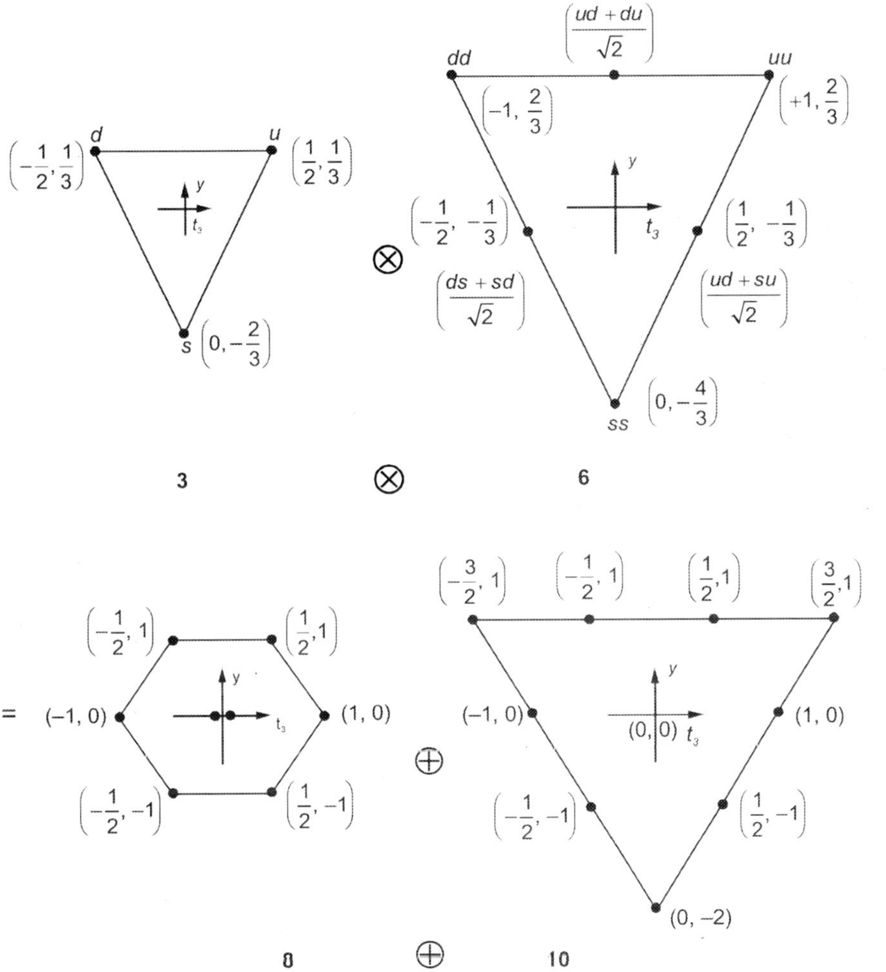

Fig. 3.14: Graphical representation of $3 \otimes 6 = 8 \oplus 10$. Since, at the sites of **8** and **10** we have three quark states, the octet (**8**) and decuplet (**10**) are baryon representations

3.5.3.4 Some Other Product Representations of SU(3)

(1)
$$\begin{aligned} \mathbf{3} \otimes \mathbf{3} \otimes \mathbf{3} &= \mathbf{3} \otimes (\mathbf{3}^* \oplus \mathbf{6}) \\ &= \mathbf{3} \otimes \mathbf{3}^* \oplus \mathbf{3} \otimes \mathbf{6} \\ &= \mathbf{1} \oplus \mathbf{8} \oplus \mathbf{8} \oplus \mathbf{10} \end{aligned} \tag{3.119}$$

Young tableau

$$\boxed{1} \otimes \boxed{2} \otimes \boxed{3} = \begin{array}{|c|}\hline 1 \\ \hline 2 \\ \hline 3 \\ \hline \end{array} \oplus \begin{array}{|c|c|}\hline 1 & 3 \\ \hline 2 \\ \cline{1-1} \end{array} \oplus \begin{array}{|c|c|}\hline 1 & 2 \\ \hline 3 \\ \cline{1-1} \end{array} \oplus \begin{array}{|c|c|c|}\hline 1 & 2 & 3 \\ \hline \end{array}$$

$$= \mathbf{1} \quad \oplus \quad \mathbf{8} \quad \oplus \quad \mathbf{8} \quad \oplus \quad \mathbf{10}$$

in which, out of the two octets (**8**) one is antisymmetric under the interchange of first two quark indices and other is symmetric under the interchange.

(2) $\mathbf{8}\otimes\mathbf{8} =$

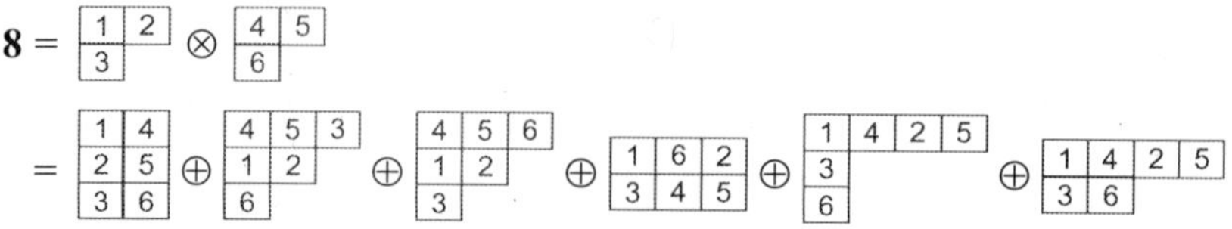

$= \mathbf{1} \oplus \mathbf{8} \oplus \mathbf{8} \oplus \mathbf{10}^* \oplus \mathbf{10} \oplus \mathbf{27}$

or $D^8(1, 1) \otimes D^8(1, 1) = D^1(0, 0) \oplus D^8(1, 1) \oplus D^8(1, 1) \oplus D^{10}(0, 3)$
$\oplus D^{10}(3, 0) \oplus D^{27}(2, 2)$ (3.120)

For demonstration, we calculate the dimension **27.**

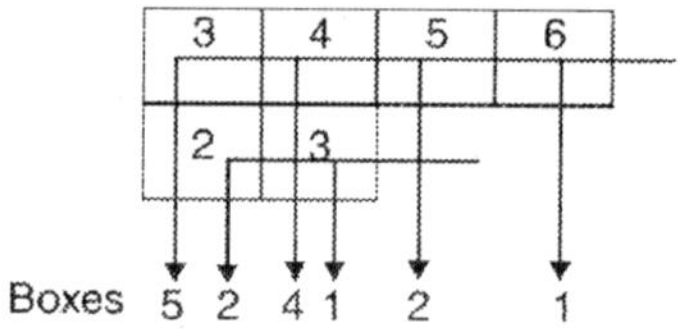

$$= \frac{3\times4\times5\times6\times2\times3}{5\times2\times4\times1\times2\times1} = \mathbf{27}$$

In the above decomposition, $\mathbf{10}^*$ is the inverted triangle of **10** (decuplet).

3.5.4 The Gell—Mann—Okubo Mass Formula

An important aspect of symmetry is connected with the relation between the degeneracy of energy eigenvalues and the symmetry. In the present context, by symmetry, we would mean that the Hamiltonian is invariant under SU(3) transformations, which implies that the operators of the theory (which cause transformations) must commute with the Hamiltonian (of strong interactions). If $|\alpha\rangle$ represents an energy eigenstate of the system, then

$$\mathrm{H}|\alpha\rangle = E_\alpha|\alpha\rangle \tag{3.121}$$

and if Ω be the operator obeying

$$[\Omega, \mathrm{H}] = 0 \tag{3.122}$$

then

$$\mathrm{H}|\Omega|\alpha\rangle\rangle = E_\alpha|\Omega|\alpha\rangle\rangle \tag{3.123}$$

implying that $\Omega|\alpha\rangle$ is also an eigenstate of H corresponding to same energy E_α to which the state $|\alpha\rangle$ belongs. If $|\alpha\rangle$ and $\Omega|\alpha\rangle$ are linearly independent, the eigenvalue E_α is degenerate (two-fold).

Now, apply above picture to the baryon octet of SU(3) shown in Fig. (3.5). Thus when the Hamiltonian is invariant under SU(3) transformations, all the eight-baryon states are degenerate in energy, i.e., belong to same energy. This means that in the rest frame, the rest masses of all the eight baryons must be same. This observation can be proved as below.

We have already discussed earlier that shift operators can transform one baryon state of the octet into other. For example, U_+ operator transforms $|\Sigma^+\rangle$ into $|p\rangle$.

Now,

$$\mathrm{H}\left|\Sigma^{+}\right\rangle = m_{\Sigma}\left|\Sigma^{+}\right\rangle \tag{3.124}$$

Since,

$$U_{+}\left|\Sigma^{+}\right\rangle = |p\rangle \tag{3.125}$$

and because of SU(3) invariance,

$$[U_{+}, \mathrm{H}] = 0 \tag{3.126}$$

we may write from eqn. (3.126)

$$\mathrm{H}U_{+}\left|\Sigma^{+}\right\rangle = U_{+}\mathrm{H}\left|\Sigma^{+}\right\rangle = m_{\Sigma}U_{+}\left|\Sigma^{+}\right\rangle$$

or

$$\mathrm{H}|p\rangle = m_{p}|p\rangle = m_{\Sigma}U_{+}\left|\Sigma^{+}\right\rangle \tag{3.127}$$

implying that $m_p = m_\Sigma$, i.e., the rest masses of proton and the sigma particle are same. Extending the process to other baryons, we would conclude that if the Hamiltonian (of strong interactions) is invariant under SU(3) transformations, the rest masses of all the particles (here baryons) belonging to a particular irreducible representation (here octet) must be equal. In reality, however, we find that the rest masses of baryons differ substantially except in case of proton and neutron, which differ slightly. Experimentally, we have $m_p, n \approx 940$ MeV, $m_\Lambda = 1115$ MeV, $m\Sigma = 1193$ MeV and $m_\Xi = 1318$ MeV. This discrepancy suggests that the strong interaction Hamiltonian is not invariant under SU(3) transformations. In other words, the SU(3) symmetry does not appear to be exact. We assume that the SU(3) symmetry is approximate and hence take the Hamiltonian of strong interactions as

$$\mathrm{H} = \mathrm{H}_0 + \mathrm{H}_{SV} \tag{3.128}$$

where H_0 is the part which is invariant under SU(3) transformations and H_{SV} is the part of the Hamiltonian which is causing the violation of SU(3) symmetry. If $\mathrm{H}_{SV} = 0$, then SU(3) symmetry would be exact and all masses would be same and since masses are not equal, the H_{SV} is not zero.

Among the shift operators, we have also isospin raising and lowering operators $T_\pm$ along with $U_\pm$ and $V_\pm$. Operators $\mathrm{T}_\pm$ change for example, p into n or Σ^+ into Σ^0 or Σ^- or Ξ^0 into Ξ^-. Since, the mass difference between the isospin states of a baryon is quite small (e.g., $m_p \approx m_n$), we can assume that as far as operators $T_\pm$ are concerned the SU(3) symmetry is exact and $T_\pm$ are assumed commuting with H_{SV} and hence with H of eqn. (3.128). However, $U_\pm$ and $U_\pm$ which transform particles of one hypercharge to other, including t_3, do not commute with H_{SV} and so SU(3) symmetry is violated. Since H_{SV} is small compared to H_0, it is like a perturbation on H_0. We also assume that H_{SV} obeys the SU(3) transformation property that it is given by the hypercharge Y, or

$$\mathrm{H}_{SV} \approx Y \approx F_8 \tag{3.129}$$

The consequence of this assumption becomes manifest when we consider that the hadrons are made of quarks. The assumption (3.129) then tells that the mass difference between the states (baryons or mesons) of a SU(3) representation (say, octet) is only due to the mass difference between the quarks constituting those states (particles). The binding energies of quarks are taken independent of flavour. Then, for the masses of the baryon octet, we may write

$$
\begin{aligned}
&\because p \to udu,\ n \to udd && \therefore m_N = m_0 + 3m_u \\
&\because \left\{\begin{matrix} \Sigma^+ \to suu \\ \Sigma^0 \to s(ud+du)/\sqrt{2} \\ \Sigma^- \to sdd \end{matrix}\right\} && \therefore m_\Sigma = m_0 + 2m_u + m_s \\
&\because \Xi^0 \to ssu,\ \Xi^- \to ssd && \therefore m_\Xi = m_0 + m_u + 2m_s \\
&\because \Lambda \to s(ud-du)/\sqrt{2} && \therefore m_\Lambda = m_0 + 2m_u + m_s
\end{aligned}
\tag{3.130}
$$

where $m_d = m_u$ has been used. m_0 is the common mass which does not depend on flavors (u, d, s). $m_u = m_d$ and m_s are flavour dependent masses.

Solving four mass equations (3.130), by eliminating m_0, m_u and m_s, we get

$$m_\Sigma + 3m_\Lambda = 2(m_N + m_\Xi) \tag{3.131}$$

which is the Gell—Mann—Okubo mass formula for the octet. The general Gell—Mann—Okubo mass formula is however,

$$m_a = m_0 + m_1 Y + m_2[T(T+1) - \tfrac{1}{4}Y^2] \tag{3.132}$$

where a is the state $|a\rangle$ of the multiplet and m_0, m_1, m_2 are the unknown constants for each multiplet.

The relation (3.131) agrees fairly well with the experimental values as we can see that

$$\{1193 + 3(1115)\}\,\text{MeV} = 4538\,\text{MeV} = \{2(940 + 1315)\}\,\text{MeV} = 4510\,\text{MeV} \tag{3.133}$$

the mass difference in the LHS and the RHS masses is small. The corresponding equation, i.e., the Gell—Mann—Okubo mass formula, for the meson octet (Fig.3.16 the charm 0 mesons) is

$$4m_K = 3m_\eta + m_\pi \tag{3.134}$$

which, using $m_K = 495$ MeV and $m_\pi = 140$ MeV and gives $m_\eta = 620$ MeV, which does not agree with its experimental value of 550 MeV. It was later discovered that the meson mass formula worked better if we use (mass)^2, i.e.

$$4m_K^2 = 3m_\eta^2 + m_\pi^2 \tag{3.135}$$

from which we get $m_\eta = 656$ MeV, a value better than 620 MeV obtained through a linear mass relation.

Therefore, an important conclusion that emerges from the Gell—Mann—Okubo mass formula for fermions and mesons (bosons) is that in the Lagrangian densities for fermionic fields, we use linear mass terms (like $m\bar{\psi}\psi$) and quadratic terms (like $\frac{1}{2}m^2\phi^2$) for bosonic fields. This seems to be justified, because the fermions obey Dirac equation in which mass is to the first order, while bosons obey Klein—Gordon equation in which mass is to the second order.

3.5.5 Applications of SU(3) Symmetry Group

Following may be deemed the applications of SU(3) group.

1. Classification of known particles and resonances can be done by SU(3) multiplets. For example, as already discussed, $\frac{1}{2}^+$ baryons form an octet; $\frac{3}{2}^+$ resonances form a decuplet, $\mathbf{1} \oplus \mathbf{8}$ represents the 0^-, 1^-, 1^+, 2^+ mesons. Here $\frac{1}{2}^+$ implies spin - $\frac{1}{2}$ and parity even (+) and so on.
2. The masses of the particles in a multiplet can be predicted by observing the breaking of SU(3) symmetry. Gell—Mann—Okubo mass formula for baryons and mesons can be obtained.
3. Using the tables of Clebsch–Gordan coefficients for SU(3), the relations between cross sections of reactions like

$$\begin{aligned} &\Delta^- \to n + \pi^- \\ &\Sigma^{*-} \to \Lambda^0 + \pi^- \\ &\Sigma^{*-} \to \Sigma^0 + \pi^- \\ &\Xi^{*-} \to \Xi^0 + \pi^- \end{aligned} \tag{3.136}$$

can be obtained. The calculated decay widths agree fairly well with the experiment.
4. The group SU(3) can predict the magnetic moments of baryons. The baryon octet ($\frac{1}{2}^+$) has been studied regarding baryon magnetic moments. On the basis of SU(3) symmetry, following relations for the magnetic moments of octet baryons can be derived.

$$\begin{aligned} &\mu_{\Sigma^0} = -\frac{\mu_n}{2}, \quad \mu_{\Lambda^0} = +\frac{\mu_n}{2}, \mu_{\Xi^0} = \mu_n \\ &\mu_{\Sigma^+} = \mu_p, \qquad \mu_{\Sigma^-} = -(\mu_n + \mu_p) = \mu_{\Xi^-} \end{aligned} \tag{3.137}$$

5. Making a comparison between the SU(3) predicted and experimentally observed magnetic moments, it is found that the predicted values of

magnetic moments of Σ^+ and Λ^0 are +2.79 and –0.85 nuclear magnetons (nm) respectively and agreeing closely with these values are the experimentally observed magnetic moments of $\Sigma^+ = 2.62$ nm and of $\Lambda^0 = -0.70$ nm. This explains the application of SU(3) in predicting the magnetic dipole moments of baryons.

We may thus summarize that the SU(3) symmetry group is considerably successful in the classification of elementary particles, in the prediction of new particles and in the prediction of their masses, magnetic moments and decay widths. These applications are also the tests of SU(3) that determine the scope of this group.

3.5.6 Limitations of SU(3) Group

The successes of SU(3) symmetry group, outlined in the previous section are in fact not total, and leave gaps where shortcomings of the group SU(3) can be seen. We mention them in the following as the limitations of SU(3).

1. Though, we can construct baryons from the **3** representation as

$$\mathbf{3}\otimes\mathbf{3}\otimes\mathbf{3}=\mathbf{1}\oplus\mathbf{8}\oplus\mathbf{8}\oplus\mathbf{10} \tag{3.138}$$

   where, however, there are two octets of baryons, one is symmetric and other antisymmetric (in the interchange of first two quark indices). The presence of two octets leaves the question that to which octet the observed eight baryons belong.

2. SU(3) predicts octets of baryons and mesons. Now, if we take the baryon octet (N, Λ, Σ, Ξ) as the basic and wish to construct a decuplet of baryons, we could do it by

$$\mathbf{8}\otimes\mathbf{8}^{*}=\mathbf{27}\oplus\mathbf{10}\oplus\mathbf{10}^{*}\oplus\mathbf{8}\oplus\mathbf{8}\oplus\mathbf{1} \tag{3.139}$$

   which though provides a decuplet **10**, but with unphysical **10***, **27** and another **8**. We can also take the octet of mesons as basic and may attempt to construct a **10** through

$$\mathbf{8}_{\mathbf{baryon}}\otimes\mathbf{8}_{\mathbf{meson}}=\mathbf{27}\oplus\mathbf{10}\oplus\mathbf{10}^{*}\oplus\mathbf{8}\oplus\mathbf{8}\oplus\mathbf{1} \tag{3.140}$$

   we get the same reduction as in eqn. (3.139), where we find the unobserved splitting of type **27** and **10***, in addition to another **8.**

3. The symmetry group SU(3), though predicts the magnetic dipole moments of octet baryons, these magnetic moments are only in terms of the magnetic dipole moments of proton (μ_p) and neutron (μ_n). SU(3) neither gives the values of μ_p and μ_n, nor their ratio.
4. The new cross section relations outside SU(3) cannot be predicted.
5. SU(3) cannot incorporate spin because of insufficient number of additive quantum numbers.
6. SU(3) cannot provide a logical basis for a regularity of level spacings for meson resonances.

3.5.7 Extension of SU(3) Group: SU(4)

The limitations of SU(3) naturally ask for symmetries higher than SU(3) so that new quantum numbers could be included. This requirement could be met by extending the SU(3) symmetry group. As already told, we have SU(3) flavour, which is equivalent to SU(3) of isospin and strangeness and there is SU(3) of color. In the former $\begin{pmatrix} u \\ d \\ s \end{pmatrix}$ or $\begin{pmatrix} p \\ n \\ \Lambda \end{pmatrix}$ is the fundamental triplet and $\begin{pmatrix} R \\ G \\ B \end{pmatrix}$ in the latter.

The simplest extension of SU(3) is to SU(4), where a fourth dimension has to be included. This can be done in two ways, one, by incorporating spin into isospin and two, by incorporating fourth flavour (charm c) of quarks. The former gives the SU(4) group of isospin and spin and latter the SU(4) flavour. The SU(4)

$$\phi_{\text{flavour}} = \begin{pmatrix} u \\ d \\ s \\ c \end{pmatrix} \tag{3.141}$$

flavour group considers as its fundamental quartet and SU(4) of isospin and spin starts with the isospin doublet $\begin{pmatrix} u \\ d \end{pmatrix}$ and incorporates two spin states, spin up $(\uparrow)$ and spin down $(\downarrow)$ to each of the members of the isospin doublet. Thus, $u^{\uparrow}$, $u_{\downarrow}$, $d^{\uparrow}$, $d_{\downarrow}$ form the four states of the SU(4) isospin-spin group.

Obviously, the SU(4) group is a 15-parameter group and its rank is 3, which is one higher than SU(3). SU(3) had two additive quantum numbers, t_3 and y the flavour SU(4) has three additive quantum numbers, t_3, y and c, where c is the charm quantum number which is similar to strangeness quantum number. A charm quark will have $c = +1$ and charm antiquark shall have $c = -1$. The values of c are assigned according to Gell—Mann Nishijima relation, which now reads

$$\frac{q}{e} = t_3 + \frac{b+s+c}{2} \tag{3.142}$$

where t_3 denotes the third component of isospin, b the baryon number, s the strangeness number and c the charm number. The eigenvalues t_3 of T_3 operator for quarks are obviously +1/2 and –1/2.

Just as the irreducible representations of SU(3) are the plots on t_3–y plane, the irreducible representations of SU(4) are three-dimensional figures plotted on t_3, y and c axes, where c is the eigenvalue similar to y and t_3. The simplest multiplet of SU(4) is the unitary singlet **1.** Next simplest are **4** and **4***. At the sites of **4** (tetrahedron, **3** was a triangle) lie the four quarks, u, d, s and c whereas at the sites of **4*** lie antiquarks $\bar{u}, \bar{d}, \bar{s}, \bar{c}$. The **4** and **4*** representations of SU(4) have been drawn in the figure (Fig. 3.15).

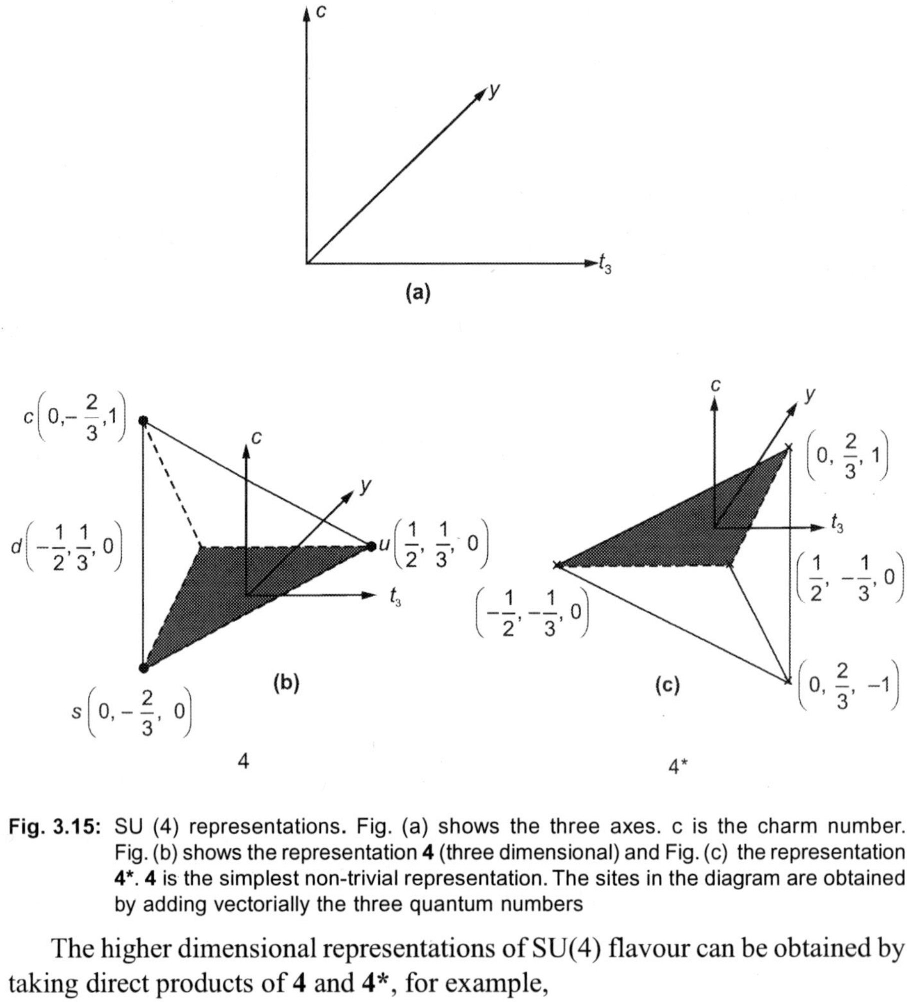

Fig. 3.15: SU (4) representations. Fig. (a) shows the three axes. c is the charm number. Fig. (b) shows the representation **4** (three dimensional) and Fig. (c) the representation **4***. **4** is the simplest non-trivial representation. The sites in the diagram are obtained by adding vectorially the three quantum numbers

The higher dimensional representations of SU(4) flavour can be obtained by taking direct products of **4** and **4***, for example,

$$\mathbf{4} \otimes \mathbf{4}^* = \mathbf{1} \oplus \mathbf{15} \tag{3.143}$$

which is a sixteen dimensional representation in which the **15**-plet denotes a multiplet of fifteen particle states and **1** denotes a singlet. In the **15**-plet, we can identify an octet **8**, a singlet **1**, a triplet **3** and an antitriplet **3***. This is the SU(3) decomposition of **15**-plet as

$$\mathbf{15} = \mathbf{3} \oplus \mathbf{8} \oplus \mathbf{1} \oplus \mathbf{3}^* \tag{3.144}$$

The following figure (Fig.3.16) whose sites are mesons expresses eqn. (3.143) in which η_c is the singlet of eqn. (3.143) and η' is the singlet of eqn. (3.144).

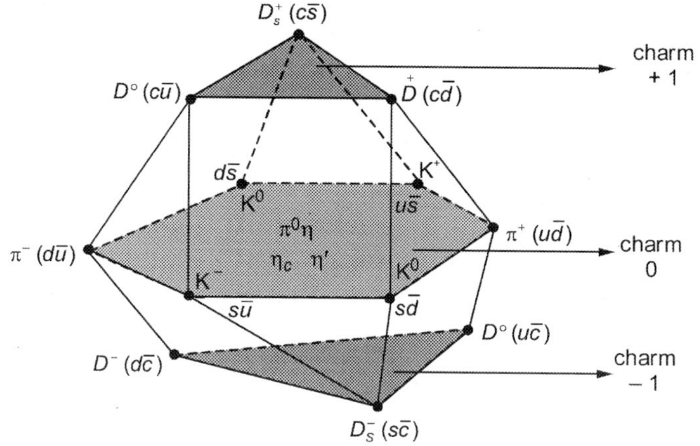

Fig. 3.16: SU(4) multiplets of **0⁻** (pseudoscalar) mesons. This is sixteen dimensional representation, where $\mathbf{4} \otimes \mathbf{4^*} = \mathbf{15} \oplus \mathbf{1}$. η_c is the singlet. The cuboctahedron (15 plet) is seen composed of a triplet (charm-1 plane), an octet plus singlet (η') (charm 0 plane) and an antitriplet (charm +1 plane)

The product (3.143) describes the charmed mesons D^0; $\bar{D}^0$; D_S^+, D_S^-; D^+, D^- in addition to uncharmed $\mathbf{1} \oplus \mathbf{8}$ mesons. Following is the list of lightest charmed mesons. Note that a meson is a quark—antiquark state and a charmed meson would contain at least one charmed quark or a charmed antiquark.

Particle	Spin	Quark structure	Mass MeV/C²
$D^0, \bar{D}^0$	0	$c\bar{u}, u\bar{c}$	1865
$D^{\pm}$	0	$c\bar{d}, d\bar{c}$	1869
$D_S^{\pm}$	0	$c\bar{s}, s\bar{c}$	1971
$D^{*0}, \bar{D}^{*0}$	1	$c\bar{u}, u\bar{c}$	
D^{*+}, D^{*-}	1	$c\bar{d}, d\bar{c}$	2010
$D_S^{*\pm}$	1	$c\bar{s}, s\bar{c}$	
η_c	0	$c\bar{c}$	2981
J/ψ	1	$c\bar{c}$	3097

Thus in $\mathbf{15} \oplus \mathbf{1}$ mesons, if the net spin of the quark and antiquark is zero, we have pseudoscalar (0^-) mesons and when the net spin is 1, we have vector (1^-) mesons. ρ^-, ρ^0, ω^0, ϕ^0, ρ^+ are the uncharmed vector (1^-) mesons.

From the representation **4** of SU(4), we can also construct charmed baryon states by taking the product

$$\mathbf{4} \otimes \mathbf{4} \otimes \mathbf{4} = \mathbf{4} \oplus \mathbf{20'} \oplus \mathbf{20'} \oplus \mathbf{20} \tag{3.145}$$

which is a sixty four dimensional representation that can be reduced into a quartet

(**4**), two **20′**-plets and a twenty-plet (**20**). By twenty-plet, we mean an irreducible representation (multiplet), which has twenty occupied particle states. The multiplets in the RHS of eqn. (3.145), though irreducible for SU(4) are reducible for SU(3) as

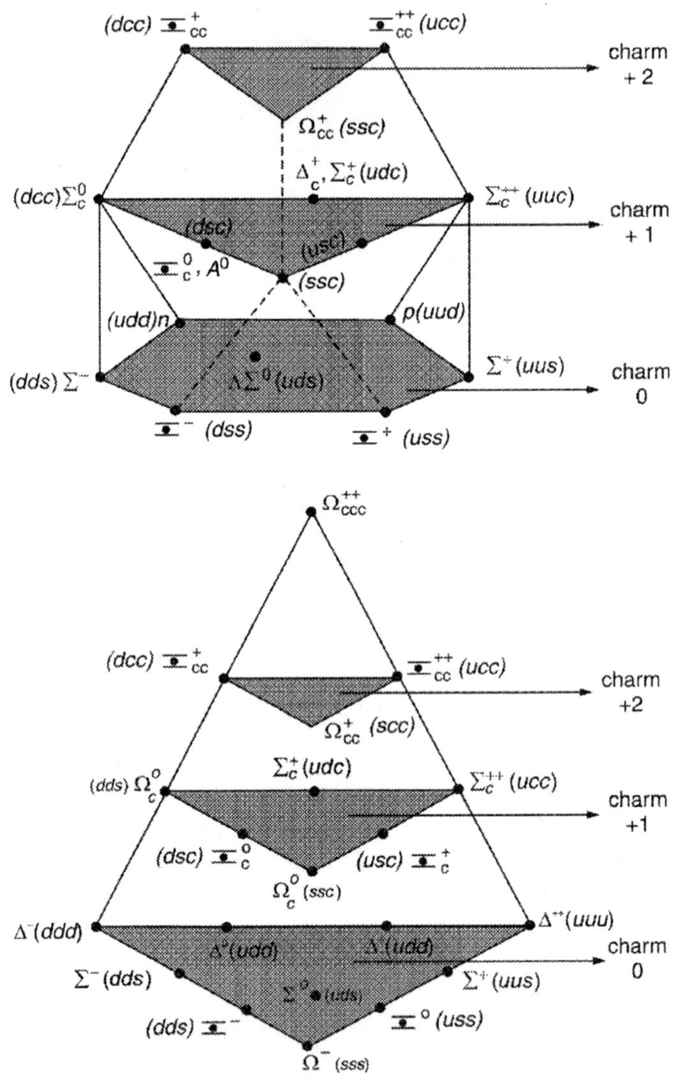

Fig. 3.17 (a, b): SU(4) multiplets of baryons. Fig. (a) $\left(\frac{1}{2}\right)^+$ baryons. Fig. (b) $\left(\frac{3}{2}\right)^+$ baryons. Both are 20 plets. In Fig. (*a*) Λ_c^0, A^0, A^+ constitute an antitriplet. Charm zero are familiar $J^P = \frac{1}{2}^+$ and $\frac{3}{2}^+$ baryons. In SU(4), we also have charm $\neq 0$ baryons.

$$\begin{aligned}\mathbf{4} &= \mathbf{1} \oplus \mathbf{3} \\ \mathbf{20'} &= \mathbf{3} \oplus \mathbf{3}^* \oplus \mathbf{6} \oplus \mathbf{8} \\ \mathbf{20} &= \mathbf{1} \oplus \mathbf{3} \oplus \mathbf{6} \oplus \mathbf{10}\end{aligned} \tag{3.146}$$

from which the difference between **20** and **20′** too becomes clear. The baryons are three quark states, their J^P (spin-parity) may be $(\frac{1}{2})^+, (\frac{3}{2})^+$, etc. The baryons corresponding to $J^P = (\frac{1}{2})^+$ belong to **20′**-plet and and $J^P = (\frac{3}{2})^+$ baryons belong to **20**-plet. Following figures (Fig. 3.17a, b) show them. From these figures the SU(3) decompositions (3.146) can be identified.

From the (Fig. 3.17a, b), e.g., in case of $J^P = (\frac{3}{2})^+$ baryons, we have in view of the decomposition (3.146)

$$\begin{aligned}1 &= charm + 3\ \ baryons \rightarrow charm + 3\ \ baryons \\ 3 &= charm + 2\ \ baryons \rightarrow\ charm + 2\ \ baryons \\ 6 &= charm + 1\ \ baryons \rightarrow\ charm + 1\ \ baryon \\ 10 &= uncharmed\ baryons\ (charm\ 0)\end{aligned} \tag{3.147}$$

which shows that no charmed quarks (only *u*, *d*, *s*) are used to construct the ten resonance states of $(\frac{3}{2})^+$ decuplet (**10**). Out of these **10** resonance states, four states namely, $\Delta^-, \Delta^0, \Delta^+, \Delta^{++}$ belong to same isospin $T = \dfrac{3}{2}$ and differ in T_3 only. These states do not use the strange quarks as well.

The SU(4) group of isospin and spin, which is

$$\text{SU(4)} = [\text{SU(2)}]_{\text{isospin}} \otimes [\text{SU (2)}]_{\text{spin}} \tag{3.148}$$

can provide the wavefunction for Δ resonance states as

$$\begin{aligned}\left|\frac{3}{2},\frac{3}{2}\right\rangle &= \Delta^{++} = uuu\ \uparrow\uparrow\uparrow \\ \left|\frac{3}{2},\frac{1}{2}\right\rangle &= \Delta^{+} = \frac{1}{\sqrt{3}}(uud + udu + duu)\ \uparrow\uparrow\uparrow \\ \left|\frac{3}{2},-\frac{1}{2}\right\rangle &= \Delta^{0} = \frac{1}{\sqrt{3}}(udd + dud + ddu)\ \uparrow\uparrow\uparrow \\ \left|\frac{3}{2},\frac{3}{2}\right\rangle &= \Delta^{++} = ddd\ \uparrow\uparrow\uparrow\end{aligned} \tag{3.149}$$

where $J = \frac{3}{2}, J_z = \frac{3}{2}$ and $\left|\frac{3}{2},\frac{3}{2}\right\rangle$, etc., imply $|T, T_3\rangle$, in which *T* is the isospin and T_3 is the third component of *T*. The *u* quark is $T_3 = +\frac{1}{2}$ and *d* is $T_3 = -\frac{1}{2}$ state. It is important to note that under the interchange of any pair of quarks the wavefunctions (3.149) are symmetric. They are also symmetric in spin.

The wavefunction of a baryon consists of a spatial part, the isospin (flavour) part, and the spin part; and later we will see that there is colour part as well. The spatial part describes the location of three quarks. In the ground state, there is no angular dependence, $l = l' = 0$; so spatial ground state wavefunction would be symmetric. The flavour component indicates, which combination of u, d, (s, c) is involved. Such combinations, similar to spin part could be both symmetric and antisymmetric.

$$\psi = \psi \text{ (space) } \psi \text{ (flavour) } \psi \text{ (spin) } [\psi\text{(colour)}] \tag{3.150}$$

Since, for a baryon ψ [LHS of eqn. (3.150)] must be antisymmetric under the interchange of any pair of quarks, eqn. (3.150) provides choices for ψ to be antisymmetric. If, out of four parts, three are symmetric, the fourth must be antisymmetric. In SU(4) group, the ψ(flavour)ψ(spin) part can be obtained by $[SU(2)]_{isospin} \otimes [SU(2)]_{spin}$. If we look at Δ^{++} baryon, in the ground state the ψ(space), is symmetric, from eqn. (3.149) we observe that *uuu*, the ψ(flavour) $\Rightarrow \psi$(isospin), is also symmetric under the interchange of two quarks and ψ(spin) of Δ^{++}, which is $\uparrow\uparrow\uparrow$, too is symmetric under the interchange of two spins. This tells that

$$\psi(\Delta^{++}) = \psi\text{(space) } \psi\text{(isospin) } \psi\text{(spin)} \tag{3.151}$$

is symmetric. The quarks are spin -$\frac{1}{2}$ objects and Δ^{++} is *uuu*, so interchange of any pair of quarks must make $\psi(\Delta^{++})$ antisymmetric. However, we are finding a symmetric $\psi(\Delta^{++})$. Similar conclusions can be derived from the wavefunctions (3.149) for ground state Δ^{+}, Δ^{0} and Δ^{-} as well. This invites the introduction of antisymmetric ψ(colour).

Thus, in SU(4) group, the SU(4) flavour can assign quark flavours to the particle (resonance) states of its multiplets and $[SU(2)]_{isospin} \otimes [SU(2)]_{spin}$ approach may be used for the wavefunctions.

As seen above, the SU(4) group can be used to classify both mesons and baryons, i.e., the hadrons, and new (higher dimensional) hadron states can be constructed. The SU(4) group also relates the magnetic moment of proton and neutron through

$$\frac{\mu_p}{\mu_n} = -1.5 \tag{3.152}$$

which is in good agreement with the experimental value of -1.46 for the above ratio. But, one of the serious drawbacks of SU(4) is that there are great mass splittings and mixings. The large masses of charmed mesons suggest a heavy charm quark and hence the masses of u, d, s and c differ a lot and hence the SU(4) symmetry is a badly broken symmetry. The masses must be equal for the symmetry to be exact. These defects in SU(4) symmetry diminish its importance in the classification of hadrons.

3.5.8 SU(6) Quark Model

We have already seen that the SU(3) symmetry group is considerably successful in the classification of elementary particles and in predicting new particles. However, the SU(3) symmetry has some shortcomings as well. Therefore, it becomes natural that SU(3) group be extended so that: (1) higher symmetry classification of particles becomes possible and (2) the possibility of removing the shortcomings of SU(3) enhances. One extension, which was discussed in the previous section, was SU(4) in which SU(2) isospin and SU(2) spin groups were combined.

Another approach made in the extension of SU(3) group is to combine the SU(3) flavor multiplets with the SU(2) spin multiplets. The enlarged group thus considered is SU(6) whose subgroup is SU(3) (flavour ⊗ SU(2) (spin). In order to obtain the SU(6) classification of particles, it is sufficient that we know the representations of SU(3) (flavour) ⊗ SU(2) (spin) subgroup of SU(6), which are used in particle (hadron) classification. Since, the SU(3) flavour group has a fundamental triplet of *u*, *d*, and s quarks and each quark has two (↑↓) spin states, we have the six states of SU(6) as {*u*(↑), *d*(↑), *s*(↑), u(↓), *d*(↓), *s*(↓), which form its fundamental sextet. The model that uses these six quark states to constitute and classify the mesons and baryons (hadrons) is called the SU(6) quark model of hadrons. Although, it is postulated that each quark has a hidden degree of freedom, called colour, we initially consider that the SU(6) quark model has no hidden degree of freedom and the quark model without colour degree of freedom is called the *simple (naïve) quark model.*

3.5.8.1 Hypotheses of SU(6) Quark Model

Following are the hypotheses on which the naïve quark model is based.

1. There exist three particles called up (*u*), down (*d*) and strange (*s*) quarks as real objects. They are fermions having spin - $\frac{1}{2}$. The characteristics of *u*, *d*, and *s* have been tabled in Chapter 1. Their internal symmetry numbers are those of **3** of SU(3).
2. To each quark *q*, there corresponds an antiquark $\overline{q}$, whose internal symmetry numbers are those of **3*** of SU(3).
3. Quark antiquark ($q\overline{q}$) combinations form the mesons.
4. Baryons are formed by three quarks (*qqq*) without any antiquarks.
5. Twelve (12) quark states (up and down spin states of d,u,s and $\overline{d},\overline{u},\overline{s}$) are required to form the elementary particles.
6. The quarks in SU(6) are static or non-relativistic.

3.5.8.2 General Features of SU(6)

The special unitary group in six dimensions SU(6) has operators that can be represented by 6×6 matrices. The lowest nontrivial representation of SU(6) is six-dimensional and is denoted by a **6**. An SU(6) group, which is six-dimensional, is made up of an SU(3) group that is three-dimensional and an SU(2) group, which is two-dimensional. Representation **3** of SU(3) represents three quark states and **3*** three antiquark states. The representation **2** of SU(2) denotes two spin states. Thus, representation **6** will represent six quark states and **6*** six antiquark states. In order to accommodate twelve quark states, we would need two SU(6) groups represented **6*** and **6***. Similar to **3** ⊗ **3*** forming the mesons and **3** ⊗ **3** ⊗ **3** forming the baryons in SU(3), we have in SU(6)

$$\mathbf{6} \otimes \mathbf{6}^* = \mathbf{36} = \mathbf{1} \oplus \mathbf{35} \tag{3.153}$$

and

$$\mathbf{6} \otimes \mathbf{6} \otimes \mathbf{6} = \mathbf{216} = \mathbf{20} \oplus \mathbf{70} \oplus \mathbf{70} \oplus \mathbf{56} \tag{3.154}$$

which respectively form mesons and baryons of the SU(6) group.

There are thirty-six generators (operators) of SU(6) each represented by 6×6 matrix. Out of these 36 operators we have

1. One identity operator,
2. Three operators, which change the spin (ordinary) only, keeping SU(3) quantum numbers unchanged.
3. Eight operators, which change SU(3) quantum numbers keeping spin unchanged.
4. Twenty four operators, which combine the spin and SU(3) quantum numbers.

3.5.8.3 The Mesons in SU(6)

In SU(6), the six quark states are

$$q = u(\uparrow)d(\uparrow)s(\uparrow)u(\downarrow)d(\downarrow)s(\downarrow) \tag{3.155}$$

and the six antiquark states are

$$\bar{q} = \bar{u}(\uparrow)\bar{d}(\uparrow)\bar{s}(\uparrow)\bar{u}(\downarrow)\bar{d}(\downarrow)\bar{s}(\downarrow) \tag{3.156}$$

From the direct product $q \otimes \bar{q}$, we get

$$\mathbf{6} \otimes \mathbf{6}^* = \mathbf{1} \oplus \mathbf{35} \tag{3.157}$$

where $\mathbf{6}^* = \overline{\mathbf{6}}$ and **1** and **35** are two irreducible representations in SU(6). We can infer, either from $q\bar{q}$ or from the comparison with **3** ⊗ **3*** = **1** ⊕ **8**, which describes mesons that **6** ⊗ **6*** would also describe 36 meson states, out of which the singlet **1** does not mix with the remaining **35.** Since, SU(3) ⊗ SU(2) is the subgroup of SU(6), the thirty-five meson states can be described in terms of their [SU(3), SU(2)] content, where SU(2) refers to the ordinary spin group. We may therefore, write

$$\mathbf{6} \equiv [\mathbf{3}\{\mathrm{SU(3)}\}, \mathbf{2}\{\mathrm{SU(2)}\}] \equiv (\mathbf{3}, \mathbf{2}) \tag{3.158}$$

and

$$\mathbf{6}^* = (\mathbf{3}^*, \mathbf{2}^*) = (\mathbf{3}^*, \mathbf{2}) \tag{3.159}$$

where due to the self-duality of SU(2) multiplets, **2*** = **2**. Therefore,

$$\begin{aligned}\mathbf{6}\otimes\mathbf{6}^* &= (\mathbf{3},\mathbf{2})\otimes(\mathbf{3}^*,\mathbf{2})\\ &= [(\mathbf{3}\otimes\mathbf{3}^*),(\mathbf{2}\otimes\mathbf{2})]\\ &= [(\mathbf{1}\oplus\mathbf{8}),(\mathbf{1}\oplus\mathbf{3})]\\ &= (\mathbf{1},\mathbf{1})\oplus(\mathbf{8},\mathbf{1})\oplus(\mathbf{1},\mathbf{3})\oplus(\mathbf{8},\mathbf{3})\end{aligned} \tag{3.160}$$

The physical content in the above decomposition can be expressed in the form of the following table, recalling that in e.g., (**1, 3**) representation of above equation **1** refers to its SU(3) part and **3** to SU(2) part.

SU(3) part	SU(2) part (Ordinary spin)		Spin of $q\bar{q}$ state	Number of states	(J^P) Multiplet
1	**1**	(singlet)	0	$1 \times 1 = 1$	0^- Singlet
8	**1**	(singlet)	0	$8 \times 1 = 8$	0^- Octet
1	**3**	(triplet)	1	$1 \times 3 = 3$	1^- Vector nonet
8	**3**	(triplet)	1	$8 \times 3 = 24$	

where $q\bar{q}$ spin zero implies, the spins of q and $\bar{q}$ are antiparallel and spin 1 implies parallel spins. Ground state of $q\bar{q}$ is also assumed $l = 0$, so that J refers to $q\bar{q}$ spin only. From the above table, we infer that in the decomposition (3.157)

$$\begin{aligned}\mathbf{1} &= (\mathbf{1},\mathbf{1})\\ \mathbf{35} &= (\mathbf{8},\mathbf{1})\oplus(\mathbf{1},\mathbf{3})\oplus(\mathbf{8},\mathbf{3})\end{aligned} \tag{3.161}$$

where the singlet **1** has a symmetric composition

$$\mathbf{1} = u(\uparrow)\bar{u}(\downarrow) + u(\downarrow)\bar{u}(\uparrow) + d(\uparrow)\bar{d}(\downarrow) + d(\downarrow)\bar{d}(\uparrow) + s(\uparrow)\bar{s}(\downarrow) + s(\downarrow)\bar{s}(\uparrow) \tag{3.162}$$

The vector (J = 1) nonet has 27 meson states, the three states form vector meson singlet and 24 states form vector meson octet. Other 8 states form the pseudoscalar octet. The nonet of vector mesons has already been drawn in Fig. (1.34). Following table recapitulates them.

Table of Vector Meson Nonet

State	J^P	T	T_3	Y	Mass MeV/C²	Main decay mode
ρ^-	1^-	1	-1	0	765	$\rho \to \pi\pi$
ρ^0	1^-	1	0	0		
ρ^+	1^-	1	1	0		
K^{*-}	1^-	$\frac{1}{2}$	$-\frac{1}{2}$	-1	893	$K^* \to K\pi$
K^{*+}	1^-	$\frac{1}{2}$	$+\frac{1}{2}$	$+1$		
K^{*0}	1^-	$\frac{1}{2}$	$-\frac{1}{2}$	$+1$		
$\bar{K}^{*0}$	1^-	$\frac{1}{2}$	$+\frac{1}{2}$	-1		

ω^0	1^-	0	0	0	784	$\omega^0 \to \pi\pi\pi$
ϕ^0	1^-	0	0	0	1019	$\phi^0 \to KK$ $\to \pi\pi\pi$

It may be mentioned again that each of the nine states from ρ^- to ϕ^0 is a vector particle ($J = 1$) state and hence, each forms a triplet [SU(2) spin]. The 0^- octet is the pseudoscalar octet of mesons of Fig. (3.13) and the 0^- singlet meson is $\eta'$$\cdot$ η' is a singlet in both SU(3) and SU(2) spin, whereas ω^0 and ϕ^0 represent singlets in SU(3) but triplet in SU(2). Thus, both ω^0 and ϕ^0 are the candidates for vector meson singlets. Because of these two candidates, the vector meson nonet, instead of an octet and a singlet, is considered.

We may then say that the mesons in SU(6) can be constructed by **6** $\otimes$ **6***, which represent themselves in the form of a pseudoscalar singlet, a pseudoscalar octet and a vector nonet of physical mesons. However, there are mesons which though are members of a separate multiplet (singlet, octet or nonet) but have identical (0, 0) quantum numbers (t_3, y). This raises the question that whether the meson states observed as physical [t_3, y) = (0, 0)] particles are pure states or mixed states. We discuss this mixing in the following.

3.5.8.3.1 The mixing of meson states

In the product representation **6** $\otimes$ **6***, we have found that we have a set of eight pseudoscalar (0^-) mesons, a single pseudoscalar meson η' and a set of nine vector mesons. At the centre of (0^-) octet lie π^0 and $\eta$$\cdot$ π^0 has $T = 1$ whereas η and η' have $T = 0$. The three however, have $t_3 = 0$, $y = 0$$\cdot$ π^0 forms the part of π^-, π^0, π^+ but η and η' tend to mix. In the above picture, it appears η being the member of the octet is a pure octet state and η' is a pure singlet state. However, this is not the case. The observed state η is a mixture of pure octet (call it η_8) and pure singlet (call η_1). Similarly, η' is a mixture of the two. We may write

$$\eta' = \eta_8 \sin\theta + \eta_1 \cos\theta$$
$$\eta = \eta_8 \cos\theta - \eta_1 \sin\theta \tag{3.162}$$

from which, we may easily infer that when θ is small, $\eta' \to \eta_1$ (the pure singlet state) and $\eta' \to \eta_8$ (the pure octet state). The value of mixing angle θ also suggests the extent of mixing. For larger θ, η' would possess a more η_8 content and η would contain more η_1 (singlet) content. It is found that the mixing angle for η, η' mixing case is (–11°), which is small hence the inference that in the formation of η and η' particles, whom we observe physically, the mixing of singlet (η_1) and octet (η_8) is quite small; η is largely an octet state and η' is largely a singlet state. The minus sign in $\theta = -11°$ implies that the mass of largely octet is less than the mass of largely singlet member. The expression for mixing angle is calculated in the following.

Define,

$$\psi = \begin{pmatrix} \eta' \\ \eta \end{pmatrix} \tag{3.163}$$

Using eqn. (3.162), we may write

$$\psi = \begin{pmatrix} \sin\theta & \cos\theta \\ \cos\theta & -\sin\theta \end{pmatrix}\begin{pmatrix} \eta_8 \\ \eta_1 \end{pmatrix} = M\psi_1 \tag{3.164}$$

where

$$M = \begin{pmatrix} \sin\theta & \cos\theta \\ \cos\theta & -\sin\theta \end{pmatrix} \tag{3.165}$$

is the mixing matrix, which is unitary. Using $M^{-1} = M$, eqn. (3.164) is

$$\psi_1 = M\psi \tag{3.166}$$

The physically observed mesons η' and η are eigenstates of the Hamiltonian operator H, with eigenvalues $m_{\eta'}^2$ and m_η^2 respectively, where $m_{\eta'}$ and m_η are the masses of η' and η. The squares of the masses have been taken as Gell—Mann—Okubo formula suggests a square mass for mesons, instead of linear. So

$$\begin{aligned} H\eta' &= m_{\eta'}^2\eta' \\ H\eta &= m_\eta^2\eta \end{aligned} \tag{3.167}$$

implying that

$$H\begin{pmatrix} \eta' \\ \eta \end{pmatrix} = \begin{pmatrix} m_{\eta'}^2 & 0 \\ 0 & m_\eta^2 \end{pmatrix}\begin{pmatrix} \eta' \\ \eta \end{pmatrix}$$

or

$$H\psi = M_1\psi \tag{3.168}$$

which on substituting for ψ from eqn. (3.164) in the RHS only, becomes

$$H\psi = M_1M\psi_1 \tag{3.169a}$$

$$\Rightarrow \quad MH\psi = MM_1M\psi_1 \tag{3.169b}$$

$$\Rightarrow \quad HM\psi = MM_1M\psi_1 \quad (\because\ M^{-1}HM = H) \tag{3.169c}$$

$$\Rightarrow \quad H\psi_1 = MM_1M\psi_1 \quad \text{[using eqn. (3.166)]} \tag{3.170}$$

In comparison to eqn. (3.167), the effect of the Hamiltonian operator on unitary-pure singlet (η_1) and pure octet (η_8) states is

$$\begin{aligned} H\eta_1 &= m_{11}^2\eta_1 + m_{18}^2\eta_8 \\ H\eta_8 &= m_{81}^2\eta_1 + m_{88}^2\eta_8 \end{aligned} \tag{3.171}$$

where $m_{11} \Rightarrow$ the mass of pure (η_1) state and m_{88} that of pure (η_8) state. m_{18} is the mass contributed to (η_1) state by (η_8) state and m_{81} is the mass to (η_8) by (η_1). However, $m_{18} = m_{81}$. We may rewrite eqn. (3.171) as

$$H\psi_1 = \begin{pmatrix} m_{88}^2 & m_{81}^2 \\ m_{18}^2 & m_{11}^2 \end{pmatrix}\psi_1 = M_2\psi_1 \tag{3.172}$$

where from eqn. (3.164)

$$\psi_1 = \begin{pmatrix} \eta_8 \\ \eta_1 \end{pmatrix} \tag{3.173}$$

Using eqn. (3.172) in eqn. (3.170), we get

$$M_2 = MM_1M \tag{3.174}$$

which may be written as

$$\begin{pmatrix} m_{88}^2 & m_{81}^2 \\ m_{18}^2 & m_{11}^2 \end{pmatrix} = \begin{pmatrix} \sin\theta & \cos\theta \\ \cos\theta & -\sin\theta \end{pmatrix} \begin{pmatrix} m_{\eta'}^2 & 0 \\ 0 & m_\eta^2 \end{pmatrix} \begin{pmatrix} \sin\theta & \cos\theta \\ \cos\theta & -\sin\theta \end{pmatrix}$$

$$= \begin{pmatrix} m_{\eta'}^2 \sin^2\theta + m_\eta^2 \cos^2\theta & \sin\theta\cos\theta(m_{\eta'}^2 - m_\eta^2) \\ \sin\theta\cos\theta(m_{\eta'}^2 - m_\eta^2) & m_{\eta'}^2 \cos^2\theta + m_\eta^2 \sin^2\theta \end{pmatrix} \tag{3.175}$$

Equating m_{88}^2 with the corresponding element in RHS, we get

$$\sin^2\theta = \frac{m_\eta^2 - m_8^2}{m_\eta^2 - m_{\eta'}^2} \tag{3.176}$$

where $m_8^2 = m_{88}^2$. Using the observed values of masses, $m_\eta \approx 549$ MeV, $m_{\eta'} \approx 958$ MeV, $m_8 \approx 276$ MeV calculated in analogy with the Gell-Mann-Okubo formula as

$$m_8^2 = \frac{1}{3}(m_k^2 - m_\pi^2) \tag{3.177}$$

where $m_K \approx 498$ MeV and $m_\pi \approx 135$ MeV, we get $\theta \approx -11°$, which is not considered a large value, implying that the singlet mixing in η octet mixing, in η′ is small and η is mainly octet and η′ mainly a singlet. These singlets and octets are with respect to SU(3) group.

In the vector meson nonet, at the centre of the $t_3 - y$ diagram are ρ^0, ω^0 and ϕ^0 mesons, in which ρ^0 ($T = 1$) forms the part of ρ^-, ρ^0, ρ^+. However, ω^0 and ϕ^0 mix. Following the approach similar to η, η′ mixing, the ω^0, ϕ^0 mixing angle can be calculated. Here ω^0 is considered a SU(3) singlet state and ϕ^0 an SU(3) octet state. However, ω^0 is partly SU(3) octet as well, and ϕ^0 is partly SU(3) singlet. The extent of mixing in ω^0, ϕ^0 is quite large compared to η, η′ mixing. Here, the angle of mixing is about 40°, which is large compared to (–)11° in η, η′ case. Assuming again, θ as the mixing angle, we write

$$\begin{aligned} \omega^0 &= \omega_8 \sin\theta + \phi_1 \cos\theta \\ \phi^0 &= \omega_8 \cos\theta - \phi_1 \sin\theta \end{aligned} \tag{3.178}$$

Here

$$\psi = \begin{pmatrix} \omega^0 \\ \phi^0 \end{pmatrix} = \begin{pmatrix} \sin\theta & \cos\theta \\ \cos\theta & -\sin\theta \end{pmatrix} \begin{pmatrix} \omega_8 \\ \phi_1 \end{pmatrix} = M\psi_1 \tag{3.179}$$

Following the similar steps as in η, η′ case, we get corresponding to eqn. (3.175)

$$\begin{pmatrix} m_{88}^2 & m_{81}^2 \\ m_{18}^2 & m_{11}^2 \end{pmatrix} = \begin{pmatrix} m_\omega^2 \sin^2\theta + m_\phi^2 \cos^2\theta & \sin\theta\cos\theta(m_\omega^2 - m_\phi^2) \\ \sin\theta\cos\theta(m_\omega^2 - m_\phi^2) & m_\omega^2\cos^2\theta + m_\phi^2 \sin^2\theta \end{pmatrix} \quad (3.180)$$

where $\omega \Rightarrow \omega^0$ and $\phi \Rightarrow \phi^0$. From the above equation, we may write

$$m_{88}^2 = m_\omega^2 \sin^2\theta + m_\phi^2\cos^2\theta = m_8^2 \quad (3.181)$$

from which we get

$$\sin^2\theta = \frac{m_\phi^2 - m_8^2}{m_\phi^2 - m_\omega^2} \quad (3.182)$$

Now the calculated value of $m_8 \approx 930$ MeV and the observed values of $m_\omega \approx 783$ MeV and that of $m_\phi \approx 1020$ MeV. Substituting these values in eqn. (3.182), we get

$$\sin\theta = 0.64$$

$$\Rightarrow \qquad \theta \approx 40° \quad (3.183)$$

This angle is large compared to 11° of η, η′ case. Had there been no mixing ($\Rightarrow \theta \to 0$), the relations (3.178) would have described $\omega^0 = \phi_1$, a pure SU(3) singlet and $\phi^0 = \omega_8$, a pure SU(3) octet. However, the angle $\theta \approx 40°$ makes $\sin^2 \approx 0.4$ and $\cos^2\theta \approx 0.6$, and for such $\theta \approx 40°$ mixing, we have from eqn. (3.178)

$$\omega^0 = \sqrt{\frac{2}{5}}\,\omega_8 + \sqrt{\frac{3}{5}}\,\phi_1$$

and

$$\phi^0 = \sqrt{\frac{3}{5}}\,\omega_8 - \sqrt{\frac{2}{5}}\,\phi_1 \quad (3.184)$$

i.e., about 60% ω^0 is in SU(3) singlet state and 40 % in octet state and reverse for ϕ^0. The ω^0, is more singlet than octet. The quark composition of pure SU(3) singlet ϕ_1 and pure SU(3) octet ω_8 are

$$\phi_1 = \frac{1}{\sqrt{3}}(u\bar{u} + d\bar{d} + s\bar{s})$$

and

$$\omega_8 = \frac{1}{\sqrt{6}}(u\bar{u} + d\bar{d} - 2s\bar{s}) \quad (3.185)$$

However, for the special case when $\theta = 35°$, we have $\sin^2\theta = 0.33$ and $\cos^2\theta = 0.67$ and the proposition (3.178) becomes

$$\omega^0 = \sqrt{\frac{1}{3}}\,\omega_8 + \sqrt{\frac{2}{3}}\,\phi_1$$

and

$$\phi^0 = \sqrt{\frac{2}{3}}\,\omega_8 - \sqrt{\frac{1}{3}}\,\phi_1 \tag{3.186}$$

i.e., when $\theta \approx 35°$, about 65% ϕ^0 is in pure SU(3) octet state and rest 35% in pure SU(3) singlet state with reverse percentages for ω^0. Such a mixing is called ideal and in this case the substitution of eqn. (3.185) into eqn. (3.186) yields

$$\omega^0 \approx \frac{1}{\sqrt{2}}(u\bar{u} + d\bar{d})$$

and

$$\phi^0 \approx s\bar{s} \tag{3.187}$$

This means that under the ideal mixing of SU(3) octet and singlet states ω^0 meson is made up of only up and down quarks antiquarks and ϕ^0 meson is made up of only strange quark antiquark. However, in reality, the mixing is not ideal (θ is 40°) so all the three quark antiquarks would make up ω^0 and ϕ^0 mesons.

3.5.8.4 The Baryons in SU(6)

In the SU(6) quark model, the baryon states can be obtained from $\mathbf{6} \otimes \mathbf{6} \otimes \mathbf{6}$ similar to SU(3), where $\mathbf{3} \otimes \mathbf{3} \otimes \mathbf{3}$ provide baryon states.

Using again the notation

$$\text{SU(6)} \rightarrow [\text{SU(3)(flavour), SU(2)(spin)}] \tag{3.188a}$$

and

$$6 = [3 \text{ of SU}(3)_f,\ 2 \text{ of SU}(2)_S] \tag{3.188b}$$

we may write

$$\begin{aligned}\mathbf{6} \otimes \mathbf{6} \otimes \mathbf{6} &= (\mathbf{3,2}) \otimes (\mathbf{3,2}) \otimes (\mathbf{3,2}) \\ &= [\mathbf{3} \otimes \mathbf{3} \otimes \mathbf{3}, \mathbf{2} \otimes \mathbf{2} \otimes \mathbf{2}]\end{aligned} \tag{3.189}$$

We may now use the relations (3.75) and (3.119) for the SU(2) and SU(3) products. We, however, mention the symmetry of the decompositions. That is

$$\mathbf{2} \otimes \mathbf{2} \otimes \mathbf{2} = \mathbf{4}_S \oplus \mathbf{2}_{MS} \oplus \mathbf{2}_{MA} \tag{3.189a}$$

where the symbols *S*, *A*, *M* stand for the symmetric, antisymmetric and mixed symmetry respectively. Under the interchange of first two quarks, the spin states of mixed symmetry doublets ($\mathbf{2}_{MS}$) will remain symmetric and in $\mathbf{2}_{MA}$ the spin states would become antisymmetric. $\mathbf{4}_S$ is totally symmetric. If χ denotes the spin state, we will have

$$\begin{aligned}\chi(S) &= \frac{1}{\sqrt{3}}(\uparrow\uparrow\downarrow + \uparrow\downarrow\uparrow + \downarrow\uparrow\uparrow) \\ \chi(M_S) &= \frac{1}{\sqrt{6}}(\uparrow\downarrow\uparrow + \downarrow\uparrow\uparrow - 2\uparrow\uparrow\downarrow) \\ \chi(M_A) &= \frac{1}{\sqrt{2}}(\uparrow\downarrow\uparrow - \downarrow\uparrow\uparrow)\end{aligned} \tag{3.190}$$

Similarly,

$$\mathbf{3} \otimes \mathbf{3} \otimes \mathbf{3} = \mathbf{10}_S \oplus \mathbf{8}_{MS} \oplus \mathbf{8}_{MA} \oplus \mathbf{1}_A \tag{3.191}$$

where for three flavours u, d, s

$$\begin{aligned}
\psi(S) &= \frac{1}{\sqrt{6}}(uds + usd + dus + dsu + sud + sdu) \\
\psi(M_S) &= \frac{1}{2}(usd + sud - dsu - sdu) \\
\psi(M_A) &= \frac{1}{\sqrt{12}}(2uds - 2dus + sdu - dsu + usd - sud) \\
\psi(A) &= \frac{1}{\sqrt{6}}(uds - usd + sud - sdu + dsu - dus)
\end{aligned} \tag{3.192}$$

It may, however, be remembered that in eqn. (3.191), there are ten different ψ_S, eight ψ_{M_S} and eight ψ_{M_A} corresponding to decuplet and two octets of eqn. (3.191).

Coming back to eqn. (3.189), we may write

$$\mathbf{6} \otimes \mathbf{6} \otimes \mathbf{6} = [\underset{S}{\mathbf{10}} \oplus \underset{M_S}{\mathbf{8}} \oplus \underset{M_A}{\mathbf{8}} \oplus \underset{A}{\mathbf{1}}, \ \underset{S}{\mathbf{4}} \oplus \underset{M_S}{\mathbf{2}} \oplus \underset{M_A}{\mathbf{2}}] \tag{3.193}$$

from which, we may write the following [SU(3), SU(2)] multiplets belonging to following categories.

$$\begin{aligned}
S &= (\mathbf{10}, \mathbf{4}) \oplus (\mathbf{8}, \mathbf{2}) \\
M_S &= (\mathbf{10}, \mathbf{2}) \oplus (\mathbf{8}, \mathbf{4}) \oplus (\mathbf{8}, \mathbf{2}) \oplus (\mathbf{1}, \mathbf{2}) \\
M_A &= (\mathbf{10}, \mathbf{2}) \oplus (\mathbf{8}, \mathbf{4}) \oplus (\mathbf{8}, \mathbf{2}) \oplus (\mathbf{1}, \mathbf{2}) \\
A &= (\mathbf{1}, \mathbf{4}) \oplus (\mathbf{8}, \mathbf{2})
\end{aligned} \tag{3.194}$$

which in terms of dimensions can be written as

$$\begin{aligned}
S &= 40 \oplus 16 = 56 \\
M_S &= 20 \oplus 32 \oplus 16 \oplus 2 = 70 \\
M_A &= 20 \oplus 32 \oplus 16 \oplus 2 = 70 \\
A &= 4 \oplus 16 = 20
\end{aligned} \tag{3.195}$$

That is,

$$\mathbf{6} \otimes \mathbf{6} \otimes \mathbf{6} = \mathbf{216} = \mathbf{56} \oplus \mathbf{70} \oplus \mathbf{70} \oplus \mathbf{20} \tag{3.196}$$

From the decomposition (3.196), we infer that the baryons in SU(6) belong to its **20, 70** and **56** irreducible representations, out of which the irreducible representation **56** represents baryons which are completely symmetric and these are the baryons which in SU(3) representation are formed as an octet of spin $\frac{1}{2}^+$

baryons [(**8** , **2**)] and [⊕] a decuplet of spin $\frac{3}{2}^+$ [(**10** , **4**)] baryons. The SU(2) dimensions of **4** and **2** refer to spin $\frac{3}{2}$ and spin $\frac{1}{2}$ states respectively. Figures (3.1) and (3.5) show the spin $\frac{1}{2}^+$ baryon octet and the baryon decuplet has been drawn in Fig. (3.2). These figures show the baryons in their ground states. We shall later note that the symmetry of the above ground state poses a problem, requiring the introduction of colour.

3.5.8.5 Applications of SU(6) Quark Model

1. Baryon Magnetic Moments

The group SU(3) was successful in predicting the magnetic dipole moments of the baryons of the octet only to the extent that, the predicted magnetic moments became known only in terms of the magnetic moments of neutron and proton. μ_n and μ_p were known in SU(3). The SU(6) quark model is able to predict not only the ratio of μ_n/μ_p, where μ denotes the magnetic (dipole) moment, but also the magnetic moments of octet baryons.

Since, baryons are three quark states, the quark model is able to predict the magnetic moments of quarks and when we know the magnetic moments of the constituents of baryons, the magnetic moments of baryons themselves can easily be predicted. It is found that the predicted magnetic moments and the observed magnetic moments of baryons agree quite well. The capability of quark model to predict the magnetic dipole moments therefore comes as one of the impressive successes of SU(6) quark model. The calculations for the magnetic moments have been carried out in the following.

Consider the octet of baryons, in which we have eight (p, n, Λ^0, Σ^+, Σ^0, Σ^-, Ξ^0 and Ξ^-) spin half ($J^P = \frac{1}{2}^+$) baryons. We assume that the baryons are in their ground state ($l = 0$) and hence the quarks constituting the baryons have zero relative orbital angular momentum and the net magnetic moment of the baryons is then the vector sum of magnetic moments of the constituent quarks or

$$\mu = \sum_{i=1}^{3} \mu_i \tag{3.197}$$

Since, for a point spin-$\frac{1}{2}$ charged particle the magnetic moment is

$$\mu = \frac{e}{2m} \tag{3.198}$$

where e is the charge on the particle and m its mass, we may write for structureless quark of charge $q_i e$ and mass m_i, the magnetic moment

$$\mu_i = \frac{q_i e}{2m_i} \tag{3.199}$$

Conventionally, the magnetic moment is not described by the vectors μ_i, but by the maximum value of $\mu_i \cdot (\sigma_3)_i$, which is $\mu_i (\sigma_3)_i$, where $(\sigma_3)_i$ is the projection of spin of ith quark along the third (z-axis). If i^{th} quark is positively charged then its μ_i is along the positive third axis. (σ_3) is + 1, when ith quark is spin up and −1, when it is spin down.

From eqn. (3.199) and the discussion below it, we may write

$$\begin{aligned} \mu_u &\approx \frac{2}{3}\frac{e}{2m_u}\sigma_{3u} \\ \mu_d &\approx -\frac{1}{3}\frac{e}{2m_d}\sigma_{3d} \\ \mu_s &\approx -\frac{1}{3}\frac{e}{2m_s}\sigma_{3s} \end{aligned} \tag{3.200}$$

where the constituent masses m_u, m_d and m_s of the three quarks in baryon are $m_\mu \approx m_d 363$ MeV and $m_s = 538$ MeV.

Let us take proton from the octet and calculate its magnetic moment. The magnetic moment of proton is given by the expectation value of the magnetic moment operator over the proton up ($\uparrow$) state.

$$\mu_p = \left\langle p(\uparrow) \middle| \sum_{i=1}^{3} \mu_i (\sigma_3)_i \middle| p(\uparrow) \right\rangle \tag{3.201}$$

Now, we require $\left| p(\uparrow) \right\rangle$. For this purpose we recall that the wave functions of $J^P = \frac{1}{2}^+$ baryons are totally symmetric and they belong to the irreducible representation **56** of SU(6). The octet of **56** was described by (**8**, **2**). Since the totally symmetric (S) octet (see eqn. 3.194) arises from the linear combination

$$\frac{1}{\sqrt{2}}\left[(\underset{M_S}{\mathbf{8}}, \underset{M_S}{\mathbf{2}}) + (\underset{M_A}{\mathbf{8}}, \underset{M_A}{\mathbf{2}}) \right] \tag{3.202}$$

the wave function for a spin up proton is

$$\left| p \uparrow \right\rangle = \frac{1}{\sqrt{2}} (p_{MS}\chi_{MS} + p_{MA}\chi_{MA}) \tag{3.203}$$

where p_{MS} and p_{MA} are mixed symmetric and mixed antisymmetric wave functions of proton. For proton (uud), we have

$$p_{MS} = \sqrt{\frac{1}{6}} (udu + duu - 2uud) \tag{3.204}$$

and

$$p_{MA} = \sqrt{\frac{1}{2}}(udu - duu) \tag{3.205}$$

χ_{MS}, χ_{MA}, the spin (up, down) wave functions are given in eqn. (3.190). Using these eqs. in eqn. (3.201), we get

$$\begin{aligned}|p\uparrow\rangle &= \frac{1}{6\sqrt{2}}(udu + duu - 2uud)(\uparrow\downarrow\uparrow + \downarrow\uparrow\uparrow - 2\uparrow\uparrow\downarrow) \\ &\quad + \frac{1}{2\sqrt{2}}(udu - duu)(\uparrow\downarrow\uparrow - \downarrow\uparrow\uparrow) \\ &= -\frac{1}{\sqrt{18}}\begin{bmatrix} uud \;\; (\uparrow\downarrow\uparrow + \downarrow\uparrow\uparrow - 2\uparrow\uparrow\downarrow) \\ +udu \;\; (\uparrow\uparrow\downarrow + \downarrow\uparrow\uparrow - 2\uparrow\downarrow\uparrow) \\ +duu \;\; (\uparrow\downarrow\uparrow + \uparrow\uparrow\downarrow - 2\downarrow\uparrow\uparrow) \end{bmatrix}\end{aligned}$$

or

$$|p\uparrow\rangle = -\frac{1}{\sqrt{18}}\begin{bmatrix} u\uparrow u\downarrow d\uparrow + \; u\downarrow u\uparrow d\uparrow - 2\; u\uparrow u\uparrow d\downarrow \\ +u\uparrow d\uparrow u\downarrow + \; u\downarrow d\uparrow u\uparrow - 2\; u\uparrow d\downarrow u\uparrow \\ +d\uparrow u\downarrow u\uparrow + \; d\uparrow u\uparrow u\downarrow - 2\; d\downarrow u\uparrow u\uparrow \end{bmatrix} \tag{3.206}$$

Now, we have to substitute eqn. (3.206) into eqn. (3.201) to obtain μ_p. After substitution, we shall have individual expectation values calculated from the terms like the *RHS* of following equation to give μ_p as

$$\mu_p = \frac{1}{18}\begin{bmatrix} \left\langle u\uparrow u\downarrow d\uparrow \left| \sum_i \mu_i(\sigma_3)_i \right| u\uparrow u\downarrow d\uparrow \right\rangle \\ +4\left\langle u\uparrow u\uparrow d\downarrow \left| \sum_i \mu_i(\sigma_3)_i \right| u\uparrow u\uparrow d\downarrow \right\rangle + \text{other terms} \end{bmatrix} \tag{3.207}$$

Now,

$$\begin{aligned}\left\langle u\uparrow u\downarrow d\uparrow \left| \sum_i \mu_i(\sigma_3)_i \right| u\uparrow u\downarrow d\uparrow \right\rangle &= \mu_u(\sigma_3\uparrow) + \mu_u(\sigma_3\uparrow) + = \mu_u(\sigma_3\uparrow) \\ &= \mu_u - \mu_u + \mu_d \\ &= \mu_d \end{aligned} \tag{3.208a}$$

and likewise

$$\begin{aligned}4\left\langle u\uparrow u\uparrow d\downarrow \left| \sum_i \mu_i(\sigma_3)_i \right| u\uparrow u\uparrow d\downarrow \right\rangle &= 4[\mu_u(\sigma_3\uparrow) + \mu_u(\sigma_3\uparrow) + \mu_d(\sigma_3\downarrow)] \\ &= 4[\mu_u + \mu_u - \mu_d] \\ &= 4(2\mu_u - \mu_d) \end{aligned} \tag{3.208b}$$

A look at the three quark states of eqn. (3.206) and the relations (3.208) show that all the six, three quark states like *u* up *u* down *d* up ($u\uparrow u\downarrow d\uparrow$ in any order) would each contribute to μ_p, a magnetic moment of μ_d, calculated as in eqn. (3.208a). The three states of type $2u\uparrow u\uparrow d\downarrow$ (disregarding the order) would each contribute to μ_p according to eqn. (3.208b). The substitution of eqn. (3.206) into eqn. (3.203) would then lead to the following expression for the magnetic moment μ_p of proton.

$$\mu_p = \frac{1}{18}[6\mu_d + 3\times 4(2\mu_u - \mu_d)]$$

or

$$\mu_p = \frac{1}{3}(4\mu_u - \mu_d) \tag{3.209}$$

The magnetic moment μ_n of neutron can be obtained by interchanging *u* quark by *d* quark in eqn. (3.209) to give

$$\mu_p = \frac{1}{3}(4\mu_u - \mu_d) \tag{3.210}$$

Since the masses of *u* and *d* quarks are nearly same $\mu_u \approx \mu_d$, we get from eqn. (3.200)

$$\mu_u = -2\mu_d \tag{3.211}$$

Substituting this relation in μ_p and μ_n relations we get

$$\mu_p = \frac{3}{2}\mu_u = -3\mu_d$$

and

$$\mu_n = -\mu_u = 2\mu_d \tag{3.212}$$

from which we get the important relation for the ratio of the magnetic moment of neutron and the magnetic moment of proton as

$$\frac{\mu_n}{\mu_p} = -\frac{2}{3} \tag{3.213}$$

which agrees excellently with experiment that gives the value

$$\frac{\mu_n}{\mu_p} = -0.685 \tag{3.214}$$

Thus the SU(6) quark model has predicted the neutron and proton magnetic dipole moment ratio quite successfully.

Since the prediction of the ratio (μ_n/μ_p) has required the relation (3.211) which itself is derived from relations (3.200), using the masses of the constituent quarks in baryons, the magnetic moments of *u*, *d*, *s* quarks may be obtained as

$$\begin{aligned} \mu_u &= 1.86 \textit{ nuclear magneton } (\mu_N) \\ \mu_d &= -0.93\,(\mu_N) \\ \mu_s &= -0.58\,(\mu_N) \end{aligned} \tag{3.215}$$

Inserting these values in equations (3.209) and (3.210), we may find

$$\mu_p = 2.79\ \mu_N$$

and
$$\mu_n = -\ 1.86\ \mu_N \tag{3.216}$$

Similar to the case of proton and neutron, relations analogous to (3.209) and (3.210) may be calculated for other octet baryons as well. The quark wave functions and the corresponding relations for their magnetic moments have been tabled as below.

Table for quark wavefunctions and magnetic moments of octet baryons

Baryon up state	Quark wavefunction	Magnetic dipole moment
p	$-\frac{1}{\sqrt{18}}(u\uparrow u\downarrow d\uparrow + u\downarrow u\uparrow d\uparrow - 2u\uparrow u\uparrow d\downarrow + per.)$	$\frac{1}{3}(4\mu_u - \mu_d)$
n	$-\frac{1}{\sqrt{18}}(d\uparrow d\downarrow u\uparrow + d\downarrow d\uparrow u\uparrow - 2d\uparrow d\uparrow u\downarrow + per.)$	$\frac{1}{3}(4\mu_u - \mu_u)$
Λ	$-\frac{1}{\sqrt{12}}(u\downarrow d\uparrow s\uparrow + d\uparrow u\downarrow s\uparrow - d\downarrow u\uparrow s\uparrow$ $- u\uparrow d\downarrow s\uparrow + per.)$	μ_s
Σ^+	$-\frac{1}{\sqrt{18}}(u\uparrow u\downarrow s\uparrow + u\downarrow u\uparrow s\uparrow - 2u\uparrow u\uparrow s\downarrow + per.)$	$\frac{1}{3}(4\mu_u - \mu_s)$
Σ^0	$-\frac{1}{\sqrt{6}}(u\uparrow d\downarrow s\uparrow + u\downarrow d\uparrow s\uparrow - 2u\uparrow d\uparrow s\downarrow + per.)$	$\frac{1}{3}(2\mu_u + 2\mu_d - \mu_s)$
Σ^-	$-\frac{1}{\sqrt{18}}(d\uparrow d\downarrow s\uparrow + d\downarrow d\uparrow s\uparrow - 2d\uparrow d\uparrow s\downarrow + per.)$	$\frac{1}{3}(4\mu_d - \mu_s)$
Ξ^0	$-\frac{1}{\sqrt{18}}(u\uparrow s\uparrow s\downarrow - u\uparrow s\downarrow s\uparrow - 2u\downarrow s\uparrow s\uparrow + per.)$	$\frac{1}{3}(4\mu_s - \mu_u)$
Ξ^-	$-\frac{1}{\sqrt{18}}(d\uparrow s\uparrow s\downarrow + d\uparrow s\downarrow s\uparrow - 2d\downarrow s\uparrow s\uparrow + per.)$	$\frac{1}{3}(4\mu_s - \mu_d)$

where per. means the permutations of the states written before them. For example, compare the proton state in the table and in eqn. (3.206). Using the expressions for the magnetic dipole moments given in the table and the magnetic moments (3.215) of the constituent quarks, the magnetic dipole moments of all the baryons of the octet can be predicted. Following table shows a fine agreement between the predicted and the observed values of magnetic moments of baryons.

Table for the predicted and the observed magnetic moments of octet baryons

Baryon	Predicted magnetic dipole moment in nuclear magneton μ_N	Observed magnetic dipole moment in μ_N
p	2.79	2.793
n	– 1.86	– 1.913
Λ	– 0.58	– 0.613
Σ^+	2.68	2.42
Σ^0	0.82	—
Σ^-	– 1.05	– 1.16
Ξ^0	– 1.40	– 1.25
Ξ^-	– 0.47	– 0.68

2. Masses of Hadrons

The Gell—Mann—Okubo formula for baryons is

$$m = m_0 + m_1 Y + m_2 [T(T+1) - \frac{Y^2}{4}] \tag{3.217}$$

and for mesons it is

$$m^2 = m_0 + m_2 [T(T+1) - \frac{Y^2}{4}] \tag{3.218}$$

where m_0, m_1 and m_2 are constants, T is the isospin and Y the hypercharge. In the case of mesons, $m_1 = 0$. Above formulae quite successfully describe the masses of the members of an irreducible representation of SU(3). In other words, the mass difference Δm, in case of baryons, and Δm^2, in case of mesons, between the members of an irreducible representation can be obtained.

We have also seen that the quark model of hadrons attributes the difference in the masses of hadrons to the difference in the masses of quarks constituting them. However, it is found that the masses of some mesons are quite different, though they have identical quark contents. Similar observation can be made for baryons as well. The examples in mesons are $\rho - \pi$, both have $u\bar{d}$ (ρ^+, π^+) quark content but the mass of ρ^+ meson is 770 MeV, whereas, the mass of π^+ meson is only 140 MeV. Similarly, K^{*+} (892) – K^+ (494) have same $u\bar{s}$ quark content. Among baryons both Δ^+ (1232) – p(938) have the uud quark content. $\Sigma - \Lambda$, $\Sigma^* - \Sigma$, are other examples. The difference in masses of quarkwise identical pair is called the mass splitting in which the mass of one member increases while the mass of other decreases from a central value.

The reason for mass splitting lies in an interaction term, which provides the additional mass, which may get added to or subtracted from the total mass of quarks constituting the hadron. The additional term is due to the spin–orbit and spin—spin interaction of quarks. If the hadrons are in their ground state ($l = 0$), the spin—spin interaction term is strong, but for higher l values, the spin–spin force due to its short range becomes weak and the spin–orbit interaction term plays its role in mass splittings. The $\rho - \pi$, $K^* - K$, $\Delta - N$, $\Sigma^* - \Sigma$, $\Sigma - \Lambda$ refer to ground state mass splittings caused by spin—spin interaction term. The mechanism of this interaction is described through gluons, which are not the part of the naïve quark model. We will explain it when we add colour to the simple quark model. We discuss here the mass formula suggested by quark model for $l > 0$ resonances. We consider the case of meson resonances.

The quark model mass formula for bosons (mesons) due to Gell—Mann is

$$m^2 = (a + bl) + c(s_1 + s_2)l \tag{3.219}$$

where a, b, c and d are constants. l is the orbital angular momentum of the quark—antiquark pair, s_1 and s_2 are the quark spins along the l axis. The first term depends on l values and the second describes the spin—orbit interaction term. A third term, which describes the radial excitation, has been neglected. If the spin—orbit interaction is not taken into account the eqn. (3.219) may be written as

$$m_0^2 = a + bl \tag{3.220}$$

The difference of m^2 and m_0^2 would give the mass splitting of mesons as

$$\Delta m^2 = c(s_1 + s_2)l \tag{3.221}$$

which on using the formula from atomic spectra can be written as

$$\Delta m^2 = a_1[j(j+1) - l(l+1) - s(s+1)] + a_2 \tag{3.222}$$

where a_1, a_2 are constants. j is the total angular momentum of the $q\bar{q}$ pair, and s is the total spin. The spin—orbit coupling term in atomic spectra gives rise to fine structure of spectral lines and each spectral line of orbital angular momentum l splits into four states, three corresponding to $s = 1$ and one corresponding to $s = 0$. The four states are given by ${}^{2s+1}l_j$, with $j = l - 1$, l and $l + 1$ for the triplet ($2s + 1 = 3$), and $j = l$ for the singlet ($2s + 1 = 1$).

Here, for each l, we must get four mass values. For example, for $l = 1$, the four states should have mass values, whose mass square difference from the central value should be

$$\Delta m^2 = a_1\ [j(j + 1) - s(s + 1)] \tag{3.223}$$

in which $2a_1$ has been balanced by a_2. This prediction of four states (for $l = 1$)has been found agreeing with the observation, where four resonance states A_2 (1310 MeV), B(1235), A_1(1100) and δ(970) have been observed. Though two more states, f(1270) and D(1285), which were not predicted by the quark model are also observed, the quark model nevertheless, predicts hadron masses and is successful in this endeavor to varying degrees.

3.5.8.6 Paradoxes of Naive Quark Model and their Resolution

The hypotheses on which the quark model (naïve) is based suffer from the following paradoxes.

1. The quarks u, d, s have fractional charges and we have yet to observe such fractional charges. This contradiction with reality is resolved by introducing the concept of quark confinement that quarks always remain inside the hadrons (see Chapter 1).
2. The quarks are assumed to be fermi particles hence the overall quark wave-function [ψ(space)ψ(isospin)ψ(spin)] must be antisymmetric. In the ground state, ($l = 0$) ψ(space) is symmetric, hence [ψ(isospin)ψ(spin)] must be antisymmetric. But we have seen in eqn. (3.151) that [ψ(isospin) ψ(spin)] for ($l = 0$) baryons is symmetric under the interchange of any two quarks. So, the Pauli's exclusion principle (here it is extended principle as isospin has been included) stands violated and this violation is a serious paradox of naïve quark model.

To resolve this paradox, quarks are associated with a new quantum number, called colour. Three colours red (R), green (G) and blue (B) are assigned to quarks and ψ(colour), given by eqn. (1.206), i.e,

$$\psi(\text{colour}) = \frac{1}{\sqrt{6}}(RGB - RBG + GBR - GRB + BRG - BGR) \qquad (3.224)$$

is antisymmetric under the interchange of two colors. Thus the antisymmetry of ψ(colour) it self requires that [ψ(space)ψ(isospin)ψ(spin)] be symmetric. However, the introduction of color raises another question that why colored particles are not observed! To this, the answer is that, all naturally occurring particles are colorless (colour singlets). Any meson, which becomes observable would contain say, red quark and antired quark and so would look colourless. Any physical baryon would contain three quarks each of different colour. For more details we refer to Chapter 1.

3. In SU(6) quark model, the quarks have been assumed static, i.e., non-relativistic. The successes of quark model are observed with static quarks. Considering quarks as relativistic makes the predictions hardly come true. In order to see that quarks are non-relativistic, the quarks must be very heavy, i.e., of the order of 5—10 GeV, whereas the predicted masses of *u, d, s* are 0.3 to 0.5 GeV. The explanation in this regard can be given that the mass greater than 5 GeV is that of a free quark, which is yet to be observed.
4. The binding energy of quarks appears to be very large and it is difficult to explain this super strong binding of quarks. The explanation for such strong binding is given that the binding could be from the exchange of a vector particle, which is coupled to the baryon current as well.
5. The mesons in the quark model are assumed to be $q\bar{q}$ and baryons $q\,q\,q$, and the hadrons are seen to be built out of only these combinations of quarks and antiquarks. There is no evidence for bound $q\,q$ and $qqqq$ states, although such states have masses comparable to the observed particles. The answer to this paradox also lies in colour degree of freedom. Since, only those states of quarks are observable, which are colourless (truly, colour singlets), the colourless states like $q\bar{q}$ (like $R\bar{R}$ etc.) and qqq (colour combinations $q_R q_B q_G$, etc.) become observable and the two or four quark states, which can't become colourless remain unobservable.

3.5.8.7 Colour Quark Model and Hadron Masses

When the quarks of the naïve quark model are endowed with colour degree of freedom, we have the colour quark model. The coloured quarks interact by exchanging coloured gluons. The theory of quark interactions in which coloured gluons participate is called the *Quantum Chromodynamics* (*QCD*).

The quark interaction is understood in terms of the potential between quarks. Just as in Quantum Electrodynamics (*QED*), we have Coulomb potential between the like and unlike charges, in *QCD* there is a potential between quark–antiquark and quark—quark. The quark—antiquark potentials are:

$$V_{q\bar{q}}(r) = -\frac{4}{3}\frac{\alpha_s}{r} \quad \text{(Colour singlet)} \tag{3.225a}$$

$$V_{q\bar{q}}(r) = \frac{1}{6}\frac{\alpha_s}{r} \quad \text{(Colour octet)} \tag{3.225b}$$

and for the quark—quark potential

$$V_{q\bar{q}}(r) = -\frac{2}{3}\frac{\alpha_s}{r} \quad \text{(Colour triplet)} \tag{3.225c}$$

$$V_{qq}(r) = \frac{1}{3}\frac{\alpha_s}{r} \quad \text{(Colour sextet)} \tag{3.225d}$$

where α_s is the strong coupling constant and the numerical coefficients denote the colour factors.

In *QCD* quarks interact via gluons, which carry a colour and an anticolour. So, with three colours, red (R), blue (B) and green (G), we can have nine kinds of gluons namely, $R\bar{R}, B\bar{B}, G\bar{G}, R\bar{B}, B\bar{R}, B\bar{G}, G\bar{B}, R\bar{G}$ and $G\bar{R}$. However, these nine gluons combine to form an octet (eight gluon states) and a singlet. The colour octet is

$$\begin{aligned}
|1\rangle &= \frac{1}{\sqrt{2}}(R\bar{B} + B\bar{R}) \quad |2\rangle = -\frac{i}{\sqrt{2}}(R\bar{B} - B\bar{R}) \\
|3\rangle &= \frac{1}{\sqrt{2}}(R\bar{R} - B\bar{B}) \quad |4\rangle = \frac{1}{\sqrt{2}}(R\bar{G} + G\bar{R}) \\
|5\rangle &= -\frac{i}{\sqrt{2}}(R\bar{G} - G\bar{R}) \quad |6\rangle = \frac{1}{\sqrt{2}}(B\bar{G} + G\bar{B}) \\
|7\rangle &= -\frac{i}{\sqrt{2}}(B\bar{G} - G\bar{B}) \quad |8\rangle = \frac{1}{\sqrt{6}}(R\bar{R} + B\bar{B} - 2G\bar{G})
\end{aligned} \tag{3.226}$$

and the colour singlet is

$$|9\rangle = \frac{1}{\sqrt{3}}(R\bar{R} + B\bar{B} + G\bar{G}) \tag{3.227}$$

it may be noted that all the three states $|3\rangle$, $|8\rangle$ and $|9\rangle$ are colourless but only $|9\rangle$ is colour singlet, $|3\rangle$ and $|8\rangle$ are not. It is the colour singlet of quarks that becomes observable. Also, for the colour singlet, the $q\bar{q}$ potential is negative, i.e., attractive. Hence, $q\bar{q}$ would make a bound state which being colour singlet would be observable, and indeed we observe such states as mesons. In quark—

quark potential, however, two quarks cannot make a singlet and an octet; we have a triplet, which is an antisymmetric combination of colours

$$\begin{aligned}&\frac{1}{\sqrt{2}}(RB - BR)\\&\frac{1}{\sqrt{2}}(BG - GB)\\&\frac{1}{\sqrt{2}}(GR - RG)\end{aligned} \tag{3.228}$$

and a sextet, which is a symmetric combination of colours

$$RR, BB, GG, \tfrac{1}{\sqrt{2}}(RB + BR), \tfrac{1}{\sqrt{2}}(BG + GB), \tfrac{1}{\sqrt{2}}(GR + RG) \tag{3.229}$$

We focus our attention on the attractive and $qq\,q\,\bar{q}$ and $q\,q$ potentials, namely, equations [3.225 (a) and (c)] and see their effect in determining the meson and baryon masses. For this purpose, we recall that the mass spectrum of mesons and baryons is caused by the mass splittings which take place due to the spin—orbit and spin—spin interaction of quarks. For $l = 0$ states, the spin–orbit interaction is absent and the spin–spin interaction would cause the mass splittings. Here we will consider the spin—spin interaction of $q\,\bar{q}$ and $q\,q$ to predict the masses of ground state (l = 0) pseudoscalar and vector mesons and also the masses of ground state octet and decuplet baryons. Since, in *QED*, e.g., in the study of the hyperfine structure of hydrogen, the proton—electron spin–spin coupling is taken into account to explain the observed energy shifts. For the ground state of hydrogen atom, the hyperfine energy shifts ΔE_{hf} are given by

$$\Delta E_{hf} = \frac{8\pi\alpha}{3m_e m_p}(\boldsymbol{S}_e.\boldsymbol{S}_p)|\psi(0)|^2 \tag{3.230}$$

where $\alpha = \dfrac{e^2}{4\pi}$ is the fine structure constant, m_p, the mass of proton and S_p its spin; m_e, S_e are for electron, $\psi(0)$ is $\psi(r_1, r_2)$ at zero separation. In *QED*, the potential (Coulomb) between proton and electron is

$$V(r) \approx -\frac{\alpha}{r} \tag{3.231}$$

In case of *QCD*, for attractive potentials [equations 3.225 a and c], we notice that the α of *QED* is substituted by $\frac{4}{3}\alpha_s$ for mesons and by $\frac{2}{3}\alpha_s$ for baryons. Therefore, the hyperfine splitting (3.230) for $q_1\,\bar{q}_2$ and $q_1\,q_2\,q_3$ cases would be

$$(\Delta E_{hfs})_{q_1\bar{q}_2} = \frac{32}{9}\pi\alpha_s|\psi(0)|^2\frac{\boldsymbol{S}_1.\boldsymbol{S}_2}{m_1 m_2} \tag{3.232}$$

and

$$(\Delta E_{hfs})_{q_1q_2q_3} = \frac{16}{9}\pi\alpha_s |\psi(0)|^2 \left[\frac{\boldsymbol{S}_1.\boldsymbol{S}_2}{m_1m_2} + \frac{\boldsymbol{S}_1.\boldsymbol{S}_3}{m_1m_3} + \frac{\boldsymbol{S}_2.\boldsymbol{S}_3}{m_2m_3}\right] \tag{3.233}$$

where m_1, m_2, m_3 and S_1, S_2, S_3 are the spins of the corresponding quarks (antiquarks). We may therefore write for the meson $(q_1\bar{q}_2)$ and baryon $(q_1q_2q_3)$ masses, the following relations

$$m(mesons) = m_1 + m_2 + \beta\frac{\boldsymbol{S}_1.\boldsymbol{S}_2}{m_1m_2} \tag{3.234}$$

and

$$m(baryons) = m_1 + m_2 + m_3 + \beta'\left[\frac{\boldsymbol{S}_1.\boldsymbol{S}_2}{m_1m_2} + \frac{\boldsymbol{S}_1.\boldsymbol{S}_3}{m_1m_3} + \frac{\boldsymbol{S}_2.\boldsymbol{S}_3}{m_2m_3}\right] \tag{3.235}$$

where β and β' coefficients in eqs. (3.234) and (3.235) respectively are constants to be adjusted to obtain the optimal fit to the data. Based on eqn. (3.234) the predictions of pseudoscalar (0^-) and vector (1^-) meson masses can be made. Likewise, the relation (3.235) can predict the octet and decuplet baryon masses. Following calculations demonstrate this.

3.5.8.7.1 Meson masses

In the coefficient

$$\beta = \frac{32}{9}\pi\alpha_s |\psi(0)|^2 \tag{3.236}$$

$|\psi(0)|^2$ cannot be calculated, therefore, we assume that it is same for all pseudoscalar and vector mesons. Now,

$$\boldsymbol{S} = \boldsymbol{S}_1 + \boldsymbol{S}_2 \tag{3.237a}$$

so

$$\boldsymbol{S}^2 = \boldsymbol{S}_1^2 + \boldsymbol{S}_2^2 + 2\boldsymbol{S}_1.\mathbf{S}_2 \tag{3.237b}$$

$$\Rightarrow \quad S(S+1) = S_1(S_1+1) + S_2(S_2+1) + 2\boldsymbol{S}_1.\boldsymbol{S}_2$$

$$= \frac{3}{2} + 2\ \boldsymbol{S}_1.\boldsymbol{S}_2 \qquad (\because S_1 = S_2 = \tfrac{1}{2})$$

$$\therefore \quad \mathbf{S}_1.\boldsymbol{S}_2 = \frac{1}{2}[S(S+1) - \frac{3}{2}] \tag{3.238}$$

where S is the spin of $q_1\,\bar{q}_2$ (meson) state. For $S = 0$, i.e., for pseudoscalar (0^-) mesons

$$\boldsymbol{S}_1.\mathbf{S}_2 = -\frac{3}{4} \tag{3.239a}$$

and for vector (1^-) mesons $S = 1$,

$$\boldsymbol{S}_1 . \boldsymbol{S}_2 = \frac{1}{4} \tag{3.239b}$$

It may be noted that these values are in the units $c = \hbar = 1$; otherwise the values (3.239) are to be multiplied by $\hbar^2$. Thus using equations (3.234) and (3.239), we may write the mass formulae for mesons

$$m(0^-) = m_1 + m_2 - \frac{3}{4}\frac{\beta}{m_1 m_2}$$

and

$$m(1^-) = m_1 + m_2 + \frac{1}{4}\frac{\beta}{m_1 m_2} \tag{3.240}$$

where m_1 and m_2 would be the masses of the quark and antiquark that constitute the meson. Taking the constituent quark masses in mesons as $m_\mu\, m_d = 310$ MeV, $m_s = 483$ MeV and $\beta = 640\, m_u^2\, \text{MeV}^3$, we obtain the following masses for pseudoscalar and vector mesons. For example,

$$m(\pi^+) = m(u\bar{d}) = m_u + m_{\bar{d}} - \frac{3}{4}\left[\frac{\beta}{m_u m_d}\right]$$
$$= 620 - 480 = 140 \text{ MeV}$$

and

$$m(\rho^+) = m(u\bar{d}) = m_u + m_{\bar{d}} + \frac{1}{4}\left[\frac{\beta}{m_u m_d}\right]$$
$$= 620 + 160 = 780 \text{ MeV} \tag{3.241}$$

Masses of other mesons may also be obtained. The following table shows an excellent agreement between the predicted and observed masses.

Table of Quark Model Masses of 0^- and 1^- Mesons

Meson	J^P	Quark content	Predicted mass MeV/C²	Observed mass MeV/C²
π^+	0^-	$u\bar{d}$	140	140
ρ^+	1^-	$u\bar{d}$	780	770
K^+	0^-	$u\bar{s}$	485	494
K^{*+}	1^-	$u\bar{s}$	896	892
η	0^-	$\frac{1}{\sqrt{6}}(u\bar{u} + d\bar{d} - 2s\bar{s})$	559	549
ω	1^-	$\frac{1}{\sqrt{2}}(u\bar{u} + d\bar{d})$	780	783
ϕ	1^-	$s\bar{s}$	1032	1019

3.5.8.7.2 Baryon masses

The mass splitting through spin—spin interaction of the constituent quarks is successfully applicable to predict the masses of the ground state ($l = 0$) baryons of the octet $(\frac{1}{2}^+)$ and decuplet $(\frac{3}{2}^+)$. Before deriving the mass equations for various baryons of these multiplets, we present the following table in which the quark contents of these baryons have been recalled and the observed masses along with the masses to be predicted on the basis of quark model have been given.

Table for the Quark Model masses of $(\frac{1}{2}^+)$ and $(\frac{3}{2}^+)$ Baryons

Baryons	J^P	Quark content	Calculated mass MeV/C²	Observed mass MeV/C²
Nucleon N $\begin{cases} p \\ n \end{cases}$	$(\frac{1}{2}^+)$	*uud* *udd*	939	939
Δ Resonances $\begin{cases} \Delta^{++} \\ \Delta^{+} \\ \Delta^{0} \\ \Delta^{-} \end{cases}$	$(\frac{3}{2}^+)$	*uuu* *uud* *udd* *ddd*	1239	1232
Ω Resonance	$(\frac{3}{2}^+)$	*sss*	1682	1672
Σ^* Resonances $\begin{cases} \Sigma^{*+} \\ \Sigma^{*0} \\ \Sigma^{*-} \end{cases}$	$(\frac{3}{2}^+)$	*uus* *uds* *dds*	1381	1384
Ξ^* Resonances $\begin{cases} \Xi^{*0} \\ \Xi^{*-} \end{cases}$	$(\frac{3}{2}^+)$	*uss* *dss*	1529	1533
Λ Particle	$(\frac{1}{2}^+)$	*uds*	1116	1114
Σ Particles $\begin{cases} \Sigma^{+} \\ \Sigma^{0} \\ \Sigma^{-} \end{cases}$	$(\frac{1}{2}^+)$	*uus* *uds* *dds*	1179	1193
Ξ Particles $\begin{cases} \Xi^{0} \\ \Xi^{-} \end{cases}$	$(\frac{1}{2}^+)$	*uss* *dss*	1327	1318

The constituent masses of u, d and s quarks in baryons are $m_u = m_d = 363$ MeV and $m_s = 538$ MeV.

Thus among the baryons of the above table, (N, Λ and Ω) belong to the category, in which the mass of each of the three quarks is same, which implies that $m_1 = m_2 = m_3$.

Σ^* and Ξ^* Resonances have two quarks of equal mass. In Σ^* first two quarks have same mass and in Ξ^*, the last two. Here all the spins are parallel.

Λ and Σ Particles too have first two quarks of equal mass and Ξ particle has last two quarks that have equal masses. However, in this case all the spins are not parallel.

The two-equal mass-quarks case comes under the category of $m_1 = m_2 \neq m_3$ along with its permutations.

We, now use spin—spin interaction to formulate mass relations for the above sets of baryons.

1. *N*, Δ, Ω *Masses*

$\boldsymbol{S}_1, \boldsymbol{S}_2, \boldsymbol{S}_3$ are the spins of individual quarks constituting the baryon. For the baryon in the ground state, its $\boldsymbol{L} = 0$, therefore

$$\begin{aligned} \boldsymbol{J} \text{ of baryon} &= \boldsymbol{S} \text{ of baryon} + \boldsymbol{L} \text{ of baryon (zero)} \\ &= \boldsymbol{S} \text{ of baryon} \end{aligned} \tag{3.242}$$

$\Rightarrow$

$$\begin{aligned} \boldsymbol{J}^2 &= j(j+1) = (\boldsymbol{S}_1 + \boldsymbol{S}_2 + \boldsymbol{S}_3)^2 \\ &= \boldsymbol{S}_1^2 + \boldsymbol{S}_2^2 + \boldsymbol{S}_3^2 + 2[\boldsymbol{S}_1.\boldsymbol{S}_2 + \boldsymbol{S}_1.\boldsymbol{S}_3 + \boldsymbol{S}_2.\boldsymbol{S}_3] \\ &= S_1(S_1+1) + S_2(S_2+1) + S_3(S_3+1) + 2[\ ] \end{aligned} \tag{3.243}$$

Since, $S_1 - S_2 - S_3 - \frac{1}{2}$ (spin of quarks), we obtain from the above equation

$$\boldsymbol{S}_1.\boldsymbol{S}_2 + \boldsymbol{S}_1.\boldsymbol{S}_3 + \boldsymbol{S}_2.\boldsymbol{S}_3 = \frac{1}{2}[j(j+1) - \frac{9}{4}] \tag{3.244}$$

in which $j = \dfrac{1}{2}$ for nucleon (N) and it is $\dfrac{3}{2}$ for Δ and Ω. So,

$$\boldsymbol{S}_1.\boldsymbol{S}_2 + \boldsymbol{S}_1.\boldsymbol{S}_3 + \boldsymbol{S}_2.\boldsymbol{S}_3 = \begin{cases} \dfrac{3}{4} \text{ for } j = \dfrac{3}{2} \text{ (decuplet)} \\ -\dfrac{3}{4} \text{ for } j = \dfrac{1}{2} \text{ (octet)} \end{cases} \tag{3.245}$$

Entering this eqn. (3.245), into eqn. (3.235), we get

$$\begin{aligned} m_N &= 3m_u - \frac{3}{4}\frac{\beta'}{m_u^2} \\ m_\Delta &= 3m_u + \frac{3}{4}\frac{\beta'}{m_u^2} \\ m_{\Omega^-} &= 3m_u + \frac{3}{4}\frac{\beta'}{m_s^2} \end{aligned} \tag{3.246}$$

where β' is a constant, whose value can be chosen. Taking

$$\beta' = 200\, m_u^2\ (\text{MeV})^3 \tag{3.247}$$

we get the optimal fit with the data. Using mass relations (3.246) and (3.247) and quark masses, we get the N, Δ and Ω^- masses as shown in the table, which agree very well with the observed masses of these particles.

2. Δ*, Ξ* *Masses*

These are the $J^P = (\frac{3}{2})^+$ baryons and in the ground state the spins of all the three quarks would be parallel and thus total spin of any two quarks would be 1. We take up S_1 and S_2, then

$$\begin{aligned} \boldsymbol{J}^2 = j(j+1) &= (\boldsymbol{S}_1 + \boldsymbol{S}_2)^2 = \boldsymbol{S}_1^2 + \boldsymbol{S}_2^2 + 2\,\boldsymbol{S}_1.\boldsymbol{S}_2 \\ &= S_1(S_1+1) + S_2(S_2+1) + 2\,\boldsymbol{S}_1.\boldsymbol{S}_2 \end{aligned} \tag{3.248}$$

where $S_1 = \frac{1}{2} = S_2$ and $j = 1$ (two quarks). So, we get from eqn. (3.248)

$$\boldsymbol{S}_1.\boldsymbol{S}_2 = \frac{1}{4}$$

and similar relation for $\boldsymbol{S}_1.\boldsymbol{S}_3$ and $\boldsymbol{S}_2.\boldsymbol{S}_3$. So, we write

$$\boldsymbol{S}_1.\boldsymbol{S}_2 = \boldsymbol{S}_1.\boldsymbol{S}_3 = \boldsymbol{S}_2.\boldsymbol{S}_3 = \frac{1}{4} \tag{3.249}$$

for the case where all the spins of three quarks are parallel ($\frac{3}{2}^+$ case).

Using eqn. (3.249) in eqn. (3.235), we get

$$m_{\Sigma^*} = 2\,m_u + m_s + \tfrac{1}{4}\beta'\left[\frac{1}{m_u^2} + \frac{2}{m_u m_s}\right] \tag{3.250}$$

and

$$m_{\Xi^*} = m_u + 2m_s + \tfrac{1}{4}\beta'\left[\frac{2}{m_u m_s} + \frac{1}{m_s^2}\right] \tag{3.251}$$

where β' is again given by eqn. (3.247). The calculated values from these relations have been shown in the table, which too agree with observed values quite excellently.

3. Λ, Σ *Masses*

These are $J^P = (\frac{1}{2})^+$ baryons, i.e., in the ground state of Λ and Σ baryons, the spins of the three quarks combine to give a net 1/2. Thus the spins of two quarks may give the total spin 1 or 0. Due to the presence of non-parallel spins, this case differs from the case of Σ^* and Ξ^*, though in both cases, two quarks have equal masses.

Now, the Σ baryon has the isospin triplet Σ^+, Σ^0, Σ^- and hence its isospin $T = 1$. On the other hand $\Lambda(\Lambda^0$ only), has isospin $T = 0$. Now, ψ(isospin) with $T = 1$ is symmetric and with $T = 0$, it is antisymmetric. Also, the baryons (here Σ and Λ) should have

$$\psi = \psi(\text{space})\psi(\text{isospin})\psi(\text{spin})\psi(\text{colour}) \tag{3.252}$$

antisymmetric, under the interchange of first two quarks. For $l = 0$, ψ(space) is symmetric, ψ(colour) already antisymmetric, hence for ψ to be antisymmetric, we must have

$$\psi(\text{isospin})\psi(\text{spin}) \longrightarrow \text{symmetric} \tag{3.253}$$

implying that for the isospin $T = 1$ (symmetric), ψ(spin) must also be symmetric. In other words, for the baryon Σ(isospin 1), the spin of first two quarks (*uu*, *ud* or *dd*, we take *ud*) must be 1. In case of Λ^0, however, because of $T = 0$ (antisymmetric); ψ(spin) too must be antisymmetric, implying that the spins of *u* and *d* quarks would add up to zero. Therefore,
for Σ

$$\boldsymbol{S}_u + \boldsymbol{S}_d = 1$$

and for Λ

$$\boldsymbol{S}_u + \boldsymbol{S}_d = 0 \tag{3.254}$$

Now,

$$\begin{aligned} \boldsymbol{J}^2 = j(j+1) &= (\boldsymbol{S}_u + \boldsymbol{S}_d)^2 = \boldsymbol{S}_u^2 + \boldsymbol{S}_d^2 + 2\,\boldsymbol{S}_u.\boldsymbol{S}_d \\ &= S_u(S_u + 1) + S_d(S_d + 1) + 2\,\boldsymbol{S}_u.\boldsymbol{S}_d \end{aligned} \tag{3.255}$$

where

$$\begin{aligned} j &= 1 \text{ for } \Sigma \\ j &= 0 \text{ for } \Lambda \end{aligned} \tag{3.256}$$

and hence

$$\boldsymbol{S}_u.\boldsymbol{S}_d = \frac{1}{4} \text{ for } \Sigma \tag{3.257}$$

and

$$\boldsymbol{S}_u.\boldsymbol{S}_d = -\frac{3}{4} \text{ for } \Lambda \tag{3.258}$$

Now, the third quark is the strange quark. So, for both Σ and Λ, the mass equation (3.235) would read as

$$m(\Sigma, \Lambda) = m_u + m_d + m_s + \beta'\left[\frac{\boldsymbol{S}_u.\boldsymbol{S}_d}{m_u m_d} + \frac{\boldsymbol{S}_u.\boldsymbol{S}_s}{m_u m_s} + \frac{\boldsymbol{S}_d.\boldsymbol{S}_s}{m_d m_s}\right] \tag{3.259}$$

which on using $m_u \approx m_d$, can be written as

$$m(\Sigma, \Lambda) = 2m_u + m_s + \beta'\left[\frac{\boldsymbol{S}_u.\boldsymbol{S}_d}{m_u^2} + \frac{1}{m_u m_s}(\boldsymbol{S}_u.\boldsymbol{S}_s + \boldsymbol{S}_d.\boldsymbol{S}_s)\right] \quad (3.260)$$

in which the numerator of the second term has to be calculated. For this, similar to eqn. (3.243), we note

$$(\boldsymbol{S}_u + \boldsymbol{S}_d + \boldsymbol{S}_s)^2 = \boldsymbol{S}_u^2 + \boldsymbol{S}_d^2 + \boldsymbol{S}_s^2 + 2(\boldsymbol{S}_u.\boldsymbol{S}_d + \boldsymbol{S}_u.\boldsymbol{S}_s + \boldsymbol{S}_d.\boldsymbol{S}_s) \quad (3.261)$$

where LHS is $J^2 = j(j+1)$. Similar to eqn. (3.244), we may then write

$$\boldsymbol{S}_u.\boldsymbol{S}_d + \boldsymbol{S}_u.\boldsymbol{S}_s + \boldsymbol{S}_d.\boldsymbol{S}_s = \frac{1}{2}[j(j+1) - \tfrac{9}{4}] = -\frac{3}{4} \text{ (for } j = \tfrac{1}{2}) \quad (3.262)$$

Note that, Σ, Λ are $j = \frac{1}{2}$ baryons. From eqn. (3.262), we may write

$$\boldsymbol{S}_u.\boldsymbol{S}_s + \boldsymbol{S}_d.\boldsymbol{S}_s = -\frac{3}{4} - \boldsymbol{S}_u.\boldsymbol{S}_d \quad (3.263)$$

which on using eqs. (3.257) and (3.258) reads

$$(\boldsymbol{S}_u.\boldsymbol{S}_s + \boldsymbol{S}_d.\boldsymbol{S}_s)_\Sigma = -\frac{3}{4} - \frac{1}{4} = -1 \quad (3.264)$$

and for Λ

$$(\boldsymbol{S}_u.\boldsymbol{S}_s + \boldsymbol{S}_d.\boldsymbol{S}_s)_\Lambda = -\frac{3}{4} + \frac{3}{4} = 0 \quad (3.265)$$

Then using equations (3.257) and (3.264) in eqn. (3.260), we get

$$m_\Sigma = 2m_u + m_s + \frac{\beta'}{4}\left[\frac{1}{m_u^2} - \frac{4}{m_u m_s}\right] \quad (3.266)$$

Similarly, from equations (3.258), (3.265) and (3.260), we would get

$$m_\Lambda = 2m_u + m_s - \frac{3\beta'}{4}\frac{1}{m_u^2} \quad (3.267)$$

From equations (3.266) and (3.267), the masses of Σ and Λ baryons can be calculated as shown in the table, which too agree very well with observed masses.

Thus, the spin—spin interaction based on colored quark model successfully predicts the masses of ground state mesons and baryons.

3.6 ASSIGNMENTS

1. Define a symmetry group. Demonstrate through an example that the symmetry group obeys the properties of closure, identity, inverse and associativity.
2. Define the basic properties of a symmetry group. Explain with examples the discrete, continuous, simple and semisimple groups.

3. What are Lie groups? Describe the properties of some important Lie groups used in particle physics. What is Lie algebra? Explain its significance.
4. Define a Lie group and its generators. Show that the generators of U(2) group are the Pauli matrices. Compute the number of generators of O(n) group.
5. Define SU(n) groups. Show that the generators of SU(n) group are $(n^2 - 1)$ traceless hermitian matrices. Mentioning the generators of SU(3) group; explain that its rank is 2.
6. Define Lie algebra and hence describe the Abelian and non-Abelian groups. Give examples. What is meant by structure constants? Explain through examples.
7. Define representation of a Lie group and briefly explain its kinds. Explain reducible and irreducible representations. Describe the condition to express a representation as the direct sum of irreducible representations. Discuss physical significance of irreducible representations.
8. Define an SU(2) group. Show that this group can describe both spin and isospin. What is meant by fundamental and adjoint representations of SU(2) groups?
9. Define fundamental representation of SU(2) group. Derive equation (3.72) andshow that (*i*) $2\otimes 2 = 1\oplus 3$ (*ii*) $2\otimes 2\otimes 2 = 4\oplus 2\oplus 2$. Explain the physical significance of above product representations. Draw these representations graphically as well as through Young's tableau technique.
10. Define an SU(3) group and show that the matrices (3.82) form the fundamental representation of SU(3) group. Show also that the diagonal generators of SU(3) group obey the commutation relation (3.84). Check it for two non-diagonal lambda matrices as well.
11. What are the shift operators of SU(3)? Deduce the commutation relations between them and explain their properties. Show that the t_3–y plot of SU(3) represents the octet representation of SU(3).
12. Define the octet representation of SU (3) group and show that the particle states of the representation can be changed into one another by shift operators.
13. Define 3 and 3* representations of SU(3) group. Show that they represent respectively the particle and antiparticle states. Derive relations (3.109) for the product representations of SU(3).
14. Define the product representations of SU(3) and show that (*i*) $3\otimes 3 = 3^*\oplus 6$ (*ii*) $3\otimes 3^* = 1\oplus 8$. Explain the physical content of the above decompositions.
15. Use Young's tableau technique to prove (*a*) $3\otimes 3\otimes 3 = 1\oplus 8\oplus 8\oplus 10$ (*b*) $8\otimes 8 = 1\oplus 8\oplus 8\oplus 10^*\oplus 10\oplus 27$. Explain the difference between 10 and 10*.

16. Deduce Gell—Mann—Okubo mass formula. How far do the baryon and meson masses agree with their masses predicted by this formula? Discuss the reasons that in the Lagrangian densities the baryon mass terms are linear but meson mass terms are quadratic in mass.
17. What are the applications of SU(3) group? Analyse them to determine the scope of the group. Point out the limitations of SU(3) group and explain how far these are removed in SU(4) group.
18. Describe an SU(4) symmetry group and discuss its scope and limitations. Write the quark content of mesons in 15 plet of $4 \otimes 4^* = 1 \oplus 15$.
19. Describe the general features of SU(6) quark model. Explain how the product representation $6 \otimes 6^*$ describes the mesons in this model.
20. Discuss the general features of SU(6) quark model and explain the mixing of meson states. Discuss and explain ideal mixing.
21. Explain how mesons and baryons can be described in SU(6) quark model.
22. Describe the SU(6) quark model and show that the model successfully predicts the neutron and proton magnetic dipole moment ratio. Estimate the magnetic dipole moments of Σ^0 and $\Sigma^{\pm}$ baryons as well.
23. Discuss the hypotheses of SU(6) quark model and explain how the model predicts the masses of hadrons. How far these predictions stand better than the predictions of Gell—Mann—Okubo mass formula?
24. Describe the SU(6) quark model hypotheses and mention its applications. Discuss its paradoxes and explain how they can be resolved.
25. Discuss the SU(6) quark model hypotheses and explain the necessity of introducing colour quantum number of quarks. Obtain the relations (3.233) and (3.234) for meson and baryon masses.
26. Describe the colour quark model and obtain the mass formulae (3.251), (3.266) and (3.267).
27. Explain, how the spin—spin interaction based on colour quark model predicts the masses of ground state mesons and baryons.

3.7 OBJECTIVE QUESTIONS

1. The family of phase transformations $e^{i\alpha}$, where α runs continuously over real numbers, forms
   (a) U(1) group (b) U(n) group
   (c) SU(n) group (d) O(n) group
2. Which among the following statements is not true?
   (a) If a group is infinite, it may be discrete or continuous.
   (b) The group of all integers in which the multiplication defines the closure property is a discrete group.
   (c) Group of all rotations in a plane is discrete group.
   (d) Under the definition of continuity, if the group manifold forms a topological space, the group is said to be a topological group.

3. A group has only two elements, the identity e and the element g, obeying $g^2 = e$. What is true among the following?
   (a) The group is finite and the element g is space-inversion.
   (b) The group is finite and the element g is charge-conjugation.
   (c) Both (a) and (b) above are true.
   (d) The element g is either space-inversion or charge-conjugation and the group is continuous.
4. The rotational symmetry related to the conservation of angular momentum is described by the group
   (a) SU(3) (b) U(3)
   (c) SO(3) (d) U(n)
5. The Lie group is a
   (a) continuous group whose parameter space is locally Euclidean.
   (b) discrete group whose parameter space is Euclidean.
   (c) finite group without any parameter space.
   (d) continuous group whose parameter space is not Euclidean.
6. SU(3) group is a generalization of
   (a) SU(2) spin group. (b) SU(2) isospin group.
   (c) O(3) group. (d) SU(n) group.
7. The four 2 × 2 hermitian matrices (Pauli matrices) are the generators of
   (a) U(1) group. (b) O(3) group.
   (c) U(2) group. (d) O(n) group.
8. The matrices $R = \begin{pmatrix} \cos\theta & -\sin\theta \\ \sin\theta & \cos\theta \end{pmatrix}$ constitute
   (a) non-Abelian O(2) group. (b) Abelian O(2) group.
   (c) non-Abelian SU(2) group. (d) Abelian U(1) group.
9. The relation $[X_i, X_j] = i\sum_{k=1}^{n} C_{ij}^{k} X_k$ represents the
   (a) ordinary algebra of a group whose generators are X_i and C_{ij}^{k} are matrices called structure constants.
   (b) Lie algebra of a group whose generators are X_i and C_{ij}^{k} are the matrices called structure constants.
   (c) Lie algebra of a group where X_i are the generators and C_{ij}^{k} the numbers called structure constants.
   (d) Lie algebra of a group where X_i are the matrices called structure constants and C_{ij}^{k} are the generators of the group.

10 The Lie algebra of SO(3) group is a commutation relation similar to
   (a) the commutation relation between angular momentum operators.
   (b) the commutation relation between linear momentum operators.
   (c) the commutation relation between the angular momentum operator and its square.
   (d) the commutation relation between the raising and lowering angular momentum operators.

11. Which of the following is true?
   (a) The group of operators to which the group of elements is mapped isomorphically is called the representation of group of elements.
   (b) The group of operators to which the group of elements is mapped homorphically is called the representation of group of elements.
   (c) The group of elements to which the group of operators is mapped isomorphically is called the representation of group of elements.
   (d) The group of elements to which the group of operators is mapped homorphically is called the representation of group of elements.

12. Match the groups with their ranks and number of generators and select the correct code.

	Group	Rank		Generator	
A	$SU(2) \approx O(3)$	A_1	5	A_2	15
B	O(3)	B_1	3	B_2	8
C	SU(3)	C_1	2	C_2	10
D	SU(4)	D_1	2	D_2	35
E	SU(6)	E_1	1	E_2	3

Codes:

	A	B	C	D	E
(a)	E_1 E_2	C_1 C_2	D_1 B_2	B_1 A_2	A_1 D_2
(b)	B_1 A_2	D_1 C_2	C_1 B_2	A_1 E_2	E_1 D_2
(c)	A_1 C_2	C_1 A_2	D_1 B_2	B_1 D_2	E_1 E_2
(d)	D_1 E_2	B_1 C_2	E_1 D_2	A_1 B_2	C_1 A_2

13. If D(*R*) is a Lie group representation and if there exists a non-singular matrix *M*, independent of group elements, such that

$$MD(R)M^{-1} = \begin{bmatrix} D^1(R) & & 0 \\ & D^2(R) & \\ 0 & & \ddots \end{bmatrix}$$ for all R in group, then $D(R)$ is called the

(a) irreducible representation (b) reducible representation
(c) defining representation (d) adjoint representation

14. The requirement that the matrices $D(R)$ be fully decomposed into irreducible matrices is that they be
(a) Hermitian (b) anti-hermitian
(c) unitary (d) orthogonal
15. Every representation is equivalent to a unitary representation for
(a) infinite groups (b) continuous groups
(c) discrete groups (d) finite groups
16. What in the following is not true about group representations?
(a) The adjoint representation is determined by the structure constants of the group.
(b) The dimensionality of adjoint representation is equal to the number of structure constants of the group.
(c) The three dimensional unitary unimodular matrices form the defining representation of SU(3).
(d) **3** is the adjoint representation of SU(2).
17. The number of Casimir operators is equal to the
(a) number of generators of the group.
(b) number of structure constants of the group.
(c) number of group elements.
(d) rank of the group.
18. The three colour charges of a quark, R, G, B form thc
(a) fundamental representation of SU(3).
(b) adjoint representation of SU(3).
(c) adjoint representation of SU(2).
(d) fundamental representation of SU(2).
19. Three spin $\frac{1}{2}$ particles group together into
(a) two triplets of spin 1 and one doublet of spin 1/2.
(b) a quartet of spin 3/2 and two doublets of spin 1/2.
(c) one triplet of spin 1, two doublets of spin 1/2 and one singlet of spin zero.
(d) one triplet of spin 1, one doublet of spin 1/2 and three singlets of spin zero.

20. The states of the spin-$\frac{1}{2}$ baryon octet can be converted into one another by
(a) I- spin and U-spin operators.
(b) I- spin and V-spin operators.
(c) U- spin and V-spin operators.
(d) I- spin, U-spin and V-spin operators.
21. Choose the correct answer for the following SU(3) shift operator

(a) $U^-|\Xi^0\rangle = |\Sigma^0\rangle$, $I^-|\Sigma^0\rangle = |p\rangle$, $I^+|p\rangle = |n\rangle$.

(b) $V^+|\Xi^-\rangle = |\Sigma^0\rangle$, $I^-|\Sigma^0\rangle = |\Sigma^-\rangle$, $U^+|\Sigma^-\rangle = |n\rangle$.

(c) $I^+|\Sigma^-\rangle = |\Lambda^0\rangle$, $V^+|\Lambda^0\rangle = |\Xi^-\rangle$, $U^-|\Xi^-\rangle = |\Sigma^-\rangle$.

(d) $I^+|n\rangle = |p\rangle$, $U^-|p\rangle = |\Sigma^+\rangle$, $V^+|\Sigma^+\rangle = |\Xi^0\rangle$.

22. Choose the correct statement for the hexagonal SU(3) irreducible representation (irrp).
(a) All the sites of the irrp are singly occupied.
(b) All the sites of the irrp are singly occupied except the sites (–1, 0), (0, 0), and (+1, 0) which are unoccupied.
(c) All the sites of the irrp are singly occupied except the sites $(-\frac{1}{2},1)$, $(+\frac{1}{2},1)$ which are doubly occupied.
(d) All the sites of the irrp are singly occupied except the site (0,0) which is doubly occupied.
23. What is true among the following?
(a) **3** ⊗ **3*** would represent a meson nonet, if **3** represents SU(3) flavour.
(b) **3** ⊗ **3*** would represent nine gluon states, if **3** represents SU(3) flavour.
(c) **3** ⊗ **3*** would represent nine gluon states, if **3** represents SU(3)colour.
(d) Statements (a) and (c) are correct.
24. **3** ⊗ **3** ⊗ **3** is decomposed into
(a) **8** ⊕ **8** ⊕ **8** ⊕ **3**, in which **8** and **3** are meson states.
(b) **8** ⊕ **8** ⊕ **8** ⊕ **3**, in which **8** and **3** are baryon states.
(c) **10** ⊕ **8** ⊕ **8** ⊕ **1**, in which **10, 8** and **1** are meson states.
(d) **10** ⊕ **8** ⊕ **8** ⊕ **1**, in which **10, 8** and **1** are baryon states.
25. Ignoring colour, on interchanging the quarks, the states of the baryon decuplet are
(a) symmetric
(b) antisymmetric
(c) five states are symmetric and five are antisymmetric
(d) three states are symmetric and seven are antisymmetric

26. In the following two mass formulae

(1) $m_\alpha + 3m_\beta = 2(m_\gamma + 3m_\delta)$

(2) $4m_\alpha^2 = 4m_\beta^2 + 3m_\gamma^2$

the Greek subscripts (α, β, γ, δ) represent
(a) mesons in (1) and baryons in (2)
(b) mesons in (2) and baryons in (1)
(c) mesons in both (1) and (2)
(d) baryons in both (1) and (2)

27. The product $4 \otimes 4^* = 1 \oplus 15$ describes
(a) strange baryons
(b) strange mesons
(c) charm baryons
(d) charm mesons

28. The six quark states of SU(6) quark model are

(a) $u(\uparrow)d(\uparrow)s(\uparrow)u(\downarrow)d(\downarrow)s(\downarrow)$

(b) $u(\uparrow)d(\uparrow)c(\uparrow)u(\downarrow)d(\downarrow)c(\downarrow)$

(c) $d(\uparrow)s(\uparrow)b(\uparrow)d(\downarrow)s(\downarrow)b(\downarrow)$

(d) $u(\uparrow)c(\uparrow)t(\uparrow)u(\downarrow)c(\downarrow)t(\downarrow)$

29. The mesons η and η' are
(a) pure octet and pure singlet states respectively.
(b) pure singlet and pure octet states respectively.
(c) both are mixture of pure octet and pure singlet states with η largely octet and η' largely singlet.
(d) both are mixture of pure octet and pure singlet states with η largely singlet and η' largely octet.

30. The ω, ϕ mixing angle is
(a) 35°　　(b) 40°
(c) –11°　　(d) 23°

31. Choose the correct answer:

(a) when ω, ϕ mixing angle is 40°, ω consists of $s\bar{s}$ and ϕ of $u\bar{u}$ and $d\bar{d}$.

(b) when ω, ϕ mixing angle is 35°, ω consists of $u\bar{u}$ and $d\bar{d}$ and ϕ of $s\bar{s}$.

(c) when ω, ϕ mixing angle is –11°, ω consists of $u\bar{u}$ and ϕ of $s\bar{s}$.

(d) when ω, ϕ mixing angle is 23°, ω consists of $d\bar{d}$ and $s\bar{s}$ and ϕ of $u\bar{u}$.

32. Match the baryons in column I with their magnetic dipole moments (μ) in column II and select the correct code.

Column I			*Column II*
A	p	1	$\frac{1}{3}(4\mu_u - \mu_s)$
B	n	2	$\frac{1}{3}(4\mu_s - \mu_d)$
C	Λ	3	$\frac{1}{3}(4\mu_u - \mu_d)$
D	Σ^+	4	μ_s
E	Ξ^-	5	$\frac{1}{3}(4\mu_d - \mu_u)$

Codes:

	A	B	C	D	E
(a)	3	4	1	2	5
(b)	3	5	1	4	2
(c)	3	5	4	1	2
(d)	1	4	2	3	5

33. The non occurrence of qq and $qqqq$ states demands that
    (a) quarks should possess colour.
    (b) quarks should be colourless.
    (c) quarks should interact via colourless gluons.
    (d) quarks should interact via photons.
34. The successful predictions of ground state masses of hadrons by the SU(6) quark model are based on
    (a) quarks carry colour and spin—spin interaction between quarks is responsible for the difference in masses of hadrons.
    (b) quarks are assumed to carry no colour and spin—spin interaction between quarks is responsible for the difference in masses of hadrons.
    (c) quarks carry colour and spin—orbit interaction between quarks is responsible for the difference in masses of hadrons.
    (d) quarks are assumed colourless and spin—orbit interaction between quarks is responsible for the difference in masses of hadrons.

ANSWERS

1a, 2c, 3c, 4c, 5a, 6b, 7c, 8b, 9c, 10a, 11b, 12a, 13b, 14c, 15d, 16b, 17d, 18a, 19b, 20d, 21b, 22d, 23d, 24d, 25a, 26b, 27d, 28a, 29c, 30b, 31b, 32c, 33a, 34a.

4

WEAK INTERACTIONS

4.1 INTRODUCTION

The weak interactions are responsible for the slow decay of unstable particles. Particles through strong interactions decay in about 10^{-23} sec.; and through electromagnetic interactions in about 10^{-16} sec. The weak interaction $\pi^- \rightarrow \mu^- \bar{\nu}_\mu$ takes a time of $\approx 10^{-8}$ sec., whereas the weak decay $\mu^- \rightarrow e^- \bar{\nu}_e \nu_\mu$ takes place in 10^{-6} sec. The neutron, which too decays through weak interaction, has a mean lifetime of about 17.3 minute. These are very long times compared to 10^{-23} sec. of strong interactions.

Weak processes can be divided into three types.

4.1.1 Purely Leptonic Interactions

In such weak interactions only the leptons participate. The examples are the decay of μ and τ leptons

$$\begin{aligned}
&\mu^- \rightarrow e^- + \bar{\nu}_e + \nu_\mu \\
&\mu^+ \rightarrow e^+ + \nu_e + \bar{\nu}_\mu \\
&\tau \rightarrow \mu + \nu_\mu + \nu_\tau \\
&\tau^+ \rightarrow e^+ + \nu_e + \bar{\nu}_\tau
\end{aligned} \tag{4.1}$$

the Leptonic weak interactions are distinguished by the fact that they obey the pure vector minus axial vector current (V–A) theory.

4.1.2 Semileptonic Interactions

The Semileptonic weak interactions involve both the leptons and hadrons. The hadrons that are involved in such processes may be both strange and non-strange. With the discovery of other flavours, the hadrons involved can also possess the charm (c) and beauty (b) (bottom ness) flavours. Depending on the involvement of non-flavoured and flavoured hadrons, this category of semileptonic weak interactions can be subdivided into

4.1.2.1 Strangeness Conserving Interactions

Such interactions involve non-strange hadrons hence, $\Delta S = 0$, between the decaying particle and the products. Examples of $\Delta S = 0$ weak processes are

1. Neutron decay, $n \rightarrow p + e^- \ \overline{\nu}_e$
2. Elastic neutrino reactions, $\nu_\mu + n \rightarrow p + \mu^-, \overline{\nu}_\mu + p \rightarrow n + \mu^+$
3. Inelastic neutrino reactions, $\pi \rightarrow e\nu$ or $\mu\nu$, $\Sigma^- \rightarrow \Lambda + e^- + \overline{\nu}_e$
4. Electron capture, $e^- + p \rightarrow n + \nu_e$
5. Muon capture, $\mu^- + p \rightarrow n + \nu_\mu$

4.1.2.2 Flavour Changing Interactions

(1) Strangeness changing $|\Delta S| = 1$

$$\begin{aligned} &\Lambda^0 \rightarrow p + e^- + \overline{\nu}_e \\ &K^+ \rightarrow \mu^+ + \nu_\mu \\ &K^+ \rightarrow \pi^0 + \mu^+ + \nu_\mu \end{aligned} \tag{4.2}$$

where only Λ^0 and K^+ are strange particles. $S(\Lambda^0) = -1$ and $S(K^+) = +1$.

(2) Charm changing $|c| = 1$, $\Delta S = 0$.

$\Lambda_c^+ \rightarrow \Lambda^0 + e^+ + \nu_e$, where $\Lambda_c^+ \Rightarrow$ charmed baryon

(3) Beauty changing $|\Delta b| = 1$, $\Delta S = 1$

$$\Lambda_b \rightarrow \Lambda_c^+ + e^- + \overline{\nu}_e \tag{4.3}$$

4.1.3. Non-Leptonic Weak Processes

These weak interactions involve only the mesons and baryons. Their subdivision is as

(1)
$$\begin{aligned} &|\Delta S| = 1 \\ &\Lambda^0 \rightarrow p + \pi^-, \ \ K^- \rightarrow \pi^- + \pi^0, K_S^0 \rightarrow \pi^+ + \pi^- \end{aligned} \tag{4.4}$$

(2)
$$\begin{aligned} &|\Delta C| = 1 \\ &D^0 \rightarrow K^- + \pi^+, \ \ \Lambda_c^+ \rightarrow \Lambda + \pi^+, \Lambda_c^+ \rightarrow \Lambda + K^+ + \pi^0 \end{aligned} \tag{4.5}$$

where D^0 is the charmed meson.

(3)
$$\begin{aligned} &|\Delta b| = 1 \\ &\Lambda_b \rightarrow \Lambda_c^+ + \pi^- \end{aligned} \tag{4.6}$$

Weak interactions violate the conservation laws of parity, charge conjugation and time reversal. Various flavours (*upness*(u), d, s, c, b) are approximately conserved by weak interactions. The color remains unaffected by weak interactions.

4.2 FERMI'S THEORY OF BETA DECAY (NON-RELATIVISTIC)

The first weak interaction that was observed is the nuclear beta decay. Following features are observed in the nuclear beta decay process.

1. Electron and positron are emitted by radioactive nuclei. The lifetimes are very long compared to nuclear time scale (10^{-23} sec).
2. The electron or positron spectrum is continuous. The total disintegration energy corresponds to the end point of the spectrum and is discrete.
3. Conservation of energy, momentum and angular momentum require the existence of neutrino and its antiparticle antineutrino. The mass of neutrino is nearly zero.
4. The β-interactions can be described by the following equations.

(a) Electron emission $n \rightarrow p + e^- + \overline{\nu}_e$

(b) Positron emission $p \rightarrow n + e^+ + \nu_e$

(c) K-capture $e^- + p \rightarrow n + \nu_e$ (4.7)

(d) Antineutrino absorption $\overline{\nu}_e + p \rightarrow n + e^+$

in which the reaction (a) occurs for free neutrons, reaction (b) is not energetically allowed for free protons but occurs only in nuclei, where the negative Q-value of the reaction is lower than the difference in the binding energy between parent and daughter nuclei. Reaction (c) occurs mainly in the heavy nuclei, where positron emission is forbidden energetically.

In all these beta interactions four fermions are involved. In Fermi's theory of beta decay, the weak decays are described by elementary interactions in which four fermions (neutron, proton, electron and antineutrino) interact at the same vertex (Fig. 4.1a). Today,

Fig. 4.1: Feynman diagrams for neutron decay. (a) Fermi's single vertex (1) interaction. (b) Two-vertex (1 and 2) interaction via W^-

weak interactions are considered to be taking place due to the mediation of W^+, W^-, or Z^0 bosons and the interaction takes place at two vertices, as shown in the Fig. (4.1b).

Thus, no intermediate vector boson (W) is involved in the original Fermi's theory of beta decay. In this theory, the decay constant λ is determined by using the Fermi's Golden Rule, i.e.,

$$\lambda = \frac{2\pi}{\hbar}\left|H_{fi}\right|^2 \rho(E_0) \tag{4.8}$$

where, $|H_{fi}|$ is the matrix element connecting the basic interactions of beta decay with wavefunctions describing the particles involved in the process. $\rho(E_0)$ is the density of states or the phase space factor. $\rho(E_0)$, is in fact the number of momentum states per unit energy at energy $E = E_0$ of the final state. The volume of final state is assumed as Ω. E is taken to be the total energy of the particles. The eqn. (4.8) also represents the rate of transition of the system from an initial state $|i\rangle$ to a final state $|f\rangle$.

Let us consider the β-decay (e^+ emission) in a pure Fermi transition

$${}^{10}\text{C} \longrightarrow {}^{10}\text{B}^* + e^+ + \nu_e \tag{4.9}$$

Since, the energy released in this reaction is ≈ 2 MeV, which is small compared to the rest energy of nuclear states, we may assume the leptons (e^+, ν_e) as *non-relativistic*. The initial and final states in eqn. (4.9) are

$$\begin{aligned} |i\rangle &= \left|{}^{10}C\right\rangle = u_i = \psi_i \\ |f\rangle &= \left|{}^{10}\text{B}^*\right\rangle\left|e^+\right\rangle\left|\nu_e\right\rangle = u_f\psi_e\psi_\nu = \psi_f \end{aligned} \tag{4.10}$$

where, for example, u_f is the final state wave function for ^{10}B*. The transition matrix element, when the system goes from $|i\rangle$ to $|f\rangle$, ^{10}B* assumed at rest, is

$$\text{H}_{fi} = \int \psi_f^* \,\text{H}(\boldsymbol{r},t)\psi_i d^3\boldsymbol{r} \tag{4.11}$$

where, H is the operator which represents the interaction and takes ψ_i to ψ_f. It destroys the initial particle and creates the final ones. Within the nucleus of ${}^{10}C$ [eqn. (4.9)], following reaction takes place

$$p \rightarrow n + e^+ + \nu_e \tag{4.12}$$

of which the LHS is destroyed by H to create its RHS.

Following Fermi, the interaction is assumed to be point like (single vertex, Fig. 4.1a) and also time independent. Hence, we may write

$$\text{H}(\boldsymbol{r}, t) = G\delta(\boldsymbol{r} - \boldsymbol{r}_0) \tag{4.13}$$

where G is the coupling constant of the Fermi interaction. The matrix element then becomes, note e represents positron [eqn. 4.9].

$$\begin{aligned} \text{H}_{fi} &= \iint \psi_f^*(\boldsymbol{r})G\delta(\boldsymbol{r}-\boldsymbol{r}_0)\psi_i(\boldsymbol{r}_0)d^3\boldsymbol{r}\, d^3\boldsymbol{r}_0 \\ &= G\int u_f^*\psi_e^*\psi_\nu^* u_i\, d^3\boldsymbol{r} \end{aligned} \tag{4.14}$$

where $u_f^* = u_f^*(\boldsymbol{r})$ and $u_i = u_i(r)$, etc. Now, since, the volume of the final state is Ω, we can assume for the purpose of normalization that it (Ω) is the volume of a large box in which the system is enclosed. Treating all particles *non-relativistically* and neglecting Coulomb interactions, we can represent the ψ_e and ψ_ν, by the plane wave forms

$$\psi_e(\boldsymbol{r}) = \frac{1}{\Omega^{\frac{1}{2}}} e^{i\boldsymbol{k}_e\boldsymbol{r}} \tag{4.15a}$$

$$\psi_\nu(\boldsymbol{r}) = \frac{1}{\Omega^{\frac{1}{2}}} e^{i\boldsymbol{k}_\nu\boldsymbol{r}} \tag{4.15b}$$

where $\boldsymbol{k}_e = \frac{\boldsymbol{p}_e}{\hbar}$ and $\boldsymbol{k}_\nu = \frac{\boldsymbol{p}_\nu}{\hbar}$. By assuming the plane wave forms, the Coulomb interaction between the nucleons and electrons has been neglected. Now,

$$\psi_e^*\psi_\nu^* = \frac{1}{\Omega} e^{-i\boldsymbol{kr}} \tag{4.16}$$

where

$$\boldsymbol{k} = \boldsymbol{k}_e + \boldsymbol{k}_\nu = \frac{\boldsymbol{p}}{\hbar} \tag{4.17}$$

We note that the wavelengths of e^+ and ν for $E = 2$ MeV is

$$\lambda\!\!\!^{-} = k^{-1} = \frac{\hbar}{p} \approx \frac{\hbar c}{E} \approx 10^{-13} \text{ m} \tag{4.18}$$

which is large compared to nuclear radius ($\approx 10^{-15}$ m). Thus, if the integration of $u_f^* u_i$ is over the nuclear volume, the expansion of

$$e^{-i\boldsymbol{kr}} = 1 - i(\boldsymbol{kr}) - \frac{(\boldsymbol{kr})^2}{2} + \ldots \tag{4.19}$$

may be approximated to unity, as the second term ($\boldsymbol{kr}$), in the nuclear region, has the value between 0.1 to 0.01 which is small compared to unity. Due to this approximation, the matrix element becomes

$$\mathrm{H}_{fi} = \frac{G}{\Omega} \int u_f^* u_i d^3\boldsymbol{r} \tag{4.20}$$

$$= G\frac{M_{fi}}{\Omega} \tag{4.21}$$

where

$$M_{fi} = \int u_f^* u_i \, d^3\boldsymbol{r} \tag{4.22}$$

is the dimensionless nuclear matrix element for single vertex interaction. Since, in the transition $^{10}C \rightarrow {}^{10}B^*$, a proton within the nucleus changes into a neutron, ^{10}C and $^{10}B^*$ become members of isotopic spin multiplet and hence, the wave-functions u_i and u_f may be taken identical, and for the transition $^{10}C \rightarrow {}^{10}B^*$, the overlap integral M_{fi} is expected to be unity.

Our next step in determining the decay constant λ (eqn. 4.8) is to calculate the density of states $\rho(E_0)$. We use an important result about the number of states that – *the number of allowed states of motion for a spinless particle in a phase space volume, which corresponds to volume Ω in ordinary space and a volume element $4\pi p^2 dp$ in momentum space, is*

$$dN = \frac{4\pi p^2 dp\Omega}{h^3} \tag{4.23}$$

The number of states per unit energy interval for a positron (e^+) and a neutrino (ν_e) in a volume Ω may be written as

$$\begin{aligned} dN_e \frac{dN_\nu}{dE_0} &= \left(\frac{4\pi p_e^2 dp_e \Omega}{h^3}\right) \frac{1}{dE_0} \left(\frac{4\pi p_\nu^2 dp_\nu \Omega}{h^3}\right) \\ &= \frac{16\pi^2\Omega^2}{h^6} p_e^2 dp_e \frac{p_\nu^2 dp_\nu}{dE_0} \end{aligned} \tag{4.24}$$

Normally in the β-decay, the neutrino (or antineutrino) is not detected. $e^+(e^-)$ is detected, therefore, we eliminate p_ν and dp_ν. E_0 is the energy difference between the initial and final states of the nucleus, so, energy conservation tells

$$E_0 = E_e + E_\nu \tag{4.25}$$

where

$$E_e^2 = p_e^2 c^2 + m_e^2 c^4 \tag{4.26}$$

and

$$E_\nu^2 = p_\nu^2 c^2 + m_\nu^2 c^4 = p_\nu^2 c^2 \tag{4.27}$$

in which m_ν, the rest mass of neutrino is taken zero. From eqs. (4.25) and (4.27), we may write

$$p_\nu^2 = \frac{(E_0 - E_e)^2}{c^2} \tag{4.28a}$$

from which

$$dp_\nu = \frac{dE_0}{c} \tag{4.28b}$$

Substituting equations (4.28) into eqn. (4.24), we get

$$\rho(E_0) = dN_e \frac{dN_\nu}{dE_0} = \frac{16\pi^2\Omega^2}{h^6 c^3} p_e^2 (E_0 - E_e)^2 dp_e \tag{4.29}$$

Now, we substitute eqs. (4.21) and (4.29) into eqn. (4.8) to get the probability of decaying the radioactive (β sensitive) nucleus to emit (e^+) of momentum between p_e and $p_e + dp_e$,

$$d\lambda(p_e) = \frac{2\pi}{\hbar c} G^2 |M_{fi}|^2 \frac{(4\pi)^2}{(2\pi\hbar)^6} \frac{(E_0 - E_e)^2}{c^2} p_e^2 dp_e \tag{4.30}$$

where we can observe that the volume Ω has disappeared. This result can be rearranged as

$$\left(\frac{d\lambda(p_e)}{p_e^2 dp_e}\right)^{\frac{1}{2}} = \frac{G|M_{fi}|}{\sqrt{2\pi^3 c^3 \hbar^7}}(E_0 - E_e) \tag{4.31}$$

A plot of the LHS versus $(E_0 - E_e)$ gives a straight line, which is called *Kurie plot*. Following graph (Fig. 4.2) shows the Kurie plot, when a neutron decays into an electron and antineutrino.

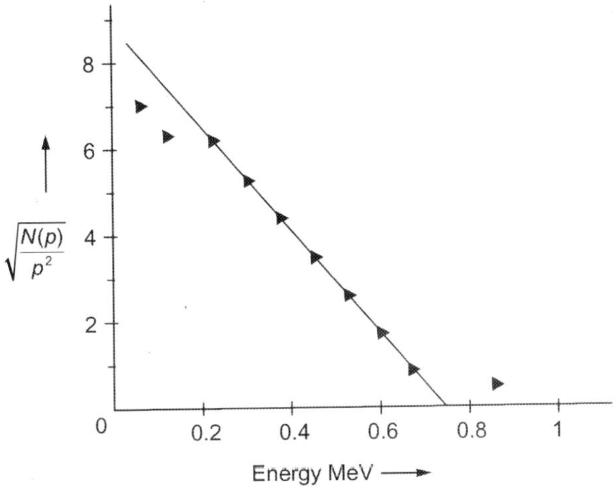

Fig. 4.2: The beta spectrum of neutron drawn as Kurie plot. *N* (*p*) denotes the number of coincidences per unit momentum interval between beta particles and protons produced in neutron decay. The plot is curved at low energies

The Kurie plot at lower energies is curved. This is due to the Coulomb interaction between nucleons and electrons, which we have neglected in the above discussion.

The total decay constant can be obtained by integrating eqn. (4.30). In the limit $E_e >> m_e$, from eqn. (4.26)

$$p_e^2 = \frac{E_e^2 - m_e^2 c^4}{c^2} \approx \frac{E_e^2}{c^2} \tag{4.32a}$$

$$dp_e = \frac{dE_e}{c} \tag{4.32b}$$

and hence from eqn. (4.30)

$$\lambda = \frac{G^2 |M_{fi}|^2}{2\pi^3 c^3 \hbar^7} \int p_e^2 (E_0 - E_e)^2 dp_e$$

$$\approx \frac{G^2 |M_{fi}|^2}{2\pi^3 c^6 \hbar^7} \int_0^{E_0} E_e^2 (E_0 - E_e)^2 dE_e$$

$$= \frac{G^2 |M_{fi}|^2}{2\pi^3 c^6 \hbar^7} \left(\frac{E_0^5}{30} \right) \tag{4.33}$$

As discussed below eqn. (4.22), taking $|M_{fi}| \approx 1$ and $\lambda = \frac{1}{\tau}$, where τ is the mean life of initial state, we may write for Fermi interaction coupling constant G

$$G = \left[\frac{\hbar^7 c^6}{\tau} \cdot \left(\frac{2\pi^3 \cdot 30}{E_0^5} \right) \right]^{\frac{1}{2}} \tag{4.34}$$

which, on using $E_0 = 2$ MeV and $\tau = 20$ sec for ${}^{10}C \rightarrow {}^{10}B^*$ transition, gives

$$G \approx 1 \times 10^{-49} \text{ erg cm}^3 \tag{4.35}$$

where, as already mentioned, the Coulomb interaction has not been included.

The Coulomb interaction between nucleons and electrons is incorporated by including a factor, which is a function of the product of ($\pm e$), ($\pm Ze$) and E, where ($\pm e$) is the charge of the emitted particle, $+Ze$ is the nuclear charge of the final nucleus when electron ($-e$) has been emitted and when the positron ($+e$) is emitted, Z is taken negative. E is the energy of the emitted particle. The factor incorporating the Coulomb interaction can then be written as $F(Z, \varepsilon)$, where $\varepsilon = \left(E_e / m_e c^2 \right)$. When this factor is included in the graph, the equation (4.31) agrees with the observed Kurie plot.

The incorporation of Coulomb interaction provides

$$\begin{aligned} G &\approx 1.39 \times 10^{-49} \text{ erg cm}^3 \\ &\approx 9 \times 10^{-50} \text{ MeV m}^3 \end{aligned} \tag{4.36}$$

In natural units $(c = \hbar = 1)$, the Fermi constant G is expressed as

$$G \approx \left(\frac{10^{-5}}{m_p^2} \right) \tag{4.37}$$

where m_p is the mass of proton.

4.2.1 The Allowed and Forbidden Transitions

In eqn. (4.11), we have seen that the matrix element H_{fi} is the overlap integral in which the operator H changes ψ_i into Hψ_i and the matrix element H_{fi} is a measure of how much Hψ_i is in ψ_f. Due to point interaction (eqn.4.13), the effect of operator H was to provide a multiplier G. Neglecting the $\boldsymbol{kr} \approx 10^{-2}$, in the expansion (4.19), only unity was retained. *If by using* $\exp(-ikr) = 1$, *the square of the matrix element* $|H_{fi}|^2$ (eqn. 4.8) *is of the order of unity, the transition* (from state $|i\rangle$ to state $|f\rangle$) *is said to be allowed.* If, by using, $\exp(-ikr) = 1$, H_{fi} becomes exactly zero, we have to consider higher terms in the expansion of $\exp(-ikr)$, viz., consider (kr), $(kr)^2$ etc. where, as seen above $kr \approx 10^{-2}$, $(kr)^2 \approx 10^{-4}$ etc. This means that, when higher order terms are required to be used, the matrix element H_{fi} would have smaller and smaller values. *Such transitions, in which* $|H_{fi}|^2$ *have small values, are called forbidden transitions.* Under such transitions of the nucleus, the probability of emission of $e^+ \nu_e$ or ($e^- \bar{\nu}_e$) becomes very small. If (kr) term has to be used to make $H_{fi} \neq 0$, the probability would decrease by a factor of 10^{-4} and so on.

Approximation $\exp(-ikr) \approx 1$, effectively means that we are taking $\psi_{e,\nu}(r)$ (eqn. 4.15) only at the centre of the nucleus, where $\boldsymbol{r} = 0$. However, the nucleus has a finite extent and ψ_e and ψ_ν have to be considered throughout the volume of the nucleus. This consideration requires higher terms in the expansion resulting in the occurrence of the forbidden transitions.

When the nucleus (state $|i\rangle$) emits the leptons (here e^+ and ν_e), its spin may or may not change. If the emitted lepton pair goes away in a singlet ($S = 0$) state that is the spins of e^+ and ν_e are antiparallel, then the spin of the nucleus ($|i\rangle$) will remain same as it was before emission. However, when the lepton pair is emitted in a triplet state, i.e., the spins of e^+ and ν_e are parallel, ($S = 1$), the spin of the nucleus ($|i\rangle$) is changed by one unit.

The transition of $|i\rangle$ to $|f\rangle$, when the emitted lepton pair is in the singlet ($S = 0$) state, is called the **Fermi (F) transition** and when the emitted lepton pair ($e^+\nu_e$) is in triplet ($S = 1$) state, the transition of the initial state $|i\rangle$ to the final state $|f\rangle$ is called the **Gamow—Teller (GT) transition.** Using the law of conservation of angular momentum, the total angular momentum of final state, in case of Fermi transition, is

$$J_f = J_i + L + 0 = J_i + L \tag{4.38}$$

and in case of GT transition, the final state has the total (orbital + spin) angular momentum

$$J_f = J_i + L + 1 \tag{4.39}$$

where J_i is the total angular momentum of initial state, L the orbital angular momentum of the lepton pair, 0 and 1 are their spin angular momenta in F and GT transitions respectively.

It may be noted that corresponding to $L = 0$, we have allowed transitions and for $L \neq 0$, the transitions are forbidden. Corresponding to $L = 1$, they are called the first forbidden transitions and to $L = 2$, the second forbidden etc.

In addition to angular momenta, the parity of the states too, has to be considered. Though, it is now known that weak interactions do not conserve parity, the conservation of parity may be considered when light particles ($e^+ \nu_e$) are emitted in beta decay. Then, *assuming* parity conservation law, the parities of the terms in the final wavefunction—the wavefunction of e^+, ν_e and final nucleus—must be the same as the parity of the initial wavefunction. Since the parity of ($e^+ \nu_e$) pair wavefunction is $(-1)^L$, the parity conservation law requires

$$\Pi_f (-1)^L = \Pi_i \tag{4.40}$$

where Π_i is the parity of the initial nuclear state and Π_f that of final nuclear state. From the parity relation (4.40), it becomes clear that the parity of the final nuclear state would change, i.e., it will be opposite, only when the ($e^+ \nu_e$) pair is emitted in $L = 1, 3\ldots$ states.

Thus taking into consideration the angular momentum and parity changes of the final state in the β-decay, we can have the following selection rules for various types of transitions.

4.2.1.1 *Fermi Transitions (FT):*

Following types of Fermi transitions would occur, when:

1. $L = 0$, $\Delta J = 0$, allowed FTs would occur without change of parity of initial and final nuclear states.
2. $L = 1$ $\Delta J = 1$, first forbidden FT is possible. The parity would change. However, for $L = 1$, $\Delta J = 0$, no first forbidden FT is possible.
3. $L = 2$, $\Delta J = 2$, second forbidden FT without change of parity. No second forbidden FT for $L = 2$, $\Delta J = 1$.
4. $L = 3$, $\Delta J = 3$, third forbidden FT with change of parity. No third forbidden FT for $L = 3$, $\Delta J = 2$.

4.2.1.2 *Gamow—Teller Transitions (GT)*

The different types (allowed, forbidden) of Gamow—Teller transitions would occur according to the following selection rules:

1. The allowed GT would occur for $L = 0$ and $\Delta J = 1$. No change of parity would take place. However, for $L = 0$, $\Delta J = 0$, there is no allowed GT.
2. The first forbidden GT is possible for $L = 1$ and $\Delta J = 0, 1, 2$ with change of parity.
3. The second forbidden GT is possible for $L = 2$ and $\Delta J = 2, 3$, without change of parity.
4. The third forbidden GT is possible for $L = 3$ and $\Delta J = 3, 4$ with change of parity.

Many beta decay processes are observed because they obey Fermi selection rules for allowed transitions. Such decays are not allowed by *GT* selection rules. Examples of such pure Fermi type transitions which are also called $0 \rightarrow 0$ transitions, implying that initial and final nucleus are spinless, are

$$\begin{aligned} {}^{10}\mathrm{C} &\longrightarrow {}^{10}\mathrm{B}^* + e^+ + \nu_e \\ {}^{14}\mathrm{O} &\longrightarrow {}^{14}\mathrm{N}^* + e^+ + \nu_e \\ {}^{26}\mathrm{Al} &\longrightarrow {}^{26}\mathrm{Mg}^* + e^+ + \nu_e \\ {}^{34}\mathrm{Cl} &\longrightarrow {}^{34}\mathrm{S}^* + e^+ + \nu_e \end{aligned} \tag{4.41}$$

There are also beta decay processes, which are observed because they obey Gamow—Teller selection rules for allowed transitions. Such decays are not allowed by Fermi selection rules. An example of such decay is

$$^{6}\mathrm{He} \longrightarrow {}^{6}\mathrm{Li} + e^- + \nu_e \; (J = 0 \rightarrow J = 1) \tag{4.42a}$$

The decay of a free neutron

$$n \longrightarrow p + e^- + \bar{\nu}_e \tag{4.42b}$$

is however, allowed by both *F* and *GT* selection rules.

Above examples suggest that both the types of interactions namely, the Fermi type and the Gamow—Teller type are in operation. In principle, these are two different types of weak interactions. For Fermi type the matrix element

$$M_{fi} = M_F = \int u_f^* u_i \, d^3\boldsymbol{r} \tag{4.43}$$

and for Gamow—Teller type of beta decay

$$M_{fi} = M_{GT} = \int u_f^* \boldsymbol{\sigma} u_i \, d^3\boldsymbol{r} \tag{4.44}$$

where the operator $\boldsymbol{\sigma}$ has Pauli-spin matrices as its components. The former is called the vector interaction and latter the axial vector interaction. We shall deal with them a little later.

4.3 HELICITY AND HANDEDNESS

The Helicity or the handedness of a particle needs to be understood, as the weak interactions are not the same for different handenesses of the particles.

Let us consider a spin $-\frac{1}{2}$ $(S = \frac{1}{2})$ particle moving with momentum $\boldsymbol{p}$ along the $+z$ axis. S_z, the component of spin S of the particle along z-axis can have $\pm\frac{1}{2}$ values, $+\frac{1}{2}$ along the $+z$ axis and $-\frac{1}{2}$ along $-z$ axis. Thus when $S_z = +\frac{1}{2}$, i.e., S_z and $\boldsymbol{p}$ are along the same direction, the particle is called *right-handed*. When $S_z = -\frac{1}{2}$, meaning that S_z and $\boldsymbol{p}$ are opposite to each other, the particle is left-handed.

The Dirac equation for a particle of mass m is

$$(-i\gamma^{\mu}\partial_{\mu} + m)\psi = 0 \tag{4.45}$$

which on using

$$i\partial_\mu = p_\mu \tag{4.46}$$

can be written as

$$(\gamma^\mu p_\mu - m)\psi = 0 \tag{4.47}$$

where

$$\gamma^0 = \begin{pmatrix} 1 & 0 \\ 0 & -1 \end{pmatrix} \text{ and } \gamma^i = \begin{pmatrix} 0 & \sigma_i \\ -\sigma_i & 0 \end{pmatrix} \tag{4.48}$$

in which 1, 0 are 2×2 unit and 2×2 null matrices respectively. σ_i $(i = 1, 2, 3)$ are three 2×2 Pauli matrices. Also introduced is the γ^5 matrix

$$\gamma^5 = i\gamma^0\gamma^1\gamma^2\gamma^3 = \begin{pmatrix} 0 & 1 \\ 1 & 0 \end{pmatrix} \tag{4.49}$$

The above representation of gamma matrices is suitable for positive and negative energy states in Dirac's theory. To explain handedness, however, following representation of gamma matrices, in which instead of γ^0, the γ^5 matrix is diagonal, is more convenient.

$$\gamma^0 = \begin{pmatrix} 0 & 1 \\ 1 & 0 \end{pmatrix}, \gamma^i = \begin{pmatrix} 0 & -\sigma_i \\ \sigma_i & 0 \end{pmatrix} \text{ and } \gamma^5 = \begin{pmatrix} 1 & 0 \\ 0 & -1 \end{pmatrix} \tag{4.50}$$

These gamma matrices can be obtained by the similarity transformation

$$U\gamma^\mu_{\text{old}}U^\dagger = \gamma^\mu_{\text{new}} \tag{4.51}$$

where the suffixes old and new refer to gamma matrices of equations (4.48) and (4.50) respectively. The unitary matrix U is

$$U = \frac{1}{\sqrt{2}}\begin{pmatrix} 1 & 1 \\ 1 & -1 \end{pmatrix} \tag{4.52}$$

Now, consider the Dirac equation for the massless particle, i.e., put $m = 0$, in eqn. (4.47) to get

$$\gamma^\mu p_\mu \psi = (\gamma^0 p_0 + \gamma^i p_i)\psi = 0 \tag{4.53}$$

Multiplying from the left by $-i\gamma^1\gamma^2,\gamma^3$, we get

$$(-i\gamma^1\gamma^2\gamma^3\gamma^0 p_0 - i\gamma^1\gamma^2\gamma^3\gamma^i p_i)\psi = 0 \tag{4.54}$$

The anticommutation relation for gamma matrices

$$\gamma^\mu\gamma^\nu + \gamma^\nu\gamma^\mu = 2g^{\mu\nu} \tag{4.55}$$

can be used to shift the position of gamma matrices in eqn. (4.54). In the first term, e.g,

$$\gamma^3\gamma^0 + \gamma^0\gamma^3 = 0 \tag{4.56a}$$

$$\Rightarrow \quad \gamma^3\gamma^0 = -\gamma^0\gamma^3 \text{ then } \gamma^2\gamma^0 = -\gamma^0\gamma^2 \text{ and } \gamma^1\gamma^0 = -\gamma^0\gamma^1 \tag{4.56b}$$

and hence

$$-i\gamma^1\gamma^2\gamma^3\gamma^0 p_0 = +i\gamma^0\gamma^1\gamma^2\gamma^3 p_0 = \gamma^5 p_0 \tag{4.57}$$

The second term of eqn. (4.54) is however,

$$\begin{aligned}-i\gamma^1\gamma^2\gamma^3(\gamma^1 p_1 + \gamma^2 p_2 + \gamma^3 p_3) &= i(\gamma^2\gamma^3 p_1 - \gamma^1\gamma^3 p_2 + \gamma^1\gamma^2 p_3)\\ &= (\sigma_1 p_1 + \sigma_2 p_2 + \sigma_3 p_3)\\ &= \sigma_i p_i = -\sigma_i p^i = -(\boldsymbol{\sigma}\cdot\boldsymbol{p})1\end{aligned} \tag{4.58}$$

where γ^1, γ^2, γ^3 matrices have been shifted and using γ^i from eqn. (4.50) we have calculated, e.g.,

$$\gamma^2\gamma^3 p_1 = -\begin{pmatrix}\sigma_2\sigma_3 & 0\\ 0 & \sigma_2\sigma_3\end{pmatrix} p_1 = -i\begin{pmatrix}\sigma_1 p_1 & 0\\ 0 & \sigma_1 p_1\end{pmatrix} \equiv -i\sigma_1 p_1.1 \tag{4.59}$$

Substitution of eqs. (4.58) and (4.59) in eqn. (4.54) yields

$$[\gamma^5 p_0 - (\boldsymbol{\sigma}\cdot\boldsymbol{p})1]\psi = 0 \tag{4.60}$$

Now, for a massless particle, from

$$E^2 = \boldsymbol{p}^2c^2 + m^2c^4 \text{ in } c = \hbar = 1 \text{ units} \tag{4.61}$$

$$E = p^0 = p_0 = |\boldsymbol{p}|\,,\ \boldsymbol{p} = p_0\hat{p} \tag{4.62}$$

Using equation (4.62) in eqn. (4.60) and canceling p_0, we get

$$(\gamma^5 - \boldsymbol{\sigma}\cdot\hat{p})\psi = 0 \tag{4.63}$$

Now, ψ is a four-component column matrix with positive energy components

$$\psi_1 = \begin{pmatrix}1\\0\\0\\0\end{pmatrix} \text{ and } \psi_2 = \begin{pmatrix}0\\1\\0\\0\end{pmatrix} \tag{4.64}$$

and negative energy components

$$\psi_3 = \begin{pmatrix}0\\0\\1\\0\end{pmatrix} \text{ and } \psi_4 = \begin{pmatrix}0\\0\\0\\1\end{pmatrix} \tag{4.65}$$

Using γ^5 from eqn. (4.50), it can be observed that

$$\begin{aligned}\gamma^5\psi_1 &= (+1)\psi_1,\ \gamma^5\psi_2 = (+1)\psi_2\\ \gamma^5\psi_3 &= (-1)\psi_3,\ \gamma^5\psi_4 = (-1)\psi_4\end{aligned} \tag{4.66}$$

which means that γ^5 has (+1) eigenvalue for positive energy states and (–1) eigenvalue for negative energy states.

Then using the observations (4.66) into eqn. (4.63), we may write,

$$\begin{aligned}\sigma\cdot\hat{p}\,\psi_{1,2} &= \gamma^5\psi_{1,2} = (+1)\psi_{1,2}\\ \sigma\cdot\hat{p}\,\psi_{3,4} &= \gamma^5\psi_{3,4} = (-1)\psi_{3,4}\end{aligned} \tag{4.67}$$

These are the eigenvalue equations for the operator $\sigma\cdot\hat{p}$. The operator $\sigma\cdot\hat{p}$ is called the helicity operator and its eigenvalue is called the *helicity of the corresponding state*.

The helicity (+1) corresponds to right-handed particles and left-handed particles have (–1) helicity. A positive helicity indicates that the spin of the particle is in the direction of its momentum $\boldsymbol{p}$ and negative helicity would mean its spin opposite to $\boldsymbol{p}$. It can also be seen from the above that the helicity has become equal to chirality (eigenvalue of γ^5).

A definite helicity can only be assigned to a massless particle. In case of massive particle, if the helicity is positive to one observer, it could be negative to another, while the helicity of massless particle would be same to all observers. This is because, the massive particles move with a speed less than c and there can be an observer who is moving with the speed greater than the speed of the particle. In the faster moving reference frame the particle's momentum would be $(-\boldsymbol{p})$ (i.e., it would appear moving backwards) and hence the helicity would change. In case of massless particle, which itself moves with the speed of light, there cannot be a frame moving faster than the particle. Therefore for massive particle, the helicity is not Lorentz invariant and while assigning a helicity to a massive particle, we will have to mention the frame of reference with respect to which the helicity has been assigned.

4.4 THE TWO-COMPONENT THEORY OF NEUTRINOS

It has been noticed from eqn. (4.67), that the positive energy components ψ_1, ψ_2 of four component

$$\psi = \begin{pmatrix}\psi_1\\ \psi_2\\ \psi_3\\ \psi_4\end{pmatrix} \tag{4.68}$$

are right-handed and the negative energy states ψ_3, ψ_4 are left-handed. ψ refers to a massless spin-$\frac{1}{2}$ particle. Denoting the right-handed part of ψ by ψ_R and the left-handed by ψ_L, we may write

$$\psi = \psi_R + \psi_L \tag{4.69}$$

where

$$\psi_R = A_R \begin{pmatrix} 1 \\ 0 \\ 0 \\ 0 \end{pmatrix} + B_R \begin{pmatrix} 0 \\ 1 \\ 0 \\ 0 \end{pmatrix} \tag{4.70}$$

and

$$\psi_L = A_L \begin{pmatrix} 0 \\ 0 \\ 1 \\ 0 \end{pmatrix} + B_L \begin{pmatrix} 0 \\ 0 \\ 0 \\ 1 \end{pmatrix} \tag{4.71}$$

where coefficients A_R, B_R and A_L, B_L are arbitrary numbers. The lower components of ψ_R and upper components of ψ_L are zero only for the particle at rest ($\boldsymbol{p} = 0$).

The general state ψ can be converted into purely right handed state or purely left handed state by the projection operators

$$P_R = \frac{1+\gamma^5}{2} \tag{4.72}$$

and

$$P_L = \frac{1-\gamma^5}{2} \tag{4.73}$$

respectively. The operator P_R projects out from $|\psi\rangle$ the eigenstate $|\psi_R\rangle$ that has the eigenvalue (+1) for γ^5. Similarly, the P_L operator projects out from $|\psi\rangle$ the eigenstate, $|\psi_L\rangle$ which has the eigenvalue (–1) for γ^5. This can be seen as below

$$\begin{aligned} P_R\psi &= P_R\psi_R + P_R\psi_L \\ &= \tfrac{1}{2}[(1+\gamma^5)\psi_R + (1-\gamma^5)\psi_L] \\ &\quad (\because \gamma^5\psi_R = (+1)\psi_R, \gamma^5\psi_L = (-1)\psi_L \\ &= \tfrac{1}{2}[(1+1)\psi_R + (1-1)\psi_L] \\ &= \psi_R \end{aligned} \tag{4.74a}$$

Similarly,

$$\begin{aligned} P_L\psi &= P_L\psi_R + P_L\psi_L \\ &= \tfrac{1}{2}[(1-\gamma^5)\psi_R + (1-\gamma^5)\psi_L] \\ &= \tfrac{1}{2}[(1-1)\psi_R + (1+1)\psi_L] \\ &= \psi_L \end{aligned} \tag{4.74b}$$

From the above relations, we can also see that

$$P_R \psi_L = 0$$

and

$$P_L \psi_R = 0 \tag{4.75}$$

whereas

$$\begin{aligned} P_R \psi_R &= \psi_R \\ P_L \psi_L &= \psi_L \end{aligned} \tag{4.76}$$

Adding eqs. (4.74a) and (4.74b), it can be seen that P_R and P_L obey the property $P^2 = P$ of a projection operator

$$P_R \psi + P_L \psi = \psi \tag{4.77}$$

from which and eqn. (4.69), we can easily see that

$$P_R^2 = P_R, P_L^2 = P_L \text{ and } P_R P_L = P_L P_R = 0 \tag{4.78}$$

Now consider that the massless particle to which ψ belongs is neutrino. Then

$$\psi_R = P_R \psi = \frac{1+\gamma^5}{2}\psi \tag{4.79}$$

would represent the right-handed neutrino whose helicity would be (+1) and

$$\psi_L = P_L \psi = \frac{1-\gamma^5}{2}\psi \tag{4.80}$$

would describe a left-handed neutrino, whose helicity would be (–1). Now, the question is, should we use both right-handed and left-handed neutrinos or use only one kind of neutrino.

Before the discovery of parity violation in β-decay (weak interactions) both kinds of neutrinos (ψ_R and ψ_L) could be used to study the beta decay. But, when it was found that weak interactions violate parity and violation of parity can be explained if only the left-handed (eqn. 4.80) neutrinos participate in weak interactions, only ψ_L was used. In fact, there exists no neutrino, which is right-handed. All neutrinos are left-handed. It may, however be noted here that the massive fermions unlike massless neutrino, have both right-handed and left handed types but only the left-handed would participate in weak interactions.

Therefore, due to the experimental fact that neutrinos are left-handed, we require only ψ_L [eqn. (4.71)] for them. This means that instead of the four components of ψ, we require only two components of ψ_L. Since in ψ_L upper two components are zero, only two, two-component states $\begin{pmatrix}1\\0\end{pmatrix}$ and $\begin{pmatrix}0\\1\end{pmatrix}$ are required to formulate the theory of left-handed particles. Here, the particles being neutrino and we require only two spinors $\begin{pmatrix}1\\0\end{pmatrix}$ and $\begin{pmatrix}0\\1\end{pmatrix}$ to formulate their theory; we call such a theory a two component theory of neutrinos.

Let us then write two component spinors as

$$\psi = \begin{pmatrix} \phi \\ \chi \end{pmatrix} \tag{4.81}$$

for which the Dirac equation (massless) [eqn. (4.53)] can be written as

$$(\gamma^i p_i)\begin{pmatrix} \phi \\ \chi \end{pmatrix} = -(\gamma^0 p_0)\begin{pmatrix} \phi \\ \chi \end{pmatrix} \tag{4.82}$$

where $i = 1, 2, 3$ and γ^0, γ^i have representation (4.50). Using eqn. (4.50), we may write eqn. (4.82) as

$$\begin{pmatrix} 0 & -\boldsymbol{\sigma}\cdot\boldsymbol{p} \\ \boldsymbol{\sigma}\cdot\boldsymbol{p} & 0 \end{pmatrix}\begin{pmatrix} \phi \\ \chi \end{pmatrix} = -\begin{pmatrix} 0 & p^0 \\ p^0 & 0 \end{pmatrix}\begin{pmatrix} \phi \\ \chi \end{pmatrix} \tag{4.83}$$

from which we obtain

$$\boldsymbol{\sigma}\cdot\boldsymbol{p}\chi = p^0\chi \tag{4.84}$$

and

$$\boldsymbol{\sigma}\cdot\boldsymbol{p}\phi = -p^0\phi \tag{4.85}$$

where $\pi^0 = E$, the energy of the massless particle (neutrino). Since, for the massless particle $p_0^2 = E^2 = |\boldsymbol{p}|^2$, there are two solutions for each of the above eqs., viz., (4.84) and (4.85), one is positive energy, $E > 0$ solution, other is $E < 0$, the negative energy solution. If both ϕ and χ are non-zero, then eqn. (4.84) tells that

$$\frac{\boldsymbol{\sigma}\cdot\boldsymbol{p}}{|\boldsymbol{p}|}\chi = (+1)\chi \text{ for } E > 0 \tag{4.86a}$$

$$\frac{\boldsymbol{\sigma}\cdot\boldsymbol{p}}{|\boldsymbol{p}|}\chi = (-1)\chi \text{ for } E < 0 \tag{4.86b}$$

whereas eqn. (4.85) would tell

$$\frac{\boldsymbol{\sigma}\cdot\boldsymbol{p}}{|\boldsymbol{p}|}\phi = (-1)\phi \text{ for } E > 0 \tag{4.87a}$$

$$\frac{\boldsymbol{\sigma}\cdot\boldsymbol{p}}{|\boldsymbol{p}|}\phi = (+1)\phi \text{ for } E < 0 \tag{4.87b}$$

Relations (4.86) mean that the helicity of state χ is (+1), that is, it is right-handed for $E > 0$ and left-handed (–1) for $E < 0$. Following hole theory of Dirac, in which the negative energy solutions represent antiparticles, the positive energy solution (4.86a) represents a right-handed (helicity +1) neutrino (particle) and the corresponding negative energy solution (4.86b) represents left-handed (helicity –1) antineutrino. Since, this observation contradicts the experimental fact that the neutrinos are left-handed and antineutrinos are right-handed, we would like χ to vanish. Conclusions derived from eqs. (4.87a, b) in which $E > 0$ solutions represent left-handed neutrinos and the negative energy solutions represent the right-handed antineutrinos agree with the experiments. *Therefore, the equation* (4.85) *is the two-component neutrino equation.*

The vanishing of χ and retaining of ϕ in $\psi = \begin{pmatrix} \chi \\ \phi \end{pmatrix}$ can be done by the operator $\left(\frac{1-\gamma^5}{2}\right)$, which we have already seen converts the general ψ into left-handed ψ. This can be checked, as done in the following. Using γ^5 of eqn. (4.50), we get

$$\frac{1-\gamma^5}{2} = \begin{pmatrix} 0 & 0 \\ 0 & 1 \end{pmatrix} \tag{4.87c}$$

so

$$\left(\frac{1-\gamma^5}{2}\right)\begin{pmatrix} \chi \\ \phi \end{pmatrix} = \begin{pmatrix} 0 & 0 \\ 0 & 1 \end{pmatrix}\begin{pmatrix} \chi \\ \phi \end{pmatrix} = \begin{pmatrix} 0 \\ \phi \end{pmatrix} \tag{4.87d}$$

which was desired. Therefore, the operator $\left(\frac{1-\gamma^5}{2}\right)$ is very important in weak interactions in which only left-handed leptons (fermions) participate. γ^5 is also called the Chirality operator and its eigenvalues (± 1), the Chirality. It may also be noted in passing that for a particle with non-zero mass, both spin states must be described by the same wavefunction, as the direction of spin is not Lorentz invariant. The same restriction, however, does not apply to neutrinos, which are massless. The spin of such a massless particle as told earlier is defined in a Lorentz covariant manner and its spin can be either parallel (+ve helicity) or antiparallel (–ve helicity) to the direction of momentum. The spin of neutrino is antiparallel and that of antineutrino parallel.

4.5 FERMI'S RELATIVISTIC THEORY OF BETA DECAY

Under section 4.2 also, the Fermi's theory of beta decay was developed. There, the matrix element was calculated by treating e^+ and ν_e non-relativistic. It may be possible to treat e^+, which has mass, non-relativistic, but it sounds unusual to treat a massless particle (neutrino) that travels with the speed of light non-relativistic. Keeping the above drawback in view, in the present section, we would get the relativistic expression for the matrix element. The decay constant would however, be obtained as in the non-relativistic case.

The matrix element H_{fi} (eqn.4.8) connects the basic interaction of beta decay with wavefunctions describing the particles participating in the process. We have already seen that in the beta decay process four fermions participate.

Fermi considered that the beta interaction was similar to electromagnetic interaction that was well known and well understood too. In an electromagnetic interaction, say, an electron and a proton, the interaction is via the photon and the Feynman diagram for this process can be drawn as (Fig.4.3). The diagram is

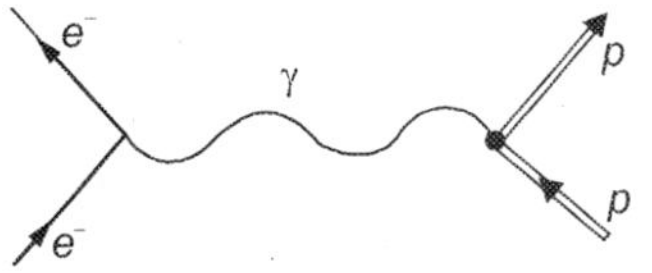

Fig. 4.3: Feynman diagram for elastic electron proton scattering

similar to that of Moller scattering (Chapter 2) except that protons replace the incoming and outgoing electrons at the vertex 2. Similar to eqn. (2.193) of Moller scattering, the matrix element for the process (Fig. 4.3) would be

$$M_{fi} = -e^2 \bar{u}_p \gamma^\mu u_p \left(\frac{g_{\mu\nu}}{q^2} \right) \bar{u}_e \gamma^\nu u_e$$
$$= -e^2 \bar{u}_p \gamma^\mu u_p \left(\frac{1}{q^2} \right) \bar{u}_e \gamma_\mu u_e \tag{4.88}$$

where q is the momentum of the interacting photon . The momenta of outgoing and incoming electron and proton have not been mentioned. Now,

$\bar{u}_p \gamma^\mu u_p$ = Electromagnetic current generated by moving charge of proton

$$= j^\mu(p) \tag{4.89a}$$

$\bar{u}_e \gamma_\mu u_e$ = Electromagnetic current generated by the moving charge of electron.

$$= j_\mu(e) \tag{4.89b}$$

in terms of which

$$M_{fi} = -\frac{e^2}{q^2} j_\mu(e) j^\mu(p) \approx j_\mu(e) A^\mu \tag{4.90}$$

Thus, M_{fi} of eqn. (4.90) denotes the familiar electromagnetic interaction in which the electromagnetic field A^μ is generated by the proton current $j^\mu(p)$. Since, $j_\mu A^\mu$ provides the interaction part of the Lagrangian density, whose volume integral gives the interaction Hamiltonian (notice, the interaction Lagrangian density does not depend on derivatives), the M_{fi} may also be equated with the interaction Hamiltonian. A similar mechanism was considered by Fermi to deal with weak interactions as well. We may consider here the beta decay $n \to p + e^- + \bar{\nu}_e$, the crossing symmetry equivalent of which is

$$n + \nu_e \to p + e^- \tag{4.91}$$

and we draw the Feynman diagram (Fig.4.4) for the reaction (4.91) without any intermediate boson, i.e., the Fig. (4.4) is the Feynman diagram for Fermi's point

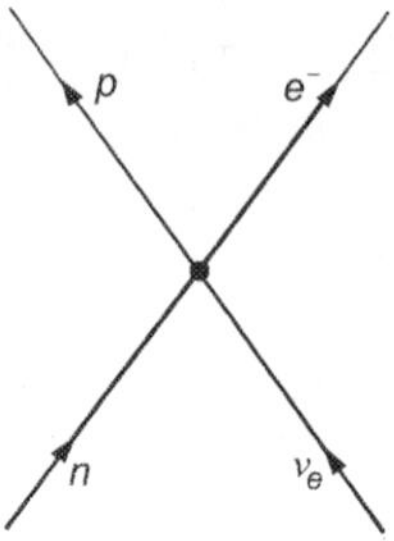

Fig. 4.4: Feynman diagram for Fermi's point interaction in the crossing symmetry equivalent process, $n + \nu_e \rightarrow p + e^-$ of beta decay

interaction. Notice that, though at the single vertex of Fig. (4.4) the weak interaction takes place, we write M_{fi} similar to one for electromagnetic interaction as in eqn. (4.88). So, for the Fig. (4.4), we have

$$M_{fi} \approx G(\bar{u}_p \gamma^\mu u_n)(\bar{u}_e \gamma_\mu u_\nu) \tag{4.92}$$

where G replaces e^2 and because of no intermediary [like the photon in Fig. (4.3)] there is no q^2 term. It may be mentioned here that, unlike eqs. (4.89), the currents in eqn. (4.92) are not proton and electron currents. We call currents like

$$\bar{u}_p \gamma^\mu u_n = \text{Weak hadron current} = j^\mu(h) \tag{4.93a}$$

and

$$\bar{u}_e \gamma_\mu u_\nu = \text{Lepton current} = j_\mu(l) \tag{4.93b}$$

Therefore, according to Fermi—the matrix element for the weak interaction can be expressed as the product of weak hadron current and lepton current. As seen in eqn. (1.108), the weak coupling constant G is related to the square of the weak charge. However, in the beta decay, two types of transitions namely, the Fermi and Gamow—Teller take place. There must be terms in $M_{fi} = H_I$ to produce both the F and GT selection rules. As proposed by Fermi, if

$$M_{fi} = G(\bar{u}_p \gamma^\mu u_n)(\bar{u}_e \gamma_\mu u_\nu) + h.c. \equiv GV \tag{4.94}$$

where $h.c.$ implies the hermitian conjugate of the first term and V stands for *vector* (current), then the interaction (4.94) would produce Fermi selection rules. The currents [eqs. (4.93)], which appear in eqn. (4.94) are four-component vectors whose $\mu = 0$ part is scalar and $\mu = i\ (1, 2, 3)$ part is a three vector. The pure scalar does not change sign under the parity (space inversion) operation and the polar vector (a three vector) changes sign. A pseudoscalar on the other hand changes sign, like $\boldsymbol{a}.(\boldsymbol{b} \times \boldsymbol{c})$ with $\boldsymbol{a}$, $\boldsymbol{b}$, $\boldsymbol{c}$ polar vectors. A pseudovector or an axial vector like $\boldsymbol{L} = \boldsymbol{r} \times \boldsymbol{p}$ does not change sign upon space inversion. Without proving, we mention here that the parity operator for spin-$\frac{1}{2}$ wavefunction is γ^0 matrix. Let us see the behavior of scalars, polar vectors, pseudoscalars and axial vectors under parity ($P = \gamma^0$) operation. $P(\bar{u}_p\gamma^\mu u_n) \Rightarrow P(\bar{u}_p\gamma^0 u_n)$ and $P(\bar{u}_p\gamma^i u_n)$, where $\bar{u}_p\gamma^0 u_n = u_p^\dagger\gamma^0\gamma^0 u_n = u_p^\dagger u_n$ is a scalar and $\bar{u}_p\gamma^i u_n$ is a three component vector.

$$P(\bar{u}_p\gamma^0 u_n) = (\bar{u}_p\gamma^0)\gamma^0(\gamma^0 u_n) = \bar{u}_p\gamma^0 u_n \ (\because \gamma^{0^2} = 1) \tag{4.95}$$
$$\Rightarrow \text{No change in sign} \Rightarrow \text{Scalar}$$

$$\begin{aligned} P(\bar{u}_p\gamma^i u_n) &= (\bar{u}_p\gamma^0)\gamma^i(\gamma^0 u_n) = \bar{u}_p\gamma^0\gamma^i\gamma^0 u_n \\ &= -\bar{u}_p\gamma^0\gamma^0\gamma^i u_n \quad (\because \gamma^i\gamma^0 = -\gamma^i\gamma^0) \\ &= -\bar{u}_p\gamma^i u_n \end{aligned} \tag{4.96}$$
$$\Rightarrow \text{Change in sign} \Rightarrow \text{Polar vector}$$

To construct four-component axial vector, γ^5 matrix is used. $i\bar{u}_p\gamma^5\gamma^\mu u_n$ is the axial four-vector whose behaviour under parity ($P = \gamma^0$) operation can be checked.

$$P(i\bar{u}_p\gamma^5\gamma^\mu u_n) \Rightarrow P(i\bar{u}_p\gamma^5\gamma^0 u_n) \text{ and } P(i\bar{u}_p\gamma^5\gamma^i u_n)$$

where

$$\begin{aligned} P(i\bar{u}_p\gamma^5\gamma^0 u_n) &= i\bar{u}_p\gamma^0 \cdot \gamma^5\gamma^0 \cdot \gamma^0 u_n \\ &= i\bar{u}_p\gamma^0\gamma^5 u_n \quad (\because \gamma^{0^2} = 1) \\ &= -i\bar{u}_p\gamma^5\gamma^0 u_n \ (\because \gamma^0\gamma^5 = -\gamma^5\gamma^0) \end{aligned} \tag{4.97}$$

Then because of change in sign on P operation $i\bar{u}_p\gamma^5\gamma^0 u_n$ is a pseudoscalar. Likewise,

$$\begin{aligned} P(i\bar{u}_p\gamma^5\gamma^i u_n) &= i\bar{u}_p\gamma^0 \cdot \gamma^5\gamma^i \cdot \gamma^0 u_n \\ &= i\bar{u}_p\gamma^5\gamma^0\gamma^0\gamma^i u_n \quad (\because \gamma^i\gamma^0 = -\gamma^0\gamma^i) \\ &= +i\bar{u}_p\gamma^5\gamma^i u_n \end{aligned} \tag{4.98}$$

which implies that because of no change in sign, $i\bar{u}_p\gamma^5\gamma^i u_n$ is a three component axial (pseudo) vector. A polar vector changes sign under parity operation. Thus a pseudoscalar ($i\bar{u}_p\gamma^5\gamma^0 u_n$) and an axial three vector ($i\bar{u}_p\gamma^5\gamma^i u_n$) form a four-component axial vector ($i\bar{u}_p\gamma^5\gamma^\mu u_n$).

Though, Fermi assumed only vector—vector interaction, as in eqs. (4.93) and (4.94), the Gamow—Teller transition would require an axial vector interaction to have a term in M_{fi} like

$$M_{fi} = G(i\bar{u}_p\gamma^5\gamma^\mu u_n)(i\bar{u}_e\gamma_5\gamma_\mu u_\nu) + h.c. \equiv GA \tag{4.99}$$

where A stands for axial four-vector.

4.5.1 The V–A Interaction

The vector current employed by Fermi in his vector—vector interaction to explain the Fermi transitions in beta decay and the axial vector current required to explain Gamow—Teller transitions could be combined to obtain the desired matrix element for the beta interaction processes. It may however, be noted that, it is not only the vector current that would give Fermi selection rules and axial, the Gamow—Teller selection rules, but a scalar can also give FTs and a tensor the GTs. These scalars and tensors are however, not used in M_{fi} because they predict a wrong correlation between the direction of neutrino and the direction of electron. In Fermi transitions the momenta of electron and anti neutrino should tend to be parallel whereas in Gamow—Teller transitions they should be opposite. The correct correlations are predicted when we use only the vector current in M_{fi} for Fermi transitions and axial vector current for the Gamow—Teller transitions. However, it is important to note that a linear combination of vector and axial vector current can always be used in the interaction Hamiltonian (M_{fi}) for beta decay. It is in this combination that vector minus axial vector current interaction emerges.

A linear combination that can be proposed is

$$M_{fi} = G\,(C_V V + C_A A) \tag{4.100}$$

where GV and GA have been defined in equations (4.94) and (4.99) respectively. C_V and C_A are coefficients. Explicitly,

$$\begin{aligned} M_{fi} &= G[C_V(\bar{u}_p\gamma^{\mu}u_n)(\bar{u}_e\gamma_{\mu}u_{\nu}) \\ &\quad + C_A(i\bar{u}_p\gamma^5\gamma^{\mu}u_n)(i\bar{u}_e\gamma_5\gamma_{\mu}u_{\nu})] + h.c. \\ &= G\sum_{i=V,A} C_i(\bar{u}_p O^i u_n)(\bar{u}_e O_i u_{\nu}) + h.c. \end{aligned} \tag{4.101}$$

where $i = V$ and $i = A$; for $i = V$, $O^i = \gamma^{\mu}$ and for $i = A$, $O_i = i\gamma_5\gamma_{\mu}$. In the combination (4.100) or (4.101) both the terms are scalars and hence do not change sign under space inversion. Consequently, the above M_{fi} would preserve the parity conservation law. However, the parity is violated in weak interactions. Thus a pseudoscalar must be added to M_{fi} of eqn. (4.101), so that the parity violation is observed. A pseudoscalar changes sign on space inversion. The product of a vector and an axial vector would construct a pseudoscalar and since, a pseudovector (axial vector) essentially contains a γ_5 matrix, we may write on the lines of eqn. (4.101) a general pseudoscalar as

$$\text{Pseudoscalar} = G\sum_{i=V,A} C_i'(\bar{u}_p O^i u_n)(\bar{u}_e O_i\gamma_5 u_{\nu}) + h.c. \tag{4.102}$$

where C_V' and C_A' are new coefficients, for $i = V$, $O^i = \gamma^{\mu}$ and for $i = A$, $O_i = i\gamma_5\gamma\mu$.

Thus the parity violating interaction Hamiltonian for the beta decay process is obtained by combining eqns. (4.101) and (4.102) to get

$$M_{fi} = G \sum_{i=V,A} [C_i(\overline{u}_p O^i u_n)(\overline{u}_e O_i u_\nu)$$
$$+C_i'(\overline{u}_p O^i u_n)(\overline{u}_e O_i \gamma_5 u_\nu)] + h.c. \tag{4.103}$$

The coefficient C_i is a measure of pure scalar and C_i' that of pseudoscalar. The time reversal invariance demands that the coefficients C_i and C_i' should be real. Now, the helicities of ν and $\overline{\nu}$ are (–1) and (+1) respectively which are definite and there is maximum parity violation. Hence, the contribution of scalar (parity preserving) and pseudoscalar (parity violating) terms in M_{fi} will be in equal measures. This would set

$$C_i' = \pm C_i \tag{4.104}$$

where $i = V$ and A. The helicity considerations of neutrino and antineutrino further require

$$C_i' = -C_i \tag{4.105}$$

which implies $C_V' = -C_V$ and $C_A' = -C_A$. Hence the eqn. (4.103) becomes

$$M_{fi} = G \sum_{i=V,A} C_i(\overline{u}_p O^i u_n)(\overline{u}_e O_i u_\nu - \overline{u}_e O_i \gamma_5 u_\nu)$$
$$= GC_V(\overline{u}_p \gamma^\mu u_n)(\overline{u}_e \gamma_\mu u_\nu - \overline{u}_e \gamma_\mu \gamma_5 u_\nu) \tag{4.106}$$
$$+ GC_A(\overline{u}_p i\gamma^5 \gamma^\mu u_n)(\overline{u}_e i\gamma_5 \gamma_\mu u_\nu + \overline{u}_e i\gamma_\mu u_\nu) + h.c.$$

where in the last term $\gamma_5\gamma_\mu = -\gamma_\mu\gamma_5$ and $\gamma_5^2 = 1$ have been used. Since, the coefficient C_V appearing in vector interaction is associated with Fermi transition and the coefficient C_A related to axial vector interaction accounts for Gamow-Teller transitions, we may consider the pure Fermi transition in e.g., the nuclear beta decay

$${}^{14}\text{O} \longrightarrow {}^{14}\text{N}^* + e^+ + \nu_e \tag{4.107}$$

and both Fermi and Gamow –Teller transitions , e.g., the neutron beta decay

$$n \rightarrow p + e^- + \overline{\nu}_e \tag{4.108}$$

to calculate the respective transition matrix elements. The calculations show that

$$\frac{C_A^2}{C_V^2} \approx 1.58 \tag{4.109}$$

which imply

$$\frac{|C_A|}{|C_V|} \approx 1.25 \approx 1 \tag{4.110}$$

From the measurements of the angular correlation of the electrons emitted in the decay of the polarized neutrons, it is concluded that the signs of C_A and C_V are opposite, hence, we use

$$C_V = -C_A \tag{4.111a}$$

and from eqn. (4.110), we may write

$$C_V \approx +1$$
$$C_A \approx -1 \quad (4.102) \tag{4.111b}$$

Using the values (4.111), eqn. (4.106) becomes

$$M_{fi} = \frac{G}{\sqrt{2}}[\bar{u}_p\gamma^\mu u_n + \bar{u}_p\gamma^5\gamma^\mu u_\nu][\bar{u}_e\gamma_\mu u_\nu + \bar{u}_e\gamma_5\gamma_\mu u_\nu]$$

$$= \frac{G}{\sqrt{2}}[\bar{u}_p\gamma^\mu(1-\gamma^5)u_n][\bar{u}_e\gamma_\mu(1-\gamma_5)u_\nu] \tag{4.112}$$

where $\gamma^5\gamma^\mu = -\gamma^\mu\gamma^5$ has been used. Equation (4.112) gives the interaction Hamiltonian for the beta decay process, which has been derived by using $C_V = -C_A$ with $C_V = 1$, and because of these opposite values of C_V and C_A, the beta decay theory described above to obtain M_{fi} is called the $V-A$ theory. Also $\bar{u}\gamma^\mu u$ is a vector current and in view of the definition $\bar{u}i\gamma^5\gamma^\mu u$ for axial vector current, (–) $\bar{u}\gamma^5\gamma^\mu u$ can be seen as minus axial vector current leading to the vector minus axial vector current concept for weak interactions.

The relation (4.112) is strictly true when strong interactions between neutron and proton are assumed absent. However, the strong interactions slightly modify the weak hadronic current as

$$\bar{u}_p\gamma^\mu(1-\gamma^5)u_n \rightarrow \bar{u}_p\gamma^\mu(1-r\gamma^5)u_n \tag{4.113}$$

where r from eqn. (4.110) is ≈ 1.25, implying that the strong interactions do not make a major change in the weak interaction Hamiltonian (4.112). When the strong interactions are absent, $C_V = -C_A$. The factor $\frac{1}{\sqrt{2}}$ is introduced to define G as in Fermi's original theory.

The equation (4.112) makes out an important point. Fermi in his theory used only *vector current* × *vector current*, where he did not consider parity violation. After many years of studies (about 25 years) on weak interactions since Fermi, the interaction Hamiltonian of Fermi (eqn. 4.92) was modified to eqn. (4.112) in which γ^μ was merely multiplied by $(1-\gamma^5)$. So, the V—A matrix element and the Fermi's matrix element differ only by a factor of $(1-\gamma^5)$, despite taking into account in the V—A theory the parity violation in weak interaction.

4.5.2 Universality of Weak Interactions

We discussed the Fermi's theory of beta decay, in which the decay

$$p \rightarrow n + e^+ + \nu_e \tag{4.114}$$

was considered and the weak coupling constant (G) was calculated as

$$G \approx 1.39 \times 10^{-62} \text{ Jm}^3 \approx 9 \times 10^{-50} \text{ MeV m}^3 \tag{4.115}$$

the neutron decay

$$n \to p + e^- + \bar{\nu}_e \tag{4.108}$$

is similar to eqn. (4.114) and hence would give the same value of G. The Fermi constant G, which is related to the square of the *weak charge,* provides the measure of the strength of the weak interactions. Notice that, above beta decays are semileptonic in which nucleons participate.

Let us consider a purely leptonic decay

$$\mu^- \to e^- + \bar{\nu}_e + \nu_\mu \tag{4.116}$$

which too is a four fermion interaction and Fermi's Golden Rule (4.8) can be applied to calculate the muon decay constant. The matrix element for the process can be calculated from the following M_{fi}, which is written similar to eqn. (4.112), from the V–A theory.

$$M_{fi} = \frac{G}{\sqrt{2}}[\bar{u}_{\nu_\mu}\gamma^\alpha(1-\gamma^5)u_\mu][\bar{u}_e\gamma_\alpha(1-\gamma_5)u_{\nu_e}] \tag{4.117}$$

where $\alpha = 0, 1, 2, 3$. Using this M_{fi}, the lifetime of muon is found to be

$$\tau_\mu = \frac{192\pi^3\hbar^7}{G^2 m_\mu^5 c^4} \tag{4.118}$$

in which substituting the experimental values of muon lifetime and mass, we get the weak coupling constant as

$$G = 1.44 \times 10^{-62} \text{ Jm}^3 \tag{4.119}$$

which is almost the same as the value of weak coupling constant [eqn. (4.115)] obtained from the semileptonic beta decay. Though, the nucleons participate in beta decay process and leptons in the muon decay, the weak coupling constant comes out to be almost same. Thus, ignoring at present the small difference between G_β (eqn. 4.115) and G_μ (eqn. 4.119), we find that the weak coupling constant is same for nucleons and leptons, indicating that both the beta and muon decay have the same physical origin. The near identical value of G for both leptons and nucleons is described as the ***universality of weak coupling constant***.

Above observations also allow us to expect that all weak interaction phenomena are described by a V—A current-current interaction with a universal coupling G. In beta decay current-current interaction was the interaction between weak hadron current $[\bar{u}_n\gamma^\mu(1-\gamma^5)u_p]$ and the lepton current $[\bar{u}_e\gamma^\mu(1-\gamma^5)u_\nu]$ and in case of muon decay between two lepton currents. For any general weak interaction, we consider a universal weak current, which interacts with itself to produce a current-current interaction. This universal weak current J^μ would also consist of a leptonic part J_e^μ and a hadronic part J_h^μ, i.e,

$$J^\mu = J_e^\mu + J_h^\mu \tag{4.120}$$

where

$$J_e^\mu = \bar{u}_{\nu_e}\gamma^\mu(1-\gamma^5)u_e + \bar{u}_{\nu_\mu}\gamma^\mu(1-\gamma^5)u_\mu \qquad (4.121)$$

and

$$J_h^\mu = \bar{u}_p\gamma^\mu(1-\gamma^5)u_n + \text{similar terms for strange particles.} \qquad (4.122)$$

Ignoring for the moment the strange hadrons, we shall have

$$J^\mu = \bar{u}_{\nu_e}\gamma^\mu(1-\gamma^5)u_e + \bar{u}_{\nu_\mu}\gamma^\alpha(1-\gamma^5)u_\mu + \bar{u}_p\gamma^\mu(1-\gamma^5)u_n \qquad (4.123)$$

The weak interaction Hamiltonian $H_I = M_{fi}$ in terms of universal weak current is then

$$M_{fi} \approx \frac{G}{\sqrt{2}} J^\mu J_\mu^\dagger \qquad (4.124)$$

where

$$J_\mu^\dagger = \bar{u}_e\gamma_\mu(1-\gamma_5)u_{\nu_e} + \bar{u}_\mu\gamma_\alpha(1-\gamma_5)u_{\nu_\mu} + \bar{u}_n\gamma_\mu(1-\gamma_5)u_p \qquad (4.125)$$

In deriving eqn. (4.125), we have used

$$\begin{aligned}[\bar{a}\gamma_\mu(1-\gamma_5)b]^\dagger &= b^\dagger(1-\gamma_5)^\dagger(\gamma_\mu)^\dagger(a^\dagger\gamma_0)^\dagger \\ &= b^\dagger(1-\gamma_5)(\gamma_0\gamma_\mu\gamma_0)(\gamma_0 a) \\ &= b^\dagger\gamma_0(1+\gamma_5)\gamma_\mu a \\ &= \bar{b}\gamma_\mu(1-\gamma_5)a\end{aligned} \qquad (4.126)$$

Then substituting equations (4.123) and (4.125) into eqn. (4.124), we would get nine terms, each of which would represent the interaction Hamiltonian for a particular weak decay. The terms will be corresponding to μ—decay, nuclear beta decay; electron and muon capture and some terms would contribute to scattering only. For example, the following terms in the product (4.124)

$$\frac{G}{\sqrt{2}}[\bar{u}_{\nu_e}\gamma^\mu(1-\gamma^5)u_e][\bar{u}_n\,\gamma_\mu(1-\gamma_5)u_p] \qquad (4.127a)$$

$$\frac{G}{\sqrt{2}}[\bar{u}_{\nu_\mu}\gamma^\alpha(1-\gamma^5)u_\mu][\bar{u}_e\,\gamma_\alpha(1-\gamma_5)u_{\nu_e}] \qquad (4.127b)$$

$$\frac{G}{\sqrt{2}}[\bar{u}_{\nu_e}\gamma^\mu(1-\gamma^5)u_e][\bar{u}_e\,\gamma_\mu(1-\gamma_5)u_{\nu_e}] \qquad (4.127c)$$

respectively represent

$$p \to n + e^+ + \nu_e \; (\textit{positron emission})$$

$$\mu^- \to e^- + \bar{\nu}_e + \nu_\mu \; (\textit{muon decay})$$

and

$$\nu_e + e \to e + \nu_e \; (\textit{elastic electron neutrino scattering})$$

Similarly, inferences from other terms can also be drawn.

In general we may write for eqn. (4.124)

$$M_{fi} \approx \frac{G}{\sqrt{2}}[J_e^{\mu} J_{\mu}^{e\dagger} + J_e^{\mu} J_{\mu}^{h\dagger} + J_e^{\mu\dagger} J_{\mu}^{h} + J_h^{\mu} J_{\mu}^{h\dagger}] \tag{4.128}$$

in which the first term represents the pure leptonic weak interactions, the second and third terms describe the semileptonic weak decay of hadrons and the last term describes the non-leptonic weak decay of hadrons. Each of the above is a current-current interaction and the weak coupling constant G is same for all. It can therefore be said that universality is a consequence of the self-interaction of the weak current.

It has however, to be noted that we have kept weak hadronic current corresponding to strange particles out of the above considerations. In fact, when strange particles are also considered, the weak hadronic current J_h^{μ} becomes a linear combination of weak hadronic currents corresponding to strangeness conserving ($\Delta S = 0$) and the strangeness changing ($\Delta S = 1$) weak interactions [eqs. (4.2)] as

$$J_{\mu}^{h} = \cos\theta_c \; J_{\mu}^{h}(\Delta S = 0) + \sin\theta_c \; J_{\mu}^{h}(\Delta S = 1) \tag{4.129}$$

where θ_c is called the ***Cabibbo angle*** which is approximately equal to

$$\theta_c \approx 0.21 \text{ radian} \approx 13° \tag{4.130}$$

We shall take up the relation (4.129) a little later.

We may thus say that the original Fermi's theory of the point interaction of four fermions has been modified by including vector and axial vector currents to account for parity violation and by Cabbibo prescription [eqn. (4.129)] to accommodate both strange and non-strange particles. In such modified form the Fermi's theory accounted for a wide range of weak interaction processes. However, as we shall see in the next section, when we go to higher energy domain, the point interaction of Fermi's theory itself requires modification.

4.5.3 The Intermediate Vector Boson

The Fermi interaction, as told in the previous section was a point interaction in which no intermediate vector boson like the photon in the electromagnetic interaction was involved. In other words, Fermi interaction was a single vertex interaction. However, problems arise when we consider the weak interaction as a single vertex interaction. An example is the elastic neutrino—electron scattering.

$$\nu_e + e^- \rightarrow \nu_e + e^- \tag{4.131}$$

which is a pure leptonic weak interaction. Assuming a point interaction for the above process, as shown in the following figure (Fig. 4.5), if the differential scattering cross section is calculated, it comes out to be

$$\left(\frac{d\sigma}{d\Omega}\right)_{\text{Fermi}}^{\nu_e e \rightarrow \nu_e e} = \frac{1}{64\pi^2 s}\overline{|M_{fi}|^2} = \frac{G^2 s}{4\pi^2} \tag{4.132}$$

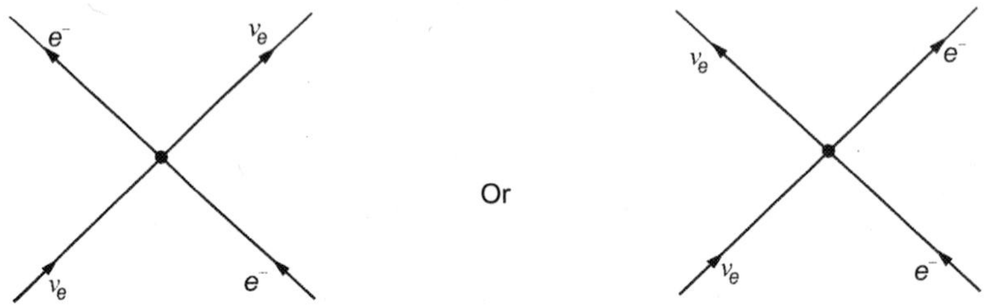

Fig. 4.5: Fermi's point interaction diagrams for elastic neutrino—electron scattering

where s is the square of the centre of mass energy of ν_e and e, i.e.,

$$s = (CMS \text{ energy})^2 = E^2 \text{ (say)} \tag{4.133}$$

Integrating eqn. (4.132), $\int d\Omega = 4\pi$, the total elastic neutrino—electron scattering cross section comes out

$$\sigma(\nu_e e \to \nu_e e) = \frac{G^2 s}{\pi} \tag{4.134}$$

from which we observe that the σ varies linearly with s, the square of the centre of mass energy. This implies that the total scattering cross section rises linearly with s without limit. Larger the square of centre of mass energy, greater should be the total scattering cross section. This observation clearly contradicts the behaviour of σ calculated on the basis of partial wave analysis, in which a de Broglie wavelength $\lambda\!\!\!^- = 1/k$ is associated with the particle in CM frame and the incident wave $\psi_i = e^{ikz}$ is expressed as the sum of partial waves of angular momentum l. From the partial wave analysis, the maximum value of σ comes out

$$\sigma_{\max} = \frac{4\pi}{k^2} \sum_l (2l+1) \tag{4.135}$$

For point like scattering (S wave), $l = 0$, i.e., in the centre of mass frame

$$\sigma_{\max} = \frac{4\pi}{k^2} = \frac{4\pi}{s} \tag{4.136}$$

which clearly decreases as the square of centre of mass energy increases. The conclusions for σ from eqn. (4.134) and eqn. (4.136) are thus opposite. Ignoring the constants in the above equations, it can be checked that for

$$\sqrt{s} = G^{-\frac{1}{2}} \approx 10^{\frac{5}{2}}\, m_p \approx 300 \text{ GeV} \tag{4.137}$$

$$\sigma_{\text{Fermi}} = \sigma_{\max} = G \tag{4.138}$$

and for

$$\sqrt{s} > G^{-\frac{1}{2}} \tag{4.139}$$

$$\sigma_{\text{Fermi}} > \sigma_{\max} \tag{4.140}$$

Thus at the centre of mass energies greater than 300 GeV σ_{Fermi} becomes unacceptable, as it will then predict that more particles would be scattered into the final state than would enter into the initial state. This is in clear violation of conservation of probability, which also means that if eqn. (4.140) becomes acceptable, the violation of unitarity takes place. The overlap *S*-matrix elements S_{fi} are the transition probability amplitudes from the initial state *i* to a particular final state *f*. The total transition probability from the state *i* to all final states is $\sum_f |S_{fi}|^2$. Probability conservation requires

$$\sum_f |S_{fi}|^2 = 1 \tag{4.141a}$$

or

$$S^*_{fi}\, S_{fi} = S^\dagger_{fi}\, S_{fi} = (S^\dagger S)_{ii} = (1)_{ii} \tag{4.141b}$$

the generalization of the above relation to all matrix elements of $S^\dagger S$ is called the ***unitarity of S***, i.e.,

$$S^\dagger S = SS^\dagger = 1 \tag{4.141c}$$

Above 300 GeV, σ_{Fermi} violates unitarity and this energy limit is called the *unitarity limit*, above which the conservation of probability would be violated.

In order to preserve unitarity at higher energies or to check the divergent behaviour of pointlike interaction cross section, the weak interaction is spread from pointlike structure by postulating an intermediate spin-1 (vector) particle, the *W*-boson, which would couple the (ν_e, *e*) and (*e*, ν_e) currents [see eqn. (4.127c)]. The diagram of Fig. (4.5) would be modified like the one drawn in

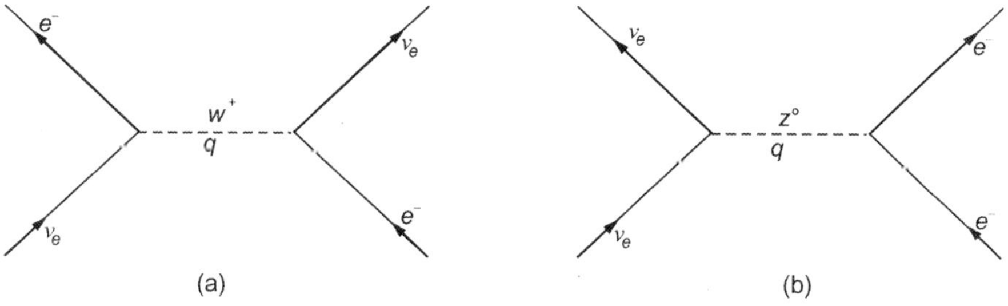

Figs. 4.6: Diagrams of Fig. 4.5 with intermediate vector bosons. (a) IVB is charged boson (W^+). (b) IVB is neutral boson (z^0)

Fig. (4.6). A propagator term like the $\left(\dfrac{1}{q^2}\right)$ of eqn. (4.88) would appear in matrix element. We shall consider at the moment only the Fig. (4.6a), where the charged vector boson W^+ is the IVB. The matrix element for the Fig. (4.6a) would be

$$M_{fi} = \left[g_W \bar{u}_e \gamma^\alpha (1-\gamma^5) u_{\nu_e}\right] \left[\frac{g_{\alpha\beta} - \dfrac{q_\alpha q_\beta}{M_W^2}}{M_W^2 - q^2}\right] \left[g_W \bar{u}_{\nu_e} \gamma^\beta (1-\gamma^5) u_e\right] \quad (4.142)$$

where the square bracket in the middle is the boson propagator. M_W is the mass of the intermediate boson and q^2 is the square of the four-momentum carried by IVB. For lower energies $q^2 << M_W$, so eqn. (4.142) becomes

$$M_{fi} = \frac{g_W^2}{M_W^2}\left(\bar{u}_e \gamma^\alpha (1-\gamma^5) u_{\nu_e}\right)\left(\bar{u}_{\nu_e} \gamma_\alpha (1-\gamma_5) u_e\right) \quad (4.143)$$

Universality demands that

$$\frac{g_W^2}{M_W^2} = \frac{G}{\sqrt{2}} \quad (4.144a)$$

where g_W is the dimensionless weak coupling constant. Notice that G too, is weak coupling constant, but from eqn. (4.37), it has the dimensions $[M]^{-2}$. Eqn. (4.144a) relates g_W and G via M_W. We may notice here, however, that conventionally in eqn. (4.142), $g_W \rightarrow \frac{g_W}{\sqrt{2}}$ and $(1-\gamma^5) \rightarrow \frac{1}{2}(1-\gamma^5)$. These replacements change eqn. (4.144a) to

$$\frac{g_W^2}{8M_W^2} = \frac{G}{\sqrt{2}} \quad (4.144b)$$

The weak interactions are weak because G is small. Now, G is small primarily because of the large mass of M_W, than due to the smallness of g_W (i.e., $g_W << e$). The electromagnetic coupling constant e is dimensionless; G has the dimensions of $(\text{MeV})^{-2}$ or $(\text{GeV})^{-2}$, so the dimensionless g_W has to be compared with e. At higher energies of the order of M_W (80 GeV), $g_W \approx e$ and hence, in the energy range of 100 GeV, the strength of the weak interaction would become comparable to electromagnetic interaction.

At lower energies, Fermi's point interaction works well, so at lower energies, the two-vertex interaction must revert to the single vertex interaction. Now, $(M_W^2 - q^2)^{-1}$ is the propagator term that exists only when q is non-vanishing. Because, $M_W \approx 80$ GeV (say, fixed), when $q^2 << M_W^2$ (lower energies), the propagator term vanishes and the two-vertex interaction reduces to the single vertex. At large energies, however, the propagator term exists and it generates the necessary moderation effect in the unphysical high-energy behaviour of the total scattering cross section.

Thus to avoid the high-energy catastrophe, which σ faced, it required the introduction of vector boson as the mediator of weak interactions. The weak interactions of the type

$$\nu_\mu + n \rightarrow \mu^- + p \tag{4.145}$$

in which leptonic (ν_μ) and hadronic (n) parts of LHS separately change the electric charge of one unit with respectively the hadronic (p) and leptonic (μ^-) parts of the RHS, are mediated by W^+ and W^-, i.e., the charged vector bosons. The weak interactions of the type (4.145) are therefore called the charged current reactions. The reactions of the type

$$\begin{aligned} \nu_\mu + p &\rightarrow \nu_\mu + p \\ \nu_\mu + e^- &\rightarrow \nu_\mu + e^- \end{aligned} \tag{4.146}$$

can also be mediated by a neutral intermediate vector boson Z^0 and such reactions are called *neutral current reactions*. The neutral current weak interactions (4.146) have been experimentally verified. The experimental verification of neutral current weak interaction is considered a landmark discovery of 1970s. It is an important step towards the unification of electromagnetic and weak interactions.

4.5.4 Cabibbo Theory

The universality of weak interactions discussed under section 4.5.2, is based on the near equality of weak coupling constants G_β and G_μ, the former from semileptonic ($\Delta S = 0$) and latter from purely leptonic weak decays. The universal current—current form to explain weak interactions was described by considering only $\Delta S = 0$ weak interactions. The strangeness changing ($\Delta S = 1$) weak interactions were ignored. The reason for excluding $\Delta S = 1$, weak decay was that they violated the universality of weak interactions. We may recall that, it is essential for weak interactions to observe universality and if it is violated, necessary changes have to be made in the theory so that universality is restored.

Keeping above points in view, in the present section we address to the following issues.

1. How the discrepancy between the values of G_β and G_μ can be removed, small though the difference between them is!
2. How the universality can be restored for the strangeness changing ($\Delta S = 1$) weak hadronic decays!

Before finding the remedy to the above problems, let us elaborate a little, on the strangeness changing weak hadronic decays. If we rewrite eqn. (4.106) by retaining C_V and C_A, we get

$$M_{fi} = [\bar{u}_p \gamma^\mu (GC_V + GC_A \gamma^5) u_n][\bar{u}_e \gamma_\mu (1 - \gamma_5) u_\nu] \tag{4.147}$$

where GC_V and GC_A are the coupling constants for vector and axial vector interactions. Eqn. (4.147) is corresponding to semileptonic weak decay. When we compare GC_V of eqn. (4.147) with G of purely leptonic decay, we get $GC_V = G$, implying $C_V = 1$, the value of C_A comes out to be approximately minus 1.25. Let us focus our attention on $C_V = 1$. In the semileptonic decay, proton and neutron are participating. Now, p and n are strongly interacting particles and have structures (implying, say, they are composed of quarks). When p and n are participating, the vector-coupling constant is GC_V, which is equal to G, which corresponds to the weak interaction involving structureless leptons (the fermions without strong interactions). Then, should it imply from $GC_V = G$, that the structure and strong interactions associated with hadrons (here p and n) do not affect the vector weak interactions! The assumption that this really happens is called the ***conserved vector current*** (***CVC***) ***hypothesis***, which postulates that the vector weak current is conserved by strong interactions, just as is the electromagnetic current. Note that electromagnetic interaction is a vector interaction and conservation law is described as the vanishing of four-divergence suggesting that the vector current in V—A

$$\partial_\mu j^\mu_{em} = 0 \tag{4.148}$$

theory would also be free from divergences. The axial vector current is hypothesized to be partially conserved (***PCAC***) in the sense that its four-divergence is slightly different from zero. This divergence is a measure of the difference of C_A (=1.25) from unity, i.e., 0.25 (small). The *CVC* hypothesis is very well verified by the decay

$$\pi^+ \to \pi^0 + e^+ + \nu_e \tag{4.149}$$

and in general the agreement of the *CVC* theory with the experiments is quite good for non-strangeness changing ($\Delta S = 0$) decays. Because of the *CVC* theory, it can be thought that a conserved (hence universal) coupling constant is associated with the vector part of weak interactions. Just as conservation of electromagnetic current implies that total charge is unaffected by the presence of strong interactions, the conservation of weak vector current implies that the total weak charge is unaffected by strong interactions. The axial vector weak charge is slightly modified by strong interactions.

It is the *CVC* theory for weak interactions, that tells that the universal weak current is expressed by eqn. (4.120), namely

$$J^\mu = J^\mu_l + J^\mu_h \tag{4.150}$$

and the weak interaction Hamiltonian is expressed by eqn. (4.124), where both the leptonic and hadronic currents have V–A form.

Now, we come to the discrepancies talked in the beginning of this section regarding G_β and G_μ values and $\Delta S = 1$ decays. Since the hadronic current J^μ_h consists of both strangeness conserving ($\Delta S = 0$) and non-strangeness conserving ($\Delta S = 1$) parts, it was believed before the Cabibbo theory that both ($\Delta S = 0$) hadronic current and ($\Delta S = 1$) hadronic current coupled with the same coupling constant G, and in this case, we could use

$$J^\mu_h = J^\mu_h(\Delta S = 0) + J^\mu_h(\Delta S = 1) \tag{4.151}$$

in the equation (4.128). Thinking that J^μ_h in the form [eqn. (4.151)] is the source of problems, Cabibbo suggested that the (ΔS = 0) hadronic current and the (ΔS = 1) hadronic current do not couple with the same coupling constant (*c.c.*), *G*, instead

$$\begin{aligned} &(\Delta S = 0) \textit{ hadronic current couples with c.c.} = G\cos\theta_c \\ &(\Delta S = 1) \textit{ hadronic current couples with c.c.} = G\sin\theta_c \end{aligned} \qquad (4.152)$$

such that the hadronic current J_h^μ can be expressed as

$$J_h^\mu = \cos\theta_c\, J_h^\mu(\Delta S = 0) + \sin\theta_c\, J_h^\mu(\Delta S = 1) \qquad (4.153)$$

where θ_c is called the ***Cabibbo angle***, whose value is to be determined by experiment. Obviously, in the form [eqn. (4.153)] the coupling constant for the whole J_h^μ would be *G*. Now, it can be said that when J_h^μ is like eqn. (4.153), both hadronic current (J_h^μ) and leptonic current (J_l^μ) couple with the same coupling constant *G*, hence universality is obeyed. The *G* in eqn. (4.152) is taken to be pure lepton decay (i.e., muon decay) coupling constant G_μ.

4.5.4.1 Consequences of Cabibbo Theory

4.5.4.1.1 The Cabibbo angle q_c

The weak interaction theory is *V—A* current–current theory and we expect both kinds of Cabibbo angles θ_V for vector coupling and θ_A for axial vector coupling. The two-particle decays

$$\begin{aligned} &\pi^\pm \to e^\pm + \nu_e(\bar{\nu}_e) \\ &\pi^\pm \to \mu^\pm + \nu_\mu(\bar{\nu}_\mu) \end{aligned} \qquad (4.154)$$

and

$$\begin{aligned} &K^\pm \to e^\pm + \nu_e(\bar{\nu}_e) \\ &K^\pm \to \mu^\pm + \nu_\mu(\bar{\nu}_\mu) \end{aligned} \qquad (4.155)$$

are the axial vector interactions. The decays (4.154) have (ΔS = 0), and the decays (4.155) have (ΔS = 1). Both pion and kaon are ($J^P = 0^-$) pseudoscalar mesons. However, the following decays in which a pseudoscalar decays into a pseudoscalar are vector interactions.

$$\left.\begin{aligned} &\pi^\pm \to \pi^0 + e^\pm + \nu_e(\bar{\nu}_e) \\ &\pi^\pm \to \pi^0 + \mu^\pm + \nu_\mu(\bar{\nu}_\mu) \end{aligned}\right. \quad (\Delta S = 0) \qquad (4.156)$$

and

$$\left.\begin{aligned} &K^\pm \to \pi^0 + e^\pm + \nu_e(\bar{\nu}_e) \\ &K^\pm \to \pi^0 + \mu^\pm + \nu_\mu(\bar{\nu}_\mu) \end{aligned}\right. \quad (\Delta S = 1) \qquad (4.157)$$

The Cabibbo angle can be found by comparing observed $\Delta S = 0$ and $\Delta S = 1$ decay rates. We take the ratio of M_{fi} for the two decays from each type of interactions (see eqs. 4.152). That is, we take the ratio

$$\frac{M_{fi}(K^+ \to \mu^+ + \nu_\mu)}{M_{fi}(\pi^+ \to \mu^+ + \nu_\mu)} = \frac{M_{fi}(\Delta S = 1)_{\text{Axial}}}{M_{fi}(\Delta S = 0)_{\text{Axial}}}$$

$$\propto \frac{G_\mu \sin\theta_A}{G_\mu \cos\theta_A} = \tan\theta_A \tag{4.158}$$

Similarly for vector interactions

$$\frac{M_{fi}(K^+ \to \pi^0 + e^+ + \nu_e)}{M_{fi}(\pi^+ \to \pi^0 + e^+ + \nu_e)} = \frac{M_{fi}(\Delta S = 1)_{\text{Vector}}}{M_{fi}(\Delta S = 0)_{\text{Vector}}}$$

$$\propto \frac{G_\mu \sin\theta_V}{G_\mu \cos\theta_V} = \tan\theta_V \tag{4.159}$$

It is found from experiments that

$$\tan\theta_A \approx 0.275 \Rightarrow \theta_A \approx 15°24'$$

and

$$\tan\theta_V \approx 0.251 \Rightarrow \theta_V \approx 14°6' \tag{4.160}$$

indicating that $\theta_A \approx \theta_V$. That, θ_A and θ_V are nearly equal, is also a prediction of Cabibbo theory.

4.5.4.1.2 Suppression of hadronic ΔS = 1 decays

Let us consider the following strangeness changing ($\Delta S = 1$) weak decays.

$$\begin{aligned} \Lambda &\to p + e^- + \bar{\nu}_e \\ \Sigma^- &\to n + e^- + \bar{\nu}_e \\ \Xi^- &\to \Lambda + e^- + \bar{\nu}_e \end{aligned} \tag{4.161}$$

where we know Λ and Σ^- have the strangeness quantum number $S = -1$, Ξ^- has $S = -2$ and p, n have $S = 0$, $\Delta S = S(\text{RHS}) - S(\text{LHS}) = 1$.

If the universal Fermi interaction is applied to these decays, it would mean the coupling constant corresponding to these ($\Delta S = 1$) weak decays would be equal to the coupling constant corresponding to ($\Delta S = 0$) weak decays. This means that the decay rate of eqn. (4.161) like ($\Delta S = 1$)decays should be equal to the decay rate of ($\Delta S = 0$) weak decays like

$$\nu \to p + e^- + \bar{\nu}_e \tag{4.162}$$

But, in reality, it is found that the decay rate of ($\Delta S = 1$) weak decays is about 20 times smaller than the ($\Delta S = 0$) decays.

Cabibbo's prescription (4.152) explains this anomaly. The matrix element M_{fi}

in the weak decays like (4.162) is obtained as the coupling of the strangeness conserving ($\Delta S = 0$) (p, n) current with lepton (e, ν_e) current and is proportional to $G \cos \theta$ [refer ($\Delta S = 0$) postulate of eqn. (4.152)]. On the other hand, ($\Delta S = 1$) decays like (4.161) would have their M_{fi} proportional to $G \sin \theta$. Therefore, noting that the decay rate Γ is proportional to the square of the coupling constant, we may write for say Σ^- decay

$$\frac{\Gamma[\Sigma^- \rightarrow n + e^- + \overline{\nu}_e](\Delta S = 1)}{\Gamma[n \rightarrow p + e^- + \overline{\nu}_e]\ (\Delta S = 0)} \propto \frac{G^2 \sin^2 \theta_c}{G^2 \cos^2 \theta_c} \propto \tan^2 \theta_c \tag{4.163}$$

where G in fact implies G_μ (the coupling constant of muon decay). The value of θ_c can be obtained by noting from eqn. (4.147) that the vector weak coupling in beta decay is GC_V with the experimental value of

$$C_V = 0.978 \tag{4.164}$$

and equating

$$GC_V = G \cos \theta_c \tag{4.165}$$

where $G \cos \theta_c\, G\cos\theta_c$ is the coupling constant for Fermi transitions (vector interactions) in beta decay. Relation (4.165) then gives

$$\cos \theta_c = 0.978,\ \sin \theta_c = 0.209 \tag{4.166a}$$

so that

$$\tan \theta_c = 0.214,\ \tan^2 \theta_c \approx 0.05 \tag{4.166b}$$

implying that

$$\theta_c \approx 12°\ 6' \approx 13° \tag{4.166c}$$

Substituting the values (4.166) in eqn. (4.163), we may write that the decay rate of ($\Delta S = 1$) [$\Sigma^- \rightarrow n + e^- + \overline{\nu}_e$]

$$\cong \left(\frac{1}{20}\right) \text{ of the decay rate of } (\Delta S = 0)[n \rightarrow p + e^- + \overline{\nu}_e] \tag{4.167}$$

which explains that as observed in practice, the ($\Delta S = 1$) hadronic decay rates are suppressed by about a factor of 20 compared to ($\Delta S = 0$) decay rates. We may write in terms of coupling constants

$$\begin{aligned} G\,(\Delta S = 0) &= (0.978)G_\mu \\ G\,(\Delta S = 1) &= (0.209)G_\mu \end{aligned} \tag{4.168}$$

which means

$$G(\Delta S = 0) \approx (4.7)\ G(\Delta S = 1) \tag{4.169}$$

The non-strangeness conserving decays like eqn. (4.161) are important because these decays inspired Cabibbo to postulate eqn. (4.153). This was because n, p, Σ^-, Λ, Ξ^- along with Σ^+, Σ^0 and Ξ^0 are the $J^P = \frac{1}{2}^+$ baryon octet of SU(3) symmetry. When the SU(3) symmetry is exact, the masses of all these eight baryons should be equal and that the weak coupling of these entire eight particle states must also be identical. Regarding the masses of these particles, we know that they are not the same, but nothing could be said about the identity of weak coupling constant. Cabibbo, then postulated, *as the masses of SU(3) octet are divided, the weak coupling constant could not be taken same* and they should be taken as

$$\left.\begin{aligned} G(\Delta S = 0) &= G\cos\theta_c \\ G(\Delta S = 1) &= G\sin\theta_c \end{aligned}\right\} G \Rightarrow G_\mu \tag{4.170}$$

4.5.4.1.3 Restoration of universality

Without further elaboration, we can say that if the hadronic current is expressed by eqn. (4.151), the coupling constants are as in eqn. (4.170), i.e., they are different and this in turn implies the violation of universality. However, when the hadronic current is expressed as eqn. (4.153), the coupling constant for both ($\Delta S = 0$) weak decays and ($\Delta S = 1$) weak decays is same $G = G_\mu$. Thus, to restore the universality, we must take the hadronic current as in eqn. (4.153), i.e,

$$J_h^\mu = \cos\theta_c\ J_h^\mu(\Delta S = 0) + \sin\theta_c\ J_h^\mu(\Delta S = 1) \tag{4.171}$$

4.5.4.1.4 Removal of G_β, G_μ difference

From equations (4.115) and (4.119), the calculated values of G_β and G_μ are

$$G_\beta = 1.39 \times 10^{-62}\ \text{Jm}^3 \tag{4.172a}$$

and

$$G_\mu = 1.44 \times 10^{-62}\ \text{Jm}^3 \tag{4.172b}$$

the difference in the above values has to be removed. Since, G_β corresponds to the beta decay for which $\Delta S = 0$, we should not expect $G_\beta = G_\mu$, instead due to the Cabibbo's postulate for $\Delta S = 0$ weak interactions

$$G_\beta \equiv G_\mu \cos\theta_c \tag{4.173}$$

Using $\cos\theta_c \approx 0.97$ and G_μ from eqn. (4.172b), we get the value of G_β as in eqn. (4.172a). Thus due to Cabibbo's assumption, that coupling constants should be according to eqs. (4.152), the discrepancy of equations (4.172) ceases to exist.

4.5.4.1.5 Quark-lepton universality

By quark-lepton universality, too, it would be meant that the weak interactions among quarks have the coupling constant equal to that of weak interactions between leptons. Here, we shall, in fact, rewrite the quark versions of consequences **2** and **3** above.

At the quark level in the beta decay process $n \to p + e^- + \bar{\nu}_e$ with $n = udd$ and $p = uud$, a d quark transforms into a u quark as

$$d \to u + e^- + \bar{\nu}_e \tag{4.174}$$

while, the remaining *ud* quarks do not participate in the decay and act just as spectators (see Fig. 1.26).

Let us also look at the weak decay of a strange particle

$$\Lambda \rightarrow p + e^- + \bar{\nu}_e \tag{4.175}$$

at the fundamental level. With $\Lambda = uds$, we may write the decay (4.174) as

$$s \rightarrow u + e^- + \bar{\nu}_e \tag{4.176}$$

where *ud* again remain spectators. The decay has been drawn in Fig. (4.7). In the

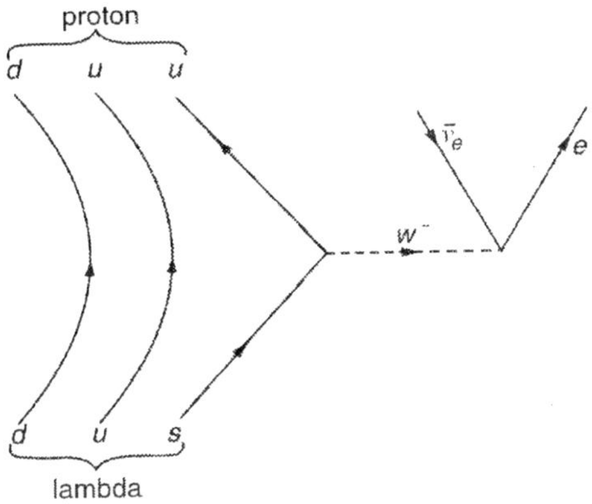

Fig. 4.7: Quark level Feynman diagram for weak decay $\Lambda \rightarrow p + e^- + \bar{\nu}_e$

weak decay of Λ, the strange quark transforms into a u quark.

Obviously, the decay (4.174) is strangeness preserving ($\Delta S = 0$) and decay (4.175), the strangeness changing ($\Delta S = 1$). Strangeness quantum number of u quark is zero and that of s quark -1. At the quark level, we may then write for the matrix elements as

$$M_{d \rightarrow u} = \frac{G^2}{\sqrt{2}}\left[\bar{u}_u \gamma^\mu (1 - \gamma^5) u_d\right]\left[\bar{u}_e \gamma_\mu (1 - \gamma_5) u_{\nu_e}\right] \tag{4.177}$$

and

$$M_{s \rightarrow u} = \frac{G^2}{\sqrt{2}}\left[\bar{u}_u \gamma^\mu (1 - \gamma^5) u_s\right]\left[\bar{u}_e \gamma_\mu (1 - \gamma_5) u_{\nu_e}\right] \tag{4.178}$$

where the same coupling constant G has been associated with both $\Delta S = 0$ and $\Delta S = 1$ decays. We recall that the decay rates of $(\Delta S = 0)$ are larger than the decay rates of $\Delta S = 1$ weak decays; hence we should have different coupling constants for $d \rightarrow u + W^-$ and $s \rightarrow u + W^-$ decays. If this happens the universality breaks down.

To restore the universality of coupling constants, Cabibbo postulated that as far as weak interactions are concerned, neither d quark nor the s quark participates instead there are quark states d' and s' which actually take part in weak interactions.

The quark states d' and s' are Cabibbo rotated quark states and from the Fig. (4.8), we may write

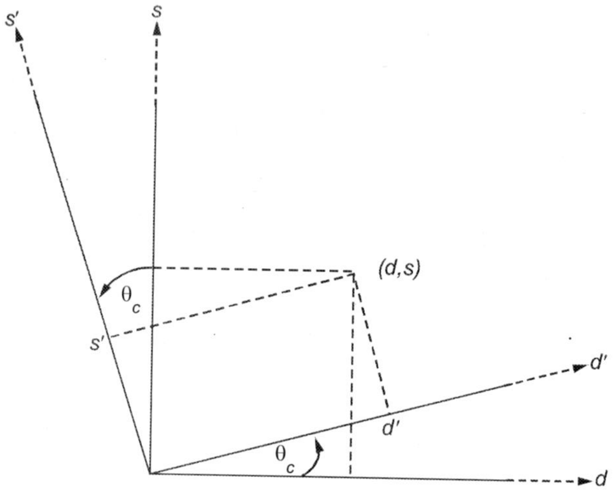

Fig. 4.8: Cabibbo rotation of quark states

$$\begin{aligned} d' &= d\cos\theta_c + s\sin\theta_c \\ s' &= -d\sin\theta_c + s\cos\theta_c \end{aligned} \quad (4.179)$$

It may be recalled that u and d quarks form an isospin doublet $\begin{pmatrix} u \\ d \end{pmatrix}$. Now, since d' has come into picture, we have another isospin doublet $\begin{pmatrix} u \\ d' \end{pmatrix}$. The former, i.e., $\begin{pmatrix} u \\ d \end{pmatrix}$ is the *strong isospin doublet* in which u and d quarks remain as such but the latter $\begin{pmatrix} u \\ d' \end{pmatrix}$ in which u remains u but d becomes d' is called the *weak isospin doublet*. In the weak interactions $\begin{pmatrix} u \\ d' \end{pmatrix}$ is used. Taking into account all the three generations of quarks, we would say that $\begin{pmatrix} u \\ d \end{pmatrix}$, $\begin{pmatrix} c \\ s \end{pmatrix}$ and $\begin{pmatrix} t \\ b \end{pmatrix}$ are three strong isospin doublets and $\begin{pmatrix} u \\ d' \end{pmatrix}$, $\begin{pmatrix} c \\ s' \end{pmatrix}$ and $\begin{pmatrix} t \\ b' \end{pmatrix}$ would be the corresponding weak isospin doublets. Eqn. (4.179) also tells that d' and s' quark states are two separate mixtures of d and s quarks. It can therefore be said that in the strong interactions quarks are in their pure states but in weak interactions they get mixed. It is also possible that instead of forming d' and s', we could Cabibbo rotate u and c quarks to get the mixed states u' and c'. However, it being a convention the d and s quarks are rotated and u and c are left unmixed. So, we may write for the first quark weak doublet

$$\begin{pmatrix} u \\ d' \end{pmatrix} = \begin{pmatrix} u \\ d\cos\theta_c + s\sin\theta_c \end{pmatrix} \tag{4.180}$$

Since, in the weak interactions only the left-handed fermions participate, the doublets

$$\begin{pmatrix} \nu_e \\ e_L^- \end{pmatrix} \text{and} \begin{pmatrix} \nu_\mu \\ \mu_L^- \end{pmatrix} \tag{4.181}$$

are the leptonic doublets and upto first generation (4.180) is the quark doublet that participates in the weak interactions. ν_e and ν_μ are already left-handed.

In the above quark model description of weak interactions, the Cabibbo hadronic current (4.153) is replaced by the quark current of the form

$$\begin{aligned} J_\mu^q &= \bar{u}_u \gamma^\mu (1-\gamma^5) u_{d'} \\ &= \bar{u}_u \gamma^\mu (1-\gamma^5) \cos\theta_c u_d + \bar{u}_u \gamma^\mu (1-\gamma^5) \sin\theta_c u_s \end{aligned} \tag{4.182}$$

The first term of this expression implies that the weak interaction vertex

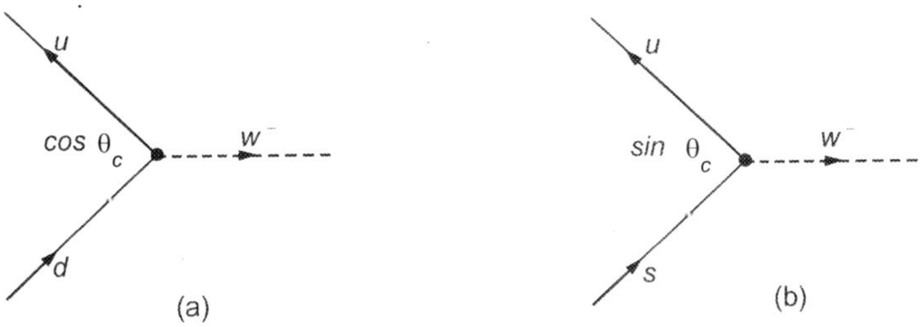

Fig. 4.9: Weak interaction vertices. Fig. a belong to $\Delta s = 0$ weak interactions and Fig. b: to $\Delta s = 1$

(Fig. 4.9a) $d \rightarrow u + W^-$ supplies additional factor of $\cos\theta_c$ and the second term implying $s \rightarrow u + W^-$ (Fig. 4.9b), provides the additional factor of $\sin\theta_c$. Obviously, in the quark current J_μ^q, the first term belongs to ($\Delta S = 0$) and the second to ($\Delta S = 1$) weak interactions.

It is to be observed that the mixing takes place between the generations. d quark belongs to the first generation and s to the second. However, there is no Cabibbo angle for leptons; implying generations of leptons do not mix. This is because the neutrinos ν_e, ν_μ and ν_τ (also called the three flavours of neutrino) are massless and any linear combination of neutrinos is also massless. The neutrino, e.g., the ν_e therefore may itself be a Cabibbo rotated state or may just be the unrotated state. The definition of ν_e (e.g.) may then be given as a massless particle which pairs with electron in weak interactions. Because of such pairings, the lepton, muon and taon numbers are conserved and we cannot have mixed lepton states.

In view of no mixing in the lepton sector, the first generation lepton current would remain

$$J_\mu^l = \bar{u}_{\nu_e}\gamma_\mu(1-\gamma_5)u_{e_L} \tag{4.183}$$

where $l \Rightarrow$ lepton and $L \Rightarrow$ left-handed. The lepton-lepton interaction Hamiltonian (see eqn. 4.128) would be

$$(M_{fi})_l = \frac{G}{\sqrt{2}} J_\mu^l J_l^{\mu\dagger} \tag{4.184}$$

and the quark–quark interaction Hamiltonian would be

$$(M_{fi})_q = \frac{G}{\sqrt{2}} J_\mu^q J_q^{\mu\dagger} \tag{4.185}$$

where J_μ^q is in the form (4.182). Both $q-q$ and $l-l$ weak interactions couple with the same coupling constant G. We may thus conclude that the Cabibbo rotated quark states [producing in turn the J_q^μ of equation (4.182)] allow us to use the same coupling constant G for the quark-quark and lepton—lepton interactions and thus establish the quark—lepton universality. Needless to say, if the quark states are not mixed, i.e., $u_{d'}$ remains u_d in eqn. (4.182), we will have to use different coupling constants $G \cos\theta_c$ and $G \sin\theta_c$ for $\Delta S = 0$ and $\Delta S = 1$ quark—quark interactions respectively. This would then mean the violation of universality of weak interactions. Since, universality is preserved, the quark states are mixed.

4.5.4.2 The Charged and Neutral Weak Currents

We have already seen that in terms of mediators, the weak interactions are of two types, one the charged weak interactions and two, the neutral weak interactions. The former are generated by the emission and absorption of charged vector bosons W^+ and W^-, the latter by the emission and absorption of neutral Z^0 bosons. We have also noticed that weak interactions are described by a *current* × *current* theory. The nature of the current, that is, whether the current is charged or neutral, is determined by its carrier. The carriers of the charged current are obviously the charged $W^\pm$ and they are neutral Z^0 for neutral currents. In other words, we may say that the currents in the *current* × *current* theory of weak interactions couple through either W^+ or W^- or Z^0. For example, as shown in the following diagrams (Fig.4.10), a W^+ is absorbed at the vertex in diagram (a) and a W^- is absorbed in

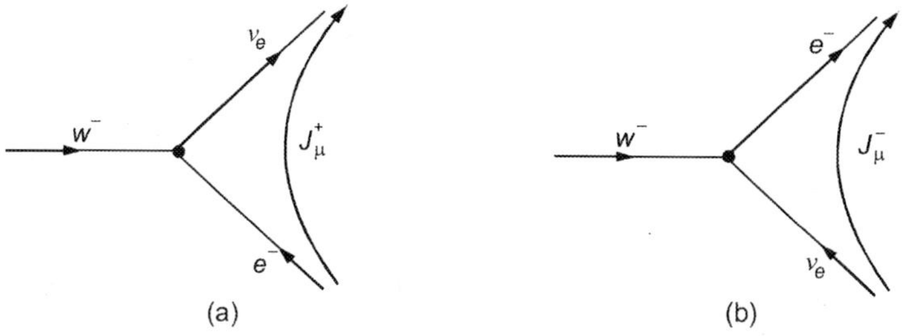

Figs. 4.10 [a, b]: The charged current J^+_μ and J^-_μ

diagram (b). In both the above diagrams, there is a change of charge ΔQ between the incoming and outgoing particles. The $\Delta Q = +1$ unit as in Fig. (4.10a) gives rise to J^+_μ current and $\Delta Q = -1$ unit [Fig. (4.10b)] causes a J^-_μ current. Remembering that the emission of $W^\pm$ is equivalent to the absorption of $W^\pm$, we have both J^+_μ and J^-_μ currcnts in a particular weak interaction (charged). For example, the weak interaction $n + \nu_e \rightarrow p + e^-$ drawn as in Fig. (4.11) depicts both the J^+_μ and J^-_μ

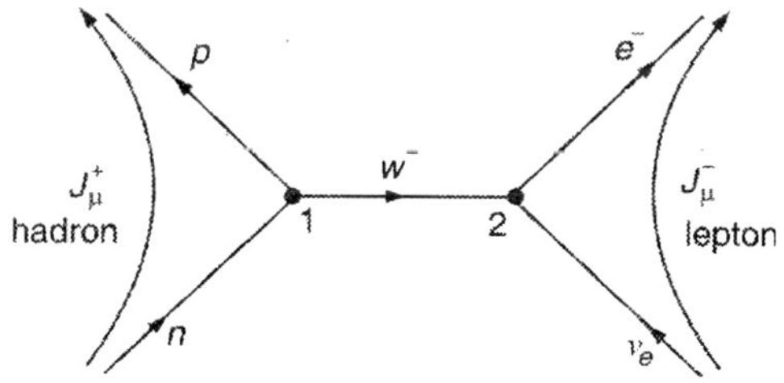

Fig. 4.11: Coupling of charged hadron and charged lepton currents

currents in which, however, J^+_μ because of $\Delta Q = +1$ between p and n is called the nucleon (in general hadron) current and J^-_μ because of $\Delta Q = -1$ between leptons (ν_e and e^-) is called the lepton current. At vertex 1, W^- has been emitted which is equivalent to the absorption of W^+, hence J^+_μ current. Fig. (4.11), then tells that the mediator W^- couples the weak currents $(J^+_\mu)_{\text{nucleon}}$ and $(J^-_\mu)_{\text{lepton}}$. Depending on the type of weak interaction (leptonic, semileptonic and non-leptonic) J^+_μ and J^-_μ could both be leptonic or hadronic or one hadronic other leptonic. In this regard reference to eqn. (4.128) can be made where however, no mediator was proposed.

Let us come back to the Fig. (4.10). In Fig. (a), the electric charge of the outgoing particle ν_e (electric charge zero) is raised by one unit from the electric charge of incoming particle electron (charge $-e$). Thus J_μ^+ is the charge raising weak current, which couples a left-handed incoming electron to the outgoing neutrino. Likewise in Fig. (4.10b), a weak charge lowering current J_μ^- couples the incoming neutrino ($Q = 0$) to the outgoing left-handed electron ($Q = -1$). Using the vector minus axial vector form for the currents, we may write for diagrams (4.10) [see eqs. (4.123–4) also]

$$J_\mu^+ = \bar{u}_{\nu_e}\gamma_\mu(1-\gamma_5)u_e \tag{4.186a}$$

and

$$J_\mu^- = \bar{u}_e\gamma_\mu(1-\gamma_5)u_{\nu_e} \tag{4.186b}$$

Using eqn. (4.126), we may find that

$$\begin{aligned}(J_\mu^+)^\dagger &= [\bar{u}_{\nu_e}\gamma_\mu(1-\gamma_5)u_e]^\dagger \\ &= \bar{u}_e\gamma_\mu(1-\gamma_5)u_{\nu_e} = J_\mu^-\end{aligned} \tag{4.187}$$

Thus, if we call the charge raising weak current

$$J_\mu^+ = J_\mu \tag{4.188}$$

then the charge lowering weak current is its hermitian conjugate

$$J_\mu^- = J_\mu^\dagger \tag{4.189}$$

Keeping in view the participation of only left-handed (L) fermions, we may incorporate a factor of $\frac{1}{2}$ in eqn. (4.186) and then using eqn. (4.80), we may write

$$J_\mu^+ = J_\mu = \bar{\nu}_L\gamma_\mu e_L \tag{4.190}$$

where

$$\nu_L = u_{\nu_e}$$

and

$$e_L = \frac{(1-\gamma_5)}{2}u_e \tag{4.191}$$

Whereas, for $J_\mu^\dagger$, we may go through the following steps

$$\begin{aligned}J_\mu^- = J_\mu^\dagger &= [\bar{u}_{\nu_e}\gamma_\mu\tfrac{1}{2}(1-\gamma_5)u_e]^\dagger \\ &= u_e^\dagger\tfrac{1}{2}(1-\gamma_5)^\dagger\gamma_\mu^\dagger(u_{\nu_e}^\dagger\gamma_0)^\dagger \\ &= [\tfrac{1}{2}(1-\gamma_5)u_e]^\dagger\gamma_0\gamma_\mu\gamma_0^2 u_{\nu_e} \\ &= e_L^\dagger\gamma_0\cdot\gamma_\mu\nu_L = \bar{e}_L\gamma_\mu\nu_L\end{aligned} \tag{4.192}$$

where $\gamma_\mu^\dagger = \gamma_0\gamma_\mu\gamma_0$ and eqs. (4.191) have been used. In the above equations ν_L

denotes the spinor and $\bar{\nu}_L$, the adjoint spinor for neutrino. Similarly, e_L and $\bar{e}_L$ are for electron. With ν_L, the subscript L, denoting left-handedness is written only to indicate symmetry with e_L. There is e_R but no ν_R.

For further discussion on charge raising (J_μ^+) and charge lowering (J_μ^-) weak currents, let us first bring them together

$$J_\mu^+ = \bar{\nu}_L \gamma_\mu e_L \tag{4.193a}$$

$$J_\mu^- = \bar{e}_L \gamma_\mu \nu_L \tag{4.193b}$$

where ν refers to the electron neutrino ν_e. The form of the above equations in which ν_L and e_L get interchanged from one to other, suggests that these equations can also be written by introducing a left-handed doublet

$$\chi_L = \begin{pmatrix} \nu_L \\ e_L \end{pmatrix} \equiv \begin{pmatrix} \nu_e \\ e \end{pmatrix}_L \tag{4.194}$$

and then writing

$$J_\mu^+ = (\bar{\nu}_L\ \bar{e}_L)\gamma_\mu \begin{pmatrix} 0 & 1 \\ 0 & 0 \end{pmatrix} \begin{pmatrix} \nu_L \\ e_L \end{pmatrix} = \bar{\nu}_L \gamma_\mu e_L \tag{4.195a}$$

and

$$J_\mu^- = (\bar{\nu}_L\ \bar{e}_L)\gamma_\mu \begin{pmatrix} 0 & 0 \\ 1 & 0 \end{pmatrix} \begin{pmatrix} \nu_L \\ e_L \end{pmatrix} = \bar{e}_L \gamma_\mu \nu_L \tag{4.195b}$$

Now

$$\begin{pmatrix} 0 & 1 \\ 0 & 0 \end{pmatrix} = \frac{1}{2}\left(\tau^1 + i\tau^2\right) = \tau^+ \tag{4.196a}$$

and

$$\begin{pmatrix} 0 & 0 \\ 1 & 0 \end{pmatrix} = \frac{1}{2}\left(\tau^1 - i\tau^2\right) = \tau^- \tag{4.196b}$$

where τ^1 and τ^2 are first two Pauli spin matrices (see σ_1 and σ_2 of eqn. 3.13). In terms of above $\tau^\pm$ matrices, we may write eqn. (4.195) in the compact form

$$J_\mu^\pm = \bar{\chi}_L \gamma_\mu \tau^\pm \chi_L \tag{4.197}$$

Since, the two Pauli spin matrices τ^1 and τ^2 have found a place in describing the $J_\mu^\pm$ weak charged currents, there must also be a current corresponding to third Pauli matrix,

$$\tfrac{1}{2}\tau^3 = \tfrac{1}{2}\begin{pmatrix} 1 & 0 \\ 0 & -1 \end{pmatrix} \tag{4.198}$$

which we may call J^3_μ and express it like eqn. (4.197) as

$$\begin{aligned} J^3_\mu &= \overline{\chi}_L \gamma_\mu \frac{\tau^3}{2} \chi_L \\ &= (\overline{\nu}_L\ \overline{e}_L)\gamma_\mu \begin{pmatrix} \frac{1}{2} & 0 \\ 0 & -\frac{1}{2} \end{pmatrix} \begin{pmatrix} \nu_L \\ e_L \end{pmatrix} \\ &= \tfrac{1}{2}\overline{\nu}_L \gamma_\mu \nu_L - \tfrac{1}{2}\overline{e}_L \gamma_\mu e_L \end{aligned} \tag{4.199}$$

Each of the term in the above equation tells that both incoming and outgoing particles are same, it is ν_e in the first term and e_L, the left-handed electron in the second term, hence no change of charge takes place between the outgoing and the incoming particles. The current that would couple the incoming and outgoing particles should be neutral (not changing the electric charge) current. However, J^3_μ is not the whole of the neutral current, because the right-handed electron e_R exists and we can have a current term $\overline{e}_R \gamma_\mu e_R$, which too would be neutral. The neutral weak current involves the right-handed components as well. We cannot, of course, have $\overline{\nu}_R \gamma_\mu \nu_R$ as ν_R does not exist. Here $\overline{\nu}_R$ is the adjoint neutrino spinor and should not be confused with anti neutrino.

The actual structure of the neutral current would be derived in the coming Chapters (Chapter 6), when we take up electroweak unification. At the moment we would call J^3_μ the third weak current, the other two being $J^\pm_\mu$.

4.5.4.2.1 Weak isospin and weak hypercharge

We know that proton and neutron form an isospin doublet $\begin{pmatrix} p \\ n \end{pmatrix}$ whose total isospin is $T = \dfrac{1}{2}$. The proton is the $T_3 = +\dfrac{1}{2}$ state and neutron the $T_3 = -\dfrac{1}{2}$ state of nucleon. Likewise we have

$$\begin{pmatrix} u \\ d \end{pmatrix}, \begin{pmatrix} c \\ s \end{pmatrix} \text{ and } \begin{pmatrix} t \\ b \end{pmatrix}$$

the three isospin ($T = \dfrac{1}{2}$) quark doublets, in which upper members u, c, t are $T_3 = +\dfrac{1}{2}$ states and lower the $T_3 = -\dfrac{1}{2}$ states. Such isospin states are familiar to us. We are also familiar with the strong force that acts between proton and neutron. All the above-mentioned isospin states are in fact the strong isospin states.

In the charged weak interactions as seen above, only the left handed fermions are involved, the doublet

$$\begin{pmatrix} \nu_e \\ e \end{pmatrix}_L$$

which has helped in giving a compact form [eqn. (4.197)] to $J_\mu^\pm$, too looks like an isospin doublet. Such left handed doublets, which participate in weak interactions, are called *weak isospin doublets*. We can have in place of eqn. (4.194) the following weak isospin doublets

$$\chi_L \to \begin{pmatrix} \nu_e \\ e \end{pmatrix}_L, \begin{pmatrix} \nu_\mu \\ \mu \end{pmatrix}_L, \begin{pmatrix} \nu_\tau \\ \tau \end{pmatrix}_L, \begin{pmatrix} u \\ d' \end{pmatrix}_L, \begin{pmatrix} c \\ s' \end{pmatrix}_L, \begin{pmatrix} t \\ b' \end{pmatrix}_L \tag{4.200}$$

in which the upper members are weak $T_3 = \frac{1}{2}$ members and the lower, weak $T_3 = -\frac{1}{2}$ members. As seen earlier d', s', b' are Cabibbo rotated quark states.

Thus the weak isospin, though applies to both leptons and quarks, involves only the left-handed ones. The right-handed leptons and quarks are singlets and are invariant as far as weak isospin is concerned. The weak isospin doublets of quarks are Cabibbo rotated. Although, the concept of weak isospin doublet does comc from the strong isospin doublet, there is no relation between the two kinds.

Similar to weak isospin, a weak hypercharge can also be defined. The Gell—Mann—Nishijima formula [see, for example, eqn. (3.142)]

$$Q = T_3 + \frac{Y}{2} \tag{4.201}$$

where T_3 and Y, the hypercharge, are for strong interactions. $Y = B + S$, with B as the baryon number and S the strangeness quantum number. Q provided the electric charge of baryons in units of e. A similar relation for the electric charge of leptons and quarks would be

$$Q = (T_3)_L + \left(\frac{Y}{2}\right)_{L,R} \tag{4.202}$$

The following table written for the quarks and leptons would explain the relation (4.202). It has to be noted that in case of neutral weak interactions, which involve right-handed components as well, it is immaterial whether we use the states d, s, b or the Cabibbo rotated states d', s', b'.

Table of weak isospin and hypercharge for fermions

State	Weak isospin		Weak hypercharge	Charge Q/e of lepton/quark
	T	T_3	Y	
ν_e, ν_μ, ν_τ	$\frac{1}{2}$	$+\frac{1}{2}$	-1	0
e_L^-, μ_L^-, τ_L^-	$\frac{1}{2}$	$-\frac{1}{2}$	-1	-1
u_L, c_L, t_L	$\frac{1}{2}$	$+\frac{1}{2}$	$+\frac{1}{3}$	$+\frac{2}{3}$
d'_L, s'_L, b'_L	$\frac{1}{2}$	$-\frac{1}{2}$	$+\frac{1}{3}$	$-\frac{1}{3}$
e_R^-, μ_R^-, τ_R^-	0	0	-2	-1
u_R, c_R, t_R	0	0	$+\frac{4}{3}$	$+\frac{2}{3}$
d'_R, s'_R, b'_R	0	0	$-\frac{2}{3}$	$-\frac{1}{3}$

Observing the Gell—Mann–Nishijima relation (4.202), we find that corresponding to T_3 there is the current J_μ^3, there is electromagnetic current J_μ^{em} corresponding to the electric charge Q of charged lepton/quark and therefore corresponding to the weak hypercharge Y, we must have a weak hypercharge current J_μ^Y, which again according to eqn. (4.202) should be

$$J_\mu^Y = 2J_\mu^{em} - 2J_\mu^3 \tag{4.203}$$

The general form of electromagnetic current is

$$J_\mu^{em} = \overline{\psi}\gamma_\mu Q\psi \tag{4.204}$$

where Q is the charge operator. If we consider only the first generation of leptons then J_μ^{em} would be due to electrons only for which $Q = -e$, where e is the electric charge. Then for an electron

$$J_\mu^{em} = -e\overline{\psi}\gamma_\mu\psi \tag{4.205}$$

In order to simplify the discussions, we omit the coupling e in the above equation and write it as

$$J_\mu^{em} = -\overline{e}\gamma_\mu e \tag{4.206}$$

where e is not the charge but the spinor and $\overline{e}$ the corresponding adjoint spinor for the electron. Since, the handedness is important in weak interactions, we must write eqn. (4.206) in terms of left-handed and right-handed electrons. Accordingly,

$$\begin{aligned} J_\mu^{em} &= -(\overline{e}_L + \overline{e}_R)\gamma_\mu(e_L + e_R) \\ &= -\overline{e}_L\gamma_\mu e_L - \overline{e}_R\gamma_\mu e_R \end{aligned} \tag{4.207}$$

where the cross terms $\overline{e}_L\gamma_\mu e_R$ and $\overline{e}_R\gamma_\mu e_L$ vanish. For example

$$\begin{aligned}\overline{e}_L\gamma_\mu e_R &= e_L^\dagger\gamma^0\gamma_\mu e_R \\ &= \left[\left(\frac{1-\gamma^5}{2}\right)e\right]^\dagger \gamma^0\gamma_\mu\left(\frac{1+\gamma^5}{2}\right)e \\ &= e^\dagger\gamma^0\left(\frac{1+\gamma^5}{2}\right)\gamma_\mu\left(\frac{1+\gamma^5}{2}\right)e \\ &= \overline{e}\gamma_\mu\left(\frac{1-\gamma^5}{2}\right)\left(\frac{1+\gamma^5}{2}\right)e \\ &= 0 \end{aligned} \tag{4.208}$$

in which eqs. (4.79) and (4.80), the anti commutation relations of gamma matrices and

$$(1-\gamma^5)(1+\gamma^5) = 1-(\gamma^5)^2 = 0 \tag{4.209}$$

have been used.

Then substituting eqn. (4.199) for J_μ^3 and eqn. (4.207) for J_μ^{em} in eqn. (4.203), we get the weak hypercharge current as

$$J_\mu^Y = -2\overline{e}_R\gamma_\mu e_R - \overline{e}_L\gamma_\mu e_L - \overline{\nu}_L\gamma_\mu\nu_L \tag{4.210}$$

Since, weak isospin does not affect the right-handed components and the left-handed part of eqn. (4.210), namely

$$\overline{\chi}_L\gamma_\mu\chi_L = \overline{e}_L\gamma_\mu e_L + \overline{\nu}_L\gamma_\mu\nu_L \tag{4.211}$$

remains unchanged on interchanging $e_L \leftrightarrow \nu_L$, we may say that the form (4.210) of the weak hypercharge current is an invariant construct for the weak isospin. Using eqn. (4.211), we may write the weak hypercharge current J_μ^Y [eqn. (4.210)] for the first generation leptons as

$$J_\mu^Y = (-2)(\overline{e}_R\gamma_\mu e_R) + (-1)(\overline{\chi}_L\gamma_\mu\chi_L) \tag{4.212}$$

where the numbers in brackets (–2) and (–1) are the hypercharges of e_R and χ_L(ν_L and e_L) respectively. Picking the hypercharge values from the table, the J_μ^Y for say, the first generation of quark multiplets would be

$$(J_\mu^Y)_{u,d'} = \left(\tfrac{4}{3}\right)(\overline{u}_R\gamma_\mu u_R) + \left(-\tfrac{2}{3}\right)(\overline{d}'_R\gamma_\mu d'_R) + \left(+\tfrac{1}{3}\right)(\overline{u}_L\gamma_\mu u_L + \overline{d}'_L\gamma_\mu d'_L) \tag{4.213}$$

and similarly we can write the hypercharge current for other generations as well.

Finally, we would say that weak isospin and weak hypercharge can be defined corresponding to their strong interaction analogues and there are three weak isospin currents and a weak hypercharge current.

4.5.5 Charged Weak Quark Current and Suppressed Weak Decays

The weak isospin doublet of first generation quarks is

$$q_L = \begin{pmatrix} u \\ d' \end{pmatrix}_L = \left(\frac{1-\gamma^5}{2}\right)\begin{pmatrix} u \\ d\cos\theta_c + s\sin\theta_c \end{pmatrix} \tag{4.214}$$

where θ_c is the Cabibbo angle and d' the Cabibbo rotated quark. The charge raising weak quark current is then (see eqn. 4.197)

$$\begin{aligned} J_\mu^+ &= \overline{q}_L \gamma_\mu \tau^+ q_L \\ &= (\overline{u}, \overline{d}\cos\theta_c + \overline{s}\sin\theta_c)\left(\frac{1+\gamma_5}{2}\right)\gamma_\mu \begin{pmatrix} 0 & 1 \\ 0 & 0 \end{pmatrix}\left(\frac{1-\gamma_5}{2}\right)\begin{pmatrix} u \\ d\cos\theta_c + s\sin\theta_c \end{pmatrix} \\ &= (\overline{u}, \overline{d}\cos\theta_c + \overline{s}\sin\theta_c)\gamma_\mu \left(\frac{1-\gamma_5}{2}\right)^2 \begin{pmatrix} d\cos\theta_c + s\sin\theta_c \\ 0 \end{pmatrix} \\ &= \cos\theta_c[\overline{u}\gamma_\mu \tfrac{1}{2}(1-\gamma_5)d] + \sin\theta_c[\overline{u}\gamma_\mu \tfrac{1}{2}(1-\gamma_5)s] \end{aligned} \tag{4.215}$$

where $\gamma_5\gamma_\mu = -\gamma_\mu\gamma_5, \overline{u} = u^\dagger\gamma^0$ and eqn. (4.196a) have been used. The first term is the current between d and u both of which have zero strangeness ($\Delta S = 0$), and hence the strangeness is preserved in the first term. In the second term current being between a strange quark s and a non-strange quark u, $\Delta S = 1$, hence strangeness conservation is violated. Replacing τ^+ by τ^-, the charge lowering weak quark current $J_\mu^-(q)$ can also be obtained. We may then write

$$J_\mu^\pm(q) = \cos\theta_c[J_\mu^\pm(q)(\Delta S = 0)] + \sin\theta_c[J_\mu^\pm(q)(\Delta S = 1)] \tag{4.216}$$

where

$$\begin{aligned} J_\mu^+(q)(\Delta S = 0) &= \overline{u}\gamma_\mu \tfrac{1}{2}(1-\gamma_5)d \\ J_\mu^+(q)(|\Delta S| = 1) &= \overline{u}\gamma_\mu \tfrac{1}{2}(1-\gamma_5)s \end{aligned} \tag{4.217}$$

$$\begin{aligned} J_\mu^-(q)(|\Delta S| = 0) &= \overline{d}\gamma_\mu \tfrac{1}{2}(1-\gamma_5)u \\ J_\mu^-(q)(|\Delta S| = 1) &= \overline{s}\gamma_\mu \tfrac{1}{2}(1-\gamma_5)u \end{aligned} \tag{4.218}$$

Equation (4.216) is the quark analogue of eqn. (4.153). As already discussed under consequences of Cabibbo's theory, the $|\Delta S| = 1$, charged weak decays are suppressed. The explanation of such suppression comes through the Cabibbo's theory that the charged weak currents must be like eqn. (4.216), so that the decay rates of $\Delta S = 0$ and $|\Delta S| = 1$ weak decays are proportional to $G^2\cos^2\theta_c$ and $G^2\sin^2\theta_c$ respectively. If Cabibbo's theory were not applied, the above proportionality for both $\Delta S = 0$ and $|\Delta S| = 1$ weak decays would be G^2. This would mean, contrary to the observed facts, equal decay rates for both $\Delta S = 0$

and $|\Delta S| = 1$ weak decays. In practice, $|\Delta S| = 1$ weak decay rates are far too small than the $\Delta S = 0$ weak decay rates. G^2, with G as the weak coupling constant is the quantity, which pure leptonic decay rates are proportional to. Specifically,

$$\Gamma(\mu^- \to e^- + \bar{\nu}_e + \nu_\mu) \propto G^2 \text{ [purely leptonic]}$$

$$\Gamma(n \to p + e^- + \bar{\nu}_e) \propto G^2 \cos^2\theta_c \text{ [}\Delta S = 0\text{, semileptonic]}$$

$$\Gamma(\Lambda^0 \to p + e^- + \bar{\nu}_e) \propto G^2 \sin^2\theta_c \text{ [}\Delta S = 1\text{, semileptonic]}$$

where Γ denotes the decay rate. From the above,

$$\begin{aligned}\Gamma(\Lambda^0 \to pe^-\bar{\nu}_e) &\approx \tan^2\theta_c\, \Gamma(n \to pe^-\bar{\nu}_e) \\ &\approx 0.05\, \Gamma(n \to pe^-\bar{\nu}_e)\end{aligned} \tag{4.219}$$

where $\theta_c \approx 13°$. The relation (4.219), which tells that $|\Delta S| = 1$ (LHS) weak decay rates are about (1/20)th of the $\Delta S = 0$ (RHS) weak decay rates, agrees with the experimental observation.

The weak decays whose decay rates are proportional to $\cos^2\theta_c$ ($M_{fi} \propto \cos\theta_c$) are called the Cabibbo favored decays and those whose decay rates are proportional to $\sin^2\theta_c$ ($M_{fi} \propto \sin\theta_c$) are called Cabibbo suppressed weak decays.

4.5.6 GIM Mechanism and Charged Quarks

In the previous section, we considered the charged weak decays through the charged weak currents $J_\mu^\pm$ (quark) and applied Cabibbo's theory to explain—why strangeness changing ($|\Delta S| = 1$) charged weak decays are suppressed $\left(\frac{1}{20}\right)$ compared to $\Delta S = 0$ decays!

The Cabibbo's theory was very successful in explaining the mutual relationships between many decay rates. However, the application of Cabibbo's theory could not provide an answer to the very small decay rate of $K_L^0 \to \mu'\mu$. The solution of this problem was provided in 1970 by Glashow, Iliopoulos and Maiani (G I M) hence, known as GIM mechanism. To understand this mechanism, let us first explain the problem itself.

To explain the problem, we consider the weak decays of two types of kaons. K^+ and K_L^0. The former mostly (64 %) decay through

$$K^+ \to \mu^+ + \nu_\mu \tag{4.220}$$

whereas, among the various modes through which K_L^0, the long lived neutral kaon, decays, the mode

$$K_L^0 \to \mu^+ + \mu^- \tag{4.221}$$

has been studied well. The similarity between these decays of K^+ and K_L^0 is that both are strangeness changing ($\Delta S = 1$) weak decays, the difference between them, however, is that the $K^+ \rightarrow \mu^+\nu_\mu$ decay is via a charged weak current and the decay $K_L^0 \rightarrow \mu^+\mu^-$ is via the neutral weak current. The branching ratio for $K_L^0 \rightarrow \mu^+\mu^-$ decay mode is a negligible 7.37×10^{-7} percent in contrast to a (64 %) for $K^+ \rightarrow \mu^+\nu_\mu$ decay mode. We express this fact by saying that the decay $K_L^0 \rightarrow \mu^+\mu^-$ is quite suppressed as compared to $K^+ \rightarrow \mu^+\nu_\mu$. In the process of explaining such suppression, let us first consider the Cabibbo's theory applied to above decays.

The Feynman diagrams for the above decays have been drawn below [Fig. 4.12 (a) and b]. The quark content of K^+ meson is $u\bar{s}$ and that of K_L^0 meson is $d\bar{s}$.

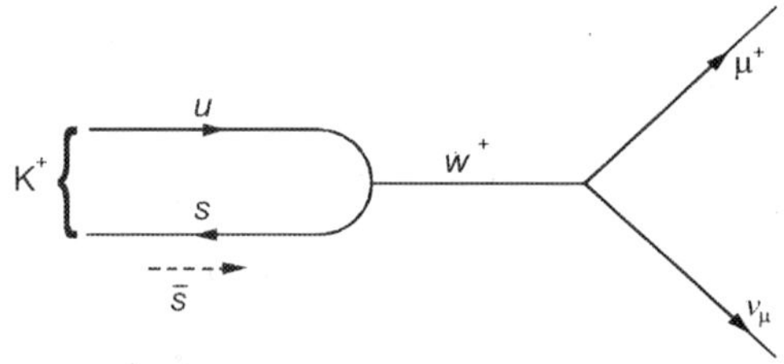

Fig. 4.12(a): The decay $K^+ \rightarrow \mu^+ \nu_\mu$. This is a charged current decay

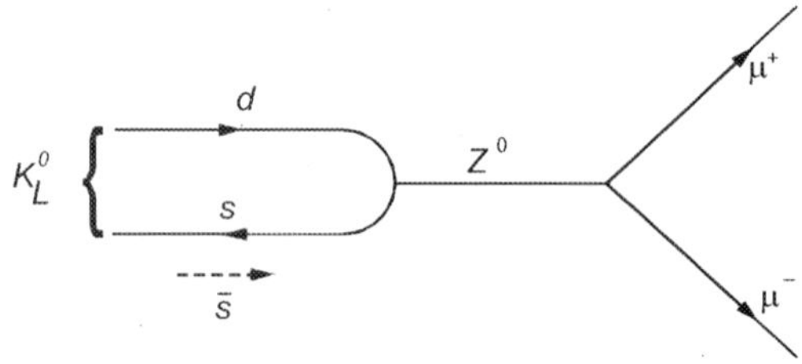

Fig. 4.12 (b): The decay $K_L^0 \rightarrow \mu^+\mu^-$. This is a neutral current decay and is highly suppressed

Though, the charmed quark was proposed in the GIM mechanism itself, we may consider here (in anticipation), the two generations of quarks, $\begin{pmatrix} u \\ d \end{pmatrix}$ and $\begin{pmatrix} c \\ s \end{pmatrix}$. Had there been no Cabibbo rotation required, the weak transitions would have been between $u \leftrightarrow d$ and $c \leftrightarrow s$, i.e., weak interactions would have been between the unrotated members of the same doublet. The decay K^+ tells that u interacts with $\bar{s}$, i.e., the first generation quark interacts with the (anti) quark of the second generation, this required the mixing of quarks of one generation with other and accordingly weak decays were not between u and d but between u and d', where $d' = d \cos \theta_c + s \sin \theta_c$ describes the generation mixing. Similarly, instead of having a weak transition between c and s, we have it between c and s' with $s' = -d \sin \theta_c + s \cos \theta_c$. Thus the decay of K^+ provides a cause for mixing the quark states.

Now, apply the above mixing, i.e., the Cabibbo theory to K_L^0 weak decay, which is mediated by a strangeness changing neutral current. Similar to eqn. (4.199), we may write for quarks, the neutral current

$$J_\mu^0(q) = \bar{q}_L \gamma_\mu \tau_3 q_L \tag{4.222}$$

which on using the eqn. (4.214) for q_L becomes

$$\begin{aligned} J_\mu^0(q) &= \begin{pmatrix} \bar{u} & \bar{d}' \end{pmatrix} \left(\frac{1+\gamma_5}{2} \right) \gamma_\mu \begin{pmatrix} 1 & 0 \\ 0 & -1 \end{pmatrix} \left(\frac{1-\gamma_5}{2} \right) \begin{pmatrix} u \\ d' \end{pmatrix} \\ &= \bar{u} \gamma_\mu \tfrac{1}{2}(1-\gamma_5) u - \bar{d}' \gamma_\mu \tfrac{1}{2}(1-\gamma_5) d' \end{aligned} \tag{4.223}$$

Substituting

$$d' = d \cos \theta_c + s \sin \theta_c \tag{4.224}$$

we may write for the above neutral current $J_\mu^0(q)$ as

$$\begin{aligned} J_\mu^0(q) \Rightarrow J^0(q) &= [\bar{u}u - \bar{d}d \cos^2 \theta_c - \bar{s}s \sin^2 \theta_c]_{\Delta S=0} \\ &\quad -[(\bar{d}s + \bar{s}d) \sin \theta_c \cos \theta_c]_{\Delta S=1} \end{aligned} \tag{4.225}$$

where for brevity the factor of $\gamma_\mu \frac{1}{2}(1-\gamma_5)$ has been dropped in between $\bar{u}$ and u, $\bar{d}$ and d, etc., for example

$$\bar{s}d \Rightarrow \bar{s} \gamma_\mu \tfrac{1}{2}(1-\gamma_5) d \tag{4.226}$$

Obviously, the first bracket of $J^0(q)$ represents the $\Delta S = 0$ neutral current and the second bracket represents the $\Delta S = 1$ neutral current.

Now, $K_L^0 \to \mu^+\mu^-$ is strangeness changing neutral current weak decay and hence it is natural to think that this decay is being mediated by the strangeness changing ($\Delta S = 1$) weak neutral current

$$J^0(q)\,(\Delta S = 1) \approx \sin \theta_c \cos \theta_c [\bar{d} \gamma_\mu (1-\gamma_5) s + \bar{s} \gamma_\mu (1-\gamma_5) d] \tag{4.227}$$

and hence the matrix element for the decay $K_L^0 \to \mu^+\mu^-$ should be proportional to $\sin \theta_c \cos \theta_c$. The calculated decay rate of $K_L^0 \to \mu^+\mu^-$ thus comes out to be far greater than the observed rate. The similar problem is faced when we deal with other strangeness changing neutral weak decays like $K^+ \to \pi^+ e^+ e^-$ and $K^+ \to \pi + \nu\bar{\nu}$.

In order to overcome the above problem, the three physicists Glashow, Iliopoulos and Maiani (GIM) proposed the existence of fourth quark, the charmed quark c and through this charmed quark showed that there should be no strangeness changing neutral current and hence no strangeness changing weak decays. This mechanism is called GIM mechanism.

According to GIM mechanism, while constructing the neutral current, not only the doublet $\begin{pmatrix} u \\ d' \end{pmatrix}$ but also the doublet $\begin{pmatrix} c \\ s' \end{pmatrix}$ must be included so that we have for $J^0_\mu(q)$, like the eqn. (4.223)

$$J^0_\mu(q) = \begin{pmatrix} \bar{u} & \bar{d}' \end{pmatrix}\left(\frac{1+\gamma_5}{2}\right)\gamma_\mu\begin{pmatrix} 1 & 0 \\ 0 & -1 \end{pmatrix}\left(\frac{1-\gamma_5}{2}\right)\begin{pmatrix} u \\ d' \end{pmatrix}$$
$$+\begin{pmatrix} \bar{c} & \bar{s}' \end{pmatrix}\left(\frac{1+\gamma_5}{2}\right)\gamma_\mu\begin{pmatrix} 1 & 0 \\ 0 & -1 \end{pmatrix}\left(\frac{1-\gamma_5}{2}\right)\begin{pmatrix} c \\ s' \end{pmatrix} \quad (4.228a)$$

which similar to equation (4.225) would become

$$\begin{aligned} J^0_\mu(q) &= [\bar{u}u + \bar{c}c - (\bar{d}d + \bar{s}s)\cos^2\theta_c - (\bar{s}s + \bar{d}d)\sin^2\theta_c] \\ &\quad +[(\bar{s}d + \bar{d}s - \bar{s}d - \bar{d}s)\sin\theta_c\cos\theta_c] \\ &= \bar{u}u + \bar{c}c - (\bar{d}d + \bar{s}s) \end{aligned} \quad (4.228b)$$

where

$$s' = -d\sin\theta_c + s\cos\theta_c \quad (4.229)$$

has been used. With the canceling of $\sin\theta_c \cos\theta_c$ term, the *strangeness changing neutral current has vanished* and remains only the $\Delta S = 0$ term in eqn. (4.228).

Thus we would say that the neutral weak current does not contain the Cabibbo angle θ_c, hence indicates that as far as neutral weak current is concerned there is no mixing of flavours (u, d, c, s). Only the strangeness preserving neutral quark currents $\bar{q}\gamma_\mu(1-\gamma_5)q$ with $q = (u, d, c, s)$ participate in such weak interactions.

The GIM mechanism can also be expressed by means of Feynman diagrams,

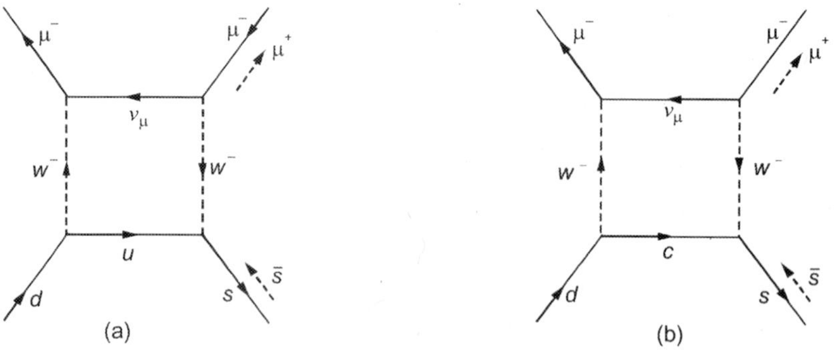

Fig. 4.13 (a, b): Diagrams for GIM mechanism

where the diagram (4.13a) shows the decay of $K_L^0 \to \mu^+ + \mu^-$ in which the virtual u quark crosses between d and $\bar{s}$ and the diagram (4.13b) replaces u quark of diagram (a) by virtual c quark. The diagram (4.13a) contributes a term proportional to $\sin \theta_c \cos \theta_c$ to the matrix element and diagram (b) contributes a term $(-\sin \theta_c \cos \theta_c)$ and thus canceling each other.

4.5.7 The Cabibbo—Kobayashi—Maskawa (CKM) Matrix

The ideas of Cabibbo and GIM can be generalized to more than four flavours (u, d, c, s) of quarks. We recall that $\begin{pmatrix} u \\ d' \end{pmatrix}$ form a weak isospin doublet and so do the $\begin{pmatrix} c \\ s' \end{pmatrix}$.

Let us imagine that there are N weak quark doublets denoted by

$$\begin{pmatrix} u_i \\ d'_i \end{pmatrix} \text{ with } i = 1, 2, 3 \ldots N \tag{4.230}$$

where, e.g.,

$$\begin{pmatrix} u_1 \\ d'_1 \end{pmatrix} \equiv \begin{pmatrix} u \\ d' \end{pmatrix} \equiv \begin{pmatrix} u \\ d \cos \theta_c + s \sin \theta_c \end{pmatrix} \tag{4.231a}$$

and

$$\begin{pmatrix} u_2 \\ d'_2 \end{pmatrix} \equiv \begin{pmatrix} c \\ s' \end{pmatrix} \equiv \begin{pmatrix} c \\ -d \sin \theta_c + s \cos \theta_c \end{pmatrix} \tag{4.231b}$$

The d'_i can be expressed by

$$d'_i = \sum_{j=1}^{N} U_{ij}\, d_j \tag{4.232}$$

where U_{ij} is a unitary $N \times N$ matrix. In total there are $2N$ quark states. Though the phases of all $2N$ quark states can be changed without altering the final observations and in turn we can remove all the phases, yet all the phases are not removed. The total removal would not leave the matrix U real. Thus there remain the essential phases, called residual phase factors, whose number is

$$\frac{1}{2}(N-1)(N-2) \tag{4.233}$$

implying that for $N = 3$, there will be one essential phase factor.

Now, we have three generations of quarks, thus $N = 3$, and each is a doublet. Using the mixing of d and s quark states [eqs. (4.224) and (4.229)], we may write

$$\begin{pmatrix} d' \\ s' \end{pmatrix} = \begin{pmatrix} \cos\theta_c & \sin\theta_c \\ -\sin\theta_c & \cos\theta_c \end{pmatrix} \begin{pmatrix} d \\ s \end{pmatrix} \tag{4.234}$$

where it is obvious that the transformation matrix

$$U = \begin{pmatrix} \cos\theta_c & \sin\theta_c \\ -\sin\theta_c & \cos\theta_c \end{pmatrix} \tag{4.235}$$

is unitary. Extending equations (4.234) to three generations of quarks, we have

$$\begin{pmatrix} d' \\ s' \\ b' \end{pmatrix} = \begin{pmatrix} U_{11} & U_{12} & U_{13} \\ U_{21} & U_{22} & U_{23} \\ U_{31} & U_{32} & U_{33} \end{pmatrix} \begin{pmatrix} d \\ s \\ b \end{pmatrix} \tag{4.236}$$

where the elements of CKM matrix are

$$\begin{aligned} & U_{11} = C_{12}C_{13}, \quad U_{12} = S_{12}C_{13}, \quad U_{13} = S_{13}\exp(-i\delta_{13}) \\ & U_{21} = -S_{12}C_{23} - C_{12}S_{23}S_{13} \ \exp(i\delta_{13}) \\ & U_{22} = C_{12}C_{23} - S_{12}S_{23}S_{13} \ \exp(i\delta_{13}) \\ & U_{23} = S_{23}C_{13} \\ & U_{31} = S_{12}S_{23} - C_{12}C_{23}S_{13} \ \exp(i\delta_{13}) \\ & U_{32} = -C_{12}S_{23} - S_{12}C_{23}S_{13} \ \exp(i\delta_{13}) \\ & U_{33} = C_{23}C_{13} \end{aligned} \tag{4.237}$$

where $C_{ij} = \cos\theta_{ij}$, and $S_{ij} = \sin\theta_{ij}$, $i, j = 1, 2, 3$ represent three generations. The angle θ_{12} is the Cabibbo angle θ_c. The phase δ_{13} is the essential phase. The magnitude of elements of U_{CKM} are determined by different experiments and in terms of magnitudes, we have

$$U_{CKM} = \begin{pmatrix} .975 & .22 & .005 \\ .22 & .97 & .045 \\ .01 & .045 & .999 \end{pmatrix} \tag{4.238}$$

It is obvious from eqs. (4.236) and (4.237) that when $\theta_{13} = \theta_{23} = 0 \Rightarrow$ absence of mixing with third generation,

$$U_{CKM} = \begin{pmatrix} \cos\theta_{12} & \sin\theta_{12} & 0 \\ -\sin\theta_{12} & \cos\theta_{12} & 0 \\ 0 & 0 & 1 \end{pmatrix} \tag{4.239}$$

which in view of eqn. (4.235) implies

$$|U_{CKM}| = |U| \tag{4.240}$$

Thus, when the third generation ($i = 3$) does not mix with other two generations

($i = 1, 2$) the CKM matrix recovers the Cabibbo—GIM picture. However, the decay of B^- meson whose quark content is $b\bar{u}$ with b for bottom (third generation) quark, does tell that there is some mixing with the third generation. Since, the GIM mechanism, in which the mixing of first two generations of quarks (d, s) is considered, is quite successful, the mixing of first two generations with the third generation of quarks must be small.

It can be observed from eqn. (4.238) that the CKM matrix is not a unit matrix. If it were a unit matrix, then d', s', b' would have been same as d, s and b, which would have implied that no cross-generational transitions could take place. The elements of CKM matrix can also be expressed in terms of couplings like U_{ud}, U_{us} etc., where U_{ud} measures the coupling u to d and U_{us} measures the coupling between u and s.

$$\begin{pmatrix} d' \\ s' \\ b' \end{pmatrix} = \begin{pmatrix} U_{ud} & U_{us} & U_{ub} \\ U_{cd} & U_{cs} & U_{cb} \\ U_{td} & U_{ts} & U_{tb} \end{pmatrix} \begin{pmatrix} d \\ s \\ b \end{pmatrix} \tag{4.241}$$

the non-zero value [eqn.(4.238)] of U_{us} permits the mixing of u and s and hence the occurrence of strangeness changing weak processes. Similar inferences can also be drawn regarding the mixing of other quark states.

4.6 ASSIGNMENTS

1. What are weak interactions? Estimate their strength and range. Giving examples mention their kinds. Define and explain the universality of weak interactions.
2. Mention the kinds of beta interaction. Develop Fermi's theory of beta decay and obtain the decay constant. How does it change when the Coulomb interaction between nucleons and electrons is incorporated?
3. Explain allowed and forbidden transitions. What are Fermi and Gamow—Teller transitions? Describe the beta decay processes allowed by these transitions.
4. Show that equation (4.48) satisfies equation (4.55). Check the same for equation (4.50).
5. Define helicity. Explain that the 'helicity' of a massive particle is frame dependent. Show that for a massless particle

$$\frac{1}{2}(1-\gamma^5)u(p) = \begin{cases} 0 \text{ if } u(p) \text{ carries helicity } +1 \\ u(p) \text{ if } u(p) \text{ carries helicity } -1. \end{cases}$$

Hence, explain that $\frac{1}{2}(1-\gamma^5)$ functions as projection operator. What change would you expect in the above results when $u(p) \rightarrow v(p)$, the antiparticle spinor?

6. Explain the two component theory of neutrinos.
7. Show that in Fermi's relativistic theory of beta decay, the demand to explain both Fermi and Gamow—Teller transitions leads to the V—A theory of weak interactions. Explain the reasons of choosing only vector minus axial vector current.
8. Describe the low energy V—A theory of beta decay. Explaining why it is called V—A theory, show that the matrix element of V—A theory differs from the Fermi's matrix element only by a factor of $(1-\gamma^5)$.
9. Show that the current current interaction of low energy V—A theory of beta decay can be universally applied to all weak interactions. Explain with reasons the modifications required in the V—A theory to include strange particles.
10. What are the difficulties, the pointlike theory of weak interactions faces at higher energies? How these difficulties are handled? Discuss the consequences of the modifications introduced.
11. Explain Cabibbo's theory of weak interactions. Elaborate on its consequences.
12. What are strangeness changing weak interactions? Explain the suppression of hadronic $\Delta S = 1$ weak decays. How the universality is restored in such decays?
13. Write notes on
    (a) Cabibbo angle.
    (b) Quark—lepton universality.
    (c) High energy catastrophe in relation to σ.
14. Explain charged and neutral weak currents. Derive equations (4.197) and (4.199). Discuss the physical significance of these equations.
15. Define weak isospin and weak hypercharge. Show that $J_\mu^Y = -2\bar{e}_R\gamma_\mu e_R - \bar{\chi}_L\gamma_\mu\chi_L$ and $J_\mu^{em} = -\bar{e}_L\gamma_\mu e_L - \bar{e}_R\gamma_\mu e_R$. Describe the physical content in these equations.
16. Describe the mixing of quark states and explain Cabibbo favoured and Cabibbo suppressed weak decays.
17. What kind of problem the strangeness changing weak decays like

    $K_L^0 \to \mu^+\mu^-$

    $K^+ \to \pi^+ e^+ e^-$

    $K^- \to \pi^+ \nu\,\bar{\nu}$

    face? Explain how the GIM mechanism solves these problems. Discuss the significance of this mechanism.
18. Describe the CKM matrix and explain its significance. Discuss the conditions under which the CKM matrix would recover the Cabibbo—GIM picture? Why the CKM matrix is not a unit matrix? Find the condition to make CKM matrix unitary.

19. Show that for the process $\nu_\mu + e^- \rightarrow \mu^- + \nu_e$, the spin averaged amplitude is

$$\sum_{\text{spin}} |M_{fi}|^2 = \frac{1}{2}\left(\frac{g_W}{M_w}\right)^2$$

$$\left[(1-\eta^2)^2 (p_1 \cdot p_4)(p_2 \cdot p_3) + (1 + 6\eta^2 + \eta^4)(p_1 \cdot p_2)(p_3 \cdot p_4)\right]$$

where the coupling has been assumed as $\gamma^\mu(1 + \eta\gamma^5)$.

4.7 OBJECTIVE QUESTIONS

1. In the following reactions (1) and (2),

   (1) $(A, Z) \rightarrow (A, Z-1)$ + emitted particles

   (2) $(A, Z) \rightarrow (A, Z+1)$ + emitted particles

   the emitted particles are

   (a) $e^+ + \nu_e$ in reaction (1) and $e^- + \overline{\nu}_e$ in reaction (2).

   (b) $e^- + \overline{\nu}_e$ in reaction (1) and $e^+ + \nu_e$ in reaction (2).

   (c) $e^+ + \overline{\nu}_e$ in reaction (1) and $e^- + \nu_e$ in reaction (2).

   (d) $e^- + \nu_e$ in reaction (1) and $e^+ + \overline{\nu}_e$ in reaction (2).

2. The parity violation in weak processes is incorporated by adding a term to the Hamiltonian density. The term has the form

   (a) $(\psi_p \Gamma^i \psi_n)(\overline{\psi}_e \Gamma_i \psi_\nu)$

   (b) $(\overline{\psi}_p \Gamma^i \gamma_5 \psi_n)(\overline{\psi}_e \Gamma_i \psi_\nu)$

   (c) $(\overline{\psi}_p \Gamma^i \gamma_5 \psi_n)(\overline{\psi}_e \Gamma_i \gamma^5 \psi_\nu)$

   (d) $\overline{\psi}_p \gamma^\mu \psi_p G_\mu$

3. Weak interactions are

   (a) same for different handednesses of particles.

   (b) same for different handednesses provided weak interactions are purely leptonic.

   (c) same for different handednesses provided weak interactions are non-leptonic.

   (d) not the same for different handednesses of particles.

4. The effective Lagrangian density that can describe the weak interactions is

   $$£_{eff}(x) = -\frac{G_F}{\sqrt{2}} J_\lambda^\dagger(x)\, J^\lambda(x), \text{ where}$$

   (a) $G_F = 10^5/m^2 p$ and J_λ is the weak current having vector minus axial vector form.

(b) $G_F = 10^{-5}/m^2p$ and J_λ is the weak current having vector minus axial vector form.
(c) $G_F = 10^{-5}/m^2p$ and J_λ is the weak current having vector multiplied by axial vector form.
(d) $G_F = 10^{5}/m^2p$ and J_λ is the weak current having vector plus axial vector form.

5. What is true?
(a) Allowed transitions are those having a small probability of occurrence and forbidden transitions are those having a zero probability of their occurrence.
(b) Allowed transitions are those having a large probability of their occurrence and forbidden transitions are those having no probability of their occurrence.
(c) Allowed transitions are those having a large probability of their occurrence and forbidden transitions are those having smaller probabilities of their occurrence.
(d) Allowed transitions are those which always take place and forbidden are those which never take place.

6. Select the correct statement among the following. A beta decay which
(a) takes place through a vector coupling is called the Gamow—Teller transition and a beta decay that takes place through an axial vector coupling is called Fermi transition.
(b) takes place through a vector coupling is called Fermi transition and a beta decay that takes place through an axial vector coupling is called Gamow—Teller transition.
(c) takes place through a vector minus axial vector coupling is called Fermi transition and a beta decay that takes place through a vector plus axial vector coupling is called Gamow—Teller transition.
(d) takes place through a vector plus axial vector coupling is called Fermi transition and a beta decay that takes place through a vector minus axial vector coupling is called Gamow—Teller transition.

7. Following expressions describe two kinds of matrix elements (1) and (2). Select the correct answer from the choices given.

1. $$|M_{fi}| = \frac{G}{V} C_V \left| \langle uud | \sum_{i=1}^{3} T_{i+} | udd \rangle \right|$$

2. $$|M_{fi}| = \frac{G}{V} C_A \left| \langle uud | \sum_{i=1}^{3} T_{i+} \boldsymbol{\sigma} \, | udd \rangle \right|$$

(a) (1) denotes the matrix element for Fermi decay and (2) the matrix element for Gamow—Teller decay with $C_V = C_A$.
(b) (1) denotes the matrix element for Fermi decay and (2) the matrix element for Gamow—Teller decay with $C_V \neq C_A$.

(c) (1) denotes the matrix element for Gamow—Teller decay and (2) the matrix element for Fermi decay with $C_V = C_A$.
(d) (1) denotes the matrix element for Gamow—Teller decay and (2) the matrix element for Fermi decay with $C_V \neq C_A$.

8. The expression $\int_{m_e c^2}^{E_0} \frac{dW}{dE_e} dE_e$ denotes the total probability per unit time.

If τ denotes the lifetime of the decaying (weakly) particle, then the above expression is equal to

(a) τ (b) 2τ

(c) $\frac{1}{\tau}$ (d) $\frac{2}{\tau}$

9. The Sargent's rule, which tells that the lifetime of a weakly decaying particle decreases as the fifth power of E_0 is expressed as

(a) $\tau \approx \frac{1}{\hbar^7 c^6}(g_V^2 + 3g_A^2)\frac{E_0^5}{60\pi^3}$

(b) $\frac{1}{\tau} \approx \frac{1}{\hbar^7 c^6}(g_V^2 + 3g_A^2)\frac{E_0^5}{60\pi^3}$

(c) $\frac{1}{\tau^2} \approx \frac{1}{\hbar^7 c^6}(g_V^2 + 3g_A^2)\frac{E_0^5}{60\pi^3}$

(d) $\tau^2 \approx \frac{1}{\hbar^7 c^6}(g_V^2 + 3g_A^2)\frac{E_0^5}{60\pi^3}$

10. Given below are three beta decays. Select the correct answer from the choices.
1. $C^{10} \rightarrow B^{10*} + e^+ + \nu_e$
2. $He^6 \rightarrow Li^6 + e^- + \overline{\nu}_e$
3. $Cl^{34} \rightarrow S^{34*} + e^+ + \nu_e$

(a) All the three are Fermi transitions.
(b) All the three are Gamow—Teller transitions.
(c) Decays (1) and (2) are Fermi transitions and (3) is Gamow—Teller.
(d) Decays (1) and (3) are Fermi transitions and (2) is Gamow—Teller.

11. The beta spectrum plotted with $K(E_e) = \sqrt{\dfrac{\frac{dN(E_e)}{dE_e}}{F(Z', E_e)E_e\sqrt{E_e^2 - m_e^2 c^4}}}$ as ordinate and the electron energy E_e as abscissa is known as

(a) Fermi plot (b) Sargent plot
(c) Kurie plot (d) Teller plot

12. If the plot of question (11) is a straight line and cuts the abscissa at the maximal energy E_0, it implies that
    (a) the beta decay process did not take place.
    (b) the beta decay process did take place but no neutrino was emitted.
    (c) the emitted neutrino of the beta decay process is massless.
    (d) the emitted neutrino of the beta decay process has a nonzero mass.
13. If the plot of beta spectrum (question 11 above), is a straight line that bends near the maximum energy (towards the origin), this implies that
    (a) no neutrino was emitted in the beta decay process.
    (b) the neutrino was emitted but its mass was not zero.
    (c) the neutrino was emitted and it was massless.
    (d) no electron was emitted, only neutrinos were emitted.
14. An experiment in the beta decay process uses an isomer of ${}_{63}Eu^{152}$ as a source from which the photons are scattered into Sm_2O_3 ring and detected in a NaI scintillation detector. The experiment is called
    (a) Goldhaber experiment to measure the helicity of neutrino from weak nuclear decays.
    (b) Frauenfelder experiment to measure the electron positron helicity.
    (c) Wu and Albert experiment to measure the momenta of electrons and positrons.
    (d) Hayward and Hudson experiment to determine the gamma anisotropy and beta asymmetry.
15. If $U\begin{pmatrix}0 & 1\\ 1 & 0\end{pmatrix}U^\dagger = \begin{pmatrix}1 & 0\\ 0 & -1\end{pmatrix}$, then U is

    (a) $\frac{1}{\sqrt{2}}\begin{pmatrix}2 & 1\\ 1 & 1\end{pmatrix}$ (b) $\frac{1}{\sqrt{2}}\begin{pmatrix}1 & 1\\ 1 & -1\end{pmatrix}$

    (c) $\frac{1}{\sqrt{2}}\begin{pmatrix}1 & 2\\ 1 & 2\end{pmatrix}$ (d) $\frac{1}{\sqrt{2}}\begin{pmatrix}2 & 1\\ -1 & 1\end{pmatrix}$
16. What is true about the following four fields?

    1. $\frac{1}{2}(1-\gamma_5)\psi$

    2. $\frac{1}{2}(1+\gamma_5)\psi$

    3. $\frac{1}{2}\gamma_5\psi$

    4. $\frac{1}{2}\gamma_0\gamma_5\psi$

(a) (1) is neutrino field and (3) is antineutrino field.
(b) (2) is antineutrino field and (4) is neutrino field.
(c) (1) is neutrino field and (2) is antineutrino field.
(d) (3) is neutrino field and (4) is antineutrino field.

17. The two component theory of neutrinos tells that
    (a) four-two component states are required to span the space of left-handed particles.
    (b) only two-four component states are required to span the space of left-handed particles.
    (c) only two-two component states are required to span the space of left-handed particles.
    (d) only one-two component state is sufficient to span the space of left-handed particles.
18. What is true about the Fermi constant?
    (a) It is related to the square of weak charge and has a value $\approx 9 \times 10^{-50}$ MeV m^3.
    (b) It is related to the square of electric charge and has a value $\approx \frac{1}{137}$.
    (c) It is related to the weak charge and has a value $\approx 9 \times 10^{-5}$ MeV m^3.
    (d) It is related to the square of colour charge and has a value $\approx 9 \times 10^{-20}$ MeV m^3.
19. The muon decay is caused by the interaction of
    (a) two weak hadron currents.
    (b) one weak hadron current and one lepton current.
    (c) two lepton currents.
    (d) a lepton current and an electromagnetic current.
20. The general weak interaction Hamiltonian for non-strange particles has the form
    (a) $\frac{G}{\sqrt{2}}\left[J_e^{\mu} J_{\mu}^{e\dagger} + J_h^{\mu} J_{\mu}^{h\dagger}\right]$
    (b) $\frac{G_1}{\sqrt{2}} J_e^{\mu} J_{\mu}^{e\dagger} + \frac{G_2}{\sqrt{2}} J_h^{\mu} J_{\mu}^{h\dagger}$
    (c) $\frac{G}{\sqrt{2}}\left[J_e^{\mu} J_{\mu}^{e\dagger} + J_e^{\mu} J_{\mu}^{h\dagger} + J_e^{\mu\dagger} J_{\mu}^{h} + J_h^{\mu} J_{\mu}^{h\dagger}\right]$
    (d) $\frac{G_1}{\sqrt{2}} J_e^{\mu} J_{\mu}^{e\dagger} + \frac{G_2}{\sqrt{2}} J_e^{\mu} J_{\mu}^{h\dagger} + \frac{G_3}{\sqrt{2}} J_e^{\mu\dagger} J_{\mu}^{h} + \frac{G_4}{\sqrt{2}} J_h^{\mu} J_{\mu}^{h\dagger}$
21. When weak interactions of strange particles are included with those of nonstrange particles
    (a) there is no change in leptonic and hadronic currents.

(b) the hadronic current is modified.
(c) the leptonic current is modified.
(d) both leptonic and hadronic currents are modified.

22. For the weak interactions of strange particles, the hadronic current is

(a) $(J_\mu^h)_{\text{strange}} = (J_\mu^h)_{\text{non-strange}}$

(b) $(J_\mu^h)_{\text{strange}} = (\cos\theta_c + \sin\theta_c)(J_\mu^h)_{\text{non-strange}}$

(c) $(J_\mu^h)_{\text{strange}} = \cos\theta_c (J_\mu^h)_{\Delta S=0} + \sin\theta_c (J_\mu^h)_{\Delta S=1}$

(d) $(J_\mu^h)_{\text{strange}} = \sin\theta_c (J_\mu^h)_{\Delta S=0} + \cos\theta_c (J_\mu^h)_{\Delta S=1}$

23. The universality of weak interactions implies that
(a) in all kinds of weak interactions neutrino must be emitted.
(b) in all kinds of weak interactions a lepton must be emitted.
(c) in all kinds of weak interactions a hadron must participate.
(d) all kinds of weak interactions must have the same coupling constant.

24. Between the $\Delta S = 0$ and $\Delta S = 1$ weak decays the decay rate of
(a) $\Delta S = 0$ weak decays is equal to the decay rate of $\Delta S = 1$ weak decays.
(b) $\Delta S = 0$ weak decays is 20 times slower than the decay rate of $\Delta S = 1$ weak decays.
(c) $\Delta S = 0$ weak decays is 20 times faster than the decay rate of $\Delta S = 1$ weak decays.
(d) $\Delta S = 0$ weak decays is 200 times faster than the decay rate of $\Delta S = 1$ weak decays.

25. If G_β be the coupling constant for beta decay and G_μ for the muon decay, what is true among the following?
(a) G_β is exactly equal to G_μ.
(b) G_β is approximately equal to G_μ. The exact relation is $G_\beta = G_\mu \cos\theta_c$ with $\theta_c \approx 13°$.
(c) G_β is approximately equal to G_μ. The exact relation is $G_\beta = G_\mu \cos\theta_c$ with $\theta_c \approx 23°$.
(d) G_β is not equal to G_μ. The exact relation is $G_\beta = 3G_\mu \tan\theta_c$ with $\theta_c \approx 13°$.

26. In weak interactions mixing of states takes place. The mixing is
(a) between the leptons of same generation.
(b) between the quarks of same generation.
(c) between the leptons of one generation and the other.
(d) between the quarks of one generation and the other.

27. What among the given choices is correct for the following relations?

$\nu'_e = \nu_e \cos\theta + \nu_\mu \sin\theta$

$\nu'_\mu = -\nu_e \sin\theta + \nu_\mu \cos\theta$

(a) Since ν'_e is different from ν_e and ν_μ, the above Cabibbo relations will exist.

(b) Since ν'_e and ν'_μ too are massless, there is no way to distinguish the Cabibbo rotated neutrinos from the unrotated ones. Above relations should not exist.
(c) Since cross generational mixing is not allowed in weak interactions, these relations should not exist.
(d) In weak interactions, left handed particles participate. Neutrinos being left handed, above relations should exist.

28. In weak interactions, the quarks paired with up (u) and charm (c) quarks are
(a) d and s respectively.
(b) $d \cos \theta_c$ and $s \sin \theta_c$ respectively.
(c) $-d \sin \theta_c$ and $s \cos \theta_c$ respectively.
(d) $d \cos \theta_c + s \sin \theta_c$ and $-d \sin \theta_c + s \cos \theta_c$ respectively.

29. What is true among the following?
(a) The carriers of charged weak currents are $\pi^\pm$ and π^0 are the carriers of neutral weak currents.
(b) $\pi^\pm$ are the carriers of charged weak currents and K^0 are the carriers of neutral weak currents.
(c) $W^\pm$ are the carriers of charged weak currents and π^0 are the carriers of neutral weak currents.
(d) $W^\pm$ are the carriers of charged weak currents and Z^0 are the carriers of neutral weak currents.

30. In case of leptons, the coupling to $W^\pm$ takes place stricitly
(a) within a particular generation.
(b) between the upper members of separate generations.
(c) between the lower members of separate generations.
(d) between the upper member of one generation and lower member of other.

31. Below are given three currents.

1. $\bar{\nu}_L \gamma_\mu e_L$ 2. $\bar{e}_L \gamma_\mu \nu_L$ 3. $-\bar{e}_L \gamma_\mu e_L - \bar{e}_R \gamma_\mu e_R$

Choose the correct answer from the following.
(a) (1) and (2) are charged currents and (3) is neutral.
(b) (1) is J^+_μ, (2) is J^-_μ and (3) is J^{em}_μ.
(c) Both (a) and (b) are correct.
(d) (1) and (2) are neutral currents and (3) is the charged current.

32. For the following table

Particles	Quantum numbers			
	α	β	γ	δ
ν_e	$\frac{1}{2}$	$\frac{1}{2}$	0	-1
e_L^-	$\frac{1}{2}$	$-\frac{1}{2}$	-1	-1
u_L	$\frac{1}{2}$	$\frac{1}{2}$	$\frac{2}{3}$	$\frac{1}{3}$
d_L	$\frac{1}{2}$	$-\frac{1}{2}$	$-\frac{1}{3}$	$\frac{1}{3}$

choose the correct answer from the alternatives given below.

(a) α is weak isospin T, β is weak T_3, γ is charge Q and δ is weak hypercharge Y.

(b) α is strong isospin T, β is strong T_3, γ is charge Q and δ is strong hypercharge Y.

(c) α is charge Q, β is weak T, γ is weak T_3, and δ is weak hypercharge Y.

(d) α is charge Q, β is strong T, γ is strong T_3 and δ is strong hypercharge Y.

33. In the following two columns Γs denote the decay rates of the weak interactions, G denotes the weak coupling constant and θ_c, the Cabibbo angle. Quantities in Column I are either proportional or equal to the quantities in Column II. Match them and select the correct code.

Column I	*Column II*
A Γ $(\mu^- \to e^- \bar{\nu}_e \nu_\mu)$	1 $(1/20)\Gamma(n \to p e^- \bar{\nu}_e)$
B Γ $(n \to p\ e^- \bar{\nu}_e)$	2 $G^2 \sin^2 \theta_c$
C Γ $(\Lambda^0 \to p\ e^- \bar{\nu}_e)$	3 $G^2 \cos^2 \theta_c$
D Γ $(\Lambda^0 \to p\ e^- \bar{\nu}_e)$	4 G^2

Codes:

	A	B	C	D
(a)	4	2	3	1
(b)	4	3	2	1
(c)	3	4	1	2
(d)	2	1	3	4

34. The GIM mechanism was required to explain the
    (a) large decay rate of $K^+ \to \mu^+ + \nu_\mu$.
    (b) small decay rates of $K_L^0 \to \mu^+ + \mu^-$.
    (c) small decay rates of $\Lambda^0 \to p + e^- + \overline{\nu}_e$.
    (d) decay rate of $K^+ \to \pi^+ + \pi^-$.
35. The current that became the cause for the introduction of GIM mechanism was
    (a) electromagnetic current.
    (b) charged weak current.
    (c) strangeness preserving neutral weak current.
    (d) strangeness changing neutral weak current.
36. In GIM mechanism
    (a) the introduction of charm quark cancels the problem creating strangeness changing neutral current.
    (b) the use of strange quark cancels the strangeness changing neutral current.
    (c) the use of first generation quarks cancels the unwanted current.
    (d) the use of leptons cancels the unwanted current

ANSWERS

1a, 2b, 3d, 4b, 5c, 6b, 7b, 8c, 9b, 10d, 11c, 12c, 13b, 14a, 15b, 16c, 17c, 18a, 19c, 20c, 21b, 22c, 23d, 24c, 25b, 26d, 27b, 28d, 29d, 30a, 31c, 32a, 33b, 34b, 35d, 36a.

5

GAUGE SYMMETRIES

5.1 INTRODUCTION

A physical system is said to be symmetric, if it is invariant under a certain transformation. The invariance of a system under certain symmetry transformation implies a conservation law. In classical physics, we have the following symmetry transformations and corresponding conservation laws.

Symmetry Transformation	Conservation Law
Translational invariance in time	Energy conservation
$t \to t + a$	$\frac{dE}{dt} = 0$
Translational invariance in space	Momentum conservation
$r^i \to r^i + b^i$	$\frac{dp^i}{dt} = 0$
Rotational invariance	Angular momentum conservation
$r^i \to R^{ij} r_j$ with $RR^T = 1$	$\frac{dJ^i}{dt} = 0$

In quantum mechanics, however, the time development is described by the Heisenberg equation of motion

$$\frac{d\Omega}{dt} = i[\mathrm{H}, \Omega] \tag{5.1}$$

where, with operator Ω is associated a physical observable. The conservation of Ω is then expressed by

$$0[\Omega, \mathrm{H}] = 0 \tag{5.2}$$

In field theory, the symmetries and conservation laws are related by Noether's theorem in which the system described by a Lagrangian is invariant under the phase transformation, which does not depend on space–time. It is already known that the field Lagrangian is expressed as

$$L = \int £ \, d^3\boldsymbol{x} \tag{5.3}$$

where the Lagrangian density is

$$£ = £ (\psi, \partial_\mu \psi) \tag{5.4}$$

with ψ denoting the field and $\partial_\mu\psi$ its four-gradient. The space-time independent phase transformation is denoted by

$$\psi \to \psi' = e^{\pm i\alpha}\psi \tag{5.5}$$

where α is a real constant. Above phase transformation is also called a gauge transformation and since it is same for *every* space–time point, we call the transformation (5.5), the *global* gauge transformation. If the parameter α is not same for every space–time point, meaning that it *depends on the space and time*, then it is denoted as $\alpha(x)$, where $x = x^\mu$, we have in place of eqn. (5.5)

$$\psi \to \psi' = e^{\pm i\alpha(x)}\psi \tag{5.6}$$

The transformations of the kind (5.6) are called the *local* gauge transformations.

The invariance of the Lagrangian under the global gauge transformations (5.5) gives rise to global gauge symmetries and the invariance of Lagrangian under local gauge transformations (5.6), to local gauge symmetries. Thus, we have two kinds of gauge symmetries, the global and the local.

5.2 GLOBAL GAUGE SYMMETRY AND NOETHER'S THEOREM

We consider a general Lagrangian density £, which as usual depends on fields ψ and their four-gradients $\partial_\mu\psi$, i.e.,

$$£ = £ (\psi, \partial_\mu \psi) \tag{5.7}$$

For example, in the case of free Dirac field, the Lagrangian density is

$$£_{\text{Dirac}} = i\bar{\psi}\gamma^\mu\partial_\mu\psi - m\bar{\psi}\psi \tag{5.8}$$

where, we see that both ψ and $\partial_\mu\psi$ are present. Of course, the adjoint of y, i.e., = ψ, i.e., $\bar{\psi} = \psi^\dagger\gamma^0$ also appears here.

Now, introduce the global gauge transformation

$$\psi \to \psi' = e^{+i\alpha}\psi(x) \tag{5.9}$$

where according to the requirement of global gauge transformations α does not depend on space-time, $x = x^\mu$. The family of global gauge transformations may be denoted by

$$U(\alpha) = e^{i\alpha} \tag{5.10}$$

Since, there is only one parameter α which may continuously vary over real numbers, the transformations $U(\alpha)$ may be seen to obey the properties of a group viz., closure, identity, inverse and associativity described in Chapter 3. For example, if we give the values α_1 and α_2, which are real numbers, to α, we shall have following two global gauge transformations, which will be the elements of the group formed by the global gauge transformations

$$U(\alpha_1) = e^{i\alpha_1}$$

and

$$U(\alpha_2) = e^{i\alpha_2} \tag{5.11}$$

Obviously, elements (5.11) commute, i.e,

$$[U(\alpha_1), U(\alpha_2)] = 0 \tag{5.12}$$

which tells that the group of global gauge transformations is an Abelian group. Because there is a single parameter to vary, the group, whose elements are formed by giving various continuous real number values to α of eqn. (5.10), is a $U(1)$ group. Accordingly, we shall call the transformations (5.10) as $U(1)$ global gauge transformations.

Taking the gradient (∂_μ) of eqn. (5.9), we see that

$$\partial_\mu \psi' = \partial_\mu (e^{i\alpha}\psi) = e^{i\alpha}(\partial_\mu \psi) \tag{5.13}$$

which tells that both ψ and $\partial_\mu \psi$ transform identically. This means that just as the new ψ i.e., ψ' (eqn. 5.9) is equal to $e^{i\alpha}$ times the old ψ, the gradient of ψ' is also equal to $e^{i\alpha}$ times the gradient of ψ. If we use the Lagrangian density (5.8), we would also need

$$\bar{\psi}' = (\psi')^\dagger \gamma^0 = (e^{i\alpha}\psi)^\dagger \gamma^0 = e^{-i\alpha}\bar{\psi} \tag{5.14}$$

Substituting equations (5.9), (5.13) and (5.14) in eqn. (5.8), we may see that

$$\begin{aligned} £'_{\text{Dirac}} &= i\bar{\psi}'\gamma^\mu \partial_\mu \psi' - m\bar{\psi}'\psi' \\ &= i(e^{-i\alpha})\bar{\psi}\gamma^\mu (e^{i\alpha})\partial_\mu \psi - m(e^{-i\alpha})\bar{\psi}(e^{i\alpha})\psi \\ &= i\bar{\psi}\gamma^\mu \partial_\mu \psi - m\bar{\psi}\psi = £_{\text{Dirac}} \end{aligned} \tag{5.15}$$

Above equation implies that the Lagrangian density (5.8) is invariant under the global gauge transformations (5.9).

The transformations (5.10) are $U(1)$ global gauge transformations forming a one dimensional Lie group $U(1)$. Since,

$$e^{i\alpha} = 1 + i\alpha + \frac{(i\alpha)^2}{2!} + \ldots \tag{5.16}$$

the infinitesimal form of global gauge transformations would be

$$U(\alpha)_{\text{infinitesimal}} \approx 1 + i\alpha \tag{5.17}$$

and for reaching at the Noether's theorem, it is sufficient that we consider the invariance of £ under infinitesimal $U(1)$ transformations. The Lie group too, being a continuous group, requires that we consider infinitesimal transformations.

We shall then see that the Lagrangian density (5.8) does not change when we carry out infinitesimal global gauge transformations

$$\psi \rightarrow \psi' = (1 + i\alpha)\psi \tag{5.18}$$

implying that we change ψ by $\partial\psi$, where

$$\delta\psi = \psi' - \psi = i\alpha\psi \tag{5.19}$$

and consequently, the four gradient $\partial_\mu \psi$ by $\delta(\partial_\mu \psi)$, where from eqn. (5.19)

$$\delta(\partial_\mu \psi) = \partial_\mu(\delta\psi) = i\alpha\partial_\mu \psi \tag{5.20}$$

The equality $\delta(\partial_\mu \psi) = \partial_\mu (\delta\psi)$ will be explained a little below. The invariance of £ due to the changes in ψ and $\partial_\mu \psi$ by $\delta\psi$ and $\delta(\partial_\mu \psi)$ respectively implies that the change in £ must be zero. That is

$$\delta £ = 0 \tag{5.21}$$

Now,

$$\begin{aligned} \delta £ &= \delta £(\psi, \partial_\mu \psi) \\ &= \frac{\partial £}{\partial \psi}\delta\psi + \frac{\partial £}{\partial(\partial_\mu \psi)}\delta(\partial_\mu \psi) \\ &= i\alpha[\frac{\partial £}{\partial \psi}\psi + \frac{\partial £}{\partial(\partial_\mu \psi)}\partial_\mu \psi] \end{aligned} \tag{5.22}$$

where eqs. (5.19) and (5.20) have been used. Since in the variation $\delta £$ [eqn. (5.22)], only ψ and $\partial_\mu \psi$ change, x does not change, we have taken

$$\delta(\partial_\mu \psi) = \partial_\mu (\delta\psi) \tag{5.23}$$

Using $\delta £ = 0$, eqn. (5.22) becomes

$$i[\frac{\partial £}{\partial \psi}\psi + \frac{\partial £}{\partial(\partial_\mu \psi)}\partial_\mu \psi] = 0 \tag{5.24}$$

where we can use

$$\partial_\mu \left(\frac{\partial £}{\partial(\partial_\mu \psi)}\psi \right) = \left[\partial_\mu \left(\frac{\partial £}{\partial(\partial_\mu \psi)} \right) \right] \psi + \frac{\partial £}{\partial(\partial_\mu \psi)}\partial_\mu \psi \tag{5.25}$$

to replace the second term and to write eqn. (5.24) as

$$i\left(\frac{\partial £}{\partial \psi} - \partial_\mu \frac{\partial £}{\partial(\partial_\mu \psi)} \right)\psi + i\partial_\mu \left(\frac{\partial £}{\partial(\partial_\mu \psi)}\psi \right) = 0 \tag{5.26}$$

The first bracket of eqn. (5.26) vanishes because of Euler–Lagrange equation [eqn. (B2)] leaving

$$\partial_\mu \left(ie\frac{\partial £}{\partial(\partial_\mu \psi)}\psi \right) = 0 \tag{5.27}$$

where, anticipating the use of the charge of electron equal to $-e$, we have multiplied the expression be e. Instead of multiplying by e here, we could also take $(ie\alpha)$ in place of $(i\alpha)$ in the transformation (5.9). Eqn. (5.27) can be written as

$$\partial_\mu j^\mu = 0 \tag{5.28}$$

where

$$j^\mu = ie\frac{\partial £}{\partial(\partial_\mu \psi)}\psi \tag{5.29}$$

From the Lagrangian density (5.8), we have

$$\frac{\partial \pounds}{\partial(\partial_\mu \psi)} = i\bar{\psi}\gamma^\mu \tag{5.30}$$

and consequently

$$j^\mu = -e\bar{\psi}\gamma^\mu\psi \tag{5.31}$$

It may now be observed that the eqn. (5.28), which is a conservation law or the continuity equation, is a consequence of invariance ($\delta\pounds = 0$) of Lagrangian density under the infinitesimal global gauge transformation (5.18). Thus $U(1)$ global gauge invariance has led to a conservation law [eqn. (5.28)]. Such a connection between symmetries (invariances) and conservation laws in field theories is called the Noether's theorem.

We may find out the quantity that is conserved in the above conservation law (5.28). The presence of $(-e)$ in eqn. (5.31) matches the four-current density j^μ with that of electromagnetic current density. From eqn. (5.31), we may write

$$j^0 = -e\bar{\psi}\gamma^0\psi = -e\psi^\dagger\psi$$

and

$$\boldsymbol{j} \equiv j^k = -e\bar{\psi}\gamma^k\psi \equiv -e\bar{\psi}\boldsymbol{\gamma}\psi \tag{5.32}$$

in which the charge density j^0 is a scalar. From eqn. (5.28), we have

$$\begin{aligned}\frac{\partial}{\partial t}\int j^0 d^3\boldsymbol{x} &= -\int \nabla.\boldsymbol{j}\, d^3\boldsymbol{x} \\ &= -e\int \nabla.(\bar{\psi}\,\gamma\,\psi)d^3\boldsymbol{x} \\ &= -e\int \bar{\psi}\,\gamma\,\psi.d\boldsymbol{S}\end{aligned} \tag{5.33}$$

where Gauss divergence theorem has been applied. If the field ψ vanishes on the boundary surface, we have

$$\frac{\partial}{\partial t}\int j^0 d^3\boldsymbol{x} = \frac{\partial Q}{\partial t} = 0 \tag{5.34}$$

implying that the charge

$$Q = \int j^0 d^3\boldsymbol{x} \tag{5.35}$$

is a constant in time. This implies that because of the $U(1)$ global gauge invariance of the Lagrangian density, the charge is conserved.

5.3 U(1) LOCAL GAUGE SYMMETRY AND QED

The gauge transformation (5.9) was global because α did not depend on space-time. Now consider that α is space–time dependent and we have local gauge transformations

$$\psi \rightarrow \psi' = e^{i\alpha(x)}\psi \tag{5.36}$$

Obviously, due to α now being differentiable, we shall have

$$\partial_\mu \psi' = \partial_\mu (e^{i\alpha(x)}\psi)$$
$$= e^{i\alpha(x)}\partial_\mu \psi + i\, e^{i\alpha(x)}\psi . \partial_\mu \alpha \qquad (5.37)$$

implying that due to the second term in eqn. (5.37), we have

$$\partial_\mu \psi' \neq e^{i\alpha(x)}\partial_\mu \psi$$

This situation of field gradient, not transforming like the field, makes this case different from that of global gauge invariance. We may also make the quite obvious observation that, if we substitute eqn. (5.36) and (5.37) in the Lagrangian density (5.8), we would see that $£'_{\text{Dirac}} \neq £_{\text{Dirac}}$, implying that the Lagrangian density (5.8) is no longer invariant under the transformation (5.36). The second term of eqn (5.37), which contains $\partial_\mu \alpha$ causes the violation.

In order to make the Lagrangian density invariant under the transformation (5.36), the term, which violates the gauge invariance, has to be canceled. For this purpose, a gauge covariant derivative

$$D_\mu = \partial_\mu + ieA_\mu \qquad (5.38)$$

is introduced. A_μ is the vector field. The origin of the covariant derivative can be seen in the quantum mechanics, where for an electron placed in an electromagnetic field, we replace in the wave equation

$$p_\mu \rightarrow p_\mu - eA_\mu \qquad (5.39)$$

which using $\partial_\mu = -ip_\mu$, implies the replacement

$$\partial_\mu \rightarrow \partial_\mu + ieA_\mu$$

With the introduction of the covariant derivative (5.38), the field gradient is now $D_\mu \psi$ and we require that the $D_\mu \psi$ transforms just like ψ transforms in eqn. (5.36). That is

$$D'_\mu \psi' = e^{i\alpha(x)} D_\mu \psi \qquad (5.40)$$

where it has to be noted that

$$D'_\mu = \partial_\mu + ieA'_\mu \qquad (5.41)$$

The requirement (5.40) would impose a condition on A'_μ which would provide the relationship between A'_μ and A_μ.

Using, from eqn. (5.36)

$$\psi = e^{-i\alpha(x)}\psi' \qquad (5.42)$$

we may write the eqn. (5.40) as

$$(\partial_\mu + ieA'_\mu)\psi' = e^{i\alpha(x)}(\partial_\mu + ieA_\mu)e^{-i\alpha(x)}\psi'$$
$$= e^{i\alpha(x)}\left(\partial_\mu\left[e^{-i\alpha(x)}\psi'\right]\right) + ieA_\mu\psi' \qquad (5.43)$$
$$= \partial_\mu\psi' - i(\partial_\mu\alpha)\psi' + ieA_\mu\psi'$$

Now peeling off ψ' from both the sides and then canceling ∂_μ, we get

$$ie A'_\mu = ie A_\mu - i\partial_\mu \alpha$$

or

$$A_\mu \to A'_\mu = A_\mu - \frac{1}{e}\partial_\mu \alpha \tag{5.44}$$

which means that eqn. (5.40) would be obeyed only when the vector field A_μ that appears in the covariant derivative transforms as eqn. (5.44). The eqn. (5.44) is called the gauge transformation for A_μ.

With the replacement of ∂_μ by D_μ, the Lagrangian density (5.8) becomes

$$£_{\text{local}} = i\bar{\psi}\gamma^\mu D_\mu \psi - m\bar{\psi}\psi \tag{5.45}$$

which now is locally gauge invariant. If we insert the eqn. (5.38), for D_μ, we may write $£_{\text{local}}$ as

$$£_{\text{local}} = i\bar{\psi}\gamma^\mu \partial_\mu \psi - m\bar{\psi}\psi - e(\bar{\psi}\gamma^\mu\psi)A_\mu \tag{5.46}$$

in which the first two terms constitute the Lagrangian density of free spin $-\frac{1}{2}$ field ψ, which was seen invariant in the previous section under global gauge transformation. The third term of $£_{\text{local}}$, in view of eqn. (5.31) is the interaction term $j^\mu A_\mu$. When we go from the global to local gauge invariance, the symmetry is enlarged and it is said that the symmetry has been gauged. In this case, the $U(1)$ symmetry is being gauged. Since, gauging requires the introduction of a vector field A_μ, we call A_μ the gauge field. The interaction term in $£_{\text{local}}$ represents the interaction of the gauge field A_μ with the Dirac field ψ through the coupling constant e. The interaction term in the form $-e\bar{\psi}\gamma^\mu\psi.A_\mu$ is the simplest term among the more complicated interaction terms and hence the interaction represented by $-e\bar{\psi}\gamma^\mu\psi.A_\mu$ is called the minimal coupling. The observation of the $£_{\text{local}}$ expression, also tells that the interaction term is introduced naturally over the free field Lagrangian when we go from global to local gauge invariance.

If we apply the transformations (5.36), (5.37) and (5.44) in the expression for $£_{\text{local}}$ [eqn. (5.46)], we find

$$\begin{aligned} £'_{\text{local}} &= i\bar{\psi}'\gamma^\mu\partial_\mu\psi' - m\bar{\psi}'\psi' - e(\bar{\psi}'\gamma^\mu\psi')A'_\mu \\ &= i\bar{\psi}\gamma^\mu\partial_\mu\psi - \bar{\psi}\gamma^\mu\psi\partial_\mu\alpha - m\bar{\psi}\psi \\ &\quad - e\bar{\psi}\gamma^\mu\psi A_\mu + \bar{\psi}\gamma^\mu\psi\partial_\mu\alpha \\ &= £_{\text{local}} \end{aligned} \tag{5.47}$$

where we observe that the gauge violating term $(-\bar{\psi}\gamma^\mu\psi.\partial_\mu\alpha)$ is cancelled by the gauge compensating term $(+\bar{\psi}\gamma^\mu\psi.\partial_\mu\alpha)$, which comes because of the gauge field A_μ. Due to this the gauge field A_μ is also called the compensating field.

In the expression for $£_{local}$ (eqn 5.45 or 5.46), it must also be noted that due to the presence of the vector field A_μ, $£_{local}$ should also contain the kinetic (gradient) and mass terms corresponding to A_μ. However, these kinetic and mass terms must be invariant under the transformation (5.44). We recall that the transformation (5.44) is required to be obeyed to make the Lagrangian density locally gauge invariant. The gauge invariant kinetic term for A_μ is

$$\text{Kinetic term} = -\frac{1}{4} F_{\mu\nu} F^{\mu\nu} \tag{5.48}$$

where

$$F_{\mu\nu} = \partial_\mu A_\nu - \partial_\nu A_\mu \tag{5.49}$$

is gauge invariant, because under the transformation (5.44)

$$\begin{aligned} F'_{\mu\nu} &= \partial_\mu A'_\nu - \partial_\nu A'_\mu \\ &= \partial_\mu (A_\nu - \frac{1}{e}\partial_\nu \alpha) - \partial_\nu (A_\mu - \frac{1}{e}\partial_\mu \alpha) \\ &= (\partial_\mu A_\nu - \partial_\nu A_\mu) - \frac{1}{e}(\partial_\mu \partial_\nu \alpha - \partial_\nu \partial_\mu \alpha) \\ &= \partial_\mu A_\nu - \partial_\nu A_\mu = F_{\mu\nu} \end{aligned} \tag{5.50}$$

where $\partial_\mu \partial_\nu \alpha = \partial_\nu \partial_\mu \alpha$ is implied. However, we cannot have a mass term for A_μ, as it would be $\frac{1}{2} m_\gamma A_\mu A^\mu$, which is not gauge invariant, because under the transformations (5.44)

$$\begin{aligned} A'_\mu A'^\mu &= (A_\mu - \frac{1}{e}\partial_\mu \alpha)(A^\mu - \frac{1}{e}\partial^\mu \alpha) \\ &\neq A_\mu A^\mu \end{aligned} \tag{5.51}$$

The absence of A_μ mass term implies that

$$\tfrac{1}{2} m_\gamma^2 A_\mu A^\mu = 0 \tag{5.52a}$$

which in view of $A_\mu \neq 0$, implies that the mass of the gauge field A_μ

$$m_\gamma = 0 \tag{5.52b}$$

Thus the gauge field A_μ is a massless vector field. The electromagnetic field is also a massless vector field. If the coupling constant e in the interaction term of $£_{local}$ were identified with the magnitude of electric charge, the gauge field A_μ would be the electromagnetic field. The interaction term would then mean the interaction between the photon field A_μ and the fermion (electron) field ψ.

Thus the requirement of local gauge invariance on the free fermion Lagrangian leads to the interacting theory of quantum electrodynamics. The complete Lagrangian density for the $U(1)$ local gauge invariant theory can then be written from eqs. (5.46) and (5.48) as

$$£_{local} = [i\bar{\psi}\gamma^\mu \partial_\mu \psi - m\bar{\psi}\psi] - \tfrac{1}{4} F_{\mu\nu} F^{\mu\nu} - e\bar{\psi}\gamma^\mu \psi A_\mu \tag{5.53}$$

5.4 SU(2) GLOBAL GAUGE SYMMETRY

Now, we apply the concept of global gauge invariance to the SU(2) isospin group. Let us now consider that we have two spin $-\frac{1}{2}$ fields ψ_1 and ψ_2. The free field Lagrangian density would be [eqn. (5.8)]

$$\begin{aligned} £ &= (£)_{\psi_1} + (£)_{\psi_2} \\ &= [i\bar{\psi}_1\gamma^\mu\partial_\mu\psi_1 - m_1\bar{\psi}_1\psi_1] + [i\bar{\psi}_2\gamma^\mu\partial_\mu\psi_2 - m_2\bar{\psi}_2\psi_2] \end{aligned} \tag{5.54}$$

where m_1 is the mass of the field ψ_1 and m_2 that of field ψ_2. We wish to write eqn. (5.54) in the form of eqn. (5.8), so we define a doublet of the spin $-\frac{1}{2}$ field as

$$\psi = \begin{pmatrix} \psi_1 \\ \psi_2 \end{pmatrix} \tag{5.55}$$

and accordingly, there is the adjoint spinor

$$\bar{\psi} = (\bar{\psi}_1 \bar{\psi}_2) \tag{5.56}$$

It has; however, to be remembered that each of the fields ψ_1 and ψ_2 itself is a four-component column matrix and as usual $\bar{\psi}_1 = \psi_1^\dagger\gamma^0$ etc. Introducing the mass matrix

$$m = \begin{pmatrix} m_1 & 0 \\ 0 & m_2 \end{pmatrix} \tag{5.57}$$

and using ψ and $\bar{\psi}$ from eqs. (5.55) and (5.56), we may write the Lagrangian density (5.54) as

$$£ = i(\bar{\psi}_1\ \bar{\psi}_2)\gamma^\mu\partial_\mu\begin{pmatrix} \psi_1 \\ \psi_2 \end{pmatrix} - (\bar{\psi}_1\ \bar{\psi}_2)\begin{pmatrix} m_1 & 0 \\ 0 & m_2 \end{pmatrix}\begin{pmatrix} \psi_1 \\ \psi_2 \end{pmatrix}$$

or

$$£ = i\bar{\psi}\gamma^\mu\partial_\mu\psi - \bar{\psi}m\psi \tag{5.58}$$

However, when $m_1 = m_2$, eqn. (5.58) assumes the familiar form of eqn. (5.8) as

$$£ = i\bar{\psi}\gamma^\mu\partial_\mu\psi - m\bar{\psi}\psi \tag{5.59}$$

The purpose of ψ as a two-element column vector (eqn. 5.55) is that, we can now introduce a more general global gauge transformation than in eqn. (5.9). In the global gauge transformation (5.9), the unitary matrix $U(\alpha)$ had only one parameter α and now we consider the unitary matrix

$$U = e^{iA} \tag{5.60}$$

in place of eqn. (5.10) of U(1). It is obvious that the condition of unitary matrix $U^\dagger U = 1$ demands that the operator A must be hermitian ($A^\dagger = A$). Since, ψ is a doublet, if it transforms as eqn. (5.9) under the unitary matrix (5.60), the unitary matrix has to be a 2 × 2 matrix, implying that the hermitian operator A be represented by a 2 × 2 hermitian matrix. Now, the most general 2 × 2 hermitian matrix can be expressed in terms of four real numbers θ_0, θ_1, θ_2, and θ_3 as

$$A = \begin{pmatrix} 1 & 0 \\ 0 & 1 \end{pmatrix}\theta_0 + \begin{pmatrix} 0 & 1 \\ 1 & 0 \end{pmatrix}\theta_1 + \begin{pmatrix} 0 & -i \\ i & 0 \end{pmatrix}\theta_2 + \begin{pmatrix} 1 & 0 \\ 0 & -1 \end{pmatrix}\theta_3$$
$$= 1.\theta_0 + \tau.\theta \tag{5.61}$$

where $\tau(\tau_1, \tau_2, \tau_3)$ are three Pauli matrices. Equation (5.61) can initially be obtained as

$$A = \begin{pmatrix} \theta_0 + \theta_3 & \theta_1 - i\theta_2 \\ \theta_1 + i\theta_2 & \theta_0 - \theta_3 \end{pmatrix} \tag{5.62}$$

then seen as eqn. (5.61), in which 1 denotes the 2 × 2 unit matrix. Using eqn. (5.61), the unitary matrix (5.60) would be

$$U = e^{i\theta_0}.e^{i\theta.\tau} \tag{5.63}$$

For SU(2) global gauge transformations both θ_0 and θ would not depend on space-time x and since, the part $e^{i\theta_0}$ only gives the U(1) gauge transformation (5.9), we can omit it here and retain only $e^{i\theta\cdot\tau}$ as the unitary matrix for SU(2). However, we are considering the internal symmetries, the SU(2) is SU(2) isospin group in which the isospin $\boldsymbol{T} = \frac{1}{2}\tau$. In other words, as discussed in the previous Chapters, here the particle states and hence the field operators ψ form an $T = \frac{1}{2}$ irreducible representation of isotopic SU(2) group, ψ is an isospinor and $\boldsymbol{T} = \frac{1}{2}\tau$, with $\tau = \tau_1, \tau_2, \tau_3$. In view of the above $\frac{1}{2}\tau$ consideration, the unitary matrix of eqn. (5.63) for SU(2) isospin group would be

$$U(\alpha_i) = e^{i\frac{\tau}{2}.\theta} = e^{i\tau.\alpha} = e^{i\tau_i\alpha_i} \tag{5.64}$$

where $i = 1, 2, 3$ and

$$\alpha = \frac{\theta}{2} \tag{5.65}$$

The set of transformation matrices $U(\alpha i)$ [equation (5.64)], form the SU(2) isospin group. The generators of this group, which are $\frac{\tau_i}{2}$, obey the commutation relation

$$\left[\frac{\tau_i}{2}, \frac{\tau_j}{2}\right] = i\varepsilon_{ijk}\frac{\tau_k}{2} \tag{5.66}$$

which implies $\tau_i\tau_j \neq \tau_j\tau_i$ and because of this, we cannot have here

$$U(\alpha_1)\, U(\alpha_2) = U(\alpha_2)\, U(\alpha_1) \tag{5.67}$$

We found the relation (5.67) to hold in U(1) case and accordingly U(1) was Abelian group. Here, because the elements of the group do not commute, the group SU(2) is non-Abelian. The determinant of matrices $U(\alpha_i)$ is obviously one, which is required by special (S) in SU(2). Now onwards we shall only use the word SU(2) that would mean the SU(2) isospin group.

Under SU(2) global gauge transformation, the field ψ transforms as

$$\psi(x) \to \psi'(x) = e^{i\tau.\alpha}\psi(x) \tag{5.68}$$

and since α is independent of x (global), we would have

$$\frac{\partial\psi}{\partial x^{\mu}} = \partial_{\mu}\psi \to \partial_{\mu}\psi' = e^{i\tau.\alpha}\partial_{\mu}\psi \tag{5.69}$$

Using the above transformations, (5.68) and (5.69), along with

$$\bar{\psi}'(x) = e^{-i\,\tau.\alpha}\bar{\psi}(x) \tag{5.70}$$

where $\tau^{\dagger} = \tau$ is implied, the Lagrangian density

$$£' = i\bar{\psi}'\gamma^{\mu}\partial_{\mu}\psi' - m\bar{\psi}'\psi' \tag{5.71}$$

can be seen to be equal to £ [eqn. (5.59)]. The exponential $e^{+i\tau\cdot\alpha}$ is simply canceled by $e^{-i\tau\cdot\alpha}$. Thus under the SU(2) global gauge transformation (5.68), the Lagrangian density is invariant.

Similar to U(1), we will bring in the continuous character of SU(2) so use infinitesimal global gauge transformation. Infinitesimally, eqn. (5.68) reads

$$\psi \to \psi' = (1 + i\tau\cdot\alpha)\psi \tag{5.72}$$

and the invariance $\delta £ = 0$ implies

$$\delta £ = \delta £(\psi, \partial_{\mu}\psi) = \frac{\partial £}{\partial\psi}\delta\psi + \frac{\partial £}{\partial(\partial_{\mu}\psi)}\delta(\partial_{\mu}\psi) = 0 \tag{5.73}$$

which using from eqn. (5.72)

$$\delta\psi = \psi' - \psi = i\tau\cdot\alpha\psi \tag{5.74}$$

and

$$\delta(\partial_{\mu}\psi) = \partial_{\mu}\psi' - \partial_{\mu}\psi = (i\tau.\alpha)\partial_{\mu}\psi \tag{5.75}$$

becomes

$$\delta £ = (i\tau.\alpha)\left[\frac{\partial £}{\partial\psi}\psi + \frac{\partial £}{\partial(\partial_{\mu}\psi)}\partial_{\mu}\psi\right] = 0 \tag{5.76}$$

Now, we may follow the steps similar to those below eqn. (5.24) and write $\tau\cdot\alpha = \tau_i\alpha_i$, to obtain similar to eqn. (5.27)

$$\partial_{\mu}\left(ig\frac{\partial £}{\partial(\partial_{\mu}\psi)}\tau_i\psi\right) = 0 \tag{5.77}$$

which on writing as

$$\partial_{\mu}j_i^{\mu} = 0 \tag{5.78}$$

suggests that there are three conserved currents corresponding to $i = 1, 2, 3$, of τ_i, which are given by

$$j_i^\mu = ig\frac{\partial £}{\partial(\partial_\mu \psi)}\tau_i\psi = -g\bar{\psi}\gamma^\mu\tau_i\psi \tag{5.79}$$

where eqn. (5.30) has been used. The eqn. (5.79) indicates that we have three (corresponding to $i = 1, 2, 3$) conserved currents. Three is the number of generators $\left(\frac{\tau_i}{2}\right)$ for the group SU(2). Had the group had n generators, we would have the number of n conserved currents like eqn. (5.79).

Thus it can be concluded that when the global gauge invariance is applied to non-Abelian groups, we still get a conserved four-current similar to one we get in U(1) (Abelian) case, but the number of these currents is not one but equal to the number of generators of the group. The form of conserved currents is like eqn. (5.79) in which g is the analogue of e in U(1) (Abelian) case. Since, SU(2) group is the SU(2) isospin group, the current j_i^μ of eqn. (5.79) may be seen as denoting the three components of isospin current (namely, j_1^μ, j_2^μ and j_3^μ) of nucleons and according to eqn. (5.78), the three components would be conserved. The field ψ in eqn. (5.79) is called the *nucleon field*.

5.5 SU(2) LOCAL GAUGE SYMMETRY AND YANG–MILLS FIELDS

The local gauge transformations demand that the parameter α of eqn. (5.68) depends on the space–time x and accordingly, we shall have the field ψ transforming as

$$\psi \to \psi' = e^{i\tau.\alpha(x)}\psi \tag{5.80}$$

However, as seen in U(1) local gauge symmetry

$$\partial_\mu\psi \to \partial_\mu\psi' \neq e^{i\tau.\alpha(x)}\partial_\mu\psi \tag{5.81}$$

instead, we shall have

$$\begin{aligned}\partial_\mu\psi' &= \partial_\mu(e^{i\tau.\alpha(x)}\psi)\\ &= e^{i\tau.\alpha(x)}[\partial_\mu\psi + i\tau.\partial_\mu\alpha(x)\psi]\end{aligned} \tag{5.82}$$

Obviously, it is the $\partial_\mu\alpha$ containing second term of eqn. (5.82), which does not allow the equality sign in eqn. (5.81) and as a result the local gauge invariance of the Lagrangian density (5.58) is violated. This means that with the transformations (5.80) and (5.82) we would not get $£' = £$.

In order to see that the Lagrangian density [eqn. (5.58)] is invariant under local gauge transformation (5.80), we have to introduce the gauge field $\boldsymbol{G}_\mu$, similar to the $U(1)$ case, where we introduced the gauge field A_μ. Remember that the gauge field A_μ was a vector field. The group being SU(2) isospin, the gauge field $\boldsymbol{G}_\mu$ must be a vector in isospin space. The group U(1) had only one generator (the charge operator); hence only one gauge field A_μ was required. The SU(2), however, has three generators $\frac{\tau_1}{2}, \frac{\tau_2}{2}$ and $\frac{\tau_3}{2}$ hence we require three gauge fields $\boldsymbol{G}_{1\mu}$, $\boldsymbol{G}_{2\mu}$, and $\boldsymbol{G}_{3\mu}$, which rotate into each other under $SU(2)$ isospin rotations. μ with values 0, 1, 2, 3 refers to the space–time (external space), which is obviously four-dimensional. Hence, with respect to the external space $G_{1\mu}$ is a four-vector and so are $\boldsymbol{G}_{2\mu}$, and $\boldsymbol{G}_{3\mu}$. We may use the notation

$$G_\mu = (G_{1\mu}, G_{2\mu}, G_{3\mu}) \tag{5.83}$$

and may call that G_μ is an isovector, i.e., a vector in isospin space and at the same time is a four-vector in external (space–time) space. When we call the four dimensional space–time as external space, the isospin space can be called the *internal space*.

Having introduced the isovector G_μ, we may also introduce the covariant derivative

$$D_\mu = \partial_\mu + ig\frac{\tau}{2}.\boldsymbol{G}_\mu \tag{5.84}$$

where g is the coupling constant for the field G_μ, and $\frac{\tau}{2} = \frac{\tau_i}{2}$ denote the three generators. Again $\tau.G_\mu$ must be an isoscalar, so $\boldsymbol{G}_\mu$ has to be an isovector to combine correctly with $\frac{\tau}{2}$, the generators of SU(2) isospin group. Having introduced the covariant derivative (5.84), we require that

$$D'_\mu \psi' = e^{i\tau.\alpha(x)} D_\mu \psi \tag{5.85}$$

where

$$D'_\mu = \partial_\mu + ig\frac{\tau}{2}.\boldsymbol{G}'_\mu \tag{5.86}$$

Substituting from eqn. (5.80)

$$\psi = e^{-i\tau.\alpha(x)}\psi' \tag{5.87}$$

into the RHS of eqn. (5.85), we get

$$D'_\mu \psi' = e^{i\tau.\alpha}(\partial_\mu + ig\frac{\tau}{2}.\boldsymbol{G}_\mu)e^{-i\tau.\alpha}\psi' \tag{5.88}$$

where we have written $\alpha(x) = \alpha$ to save the space. Writing for D'_μ from eqn. (5.86), we write eqn. (5.88) as

$$(\partial_\mu + ig\frac{\tau}{2}.\boldsymbol{G}'_\mu)\psi' = e^{i\,\tau.\alpha}\partial_\mu[e^{-i\tau.\alpha}\psi'] + ige^{i\tau.\alpha}(\frac{\tau}{2}.\boldsymbol{G}_\mu)e^{-i\tau.\alpha}\psi' \tag{5.89}$$

$$=[\partial_\mu - i\tau.\partial_\mu\alpha + ige^{i\tau.\alpha}(\frac{\tau}{2}.\boldsymbol{G}_\mu)e^{-i\tau.\alpha}]\psi'$$

Peeling off y¢ from both the sides and also canceling ∂_μ, we may write eqn. (5.89) as

$$\tau.\boldsymbol{G}'_\mu = e^{i\tau.\alpha}(\tau.\boldsymbol{G}_\mu)e^{-i\tau.\alpha} - \frac{2}{g}\tau.\partial_\mu\alpha \tag{5.90}$$

The continuous nature of SU(2) groups (rotations in isospin space) demands the use of infinitesimal form of transformation (5.80). When α is infinitesimal, we may write

$$e^{\pm i\tau.\alpha} \approx 1 \pm i\tau.\alpha \tag{5.91}$$

Accordingly the eqn. (5.90) becomes

$$\begin{aligned}
\tau.\boldsymbol{G}'_\mu &= (1+i\tau_i\alpha^i)(\tau_j G^j_\mu)(1-i\tau_i\alpha^i) - \frac{2}{g}\tau.\partial_\mu\alpha \\
&= \tau_j G^j_\mu + i\tau_i\tau_j\alpha^i G^j_\mu - i\tau_j\tau_i\alpha^i G^j_\mu - \frac{2}{g}\tau.\partial_\mu\alpha \\
&= \tau.\boldsymbol{G}_\mu + i[\tau_i, \tau_j]\alpha^i G^j_\mu - \frac{2}{g}\tau.\partial_\mu\alpha \\
&= \tau.\boldsymbol{G}_\mu - 2\varepsilon_{ijk}\tau_k\alpha^i G^j_\mu - \frac{2}{g}\tau.\partial_\mu\alpha \\
&= \tau.\boldsymbol{G}_\mu - 2\tau.(\alpha\times\boldsymbol{G}_\mu) - \frac{2}{g}\tau.\partial_\mu\alpha
\end{aligned} \tag{5.92}$$

where the commutation relation (5.66) and

$$\tau_k\varepsilon_{ijk}\alpha^i G^j_\mu = \tau.(\alpha\times\boldsymbol{G}_\mu) \tag{5.93}$$

have been used. Deleting τ from both the sides of eqn. (5.92), we get the transformation law for SU(2) gauge field $\boldsymbol{G}_\mu$ as

$$\boldsymbol{G}'_\mu = \boldsymbol{G}_\mu - 2(\alpha\times\boldsymbol{G}_\mu) - \frac{2}{g}\partial_\mu\alpha \tag{5.94}$$

This equation can be compared with eqn. (5.44), namely

$$A'_\mu = A_\mu - \frac{1}{e}\partial_\mu\alpha \tag{5.95}$$

The comparison shows that there is an additional term (α × $\boldsymbol{G}_\mu$) in the transformation law for $\boldsymbol{G}_\mu$. This additional term signifies the rotation of $\boldsymbol{G}_\mu$. The term $\frac{2}{g}\partial_\mu\alpha$ denotes the shifting of $\boldsymbol{G}_\mu$. Thus we conclude that

1. The SU(2) gauge field $\boldsymbol{G}_\mu$ must transform as eqn. (5.94) so that the field ψ and the gradient of ψ, i.e., $D_\mu\psi$ transform identically, i.e., the eqn. (5.85) is obeyed.

2. Whereas in case of U(1) group, the realization of local gauge symmetry requires that the gauge field A_μ should merely be shifted by $\left(-\frac{1}{e}\partial_\mu\alpha\right)$, the realization of SU(2) local gauge invariance demands that the SU(2) gauge field $\boldsymbol{G}_\mu$ should not only be shifted by $\left(-\frac{2}{g}\partial_\mu\alpha\right)$ but it must also be rotated $(\alpha \times \boldsymbol{G}_\mu)$. The rotation is in the internal (isospin) space. The rotation of the gauge field in the internal space is the additional feature which non-Abelian gauge fields possess, vis-à-vis the Abelian gauge fields.

The locally gauge invariant Lagrangian density, from eqn. (5.59) becomes $\partial_\mu \to D_\mu$

$$\pounds_{\text{local}} = i\bar{\psi}\gamma^\mu D_\mu\psi - m\bar{\psi}\psi \tag{5.96}$$

where we rewrite D_μ from eqn. (5.84)

$$D_\mu = \partial_\mu + ig\frac{\tau}{2}.\boldsymbol{G}_\mu \tag{5.97}$$

Since, D_μ contains the gauge field $\boldsymbol{G}_\mu$, the Lagrangian density (5.96) should also contain the kinetic and mass parts of field $\boldsymbol{G}_\mu$. m [already present in eqn. (5.96)] is the mass of ψ field.

In the case of SU(2), it is important to observe the construction of kinetic part of $\boldsymbol{G}_\mu$. We can expect from U(1) gauge theory that the kinetic part of $\boldsymbol{G}_\mu$ could also be like the kinetic part of A_μ, i.e., like eqn. (5.48) and we could write

$$\boldsymbol{G}_{\mu\nu} = \partial_\mu\boldsymbol{G}_\nu - \partial_\nu\boldsymbol{G}_\mu \tag{5.98}$$

where it has to be noted that because of three gauge fields $G_{1\mu}$, $G_{2\mu}$, $G_{3\mu}$, we would have three kinetic terms

$$-\frac{1}{4}G_{1\mu\nu}G^{1\mu\nu} - \frac{1}{4}G_{2\mu\nu}G^{2\mu\nu} - \frac{1}{4}G_{3\mu\nu}G^{3\mu\nu} = -\frac{1}{4}\boldsymbol{G}_{\mu\nu}\boldsymbol{G}^{\mu\nu} \tag{5.99}$$

in which $G_{\mu\nu}$ has not the form (5.98), we have to find it in the following.

If $G_{\mu\nu}$ is like eqn. (5.98), it must be invariant under the gauge transformations of $\boldsymbol{G}_\mu$ given by eqn. (5.94). Substituting eqn. (5.94) into eqn. (5.98), we get

$$\begin{aligned}
\boldsymbol{G}'_{\mu\nu} &= \partial_\mu\boldsymbol{G}'_\nu - \partial_\nu\boldsymbol{G}'_\mu \\
&= \partial_\mu[\boldsymbol{G}_\nu - 2(\alpha\times\boldsymbol{G}_\nu) - \frac{2}{g}\partial_\nu\alpha] - \partial_\nu[\boldsymbol{G}_\mu - 2(\alpha\times\boldsymbol{G}_\mu) - \frac{2}{g}\partial_\mu\alpha] \\
&= (\partial_\mu\boldsymbol{G}_\nu - \partial_\nu\boldsymbol{G}_\mu) - 2\partial_\mu(\alpha\times\boldsymbol{G}_\nu) + 2\ \ \partial_\nu(\alpha\times\boldsymbol{G}_\mu) \\
&= (\partial_\mu\boldsymbol{G}_\nu - \partial_\nu\boldsymbol{G}_\mu) - 2\alpha\times(\partial_\mu\boldsymbol{G}_\nu - \partial_\nu\boldsymbol{G}_\mu) - 2[\partial_\mu\alpha\times\boldsymbol{G}_\nu - \partial_\nu\alpha\times\boldsymbol{G}_\mu] \\
&\neq \boldsymbol{G}_{\mu\nu}
\end{aligned} \tag{5.100}$$

where $\partial_\mu \partial_\nu = \partial_\nu \partial_\mu$, has canceled the (2/*g*) terms. Eqn. (5.100) tells that the form (5.98) in not invariant under eqn. (5.94). In order to find a SU(2) gauge invariant form for $\boldsymbol{G}_{\mu\nu}$, we try a term $g(\boldsymbol{G}_\mu \times \boldsymbol{G}_\nu)$ and check its transformation when $\boldsymbol{G}_\mu \to \boldsymbol{G}'_\mu$, i.e,

$$\begin{aligned} g(\boldsymbol{G}'_\mu \times \boldsymbol{G}'_\nu) &= g[\boldsymbol{G}_\mu - 2(\alpha \times \boldsymbol{G}_\mu) - \frac{2}{g}\partial_\mu \alpha] \times [\boldsymbol{G}_\nu - 2(\alpha \times \boldsymbol{G}_\nu) - \frac{2}{g}\partial_\nu \alpha] \\ &= g(\boldsymbol{G}_\mu \times \boldsymbol{G}_\nu) - 2g\boldsymbol{G}_\mu \times (\alpha \times \boldsymbol{G}_\nu) - 2g(\alpha \times \boldsymbol{G}_\mu) \times \boldsymbol{G}_\nu \\ &\quad -2\boldsymbol{G}_\mu \times \partial_\nu \alpha - 2\partial_\mu \alpha \times \boldsymbol{G}_\nu \end{aligned} \tag{5.101}$$

where we have retained in the RHS the terms upto the order α only. From the property of the vector triple product, we can see that

$$\boldsymbol{G}_\mu \times (\alpha \times \boldsymbol{G}_\nu) + (\alpha \times \boldsymbol{G}_\mu) \times \boldsymbol{G}_\nu = \alpha \times (\boldsymbol{G}_\mu \times \boldsymbol{G}_\nu) \tag{5.102}$$

Using eqn. (5.102) in the RHS of eqn. (5.101), we get

$$\begin{aligned} -g(\boldsymbol{G}'_\mu \times \boldsymbol{G}'_\nu) &= -g(\boldsymbol{G}_\mu \times \boldsymbol{G}_\nu) + 2g\alpha \times (\boldsymbol{G}_\mu \times \boldsymbol{G}_\nu) \\ &\quad + 2[\partial_\mu \alpha \times \boldsymbol{G}_\nu - \partial_\nu \alpha \times \boldsymbol{G}_\mu] \end{aligned} \tag{5.103}$$

Now, we add eqs. (5.100) and (5.103) to write

$$\begin{aligned} \partial_\mu \boldsymbol{G}'_\nu - \partial_\nu \boldsymbol{G}'_\mu - g(\boldsymbol{G}'_\mu \times \boldsymbol{G}'_\nu) &= [\partial_\mu \boldsymbol{G}_\nu - \partial_\nu \boldsymbol{G}_\mu - g(\boldsymbol{G}_\mu \times \boldsymbol{G}_\nu)] \\ &\quad -2\alpha \times [\partial_\mu \boldsymbol{G}_\nu - \partial_\nu \boldsymbol{G}_\mu - g(\boldsymbol{G}_\mu \times \boldsymbol{G}_\nu)] \end{aligned} \tag{5.104}$$

where the third term in the RHS of $\boldsymbol{G}'_{\mu\nu}$ [eqn.(5.100)] has canceled with the third term in the RHS of eqn. (5.103). Equation (5.104) suggests that, if we define

$$\boldsymbol{G}_{\mu\nu} = \partial_\mu \boldsymbol{G}_\nu - \partial_\nu \boldsymbol{G}_\mu - g(\boldsymbol{G}_\mu \times \boldsymbol{G}_\nu) \tag{5.105}$$

we would, under the gauge transformation for $\boldsymbol{G}_\mu$ [eqn. (5.94)] get, upto the order α

$$\boldsymbol{G}'_{\mu\nu} = \boldsymbol{G}_{\mu\nu} - 2(\alpha \times \boldsymbol{G}_{\mu\nu}) \tag{5.106}$$

Equation (5.106) is the expected transformation law for SU (2) field tensor $\boldsymbol{G}_{\mu\nu}$.

In U(1), since, A_μ was merely shifted by $\left(-\frac{1}{e}\partial_\mu \alpha\right)$, we got $F'_{\mu\nu} = F_{\mu\nu}$. But, in SU(2), $\boldsymbol{G}_\mu$ shifts as well as rotates, we must expect a rotation of $\boldsymbol{G}_{\mu\nu}$ as well. The rotation term $(\alpha \times \boldsymbol{G}_{\mu\nu})$ has to be present in the transformation of SU(2) field tensor $\boldsymbol{G}_{\mu\nu}$ under the change (5.94), i.e., $\boldsymbol{G}_{\mu\nu}$ also undergoes isorotation.

Since, eqn. (5.106) is obeyed by eqn. (5.105), we have the field tensor for SU(2) group as

$$\boldsymbol{G}_{\mu\nu} = \partial_\mu \boldsymbol{G}_\nu - \partial_\nu \boldsymbol{G}_\mu - g(\boldsymbol{G}_\mu \times \boldsymbol{G}_\nu) \tag{5.107}$$

and the kinetic term would be $\left(-\frac{1}{4}\boldsymbol{G}_{\mu\nu}\boldsymbol{G}^{\mu\nu}\right)$. The field tensor $\boldsymbol{G}_{\mu\nu}$ in the form

(5.107) is called the Yang–Mills field tensor. The last term of $\boldsymbol{G}_{\mu\nu}$ denotes the self-interaction of gauge field.

Thus the kinetic term of the gauge field $\boldsymbol{G}_{\mu}$ would be the RHS of eqn. (5.99) with $\boldsymbol{G}_{\mu\nu}$ as eqn. (5.107). The mass term corresponding to the gauge field $\boldsymbol{G}_{\mu}$ would again be

$$\frac{1}{2}m_G^2\boldsymbol{G}_{\mu}\boldsymbol{G}^{\mu}$$

which would again be not invariant under the transformation (5.94), i.e., we get

$$\frac{1}{2}m_G^2\boldsymbol{G}'_{\mu}\boldsymbol{G}'^{\mu} \neq \frac{1}{2}m_G^2\boldsymbol{G}_{\mu}\boldsymbol{G}^{\mu} \tag{5.108}$$

implying similar to the U(1) case that the mass of SU(2) gauge field

$$m_G = 0 \tag{5.109}$$

In fact above description suggests that the *requirement of local gauge invariance demands a massless gauge field.*

The complete locally gauge invariant Lagrangian density for the SU(2) isospin group is [eqs. (5.96) and (5.99)]

$$\begin{aligned}\pounds_{\text{local}} &= i\bar{\psi}\gamma^{\mu}D_{\mu}\psi - m\bar{\psi}\psi - \frac{1}{4}\boldsymbol{G}_{\mu\nu}\boldsymbol{G}^{\mu\nu}\\ &= [i\bar{\psi}\gamma^{\mu}\partial_{\mu}\psi - m\bar{\psi}\psi] - \frac{1}{4}\boldsymbol{G}_{\mu\nu}\boldsymbol{G}^{\mu\nu} - g\bar{\psi}\gamma^{\mu}\boldsymbol{T}\psi.\boldsymbol{G}_{\mu}\end{aligned} \tag{5.110}$$

where the terms in the square bracket constitute the globally gauge invariant SU(2) Lagrangian density (5.59). Thus in SU(2) case also, we may conclude that when we enlarge the gauge symmetry from global to local an isovector field $\boldsymbol{G}_{\mu}$ has to be introduced, which automatically introduces an interaction term $g\bar{\psi}\gamma^{\mu}\boldsymbol{T}\psi.\boldsymbol{G}_{\mu}$ with $\boldsymbol{T} = \frac{\tau}{2}$ as the generators of the group. Eqn. (5.110) is comparable to the eqn. (5.53). In eqn. (5.53) we had the gauge field A_{μ}, which was called the electromagnetic field and here in eqn. (5.110), we have the gauge field $\boldsymbol{G}_{\mu}$, which is called the Yang–Mills field, and accordingly the Lagrangian density (5.110) is also called the Yang–Mills Lagrangian density. It may also be noted that U(1) locally gauge invariant Lagrangian density is $T \to 1$ and $\boldsymbol{G}_{\mu} \to A_{\mu}$ version of the Yang–Mills Lagrangian density.

5.6 PROPERTIES OF YANG–MILLS FIELDS

5.6.1 Adjoint Representation

In Chapter 3, we discussed various representations of symmetry groups and briefly described the adjoint representation. Here we discuss the adjoint representation in some detail. We know there are three generators of SU(2) group and 8 for

SU(3) group. The triplet or the **3** representation is called the adjoint representation for SU(2). Remember, as told in Chapter 3, the fundamental representation of SU(2) is **2**. Similarly, **8** is the adjoint representation of SU(3), whereas **3** is its fundamental representation.

The definition of adjoint representation is given as; *it is that irreducible representation which is completely determined by the structure constants of the group and the dimensionality of adjoint representation is equal to the generators of the group.*

For SU(2), the three generators T_i are equal to $\frac{\tau_i}{2}$ in its fundamental representation, where $i = 1, 2, 3$ and τ_i are (2×2) Pauli matrices. However, for SU(2) if we express the three generators T_i as

$$T_1 = \begin{pmatrix} 0 & 0 & 0 \\ 0 & 0 & -i \\ 0 & i & 0 \end{pmatrix}, T_2 = \begin{pmatrix} 0 & 0 & i \\ 0 & 0 & 0 \\ -i & 0 & 0 \end{pmatrix}, T_3 = \begin{pmatrix} 0 & -i & 0 \\ i & 0 & 0 \\ 0 & 0 & 0 \end{pmatrix} \quad (5.111)$$

we find that the matrix elements of above three matrices can be obtained from

$$(T_i)_{jk} = -i\varepsilon_{ijk} \quad (5.112)$$

where, ε_{ijk} is completely antisymmetric Levi-Civita tensor. ε_{ijk} is $+1$ if the order of i, j, k is a cyclic permutation of 1, 2, 3; ε_{ijk} is -1, if i, j, k are anticyclic and $\varepsilon_{ijk} = 0$, if any two indices are same. It can also be seen from the above matrices (5.111) that they obey

$$[T_i, T_j] = i\varepsilon_{ijk} T_k \quad (5.113)$$

Now, for an SU(n) group, we have $n^2 - 1$ generators denoted by T_a with $a = 1, 2, 3 \ldots n^2 - 1$. These generators obey

$$[T_a, T_b] = i\, f_{abc} T_c \quad (5.114)$$

where, f_{abc} are called the structure constants of the group SU(n). The structure constants f_{abc} are taken to be real and completely antisymmetric with respect to a, b, c. Comparing eqn. (5.113) with eqn. (5.114), we may say that ε_{ijk} are the structure constants of SU(2) group and from eqn. (5.112), we find that ε_{ijk}, now the structure constants, define the generators of SU(2) group. Thus in the adjoint representation of SU(2), its generators are given by eqn. (5.112).

Now, recall the equation (5.92), which on using $\boldsymbol{T} = \frac{\tau}{2}$ (fundamental representation) can be written as

$$\boldsymbol{T}.\boldsymbol{G}'_\mu = \boldsymbol{T}.\boldsymbol{G}_\mu - \boldsymbol{T}.(\alpha \times \boldsymbol{G}_\mu) - \frac{1}{g}\boldsymbol{T}.\partial_\mu \alpha \quad (5.115)$$

that must be valid for the adjoint representation as well. So we do not delete $\boldsymbol{T}$. Eqn. (5.115) is then the general transformation law for $\boldsymbol{G}_\mu$. This equation suggests that $G_{\mu\nu}$ of eqn. (5.107) must have the general form

$$(\boldsymbol{T}.\boldsymbol{G})_{\mu\nu} = \partial_\mu(\boldsymbol{T}.\boldsymbol{G}_\nu) - \partial_\nu(\boldsymbol{T}.\boldsymbol{G}_\mu) - g[(\boldsymbol{T}.\boldsymbol{G}_\mu)\times(\boldsymbol{T}.\boldsymbol{G}_\nu)] \tag{5.116}$$

Thus retaining the generator $\boldsymbol{T}$, we may write

$$\begin{aligned} \boldsymbol{G}_{\mu\nu} &\equiv \boldsymbol{G}^i_{\mu\nu} T^i \\ \boldsymbol{G}_\mu &\equiv \boldsymbol{G}^i_\mu T^i \end{aligned} \tag{5.117}$$

We shall make use of eqn. (5.117) to understand the properties of Yang–Mills fields. Expressing eqn. (5.107) as

$$G^i_{\mu\nu} = \partial_\mu G^i_\nu - \partial_\nu G^i_\mu - g\varepsilon_{ijk} G^j_\mu G^k_\mu \tag{5.118}$$

with ε_{ijk} as structure constants, we say that $G^i_{\mu\nu}$ and G^i_μ each form an adjoint representation of the gauge group SU(2). Replacing ε_{ijk} by f_{abc}, with a, b, c running from 1 to the number of generators of the gauge group, eqn. (5.118) can be written for any SU(n) gauge group.

It is important to note that in all non-Abelian gauge groups SU(2), SU(3) etc., [U(1) is Abelian], the transformations of $\boldsymbol{G}_{\mu\nu}$ and $\boldsymbol{G}_\mu$ are

$$G'^a_{\mu\nu} = G^a_{\mu\nu} - 2 f_{abc}\alpha^b G^c_{\mu\nu} \tag{5.119a}$$

and

$$G'^a_\mu = G^a_\mu - 2 f_{abc}\alpha^b G^c_\mu - \left(\frac{2}{g}\right)\partial_\mu \alpha^a \tag{5.119b}$$

which are quite different from the corresponding U(1) transformations. Due to this and in fact due to the presence of second term (rotation) of eqn. (5.119b) the gauge fields $\boldsymbol{G}_\mu$ of non-Abelian gauge groups are not neutral they carry charges. For example, the gauge fields of SU(2) group carry isocharges and the gauge fields of SU(3) colour group carry colour charges. On the contrary, A_μ the gauge field of U(1) group, which is the electromagnetic field is neutral. Combining the above features, we may say that a non-Abelian gauge group possesses the following properties:

(a) The gauge particles (i.e., the quanta of gauge fields) are massless.
(b) The gauge fields form an adjoint representation of the gauge group. This implies that the gauge fields possess 'charge'.
(c) The gauge fields have vertices (in Feynman diagrams) with three gauge particles interacting and also the vertices where four gauge particles interact.

Needless to mention, property (a) is applicable to Abelian [U(1)] gauge group as well.

5.6.2 Electric and Magnetic Fields

From the Yang–Mills field tensor (5.118), which we express as

$$G^{\mu\nu}_a(x) = \partial^\mu G^\nu_a(x) - \partial^\nu G^\mu_a(x) - g f_{abc} G^\mu_b(x) G^\nu_c(x) \tag{5.120}$$

where a, b, c each running from 1 to 3 are indices for internal (isospin) space; μ, $\nu = 0, 1, 2, 3$ are for external space. We can notice that for $\nu = \mu$

$$G_a^{\mu\nu}(x) = 0 \tag{5.121}$$

because when $\nu = \mu$

$$G_a^{\mu\mu}(x) = -f_{abc} G_b^{\mu} G_c^{\mu}$$

in which f_{abc} being antisymmetric and $G_b^{\mu} G_c^{\mu} = G_c^{\mu} G_b^{\mu}$, the contraction vanishes. Taking $f_{abc} = \varepsilon_{abc}$ (Levi–Civita antisymmetric tensor), the vanishing can also be seen explicitly. This situation is similar to electromagnetic case, where the diagonal elements of

$$F^{\mu\nu} = \partial^{\mu} A^{\nu} - \partial^{\nu} A^{\mu} \tag{5.122}$$

vanish. In the electromagnetic case we have

$$F^{\mu\nu} = \begin{pmatrix} 0 & -E^1 & -E^2 & -E^3 \\ E^1 & 0 & -B^3 & B^2 \\ E^2 & B^3 & 0 & -B^1 \\ E^3 & -B^2 & B^1 & 0 \end{pmatrix} \tag{5.123}$$

where

$$E^i = F^{i0} = \partial^i A^0 - \partial^0 A^i \tag{5.124a}$$

and

$$B^i = -\frac{1}{2} \varepsilon^{ijk} F^{jk} \tag{5.124b}$$

Corresponding to Yang–Mills field tensor (5.120) as well, we can make the similar identification of electric and magnetic fields. In view of eqn. (5.117), we shall however, have

$$\boldsymbol{E} \equiv \boldsymbol{E}_a T_a$$

and

$$\boldsymbol{B} \equiv \boldsymbol{B}_a T_a \tag{5.125}$$

in which $\boldsymbol{E}_a$ and $\boldsymbol{B}_a$ can be obtained by replacing $F^{\mu\nu} \to G_a^{\mu\nu}$. Accordingly, we obtain for the electric and magnetic fields from the Yang–Mills field tensor

$$E_a^i = G_a^{i0} = \partial^i G_a^0 - \partial^0 G_a^i - g\, f_{abc} G_b^i G_c^0 \tag{5.126}$$

and

$$\begin{aligned} B_a^i &= -\frac{1}{2} \varepsilon^{ijk} G^{jk} \\ &= -\frac{1}{2} \varepsilon^{ijk} (\partial^j G_a^k - \partial^k G_a^j - g\, f_{abc} G_b^j G_c^k) \\ &= (\partial^k G_a^j - \partial^j G_a^k)_i + \frac{1}{2} g\, f_{abc} (G_b^j G_c^k - G_c^j G_b^k)_i \end{aligned} \tag{5.127}$$

The above expressions for E_a^i and B_a^i may be written in a more familiar form by using

$$\boldsymbol{E}_a = (E_a^1, E_a^2, E_a^3)$$

and

$$\boldsymbol{B}_a = (B_a^1, B_a^2, B_a^3) \tag{5.128}$$

to get

$$\boldsymbol{E}_a = -\nabla G_a^0 - \frac{\partial \boldsymbol{G}_a}{\partial t} - g f_{abc} \boldsymbol{G}_b G_c^0 \tag{5.129}$$

and

$$\boldsymbol{B}_a = (\nabla \times \boldsymbol{G}_a) + \frac{1}{2} g\, f_{abc} (\boldsymbol{G}_b \times \boldsymbol{G}_c) \tag{5.130}$$

which can be substituted in eqn. (5.125) to obtain $\boldsymbol{E}$ and $\boldsymbol{B}$ for Yang–Mills case.

$$-\partial^i = -\nabla$$

An important observation can be made from eqn. (5.130) by taking its three-divergence. We obtain

$$\nabla.\boldsymbol{B}_a = \frac{1}{2} g f_{abc} \nabla.(\boldsymbol{G}_b \times \boldsymbol{G}_c) \neq 0 \tag{5.131}$$

Comparing eqn. (5.131) with the Maxwell's field equation which is for the Abelian gauge field (e.m.field)

$$\nabla.\boldsymbol{B} = 0 \tag{5.132}$$

where RHS is zero because no magnetic charge density is assumed to exist, we find that the non-Abelian gauge fields show that the magnetic charge density can exist $(\nabla \cdot \boldsymbol{B}_a \neq 0)(\nabla.\boldsymbol{B}_a \neq 0)$. The total magnetic charge is a surface integral at spatial infinity and it does not vanish for special solutions called the *monopole solutions*.

5.7 SU(3) GLOBAL GAUGE INVARIANCE

This case of SU(3) is a natural extension of SU(2) group. The group SU(3) is a group of unitary unimodular transformation on three objects. The three objects may be

$$\begin{pmatrix} p \\ n \\ \Lambda \end{pmatrix} \equiv \begin{pmatrix} u \\ d \\ s \end{pmatrix} \text{ or } \begin{pmatrix} R \\ B \\ G \end{pmatrix} \tag{5.133}$$

If (u, d, s), the up, down and strange flavours of quarks are the three objects, the group is called the SU(3) flavour group. Since each quark comes in three colours red, blue and green and if (R, B, G) are the three objects, we have SU(3) colour group. We choose here SU(3) colour group. The reason is that in SU(3) flavour, the masses of different flavours (i.e., the quarks of different flavours) are not same, whereas in case of (*R*, *B*, *G*) a quark of a particular flavour has three colours. This means the same mass is associated with different colours (*R*, *B*, *G*).

The Lagrangian density for SU(3) colour would then be the sum of Lagrangian densities of quarks of different colours. Remember, quark is spin $-\frac{1}{2}$ object.

$$\begin{aligned} £ &= £_{\text{red}} + £_{\text{blue}} + £_{\text{green}} \\ &= (i\bar{\psi}_R\gamma^\mu\partial_\mu\psi_R - m\bar{\psi}_R\psi_R) + (i\bar{\psi}_B\gamma^\mu\partial_\mu\psi_B - m\bar{\psi}_B\psi_B) \\ &\quad + (i\bar{\psi}_G\gamma^\mu\partial_\mu\psi_G - m\bar{\psi}_G\psi_G) \end{aligned} \tag{5.134}$$

Now, introducing

$$\psi = \begin{pmatrix} \psi_R \\ \psi_B \\ \psi_G \end{pmatrix}, \ \bar{\psi} = (\bar{\psi}_R, \bar{\psi}_B, \bar{\psi}_G) \tag{5.135}$$

we may write eqn. (5.134), similar to SU(2) case, as

$$£ = i\bar{\psi}\gamma^\mu\partial_\mu\psi - m\bar{\psi}\psi \tag{5.136}$$

The unitary matrix is now

$$U = e^{iA}, \ A^\dagger = A \tag{5.137}$$

with (see eqn. 5.61)

$$A = 1.\theta_0 + \lambda.\theta \tag{5.138}$$

where 1 is 3 × 3 unit matrix, λ_1, λ_2, ...λ_8 are the eight Gell–Mann matrices [eqn. (3.15)]

$$\lambda.\theta = \lambda_1\theta_1 + \lambda_2\theta_2 + \lambda_3\theta_3 + \ldots + \lambda_8\theta_8 \tag{5.139}$$

Hence,

$$U = e^{i\theta_0}.e^{i\lambda.\theta} \tag{5.140}$$

The first part of *U*, namely, $e^{i\theta_0}$ has already been discussed under U(1) transformations. $e^{i\lambda\theta}$ is the new term, which provides the unitary matrix

$$U = e^{i\frac{\lambda}{2}.\theta} \equiv e^{i\lambda.\alpha} \tag{5.141}$$

with $\alpha = \frac{\theta}{2}$, for SU(3) group. In terms of its generators

$$T_a = \frac{\lambda_a}{2} \tag{5.142}$$

the unitary matrix is

$$U = e^{iT_a\theta_a} \tag{5.143}$$

The matrices λ_a obey

$$\left[\frac{\lambda_a}{2}, \frac{\lambda_b}{2}\right] = i\, f_{abc} \frac{\lambda_c}{2}$$

or

$$[T_a, T_b] = i\, f_{abc} T_c \tag{5.144}$$

where f_{abc} are the real constants, called the structure constants of SU(3) group. We may follow the steps similar to SU(2) case i.e., the steps below eqn. (5.68) and arrive at the following equation, which is the corresponding eqn. of (5.77).

$$\partial_\mu \left[i g_s \frac{\partial \pounds}{\partial(\partial_\mu \psi)} \lambda_a \psi \right] = 0 \tag{5.145}$$

which on writing as

$$\partial_\mu j_a^\mu = 0 \tag{5.146}$$

with

$$\begin{aligned} j_a^\mu &= i g_s \frac{\partial \pounds}{\partial(\partial_\mu \psi)} \lambda_a \psi \\ &= -g_s \bar{\psi} \gamma^\mu \lambda_a \psi \\ &\equiv -g_s \bar{\psi} \gamma^\mu T_a \psi \end{aligned} \tag{5.147}$$

suggests that the global gauge invariance of SU(3) colour group projects out the eight (a = 1, 2, 3 ..., 8) conserved colour currents.

5.8 SU(3) LOCAL GAUGE SYMMETRY AND QCD

When the parameters α of eqn. (5.141) is space–time dependent, the Lagrangian density [eqn. (5.136)], is no longer invariant under

$$\psi(x) \to \psi'(x) = e^{i\lambda.\alpha\,(x)} \psi(x) \tag{5.148}$$

As usual, the restoration of gauge invariance is through the introduction of a covariant derivative

$$D_\mu = \partial_\mu + i g \frac{\lambda}{2} . \boldsymbol{G}_\mu \tag{5.149}$$

where $g \Rightarrow g_s$ of eqn. (5.147), the SU(3) analogue of e of U(1). e, g and g_s are identified as coupling constants of electromagnetic [U(1)], weak [SU(2)] and strong [SU(3) colour] interactions. The field $\boldsymbol{G}_\mu$ in the covariant derivative (5.149) is now the gauge field of the SU(3) group. In fact, we have eight gauge fields $G_\mu^a (a = 1, 2, 3..., 8)$ and it is better to write the covariant derivative (5.149) as

$$D_\mu = \partial_\mu + i g T_a G_\mu^a \tag{5.150}$$

The locally gauge invariant SU(3) Lagrangian density would be from eqn. (5.136)

$$\pounds_{\text{local}} = i\bar{\psi}\gamma^{\mu}D_{\mu}\psi - m\bar{\psi}\psi - \frac{1}{4}G^{a}_{\mu\nu}G^{\mu\nu}_{a}$$

$$= [i\bar{\psi}\gamma^{\mu}\partial_{\mu}\psi - m\bar{\psi}\psi] - g(\bar{\psi}\gamma^{\mu}T_{a}\psi)G^{a}_{\mu} - \frac{1}{4}G^{a}_{\mu\nu}G^{\mu\nu}_{a} \tag{5.151}$$

where the naturally occurring interaction term is $\left[-g(\bar{\psi}\gamma^{\mu}T_{a}\psi)G^{a}_{\mu}\right]$ and $[-\frac{1}{4}G^{a}_{\mu\nu}G^{\mu\nu}_{a}]$ is the kinetic term for the gauge fields [G^{a}_{μ}]. Proceeding in a manner of SU(2) local gauge symmetry, it can be easily found that the gauge fields G^{a}_{μ} transform as eqn. (5.94)

$$\boldsymbol{G}'_{\mu} = \boldsymbol{G}_{\mu} - 2(\alpha \times \boldsymbol{G}_{\mu}) - \frac{2}{g}\partial_{\mu}\alpha \tag{5.152}$$

which on using for SU(3)

$$(\alpha \times \boldsymbol{G}_{\mu})_{a} = \sum_{b,c=1}^{8} f_{abc}\alpha_{b}G^{c}_{\mu} \tag{5.153}$$

can be expressed as

$$G'^{a}_{\mu} = G^{a}_{\mu} - \frac{2}{g}\partial_{\mu}\alpha_{a} - 2f_{abc}\alpha_{b}G^{c}_{\mu} \tag{5.154}$$

and the SU(3) field tensor is

$$G^{a}_{\mu\nu} = \partial_{\mu}G^{a}_{\nu} - \partial_{\nu}G^{a}_{\mu} - g\, f_{abc}G^{b}_{\mu}G^{c}_{\nu} \tag{5.155}$$

Obviously, the eight gauge fields G^{a}_{μ} are massless. The quanta associated with these eight fields are eight gluons and are massless because G^{a}_{μ} are massless. As discussed in SU(2) case, the third term in the RHS of eqn. (5.154) describes that the gauge fields $\boldsymbol{G}_{\mu}$ carry charge, here the colour charge.

Now, let us describe the Lagrangian density (5.151). This is the Lagrangian density for a particular flavour of quark (mass m), which is invariant under SU(3) local gauge transformations. This Lagrangian density describes three equal mass Dirac fields (three colours of a particular quark flavour) which are in interaction with eight massless vector fields (quanta gluons). The Dirac fields constitute eight colour currents

$$j^{\mu}_{a} = (-)g\bar{\psi}\gamma^{\mu}T_{a}\psi \tag{5.156}$$

which act as the sources of colour fields G^{a}_{μ}, in the same way the electric currents act as the source of electromagnetic field.

Just as the U(1) local gauge invariance led us to the interacting field theory of quantum electrodynamics, the local gauge symmetry of SU(3) colour group leads us to the interacting theory of Quantum Chromodynamics (*QCD*). The Lagrangian density (eqn. 5.151) is then the gauge invariant *QCD* Lagrangian density. It should be remembered that there would be six of such Lagrangian densities corresponding to each of the six flavours of quarks. The distinguishing feature would be the flavor mass m_u, m_d, m_s, m_c, m_t, and m_b.

The *QCD* theory is very close in structure with Yang–Mills theory. However, SU(2) theory is the theory of weak interactions and *QCD* is the theory of strong interactions.

5.9 SPONTANEOUS SYMMETRY BREAKING

We have seen that the vector gauge field, which is introduced to achieve the local gauge invariance, has to be massless. This is due to the fact that the mass term having the form $\frac{1}{2}m^2\boldsymbol{G}_\mu\boldsymbol{G}^\mu$ is not invariant under the transformations of G_μ's. In case of U(1) local gauge invariance the gauge field is the electromagnetic field. The masslessness of electromagnetic field agrees with the observation as this field has infinite range for which we do need a massless quantum (photon). In case of colour SU(3) local gauge invariance, the gauge fields are eight colour fields whose quanta are gluons, which too have to be massless; otherwise local gauge invariance would be violated. The SU(3) colour gauge theory (*QCD*) is the theory of strong interactions, which, however, have a short range. The massless gluon would provide an infinite range to strong interactions and such an inference would be against the observed facts. The explanation to this contradiction is that the strong interaction acting between say, proton and neutron is via a pion which is massive hence accounts for short range. The proton, neutron and pion are colour singlets meaning that their states remain unaffected by colour SU(3) transformations. Only a singlet (pion) can be exchanged between two singlets (proton and neutron). Since singlet gluons do not exist, the gluons cannot be exchanged between proton and neutron. The manifestation of strong interactions is only through the pions (massive), which explains their short range. The quark confinement and absence of singlet gluon save the strong interactions from having an infinite range.

The SU(2) local gauge invariance too introduces a massless gauge field and accordingly, the weak interactions [gauge group $SU(2)_L$] too should have an infinite range but the reality is that they too have a quite short range. We do not have an explanation for this short range similar to one for SU(3). Therefore the masslessness of gauge fields does create a problem in explaining the short range of weak interactions. The problem can be tackled by manipulating the gauge theory in such a way that it can accommodate massive gauge fields. The procedure to achieve this is known as spontaneous symmetry breaking. We would consider the spontaneous breaking of both global gauge symmetry and local gauge symmetry. The mechanism responsible for the spontaneous breaking of local gauge symmetry is known as Higgs mechanism. The gauge fields, which are initially massless, acquire mass due to spontaneous breaking of gauge symmetry. In order to know that the gauge field has acquired mass, we must look at the mass term in the Lagrangian density, about which we talk in the following.

5.9.1 The Mass Term

Let us look at the Lagrangian density for a neutral scalar (Klein–Gordon) field (Appendix B)

$$£ = \frac{1}{2}(\partial_\mu \phi \partial^\mu \phi - m^2\phi^2) \tag{5.157}$$

where the first term is the kinetic term and the second term which contains the square of the field (i.e. ϕ^2) is the mass term. It is important to note that the relative sign between the kinetic and mass term is negative. However, the Lagrangian may not show the mass term as clearly as the eqn. (5.157) shows. For instance, from

$$£ = \frac{1}{2}\partial_\mu \phi \partial^\mu \phi + e^{-(\beta\phi)^2} \tag{5.158}$$

we cannot just see the mass term. However, expanding the second term, we may write

$$£ = \frac{1}{2}\partial_\mu \phi \partial^\mu \phi + 1 - \beta^2\phi^2 + \frac{1}{2}\beta^4\phi^4 - .... \tag{5.159}$$

from which we may observe that there is a relative minus sign between the kinetic term $\frac{1}{2}\partial_\mu \phi \partial^\mu \phi$ and the term $\beta^2\phi^2$. So comparing the Lagrangian density (5.159) with eqn. (5.157), we find

$$\beta^2 = \frac{1}{2}m^2 \tag{5.160}$$

from which we infer that the field ϕ of eqn. (5.158) has mass equal to $\beta\sqrt{2}$. Depending on the field, i.e., whether the field is scalar, vector or spinor, the Lagrangian densities given in Chapter 2, can be taken as reference for comparing with a given Lagrangian density. In the above example, we see that by mere expanding the exponential term the mass term that was hidden in eqn. (5.158) form of the Lagrangian density gets revealed in its eqn. (5.159) form. The factor 1 is a constant and does not change the field equations hence meaningless.

5.9.2 Mass Generation through Spontaneous Symmetry Breaking

The spontaneous symmetry breaking provides a way to generate mass of a particle. To understand the manner in which a particular symmetry is broken, we consider a simple system made of only the scalar (spin zero and neutral) field with an interaction term proportional to the fourth power of field. Let such a system be described by a Lagrangian density

$$\pounds \equiv T - V = \frac{1}{2}\partial_\mu \phi \partial^\mu \phi - (\tfrac{1}{2}\mu^2\phi^2 + \tfrac{1}{4}\lambda\phi^4) \tag{5.161}$$

This Lagrangian density is symmetric under $\phi \to -\phi$. This means the Lagrangian density (5.161) has the reflection symmetry. We expect that this reflection symmetry is the one that will be spontaneously broken. μ and λ are two important parameters in eqn. (5.161). From the outset λ is taken positive ($\lambda > 0$), however for μ, we can have both the conditions (I) $\mu^2 > 0$ and (II) $\mu^2 < 0$. For ($\lambda < 0$), the Hamiltonian of the system does not have a minimum and hence no ground state.

Case I: $\mu^2 > 0$

When the parameter μ is positive, the potential energy is just

$$V(\phi) = \tfrac{1}{2}\mu^2\phi^2 + \tfrac{1}{4}\lambda\phi^4 \tag{5.162}$$

which when plotted against ϕ has the shape of the Fig. (5.1).The minimum of the

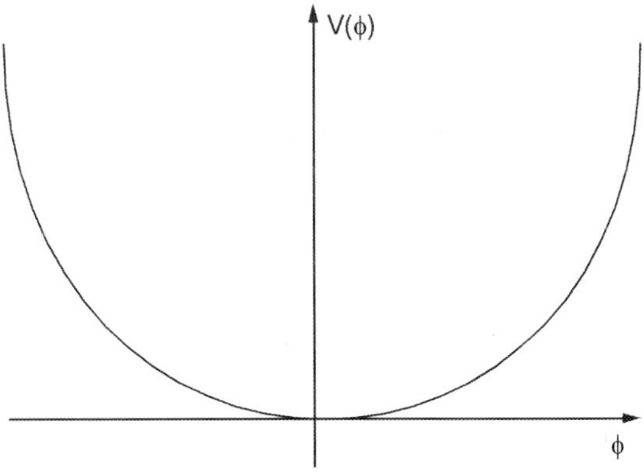

Fig. 5.1: The potential $V(\phi) = \frac{1}{2}\mu^2\phi^2 + \frac{1}{4}\lambda\phi^2$ for $\lambda > 0$, $\mu^2 > 0$

potential energy is obtained from

$$\frac{dV(\phi)}{d\phi} = 0 \text{ and } \frac{d^2V(\phi)}{d\phi^2} > 0 \tag{5.163}$$

$$\frac{dV(\phi)}{d\phi} = \phi(\mu^2 + \lambda\phi^2) = 0$$

implying that

$$\phi = 0 \text{ and } \phi^2 = -\frac{\mu^2}{\lambda} \tag{5.164}$$

$$\frac{d^2V(\phi)}{d\phi^2} = \mu^2 + 3\lambda\phi^2 \tag{5.165}$$

For $\phi = 0, \quad \frac{d^2V(\phi)}{d^2\phi^2} = \mu^2 \text{(positive)}$

For $\phi^2 = -\frac{\mu^2}{\lambda}, \quad \frac{d^2V(\phi)}{d\phi^2} = \mu^2 - 3\mu^2 = -2\mu^2 \text{ (negative)} \quad (5.166)$

Therefore for $\mu^2 > 0$ the minimum of potential energy $V(\phi)$ is obtained when $\phi = 0$. Corresponding to $\phi = 0$, the kinetic term $\partial_\mu\phi\partial^\mu\phi$ is obviously zero, hence at $\phi = 0$; the total energy is equal to the minimum of potential energy. The minimum energy state is the vacuum state, which here corresponds to $\phi = 0$. At $\phi = 0$, $V(\phi) = 0$, implying that $V(-\phi) = 0$. This means the ground state of the system retains the reflection ($\phi \rightarrow -\phi$) symmetry. When the symmetry of the Lagrangian is maintained by the ground state as well, there is no breaking of symmetry. Thus in the case of $\mu^2 > 0$, the symmetry (reflection) of the Lagrangian density (5.161) is not broken. The Lagrangian density (5.161) then ($\mu^2 > 0$) describes a scalar (neutral) particle of mass μ and $\frac{1}{4}\lambda\phi^4$ term shows that the ϕ field interacts with itself and there is a four-particle vertex with coupling constant λ.

Case II: $\mu^2 > 0$

This case of negative μ^2 is interesting. In this case also, the extremum would be at

$$\phi = 0 \text{ or } \phi^2 = -\frac{\mu^2}{\lambda} \tag{5.167}$$

where μ^2 now is negative. Substituting $\phi = 0$ in

$$\frac{d^2V(\phi)}{d\phi^2} = \mu^2 + 3\lambda\phi^2$$

we get

$$\frac{d^2V(\phi)}{d\phi^2} = \mu^2 \text{(negative)} \tag{5.168}$$

So at $\phi = 0$, the potential energy $V(\phi)$ is not minimum. However, for $\phi^2 = -\frac{\mu^2}{\lambda}$,

$$\frac{d^2V(\phi)}{d\phi^2} = -2\mu^2 \text{ (positive)} \tag{5.169}$$

Let

$$\phi^2 = -\frac{\mu^2}{\lambda} = v^2 \text{ say}$$

or

$$\phi = \pm v = \pm\sqrt{\frac{-\mu^2}{\lambda}} \tag{5.170}$$

Thus, in the $\mu^2 < 0$ case, the potential energy of the system is minimum at two values of ϕ. One at $+\nu$ and other at $-\nu$. Since ν has a constant value, the kinetic energy term $\partial_\mu\phi\partial^\mu\phi$ corresponding to $\phi = \pm\nu$ (constant) would again be zero and hence the minimum energy state, i.e., the ground state would be the minimum potential energy state. Therefore, for $\mu^2 < 0$, the ground state corresponds to $\phi = \pm\nu$. The ground state now does not belong to $\phi = 0$. The potential energy graph for this case has been drawn in figure (5.2). With the vacuum state corresponding to two values of ϕ, the system will choose one of these values. The choice of $\phi = \pm\nu$ or $\phi = -\nu$ but not both, results in the spontaneous breaking of the symmetry of the Lagrangian density. We choose $\phi = +\nu$. The Lagrangian density is invariant under $\phi \rightarrow -\phi$ but the ground state $\phi = +\nu$ because $+v \neq -v$, is not similarly invariant.

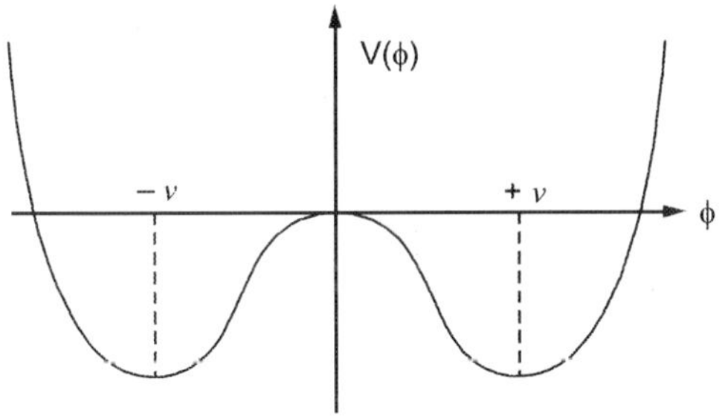

Fig. 5.2: The potential $V(\phi) = \frac{1}{2}\mu^2\phi^2 + \frac{1}{4}\lambda\phi^4$ for $\lambda > 0$, $\mu^2 < 0$, $\nu = \sqrt{\frac{-\mu^2}{\lambda}}$

When μ^2 is negative, the Lagrangian density (5.161) is not able to convey the mass of the field ϕ. If we read it as such, because of negative μ^2, the term $(-\frac{1}{2}\mu^2\phi^2)$ suggests an imaginary mass and thus makes the system unphysical. In order to interpret the Lagrangian density (5.161) correctly for $\mu^2 < 0$ case, we treat the field (ϕ) as the fluctuations around the state of minimum energy. Such a concept of generating a field is known as Feyman's perturbative technique in which fluctuations are perturbations around the ground state. Since, we have chosen $\phi = +\nu$ as the ground state and let $\eta(x)$ denote the fluctuations (a new field) about $\phi = +\nu$, we may write

$$\phi(x) = \nu + \eta(x) \tag{5.171}$$

Since, $\nu = \sqrt{\frac{-\mu^2}{\lambda}}$ is a constant, we may write

$$\partial_\mu\phi = \partial_\mu\eta \tag{5.172}$$

Substituting equations (5.171) and (5.172) into the Lagrangian density (5.161), we may write it as

$$£_{new} = \frac{1}{2}\partial_\mu \eta \partial^\mu \eta - \frac{1}{2}\mu^2(v^2 + 2v\eta + \eta^2) - \frac{1}{4}\lambda(v^2 + 2v\eta + \eta^2)^2 \qquad (5.173)$$

Replacing $\mu^2 = -\lambda v^2$, above equation may be simplified to

$$£_{new} = \frac{1}{2}\partial_\mu \eta \partial^\mu \eta - \lambda v^2 \eta^2 - \lambda v \eta^3 - \frac{1}{4}\lambda \eta^4 + \frac{1}{4}\lambda v^4 \qquad (5.174)$$

Remember μ^2 is negative implying that v^2 is positive. Now we may compare £ and $£_{new}$. We rewrite £

$$£ = \frac{1}{2}\partial_\mu \phi \partial^\mu \phi - \frac{1}{2}\mu^2 \phi^2 - \frac{1}{4}\lambda \phi^4 \qquad (5.175)$$

and note the following points

1. £ is the Lagrangian density of field $\phi(x)$.
2. $£_{new}$ is the Lagrangian density of field $\eta(x)$.
3. Whereas £ **is symmetric** under $\phi \to -\phi$, $£_{new}$ is **not symmetric** under $\eta \to -\eta$. The term $\lambda v \eta^3$ ***destroys*** this symmetry.
4. Since $\mu^2 < 0$, the mass of the ϕ field from £ is unphysical (imaginary), whereas v^2 being positive the term $\lambda v^2 \eta^2$ in $£_{new}$ has the correct minus sign relative to the kinetic $\frac{1}{2}\partial_\mu \eta \partial^\mu \eta$ term, hence the η field possesses a real (physical) mass given by

$$\frac{1}{2}m_\eta^2 = \lambda v^2$$

or

$$m_\eta = \sqrt{2\lambda v^2} = \sqrt{-2\mu^2} \ (\mu^2 < 0) \qquad (5.176)$$

Thus we find that choosing the ground state $\phi = v$ and taking $\phi = v + \eta$, £ became $£_{new}$ in which £ was symmetric and $£_{new}$ not. The symmetry of the Lagrangian density £ was spontaneously broken when we made the change $\phi = v + \eta$. Since, no external agency is responsible to cause the symmetry breaking; we call it spontaneous symmetry breaking. An example where external agency breaks the symmetry is the three dimensional symmetry which is broken by gravity. The up and down determined by external agency (gravity) are quite different from left and right.

The Lagrangian density £ and $£_{new}$ are equivalent because $£_{new}$ is obtained from £ by shifting ϕ from v to $v + \eta$. The physics from both the densities is same. The mass of the scalar particle which was unphysical (hidden) in £ got revealed in $£_{new}$, where the reflection symmetry was broken by our choice of the ground state, $\phi = v$. When a particular ground state is selected, the symmetry in fact becomes hidden and is no longer manifest.

5.9.3 Spontaneous Breaking of Global Gauge Symmetry

In the example that was considered in the previous section, we had only two ground states $\phi = \pm \nu$. A physical example of two ground states is buckling of a thin strip. If we press the ends of a thin plastic strip, e.g., a short ruler, it will snap into a curved configuration, but it can buckle either left or to the right–both are ground states of the system. The moment the strip buckles to one of the two positions, the left-right symmetry is broken. Since, we have in the above example only two ground states, the section 5.5.2 described the case of spontaneous breaking of a discrete symmetry. In the present section, we consider the spontaneous breaking of a continuous symmetry (global gauge), where it will be seen that the new Lagrangian density provides the example of the Goldstone theorem.

In this case, we consider a complex scalar field

$$\phi = \frac{\phi_1 + i\phi_2}{\sqrt{2}} \tag{5.177}$$

where ϕ_1 and ϕ_2 are real. The Lagrangian density for a complex scalar field is

$$£ = (\partial_\mu \phi)^* (\partial^\mu \phi) - \mu^2 \phi^* \phi - \lambda(\phi^* \phi)^2 \tag{5.178}$$

Obviously, under the global gauge transformation

$$\phi(x) \to \phi'(x) = e^{i\alpha} \phi(x)$$

implying also that

$$\partial_\mu \phi(x) \to \partial_\mu \phi'(x) = e^{i\alpha} \partial_\mu \phi(x) \tag{5.179}$$

£ is invariant. Since, eqn. (5.179) is U(1) global gauge transformation, the Lagrangian density £ is symmetric under U(1) global gauge transformation. In other words, the U(1) global gauge symmetry is the symmetry of the £ [eqn. (5.178)] and this U(1) global gauge symmetry would be seen spontaneously broken. In the previous section of discrete symmetry, the £ had reflection symmetry that was broken.

Here too, we will consider ($\lambda > 0$). As before, we shall have two cases: (1) $\mu^2 > 0$ and (2) $\mu^2 < 0$. Without repeating the details of the previous section, we can say that for $\mu^2 > 0$, the lowest state is $\phi = 0$ or $\phi^* = 0$ and there is no symmetry breaking. In this case, the Lagrangian density (5.178) represents a complex scalar field of mass μ (i.e., the charged scalar particles of mass μ) with its interaction denoted by λ term.

To deal with the $\mu^2 < 0$ case, we rewrite the Lagrangian density (5.178) in terms of two real scalar fields ϕ_1 and ϕ_2. Using eqn. (5.177) and its complex conjugate, we may write

$$\begin{aligned} £ = &\frac{1}{2}(\partial_\mu \phi_1)(\partial^\mu \phi_1) + \frac{1}{2}(\partial_\mu \phi_2)(\partial^\mu \phi_2) \\ &- \frac{1}{2}\mu^2(\phi_1^2 + \phi_2^2) - \frac{1}{4}\lambda(\phi_1^2 + \phi_2^2)^2 \end{aligned} \tag{5.180}$$

The potential energy term $V(\phi_1, \phi_2)$ is

$$V(\phi_1,\phi_2)=\frac{1}{2}\mu^2(\phi_1^2+\phi_2^2)+\frac{1}{4}\lambda(\phi_1^2+\phi_2^2)^2 \tag{5.181}$$

The extremum condition

$$\frac{\partial V}{\partial \phi_1}=0=\frac{\partial V}{\partial \phi_2}$$

gives

$$\phi_1[\mu^2+\lambda(\phi_1^2+\phi_2^2)]=0=\phi_2[\mu^2+\lambda(\phi_1^2+\phi_2^2)] \tag{5.182}$$

from which the extremum occurs at

(a) $\phi_1 = 0$

(b) $\phi_2 = 0$

(c) $\phi_1^2+\phi_2^2=-\dfrac{\mu^2}{\lambda}=v^2$ (5.183)

Now from

$$\frac{d^2V(\phi)}{d\phi^2}=\mu^2+3\lambda\phi^2(\phi\Rightarrow\phi_1,\phi_2) \tag{5.184}$$

we find that for $\phi_1 = 0$ and $\phi_2 = 0$

$$\frac{d^2V}{d\phi_1^2}=\mu^2=\frac{d^2V}{d\phi_2^2}=\text{negative }(\mu^2<0) \tag{5.185}$$

which imply that $V(\phi_1, \phi_2)$ is not minimum at $\phi_1 = 0$ and $\phi_2 = 0$. The only possibility of becoming $V(\phi_1, \phi_2)$ minimum is when [eqn. (5.183c)]

$$\phi_1^2+\phi_2^2=v^2\ (v^2=-\frac{\mu^2}{\lambda}) \tag{5.186}$$

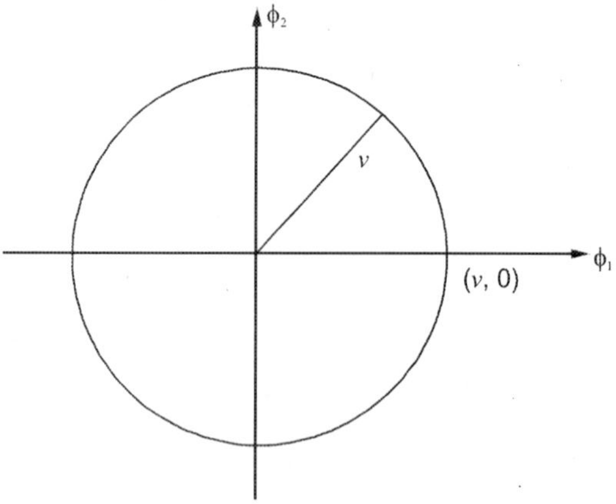

Fig. 5.3: The circle $\phi_1^2+\phi_2^2=v^2$

We observe that the eqn. (5.186) is the equation of a circle with radius ν. If the equation (5.186) makes the $V(\phi_1, \phi_2)$ minimum, it means at every point on the circle the $V(\phi_1, \phi_2)$ must be minimum. We may now choose on the circle the point $\phi_2 = 0$ and $\phi_1 = \nu$ (Fig. 5.3) at which too $V(\phi_1, \phi_2)$ must be minimum. Then

$$(\phi_1, \phi_2) = (v, 0) \tag{5.187}$$

$$\frac{\partial^2 V(\phi)}{\partial \phi_1^2} = \mu^2 + 3\lambda\phi_1 = \mu^2 + 3\lambda v \; (v = -\frac{\mu^2}{\lambda})$$

$$= -2\mu^2 = \text{positive} \tag{5.188a}$$

For the point $(\phi_1, \phi_2) = (0, \nu)$ on the circle, we would also get

$$\frac{\partial^2 V(\phi)}{\partial \phi_2^2} = \text{positive} \tag{5.188b}$$

Since the second differentials have come out to be positive, we will call the circle described by eqn. (5.186) as the circle of minima. $V(\phi_1, \phi_2)$ and the circle of minima have been plotted in Fig. (5.4). In the previous section of discrete symmetry, we had only two states, corresponding to which, $V(\phi_1)$ was minimum. Here we have a circle of minima, implying that we have infinite states (infinitesimally close to each other), where $V(\phi_1)$ is minimum. The perturbation calculation will have to be about any one of these infinite states and we choose that one state as given by eqn. (5.187), i.e., the state $\phi_1 = \nu$, $\phi_1 = 0$. Since, v and zero are the minimum values of ϕ_1 and ϕ_2 respectively, we may express

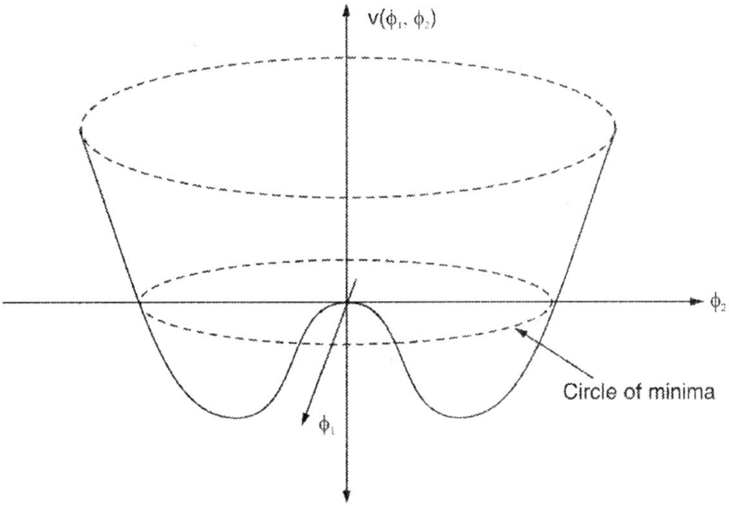

Fig. 5.4: The potential function $V(\phi_1, \phi_2)$, and the circle of minima

$$\phi_{1\min} = \nu$$

and

$$\phi_{2\min} = 0 \tag{5.189}$$

Choosing one state (5.189) out of infinite ground states, breaks the symmetry of the Lagrangian spontaneously because eqn. (5.189) is not invariant under the transformation that keeps £ invariant.

Now, the field $\phi_1(x)$ can be treated as the fluctuations about $\phi_{1\ \min}$ and $\phi_2(x)$ can be treated as fluctuations about $\phi_{2\ \min}$. Let $\eta(x)$ denote the fluctuations about $\phi_{1\ \min} = \nu$ and $\xi(x)$ the fluctuations about $\phi_{2\ \min} =$ zero. Obviously $\eta(x)$ and $\xi(x)$ are also real fields. Thus we may write

$$\phi_1(x) = v + \eta(x)$$

and

$$\phi_2(x) = 0 + \xi(x) = \xi(x) \tag{5.190}$$

Now we bring back the field $\phi(x)$ and substitute eqn. (5.190) into eqn. (5.177) to write

$$\phi(x) = \frac{1}{\sqrt{2}}[v + \eta(x) + i\xi(x)] \tag{5.191}$$

and then substitute it (eqn.5.191) and its complex conjugate into the Lagrangian density (5.178). Since, v is a constant, we would write

$$(\partial_\mu \phi)^* = \frac{1}{\sqrt{2}}(\partial_\mu \eta - i\partial_\mu \xi)$$

and

$$(\partial^\mu \phi) = \frac{1}{\sqrt{2}}(\partial^\mu \eta + i\partial^\mu \xi) \tag{5.192}$$

The substitutions yield

$$£_{\text{new}} = \frac{1}{2}\left(\partial_\mu \eta - i\partial_\mu \xi\right)\left(\partial^\mu \eta + i\partial^\mu \xi\right)$$

$$-\frac{1}{2}\mu^2\left[(v+\eta)^2 + \xi^2\right] - \frac{1}{4}\lambda\left[(v+\eta)^2 + \xi^2\right]^2 \tag{5.193a}$$

which on using $\mu^2 = -\nu^2\lambda$ and simplifying becomes

$$£_{\text{new}} = \left[\frac{1}{2}\partial_\mu \eta \partial^\mu \eta - \lambda v^2 \eta^2\right] + \left[\frac{1}{2}\partial_\mu \xi \partial^\mu \xi\right]$$

$$-\left[\lambda v(\eta^3 + \eta\xi^2) + \frac{\lambda}{4}\left(\eta^4 + \xi^4 + 2\eta^2\xi^2\right)\right] + \left(\frac{\lambda}{4}\right)v^4 \tag{5.193b}$$

where the last term, being a constant does not effect the field equations hence has no significance. Recall that due to $\mu^2 < 0$, ν^2 is positive. The $£_{\text{new}}$ in the form (5.193) does not look symmetrical at all, the symmetry of eqn. (5.180) has been spontaneously broken by the selection of a particular vacuum state.

Let us now analyse $£_{new}$ (eqn. 5.193b). We observe the following points.

1. The first square bracket denotes the Lagrangian density of a real scalar field η, which has acquired a mass
$$m_\eta = \sqrt{2\lambda v^2} = \sqrt{-2\mu^2} \qquad (\mu^2 < 0) \tag{5.194}$$
2. The second square bracket contains only the kinetic part of scalar ξ field and there is no corresponding mass term. This means the ξ field is a massless field. Nothing of this kind was observed in the spontaneous breaking of discrete symmetry.
3. The third square bracket denotes various cubic and quartic field interactions shown in the following figure (Fig. 5.5).

The point (2) above requires a little more elaboration. It is only in the spontaneous breaking of a continuous symmetry that we get a massless scalar field ξ. The quantum of ξ field would be a massless scalar boson. *It was proved by J.Goldstone that the spontaneous breaking of a continuous global*

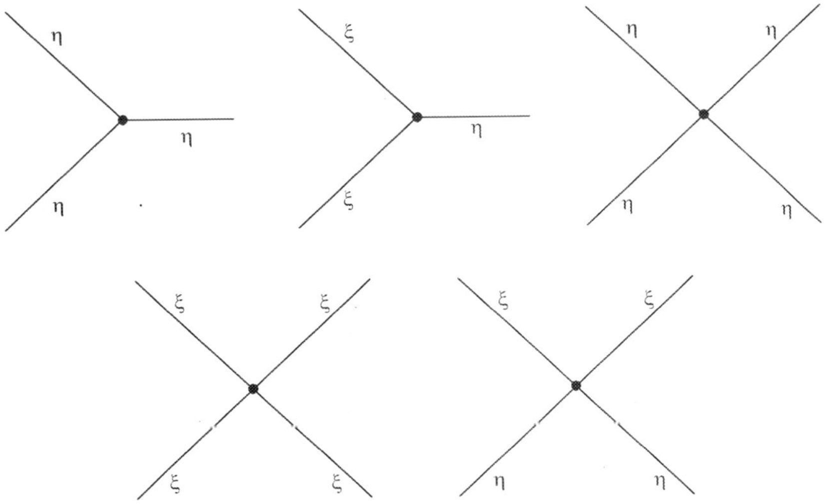

Fig. 5.5: Five kinds of couplings

symmetry always generates one or more massless scalar bosons and such bosons are called Goldstone bosons. In the above example, the ξ bosons are the Goldstone bosons. However, the appearance of massless field ξ is not a welcome situation. The spontaneous symmetry breaking was supposed to account for the mass of the gauge fields of weak interactions but the process introduces a massless scalar boson. Such a boson is not found in the list of particles (elementary). Therefore, a mechanism has to be sought which would remove these Goldstone bosons. The mechanism, which can remove them, is called the *Higgs mechanism*, which we describe in the next section.

5.9.4 The Spontaneous Breaking of U(1) Local Gauge Symmetry

The Higgs Mechanism

In order to eliminate the massless field ξ, which appears when a continuous U(1) global gauge symmetry is broken, we move to the spontaneous breaking of the U(1) local gauge symmetry.

We may again start with the Lagrangian density (5.178), which we rewrite

$$£ = (\partial_\mu \phi)^* (\partial^\mu \phi) - \mu^2 \phi^* \phi - \lambda (\phi^* \phi)^2 \tag{5.195}$$

which is obviously not invariant under the local gauge transformation of ϕ field, viz.,

$$\phi'(x) = e^{i\alpha(x)} \phi(x) \tag{5.196}$$

because

$$\partial_\mu \phi'(x) \neq e^{i\alpha(x)} \partial_\mu \phi(x) \tag{5.197}$$

In order to make £ invariant under the transformation (5.196), we introduce the covariant derivative [section (5.3)]

$$D_\mu = \partial_\mu + ieA_\mu \tag{5.198}$$

where the gauge field transforms as [eqn. (5.44)]

$$A_\mu \to A'_\mu = A_\mu - \frac{1}{e} \partial_\mu \alpha \tag{5.199}$$

The locally gauge invariant Lagrangian density, then is

$$£_{\text{local}} = (D_\mu \phi)^* (D^\mu \phi) - \mu^2 \phi^* \phi - \lambda (\phi^* \phi)^2 - \frac{1}{4} F_{\mu\nu} F^{\mu\nu} \tag{5.200}$$

in which

$$(D_\mu \phi)^* (D^\mu \phi) = [(\partial_\mu - ieA_\mu)\phi^*][(\partial^\mu + ieA_\mu)\phi] \tag{5.201}$$

Again, we have two cases (1) $\mu^2 > 0$, and (2) $\mu^2 < 0$. For $\mu^2 > 0$, the Lagrangian density (5.200) is a quantum electrodynamic Lagrangian density for a charged scalar particle of mass μ with interaction expressed by the λ term. No mass generation takes place in $\mu^2 > 0$ case.

For spontaneous breaking of U(1) local gauge symmetry, we therefore consider $\mu^2 < 0$ in eqn. (5.200). Similar to the method adopted in the previous section, we convert ϕ in $£_{\text{local}}$ into and ϕ_1 and ϕ_2 by using eqn. (5.177) to write it as

$$£_{\text{local}} = \frac{1}{2} (D_\mu \phi_1)(D^\mu \phi_1) + \frac{1}{2} (D_\mu \phi_2)(D^\mu \phi_2)$$
$$- \frac{1}{2} \mu^2 (\phi_1^2 + \phi_2^2) - \frac{1}{4} \lambda (\phi_1^2 + \phi_2^2)^2 - \frac{1}{4} F_{\mu\nu} F^{\mu\nu} \tag{5.202}$$

Since, the potential energy term is same as in the previous section, it would be minimum when

$$\phi_1^2 + \phi_2^2 = -\frac{\mu^2}{\lambda} = \nu^2 \tag{5.203}$$

Out of the infinite states of minimum potential energy described by the circle of minima [eqn. (5.203)], we pick up the state for which

$$\phi_{1\,\min} = \nu$$

and

$$\phi_{2\,\min} = 0 \tag{5.204}$$

It can be seen that for these values (5.204) of ϕ_1 and ϕ_2, the kinetic part $(D_\mu\phi)^*(D^\mu\phi)$ would vanish because $\partial_\mu\phi_1 = 0$ for $\phi_1 = \nu$ and $\partial_\mu\phi_2 = 0$ for $\phi_2 = 0$. So, the minimum energy state (the ground state) would be the state of minimum potential energy. When we choose the ground state as eqn. (5.204), which is not invariant under the transformation $£_{local}$ is, the local gauge symmetry of $£_{local}$ is spontaneously broken.

Then as usual defining the fields ϕ_1 and ϕ_2 as fluctuations around their minimum values, we introduce like eqn. (5.190)

$$\phi_1(x) = v + \eta(x)$$

and

$$\phi_2(x) = 0 + \xi(x) = \xi(x) \tag{5.205}$$

so that we can rewrite

$$\begin{aligned}\phi(x) &= \frac{1}{\sqrt{2}}(\phi_1 + i\phi_2) \\ &= \frac{1}{\sqrt{2}}[v + \eta(x) + i\xi(x)]\end{aligned} \tag{5.206}$$

Substituting eqn. (5.206) in eqn. (5.200), we may write

$$\begin{aligned}£_{local}^{new} = &\frac{1}{2}\left[(\partial_\mu - ieA_\mu)(v+\eta-i\xi)\right]\left[(\partial^\mu + ieA^\mu)(v+\eta+i\xi)\right] \\ &-\frac{1}{2}\mu^2\left[(v^2+\eta^2)^2+\xi^2\right]-\frac{1}{4}\lambda\left[(v^2+\eta^2)^2+\xi^2\right]^2 \\ &-\frac{1}{4}F_{\mu\nu}F^{\mu\nu}\end{aligned} \tag{5.207}$$

Simplifying eqn. (5.207), we find that all the imaginary terms cancel and we get on using $\mu^2 = -\nu^2\lambda$ and rearranging the terms

$$\begin{aligned}
\pounds_{\text{local}}^{\text{new}} &= \left[\frac{1}{2}\partial_\mu \eta \partial^\mu \eta - \lambda v^2 \eta^2\right] + \left[\frac{1}{2}\partial_\mu \xi \partial^\mu \xi\right] \\
&+ \left[-\frac{1}{4}F_{\mu\nu}F^{\mu\nu} + \frac{1}{2}e^2 v^2 A_\mu A^\mu\right] + e v A_\mu \partial^\mu \xi \\
&+ \Big\{e\left[\eta(\partial_\mu \xi) - \xi(\partial_\mu \eta)\right] A^\mu + e^2 v \eta A_\mu A^\mu \\
&+ \frac{1}{2}e^2(\eta^2 + \xi^2) A_\mu A^\mu - \lambda v(\eta^3 + \eta\xi^2) \\
&- \frac{\lambda}{4}(\eta^4 + \xi^4 + 2\eta^2\xi^2)\Big\} + \frac{\lambda}{4}v^4
\end{aligned} \tag{5.208}$$

where the use of

$$\partial_\mu (v + \eta - i\xi) = \partial_\mu \eta - i\partial_\mu \xi, \text{ etc.,} \tag{5.209}$$

is implied.

The analysis of $\pounds_{\text{local}}^{\text{new}}$ is important. We can note the following points. Obviously, this form of Lagrangian density is not symmetric as compared to $\pounds_{\text{local}}$ [eqn. (5.207)]. The local gauge symmetry of eqn. (5.207) has spontaneously broken because we chose a particular ground state [eqn. (5.204)] and this choice resulted in $\pounds_{\text{local}}^{\text{new}}$.

1. The real scalar field η has acquired mass
$$m_\eta = \sqrt{2\lambda v^2} = \sqrt{-2\mu^2} \tag{5.210}$$
This observation is similar to one we had in the global gauge of previous section
2. The second square bracket tells the massless ξ field is still there. Though we had in the beginning told, we are attempting to get rid of ξ field but in the form (5.208), it is still there.
3. The third square bracket tells that the A_μ field has acquired a mass
$$m_{A_\mu} = ev \tag{5.211}$$
It has to be remembered that the vector gauge field A_μ was introduced via the covariant derivative (5.198) and since all gauge fields are required to be massless, A_μ too was massless in the Lagrangian density $\pounds_{\text{local}}$ (eqn.5.202). When the local gauge symmetry of $\pounds_{\text{local}}$ broke, the A_μ field became massive. Since, the gauge fields (like A_μ) are introduced only when we consider local gauge invariance, there were obviously no such terms in the previous section of global gauge symmetry breaking.
4. The term $[evA_\mu(\partial^\mu\xi)]$ is quite a significant interaction term that would be found helping the elimination of ξ field.

5. The various terms in the curly bracket {} also represent the interaction terms of ξ, η and A_μ fields.
6. The last term $\lambda v^4/4$ is a constant, hence of no significance as it cannot change the field equations obtained from $£_{\text{local}}^{\text{new}}$ without this term.

The analysis of $£_{\text{local}}^{\text{new}}$ points out that the Goldstone bosons have not yet been eliminated. To eliminate the massless ξ field, we now discuss the term $[evA_\mu(\partial^\mu\xi)]$, which is read as an interaction term. This is a term, which is bilinear in two different fields A_μ and ξ. The interpretation of such bilinear interaction terms is that in such an interaction, one field converts into other and neither one exists as an independent field. Thus $[A_\mu(\partial^\mu\xi]$ interaction term would suggest that neither the field A_μ nor ξ is independent. When interaction of these two fields takes place, A_μ converts itself into ξ or ξ into A_μ. The interaction vertex is drawn in Fig. (5.6), where it is shown that at the vertex the ξ field has turned into A field. Since, we do

Fig. 5.6: The interaction vertex of A_μ and the ξ fields

not require the ξ field, we would like that ξ field transforms into A_μ field. When such a transformation takes place, the massless ξ field would be eliminated from the Lagrangian density $£_{\text{local}}^{\text{new}}$ [eqn. (5.208)]. We, therefore have to find the condition that the ξ field would disappear.
In order to remove ξ, we notice the local gauge transformation for ϕ, namely,

$$\phi(x) \to \phi'(x) = e^{i\alpha(x)}\phi(x) \tag{5.212}$$

and also recall eqn. (5.206), viz.,

$$\phi(x) = \frac{1}{\sqrt{2}}[v + \eta(x) + i\xi(x)] \tag{5.213}$$

Now, we will make the eqn. (5.213) look like eqn. (5.212). To achieve this, we notice that

$$\phi(x) = \frac{1}{\sqrt{2}}[v + \eta(x)]e^{i\frac{\xi(x)}{v}} \approx \frac{1}{\sqrt{2}}[v + \eta(x) + i\xi(x)] \tag{5.214}$$

provided that only the terms to the lowest order in ξ are taken into account. Let the real fields

$$\eta(x) = h(x)$$

and

$$\xi(x) = \theta(x) \tag{5.215}$$

provide the exponential form to $\phi(x)$, i.e., allow us to write

$$\phi(x) = \frac{1}{\sqrt{2}}[v + h(x)]e^{i\frac{\theta(x)}{v}} \tag{5.216}$$

Now we want that when $\phi(x)$, having the form of eqn. (5.216), transforms to $\phi'(x)$ [according to local gauge transformation (5.212)], the massless field must get eliminated.

Let, corresponding to $\phi'(x)$, the massless field is $\theta'(x)$. Therefore, from eqn. (5.216), we may write

$$\phi'(x) = \frac{1}{\sqrt{2}}[v + h(x)]e^{i\frac{\theta'(x)}{v}} \tag{5.217}$$

The same $\phi'(x)$ as obtained by substituting eqn. (5.216) into eqn. (5.212) gives

$$\phi'(x) = \frac{1}{\sqrt{2}}[v + h(x)]e^{i\left[\alpha(x) + \frac{\theta(x)}{v}\right]} \tag{5.218}$$

Equating equations (5.217) and (5.218), we may write

$$\frac{\theta'(x)}{v} = \alpha(x) + \frac{\theta(x)}{v} \tag{5.219}$$

Now, $\theta'(x)$ is the massless field in ϕ' (i.e., transformed ϕ), which must vanish. Hence, putting $\theta'(x) = 0$ in eqn. (5.219), we get

$$\alpha(x) = -\frac{\theta(x)}{v} \tag{5.220}$$

Thus when the parameter $\alpha(x)$ in the local gauge transformation

$$\phi(x) = e^{i\alpha(x)}\phi(x)$$

takes the value (5.220), the massless fields (ξ fields) would disappear from the Lagrangian density and in that case the transformation of the gauge field A_μ which was [eqn. (5.44)]

$$A'_\mu = A_\mu - \frac{1}{e}\partial_\mu \alpha(x)$$

would become

$$A'_\mu = A_\mu + \frac{1}{ev}\partial_\mu \theta(x) \tag{5.221}$$

Thus we can conclude that when the massless fields have disappeared ($\theta' = 0$, eqn. (5.217) suggests that the complex scalar field should be expressed [$h'(x) = h(x)$] as

$$\phi(x) \to \frac{1}{\sqrt{2}}[v + h(x)] \tag{5.222}$$

where $h(x)$ is a real scalar field and is commonly called the *Higgs field*. Making the above replacement of $\phi(x)$ by $\frac{1}{\sqrt{2}}[v + h(x)]$ in the Lagrangian density is the Higgs mechanism for U(1) gauge symmetry. As we shall see in the next section for SU(2), the Higgs mechanism would make the replacement

$$\phi(x) \to \frac{1}{\sqrt{2}}\begin{pmatrix} 0 \\ v + h(x) \end{pmatrix}$$

in the SU(2) Lagrangian density. At the moment for the U(1) case, we replace both $\phi(x)$ and $\phi^*(x)$ by the RHS of eqn. (5.222) in the Lagrangian density (5.200) to obtain the final Lagrangian density as

$$\begin{aligned} £^{\text{final}}_{\text{local}} = & \left[\frac{1}{2}(\partial_\mu h)(\partial^\mu h) - \lambda v^2 h^2\right] \\ & + \left[-\frac{1}{4}F_{\mu\nu}F^{\mu\nu} + \frac{1}{2}e^2 v^2 A_\mu A^\mu\right] \\ & + \left\{e^2 v h A_\mu A^\mu + \frac{1}{2}e^2 h^2 A_\mu A^\mu - \lambda v h^3 - \frac{1}{4}\lambda h^4\right\} \\ & + \frac{1}{4}\lambda v^4 \end{aligned} \tag{5.223}$$

where again the last term being constant may not be taken into account.

The final Lagrangian density then gives us the following picture.

1. There is no massless field hence Goldstone bosons have been eliminated.
2. The Higgs field h (the real scalar) has mass.

$$\frac{1}{2}m_h^2 = \lambda v^2$$

or

$$m_h = \sqrt{2\lambda v^2} = \sqrt{-2\mu^2}\,(\mu^2 < 0) \tag{5.224}$$

3. The gauge field A_μ continues to have mass (as in $£^{\text{new}}_{\text{local}}$)

$$\begin{aligned} \frac{1}{2}m_{A_\mu}^2 &= \frac{1}{2}e^2 v^2 \\ m_{A_\mu} &= ev \end{aligned} \tag{5.225}$$

4. The terms in the curly bracket now denote the various kinds of couplings between the gauge fields A_μ and the Higgs field h.

The physical interpretation of the above mechanism is significant. The massless fields, which appeared in the previous section, are in fact the oscillations in the parameter that distinguishes different vacuum states. In the present case this parameter is the phase $\alpha(x)$. The quanta of θ field are the Goldstone bosons. We have made $\theta' = 0$ by choosing the gauge as $[\alpha(x) = -\theta/v]$, so that the Goldstone bosons have disappeared.

The final Lagrangian density describes two interacting massive particles namely the Higgs boson (h) and a vector gauge boson (A_μ). A massive vector boson has three degrees of freedom corresponding to three polarization states, two being transverse and third longitudinal. This massless vector particle like photon has two (transverse) degrees of freedom. The massless scalar boson has only one degree of freedom. Thus when massless A_μ became massive A_μ, it required one more degree of freedom. A_μ, acquired it, by swallowing the massless scalar field and naturally the massless field disappeared. The massless scalar field provided the vital third (longitudinal) polarization required by erstwhile massless A_μ to become massive A_μ.

The significance of the interaction term $A_\mu(\partial^\mu\xi)$ also lies in the number of degrees of freedom. The Lagrangian density $£_{local}$ [eqn. (5.200)] had two scalar fields ϕ_1 and ϕ_2 and one massless vector field A_μ. With two degrees of freedom for A_μ (two helicity states of photon) and one each for ϕ_1 and ϕ_2, we had in total four degrees of freedom corresponding to $£_{local}$ of eqn. (5.200). After this, we simply chose a particular ground state [eqn. (5.204)] and obtained $£^{new}_{local}$ [eqn. (5.208)]. The total number of degrees of freedom in eqn. (5.208) is five; A_μ, now massive, has three, η (one and ξ (one). In this process one degree of freedom appeared to have been created by simply changing the variables. Such a thing cannot happen and hence an unphysical field (ξ) had to be identified and eliminated. The final Lagrangian density [eqn. (5.223)] after eliminating ξ (θ) field has again four degrees of freedom h (one) and A_μ (three) and hence the Lagrangian density (5.223) correctly interprets the particle spectrum of the theory.

In the above-considered case of spontaneous breaking of U(1) local gauge symmetry, the complex scalar field was replaced according to eqn. (5.222) by $(v+h)/\sqrt{2}$ and this replacement finally led to the Lagrangian density (5.223) in which the unwanted massless Goldstone boson was turned into the all important longitudinal degree of freedom of massive gauge particle (A_μ). This whole process is in fact given the name the **Higgs Mechanism**.

5.9.5 Spontaneous Breaking of Local SU(2) Gauge Symmetry

In this section we extend the concept of spontaneous symmetry breaking to non-Abelian gauge groups and consider the SU(2) group. We consider the all-important

Higgs mechanism applied to SU(2) gauge group or in other words consider the spontaneous breaking of local SU(2) gauge symmetry.

Let us start with the Lagrangian density

$$£ = (\partial_\mu \phi)^\dagger (\partial^\mu \phi) - \mu^2 \phi^\dagger \phi - \lambda(\phi^\dagger \phi)^2 \tag{5.226}$$

This Lagrangian density is similar to the Lagrangian density of U(1) case except that here ϕ is an SU(2) doublet of complex scalar fields. The doublet has required a dagger instead of just the complex conjugation in U(1) case. Then ϕ will be denoted as

$$\phi = \begin{pmatrix} \phi_\alpha \\ \phi_\beta \end{pmatrix} = \frac{1}{\sqrt{2}} \begin{pmatrix} \phi_1 + i\phi_2 \\ \phi_3 + i\phi_4 \end{pmatrix} \tag{5.227}$$

It can easily be verified that under the global SU(2) gauge transformations [eqn. (5.64)]

$$\phi(x) \to \phi'(x) = e^{i\,\tau_a \alpha_a} \phi(x) \equiv e^{i\,T_a \theta_a} \phi(x) \tag{5.228}$$

the Lagrangian density (5.226) remains invariant. Because of the presence of $\partial_\mu \phi$, this Lagrangian density would not remain invariant under the local SU(2) gauge transformations

$$\phi(x) \to \phi'(x) = e^{i\frac{\tau_a}{2}\theta_a} \phi(x) \tag{5.229a}$$

where

$$\alpha(x) = \frac{\theta(x)}{2} \tag{5.229b}$$

is implied. To achieve local gauge invariance, we have to replace ∂_μ by the covariant derivative [eqn. (5.84)]

$$D_\mu = \partial_\mu + ig\frac{\tau_a}{2} G^a_\mu \tag{5.230}$$

where G^a_μ, with $a = 1, 2, 3$ denote three gauge fields. We have already seen in section (5.5) that $\boldsymbol{G}_\mu$ transforms as [eqn. (5.94)]

$$\boldsymbol{G}_\mu \to \boldsymbol{G}'_\mu = \boldsymbol{G}_\mu - 2(\alpha \times \boldsymbol{G}_\mu) - \frac{2}{g}\partial_\mu \alpha \tag{5.231}$$

in which the factor of 2 will disappear if we use from eqn. (5.229) $\theta(x)$ in place of $\alpha(x)$.

Replacing ∂_μ by D_μ in the Lagrangian density (5.226) and incorporating the kinetic part of the gauge field $\boldsymbol{G}_\mu$, the local SU(2) gauge invariant Lagrangian density would be

$$£_{\text{local}} = [(\partial_\mu + i\frac{g}{2}\tau.\boldsymbol{G}_\mu)\phi]^\dagger [(\partial^\mu + i\frac{g}{2}\tau.\boldsymbol{G}^\mu)\phi] \\ -\mu^2 \phi^\dagger \phi - \lambda(\phi^\dagger \phi)^2 - \frac{1}{4}\boldsymbol{G}_{\mu\nu}\boldsymbol{G}^{\mu\nu} \tag{5.232}$$

where $\boldsymbol{G}_{\mu\nu}$ already obtained in section (5.5) has the form

$$\boldsymbol{G}_{\mu\nu} = \partial_\mu \boldsymbol{G}_\nu - \partial_\nu \boldsymbol{G}_\mu - g(\boldsymbol{G}_\mu \times \boldsymbol{G}_\nu) \tag{5.233}$$

where the cross product is in the internal space of SU(2) group, i.e., the isospin space.

After the recapitulation of above-mentioned features of SU(2) local gauge invariance, we may undertake the process of spontaneous breaking of SU(2) local gauge symmetry. The starting point would be the Lagrangian density $£_{\text{local}}$ [eqn. (5.232)]. As usual, we have two cases corresponding to (1) $\mu^2 > 0$ and (2) $\mu^2 < 0$. We have $\lambda < 0$ in both the cases. When $\mu^2 > 0$, the Lagrangian density (5.232) describes a system of four particles (corresponding to ϕ_1, ϕ_2, ϕ_3, ϕ_4, of eqn. (5.227), each having the mass μ and interacting with three massless gauge bosons of fields G_μ^a. The minimum of potential energy

$$V(\phi) = \mu^2 \phi^\dagger \phi + \lambda(\phi^\dagger \phi)^2 \tag{5.234}$$

can be obtained by writing $\phi^\dagger\phi$ [eqn.(5.227)] as

$$\phi^\dagger \phi = \frac{1}{2}(\phi_1^2 + \phi_2^2 + \phi_3^2 + \phi_4^2) = \frac{1}{2}\sum_{j=1}^{4} \phi_j^2 \tag{5.235}$$

We then have

$$V(\phi) = \sum_{j=1}^{4} \left[\frac{1}{2}\mu^2 \phi_j^2 + \frac{1}{4}\lambda \phi_j^4 \right] \tag{5.236}$$

and the condition of extremum

$$\frac{\partial V(\phi)}{\partial \phi_j} = 0 \tag{5.237a}$$

gives

$$\phi_j \left(\mu^2 + \lambda \sum_{j=1}^{4} \phi_j^2 \right) = 0 \tag{5.237b}$$

implying that the extremum of $V(\phi)$ would occur at

$$\phi_j = 0 (j = 1, 2, 3, 4) \tag{5.238a}$$

or at

$$\sum_{j=1}^{4} \phi_j^2 = -\frac{\mu^2}{\lambda} (\mu^2 > 0) \tag{5.238b}$$

The second order differential

$$\frac{\partial^2 V(\phi)}{\partial \phi_j^2} = \mu^2 + 3\lambda \sum_{j=1}^{4} \phi_j^2 \tag{5.239}$$

which for equation (5.238a) becomes

$$\frac{\partial^2 V(\phi)}{\partial \phi_j^2} = \mu^2 = \text{positive} \tag{5.240}$$

and for the values of eqn. (5.238b)becomes

$$\frac{\partial^2 V(\phi)}{\partial \phi_j^2} = -2\mu^2 = \text{ negative} \tag{5.241}$$

Therefore for the case $\mu^2 > 0$, the minimum of the potential energy is for $\phi_j = 0$, and the symmetry of the Lagrangian density (5.232) is not broken at $\phi_j = 0$ which implies from eqn. (5.227)

$$\phi \equiv \begin{pmatrix} 0 \\ 0 \end{pmatrix} \tag{5.242}$$

In the second case of $\mu^2 > 0$, λ already positive, we find from eqn. (5.241) that $V(\phi)$ would be minimum for

$$\sum_{j=1}^{4} \phi_j^2 = -\frac{\mu^2}{\lambda} (\mu^2 < 0) \tag{5.243}$$

which when expanded is

$$(\phi_1^2 + \phi_2^2 + \phi_3^2 + \phi_4^2) = -\frac{\mu^2}{\lambda} = v^2 (\mu^2 < 0) \tag{5.244}$$

From eqn. (5.244) we may note that

$$\partial_\mu \phi_j = 0$$

hence the kinetic part corresponding to eqn. (5.244) would vanish and consequently the minimum of potential energy state would be the ground state or the vacuum state. Since,

$$\sum_{j=1}^{4} \phi_j^2 = 2(\phi^\dagger \phi) \tag{5.245}$$

the minimum of potential energy occurs [eqn.(5.244)] corresponding to the manifold of points represented by

$$\phi^\dagger \phi = \frac{v^2}{2} \tag{5.246}$$

which obviously is invariant under the SU(2) local gauge transformations [eqn. (5.228)]. Therefore, as long as eqn. (5.244) is obeyed, the Lagrangian density $£_{\text{local}}$ and the ground state shall have the same symmetry and there will be no spontaneous symmetry breaking. Therefore, we choose one particular point in the manifold as the minimum of energy and about that minimum create fluctuations to generate the field.

Let that particular minimum correspond to

$$\phi_1 = \phi_2 = \phi_4 = 0$$

and

$$\phi_3^2 = v^2 = -\frac{\mu^2}{\lambda}(\mu^2 < 0) \tag{5.247}$$

Then we take

$$\phi_{1\min} = \phi_{2\min} = \phi_{4\min} = 0$$

and

$$\phi_{3\min} = +v \tag{5.248}$$

so that the vacuum (the lowest energy state) corresponds from eqn. (5.227) to

$$\phi_0 = \frac{1}{\sqrt{2}}\begin{pmatrix} 0 \\ v \end{pmatrix} \tag{5.249}$$

Now we treat fields ϕ_1, ϕ_2, ϕ_3 and ϕ_4 as the fluctuations about their respective minima. That is $\phi_1(x)$ would be the consequence of fluctuations about $\phi_{1\,\min}$, $\phi_2(x)$ due to the deviations around $\phi_{2\,\min}$ and so on.

Let $\eta(x)$ denote the fluctuations (field) about $\phi_{3\,\min}$ and let $\xi_1(x)$, $\xi_2(x)$ and $\xi_3(x)$ denote the fluctuations about respectively the $\phi_{1\,\min}$, $\phi_{2\,\min}$, and $\phi_{4\,\min}$. We define the fluctuations as

$$\begin{aligned}
\phi_1(x) &= \phi_{1\,\min} + \xi_1(x) = \xi_1(x) \\
\phi_2(x) &= \phi_{2\,\min} + \xi_2(x) = \xi_2(x) \\
\phi_3(x) &= \phi_{3\,\min} + \eta(x) = v + \eta(x) \\
\phi_4(x) &= \phi_{4\,\min} + \xi_3(x) = \xi_3(x)
\end{aligned} \tag{5.250}$$

Hence

$$\phi(x) = \frac{1}{\sqrt{2}}\begin{pmatrix} \xi_1 + i\xi_2 \\ v + \eta + i\xi_3 \end{pmatrix} \tag{5.251}$$

We recall that in the case of spontaneous breaking of U(1) local gauge symmetry there was only one massless field ξ, which was removed (gauged away), here we have three massless fields ξ_1, ξ_2 and ξ_3 implying that here in SU(2), we have to remove three Goldstone bosons. As in U(1) case, we will try to take the three ξ fields into the exponential. For this purpose, we take note of SU(2) local gauge transformations

$$\phi(x) \to \phi'(x) = e^{i\tau.\alpha\,(x)}\phi(x) \tag{5.252}$$

and try to express $\phi(x)$ of eqn. (5.251) in the form of above mentioned eqn. (5.252). Anticipating that ξ fields may be taken to the exponential of $i[\tau.\xi(x)$, we provide the following real values to η and ξ fields so that $\phi(x)$ of eqn. (5.251) can be expressed as exponential. We parameterize the fields as

$$\begin{aligned} \eta(x) &= h(x) \\ \xi_1(x) &= \beta_2(x) \\ \xi_2(x) &= \beta_1(x) \\ \xi_3(x) &= -\beta_3(x) \end{aligned} \tag{5.253}$$

In terms of above real fields, eqn. (5.251) becomes

$$\phi(x) = \frac{1}{\sqrt{2}}\begin{pmatrix} \beta_2 + i\beta_1 \\ v + h(x) - i\beta_3 \end{pmatrix} \tag{5.254}$$

Now, it can be checked that

$$e^{i\tau.\frac{\beta(x)}{v}}\begin{pmatrix} 0 \\ \dfrac{v+h(x)}{\sqrt{2}} \end{pmatrix} \approx \frac{1}{\sqrt{2}}\begin{pmatrix} \beta_2 + i\beta_1 \\ v + h(x) - i\beta_3 \end{pmatrix} \tag{5.255}$$

in which

$$\begin{aligned} \exp\left[i\tau.\frac{\beta(x)}{v}\right] &\approx 1 + i\tau.\frac{\beta}{v} \\ &= 1 + i\tau_1\frac{\beta_1}{v} + i\tau_2\frac{\beta_2}{v} + i\tau_3\frac{\beta_3}{v} \end{aligned} \tag{5.256}$$

or

$$\exp\left[i\tau.\frac{\beta(x)}{v}\right] = \begin{pmatrix} 1 + i\beta_3/v & i(\beta_1 - i\beta_2)/v \\ i(\beta_1 + i\beta_2)/v & 1 - i\beta_3/v \end{pmatrix} \tag{5.257}$$

where in eqn. (5.256), 1 is 2×2 unit matrix and τ_1, τ_2, τ_3 are three Pauli matrices. In the matrix multiplication of LHS of eqn. (5.255), only the lowest order terms have been retained, (h/v) terms neglected.

Thus combining equations (5.254) and (5.255), we may write the $\phi(x)$ of eqn. (5.251) as

$$\phi(x) = e^{i[\tau.\beta(x)]/v}\begin{pmatrix} 0 \\ \dfrac{v+h(x)}{\sqrt{2}} \end{pmatrix} \tag{5.258}$$

Now, let the $\phi'(x)$ of eqn. (5.252) is obtained from the $\phi(x)$ of eqn. (5.258) as

$$\phi'(x) = e^{i\tau.\frac{\beta'(x)}{v}}\begin{pmatrix} 0 \\ \dfrac{v+h(x)}{\sqrt{2}} \end{pmatrix} \tag{5.259}$$

The $\phi'(x)$ of eqn. (5.252) on substituting eqn. (5.258) for $\phi(x)$ becomes

$$\phi'(x) = e^{i\tau.[\alpha(x)+\frac{\beta(x)}{v}]}\begin{pmatrix} 0 \\ \frac{v+h(x)}{\sqrt{2}} \end{pmatrix} \tag{5.260}$$

Equating above two equations of $\phi'(x)$ [i.e., equations (5.259) and (5.260)], we get

$$\frac{\beta'(x)}{v} = \alpha(x) + \frac{\beta(x)}{v} \tag{5.261}$$

Now $\beta'(x)$ represent massless fields (Goldstone bosons), which must vanish when, $\phi(x)$ transforms to $\phi'(x)$. So, putting $\beta'(x) = 0$ we get

$$\alpha(x) = -\frac{\beta(x)}{v} \tag{5.262}$$

Conversely, when the condition (5.262) is obeyed, $\beta'(x) = 0$ and we can write from eqn. (5.259) that $\phi(x)$ expressed as

$$\phi(x) = \frac{1}{\sqrt{2}}\begin{pmatrix} 0 \\ v+h(x) \end{pmatrix} \tag{5.263}$$

is independent of the unphysical massless fields. Thus when the factor $\alpha(x)$ of SU(2) local gauge transformations obeys the condition (5.262), the Goldstone bosons would be removed.

The prescription of Higgs mechanism is then that we must simply replace $\phi(x)$ by eqn. (5.263), where $h(x)$ is real scalar field, the Higgs field.

We may now substitute eqn. (5.263) into the Lagrangian density (5.232) to obtain the final local SU(2) gauge invariant Lagrangian density, which would be free from the unphysical massless fields. Let us separately calculate the kinetic terms of eqn. (5.232).

$$[\partial_\mu\phi + i\frac{g}{2}\tau.\boldsymbol{G}_\mu\phi]^\dagger[\partial^\mu\phi + i\frac{g}{2}\tau.\boldsymbol{G}^\mu\phi] = (\partial_\mu\phi^\dagger)(\partial^\mu\phi) + i\frac{g}{2}(\partial_\mu\phi^\dagger)\tau.\boldsymbol{G}^\mu]\phi$$

$$-i\frac{g}{2}\phi^\dagger(\tau.\boldsymbol{G}_\mu)^\dagger\partial^\mu\phi + \frac{g^2}{4}\phi^\dagger(\tau\cdot G_\mu)^\dagger(\tau\cdot\boldsymbol{G}^\mu)\phi \tag{5.264}$$

Now,

$$(\tau.\boldsymbol{G}_\mu) = \tau_1 G_\mu^1 + \tau_2 G_\mu^2 + \tau_3 G_\mu^3$$

$$= \begin{pmatrix} G_\mu^3 & G_\mu^1 - iG_\mu^2 \\ G_\mu^1 + iG_\mu^2 & -G_\mu^3 \end{pmatrix} \tag{5.265}$$

Since, the second term of eqn. (5.264) is

$$= -i\frac{g}{4}(v+h)\partial_\mu h.G_3^\mu$$

and the third term is

$$= +i\frac{g}{4}(v+h)G_{\mu}^{3}\partial^{\mu}h$$

both cancel each other. The last term (*LT*) of eqn. (5.264) can be calculated as

$$LT = \frac{g^2}{8}[0 \quad v+h]\begin{bmatrix} G_{\mu}^{3} & G_{\mu}^{1}-iG_{\mu}^{2} \\ G_{\mu}^{1}+iG_{\mu}^{2} & -G_{\mu}^{3} \end{bmatrix}\begin{bmatrix} G_{3}^{\mu} & G_{1}^{\mu}-iG_{2}^{\mu} \\ G_{1}^{\mu}+iG_{2}^{\mu} & -G_{3}^{\mu} \end{bmatrix}\begin{bmatrix} 0 \\ v+h \end{bmatrix}$$

or

$$LT = \frac{g^2}{8}(v+h)^2(G_{\mu}^{1}G_{1}^{\mu} + G_{\mu}^{2}G_{2}^{\mu} + G_{\mu}^{3}G_{3}^{\mu}) \tag{5.266}$$

Likewise, following terms of eqn. (5.232) can be calculated as

$$-\mu^2\phi^{\dagger}\phi = \frac{1}{2}\lambda v^2\,(0 \ v+h)\begin{pmatrix} 0 \\ v+h \end{pmatrix}$$

$$= \frac{1}{2}\lambda v^4 + \lambda v^3 h + \frac{1}{2}\lambda v^2 h^2 \tag{5.267}$$

and

$$-\lambda(\phi^{\dagger}\phi)^2 = -\frac{1}{4}\lambda(v^2+2vh+h^2)^2 \tag{5.268}$$

The first term of eqn. (5.264) becomes

$$(\partial_{\mu}\phi^{\dagger})(\partial^{\mu}\phi) = \frac{1}{2}\begin{pmatrix} 0 & \partial_{\mu}h \end{pmatrix}\begin{pmatrix} 0 \\ \partial^{\mu}h \end{pmatrix} = \frac{1}{2}(\partial_{\mu}h)(\partial^{\mu}h) \tag{5.269}$$

Substituting above terms in the Lagrangian density (5.232), we finally get

$$\begin{aligned} £_{\text{local}}^{\text{new}} = & \left[\frac{1}{2}(\partial_{\mu}h)(\partial^{\mu}h) - \lambda v^2 h^2\right] \\ & + \left[-\frac{1}{4}G_{\mu\nu}^{a}G_{a}^{\mu\nu} + \frac{g^2 v^2}{8}(G_{\mu}^{1}G_{1}^{\mu} + G_{\mu}^{2}G_{2}^{\mu} + G_{\mu}^{3}G_{3}^{\mu})\right] \\ & + \left\{\frac{g^2 h}{4}(v+\tfrac{1}{2}h)(G_{\mu}^{1}G_{1}^{\mu} + G_{\mu}^{2}G_{2}^{\mu} + G_{\mu}^{3}G_{3}^{\mu}) - \lambda v h^3 - \frac{\lambda h^4}{4}\right\} + \frac{1}{4}\lambda v^4 \end{aligned} \tag{5.270}$$

in which the first bracket denotes the Lagrangian density of free Higgs scalar field of mass

$$m_h = \sqrt{2\lambda v^2} = \sqrt{-2\mu^2} \ (\mu^2 < 0) \tag{5.271}$$

The second square bracket denotes the Lagrangian density of free massive vector field in which each of the three fields has equal mass, i.e.,

$$m_{G_{\mu}^{1}} = m_{G_{\mu}^{2}} = m_{G_{\mu}^{3}} = \frac{gv}{2} \tag{5.272}$$

The curly bracket denotes various interaction terms and the last term being a constant is insignificant. In the Lagrangian density (5.232) each of the gauge field was massless and they became massive in eqn. (5.270) by eating the three massless fields, one by each gauge field. In this way the three massless Goldstone bosons provided the vital longitudinal degree of freedom to each of the three vector gauge particles of the SU(2) group. This is the **Higgs Mechanism for SU(2)** (non-Abelian) gauge theory.

5.10 ASSIGNMENTS

1. What is gauge symmetry? How Noether's theorem is related to it? State and prove Noether's theorem.
2. What is U(1) global gauge symmetry? Show that U(1) global gauge invariance leads to a conservation law. Identify the conserved quantity
3. The field variable ψ_α is subjected to an infinitesimal global transformation $\delta\psi_\alpha$. Show that the Lagrangian $\pounds(\psi_\alpha, \partial_\mu\psi_\alpha)$ undergoes change by an amount

   $\delta\pounds = \partial_\mu\left(\dfrac{\partial\pounds}{\partial(\partial_\mu\psi_\alpha)}\delta\psi_\alpha\right)$. Now proceed to obtain the conservation law.
4. Define U(1) local gauge symmetry. Show that when U(1) global gauge symmetry is extended to U(1) local gauge symmetry, an interaction term appears naturally in the Lagrangian density.
5. What is meant by gauge invariance? Explain global and local gauge invariances. Show that the requirement of local gauge invariance on the free fermion field Lagrangian leads to the interacting theory of quantum electrodynamics.
6. Explain the necessity of introducing a covariant derivative while enlarging the global gauge symmetry to local. Why the additional field of the covariant derivative is bosonic and massless? Comment on the origin of the covariant derivative and explain the concept of minimal coupling.
7. Discuss the origin of the term gauge and then explain global and local gauge invariances. Why it is called the enlargement of gauge symmetry, when we go from global to local gauge invariance? Explain that minimal coupling is a device to achieve local gauge invariance. Give examples of Lagrangian densities, which are gauge invariant (1) globally (2) locally.
8. Yukawa Lagrangian is expressed as

   $$\pounds = [i\bar{\psi}\gamma^\mu\partial_\mu\psi - m_1\bar{\psi}\psi] + [\frac{1}{2}(\partial_\mu\phi)\partial^\mu\phi) - \frac{1}{2}m_2^2\phi^2] + \alpha_Y\bar{\psi}\psi\phi.$$

   Obtain/Identify

   (i) Field equations

   (ii) Masses and spins of fields ψ and ϕ.

   (iii) The free field Lagrangian densities.

(iv) Write the gauge transformations (both global and local) for ψ and ϕ fields and check the invariances of fields' Lagrangian densities. Suggest modifications, if any, in the Lagrangian density to make it gauge invariant.

9. Explain SU(2) global gauge transformations and show that these transformations cause a change of $\delta£$ in £, where

$\delta £ = (i\tau \cdot \alpha)\left(\frac{\partial £}{\partial \psi}\psi + \frac{\partial £}{\partial(\partial_\mu \psi)}\partial_\mu \psi\right)$. Hence obtain the conservation laws and compare them with corresponding U(1) conservation law. Discuss the difference.

10. Explain global and local gauge invariances. Show that the application of global gauge invariance to non-Abelian gauge groups leads to the conserved currents whose number is equal to the number of generators of the group.

11. What is SU(2) local gauge invariance? Explain that the SU(2) gauge field $\boldsymbol{G}_\mu$ is an isovector and show that it transforms as $\boldsymbol{G}'_\mu = \boldsymbol{G}_\mu - 2(\alpha \times \boldsymbol{G}_\mu) - \frac{2}{g}\partial_\mu \alpha$. Compare this transformation with the Abelian gauge field transformation and explain the difference.

12. Show that $\boldsymbol{G}_{\mu\nu} = \partial_\mu \boldsymbol{G}_\nu - \partial_\nu \boldsymbol{G}_\mu$ is not the correct field tensor of SU(2) group and hence obtain the correct field tensor of SU(2) gauge group. What this field tensor is generally known as? Obtain the transformation law $\boldsymbol{G}'_{\mu\nu} = \boldsymbol{G}_{\mu\nu} - 2(\boldsymbol{\alpha} \times \boldsymbol{G}_{\mu\nu})$. Compare it with its Abelian counterpart and explain the difference.

13. Proceeding from the SU(2) local gauge invariance; obtain the expression for the Yang–Mills field tensor. Show that due to the SU(2) local gauge transformation, this field tensor rotates. What should be the mass of Yang–Mills field?

14. Define and explain the adjoint representation of non-Abelian gauge groups and show that non-Abelian gauge fields carry charges. What these charges are known as, for SU(2) and SU(3) groups?

15. Distinguish adjoint representation from fundamental representation and obtain the expression for Yang–Mills field tensor. Show that Yang–Mills field shows magnetic monopole solutions.

16. Taking $V(\phi^*\phi) = \lambda[\phi^*\phi - \phi_0^2](\phi_0 \neq 0)$ write down the Lagrangian density for local gauge invariant scalar electrodynamics and mention the transformations. Obtain the field equations and canonical conjugates.

17. Explain global and local gauge invariances and show that the global gauge invariance of SU(3) colour group leads to eight conserved colour currents.

18. Explain global and local gauge symmetries. Show that the local SU(3) colour gauge symmetry leads to the interacting theory of quantum chromodynamics. Mention a few significant features of quantum chromodynamics.
19. Explain spontaneous breaking of gauge symmetry. Why is it desired that the symmetry should break? Explain through examples.
20. Illustrate that mass is generated by symmetry breaking. Explain it by considering the analogy of ferromagnetism. [Above Curie temperature, iron is paramagnetic and spins of valence electrons are distributed isotropically. The spins change without any force being applied. The fields, which carry magnetic interaction, are massless. When the temperature lowers to the Curie point, the iron becomes ferromagnetic. This is a phase transition. The spins (magnetic moments) of the valence electrons turn spontaneously into a common direction. It could be any direction. The space within the ferromagnetic ceases to be isotropic; instead, it has a preferred direction. We would require force to turn the spins from this definite preferred direction. As far as rotations are concerned, the carriers of magnetic interactions now have mass. This process is spontaneous symmetry breaking.] Explain the Meissner effect in superconductors as well. [Above the transition temperature of the superconductor, magnetic fields propagate freely within the superconductor. When there is transition to the superconducting phase, the magnetic field is expelled from the superconductor. An observer within the superconductor would find a finite range of magnetic field in the superconductor, implying that he would discover that the photon has acquired mass]. Search mathematical details.
21. .... ξ is expressed as

$$£ = \frac{1}{2}(\partial_\mu \xi)\partial^\mu \xi) - \alpha\xi^2 + e^{(\alpha+\beta)\xi^2},$$ what is the mass of field ξ for

(*i*) $\alpha = 0$, (*ii*) $\alpha > 0$, (*iii*) $\alpha < 0$, (*iv*) $\alpha = 0$. Assume $\beta > 0$.
22. Proceeding from the Lagrangian density of a neutral scalar field with quartic interaction, derive equation (5.174) and explain that symmetry has broken and mass has been generated. Identify the symmetry that was broken.
23. Starting from a suitable Lagrangian density, explain the spontaneous breaking of global gauge symmetry and show that a Goldstone boson has emerged. Why this Goldstone boson is undesirable and mention how it can be removed?
24. Show that a Goldstone boson appears when global gauge symmetry is broken and explain the mechanism of its removal.
25. Describe the spontaneous breaking of U(1) local gauge symmetry. What the spontaneous breaking of local gauge symmetry is generally known as? Explain, how the unphysical gauge boson has disappeared? Obtain the condition for its elimination.

26. Define and explain the Higgs mechanism. Take the example of U(1) gauge symmetry. Explain how the Goldstone boson disappears through this mechanism. Derive the condition of its disappearance. Give a physical interpretation of Higgs mechanism.
27. State Goldstone theorem. How the unphysical Goldstone bosons can be eliminated by Higgs mechanism? Consider the case of SU(2) gauge symmetry. Show that the SU(2) gauge fields acquire a mass of $(gv/2)$ and the mass of Higgs field is $\sqrt{2\lambda v^2}$. Obtain the condition that eliminated the Goldstone boson.
28. Show that
    (a) The electromagnetic field is a gauge field.
    (b) To apply Higgs mechanism we replace the field

    ϕ by (1) $\frac{1}{\sqrt{2}}[v+h(x)]$ in U(1) gauge theory

    and by

    (2) $\frac{1}{\sqrt{2}}\begin{pmatrix} 0 \\ v+h(x) \end{pmatrix}$ in SU(2) gauge theory

    (c) The condition to eliminate Goldstone boson(s) is

    (1) $\alpha(x) = -\frac{\theta(x)}{v}$ in U(1) gauge theory

    and

    (2) $\alpha(x) = -\frac{\beta(x)}{v}$ in SU(2) gauge theory

    (d) The massless Goldstone boson provides the longitudinal degree of freedom to massless gauge particle to make it massive.
    (e) The Lagrangian densities before and after spontaneous symmetry breaking have equal number of degrees of freedom.

5.11 OBJECTIVE QUESTIONS

1. In Lagrangian field theory the symmetries and conservation laws are related by
   (a) Noether's theorem (b) Goldstone's theorem
   (c) Wick's theorem (d) Coleman's theorem
2. The Lagrangian density of a free Dirac field is
   (a) always invariant under the space time dependent phase transformations of fields.
   (b) always invariant under the space time independent phase transformations of fields.

(c) never invariant under the space time independent phase transformations of fields.
(d) always invariant under any kind of phase transformations of fields.

3. For the field ψ, the transformation $[\psi' = e^{-i\alpha}\psi]$
(a) is a global gauge transformation if α depends on space time.
(b) is a local gauge transformation if α is independent of space time.
(c) both (a) and (b) are correct.
(d) is a global gauge transformation if α is space time independent and local if α is space time dependent.

4. The phase transformations are called the
(a) global gauge transformations if the phase parameter is same at every space–time point.
(b) local gauge transformations if the phase parameter is not same at every space–time point.
(c) global and local as defined in (a) and (b) above.
(d) global when the phase parameter is different and local when the phase parameter is same at every space–time point.

5. The gauging of symmetries implies
(a) making local gauge symmetries global.
(b) making global gauge symmetries local.
(c) making no change in the gauge symmetry.
(d) introducing additional global gauge symmetry.

6. The gauging of symmetry requires the introduction of
(a) a covariant derivative $D_\mu = \partial_\mu + ieA_\mu$, where A_μ is a real scalar field.
(b) a covariant derivative $D_\mu = \partial_\mu + ieA_\mu$, where A_μ is a complex vector field.
(c) a covariant derivative $D_\mu = \partial_\mu + ieA_\mu$, where A_μ is a real vector field.
(d) a covariant derivative $D_\mu = \partial_\mu + ieA_\mu$, where A_μ is a complex scalar field.

7. The Lagrangian density $£ = \partial_\mu\phi^*\partial^\mu\phi - \mu^2\phi^*\phi - \lambda(\phi^*\phi)^2$ is invariant under
(a) U(1) global gauge transformations.
(b) U(1) local gauge transformations.
(c) both U(1) global and local gauge transformations.
(d) none of the above.

8. The Lagrangian density

$$£ = -\frac{1}{4}(\partial_\mu A_\nu - \partial_\nu A_\mu)(\partial^\mu A^\nu - \partial^\nu A^\mu) + (\partial_\mu - ieA_\mu)\phi^*(\partial^\mu + ieA^\mu)\phi$$
$$-\mu^2\phi^*\phi - \lambda(\phi^*\phi)^2$$

is invariant under
(a) U(1) global gauge transformations.

(b) U(1) local gauge transformations.
(c) both U(1) global and local gauge transformations.
(d) none of the above.

9. The free Dirac field Lagrangian density would be invariant under local gauge transformations provided
(a) ∂_μ is replace by $D_\mu = \partial_\mu + ieA_\mu$ with A_μ a massive vector field transforming as $A'_\mu = A_\mu - \frac{1}{e}\partial_\mu \alpha$.
(b) ∂_μ is replace by $D_\mu = \partial_\mu + ieA_\mu$ with A_μ a massless vector field transforming as $A'_\mu = A_\mu - \frac{1}{e}\partial_\mu \alpha$.
(c) a massless vector field A_μ is introduced in the Lagrangian density without changing the gradient terms.
(d) a massive vector field A_μ is introduced in the Lagrangian density without changing the gradient terms.

10. The Lagrangian density [$£ = i\bar{\psi}\gamma^\mu\partial_\mu\psi - m\bar{\psi}\psi - e\bar{\psi}\gamma^\mu\psi A_\mu$] is
(a) globally gauge invariant.
(b) locally gauge invariant.
(c) gauge invariant neither globally nor locally.
(d) globally as well as locally gauge invariant.

11. The U(1) phase invariance of the Lagrangian density $£ = \partial_\mu\phi^*\partial^\mu\phi - m^2\phi^*\phi$ implies the existence of a
(a) conserved current $[-ie(\phi^*\partial^\mu\phi - \phi\partial^\mu\phi^*)]$.
(b) non-conserved current $[-ie(\phi^*\partial^\mu\phi - \phi\partial^\mu\phi^*)]$.
(c) conserved current $[-ie\phi^*\partial^\mu\phi]$.
(d) non-conserved current $[-ie\phi^*\partial^\mu\phi]$.

12. The conditions imposed on the gauge field are obeyed by
(a) photons, gluons and $W^\pm$, Z^0 bosons.
(b) photons and $W^\pm$, Z^0 bosons.
(c) photons and gluons.
(d) gluons and $W^\pm$, Z^0 bosons.

13. The requirement of local gauge invariance demands that the quanta of weak field must have
(a) no mass.
(b) a mass of 81 GeV/C^2.
(c) a mass of 92 GeV/C^2.
(d) masses of 81 and 92 GeV/C^2.

14. The requirement of U(1) local gauge invariance on the fermion Lagrangian leads to
(a) an interacting theory of scalar electrodynamics.
(b) an interacting theory of quantum electrodynamics.
(c) an interacting theory of quantum flavordynamics.
(d) an interacting theory of quantum chromodynamics.

15. If τ_3 be the third Pauli matrix, $e^{i\tau_3\alpha}$ is equal to
(a) $\cos\tau_3\alpha + i\sin\tau_3\alpha$
(b) $\cos\alpha + i\sin\alpha$
(c) $\cos\alpha + i\sin\tau_3\alpha$
(d) $\cos\tau_3\alpha + i\sin\alpha$

16. The SU(2) local gauge symmetry demands that the four gradient ∂_μ in the Lagrangian density be replaced by
(a) $D_\mu = \partial_\mu + igG_\mu$ with G_μ a massless vector field.
(b) $D_\mu = \partial_\mu + ig\frac{\tau}{2}.G_\mu$ with $\frac{\tau}{2}$ as the generators and G_μ a massless vector field.
(c) $D_\mu = \partial_\mu + ig\frac{\tau}{2}.G_\mu$ with $\frac{\tau}{2}$ as the generators and G_μ a massless isovector field.
(d) $D_\mu = \partial_\mu + ig\frac{\tau}{2}.G_\mu$ with $\frac{\tau}{2}$ as the generators and G_μ a massive isovector field.

17. The gauge field $\boldsymbol{G}_\mu$ of SU(2) group is called
(a) the electromagnetic field.
(b) the Dirac field.
(c) the Proca field.
(d) the Yang–Mills field.

18. The transformation law for SU(2) gauge field $\boldsymbol{G}_\mu$ is
(a) $\boldsymbol{G}'_\mu = \boldsymbol{G}_\mu - \frac{1}{g}\partial_\mu\alpha$
(b) $\boldsymbol{G}'_\mu = \boldsymbol{G}_\mu - 2\alpha.\boldsymbol{G}_\mu - \frac{2}{g}\partial_\mu\alpha$
(c) $\boldsymbol{G}'_\mu = \boldsymbol{G}_\mu - 2\alpha\times\boldsymbol{G}_\mu - \frac{2}{g}\partial_\mu\alpha$
(d) $\boldsymbol{G}'_\mu = \boldsymbol{G}_\mu - \frac{2}{g}(\alpha\times\boldsymbol{G}_\mu)\ \times\partial_\mu\alpha$

19. The U(1) gauge field tensor $F_{\mu\nu}$ and SU(2) gauge field tensor $G_{\mu\nu}$ transform as

(a) $F'_{\mu\nu} = F_{\mu\nu} - \alpha \times F_{\mu\nu}$ and $\boldsymbol{G}'_{\mu\nu} = \boldsymbol{G}_{\mu\nu}$

(b) $F'_{\mu\nu} = F_{\mu\nu}$ and $\boldsymbol{G}'_{\mu\nu} = \boldsymbol{G}_{\mu\nu} - 2\alpha \times \boldsymbol{G}_{\mu\nu}$

(c) $F'_{\mu\nu} = F_{\mu\nu} - 2\alpha \times F_{\mu\nu}$ and $\boldsymbol{G}'_{\mu\nu} = \boldsymbol{G}_{\mu\nu} - 2\alpha \times \boldsymbol{G}_{\mu\nu}$

(d) $F'_{\mu\nu} = F_{\mu\nu}$ and $\boldsymbol{G}'_{\mu\nu} = \boldsymbol{G}_{\mu\nu}$

20. The electromagnetic field tensor $F_{\mu\nu}$ and the Yang–Mills field tensor $G_{\mu\nu}$ are

(a) different because $F_{\mu\nu}$ has an additional self-interaction term $[-e(A_\mu \times A_\nu)]$.

(b) different because $G_{\mu\nu}$ has an additional self-interaction term $[-g(G_\mu \times G_\nu)]$.

(c) similar, because both have similar self-interaction terms.

(d) similar because there is no self-interaction term in either of the field tensors.

21. The gauge fields having self-interaction terms in their field tensors are called

(a) Maxwell field
(b) Abelian fields
(c) matter fields
(d) non-Abelian fields

22. From the free Yang–Mills field Lagrangian density $-\frac{1}{4}G^a_{\mu\nu}G^{\mu\nu}_a$, the field equations can be obtained as

(a) $\partial_\mu G^{\mu\nu}_a = 0$

(b) $\partial^\mu \partial_\mu G^{\mu\nu}_a - g^2 \varepsilon_{abc} G_{b\mu} G^{\mu\nu}_c = 0$

(c) $\partial_\mu G^{\mu\nu} - g\varepsilon_{abc} G_{b\mu} G^{\mu\nu}_c = 0$

(d) $\partial_\mu (G^{\mu\nu} - g^2 \varepsilon_{abc} G_{b\mu} G^{\mu\nu}_c) = 0$

23. The Maxwell's equations $[\nabla.\boldsymbol{E} = j^0$ and $\nabla \times \boldsymbol{B} - \frac{\partial \boldsymbol{E}}{\partial t} = \boldsymbol{j}]$ are generalized in Yang–Mills theory to

(a) $\nabla.\boldsymbol{E}_a + g\varepsilon_{abc}\boldsymbol{A}_b.\boldsymbol{E}_c = j^0_a$

$$\nabla \times \boldsymbol{B}_a - \frac{\partial \boldsymbol{E}_a}{\partial t} + g\varepsilon_{abc}(\boldsymbol{A}^0_b.\boldsymbol{E}_c + \boldsymbol{A}_b \times \boldsymbol{E}_c) = \boldsymbol{j}_a$$

(b) $\nabla.\boldsymbol{E}_a = j^0_a$

$$\nabla \times \boldsymbol{B}_a - \frac{\partial \boldsymbol{E}_a}{\partial t} + g\varepsilon_{abc}(\boldsymbol{A}^0_b.\boldsymbol{E}_c + \boldsymbol{A}_b \times \boldsymbol{E}_c) = \boldsymbol{j}_a$$

(c) $\nabla.(\boldsymbol{E}_a + g\varepsilon_{abc}\boldsymbol{A}_b.\boldsymbol{E}_c) = j_a^0$

$$\nabla \times \boldsymbol{B}_a - \frac{\partial \boldsymbol{E}_a}{\partial t} + g\varepsilon_{abc}\boldsymbol{A}_b^0.\boldsymbol{E}_c = \boldsymbol{j}_a$$

(d) $\nabla.\boldsymbol{E}_a + g\varepsilon_{abc}\boldsymbol{A}_b.\boldsymbol{E}_{cc} = j_a^0$

$$\nabla \times \boldsymbol{B}_a - \frac{\partial \boldsymbol{E}_a}{\partial t} + g\varepsilon_{abc}\boldsymbol{A}_b \times \boldsymbol{E}_c = \boldsymbol{j}_a$$

24. The divergence of magnetic field, namely $\nabla\cdot B$ or $\nabla\cdot B_a$ is
    (a) non zero in Maxwell field and hence implies the existence of magnetic charge density.
    (b) zero in Yang–Mills field and hence implies the existence of magnetic charge density.
    (c) non zero in Yang–Mills field and hence implies the existence of magnetic charge density.
    (d) zero in Maxwell field and hence implies the existence of magnetic monopoles.
25. The conserved currents $[-g_s\bar{\psi}\gamma^\mu\lambda_a\psi]$ are
    (a) eight in number and are the consequence of global gauge invariance of SU(3) flavour group.
    (b) three in number and are the consequence of global gauge invariance of SU(3) colour group.
    (c) eight in number and are the consequence of global gauge invariance of SU(3) colour group.
    (d) three in number and are the consequence of local gauge invariance of SU(3) colour group.
26. Local SU(3) colour symmetry describes that
    (a) three Dirac fields of different mass interact with eight massless gluons.
    (b) three Dirac fields of equal mass interact with eight massless gluons.
    (c) eight Dirac fields of equal mass interact with three massless gluons.
    (d) eight Dirac fields of unequal mass interact with three massless gluons.
27. The principle of local gauge invariance works well for
    (a) all the three interactions, weak, em and strong.
    (b) only weak and em interactions.
    (c) only em and strong interactions.
    (d) only weak and strong interactions.
28. What is true among the following?
    (a) Conservation of charge, isospin, strangeness, baryon number and lepton number is described by global gauge symmetry.
    (b) Conservation of charge is described by local gauge symmetry whereas, the conservation of isospin, strangeness, baryon number and lepton number is described by global gauge symmetry.

(c) Conservation of charge, isospin and strangeness is described by local gauge symmetry whereas, the conservation of baryon number and lepton number is described by global gauge symmetry.
(d) Conservation of charge, baryon number and lepton number is described by local gauge symmetry whereas, the conservation of isospin and strangeness is described by global gauge symmetry.

29. Following are two Jacobi identities (JI).

(1) $\partial_\rho F_{\mu\nu} + \partial_\mu F_{\nu\rho} + \partial_\nu F_{\rho\mu} = 0$

(2) $D_\rho G_{\mu\nu} + D_\mu G_{\nu\rho} + D_\nu G_{\rho\mu} = 0$

What is true about them?
(a) (1) is Abelian JI with $F_{\mu\nu}$ as Abelian field tensor and (2) is non-Abelian JI with $G_{\mu\nu}$ as non-Abelian field tensor.
(b) (1) is Abelian JI with $F_{\mu\nu}$ as non-Abelian field tensor and (2) is non-Abelian JI with $G_{\mu\nu}$ as Abelian field tensor.
(c) (1) and (2) are Abelian JIs if $F_{\mu\nu}$ and $G_{\mu\nu}$ are Abelian field tensors.
(d) (1) and (2) are non-Abelian JIs if $F_{\mu\nu}$ and $G_{\mu\nu}$ are non-Abelian field tensors.

30. What is true among the following?
(a) Gauge fields are massless. To accommodate massive gauge fields in the gauge theory, we adopt a procedure that is called spontaneous symmetry breaking.
(b) From the very inception, some gauge fields are massless and some are massive. The right choice of the gauge field takes care of massive gauge fields.
(c) Abelian gauge fields are massless. To accommodate massive gauge fields we consider non-Abelian gauge fields.
(d) non-Abelian gauge fields are massless. To accommodate massive gauge fields we consider Abelian gauge fields.

31. The spontaneous symmetry breaking takes place when
(a) the reformulated Lagrangian has the same symmetry as the original one.
(b) the reformulated Lagrangian does not have the symmetry of the original Lagrangian.
(c) the reformulated Lagrangian contains only the kinetic terms.
(d) the reformulated Lagrangian does not contain the kinetic terms.

32. A straight rod is initially held vertically on a table. Now the rod is pressed andit goes into a bent position. What is true among the following statements?
(a) Both initial and bent positions exhibit no symmetry breaking.
(b) Initial position exhibits spontaneous symmetry breaking and the bent position exhibits restoration of symmetry.

(c) Both positions exhibit spontaneous breaking of symmetry.
(d) The bent position exhibits spontaneous symmetry breaking.

33. If $\pounds = \frac{1}{2}(\partial_\mu \eta)(\partial^\mu \eta) - \mu^2\eta^2 \pm \mu\lambda\eta^3 - \frac{1}{4}\lambda^2\eta^4 \pm \frac{1}{4}\left(\frac{\mu^2}{\lambda}\right)^2$, the mass of η field is

(a) zero
(b) $\sqrt{2}\frac{\mu\hbar}{c}$
(c) $\sqrt{2\mu\lambda}\ \frac{\hbar}{c}$
(d) $\frac{\lambda}{\sqrt{2}}\frac{\hbar}{c}$

34. The spontaneously broken symmetry is
(a) discrete if the system has two ground states and continuous if it has four ground states.
(b) discrete if it has four ground states and continuous if it has only one ground state.
(c) discrete if the system has small finite number of ground states and continuous if it has large finite number of ground states.
(d) discrete, if the system has finite number of ground states and continuous if the system has infinite ground states.

35. The symmetry of a system is spontaneously broken because
(a) one of the ground states does not possess the symmetry of the Lagrangian.
(b) the collection of all the ground states does not possess the symmetry of Lagrangian.
(c) one of the excited states of the system does not possess the symmetry of the Lagrangian.
(d) all the excited states of the system do not possess the symmetry of the Lagrangian.

36. Goldstone bosons are
(a) massless spin-1 particles.
(b) massive spin-1 particles.
(c) massless spin-0 particles.
(d) massive spin-0 particles.

37. Goldstone bosons are produced when
(a) continuous global symmetry is spontaneously broken.
(b) discrete global symmetry is spontaneously broken.
(c) continuous local symmetry is spontaneously broken.
(d) discrete local symmetry is spontaneously broken.

38. Goldstone bosons are
(a) desired particles because they are produced in SSB.
(b) desired particles because they are massless.

(c) undesired particles because we have no particles of their kind in the list of elementary particles.
(d) undesired particles because more than one Goldstone bosons are produced.

39. The spontaneous breaking of local gauge symmetry is required to
(a) provide mass to massless gauge particles without having Goldstone bosons.
(b) provide mass to massless gauge particles and having Goldstone bosons.
(c) produce massive particles along with Goldstone bosons.
(d) separate massive and massless particles.

40. The Higgs mechanism is the
(a) spontaneous breaking of global and local gauge symmetries.
(b) spontaneous breaking of global gauge symmetry.
(c) spontaneous breaking of local gauge symmetry.
(d) dynamical breaking of global and local gauge symmetries.

41. To apply Higgs mechanism to SU(2) gauge theory, the field $\phi(x)$ is replaced by

(a) $\phi(x) = \frac{1}{\sqrt{2}}(v + h(x))$ (b) $\phi(x) = \frac{1}{\sqrt{2}}\begin{pmatrix} 0 \\ v + h(x) \end{pmatrix}$

(c) $\phi(x) = \frac{1}{\sqrt{2}}(v + h(x))\begin{pmatrix} 1 \\ 1 \end{pmatrix}$ (d) $\phi(x) = \frac{1}{\sqrt{2(v + h(x))}}$

42. The mass of the Higgs particle is

(a) $\sqrt{-2\mu^2}$ with $\mu^2 < 0$ (b) $\sqrt{-2\mu^2}$ with $\mu^2 = -\frac{1}{4}$

(c) $\sqrt{\frac{\mu^2}{\lambda}}$ with $\mu^2 > 0, \lambda > 0$ (d) $\sqrt{-\frac{\mu^2}{\lambda}}$ with $\mu^2 < 0, \lambda > 0$

43. The Meissner effect in a superconductor can also be expressed by stating that
(a) photons have vacated the superconductor.
(b) as in Higgs phenomenon, the photons have become massive.
(c) photons are massless but they have formed pairs.
(d) the electrons have fallen into one quantum state.

ANSWERS

1.a, 2.b, 3.d, 4.c, 5.b, 6.c, 7.a, 8.c, 9.b, 10.d, 11.a, 12.c, 13.a, 14.b, 15.c, 16.c, 17.d, 18.c, 19.b, 20.b, 21.d, 22.c, 23.a, 24.c, 25.c, 26.b, 27.c, 28.b, 29.a, 30.a, 31.b, 32.d, 33.b, 34.d, 35.a, 36.c, 37.a, 38.c, 39.a, 40.c, 41.b, 42.a, 43.b.

6

THE STANDARD MODEL

6.1 INTRODUCTION

According to our present knowledge of matter its basic patterns are the subatomic particles and all forces between them arise out of four types of basic interactions, strong, electromagnetic, weak and gravitational. However, at atomic and subatomic levels the gravitational interaction is extremely feeble and can be ignored. The Standard Model provides a successful theory of the strong, the electromagnetic and the weak interactions. The electromagnetic and weak interactions have been unified into the electroweak interaction that is described by the Glashow–Weinberg–Salam (GWS) model. The Quantum Chromody-namics (QCD) describes the theory of strong interactions. The GWS model and the QCD together form the Standard Model.

All these three interactions are described by gauge theories; the particles (quanta), which transmit the interactions, are vector (spin-1) gauge bosons. Electromagnetic interaction is transmitted by photons, the strong interaction by massless spin-1 gluons and the weak interactions by spin-1 W^+, W^- and Z bosons. The W and Z bosons were discovered in 1983 at CERN and the gluons at Deutsches Elektronensynchrotron in Hamburg. These discoveries definitely show the amazing success of the Standard Model. Since, the Standard Model describes all the three interactions, we ın the following first describe the GWS model of electroweak interaction and then incorporate the QCD.

6.2 THE GWS MODEL

In 1961, S.L. Glashow aimed at the unification of electromagnetic and weak interactions. He tried to combine these two interactions into one fundamental electroweak interaction. However, there was a great disparity in the strength of these two interactions and naturally Glashow and others realized that the disparity could be accounted for, if the weak interactions are mediated by particles quite massive. This obviously raised the question, if both the interactions are one why photon is massless and weak mediators heavy? S.Weinberg and A.Salam in the form of Higgs mechanism in which the local gauge symmetry was spontaneously broken to generate the masses provided the solution to this problem in 1967. The gauge theory of weak interactions, which unifies the electromagnetic, and weak interactions is therefore called the *GWS model*. From this model it is natural to expect that while the mediators of weak interactions become massive, the photon must remain massless.

The general idea of this model is that the weak interactions are mediated by gauge bosons, which initially are massless. The Lagrangian of the theory also contains massless fermions (leptons and quarks) and the Lagrangian (density) is invariant under an internal symmetry group, which is the gauge symmetry. A scalar field (Higgs field) whose vacuum expectation value is not zero $[\langle 0|\phi_0|0\rangle \neq 0]$ is then introduced which causes the spontaneous breakdown of gauge symmetry (local gauge). As a result of this Higgs mechanism, not only the mediators of weak interactions become massive, the leptons (massive) and quarks too acquire mass. The photon and neutrino remain massless.

In the following, we will mathematically realize the above features of the GWS model for which we will be referring to Chapter 4 on weak interactions. We note that weak interactions can be described by a Yang–Mills type of gauge theory and that only left-handed leptons and quarks participate in weak interactions. These left-handed particles form weak isospin doublets. Neutrinos are massless hence there are no right-handed neutrinos. All other massive fermions have both the left-handed and right-handed companions. Right-handed are singlets (single member) and do not participate in weak interactions. The weak interactions have a $V-A$ coupling and only the left-handed particle operators should be affected by SU(2) (Yang–Mills) transformations and hence we denote the gauge group for weak interactions as SU_L (2), where L implies left-handed. The right-handed are invariant (singlet) under this SU_L (2). Whereas the weak interaction notices the handedness, the electromagnetic interaction does not take handedness into account. Keeping the above background in view, the GWS model may be described under the following subtitles.

6.2.1 Electromagnetic and Weak Embedding

We recall that the weak interaction is described by

$$J^\mu_{\text{weak}} W_\mu$$

in analogy with $J^\mu A_\mu$ of *QED*. The J^μ_{weak} has vector minus axial vector form

$$\begin{aligned} J^\mu_{\text{weak}} &= \overline{\psi}_{1L}\gamma^\mu\psi_{2L} \\ &= \frac{1}{2}\left[\overline{\psi}_1\gamma^\mu\psi_2 - \overline{\psi}_1\gamma^\mu\gamma^5\psi_2\right] \end{aligned} \tag{6.1}$$

where

$$\psi_L = \frac{(1-\gamma^5)}{2}\psi$$

and

$$\gamma^5\gamma^\mu = -\gamma^\mu\gamma^5 \tag{6.2}$$

have been used. First term of eqn. (6.1) is vector and other the axial vector. We have seen under section (4.6.4.2) that the weak interactions are mediated by $W^\pm$ and Z^0 bosons. However, before the discovery of neutral current weak decays, all the weak decays were attributed to the exchange of $W^\pm$ bosons and hence J^μ_{weak} of eqn. (6.1) was just charged weak current written as $J^{\mu\pm}_{\text{weak}}$, in which the plus (+) sign implied (Fig.6.1)

$$q_2 - q_1 = +1 \quad W \to W^+ \tag{6.3a}$$

and minus (–) sign implied

$$q_2 - q_1 = -1 \quad W \to W^- \tag{6.3b}$$

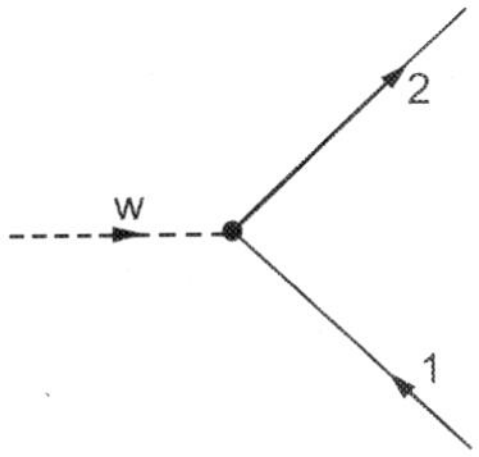

Fig. 6.1: A weak interaction vertex

where q_2 and q_1 are the electric charges of outgoing and incoming particles.

The covariant derivative (eqn. 5.84) for SU(2) gauge theory is

$$D_\mu = \partial_\mu + ig\boldsymbol{T}.\boldsymbol{W}_\mu \tag{6.4}$$

where the gauge field $\boldsymbol{G}_\mu$ has been replaced by $\boldsymbol{W}_\mu$ and $\boldsymbol{T}$ denotes the isospin operator whose three components (T_1, T_2, T_3) are the generators of SU(2). In the covariant derivative

$$\boldsymbol{T}.\boldsymbol{W}_\mu = \left(T_1W^1_\mu + T_2W^2_\mu\right) + T_3W^3_\mu \tag{6.5}$$

there is a neutral gauge field W^3_μ whose quanta could be identified with photons. If this happened, the SU(2) gauge theory with the gauge field

$$\boldsymbol{W}_\mu = (W^1_\mu, W^2_\mu, W^3_\mu) \tag{6.6}$$

would describe, both weak and electromagnetic interactions. Thus, it was thought some fifty years ago that the vector current operators $J^{\mu\pm}_{\text{weak}}$ are the charged members of an $T = 1$ operator and its neutral member is related to the electromagnetic current. We may mention that

$$W^\pm_\mu = \frac{W^1_\mu \pm iW^2_\mu}{\sqrt{2}} \tag{6.7}$$

are the charged gauge fields coupled to $J^{\mu\pm}_{\text{weak}}$. The neutral gauge field W^3_μ appears in the covariant derivative in the term $igT_3W^3_\mu$. Now, we look this term with respect to a weak lepton doublet say $\begin{pmatrix}\nu_e \\ e^-\end{pmatrix}_L$ [eqn. (4.199)] in which $T_3 = +\frac{1}{2}$ and $-\frac{1}{2}$ for ν_e and e^- respectively. Thus $igT_3W^3_\mu$ suggests that W^3_μ couples with both the neutral member, the neutrino and the charged member, the electron with the same strength $\frac{g}{2}$ (magnitude). If W^3_μ represents photon, it will have its coupling with neutrino, whereas in reality a photon interacts with only charged particles. Hence the quantum of the neutral gauge field W^3_μ cannot be the photon and the group SU(2) is insufficient to describe the electromagnetic interactions. We have to extend this group. As far as the weak interaction group SU(2) is concerned, we will write it $\text{SU}_\text{L}(2)$ because only left-handed particles are involved in the weak interactions (charged).

In order to extend the group to include electromagnetic interactions, we note that U(1) is the group related to electromagnetic interactions and the electric charge operator Q is its generator. The group of weak interactions $\text{SU}_\text{L}(2)$, has $T_a\,(a = 1, 2, 3)$, the components of isospin $\boldsymbol{T}$, as its generators. Now, the following equation [eqn. (4.200)] relates Q and isospin (T_3)

$$Q = T_3 + \frac{Y}{2} \tag{6.8}$$

where Y is the hypercharge of the particle whose electric charge is Q. Thus in order to embedd electromagnetic interactions, hypercharge must be included with isospin and hence in terms of groups, to $\text{SU}_\text{L}(2)$ must be included a group whose generator is $\frac{Y}{2}$. We therefore consider like U(1) a group of transformations

$$\psi(x) \to \psi'(x) = e^{i(\frac{Y}{2})\theta}\psi(x) \tag{6.9}$$

and call such a group $\text{U}_\text{Y}(1)$. The hypercharge values of leptons and quarks can be noted from the table (Chapter 4) as

$$\begin{aligned} &\text{Leptons} \quad Y_L = -1, \quad Y_R = -2 \\ &\text{Quarks} \quad Y_L = \frac{1}{3}, \quad Y_R = \frac{4}{3} \quad (u, c, t) \\ &\qquad\qquad\qquad\quad\; Y_R = -\frac{2}{3} \quad (d, s, b) \end{aligned} \tag{6.10}$$

Thus, the gauge group that would describe both the weak and electromagnetic interactions would be

$$\mathrm{SU_L}\,(2) \otimes \mathrm{U_Y}(1) \tag{6.11}$$

where $\otimes$ denotes that both the groups are independent. Just as we have U(1), SU(2), SU(3), etc., gauge symmetries, here we have $[\mathrm{SU_L}(2) \otimes \mathrm{U_Y}(1)]$ gauge symmetry.

6.2.2 The Combined Covariant Derivative

Now, since the gauge group is the direct product of two (independent)—gauge groups $\mathrm{SU_L}(2)$ and $\mathrm{U_Y}(1)$, both individually have their own covariant derivatives with different coupling constants. Let B_μ be the gauge field for $\mathrm{U_Y}(1)$ gauge group and g_2 be the corresponding coupling constant. Let g_1 be the coupling constant for $\mathrm{SU_L}(2)$ gauge group and its gauge field is $\boldsymbol{W}_\mu$. The $\mathrm{U_Y}(1)$ group would have its U(1) like [eqn. (5.38)] covariant derivative and $\mathrm{SU_L}(2)$ has its covariant derivative in eqn. (6.4). The covariant derivative, which contains both the gauge fields $\boldsymbol{W}_\mu$ and B_μ can be written as

$$D_\mu = \partial_\mu + ig_1 \boldsymbol{T}.\boldsymbol{W}_\mu + ig_2\left(\frac{Y}{2}\right)B_\mu \tag{6.12}$$

where we recall that $\boldsymbol{T} = \dfrac{\tau}{2}$ are the generators of $\mathrm{SU_L}(2)$ and $\dfrac{Y}{2}$, the generator of $\mathrm{U_Y}(1)$ gauge group.

Since, only the left-handed particles (which form isospin doublets) participate in weak interactions, the covariant derivative for them would be

$$(D_\mu)_L = \partial_\mu + ig_1 T_a W_\mu^a + ig_2 \frac{Y_L}{2} B_\mu \tag{6.13}$$

where Y_L are determined from eqn. (6.10), $a = 1, 2, 3$. For right-handed particles, due to their non-participation in weak interactions, we will have

$$(D_\mu)_R = \partial_\mu + ig_2 \frac{Y_R}{2} B_\mu \tag{6.14}$$

where again eqn. (6.10) determine Y_R. We may also mention here that under the $\mathrm{SU_L}(2)$ local gauge transformations, the fermion (leptons and quarks) fields transform as [eqn. (5.229)]

$$\begin{aligned} \psi_L(x) &\to \psi'_L(x) = e^{i\,\boldsymbol{T}.\boldsymbol{\theta}\,(x)}\psi_L(x) \\ \psi_R(x) &\to \psi'_R(x) = \psi_R(x) \end{aligned} \tag{6.15}$$

and under these $\mathrm{SU_L}(2)$ transformations, the gauge fields W_μ and B_μ transform as [eqn.(5.231)]

$$\begin{aligned} \boldsymbol{W}_\mu &\to \boldsymbol{W}'_\mu = \boldsymbol{W}_\mu - (\boldsymbol{\theta}\times\boldsymbol{W}_\mu) - \frac{1}{g_1}\partial_\mu \boldsymbol{\theta}(x) \\ B_\mu &\to B'_\mu = B_\mu \end{aligned} \tag{6.16}$$

On the other hand under the $U_Y(1)$ local gauge transformations the fermion fields transform as eqn. (6.9), which can be written as

$$\begin{aligned} \psi_L(x) &\rightarrow \psi'_L(x) = e^{i\,Y_L\,\alpha(x)}\psi_L(x) \\ \psi_R(x) &\rightarrow \psi'_R(x) = e^{i\,Y_R\,\alpha(x)}\psi_R(x) \end{aligned} \tag{6.17}$$

and under these $U_Y(1)$ transformations the gauge fields W_μ and B_μ transform as [eqn. (5.44)]

$$\boldsymbol{W}_\mu \rightarrow \boldsymbol{W}'_\mu = \boldsymbol{W}_\mu$$

and

$$B_\mu \rightarrow B'_\mu = B_\mu - \frac{1}{g_2}\partial_\mu \alpha(x) \tag{6.18}$$

The covariant derivative, [eqn. (6.12)], in which gauge fields transform as eqns. (6.16) and (6.18) would maintain the local gauge invariance of the electroweak Lagrangian density.

6.2.3 Lagrangian Density and Mass Term

Let ψ_f denote the fermion fields. By ψ_f, we would mean any of the $SU_L(2)$ multiplet field of leptons and quarks. These multiplets have been listed in the following table, where the mixing of quarks has been ignored.

Table 6.1 ***Lepton and Quark $SU_L(2)$ Multiplets***

Generations	1	2	3
Lepton doublets	$\begin{pmatrix} \nu_e \\ e^- \end{pmatrix}_L$	$\begin{pmatrix} \nu_\mu \\ \mu^- \end{pmatrix}_L$	$\begin{pmatrix} \nu_\tau \\ \tau^- \end{pmatrix}_L$
Lepton singlets	e_R^-	μ_R^-	τ_R^-
Quark doublets	$\begin{pmatrix} u \\ d \end{pmatrix}_L$	$\begin{pmatrix} c \\ s \end{pmatrix}_L$	$\begin{pmatrix} t \\ b \end{pmatrix}_L$
Quark singlets	u_R, d_R	c_R, s_R	t_R, b_R

The fields ψ_f for say the first generation of leptons would mean

$$\psi_{fL} = \begin{pmatrix} \nu_e \\ e^- \end{pmatrix} \equiv \begin{pmatrix} \nu_L \\ e_L^- \end{pmatrix}$$

and

$$\psi_{fR} = e_R \tag{6.19}$$

Correspondingly

$$\overline{\psi}_{fL} = \begin{pmatrix} \nu_L^\dagger & e_L^{-\dagger} \end{pmatrix}\gamma^0$$

and also

$$\overline{\psi}_{fR} = \overline{e}_R \tag{6.20}$$

where ν_L, e_L, e_R now represent the corresponding fields ($\nu_L \Rightarrow$ neutrino field, etc.,). All the ψ_{fL}'s and ψ_{fR}'s given in the above table have to be incorporated in the Dirac Lagrangian density. We would write $\sum_f \psi_f$ for all the $SU_L(2)$ multiplet fields. The fields being fermionic, we write the Lagrangian density $£_0$ as

$$£_0 = i\sum_f \overline{\psi}_f \gamma^\mu \partial_\mu \psi_f - \sum_f m_f \overline{\psi}_f \psi_f \tag{6.21}$$

Before proceeding further, let us discuss the mass term. If handedness were not considered such a term would be invariant under the local (global as well) gauge transformations. But weak interactions do notice the handedness, so difficulty occurs if mass term is retained in the Lagrangian density (6.21). Following calculations would make this point clear. We may note that the mass term of eqn. (6.21) is $m\overline{\psi}\psi$ and

$$\begin{aligned} m\overline{\psi}\psi &= m\overline{\psi}\left(\frac{1-\gamma^5}{2}+\frac{1+\gamma^5}{2}\right)\psi \\ &= m\left[\overline{\psi}\frac{1-\gamma^5}{2}+\overline{\psi}\frac{1+\gamma^5}{2}\right]\psi \\ &= m\left(\overline{\psi}_R+\overline{\psi}_L\right)\left(\psi_R+\psi_L\right) \\ &= m\overline{\psi}_R\psi_R + m\overline{\psi}_R\psi_L + m\overline{\psi}_L\psi_R + m\overline{\psi}_L\psi_L \\ &= m\left(\overline{\psi}_R\psi_L+\overline{\psi}_L\psi_R\right) \end{aligned} \tag{6.22}$$

where eqns. (4.68), (4.78), (4.79) and

$$\overline{\psi}_R\psi_R = \left(\overline{\psi}\frac{1-\gamma^5}{2}\right)\left(\frac{1+\gamma^5}{2}\psi\right) = 0 = \overline{\psi}_L\psi_L \tag{6.23}$$

have been used. Looking at the RHS (last line) of eqn. (6.22) we find that whereas ψ_L is a doublet ψ_R is a singlet under $SU_L(2)$. In other words, the isospin T of ψ_L is $\frac{1}{2}$ and that of ψ_R is zero. When $T = \frac{1}{2}$ and $T = 0$ objects combine, we have an $T = \frac{1}{2}$ object, which is not an invariant. Singlet $(T = 0)$ is an invariant object. In view of this non-invariance, we cannot retain the mass term in the Lagrangian density (6.21). In order to get an $(T = 0)$ object, $T = \frac{1}{2}$ should couple with $T = \frac{1}{2}$, such situation we will create later on, where employing the Higgs mechanism masses will be generated.

Thus the fields ψ_f should be massless and accordingly, the Lagrangian density (6.21) for such Dirac fields ψ_f would become

$$\pounds_0 = i\sum_f \overline{\psi}_f \gamma^\mu \partial_\mu \psi_f \tag{6.24}$$

Now, we replace ∂_μ by D_μ to incorporate the requirement of local gauge invariance. The covariant derivative D_μ [eqn. (6.12)] contains the gauge fields W_μ and B_μ and hence, while replacing ∂_μ by D_μ in eqn. (6.24), we will also include the kinetic terms corresponding to both W_μ and B_μ. The kinetic terms (*K.T.*) would be

$$K.T. = -\frac{1}{4}\boldsymbol{W}_{\mu\nu}\boldsymbol{W}^{\mu\nu} - \frac{1}{4}B_{\mu\nu}B^{\mu\nu} \tag{6.25}$$

where $W_{\mu\nu}$ is the SU(2) like tensor and $B_{\mu\nu}$, U(1) like, and hence have following forms.

$$\boldsymbol{W}_{\mu\nu} = \partial_\mu \boldsymbol{W}_\nu - \partial_\nu \boldsymbol{W}_\mu - g_1(\boldsymbol{W}_\mu \times \boldsymbol{W}_\nu) \tag{6.26}$$

and

$$B_{\mu\nu} = \partial_\mu B_\nu - \partial_\nu B_\mu \tag{6.27}$$

Thus the local gauge invariant electroweak Lagrangian density for the massless field ψ_f, from eqn. (6.24) would be

$$\begin{aligned}\pounds_1 &= i\sum_f \overline{\psi}_f \gamma^\mu D_\mu \psi_f + K.T. \\ &= i\sum_f \overline{\psi}_{fL}\gamma^\mu (D_\mu)_L \psi_{fL} + i\sum_f \overline{\psi}_{fR}\gamma^\mu (D_\mu)_R \psi_{fR} + K.T.\end{aligned} \tag{6.28}$$

where *K.T.* are given by eqn. (6.25) and $(D_\mu)_L$ and $(D_\mu)_R$ have been defined in equations (6.13) and (6.14) respectively.

6.2.4 Calculation of Interactions

From the various fundamental fermions (Table 6.1), we choose the first generation leptons and write $\pounds_1$ (eqn. 6.28) for them. The first generation leptons are

$$\begin{pmatrix} \nu_e \\ e^- \end{pmatrix}_L \Rightarrow \begin{pmatrix} \nu_L \\ e_L \end{pmatrix} \text{ and } e_R^- \Rightarrow e_R \tag{6.29}$$

where the subscript L with neutrino is written only to match with e_L. Neutrinos are always left-handed. Then

$$\psi_{fL} = \begin{pmatrix} \nu_L \\ e_L \end{pmatrix} \text{ and } \psi_{fR} = e_R$$

and also

$$\overline{\psi}_{fL} = \begin{pmatrix} \overline{\nu}_L & \overline{e}_L \end{pmatrix} \text{ with } \overline{\psi}_{fR} = \overline{e}_R \tag{6.30}$$

Hence,

$$\pounds_1 = i(\overline{\nu}_L \quad \overline{e}_L)\gamma^\mu\left(\partial_\mu + ig_1\frac{\tau_a}{2}W^a_\mu + ig_2\frac{Y_L}{2}B_\mu\right)\begin{pmatrix}\nu_L\\ e_L\end{pmatrix}$$
$$+i\overline{e}_R\gamma^\mu\left(\partial_\mu + ig_2\frac{Y_R}{2}B_\mu\right)e_R + K.T. \tag{6.31}$$

where T_a in $(D_\mu)_L$ has been replaced by $\frac{\tau_a}{2}$. It may, however, be noted that similar left handed and right handed terms written for other fundamental fermions of Table 6.1, are required to be included in $\pounds_1$. Now, in eqn. (6.31) the interaction terms are those, which contain W_μ and B_μ. The factor $\tau_a W^a_\mu$ may be expressed in the matrix form according to eqn. (5.265) as

$$\tau_a W^a_\mu = \begin{pmatrix} W^3_\mu & W^1_\mu - iW^2_\mu \\ W^1_\mu + iW^2_\mu & -W^3_\mu \end{pmatrix} \tag{6.32}$$

which on using eqn. (6.7) is written as

$$\tau_a W^a_\mu = \begin{pmatrix} W^3_\mu & \sqrt{2}W^-_\mu \\ \sqrt{2}W^+_\mu & -W^3_\mu \end{pmatrix} \tag{6.33}$$

Substituting the matrix (6.33) for $\tau_a W^a_\mu$ in eqn. (6.31) and noting this term down along with all B_μ terms, we write for the *interaction terms* (*I.T.*)

$$(-)\ I.T. = \frac{g_1}{2}(\overline{\nu}_L \quad \overline{e}_L)\gamma^\mu\begin{pmatrix} W^3_\mu & \sqrt{2}\ W^-_\mu \\ \sqrt{2}\ W^+_\mu & -W^3_\mu \end{pmatrix}\begin{pmatrix}\nu_L\\ e_L\end{pmatrix}$$
$$+\frac{g_2}{2}(\overline{\nu}_L \quad \overline{e}_L)\gamma^\mu Y_L B_\mu\begin{pmatrix}\nu_L\\ e_L\end{pmatrix}$$
$$+\frac{g_2}{2}\overline{e}_R\gamma^\mu Y_R B_\mu e_R \tag{6.34}$$

where the minus sign has been shifted to LHS. Using $Y_L = -1$ and $Y_R = -2$ for leptons and simplifying, the eqn. (6.34) becomes

$$(-)I.T. = g_1\left[\frac{1}{\sqrt{2}}(\overline{e}_L\gamma^\mu\nu_L)\ W^+_\mu + \frac{1}{\sqrt{2}}(\overline{\nu}_L\gamma^\mu e_L)\ W^-_\mu\right.$$
$$\left.+\frac{1}{2}(\overline{\nu}_L\gamma^\mu\nu_L - \overline{e}_L\gamma^\mu e_L)W^3_\mu\right]$$

$$+g_2\left[-\frac{1}{2}(\overline{\nu}_L\gamma^\mu\nu_L+\overline{e}_L\gamma^\mu e_L)-\overline{e}_R\gamma^\mu e_R\right]B_\mu \tag{6.35}$$

Let us analyse the above equation. $W_\mu^\pm$ are charged gauge fields and W_μ^3 and B_μ are neutral. The coefficients of $W_\mu^\pm$, W_μ^3 and B_μ are currents. The coefficients of W_μ^+ and W_μ^- are respectively the charged weak currents $J^{\mu+}$ and $J^{\mu-}$. Recalling eqn. (4.198), the coefficient of W_μ^3 is J_3^μ, the third weak current, which is of course neutral. The coefficient of B_μ from eqn. (4.209) is $\frac{1}{2}J_\mu^Y$, the weak hypercharge current, which is also neutral. It may be recalled that the electromagnetic current [eqn. (4.206)] is

$$J_{em}^\mu = J_3^\mu + \frac{J_Y^\mu}{2} \quad (\text{from } Q = T_3 + \frac{Y}{2})$$

$$= -\overline{e}_L\gamma^\mu e_L - \overline{e}_R\gamma^\mu e_R \tag{6.36}$$

which is also neutral. In electromagnetic current, there are both left-handed and right-handed components of electron. In J_3^μ there are only left-handed components.

Among the various current and gauge field interaction terms of eqn. (6.35), the terms involving charged weak currents represent the already well-known beta decay process. So, we will not consider $J^{\mu\pm}W_\mu^\pm$ interaction terms. We would take up the, neutral weak current-gauge field, interaction terms. For this we look at J_3^μ which contains both the neutrino current $\overline{\nu}_L\gamma^\mu\nu_L$ and the electron current $\overline{e}_L\gamma^\mu e_L$. Since, the source of electron current is charged particles, i.e., the electrons that are also the source of electromagnetic current; weak interactions involving electron current are difficult to differentiate from the electromagnetic interactions involving electron (e_L, e_R) current.

6.2.5 Neutral Gauge Fields Z_μ and A_μ

In view of the above, we take up the current $\overline{\nu}_L\gamma^\mu\nu_L$, the weak neutrino (neutral) current and pick up the interaction terms of this current with gauge fields. We denote these terms by *NCIT* and write from eqn. (6.35)

$$NCIT = \left(\overline{\nu}_L\gamma^\mu\nu_L\right)\frac{1}{2}\left(g_1W_\mu^3 - g_2B_\mu\right) \tag{6.37}$$

Since, the second bracket on the RHS denotes a linear combination of neutral gauge fields, we can replace it by a single neutral gauge field denoted by Z_μ and may write

$$Z_\mu \propto \left(g_1 W_\mu^3 - g_2 B_\mu\right) \tag{6.38}$$

Realizing that W_μ^3 and B_μ are orthogonal fields, we may write for the normalized Z_μ as

$$Z_\mu = \frac{g_1 W_\mu^3 - g_2 B_\mu}{\sqrt{g_1^2 + g_2^2}} \tag{6.39}$$

where $1\Big/\sqrt{g_1^2 + g_2^2}$ is the normalization constant. From the coefficients of W_μ^3 and B_μ in eqn. (6.39), we can see that

$$\frac{g_1^2}{g_1^2 + g_2^2} + \frac{g_2^2}{g_1^2 + g_2^2} = 1 \tag{6.40}$$

from which we may define

$$\cos\theta_W = \frac{g_1}{\sqrt{g_1^2 + g_2^2}} \quad \text{and} \quad \sin\theta_W = \frac{g_2}{\sqrt{g_1^2 + g_2^2}} \tag{6.41}$$

such that the eqn. (6.39) is

$$Z_\mu = (\cos\theta_W)W_\mu^3 - (\sin\theta_W)B_\mu \tag{6.42}$$

Above relation can be drawn as Fig. (6.2). The angle θ_W is called the Weinberg angle or the weak-mixing angle.

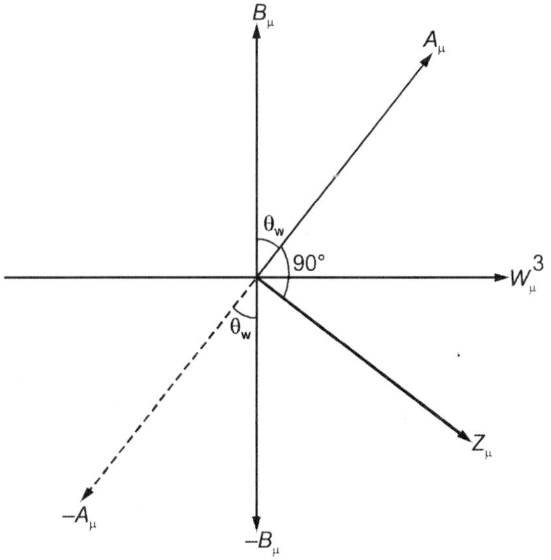

Fig. 6.2: Orthogonality of A_μ and Z_μ

$$Z_\mu = (\cos\theta_W)W_\mu^3 - (\sin\theta_W)B_\mu$$

$$A_\mu = (\cos\theta_W)B_\mu + (\sin\theta_W)W_\mu^3$$

$$\theta_W = \text{Weinberg angle} \simeq 28°$$

Since, the above theory is desired to describe electromagnetic interactions as well, we could expect that Z_μ can be identified with A_μ, which is also a neutral gauge field. However, Z_μ cannot be A_μ because Z_μ couples with electrically neutral particles ν_L and A_μ couples only with the charged particles. Thus the two neutral gauge fields W_μ^3 and B_μ should combine to construct A_μ as well, but they should combine in a way different from they did for Z_μ. Since A_μ does not couple to the neutrino current, it must be orthogonal to Z_μ. The combination of W_μ^3 and B_μ, orthogonal to Z_μ is

$$A_\mu = (\cos\theta_W)B_\mu + (\sin\theta_W)W_\mu^3 \tag{6.43}$$

Thus we get two distinct neutral gauge fields Z_μ and A_μ. Experiments have confirmed them. The neutral gauge field A_μ produces electromagnetic interactions and the neutral gauge field Z_μ produces the neutral weak interactions. It may be noted here that before such a theory, the neutral weak interactions were not known and we knew only charged weak interactions produced by charged gauge fields W^+ and W^-. The discovery of neutral weak interactions was new. The reasons why such (neutral) weak interactions were not experimentally observed before may be stated as:

1. The detection of neutrinos is difficult and we need neutrinos to have the weak current $\bar{\nu}_L\gamma^\mu\nu_L$ which couples with Z_μ.
2. The weak interaction of electron current with Z_μ is difficult to be distinguished from the interaction (electromagnetic) of electron current with A_μ.

6.2.6 Electromagnetic Coupling Constant *e*

Solving the two equations (6.42) and (6.43), which we rewrite

$$Z_\mu = (\cos\theta_W)W_\mu^3 - (\sin\theta_W)B_\mu$$

$$A_\mu = (\cos\theta_W)B_\mu + (\sin\theta_W)W_\mu^3 \tag{6.44}$$

we can express W_μ^3 and B_μ in terms of Z_μ and A_μ as

$$B_\mu = (\cos\theta_W)A_\mu - (\sin\theta_W)Z_\mu \tag{6.45}$$

and

$$W_\mu^3 = (\sin\theta_W)A_\mu + (\cos\theta_W)Z_\mu \tag{6.46}$$

Now we pick up from eqn. (6.35), the interaction terms of electron currents ($\bar{e}_L\gamma^\mu e_L$ and $\bar{e}_R\gamma^\mu e_R$) with neutral gauge fields. We denote them by *ECIT* (electron current interaction terms)

$$ECIT = \bar{e}_L\gamma^\mu e_L\left(-\frac{g_1}{2}W_\mu^3 - \frac{g_2}{2}B_\mu\right) - \bar{e}_R\gamma^\mu e_R . g_2 B_\mu \quad (6.47)$$

which on substituting the eqns. (6.45) and (6.46) becomes

$$ECIT = -[\bar{e}_L\gamma^\mu e_L(\tfrac{1}{2})(g_1 \sin\theta_W + g_2\cos\theta_W) + \bar{e}_R\gamma^\mu e_R . g_2 \cos\theta_W]A_\mu$$
$$+ [\bar{e}_L\gamma^\mu e_L(\tfrac{1}{2})(-g_1\cos\theta_W + g_2\sin\theta_W) + \bar{e}_R\gamma^\mu e_R . g_2 \sin\theta_W]Z_\mu \quad (6.48)$$

At the moment, we focus our attention at the A_μ term of eqn. (6.48). From eqn. (6.41) we can easily see that

$$g_1 \sin\theta_W = g^2 \cos\theta_W \quad (6.49)$$

Substituting this equation in the first bracket (A_μ term) of eqn. (6.48), we get

$$A_\mu \text{ term} = g_2\cos\theta_W(-\bar{e}_L\gamma^\mu e_L - \bar{e}_R\gamma^\mu e_R)A_\mu \quad (6.50)$$

which on using eqn. (6.36) can be read as

$$A_\mu \text{ term } = g_2\cos\theta_W J^\mu_{em} A_\mu \quad (6.51)$$

where J^μ_{em} is the electromagnetic current. Naturally the interaction term (6.51) denotes the electromagnetic interaction and $g^2 \cos\theta_W$ is the coupling constant for this (electromagnetic) interaction. It is well-known that the electromagnetic interaction is expressed by

$$e\, J^\mu_{em}\, A_\mu \quad (6.52)$$

where *e* is the magnitude of the charge of electron. Hence we get from eqns. (6.51) and (6.52)

$$g^2 \cos\theta_W = e = g_1 \sin\theta_W \quad (6.53)$$

from which we get

$$g_1 = \frac{e}{\sin\theta_W} \quad \text{and} \quad g_2 = \frac{e}{\cos\theta_W} \quad (6.54)$$

We recall that g_1 is the coupling constant for weak interactions and g_2 couples B_μ. The relation (6.54) shows that the weak coupling constant g_1 and the electromagnetic coupling constant *e* are not independent. In order to determine g_1 and g_2, we must know θ_W and various weak interaction experiments give the value (almost same)

$$\sin^2\theta_W = 0.2325$$

implying

$$\theta_W = 28° \quad (6.55)$$

Substituting this value of θ_W in eqn. (6.54), we get

$$g_1 \approx 2.07\, e \quad \text{and} \quad g_2 \approx 1.14\, e \quad (6.56)$$

The result (6.56) *implies that the weak interaction coupling constant* g_1 *is twice as strong the electromagnetic coupling constant e. Such a result contradicts the* ***observed facts***. The experiments suggest that **(a) the weak interactions are quite weaker** (have a much smaller coupling constant than the electromagnetic) and **(b) have a short range**. Here the gauge fields W_μ and Z_μ are massless and hence their gauge particles suggest the long-range interactions (weak). Of course, the massless neutral gauge field A_μ has been found which represents the electromagnetic interactions. The fields ψ_f, which continue to be massless, also require mass.

The discrepancies (a) and (b) mentioned above regarding the weak interactions have to be removed. The removal is through the incorporation of spontaneous breaking of gauge symmetry. However, before talking about the gauge symmetry breaking, we explain the structure of neutral current in the following subsection.

6.2.7 Structure of Neutral Current

Now, since we have neutral gauge fields Z_μ and A_μ, the current that couples with Z_μ is the weak neutral current and the current coupling with A_μ is the electromagnetic current. Equation (6.48) provides the Z_μ coupling through the Z_μ term, which is

$$\begin{aligned} Z_\mu \text{ term} = & [(\tfrac{1}{2})\bar{e}_L\gamma^\mu e_L(-g_1\cos\theta_W + g_2\sin\theta_W) \\ & + \bar{e}_R\gamma^\mu e_R . g_2 \sin\theta_W]Z_\mu \end{aligned} \tag{6.57}$$

But, eqn. (6.37) also gives the neutrino current (which is also neutral) interaction, where neutrino current couples with the neutral gauge fields W_μ^3 and B_μ. Using equations (6.45) and (6.46) we express *NCIT* in Z_μ form (A_μ term cancels) and then add the term to eqn. (6.57). Equations (6.37), (6.45) and (6.46), provide us

$$NCIT = \frac{1}{2}\bar{\nu}_L\gamma^\mu\nu_L(g_1\cos\theta_W + g_2\sin\theta_W)Z_\mu \tag{6.58}$$

in which g_1 ($\sin\theta_W$)A_μ has canceled with $-g_2$ ($\cos\theta_W$) A_μ because of eqn. (6.49). Adding equations (6.57) and (6.58), we get the Z_μ interaction term $Z_\mu(IT)$ as

$$\begin{aligned} Z_\mu(IT) = & \Big[\frac{1}{2}g_1\cos\theta_W(\bar{\nu}_L\gamma^\mu\nu_L - \bar{e}_L\gamma^\mu e_L) \\ & + \frac{1}{2}g_2\sin\theta_W(\bar{\nu}_L\gamma^\mu\nu_L + \bar{e}_L\gamma^\mu e_L) \\ & + g_2\sin\theta_W(\bar{e}_R\gamma^\mu e_R)\Big]Z_\mu \end{aligned} \tag{6.59}$$

Now we introduce J^μ_{em} to replace the right-handed components, i.e., we use eqn. (6.36) as

$$\bar{e}_R\gamma^\mu e_R = -\bar{e}_L\gamma^\mu e_L - J^\mu_{em} \tag{6.60}$$

in eqn. (6.59). This substitution and the use of

$$g_1\cos\theta_W = (g_1^2 + g_2^2)^{\frac{1}{2}}\cos^2\theta_W$$

and

$$g_2\sin\theta_W = (g_1^2 + g_2^2)^{\frac{1}{2}}\sin^2\theta_W \tag{6.61}$$

which follow from eqn. (6.41) lead to

$$Z_\mu(IT) = \left(g_1^2 + g_2^2\right)^{\frac{1}{2}}\left(J_3^\mu - \sin^2\theta_W J^\mu_{em}\right)Z_\mu \tag{6.62}$$

in which

$$J_3^\mu = \frac{1}{2}\left(\bar{\nu}_L\gamma^\mu\nu_L - \bar{e}_L\gamma^\mu e_L\right) \tag{6.63}$$

has been used. From eqn. (6.62), we write the neutral weak current as

$$J^\mu_{NC} = J_3^\mu - \sin^2\theta_W\, J^\mu_{em} \tag{6.64}$$

The experimental fact is that the neutral weak current J^μ_{NC} is predominantly *V–A* type, i.e., dominantly left-handed, but there is right-handed component as well, hence J^μ_{NC} is not pure *V–A*. The structure of J^μ_{NC} in relation (6.64) is such that the right-handed component in the LHS is canceled by the right-handed component in $\sin^2\theta_W\, J^\mu_{em}$ leaving a pure left-handed J_3^μ of $SU_L(2)$.

6.2.8 Spontaneous Breaking of Local [$SU_L(2) \otimes U_Y(1)$] Gauge Symmetry

The relation (6.64) shows an intimate connection between electromagnetic and weak interactions. Similar inference is derived from the relation between coupling constants, namely, eqn. (6.53). However, problem remains regarding the masses of gauge bosons. In practice, the masses of $W^\pm$ and Z^0 bosons, the quanta of W^+ and Z_μ fields, are 80 and 91 GeV respectively, whereas the local gauge invariance of electroweak interaction demands them to be massless. To generate masses for $W^\pm$ and Z^0 bosons and retaining photon (quantum of A_μ) massless, we apply Higgs mechanism to undertake the spontaneous breaking of local gauge symmetry. The gauge symmetry is [$SU_L(2) \otimes U_Y(1)$]. As described earlier, the Higgs boson is a spin zero electrically neutral particle, which carries weak isospin and weak hypercharge. The $W^\pm$ and Z^0 bosons interact with Higgs boson and therefore they acquire mass. The photon and also gluons do not interact with the Higgs particle and hence remain massless.

In order to explain the Higgs mechanism in this case, we recall the local gauge invariant electroweak Lagrangian density [eqn. (6.28)], which we write as

$$£_1 = \overline{\psi}_L \gamma^\mu (i\partial_\mu - g_1 T_a W_\mu^a - g_2 \frac{Y_L}{2} B_\mu) \psi_L$$
$$+\overline{\psi}_R \gamma^\mu (i\partial_\mu - g_2 \frac{Y_R}{2} B_\mu)\psi_R + K.T. \text{ for } W \text{ and } B \text{ fields} \tag{6.65}$$

where the minus sign in covariant derivative is due to (i) taken inside and ψ_L and ψ_R imply

$$\psi_L = \sum_f \psi_{fL} \quad \text{and} \quad \psi_R = \sum_f \psi_{fR}$$

with similar substitutions for adjoints. The fermions (ψ_f) and bosons $W^\pm$, Z^0 in the above Lagrangian density are massless. To generate masses through the Higgs mechanism, we introduce a Higgs field ϕ. The choice of the Higgs field has been described in the following subsection.

6.2.8.1 Choice of Higgs Field

In order to see the desired mass spectrum for $W^\pm$, Z^0 and the photon, a Higgs scalar field ϕ has to be introduced. The Lagrangian density for this scalar field is

$$£_\phi = (\partial_\mu \phi)^\dagger (\partial^\mu \phi) - \mu^2 \phi^\dagger \phi - \lambda(\phi^\dagger \phi)^2 \tag{6.66}$$

in which $\mu^2 < 0$ leads to spontaneous breaking of the gauge symmetry. λ is positive. For the local gauge invariance ∂_μ is replaced by D_μ, which now [eqn.6.12)] is

$$D_\mu = \partial_\mu + ig_1 T_a W^{\mu a} + ig_2 \frac{Y}{2} B_\mu \tag{6.67}$$

and hence $£_\phi$ becomes $£_2$, expressed as

$$£_2 = (D_\mu \phi)^\dagger (D^\mu \phi) - \mu^2 \phi^\dagger \phi - \lambda(\phi^\dagger \phi)^2 \tag{6.68}$$

where the kinetic terms have not been included as they will come from $£_1$. This Lagrangian density will have to be added to $£_1$ to get masses for $W^\pm$ and Z^0 mediators of weak interaction. To keep the Lagrangian density $£_2$ gauge invariant, Weinberg suggested that the Higgs field ϕ must not be one but four fields ϕ_j, which must belong to $SU_L(2) \otimes U_Y(1)$ multiplets. *The most suitable choice was to arrange these ϕ_j into an isospin doublet with weak hypercharge $Y = 1$.* Therefore it must be understood that in the covariant derivative (6.67), $Y = 1$.

Following Weinberg, we may therefore write

$$\phi = \frac{1}{\sqrt{2}} \begin{pmatrix} \phi_1 + i\phi_2 \\ \phi_3 + i\phi_4 \end{pmatrix} = \begin{pmatrix} \phi^+ \\ \phi^0 \end{pmatrix} \tag{6.69}$$

where ϕ^+ have the electric charge (+1) and ϕ^0 have zero electric charge. Since, ϕ of eqn. (6.69) is similar to ϕ of eqn. (5.227), the steps may be retraced.

$$V(\phi) = \mu^2 \phi^\dagger \phi + \lambda(\phi^\dagger \phi)^2 \tag{6.70}$$

with

$$\phi^\dagger\phi = \frac{1}{2}\sum_{j=1}^{4}\phi_j^2 \tag{6.71}$$

When $\lambda > 0$, $\mu^2 < 0$; $V(\phi)$ would be minimum for

$$\sum_{j=1}^{4}\phi_j^2 = \left(\phi_1^2 + \phi_2^2 + \phi_3^2 + \phi_4^2\right) = -\frac{\mu^2}{\lambda} = v^2 \tag{6.72}$$

Thus the ground state corresponds to eqn. (6.72). In order to break the gauge symmetry spontaneously, we choose

$$\phi_2 = \phi_2 = \phi_4 = 0 \tag{6.73}$$

and

$$\phi_3^2 = \nu^2 = -\frac{\mu^2}{\lambda} \qquad (\mu^2 < 0) \tag{6.74}$$

Therefore, from eqn. (6.69), the lowest energy state corresponds to $\phi = \phi^0$, where

$$\phi_0 = \frac{1}{\sqrt{2}}\begin{pmatrix} 0 \\ \nu \end{pmatrix} \tag{6.75}$$

This ϕ_0 has only lower component non-vanishing. This implies that ϕ_0 has $T = \frac{1}{2}, T_3 = -\frac{1}{2}$ and $Y = 1$. Hence, from the relation

$$Q = T_3 + \frac{Y}{2} \tag{6.76}$$

the generator Q of the electromagnetic interactions vanishes for above values of T_3 and Y. This implies that

$$Q\phi_0 = 0 \tag{6.77}$$

and hence

$$\phi_0 \to \phi_0' = e^{i\alpha(x)Q}\phi_0 = \phi_0 \tag{6.78}$$

The above transformation from ϕ_0 to ϕ_0', where the generator of transformation is Q (the electric charge operator), is the transformation of gauge group $U(1)_{em}$. Thus $\phi_0(x)$ remains invariant under $U(1)_{em}$ symmetry.

The vacuum ϕ_0 chosen as eqn. (6.75) breaks both $SU_L(2)$ and $U_Y(1)$ gauge symmetries but does not break the $U(1)_{em}$ gauge symmetry. Hence due to the breaking of above gauge symmetries $W^{\pm}$ and Z^0 acquire mass and due to invariance of $U(1)_{em}$ gauge symmetry, photon remains massless.

Having obtained the correct vacuum ϕ_0, the fluctuations over ϕ_0 may generate any arbitrary $\phi(x)$. The steps from eqn. (5.249) to eqn. (5.262) may be repeated to finally write the Higgs doublet of eqn. (6.69) as

$$\phi(x) = \frac{1}{\sqrt{2}}\begin{pmatrix} 0 \\ v + h(x) \end{pmatrix} \tag{6.79}$$

Above is therefore the Higgs field, which has to be used in $£_2$ [eqn. (6.68)].

6.2.8.2 Massesn of Gauge Bosons

To substitute eqn. (6.79) for $\phi(x)$ in $£_2$, we first write $£_2$ as (using $T_a = \frac{\tau_a}{2}$)

$$£_2 = (\partial_\mu \phi^\dagger + \phi^\dagger S_\mu^\dagger)(\partial^\mu \phi + S^\mu \phi) - \mu^2 \phi^\dagger \phi - \lambda(\phi^\dagger \phi)^2 \tag{6.80}$$

where

$$S_\mu = ig_1 \frac{\tau_a}{2} W_\mu^a + ig_2 \frac{Y}{2} B_\mu \tag{6.81}$$

Expanding the brackets, we get

$$\begin{aligned} £_2 &= (\partial_\mu \phi^\dagger)(\partial^\mu \phi) + (\partial_\mu \phi^\dagger)(S^\mu \phi) + (\phi^\dagger S_\mu^\dagger)(\partial^\mu \phi) \\ &\quad + (\phi^\dagger S_\mu^\dagger)(S^\mu \phi) - \mu^2 \phi^\dagger \phi - \lambda(\phi^\dagger \phi)^2 \end{aligned} \tag{6.82}$$

Using ϕ from eqn. (6.79) and obtaining S_μ as

$$S_\mu = \frac{i}{2}\begin{bmatrix} g_1 W_\mu^3 + g_2 Y B_\mu & g_1 (W_\mu^1 - i W_\mu^2) \\ g_1 (W_\mu^1 + i\ W_\mu^2) & -g_1 W_\mu^3 + g_2 Y B_\mu \end{bmatrix} \tag{6.83}$$

we can calculate

$$\begin{aligned}
(\partial_\mu \phi^\dagger)(\partial^\mu \phi) &= \frac{1}{2}(\partial_\mu h)(\partial^\mu h) \\
(\partial_\mu \phi^\dagger)(S^\mu \phi) &= \frac{i}{4}(v + h)(-g_1 W_3^\mu + g_2 Y B^\mu)\partial_\mu h \\
\phi^\dagger (S_\mu^\dagger \partial^\mu \phi) &= -\frac{i}{4}(v + h)(-g_1 W_\mu^3 + g_2 Y B_\mu)\partial^\mu h \\
\phi^\dagger S_\mu^\dagger S^\mu \phi &= \frac{(v + h)^2}{8}\Big[g_1^2 (W_\mu^1 W_1^\mu + W_\mu^2 W_2^\mu) \\
&\quad + (g_1 W_\mu^3 - g_2 Y B_\mu)(g_1 W_3^\mu - g_2 Y B^\mu)\Big] \\
&\equiv \frac{(v + h)^2}{8} D \\
-\mu^2 \phi^\dagger \phi &= \frac{1}{2}\lambda v^2 (v + h)^2 \\
-\lambda(\phi^\dagger \phi)^2 &= -\frac{\lambda}{4}(v + h)^4
\end{aligned} \tag{6.84}$$

Substituting these values (second and third cancel) in eqn. (6.82), we may write

$$\begin{aligned}£_2 = & \left[\frac{1}{2}(\partial_\mu h)(\partial^\mu h) - \lambda v^2 h^2\right] + g_1^2 \frac{v^2}{8}(W_\mu^1 W_1^\mu + W_\mu^2 W_2^\mu) \\ & + \frac{v^2}{8}(g_1 W_\mu^3 - g_2 Y B_\mu)(g_1 W_3^\mu - g_2 Y B^\mu) + \frac{h^2}{8} D + \frac{vh}{4} D \\ & - \lambda v h^3 - \lambda \frac{h^4}{4} + \lambda \frac{v^4}{4} - \frac{1}{4}\boldsymbol{W}_{\mu\nu}\boldsymbol{W}^{\mu\nu} - \frac{1}{4} B_{\mu\nu} B^{\mu\nu} \end{aligned} \tag{6.85}$$

where we have included the $K.E$ terms from $£_1$. The hypercharge $Y = 1$. Since,

$$\boldsymbol{W}_{\mu\nu} = \partial_\mu \boldsymbol{W}_\nu - \partial_\nu \boldsymbol{W}_\mu - g(\boldsymbol{W}_\mu \times \boldsymbol{W}_\nu) \tag{6.86}$$

in which the last term in the RHS denotes the self-interaction of W_μ fields. The kinetic term of W_μ field is therefore

$$\begin{aligned}-\frac{1}{4}\boldsymbol{W}_{\mu\nu}\boldsymbol{W}^{\mu\nu} = & -\frac{1}{4}(\partial_\mu \boldsymbol{W}_\nu - \partial_\nu \boldsymbol{W}_\mu)(\partial^\mu \boldsymbol{W}^\nu - \partial^\nu \boldsymbol{W}^\mu) \\ & - \frac{g}{2}(\boldsymbol{W}_\mu \times \boldsymbol{W}_\nu)(\partial^\mu \boldsymbol{W}^\nu - \partial^\nu \boldsymbol{W}^\mu) \\ & + \frac{g^2}{4}(\boldsymbol{W}_\mu \times \boldsymbol{W}_\nu)(\boldsymbol{W}^\mu \times \boldsymbol{W}^\nu) \end{aligned} \tag{6.87}$$

in which the second and third terms are the self-interaction terms. We may therefore write

$$-\frac{1}{4}\boldsymbol{W}_{\mu\nu}\boldsymbol{W}^{\mu\nu} = \left[-\frac{1}{4}\boldsymbol{W}_{\mu\nu}\boldsymbol{W}^{\mu\nu}\right]_{\text{free}} + \text{self-interaction terms} \tag{6.88}$$

where

$$\left[-\frac{1}{4}\boldsymbol{W}_{\mu\nu}\boldsymbol{W}^{\mu\nu}\right]_{\text{free}} = -\frac{1}{4}(\partial_\mu \boldsymbol{W}_\nu - \partial_\nu \boldsymbol{W}_\mu)(\partial^\mu \boldsymbol{W}^\nu - \partial^\nu \boldsymbol{W}^\mu) \tag{6.89}$$

Taking only the free field part, we may write

$$\left[-\frac{1}{4}\boldsymbol{W}_{\mu\nu}\boldsymbol{W}^{\mu\nu}\right]_{\text{free}} = -\frac{1}{4}W_{\mu\nu}^1 W_1^{\mu\nu} - \frac{1}{4}W_{\mu\nu}^2 W_2^{\mu\nu} - \frac{1}{4}W_{\mu\nu}^3 W_3^{\mu\nu} \tag{6.90}$$

Now we substitute eqn. (6.90) in eqn. (6.85) and find that we have terms

$$\left[-\frac{1}{4}W_{\mu\nu}^1 W_1^{\mu\nu} + \frac{g_1^2 v^2}{8} W_\mu^1 W_1^\mu\right] + \left[-\frac{1}{4}W_{\mu\nu}^2 W_2^{\mu\nu} + \frac{g_1^2 v^2}{8} W_\mu^2 W_2^\mu\right] \tag{6.91}$$

for gauge fields W_μ^1 and W_μ^2 but we do not have similar terms for W_μ^3 and B_μ. Above square brackets indicate that the fields W_μ^1 and W_μ^2 have acquired masses

$$m(W_\mu^1) = \frac{g_1 v}{2} = m(W_\mu^2) \tag{6.92}$$

However, no conclusion about the masses of W_μ^3 and B_μ can be derived due to the absence of eqn. (6.91) type terms for W_μ^3 and B_μ. In fact, this is true that we will not get masses for W_μ^3 and B_μ because they are related to A_μ and Z_μ by eqns. (6.45) and (6.46). Instead, we will obtain the masses of A_μ and Z_μ fields. The third term of eqn. (6.85) also indicates that W_μ^3 and B_μ are mixed. Now using equations (6.45) and (6.46), we can check that [eqn. (6.85) with $Y = 1$]

$$\begin{aligned}
&\frac{v^2}{8}(g_1 W_\mu^3 - g_2 B_\mu)(g_1 W_3^\mu - g_2 B^\mu) \\
&= \frac{v^2}{8}(g_1 \cos\theta_W + g_2 \sin\theta_W)^2 Z_\mu Z^\mu \\
&= (g_1^2 + g_2^2)\frac{v^2}{8} Z_\mu Z^\mu
\end{aligned} \tag{6.93}$$

where ($g_1 \sin\theta_W = g_2 \cos\theta_W$) has canceled the A_μ terms and from eqn. (6.54) $g_1^2 g_2^2 = e^2(g_1^2 + g_2^2)$ has been used in the last step. Similarly, using equations (6.45) and (6.46), we can also see that

$$\begin{aligned}
-\frac{1}{4}W_{\mu\nu}^3 W_3^{\mu\nu} - \frac{1}{4}B_{\mu\nu}B^{\mu\nu} &= -\frac{1}{4}(\partial_\mu A_\nu - \partial_\nu A_\mu)(\partial^\mu A^\nu - \partial^\nu A^\mu) \\
&\quad -\frac{1}{4}(\partial_\mu Z_\nu - \partial_\nu Z_\mu)(\partial^\mu Z^\nu - \partial^\nu Z^\mu) \\
&= -\frac{1}{4}A_{\mu\nu}A^{\mu\nu} - \frac{1}{4}Z_{\mu\nu}Z^{\mu\nu}
\end{aligned} \tag{6.94}$$

where

$$W_{\mu\nu}^3 = \partial_\mu W_\nu - \partial_\nu W_\mu \tag{6.95}$$

$$A_{\mu\nu} = \partial_\mu A_\nu - \partial_\nu A_\mu \tag{6.96a}$$

and

$$Z_{\mu\nu} = \partial_\mu Z_\nu - \partial_\nu Z_\mu \tag{6.96b}$$

are implied. Thus substituting equations (6.88), (6.91), (6.93) and (6.96) in $£_2$ [eqn. (6.85)], we get

$$£_2 = \left[\frac{1}{2}(\partial_\mu h)(\partial^\mu h) - \lambda v^2 h^2\right]$$

$$+\left[-\frac{1}{4}W^1_{\mu\nu}W_1^{\mu\nu}+\frac{g_1^2v^2}{8}W^1_\mu W_1^\mu\right]$$

$$+\left[-\frac{1}{4}W^2_{\mu\nu}W_2^{\mu\nu}+\frac{g_1^2v^2}{8}W^2_\mu W_2^\mu\right]$$

$$+\left[-\frac{1}{4}Z_{\mu\nu}Z^{\mu\nu}+\frac{(g_1^2+g_2^2)v^2}{8}Z_\mu Z^\mu\right]$$

$$+\left[-\frac{1}{4}A_{\mu\nu}A^{\mu\nu}+(\text{zero})A_\mu A^\mu\right]+\text{self-interaction terms}$$

$$+\frac{h}{8}(h+v)D-\lambda vh^3-\lambda\frac{h^4}{4}+\lambda\frac{v^4}{4} \tag{6.97}$$

from which we can easily read the masses of gauge bosons. The square brackets are the Lagrangian densities of various fields. We read that there is (are)

One Higgs boson with mass

$$m(h)=\sqrt{2\lambda v^2}=\sqrt{-2\mu^2}\ (\mu^2<0)$$

Two W bosons each with mass

$$m(W)=\frac{1}{2}g_1\, v$$

One Z boson with mass

$$m(Z)=\frac{v}{2}(g_1^2+g_2^2)^{\frac{1}{2}}=\frac{m_W}{\cos\theta_W}$$

and a (massless) photon

$$m(\gamma)=0 \tag{6.98}$$

From the above we see that when we add £$_2$ to £$_1$, the gauge bosons ($W^\pm$, Z) of weak interactions would acquire mass and the photon would remain massless. Our next step is to provide masses to ψ_f fields, i.e., to leptons and quarks or to fermions.

6.2.8.3 Masses of Leptons

Due to chirality (handedness), the fermions (leptons and quarks) in the Lagrangian density (6.21) were required to be massless. However, in reality except neutrinos, all leptons and all quarks have mass. Therefore the GWS model must be able to explain how leptons and quarks acquire mass. We first take up the case of leptons.

It is again the Higgs mechanism, which will provide mass to leptons (also to quarks). Since the left-handed leptons form isospin doublets and upper $\left(T_3=\frac{1}{2}\right)$ members (neutrinos) of these doublets are massless hence the Higgs mechanism should provide mass only to the lower $\left(T_3 = -\frac{1}{2}\right)$ members of the weak isospin doublets and leave upper members massless.

We again choose the first generation leptons $\begin{pmatrix} \nu_L \\ e_L \end{pmatrix}$ and e_R. We have seen under the subsection *choice of the Higgs field* that its quantum, the Higgs boson, has isospin one half and hypercharge one. We table these quantum numbers as below (Table 6.2):

Table 6.2

Particle	T	T_3	Y
$\nu_e(\nu_L)$	$\frac{1}{2}$	$+\frac{1}{2}$	-1
e_L	$\frac{1}{2}$	$-\frac{1}{2}$	-1
e_R	0	0	-2
$h^0(\phi)$	$\frac{1}{2}$	$-\frac{1}{2}$	1

Recalling the arguments below the eqn. (6.23), we have to construct a $T = 0$ combination, so that the combination is invariant under $SU_L(2)$ transformations. Now for the first generation leptons

$$\psi_L = \begin{pmatrix} \nu_L \\ e_L \end{pmatrix}, \quad \psi_R = e_R$$

$$\bar{\psi}_L = (\bar{\nu}_L \quad \bar{e}_L) \text{ and } \bar{\psi}_R = \bar{e}_R \tag{6.99}$$

ψ_L implies ψ_{fL} etc. We have noted that a mass term like [see eqn. (6.21)]

$$m\bar{\psi}_L\psi_R + m\bar{\psi}_R\psi_L \tag{6.100}$$

cannot be included in the Lagrangian density because ψ_L is a doublet and ψ_R a singlet. A doublet should combine with a doublet and singlet with a singlet to provide a $T = 0$ (gauge invariant) combination. Therefore, using the Higgs field ϕ and ψ_R, we construct a combination $\phi\psi_R$, whose $T = \frac{1}{2}, Y = -1$ (Fig. 6.3)

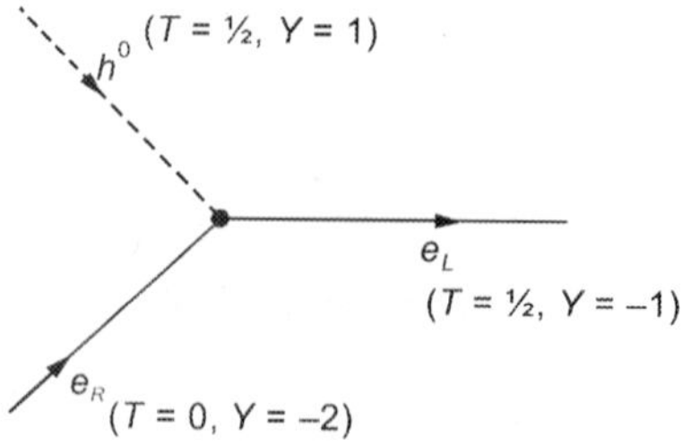

Fig. 6.3: Coupling of Higgs scalar and electron

and hence behaves as a doublet. Using such a doublet, a term $£_3$ is added to the Lagrangian density $£_1$ to which $£_2$ is already added. First we will write $£_3$ which is invariant under $[SU_L(2) \otimes U_Y(1)]$ symmetry, and then break the symmetry through Higgs mechanism that results in the generation of electron mass. The gauge invariant $£_3$ is

$$£_3 = -g_e\left(\overline{\psi}_L \phi \psi_R + \overline{\psi}_R \phi^\dagger \psi_L\right) \tag{6.101}$$

where g_e is the arbitrary coupling constant and within the bracket, we have products of two isospinors. $\phi\psi_R$ is another isospinor. Now the Higgs field ϕ is

$$\phi = \frac{1}{\sqrt{2}}\begin{pmatrix} \phi_1 + i\phi_2 \\ \phi_3 + i\phi_4 \end{pmatrix} = \begin{pmatrix} \phi^+ \\ \phi^0 \end{pmatrix} \tag{6.102}$$

which along with eqn. (6.99) gives $£_3$ as

$$£_3 = -g_e\left[\left[\overline{\nu}_L \quad \overline{e}_L\right]\begin{pmatrix} \phi^+ \\ \phi^0 \end{pmatrix} e_R + \overline{e}_R\left(\phi^- \quad \overline{\phi}^0\right)\begin{pmatrix} \nu_L \\ e_L \end{pmatrix}\right] \tag{6.103}$$

To break spontaneously the $[SU_L(2) \otimes U_Y(1)]$ symmetry of eqn. (6.103), we apply Higgs mechanism, according to which we must replace ϕ of eqn. (6.102) by

$$\phi(x) = \frac{1}{\sqrt{2}}\begin{pmatrix} 0 \\ v + h(x) \end{pmatrix} \tag{6.104}$$

It is interesting to note that 'same' ϕ [eqn. (6.104)] was used to generate the masses of $W^\pm$ and Z bosons. Thus a single Higgs doublet [eqn. (6.104)] would provide mass to gauge bosons as well as to leptons. Only thing we will require is to add eqn. (6.101) [$£_3$] to the Lagrangian densities.

When eqn. (6.104) is substituted in eqn. (6.101) the symmetry is broken and $£_3$ becomes $£_3^{\text{new}}$ as

$$£_3^{\text{new}} = -\frac{g_e}{\sqrt{2}}\left[\left(\overline{\nu}_L \quad \overline{e}_L\right)\begin{pmatrix} 0 \\ v + h \end{pmatrix} e_R + \overline{e}_R\left(0 \quad v + h\right)\begin{pmatrix} \nu_L \\ e_L \end{pmatrix}\right]$$

$$= -\frac{g_e v}{\sqrt{2}}\left(\overline{e}_L e_R + \overline{e}_R e_L\right) - \frac{g_e}{\sqrt{2}}\left(\overline{e}_L e_R + \overline{e}_R e_L\right)h \tag{6.105}$$

which on using the eqn. (6.22) can be expressed as

$$£_3^{\text{new}} = -\frac{g_e v}{\sqrt{2}}\,\overline{e}e - \frac{g_e}{\sqrt{2}}\overline{e}eh \tag{6.106}$$

The first term of eqn. (6.106) is like $m\overline{\psi}\psi$ and hence the electron acquires a mass

$$m_e = \frac{g_e v}{\sqrt{2}} \tag{6.107}$$

The second term indicates the interaction of Higgs particle with electron. The strength with which the Higgs particle couples to the electron is

$$g_{h-e} = \frac{g_e}{\sqrt{2}} = \frac{m_e}{v} \tag{6.108}$$

We can generalize this relation for all leptons as

$$g_{h-lepton} = \frac{mass\ of\ the\ lepton}{v} \tag{6.109}$$

Since $v = \sqrt{-\mu^2/\lambda}$ $(\mu^2 < 0)$ is a constant, eqn. (6.109) tells that the Higgs-lepton (charged) coupling strength is proportional to the mass of charged lepton. The experimental consequence of such proportionality is that the Higgs particle can be produced more easily by heavy leptons (fermions) and will decay predominantly into heavy fermions.

From eqn. (6.98), the mass of Higgs particle is

$$m_h = \sqrt{2\lambda v^2} \tag{6.110}$$

and $v(\approx 150$ GeV) is known. However, the quartic (ϕ^4) coupling λ is not known. This makes the estimation of m_h difficult. Anticipating that $\lambda < 1$, the upper limit of m_h is estimated below 350 GeV.

In the above analysis we saw that the electron (charged lepton) acquired mass and because in $£_3$ we cannot have terms like

$$\bar{\nu}_R \nu_L + \bar{\nu}_L \nu_R \tag{6.111}$$

the neutrino remains massless. There is no right-handed neutrino (ν_R), so eqn. (6.111) cannot be included in the Lagrangian density.

6.2.8.4 Masses of Quarks

The difference between quark doublets and lepton doublets is that in the case of leptons, the upper $(T_3 = -\frac{1}{2})$ member of the weak isospin doublet was neutrino (ν_e, ν_μ or ν_τ) and was massless. But in the case of weak isospin doublet of quarks

$$\begin{pmatrix} u \\ d' \end{pmatrix}_L, \begin{pmatrix} c \\ s' \end{pmatrix}_L, \begin{pmatrix} t \\ b' \end{pmatrix}_L$$

both the $(T_3 = +\frac{1}{2})$ and $(T_3 = -\frac{1}{2})$ members are massive. So we have to generate masses of $(T_3 = +\frac{1}{2})$ members as well. For this purpose, we construct a new Higgs doublet ϕ_c from the old Higgs doublet ϕ as

$$\phi_c = -i\tau_2\phi^* = \begin{pmatrix} -\bar{\phi}^0 \\ \phi^- \end{pmatrix} \tag{6.112}$$

We already know that

$$\phi = \begin{pmatrix} \phi^+ \\ \phi^0 \end{pmatrix} \tag{6.113}$$

Taking

$$\psi_L = \begin{pmatrix} u_L \\ d'_L \end{pmatrix}, \qquad \psi_R = u_R, d_R \tag{6.114}$$

the Lagrangian density $£_4$ that would be $[SU_L(2) \otimes U_Y(1)]$ symmetric would be

$$\begin{aligned} £_4 &= -g_d[\overline{\psi}_L \phi d_R + \overline{d}_R \phi^\dagger \psi_L] \\ &\quad - g_u[\overline{\psi}_L \phi_c u_R + \overline{u}_R \phi_c^\dagger \psi_L] \end{aligned} \tag{6.115}$$

Substituting from the above equations, we may write for

$$\begin{aligned} £_4 = -g_d &\left[\begin{pmatrix} \overline{u}_L & \overline{d}'_L \end{pmatrix} \begin{pmatrix} \phi^+ \\ \phi^0 \end{pmatrix} d_R + \overline{d}_R \begin{pmatrix} \phi^- & \overline{\phi}^0 \end{pmatrix} \begin{pmatrix} u_L \\ d'_L \end{pmatrix} \right] \\ -g_u &\left[\begin{pmatrix} \overline{u}_L & \overline{d}'_L \end{pmatrix} \begin{pmatrix} -\overline{\phi}^0 \\ \phi^- \end{pmatrix} u_R + \overline{u}_R \begin{pmatrix} -\phi^0 & \phi^+ \end{pmatrix} \begin{pmatrix} u_L \\ d'_L \end{pmatrix} \right] \end{aligned} \tag{6.116}$$

To break the symmetry of eqn. (6.116), we apply Higgs mechanism according to which we will replace

$$\phi(x) = \frac{1}{\sqrt{2}} \begin{pmatrix} 0 \\ v+h \end{pmatrix}$$

and

$$\phi_c(x) = \frac{1}{\sqrt{2}} \begin{pmatrix} v+h \\ 0 \end{pmatrix} \tag{6.117}$$

and hence

$$\begin{aligned} £_4^{\text{new}} &= -\frac{g_d}{\sqrt{2}} \left[\begin{pmatrix} \overline{u}_L & \overline{d}'_L \end{pmatrix} \begin{pmatrix} 0 \\ v+h \end{pmatrix} d_R + \overline{d}_R \begin{pmatrix} 0 & v+h \end{pmatrix} \begin{pmatrix} u_L \\ d'_L \end{pmatrix} \right] \\ &\quad -\frac{g_u}{\sqrt{2}} \left[\begin{pmatrix} \overline{u}_L & \overline{d}'_L \end{pmatrix} \begin{pmatrix} v+h \\ 0 \end{pmatrix} u_R + \overline{u}_R \begin{pmatrix} v+h & 0 \end{pmatrix} \begin{pmatrix} u_L \\ d'_L \end{pmatrix} \right] \\ &= -\frac{g_d}{\sqrt{2}} [v\overline{d}d + h\overline{d}d] - \frac{g_u}{\sqrt{2}} [v\overline{u}u + h\overline{u}u] \end{aligned} \tag{6.118}$$

where

$$\overline{d}d = \overline{d}'_L d_R + \overline{d}_R d'_L$$

and

$$\bar{u}u = \bar{u}_L u_R + \bar{u}_R u_L \tag{6.119}$$

have been used. Similar to eqn. (6.107), we may write

$$m_d = \frac{g_d v}{\sqrt{2}}, \frac{g_d}{\sqrt{2}} = \frac{m_d}{v} \tag{6.120}$$

and also

$$m_u = \frac{g_u v}{\sqrt{2}}, \quad \frac{g_u}{\sqrt{2}} = \frac{m_u}{v} \tag{6.121}$$

where m_d and m_u are masses of d and u quarks, g_d and g_u are arbitrary coupling constants. Substituting equations (6.120) and (6.121) in eqn. (6.118) we write

$$£_4^{\text{new}} = -m_d\bar{d}d - m_u\bar{u}u - \frac{m_d}{v}\bar{d}dh - \frac{m_u}{v}\bar{u}uh \tag{6.122}$$

The first two terms are like $m\bar{\psi}\psi$ hence d and u quarks acquire masses m_d and m_u respectively expressed in terms of coupling constants g_d and g_u through relations (6.120) and (6.121). Again the last two terms of eqn. (6.122) denote the coupling of Higgs boson with d and u quarks. The strength of this coupling is also proportional to the mass of the quarks. This again implies that Higgs particle can be produced by heavier flavours of quarks. We will include $£_4$ with other three Lagrangian densities.

Thus the spontaneous breaking of the $[SU_L(2) \otimes U_Y(1)]$ local gauge symmetry has been able to provide masses to $W^{\pm}$ and Z^0 gauge bosons, to leptons and to quarks and has left photon and the neutrinos massless. This in fact completes the GWS model. However, *the discrepancies pointed out in section* 6.2.6 *need to be explained.* We do that in the following subsection.

6.2.8.5 Strength of Weak Interactions

In section 6.2.6, the discrepancies regarding the strength of the coupling constant (g_1) and the range of weak interactions were observed. Both the discrepancies get removed when we take into consideration the largeness of mass of weak interaction gauge bosons. For this let us consider the exchange of a $Z \Rightarrow Z^0$ boson as shown in Fig. (6.4).

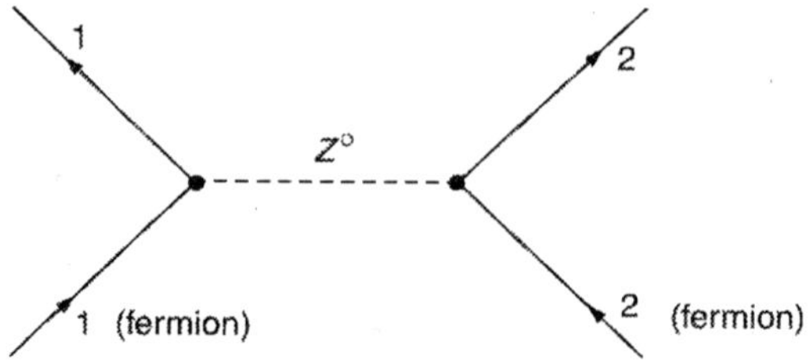

Fig. 6.4: Neutral current interaction 1 + 2 → 1 + 2

The propagator, which exchanges the boson Z^0 between the fermions 1 and 2, can be written as

$$\frac{g_Z^2}{k^2 - M_Z^2} \tag{6.123}$$

where k and M_Z are the four-momentum and mass of Z boson and g_Z is the coupling constant formed by the coupling constants g_1 and g_2 in the following manner. From eqn. (6.48), we can check by observing the coefficient of Z_μ that for the left-handed electron

$$g_L^Z = -g_1 \cos\theta_W + g_2 \sin\theta_W$$

$$= +\frac{(g_2^2 - g_1^2)}{\sqrt{g_1^2 + g_2^2}} \tag{6.124}$$

and for the right-handed electron

$$g_R^Z = g_2 \sin\theta_W = \frac{g_2^2}{\sqrt{g_1^2 + g_2^2}} \tag{6.125}$$

From the above relations, we infer that g_Z, the coupling constant for Z^0 with fermions is a combination of weak coupling constants g_1 and g_2, which couples B_μ gauge field.

In the energy range that is quite smaller than $M_Z (\approx 92$ GeV), the square of four-momentum k^2 in eqn. (6.123) can be neglected $(k^2 << M_Z^2)$ and the eqn. (6.123) then yields the effective interaction strength G_Z, where

$$G_Z \approx \frac{g_Z^2}{M_Z^2} \tag{6.126}$$

To make G_Z dimensionless, we may use $m_p \approx 1$ GeV to write

$$G_Z \approx \frac{g_Z^2 m_p^2}{M_Z^2} \tag{6.127}$$

Since,

$$\left(\frac{m_p^2}{M_Z^2}\right) \approx \left(\frac{1}{92}\right)^2 \approx 10^{-4} \tag{6.128}$$

the dimensionless weak coupling constant

$$G_Z \approx 10^{-4} g_Z^2 \tag{6.129}$$

Thus the G_Z, which now is the weak coupling constant, is about 10^{-4} times smaller than g_Z^2 (a combination of g_1 and g_2). From Chapter 1 [eqn (1.107)], the range of weak interaction

$$R_{\text{weak}} \approx \frac{v\hbar}{M_Z c^2} \tag{6.130}$$

which due to the large mass of M_Z gives R_{weak} » 10^{-19} m. In this way the largeness of mass of the mediators of weak interaction explains that the weak interactions have small coupling strength and that their range is short.

6.3 RENORMALIZABILITY OF ELECTROWEAK GAUGE THEORY

A well-known technique of performing calculations in Quantum Electrodynamics (QED) is the perturbation theory. When this technique is applied to QED, for example, to the calculation of Lamb Shift in the energy spectrum of hydrogen atom, it is found that the perturbative corrections to the Coulomb potential arise from the production of virtual electron–positron pairs by the electromagnetic field. When the effects of these pairs are calculated they tend to take the measurable quantities to infinity. The mathematical procedure, which eliminates these infinities, is known as renormalization. The process of renormalization can be undertaken only in a perturbative technique, because in this technique the subtraction of one infinite term from the other can be done in each step in the perturbative calculation. Since, in the perturbative theory a small correction term is added to the known solution of a simpler problem, the smallness of the electromagnetic coupling constant allows the adoption of perturbative approach for *QED* calculations and it is possible in *QED* that the renormalization procedure can remove all the infinities.

In the strong interactions, the coupling constant is large compared to unity hence the perturbative techniques cannot be applied. Therefore the calculations for strong interactions lack in accuracy, as they remain unrenormalized.

For the weak interactions, the coupling constant is even smaller than the electromagnetic, hence it is natural to expect that the perturbative approach would be applicable to them and the calculations of weak interactions would be renormalizable. However, the renormalization procedure breaks down in weak interactions because of the appearance of new type of infinities, which cannot be canceled.

In section 4.5.3, we have briefly discussed the neutrino and electron scattering, where the scattering cross section without introducing the vector bosons has been shown to be

$$\sigma(\nu_e e \rightarrow \nu_e e) = \frac{G^2 s}{\pi} \tag{6.131}$$

where s is the square of center of mass energy of ν_e and e. The scattering cross section σ, increases with s, without any upper limit. At energies greater than 300

GeV, σ violates unitarity. The conservation of probability is restored by the introduction of intermediate vector boson W as in the following Figure (Fig. 6.5). The total scattering cross section for Figure (6.5) is

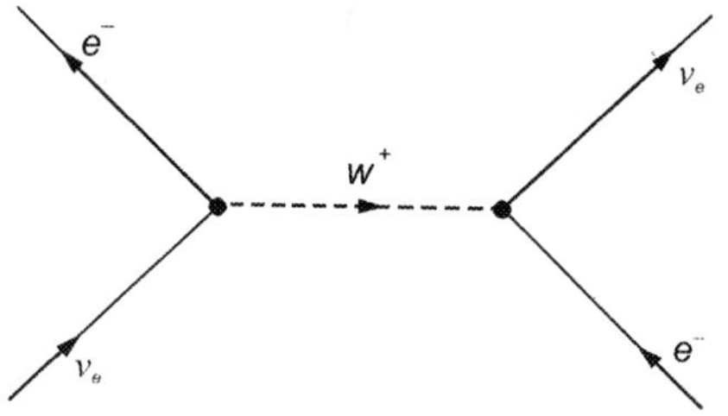

Fig. 6.5: Diagram for the scattering $\nu_e e^- \xrightarrow{W^+} e^- \nu_e$

$$\sigma = \frac{G^2 M_W^2}{\pi} \tag{6.132}$$

which is free from divergence. However, when W bosons have been introduced, we have certain other processes in which $W^\pm$ bosons are involved and they are not free from divergences. An example is $\nu_e \bar{\nu}_e$ annihilation.

$$\nu_e + \bar{\nu}_e \rightarrow W^+ + W^- \tag{6.133}$$

The first order diagram for this process is shown in Fig. (6.6)

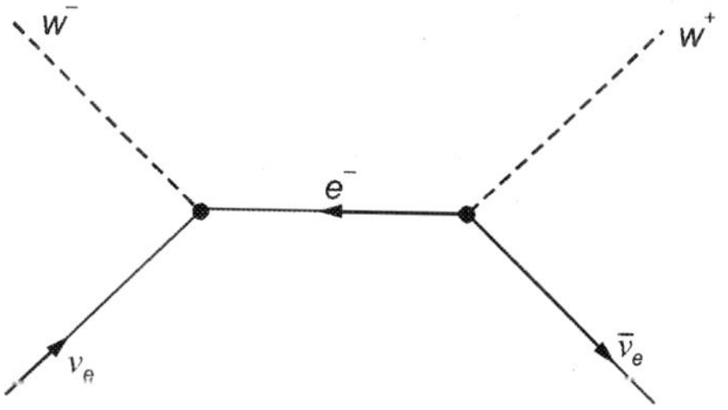

Fig. 6.6: Diagram for neutrino-antineutrino annihilation $\nu_e + \bar{\nu}_e \rightarrow W^+ + W^-$

The scattering cross section for the above annihilation process [eqn. (6.133)] is

$$\sigma(\nu\bar{\nu} \rightarrow W^+ W^-) = \frac{G^2 s}{3\pi} \tag{6.134}$$

which obviously goes to infinity when s does so. It is here we notice the remarkable structure of electroweak theory, whose most significant aspect is presented by the neutral weak interactions mediated by Z^0 bosons. Introducing Z^0 boson, a diagram as in Figure (6.7) can be constructed, which cancels the divergence of diagram (6.6) and brings

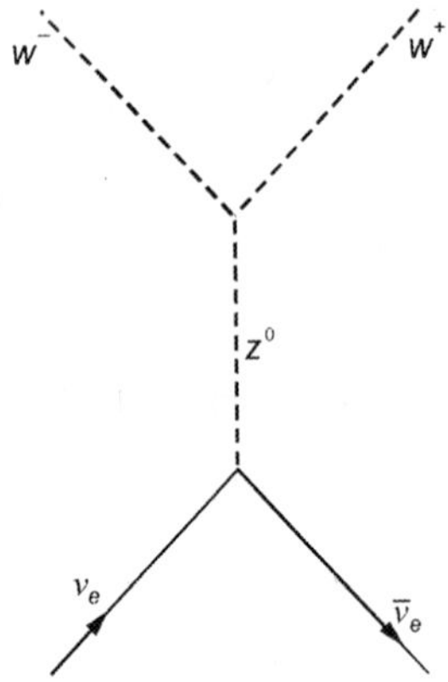

Fig. 6.7: Diagrams for neutrino-antineutrino annihilation through $\nu_e + \bar{\nu}_e \xrightarrow{Z^0} W^- W^+$

about the renormalization of $\nu_e\bar{\nu}_e$ annihilation process. Another example, where renormalization in weak interactions becomes manifest is the process

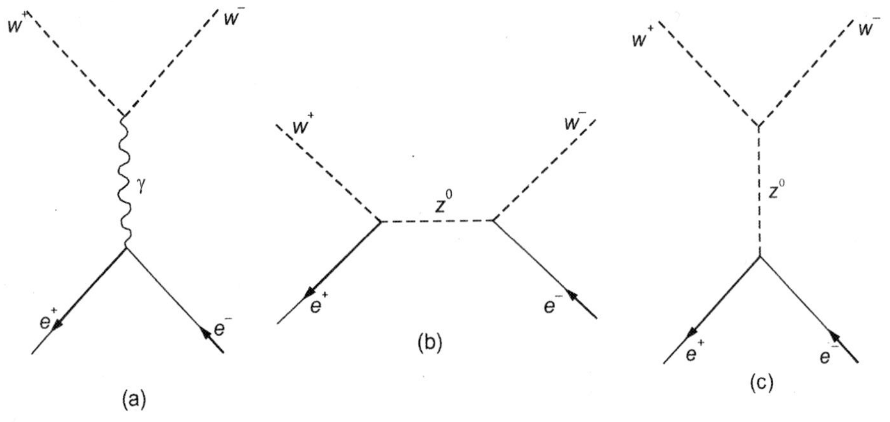

Fig. 6.8: Diagrams for electron–positron annihilation. Fig. a: annihilation through electromagnetic interaction. Fig. b, c: e⁺ e⁻ annihilation through weak interactions

$$e^+ + e^- \rightarrow W^+ + W^- \tag{6.135}$$

The three first order diagrams for the above process are shown in Fig. (6.8). The first diagram is due to electromagnetic interactions, whereas the diagrams (b) and (c) arise from weak interactions. The individual diagrams show divergence but the sum of diagrams a, b and c is finite.

A quite interesting example, where even Z^0 diagrams fail to cancel the divergence is the scattering of charged bosons $W^{\pm}$ shown as in the three diagrams of Fig. (6.9).

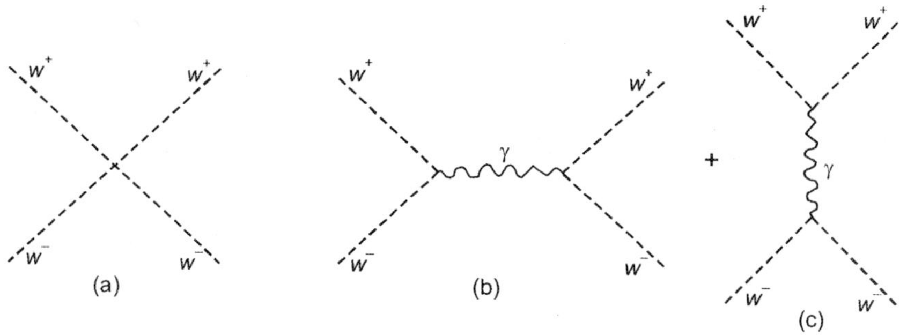

Fig. 6.9: Diagrams for the scattering of charged W bosons

Individual diagrams diverge as s^2/M_W^2, but their sum is s/M_W^2, which relatively is less divergent. The divergence does not cancel even if we include diagram with Z^0 exchange. It still diverges as s/M_W^2. The cancellation of this diagram comes from the diagram with Higgs boson exchange (Fig. 6.10). Thus the Higgs scalar

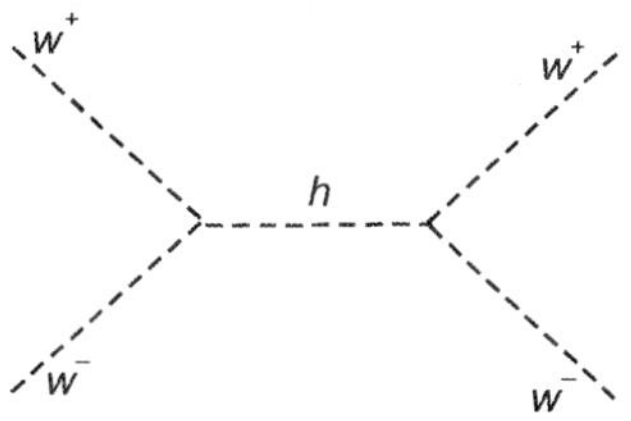

Fig. 6.10: Diagrams for the scattering of W bosons through Higgs particle h

particle becomes essential for the renormalization. The Higgs boson in fact was meant for spontaneous symmetry breaking. If somehow the symmetry breaking had not required the Higgs boson, it would have been essential for renormalization of electroweak gauge theory. In fact, for a theory to be renormalizable, it must be a Yang–Mills theory. The GWS model is thus the simplest model that fulfills the vital requirement of renormalizability.

6.4 QUANTUM CHROMODYNAMICS

In order to complete the description of gauge theories of three interactions, we take up the strong interactions of coloured quarks. As described in section (5.8), the local gauge theory of colour SU(3) group leads us to the theory of Quantum Chromodynamics (QCD). The $SU_c(3)$ colour gauge invariant Lagrangian density has already been obtained in eqn. (5.151). Here we will briefly retrace the steps.

We consider the strong interactions of coloured quarks mediated by the exchange of coloured gluons. Since, the gluons are massless; we need not incorporate the mechanism of mass generation, which we required in electroweak theory. The quarks appear in six flavours and each flavour has three colours.

Thus denoting flavour by index q and colour indices by j, k we may write the Lagrangian density for the free quark (assumed) field as

$$\pounds_q = i\sum_q \bar{\psi}_q^j \gamma^\mu \partial_\mu \psi_q^k - \sum_q m_q \bar{\psi}_q^j \psi_q^j \tag{6.136}$$

where $\sum_q$ means, we have to include all the six flavours, u, d, c, s, t, b of quarks.

$j, k = 1, 2, 3$ refer to three colours. ψ_q^j is the Dirac spinor for a given flavour q and colour j of quark. The local gauge invariance of $\pounds_q$ is obtained as usual by replacing ∂_μ by D_μ, which we express because of colour indices as (see eqn. 5.150)

$$(D_\mu)_{jk} = \delta_{jk}\partial_\mu + ig_s (T_a)_{jk} G_\mu^a \tag{6.137}$$

where g_s is the strong coupling constant, and G_μ^a represent the colour gauge fields. Since there are eight generators for SU(3) group, a will run from 1 to 8.

Thus replacing ∂_μ of eqn. (6.136) by D_μ of eqn. (6.137) and including kinetic term of G_μ^a as

$$-\frac{1}{4} G_{\mu\nu}^a G_a^{\mu\nu} \tag{6.138}$$

with

$$G_{\mu\nu}^a = \partial_\mu G_\nu^a - \partial_\nu G_\mu^a - g_s f_{abc} G_\mu^b G_\nu^c \tag{6.139}$$

where f_{abc} are the structure constants of SU(3) and $a, b, c = 1, 2, 3, ..., 8$, we get the local gauge invariant $\pounds_q$, the QCD Lagrangian density, as

$$\pounds_q = i\sum_q \bar{\psi}_q^j \gamma^\mu (D_\mu)_{jk} \psi_q^k - \sum_q m_q \bar{\psi}_q^j \psi_q^j - \frac{1}{4} G_{\mu\nu}^a G_a^{\mu\nu} \tag{6.140}$$

We expand the first term by using eqn. (6.137) for the covariant derivative and write $\pounds_q$ as

$$\pounds_q = i\sum_q \delta_{jk} \bar{\psi}_q^j \gamma^\mu \partial_\mu \psi_q^k - \sum_q m_q \bar{\psi}_q^j \psi_q^j$$

$$-\sum_q g_s \bar{\psi}_q^j \gamma^\mu (T_a)_{jk} \psi_q^k G_\mu^a - \frac{1}{4} G_{\mu\nu}^a G_a^{\mu\nu} \tag{6.141}$$

where m_q is the mass of a quark flavour. It is expected that this Lagrangian density $\pounds_q$ would be added to the Lagrangian densities $\pounds_1$ to $\pounds_4$ of GWS model to include QCD. However, the first term of eqn. (6.141) is already in $\pounds_1$, where quark flavours are included in ψ_f and since the quarks' masses m_q are being generated by $\pounds_4$, we have only last two terms of eqn. (6.141) that need be added to GWS model to include QCD. We denote these terms by $\pounds_5$ with

$$\pounds_5 = -\frac{1}{2} g_s \sum_q (\bar{\psi}_q^j \gamma^\mu \lambda_{jk}^a \psi_q^k) G_\mu^a - \frac{1}{4} G_{\mu\nu}^a G_a^{\mu\nu} \tag{6.142}$$

where $T_a = \frac{\lambda_a}{2}$ has been used. The first term in £$_5$ denotes quark gluon couplings and second the kinetic term, which includes self-interaction of colour gauge fields G_a^μ.

6.5 THE STANDARD MODEL LAGRANGIAN

Since in the Standard Model, SU(3) colour group is included in the gauge group of GWS model, we call it the $SU_c(3) \otimes SU_L(2) \otimes U_Y(1)$ Standard Model, whose Lagrangian density would be

$$£ = £_1 + £_2 + £_3 + £_4 + £_5 \tag{6.143}$$

Substituting the previously obtained Lagrangian densities £$_1$ to £$_5$ and rearranging, the Standard Model Lagrangian density is written as

$$£ = \left[-\frac{1}{4} W_{\mu\nu}.W^{\mu\nu} - \frac{1}{4} B_{\mu\nu}B^{\mu\nu} - \frac{1}{4} G^a_{\mu\nu}G^{\mu\nu}_a \right] \begin{cases} W^\pm, Z^0, \gamma \\ \text{kinetic energies} \\ \text{and self-} \\ \text{interactions} \end{cases}$$

$$+ \begin{bmatrix} i\overline{\psi}_L\gamma^\mu(\partial_\mu + \frac{i}{2}g_1\boldsymbol{\tau}.\boldsymbol{W}_\mu + \frac{i}{2}g_2Y_LB_\mu)\psi_L \\ +i\overline{\psi}_R\gamma^\mu(\partial_\mu + \frac{i}{2}g_2Y_RB_\mu)\psi_R \end{bmatrix} \begin{cases} \text{lepton and quark} \\ \text{kinetic energies} \\ \text{and their} \\ \text{interactions} \\ \text{with} \\ W^\pm, Z^0, \gamma \end{cases}$$

$$+ \begin{bmatrix} \left| (\partial_\mu + \frac{i}{2}g_1\boldsymbol{\tau}.\boldsymbol{W}_\mu + \frac{i}{2}g_2YB_\mu)\phi \right|^2 \\ -\mu^2|\phi|^2 - \lambda|\phi|^4 \end{bmatrix} \begin{cases} W^\pm, Z, \gamma \text{ and Higgs} \\ \text{boson masses and} \\ \text{couplings} \end{cases}$$

$$+ \begin{bmatrix} -G_1\left[\overline{\psi}_L\phi\psi_R + \overline{\psi}_R\phi^\dagger\psi_L\right] \\ -G_2\left[\overline{\psi}_L\phi_c\psi_R + \overline{\psi}_R\phi_c^\dagger\psi_L\right] \end{bmatrix} \begin{cases} \text{lepton and quark} \\ \text{masses and couplings} \\ \text{to Higgs boson} \end{cases}$$

$$-\left[\frac{1}{2}g_s(\overline{\psi}^j_q\gamma^\mu\lambda^a_{jk}\psi^k_q)G^a_\mu\right] \begin{cases} \text{quark–gluon} \\ \text{couplings} \end{cases} \tag{6.144}$$

where ψ_L denotes a left-handed lepton or quark doublet and ψ_R denotes right-handed fermion (lepton or quark) singlet.

6.6 EXPEIMENTAL EVIDENCES FOR THE STANDARD MODEL

The fundamental fermions may be arranged in the following families

$$\begin{bmatrix} \nu_e \\ e \\ u \\ d \end{bmatrix} \quad \begin{bmatrix} \nu_\mu \\ \mu \\ c \\ s \end{bmatrix} \quad \begin{bmatrix} \nu_\tau \\ \tau \\ t \\ b \end{bmatrix}$$

The interactions between them are described by the gauge theories. Since the Standard Model puts together all the three interactions, *electromagnetic* (QED), *Weak and Strong* (QCD), we here mention the experimental evidence for each of the above theories.

6.6.1 Experimental Tests of QED

The *QED*, which is the U(1) local gauge theory, has been verified experimentally to a fair degree of accuracy. Following are the examples where *QED* has been verified quite accurately.

6.6.1.1 Energy Levels of Atomic Hydrogen

Each level of hydrogen spectrum is a doublet. The transition between two sublevels of $1\,^2S_{\frac{1}{2}}$ state gives rise to well known 21cm hydrogen line, which corresponds to the frequency of 1420.4 MHz. Assuming electron a Dirac particle having a magnetic moment of 1 Bohr magneton and taking into account the higher order diagrams, the agreement between theoretical and experimental values of $\Delta\nu$ is found to be quite good.

$$\begin{aligned} \Delta\nu\ (\text{QED theory}) &= 1420.404\ \text{MHz} \\ \Delta\nu\ (\text{Experiment}) &= 1420.405\ \text{MHz} \end{aligned} \tag{6.145a}$$

The agreement between the theoretical and experimental values of Lamb shift

$$\begin{aligned} \Delta\nu\ (\text{theory}) &= 1057.898\ \text{MHz} \\ \Delta\nu\ (\text{Experiment}) &= 1057.878\ \text{MHz} \end{aligned} \tag{6.145b}$$

also provides a sound experimental test of QED.

6.6.1.2 Electron and Muon Magnetic Moments

The electron and muon are charged particles. A charged particle in orbital motion generates a magnetic dipole moment whose magnitude is proportional to orbital angular momentum. A particle with intrinsic spin has intrinsic magnetic moment. If it is assumed that the electron or the muon is in $l = 0$ state, the orbital magnetic moment will be zero and only spin magnetic moment would exist.

According to Dirac theory, the electron or the muon is a spin $\frac{1}{2}$ particle and the magnetic moment (μ) corresponding to its spin is equal to one Bohr magneton (μ_B). The value $\mu = \mu_B$ is based on the value of Lande's splitting factor g equal to 2. However, these charged leptons are not just bare charges but possess e^+e^- pairs and clouds of virtual photons surrounding them, hence their magnetic moment changes. The corrections to the magnetic moment due to e^+e^- pairs, etc., is expressed by

$$\mu = \mu_B (1 + y) \tag{6.146}$$

where y representing the correction is

$$y = \frac{g-2}{2} \tag{6.147}$$

which obviously does not exist for $g = 2$ and therefore, for y to exist, $(g-2)$ should not be zero.

The experimental technique now makes the very precise measurements of **$(g-2)$** for both electron and muon from which we get y values, which from eqn. (6.146) provide us the more precise values of their magnetic moments (μ). The theoretically calculated values of y for both electron and muon agree very well with corresponding experimental values.

For electron

$$\begin{aligned} y\,(\text{Theory}) &= (1159652133 \pm 29) \times 10^{-12} \\ y\,(\text{Experiment}) &= (1159652188 \pm 4) \times 10^{-12} \end{aligned} \tag{6.148}$$

For muon

$$\begin{aligned} y\,(\text{Theory}) &= (11659194 \pm 14) \times 10^{-10} \\ y\,(\text{Experiment}) &= (11659230 \pm 84) \times 10^{-10} \end{aligned} \tag{6.149}$$

The above agreements provide the very important test for the validity of QED.

6.6.1.3 QED at High Energies

The QED at high energies shows that the electron is a point particle and has no structure down to a distance of about 2×10^{-21} m. Such a conclusion was derived by studying the Bhabha scattering $e^+e^- \to e^+e^-$, where beam machines of e^+ and e^- were used. The PETRA (DESY, Hamburg, Germany) uses e^+ and e^- colliding beams of 23 GeV each and the e^+e^- scattering at such energies show that the electron is a point particle even at higher energies.

6.6.2 Evidences for Weak Interactions

The weak interactions involving charged weak currents have been studied quite extensively since 1930s. In addition to charged weak decays, the handedness, violation of parity, conservation of CP and its small violation in weak interactions too have been experimentally studied. All the above studies provide experimental evidence for weak interactions. To above studies, however, should be added the experimental evidence for neutral weak interactions.

6.6.2.1 The Experimental Verification of Weak Neutral Current

At CERN (European Center for Nuclear Research, Geneva, Switzerland) the bubble chamber was exposed to a beam of muon antineutrinos. The observed scattering was (Fig. 6.11)

$$\overline{\nu}_\mu + e^- \to \overline{\nu}_\mu + e^- \tag{6.150}$$

Though, the scattering cross section (σ) for the above scattering was small, the $\overline{\nu}_\mu e^- \to \overline{\nu}_\mu e^-$ scattering, a neutral weak interaction, ***was*** observed. After the above process, the more common neutral processes

$$\begin{aligned} \nu_\mu + N &\to \nu_\mu + \text{hadrons} \\ \overline{\nu}_\mu + N &\to \overline{\nu}_\mu + \text{hadrons} \end{aligned} \tag{6.151}$$

were also observed in the same experiment.

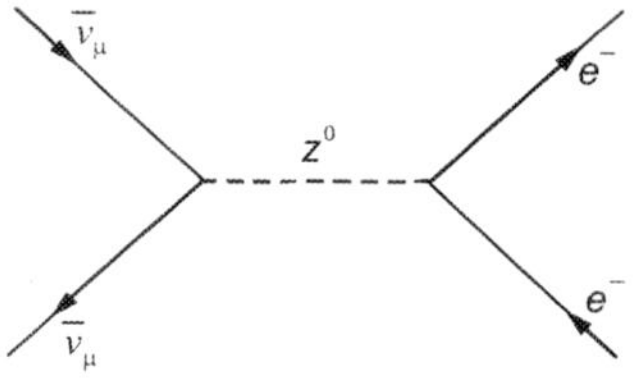

Fig. 6.11: Diagram for muon antineutrino-electron scattering

These scatterings observed in 1973 provided the vital experimental tests for the electroweak theory.

6.6.2.2 Sin² θ_W Measurements

Another significant test of electroweak theory is the measurement of $\sin^2\theta_W$. First, we obtain an expression for $\sin^2\theta_W$ in terms of known quantities. From eqn. (4.143b), we may write

$$\frac{g_1^2}{8m_W^2} = \frac{G}{\sqrt{2}} \tag{6.152}$$

where m_W from eqn. (6.92) is

$$m_W = \frac{1}{2} g_1 v \tag{6.153}$$

Equating these equations and using $G = 1.17 \times 10^{-5}$ GeV^{-2}, we get

$$v = (G\sqrt{2})^{-\frac{1}{2}} \approx 246 \text{ GeV} \tag{6.154}$$

Using eqn. (6.53), viz., $g_1 \sin\theta_W = e$, and eqn. (6.152), we may write

$$m_W^2 \sin^2\theta_W = \frac{\pi\alpha}{G\sqrt{2}} \tag{6.155}$$

where $\alpha = e^2/4\pi$ $(c = \hbar = 1)$ is the fine structure constant. The third of eqn. (6.98) allows us to write

$$\sin^2\theta_W = 1 - \frac{m_W^2}{m_Z^2} \tag{6.156}$$

Combining equations (6.155) and (6.156), we get a quadratic equation for $\sin^2\theta_W$ whose solutions can be written as

$$\sin^2\theta_W = \frac{1}{2} \pm \frac{1}{2}\sqrt{1 - \frac{4\pi\alpha}{m_Z^2 G\sqrt{2}}} \tag{6.157}$$

In eqn. (6.157), α and G have accurate experimental values and since, the value of m_Z is also known , the value of $\sin^2\theta_W$ must have a fixed value. If we get a fixed value of $\sin^2\theta_W$ for different weak interaction processes, then this becomes the test of electroweak theory. We consider three processes in the following subsections, which show that the value of $\sin^2\theta_W$ is same.

6.6.2.2.1 Elastic neutrino electron scattering

The elastic muon neutrino electron scattering shown in Fig. (6.12) is the one

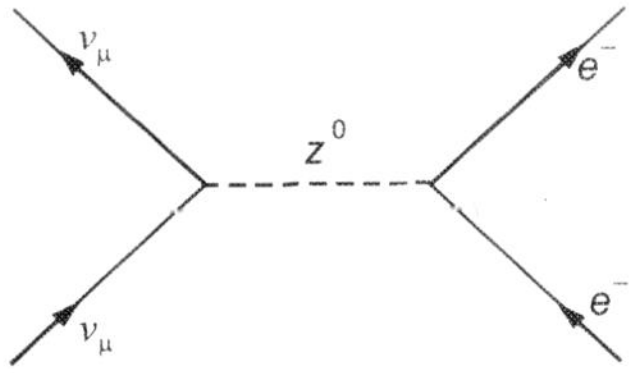

Fig. 6.12: Diagram for elastic $\nu_\mu e^-$ scattering

which proceeds only via a weak neutral current interaction. From eqn. (6.62), the weak neutral current interaction is

$$Z_\mu(IT) = \frac{g_1}{\cos\theta_W}\left(J_3^\mu - \sin^2\theta_W\, J_{em}^\mu\right)Z_\mu \tag{6.158}$$

where eqn. (6.41) has been used. The bracket [see eqn. (6.64)] denotes the weak neutral current J_{NC}^μ. A general neutral current is customarily written as

$$J_{NC}^\mu(f) = \overline{\psi}_f\gamma^\mu\frac{1}{2}(C_V^f - C_A^f\gamma^5)\psi_f \tag{6.159}$$

where ψ_f is a fermion spinor and C_V^f, C_A^f are the vector and axial vector coupling constants for fermion f. The neutrino is left-handed so that

$$C_V^\nu = C_A^\nu = \frac{1}{2} \tag{6.160}$$

and

$$J^{\mu}_{NC}(\text{neutrino}) = \frac{1}{2}\left[\bar{\psi}_{\nu}\gamma^{\mu}\tfrac{1}{2}(1-\gamma^{5})\psi_{\nu}\right] \tag{6.161}$$

However, for quarks

$$C^{q}_{V} \neq C^{q}_{A} \tag{6.162}$$

hence

$$J^{\mu}_{NC}(quark) = \bar{\psi}_{q}\gamma^{\mu}\frac{1}{2}(C^{q}_{V} - C^{q}_{A}\gamma^{5})\psi_{q} \tag{6.163}$$

This means the neutral current for neutrino is pure V–A, while for the quark it is not pure V–A. Now, from eqn. (6.63), we may write [see eqn. (4.198)]

$$\begin{aligned} J^{\mu}_{3} &= \bar{\psi}_{fL}\gamma^{\mu}T_{3}\psi_{fL} \\ &= \bar{\psi}_{f}\gamma^{\mu}\tfrac{1}{2}(1-\gamma^{5})T_{3}\psi_{f} \end{aligned} \tag{6.164}$$

where from eqn. (6.30), $T_3 = +\frac{1}{2}$ for $\psi_{fL} = \nu_L$ and $T_3 = -\frac{1}{2}$ for $\psi_{fL} = e_L$. The left-handed character is incorporated by the factor $\frac{1}{2}(1-\gamma^5)$. Now, the J^{μ}_{em} from eqn. (6.60) can be written as

$$J^{\mu}_{em} = \bar{\psi}_{f}\gamma^{\mu}Q\psi_{f} \tag{6.165}$$

where ψ_f for electron is $e_L + e_R$; $Q = -1$ for electron (e_L, e_R); $Q = 0$ for ν_L; $Q = \frac{2}{3}$ for u_L, u_R and so on. Substituting equations (6.164) and (6.165) in eqn. (6.158), we can write [see eqn. (6.64)]

$$\begin{aligned} J^{\mu}_{NC} &= J^{\mu}_{3} - \sin^2\theta_W\, J^{\mu}_{em} \\ &= \bar{\psi}_{f}\,\gamma^{\mu}\tfrac{1}{2}(1-\gamma^{5})T_{3}\,\psi_{f} - \sin^2\theta_W\bar{\psi}_{f}\,\gamma^{\mu}Q\,\psi_{f} \end{aligned} \tag{6.166}$$

or

$$J^{\mu}_{NC} = \bar{\psi}_{f}\,\gamma^{\mu}\left(\frac{T_3}{2} - \sin^2\theta_W\,Q\right)\psi_{f} - \frac{1}{2}\bar{\psi}_{f}\gamma^{\mu}\gamma^{5}T_{3}\,\psi_{f} \tag{6.167}$$

Comparing eqn. (6.167) with eqn. (6.159), we may write

$$C^{f}_{V} = T_3 - 2\sin^2\theta_W Q$$

and

$$C^{f}_{A} = T_3 \tag{6.168}$$

The total scattering cross section for $\nu_\mu e^- \to \nu_\mu e^-$ scattering is obtained as

$$\sigma(\nu_\mu e^-) = \frac{G^2 s}{3\pi}\left(C^2_V + C_V C_A + C^2_A\right) \tag{6.169}$$

where G being the Fermi's coupling constant and s the square of center of mass energy. The $(\bar{\nu}_\mu e^-)$ scattering cross section is $\sigma(\nu_\mu e^-)$ with $C_A \to -C_A$, i.e.

$$\sigma(\bar{\nu}_\mu e^-) = \frac{G^2 s}{3\pi}\left(C_V^2 + C_V C_A + C_A^2\right) \qquad (6.170)$$

Usually, instead of measuring $\sigma(\nu_\mu e^-)$ or $\sigma(\bar{\nu}_\mu e^-)$, the ratio

$$R = \frac{\sigma(\nu_\mu e^- \to \nu_\mu e^-)}{\sigma(\bar{\nu}_\mu e^- \to \bar{\nu}_\mu e^-)} = \frac{C_V^2 + C_V C_A + C_A^2}{C_V^2 - C_V C_A + C_A^2} \qquad (6.171)$$

is measured. Substituting equations (6.168) in eqn. (6.171), we can obtain the value of $\sin^2 \theta_W$, which from the above experiment (R is obtained from the experiments) is found to be

$$\sin^2 \theta_W = 0.231 \pm 0.010 \qquad (6.172)$$

6.6.2.2.2 Deep inelastic neutrino quark scattering

The nucleon has a structure; it is a three-quark state. The electrons, photons and neutrinos are considered to be structureless. These structureless particles are used as probes in various experiments. In order to have a Rutherford kind of scattering between neutrino and quark, the neutrino must be highly energetic. Such experiments are called deep inelastic scatterings. At low energies, the nucleon as a whole interacts with the incident particle. In deep inelastic scatterings, the probe particle scatters from a charged structureless point constituent of the nucleon. Earlier, this pointlike charged constituent was known as parton, which was later identified as quark. In such deep inelastic scatterings, the probe is scattered by one quark and remaining quarks behave just as spectators. The neutrinos used in deep inelastic process scatter off the target nucleon via the weak interaction. The weak interaction could be both charged current mediated by $W^\pm$ and neutral current mediated by Z^0 as shown in the following Figures (Figs. 6.13 a, b)

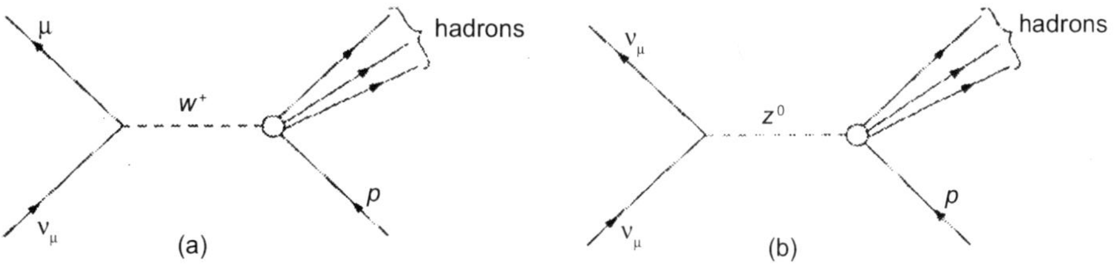

Fig. 6.13: Deep inelastic–neutrino-quark scattering diagrams. For highly energetic neutrinos the quark in proton acts as a scattering centre. The scattering is via the weak interactions. Fig. a- the scattering is due to charged weak interaction and in Fig. b- it is due to neutral weak current

The differential scattering cross section for the above charged and neutral weak interactions can be obtained. In the experiment, the ratio of above two, total scattering cross sections is measured. If it is assumed that in the target nucleon there are only u and d quarks, the ratio R and $\overline{R}$ are obtained as

$$R = \frac{\sigma^{NC}(\nu)}{\sigma^{CC}(\nu)} = \frac{1}{2} - \sin^2\theta_W + \frac{20}{27}\sin^4\theta_W \tag{6.173}$$

$$\overline{R} = \frac{\sigma^{NC}(\overline{\nu})}{\sigma^{CC}(\overline{\nu})} = \frac{1}{2} - \sin^2\theta_W + \frac{20}{9}\sin^4\theta_W \tag{6.174}$$

where $CC \Rightarrow$ charged current, $NC \Rightarrow$ neutral current. ν and $\overline{\nu}$ imply the incident particles on the target (nucleon).

Measuring R and $\overline{R}$ by experiments, it is found that

$$\sin^2\theta_W = 0.230 \pm 0.004 \tag{6.175}$$

This value of $\sin^2\theta_W$ is quite close to the value obtained in eqn. (6.172) for elastic electron neutrino scatterings.

6.6.2.2.3 Asymmetry measurements in cross sections

Measuring the asymmetries in the differential cross sections of electron-positron annihilation and scattering of charged leptons from quarks, the value of $\sin^2\theta_W$ can also be obtained.

The annihilation process e^+e^- can take place as in Figs. (6.14(a, b)), where

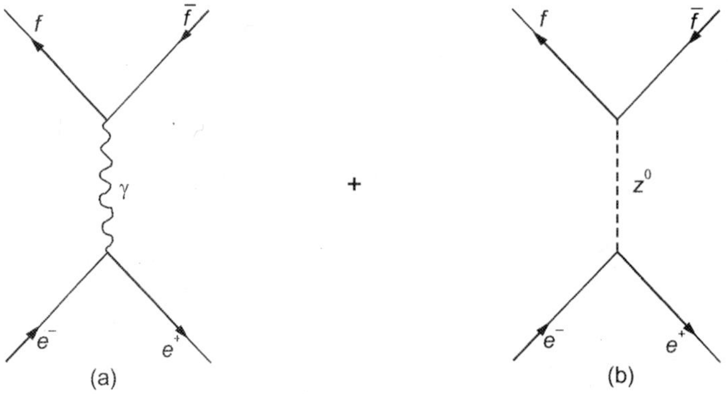

Fig. 6.14: Electron – positron annihilation diagrams. The exchange in Fig. a is due to photon and in Fig. b is due to Z^0 boson

f represents the fermion e, μ, τ or q (quark). The contributions to the scattering amplitude come from both the photon exchange (Fig.a) and Z^0 exchange (Fig.b) processes. The interference between the partial amplitudes corresponding to (a) and (b) processes creates an asymmetry in the differential cross section of e^+e^-annihilation process. The measured asymmetry leads to

$$\sin^2\theta_W = 0.210 \pm 0.019 \tag{6.176}$$

The scattering of spin-polarized fermions can take place from the quarks in the nuclei and contributions may again come from γ exchange and Z^0 exchange

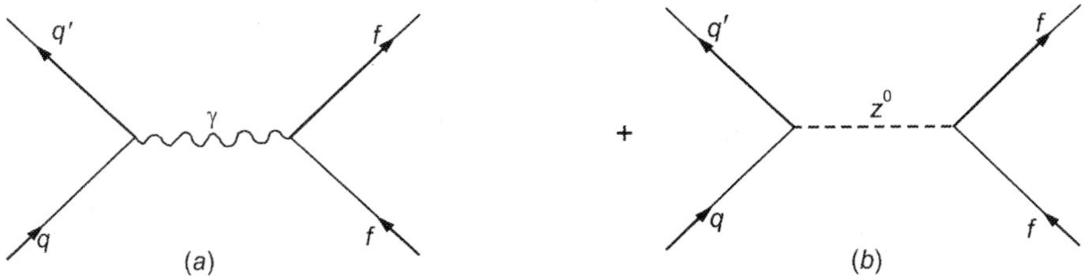

Fig. 6.15: Quark–fermion scattering diagrams through photons Fig. a and Z^0 bosons Fig. b. The fermion is spin polarized

processes (Fig.6.15(a, b)). The interference between the electromagnetic and weak interaction contributions to the differential cross-section of the above scattering leads to the asymmetry in the scattering. The asymmetry measurements show that

$$\sin^2 \theta_W = 0.23 \pm 0.08 \tag{6.177}$$

Above obtained two values of $\sin^2 \theta_W$ agree with the values obtained in eqn. (6.175) and eqn. (6.172). Thus the same value of $\sin^2 \theta_W$ is obtained in different processes, which verifies that the electroweak theory stands the experimental tests.

6.6.2.3 Discovery of W and Z Bosons

The equations (6.155) and (6.156) provide us

$$m_W^2 = \frac{\pi\alpha}{(G\sqrt{2})\sin^2 \theta_W} \tag{6.178}$$

and

$$m_Z^2 = \frac{m_W^2}{1-\sin^2 \theta_W} \tag{6.179}$$

Substituting, α 1/137, $G = 1.17 \times 10^{-5}$ GeV^{-2}, $\sin^2 \theta_W = 0.23$, we get

$$m_W \approx 78 \text{ GeV}$$

and

$$m_Z \approx 89 \text{ GeV} \tag{6.180}$$

However, when radiative corrections (inclusion of loop diagrams in Feynman formalism) are incorporated, the predicted masses are

$$m_W \approx 80.2 \text{ GeV}$$

and

$$m_Z \approx 91.6 \text{ GeV} \tag{6.181}$$

A crucial test of electroweak theory lies therefore in the experimental observation of the $W^{\pm}$ and Z^0 gauge bosons. In order to produce these very massive particles ($W^{\pm}$, Z^0 masses are comparable to the mass of a gold atom) an accelerator must have such large energy in the center of mass frame. For the fixed target accelerators this is very difficult and therefore colliders are used. In the year 1983, at the, *$Sp\bar{p}S$ CERN Collider* the $W^{\pm}$, Z^0 bosons were found at its two different detector facilities *UA*1 and *UA*2. At the detector *UA*1, out of 10^9 recorded events, ***six*** events were of *W* boson. ***Five*** *Z* events were recorded at *UA*1 and ***eight*** at *UA*2. Above events convinced that $W^{\pm}$ and Z^0 were found at the predicted masses. Thus the discovery of *W* and *Z* bosons provided an important test of electroweak theory.

6.7 EVIDENCES FOR QUANTUM CHROMODYNAMICS

Quantum Chromodynamics (QCD) is a SU(3)-colour gauge theory. The quark structure of hadrons predicts six flavours of quarks. The flavors *u* and *d* form the first generation of quarks. The experimental evidence for the first generation of quarks lies in the magnitude of magnetic moments of proton and neutron. If proton and neutron had no substructure, the magnetic moment of proton should have been one nuclear magneton and that of neutron zero. But, observed values are $\mu_p = +2.79\ \mu_N$ and $\mu_n = -1.91\mu_N$. The ratio is $\mu_n/\mu_p = -0.685$. From the quark model, we have already seen in Chapter 3 that $\mu_n/\mu_p = -2/3$. This successful explanation of neutron-proton magnetic moment ratio is a strong evidence for first generation (*u*, *d*) quarks. The deep inelastic scatterings have further confirmed that the nucleon is not a fundamental (point) particle and have shown that there are point-like constituents in the proton or the neutron, which scatter at high energies, as if they were free particles. The results of scatterings have shown that the charged constituents of proton or neutron have spin-$\frac{1}{2}$.

The discovery of Ω^- particle (mass 1.672 GeV, quark content *sss*) provided the evidence of strange (*s*) flavour, which was heavier than the up and down flavours. The ϕ meson was found to be a nearly pure $s\bar{s}$ state. The discovery of J/ψ particle (3.1 GeV) provided the evidence for the fourth flavour, charm. J/ψ is a bound state of $c\bar{c}$. The *s* and *c* quarks comprise the second generation.

The discovery of upsilon (γ) particle (mass 9.4 GeV) as e^+e^- resonance provided the evidence for fifth flavour, the bottom. The sixth flavour, the top quark was discovered in 1995. The top and bottom flavours constitute the third generation of quarks.

Thus, there exists sufficient evidence for five flavours of quarks. The existence of the sixth flavour has also been established. Each flavour comes in three colours. The evidence for colour would therefore be the first evidence for Quantum Chromodynamics.

6.7.1 Evidence for Colour

The evidence for three colours of quarks has a connection with the discovery of

J or ψ resonance. The J/ψ resonance was discovered at the *SPEAR* e^+e^- collider of Stanford Linear Accelerator Centre Stanford (USA). At the center of mass energy of 3.1 GeV, a large narrow peak was observed in the cross sections for the processes

$$e^+ + e^- \rightarrow \mu^+ + \mu^- \tag{6.182a}$$

and

$$e^+ + e^- \rightarrow \text{hadrons} \tag{6.182b}$$

The observed peak was called the J/ψ resonance having mass 3.1 GeV. A second peak at 3.7 GeV was also observed, which was however, called ψ'. J/ψ was also produced in the fixed target experiment at the Brookhaven Alternating-Gradient Synchrotron (AGS).

The process (6.182a) may take place in two ways. One is pure *QED* in which a virtual photon mediates. In the other, e^+ and e^- first produce a resonance J/ψ, which then decays into μ^+ and μ^-.

In the process (6.182b), hadrons are produced in the final state. The mechanism of their production is that; e^+ and e^- interact via the virtual photon, which then produces the quark–antiquark pairs $(q, \overline{q})$. For a brief moment, the quarks fly apart as free particles but on reaching a separation of about the diameter of hadron ($\approx 10^{-15}$ m), their strong interaction is strong enough to produce new quark antiquark pairs. Since, strong interaction is via gluons, the new quark antiquark pairs are produced mainly by gluons. The bound $(q\overline{q})$ pairs are hadrons, which may themselves decay into multihadronic states. When the beam energy is increased, initially muons and light quarks are produced. At about 1.5 GeV, the charmed (c) quarks, at 1.784 GeV, the (τ) leptons and at about 4.7 GeV, the bottom (b) quarks are produced. J/ψ and ψ' are the lowest bound states ($c\overline{c}$) of charm quark. $\phi = s\overline{s}$, $\Upsilon = b\overline{b}$ are other significant bound states.

To get an insight into the above process, the ratio (R) of the rate of hadron production and the rate of muon pair production is examined.

$$R = \frac{\sigma(e^+e^- \rightarrow \text{hadrons})}{\sigma(e^+e^- \rightarrow \mu^+\mu^-)} \tag{6.183}$$

where the numerator includes all the quark antiquark events. The calculations lead to the following relation for the ratio $R(R_i)$

$$R_i = \sum_{i=1}^{N} \frac{Q_i^2}{e^2} \tag{6.184}$$

where N is the total number of quark flavours and Q_i is the electric charge of ith quark flavour. For example

$$Q_1 = Q_u = +\frac{2}{3}e$$

$$Q_2 = Q_d = -\frac{1}{3}e \quad \text{etc.}$$
$$Q_3 = Q_s = -\frac{1}{3}e \tag{6.185}$$

From eqn. (6.184), the value of R^i can be calculated for different number of flavors. If, only three flavours are considered, we will have

$$R_3 = \frac{Q_u^2}{e^2} + \frac{Q_d^2}{e^2} + \frac{Q_s^2}{e^2} = \left(\frac{2}{3}\right)^2 + \left(-\frac{1}{3}\right)^2 + \left(-\frac{1}{3}\right)^2 = \frac{2}{3} \tag{6.186}$$

If u, d, s and c (four flavours) are considered, we would get

$$R_4 = \left(\frac{2}{3}\right)^2 + \left(-\frac{1}{3}\right)^2 + \left(-\frac{1}{3}\right)^2 + \left(\frac{2}{3}\right)^2 = \frac{10}{9} \tag{6.187}$$

Likewise, for five flavours (u, d, s, c and b), we would get $(Q_b = -\frac{1}{3}e)$

$$R_5 = \frac{11}{9} \tag{6.188}$$

Now, we compare the above-calculated ratios R_i with their experimentally obtained values. Following table demonstrates the comparison

Table 6.3 $R = \dfrac{\sigma(e^+e^- \to hadrons)}{\sigma(e^+e^- \to \mu^+\mu^-)}$

e^+e^- Collider Center of mass energy GeV	Flavours excited	Expected Ratio R_{ex}	Observed Ratio R_{ob}	Conclusion
2–3	u, d, s	$\frac{2}{3}$	2	$R_{ob} = 3R_{ex}$
> 3–4	u, d, s, c	$\frac{10}{9}$	$\frac{10}{3}$	$R_{ob} = 3R_{ex}$
> 10	u, d, s, c, b	$\frac{11}{9}$	$\frac{11}{3}$	$R_{ob} = 3R_{ex}$

From the above table, it is easily observed that if $\left(\frac{Q_i}{e}\right)^2$ for each flavour is *multiplied by* **three,** the expected R becomes equal to observed R. The factor 3 is for three colours in which each flavour should occur. This observation provides a very strong evidence for the existence of three colours of quarks.

It may also be observed that below 3.1 GeV (J/ψ resonance) only three flavours of quarks (u, d, s) are produced. Between 3 GeV and 4 GeV, the fourth flavour (charm) gets excited and above 10GeV the fifth flavour (bottom) also appears.

An equally convincing evidence for the existence of three colours comes from the decay $\pi^0 \to 2\gamma$. We know that the transition (decay) rate of a process is proportional to the square of the matrix element between the initial and final states. If the matrix element for the process $\pi^0 \to 2\gamma$ is calculated by considering quark composition of π^0, but without considering colour, the matrix element is obtained smaller by a factor 3, than the matrix element which would have given the observed decay rate. By considering a factor of 3 corresponding to three colours, the observed decay rate is found to be in excellent agreement with the calculated decay rate. Thus we get another evidence for three colours for each flavour.

6.7.2 Variation of Strong Coupling Constant (α_s) with Energy

In Chapter 1, the concept of QED and QCD running coupling constant was explained qualitatively. From eqn. (1.80), the strong coupling constant (α_s) is

$$\alpha_s = \frac{g_3^2}{4\pi\hbar c} = \frac{g_3^2}{4\pi} \quad (c = \hbar = 1) \tag{6.189}$$

The concept of running coupling constant tells that α_s is a function of energy Q. It is defined by

$$\log \frac{Q^2}{\mu^2} = \int_{\alpha_s(\mu^2)}^{\alpha_s(Q^2)} \frac{d\alpha_s(Q^2)}{\beta(\alpha_s)} \tag{6.190}$$

where μ is a particular (chosen) value of Q at which the initial value of $\alpha_s(Q)$ is specified. The function β is calculated using perturbation theory and it depends on α_s, as well as on the number of quark flavors. Taking n as the number of quark flavors

$$\beta(\alpha_s) \approx -b_n \alpha_s^2 \tag{6.191}$$

where

$$b_n = \frac{33 - 2n}{12\pi} \tag{6.192}$$

and hence

$$\log\left(\frac{Q^2}{\mu^2}\right) \approx -\int \frac{d\alpha_s}{b_n \alpha_s^2} = \frac{1}{b_n}\left[\frac{1}{\alpha_s(Q)} - \frac{1}{\alpha_s(\mu)}\right] \tag{6.193}$$

implying that

$$\alpha_s(Q) = \frac{\alpha_s(\mu)}{1 + b_n \alpha_s(\mu) \log\left(\frac{Q^2}{\mu^2}\right)} \tag{6.194}$$

Since, $\alpha_s(Q)$ decreases with increasing $Q(Q^2 \to \infty, \alpha_s = 0$ is asymptotic freedom), we can use perturbative calculations in QCD for small values of α_s. The coupling constant α_s will become large for sufficiently low Q^2. Let α_s becomes large enough at $Q^2 = \Lambda^2$, where we cannot allow perturbative calculations, and then Λ is called the ***QCD scale parameter***. Only for $Q^2 > \Lambda^2$, the QCD can be described perturbatively. Λ can be defined from eqn. (6.194) by demanding that for $Q^2 \to \Lambda^2$, $\alpha_s(Q^2) \to \infty$, which implies

$$1 + b_n \alpha_s(\mu) \log\left(\frac{\Lambda^2}{\mu^2}\right) = 0 \tag{6.195}$$

or

$$\frac{1}{\alpha_s(\mu)} = b_n \log\left(\frac{\mu^2}{\Lambda^2}\right) \tag{6.196}$$

Substituting eqn. (6.196) into eqn. (6.194), we get

$$\alpha_s(Q) = \frac{1}{b_n \log\left(\frac{Q^2}{\Lambda^2}\right)} \tag{6.197}$$

where b_n is defined by eqn. (6.192). Eqn. (6.197) is used to determine the Λ, the fundamental parameter of QCD. LHS of eqn. (6.197) is determined experimentally. Λ has a value between about 100 MeV to 500 MeV. If we get a constant value of Λ from the analyses of different processes, that would mean that the QCD has been verified. For example, for the process

$$e^+ e^- \to \text{hadrons} \tag{6.198}$$

the theoretically calculated $\alpha_s(Q^2)$ values for particular Λ^2 values have been found to correspond with the experimentally observed values of $\alpha_s(Q^2)$. Such an agreement provides another test of QCD.

In addition to above tests of QCD, the detailed studies of resonance states of quarkonium ($s\bar{s}$, $c\bar{c}$, $b\bar{b}$) systems, studies of glueballs (particles consisting of gluons only), studies of non-$q\bar{q}$ states have also indicated the experimental verification of QCD. However, the verification has been in the high-energy range, where QCD is perturbative. At smaller energies ($Q^2 < \Lambda^2$), the QCD becomes non-perturbative. As far as the Standard Model is concerned, the non-perturbative QCD may or may not be considered its part.

Therefore, after presenting experimental evidences for QED, electroweak theory and perturbative QCD, we may say that there are enough evidences, which prove that the Standard Model stands the experimental tests.

6.8 GRAND UNIFICATION

The Standard Model, which comprises of the GWS model and the QCD, combines the $SU_c(3)$ gauge group of QCD with the $SU_L(2) \otimes U_Y(1)$ gauge group of electroweak theory. The gauge group of the Standard Model is therefore $SU_c(3) \otimes SU_L(2) \otimes U_Y(1)$. However, this gauge group, instead of being a single gauge group, is a product of three disconnected groups. There are three coupling constants, one for each group and $\sin\theta_W$ has to be measured experimentally. Such a situation does not provide a true unification of three interactions. The true unification of electromagnetic, weak and strong interactions could be realized if we have a single gauge group with one coupling constant possessing the property of manifesting itself as strong, weak and electromagnetic coupling constants. Such conjectured unification of these three interactions is called the *Grand Unification*.

Let us assume that there exists a grand unified group G having $SU_c(3)$, $SU_L(2)$ and $U_Y(1)$ as its subgroups, i.e.,

$$G \supset SU_c(3) \otimes SU_L(2) \otimes U_Y(1) \tag{6.199}$$

and let g_G be the single coupling constant corresponding to the group G. As described in GWS model g_1 is the coupling constant for $SU_L(2)$ and g^2 for $U_Y(1)$ and we replace g_s of section (6.4) by g_3 as the coupling constant for $SU_c(3)$. We may then denote $g_i(Q)$ as the coupling constant for $SU_L(2)$ ($i = 1$), $U_Y(1)$ ($i = 2$), and $SU_c(3)$ ($i = 3$) gauge groups. The coupling constants $g_1(Q)$ and $g_3(Q)$ are for non-Abelian gauge groups where the gauge bosons carry charges. The W and Z bosons carry weak charge and gluons the colour charge. Therefore, due to these charges the self-interaction of gauge bosons takes place giving rise to antiscreening effects, as a result of which the coupling constants corresponding to both $SU_L(2)$ and $SU_L(3)$ decrease with increasing Q^2. Of course the decrease in both the cases is not identical. The coupling constant for $SU_L(2)$ decreases slowly, whereas the coupling constant for $SU_L(3)$ decreases more rapidly. The coupling constants are defined through the following relations [see eqn. (6.54)] in natural units

$$g_1 = \frac{e}{\sin\theta_W}, \quad g_2 = \frac{e}{\cos\theta_W}$$

$$\alpha = \frac{e^2}{4\pi} \approx \frac{1}{137}, \quad \alpha_1 = \frac{g_1^2}{4\pi} \approx \frac{1}{30}$$

$$\alpha_2 = \frac{g_2^2}{4\pi} \approx \frac{1}{100}, \quad \alpha_3 = \frac{g_3^2}{4\pi} \approx 0.4 \to 0.1 \tag{6.200}$$

where e is the electromagnetic coupling constant. The rationalized coupling constants α, α_1, α_2 and $\alpha_3$belong to gauge groups $U_{em}(1)$, $SU_L(2)$, $U_Y(1)$ and $SU_c(3)$ respectively. As we already know these coupling constants depend on Q, above values of α, α_1, α_2 are for Q equal to few GeVs and α_3 is for $Q \approx 100$ GeV. For the variation of $\alpha_i(Q)$ with Q, a relation similar to eqn. (6.194) can be written as

$$\alpha_i(Q) = \frac{\alpha_i(\mu)}{1 + b_i\alpha_i(\mu)\log\left(\frac{Q^2}{\mu^2}\right)} \tag{6.201}$$

where the coefficients b_i of the logarithm term are different for three [$SU_L(2)$, $U_Y(1)$ and $SU_c(3)$] groups. The characteristic values of b_i are

$$b_i = \frac{1}{12\pi}(11n_b - 4n_f) \tag{6.202}$$

where n_b is the number of vector bosons and n_f, the number of fermion generations, which make contributions to one loop vacuum polarization diagrams. $n_b = 0$ for $U_Y(1)$, 2 for $SU_L(2)$ and 3 for $SU_c(3)$. $n_f = 3$. From eqn. (6.202), we can see that

$$\begin{aligned} b_1 &= b[SU_L(2)] = +\frac{5}{6\pi} \\ b_2 &= b[U_Y(1)] \;\; = -\frac{1}{\pi} \\ b_3 &= b[SU_c(3)] \;\; = +\frac{7}{4\pi} \end{aligned} \tag{6.203}$$

where each b_i has magnitude less than one. Due to a negative b and slowly changing logarithm, the $\alpha_2[U_Y(1)]$ increases very slowly and b_i being positive with different magnitudes, $\alpha_1[SU_L(2)]$ decreases slowly compared to $\alpha_3[SU_c(3)]$, which decreases more rapidly. These variations of α_1, α_2, α_3 with Q have been shown in Fig. (6.16).

The variations of α_1, α_2 and α_2 show that at low energies all the coupling constants are different but the extrapolation of these running coupling constants to very high energies indicate that at the energy $M_x \approx 10^{15}$ GeV, all the coupling constants would merge into one, i.e., then only one coupling constant α_G would exist. The estimation of the grand unification energy M_x can be made as below:

To estimate M_x, we recall that $g_1(Q)$, $g_2(Q)$ and $g_1(Q)$ are the coupling constants, which we have so far associated with $SU_L(2)$, $U_Y(1)$ and $SU_c(3)$ groups respectively. Assuming now that, there exists a group G of eqn. (6.199) with its Clebsch–Gordon coefficient as C, we introduce new coupling constants, k_1, k_2 and k_3, where

$$
\begin{aligned}
k_1(Q) &= g_1(Q) \\
k_2(Q) &= Cg_2(Q) \\
k_3(Q) &= g_3(Q)
\end{aligned}
\tag{6.204}
$$

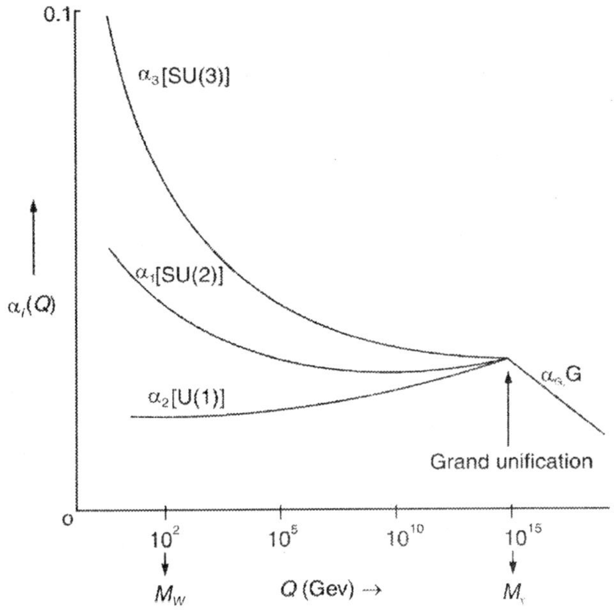

Fig. 6.16: The variation of three coupling constants with Q. The coupling constants are $\alpha_i = \frac{k_i^2}{4\pi}$. α_1 and α_3 decrease with Q, while α_2 increases slowly with Q. The decrease of α_3 is sharper. All the three coupling constants merge into one at $Q \simeq M_X$. The graph thus shows that at very short distances $\frac{1}{Q} - \frac{1}{M_X}$, the strong [$SU_c(3)$] and electroweak [$SU_L(2) \otimes U_Y(1)$] interactions would unify. The gauge group beyond grand unification is G with coupling constants α_G

We now rewrite eqn. (6.201) as

$$
\frac{1}{\alpha_i(\mu)} = \frac{1}{\alpha_i(Q)} - 2b_i \log\frac{Q}{\mu}
\tag{6.205}
$$

where b_i are given by eqn. (6.203). Since, due to the new coupling constants $\alpha_i = \frac{k_i^2}{4\pi}$ and for $Q \geq M_x$, $k_i(Q) = g_G$, we may write from eqn. (6.205)

$$
\frac{1}{k_i^2(\mu)} = \frac{1}{g_G^2} - \frac{2b_i}{4\pi}\log\left(\frac{M_x}{\mu}\right)
\tag{6.206}
$$

where $i = 1$, 2 and 3 are respectively for $SU_L(2)$, $U_Y(1)$ and $SU_c(3)$. Using eqn. (6.206), we construct a linear combination

$$\frac{1}{k_1^2}+\frac{C^2}{k_2^2}-\frac{(1+C^2)}{k_3^2}=-\frac{2}{4\pi}\left[b_1+C^2b_2-(1+C^2)b_3\right]\log\left(\frac{M_x}{\mu}\right) \tag{6.207}$$

where $\frac{1}{g_G^2}$ terms have canceled. In the LHS of equation (6.207) $k_i^2 = k_i^2(\mu)$.

From eqn. (6.204), we can see that

$$\frac{1}{k_1^2}+\frac{C^2}{k_2^2}=\frac{1}{g_1^2}+\frac{1}{g_2^2}=\left(\frac{1}{g_2\cos\theta_W}\right)^2=\frac{1}{e^2} \tag{6.208}$$

where equations (6.41) and (6.53) have been used. Now substituting eqn. (6.203) in the RHS and eqn. (6.208) in the LHS of eqn. (6.207), we may write

$$\frac{1}{e^2}-\frac{(1+C^2)}{k_3^2}=\frac{11(1+3C^2)}{(6\pi)(4\pi)}\log\left(\frac{M_x}{\mu}\right) \tag{6.209}$$

from which, we get

$$\log\left(\frac{M_x}{\mu}\right)=\frac{6\pi}{11(1+3C^2)}\left[\frac{1}{\alpha}-\frac{(1+C^2)}{\alpha_3}\right] \tag{6.210}$$

where $\alpha = \frac{e^2}{4\pi}$ and $\alpha_3 = \frac{k_3^2}{4\pi}$ have been used. Eqn. (6.210) can finally be used to estimate M_x. We know that for $\mu = 10$ GeV, $\alpha \approx \frac{1}{137}$, $\alpha_3 = \alpha_s \approx 0.1$ and we take $C^2 = \frac{5}{3}$. Substituting these values in eqn. (6.210), we get

$$M_x \approx 5\times 10^{14}\ \text{GeV} \tag{6.211}$$

The value of $C^2 = \frac{5}{3}$ has been taken on the basis that the smallest group having SU(3), SU(2) and U(1) as its sub groups is SU(5) and the SU(5) value of the Clebsch–Gordon coefficient relating g_1 and g_2 is $C = \sqrt{\frac{5}{3}}$.

Thus at the energy $M_x \approx 10^{15}\,\text{GeV}$, the electroweak and strong interactions would merge into a single interaction which may be called the grand unified interaction. A significant consequence of the grand unification is that the weak mixing angle θ_W can be predicted by the theory rather than being a parameter whose value is determined by experiment. This is because, from eqn. (6.53), we have

$$\frac{g_2}{g_1} = \tan\theta_W \tag{6.212}$$

and using the conversions (6.204), this relation becomes

$$\frac{k_2(Q)}{Ck_1(Q)} = \tan\theta_W \tag{6.213}$$

Now, at the unification scale $Q \approx M_x$, $k_2(M_x) = k_1(M_x)$ and hence C alone determines the Weinberg angle θ_W. Using $C = \sqrt{\frac{5}{3}}$ for SU(5), we get at the unification scale

$$\sin^2\theta_W = \frac{1}{1+C^2} = \frac{3}{8} \tag{6.214}$$

However, above value of $\sin^2\theta_W$ is required to be 0.205 so that the couplings at lower energy ($Q = M_W \approx 10^2$GeV) when evolved, produce a value of coupling constant that coincides with that at $Q = M_x$. The value of 0.205 is lower than the measured value of 0.2325, which indicates that SU(5) group may not be the correct grand unifying group. Other groups, which have been considered for grand unification, are SO(10) and E_6.

6.8.1 Implications of Grand Unification

Above the grand unified energy of 10^{15} GeV, the difference between electromagnetic, weak and strong forces would cease to exist. In this energy range only one unified interaction would exist. For experiments, this energy is too large to be achieved but is supposed to be there in the early stages of the creation of universe, the Big Bang.

The strong force operates between quarks and weak force between quarks and leptons, therefore, when strong and weak interactions get combined, the quarks and leptons should be the different aspects of one more fundamental quantity. This implies that the GUT force should be able to transform quarks into leptons and vice versa.

The unified interaction of grand unified theories would be mediated by twenty-four (24) particles shown in the following table.

Table 6.4 ***Grand Unified Family of Mediating Particles***

Force Mediated	By the particle(s)	Having the number
Electromagnetic	Photon	1
Weak	W^+, W^-, Z^0 (Vector bosons)	3
Strong	Gluons	8
Strong-Electroweak	X particles	12

These 24 particles constitute one family of particles that mediate the various interactions. The new *X* particles mediate between the electroweak and strong force and naturally these particles would be responsible for changing hadrons into leptons by converting quarks (strong) into leptons and vice versa. These X particles (bosons), sometimes called lepto-quark particles or progenitors, come in four different electrical charges $\pm\frac{1}{3}e$ and $\pm\frac{4}{3}e$. They are written as $X^{\frac{1}{3}}, X^{-\frac{1}{3}}, X^{\frac{4}{3}}$ and $X^{-\frac{4}{3}}$. The particles $X^{\pm\frac{1}{3}}$ are sometimes called the *Y* particles. Since, these particles have spin 1, they are bosons. Because of each particle appearing in three colours, the total number of X particles would be twelve (12). Since, the X particles cause the conversion of quark and leptons into each other and the simple SU(5) model of GUT predicts the dominant decays as

$$\begin{aligned} X &\to \bar{q} + \bar{l} \\ &\to q + q \end{aligned} \tag{6.215}$$

$$\begin{aligned} \bar{X} &\to q + l \\ &\to \bar{q} + \bar{l} \end{aligned} \tag{6.216}$$

the violation of baryon and lepton number takes place. Above processes obviously take place only at GUT energies, the mass of mediating (virtual) X particle should be $\approx 10^{15}$ GeV. Because of the introduction of quark–lepton mediating particles, the grand unified theories also imply that an isolated proton should also decay. The following diagrams (Fig. 6.17) show the proton decay. Both the Figs. (6.17)

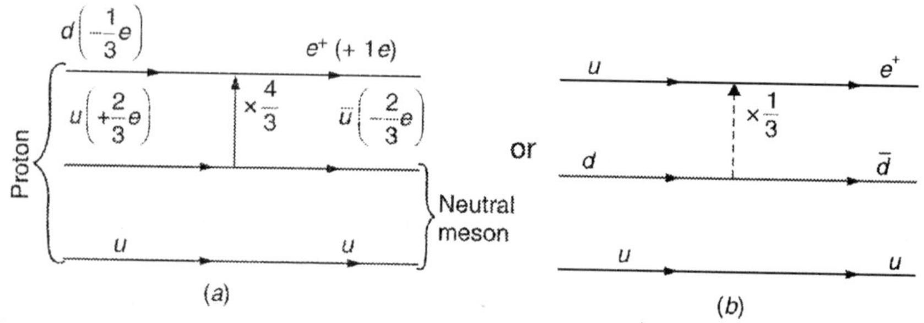

Fig. 6.17: Diagrams showing the possible mechanism for proton decay. In the final state, the quark and antiquark combine to from the neutral mesons and we may have the decay modes as $p \to e^+\pi^0$, $e^+\rho^0$, $e^+\omega^0$ etc. The mode $p \to e^+\pi^0$ is expected to be a dominant decay mode of proton in SU(5)

denote proton *uud* decaying into a neutral meson and positron. In figure (a), the mediator is the super heavy boson $X^{\frac{4}{3}}$ (electric charge $+\frac{4}{3}e$) and in Fig. (b), it is

$X^{\frac{1}{3}}$ (electric charge $+\frac{1}{3}e$). The neutral meson could be π^0, ω^0, ρ^0, etc. The gauge group SU(5) expects the dominant decay mode of proton as

$$p \rightarrow e^{+} + \pi^{0} \tag{6.217}$$

The decay rate of this process comes out to be proportional to M_x^{-4}. Taking $M_x = 2\times 10^{14}$ GeV, the lifetime of proton comes

$$\tau_p \approx 10^{31} \text{ years} \tag{6.218}$$

The conclusive experimental evidence of proton decay is however, waited. Two candidate events were reportedly observed in the Kolar Gold Field Laboratory corresponding to a lifetime of $(1.7 \pm 1.2)\times 10^{30}$ years. Based on larger grand unified groups than SU(5), longer proton lifetimes can be predicted.

6.8.2 Gravitational Interaction and the Planck Scales

The gravitational force between two subatomic particles m_1 and m_2 is

$$F_g = G_N \frac{m_1 m_2}{r^2} \tag{6.219}$$

where $G_N = 6.67 \times 10^{-11}$ Nm^2 kg^{-2} is the Newton's gravitational constant. At the subatomic level, the gravitational force would become significant only when the separation between two particles becomes far too small [$< 10^{-30}$ m]. The gravitational effects become of the order of unity when the gravitational potential energy at a distance r from mass M becomes equal to the rest energy of mass M. That is

$$\frac{\text{Gravitational potential energy}}{\text{Rest energy}} \approx 1$$

or

$$\frac{G_N M^2/r}{Mc^2} = 1 \tag{6.220}$$

If the masses are separated by natural unit of length

$$r = \frac{\hbar}{Mc} \tag{6.221}$$

then from eqn. (6.220)

$$Mc^2 = \sqrt{\frac{\hbar c^5}{G_N}} \tag{6.222}$$

Substituting the values, we get

$$Mc^2 = 1.20\times 10^{19}\,\text{GeV} \tag{6.223}$$

The energy [eqn. (6.223)] is called the Planck energy or taking $c = 1$, it is also called Planck mass.

The distance of one natural unit of length [eqn. (6.221)] is also called the Planck length L_p

$$L_p = \frac{\hbar}{Mc} = \sqrt{\frac{\hbar G_N}{c^3}}$$
$$= 1.61 \times 10^{-35} \text{ m} \tag{6.224}$$

The quantity (L_p/c) is called the Planck time T_p

$$T_p = \frac{L_p}{c} = 5.37 \times 10^{-44} \text{ s} \tag{6.225}$$

Thus we have the Planck scales as

$$\begin{aligned} \text{Planck energy} &= 1.20 \times 10^{19} \text{ GeV} \\ \text{Planck length} &= 1.61 \times 10^{-35} \text{ m} \\ \text{Planck time} &= 5.37 \times 10^{-44} \text{ s} \end{aligned} \tag{6.226}$$

From the above we observe that the gravitational interactions become significant in the energy range of 10^{19} GeV, which is about 10^4 times above the *GUT* energy. Therefore, at *GUT* energies, the gravitational interaction can still be ignored.

The unification of gravitational interaction with strong and electroweak interactions can be expected only around the Planck energy. The Planck energy also represents the energy of graviton, the quantum of gravitational field. The mass associated with 10^{19} GeV (Planck energy) is 2×10^{-8} kg., which though is small from the point of view of ordinary gravity, is very large compared to the mass energy of all other elementary particles. However, for any particle of this size, both quantum theory and gravitation should be taken into account. For all other elementary particles, the gravity is ignored.

The fusion of quantum mechanics and the theory of general relativity (gravitation) is called the *quantum theory of gravity*. On a microscopic level the gravitational field should be quantized into gravitons. Since the gravitational field is nothing but the warping of space–time, the graviton would appear to be a quantum of spacetime. The quantum fluctuations of spacetime, which manifest as gravitational waves, start to occur at distances of the order of the Planck length (1.61×10^{-35} m). The intervals of space and time given by Planck length and Planck time (5.37×10^{-44} S) are the regions in which quantum gravity has to be considered. These regions are much smaller than even the size of an electron ($\approx 10^{-19}$ m). Thus Planck scales [eqn. (6.226)] determine and open up a new aspect of gravity, the quantum gravity.

6.9 GRAND UNIFICATION AND THE BIG BANG

Big Bang theory is one of the theories, which describes the creation of universe.

Hubble in 1929 observed that the universe is expanding and different galaxies are moving away from us. This observation suggests that there was a time some ten or twenty thousand million years ago that all the matter of the universe was exactly at the same place and therefore the density of the universe was infinite.

Hubble's observation suggested that there was a time called Big Bang when the universe was infinitesimally small and infinite dense. Under such condition all laws of science and therefore all ability to predict future would breakdown. If there were events earlier than this time they would not effect what happens at the present time. Thus the time had a beginning at Big Bang.

The world in which we live today is low energy and cold universe in which forces and matter appear stable and disconnected. We can produce energies only about 10^5 GeV with accelerators and for the test of the *GUT* unification and beyond; the energies as high as 10^{-19} GeV are required. It can be speculated that such energies were commonplace just after the Big Bang. The early universe was just like a high energy physics experiment in which events starting from time $t = 0$ took place. The Big Bang model is considered the standard model of the origin of the universe and the detection of 3K Microwave Background Radiation (MBR) in 1965 by Penzias and Wilson was a crucial evidence for this model. In the following we summarize the events starting from $t = 0$ (about 15 million years ago)

6.9.1 $t = 0$

Big Bang took place. It was not an explosion of matter into already existing space and time; instead it was the very creation of space-time and matter themselves.

6.9.2 From $t = 0$ to $t = 10^{-43}$ sec.

This was the duration between the creation and the Planck time. During this time the energy or the temperature was too high. Energy was greater then 10^{19} GeV and the temperature of the universe was greater than 10^{33} K. The energy dropped to 10^{19} GeV (10^{33} K) at Planck time. The quantum gravitational effects became strong. This stage of universe can only be discussed by *quantum gravity* as for this stage the classical theory because of the singularity does not provide a good description of universe. However, relatively little is known about this era (duration). The extremely high energies would have caused all the four forces viz.; gravitational, strong, weak and electromagnetic to merge into one force called the *superforce*. The Physicists working in the fields of supersymmetry and supergravity theories are keenly researching this era.

6.9.3 From 10^{-43} to 10^{-35} sec.

Recall at the Planck time $\approx 5.37 \times 10^{-44}$ sec., the energy of the universe was 1.20×10^{19} GeV. As the universe expanded, the energy dropped below 10^{19} GeV and the temperature to 10^{32} K (1GeV $\equiv$ 10^{13} K). At this stage the *gravitational force broke bond* with the superforce and now there were two forces in the nature, the gravitational force and the *GUT* force. This era was governed by grand unified theories. There was an abundance of *X* particles and their antiparticles $\overline{X}$. The *X* particles decayed into quarks and leptons and $\overline{X}$ into antiquarks and antileptons. Although the densities of *X* and $\overline{X}$ before 10^{-35} sec. were equal, the decay rates of *X* and $\overline{X}$ were not same. This resulted into creation of more particles than antiparticles. This unequal number of particles and antiparticles later caused the asymmetry of matter and antimatter. The *X* particles continually converted into quarks and leptons and vice versa. In this era at 10^{-35} sec. the size of the universe is estimated to be only 10^{-24} cm and in it were plenty of quarks, electrons, neutrinos, photons, gluons, *X*-particles and their antiparticles. However, all these particles effectively had no individuality.

6.9.4 From 10^{-35} to 10^{-10} sec.

The further expansion of universe continued. The energy dropped to $\approx 10^{15}$ GeV and the temperature to 10^{27} K. This energy became low for the existence of *X* particles and they disappeared. The quarks and leptons started to have an individual identity. Now they could not be converted into each other. At the energy 5×10^{14} GeV (grand unification scale) the *strong force separated* from the *GUT* force and left the electroweak force as the only unified force. During the era 10^{-35} sec. to 10^{-10} sec., there were three interactions in the universe, the gravitational, and strong and electroweak. The leptons, quarks, neutrinos, photons, gluons, $W^{\pm}$ and Z^0 particles were present during the above era. However, it was still too hot for quarks to combine to form baryons and mesons. According to S.W. Hawking, during this era, at about 10^{-20} sec., the mini black holes must have formed.

6.9.5 From 10^{-10} to 10^{-3} sec.

The expansion of universe continued and energy at about 10^{-10} sec., dropped to 10^2 GeV and the temperature came down to 10^{15} K level. We may recall that at the *CERN* $p\overline{p}$ collider the effective energy available to produce *W* and *Z* bosons was about 90 GeV($\approx 10^2$ GeV). This experiment, which proved the existence of W^+ and Z^0 bosons led to the award of 1984 Nobel Prize for Physics to Carlo Rubbia and Simon van der Meer. In the case of universe, the energy has dropped to 10^2 GeV and when it dropped below 10^2 GeV, the energy became insufficient for the formation of *W* and Z^0 particles and they disappeared. At this stage of the universe, the *weak force* also *broke away* from the unified electroweak force. Now there were four familiar forces in the universe, namely, the gravitational, strong, weak and the electromagnetic. At about 10^{-4} sec., the cooling was sufficient enough for quarks to combine to form qqq, the baryons and $q\bar{q}$, the mesons. The proton and neutron found their existence. Due to the abundance of quarks over antiquarks, there were excess of protons and neutrons over their antiparticles. The formation of protons and neutrons is an important event in the evolution of the universe.

6.9.6 From 10^{-3} to 35 min.

At about 10^{-2} sec., the matter in the universe consisted of protons, neutrons, electrons, positrons, neutrinos and antineutrinos. The density of the matter was sufficiently high that even the neutrinos were trapped within the size of the universe. In such a cosmic fluid, light particles dominated. For every 10^9 photons, electrons or neutrinos there was only about one proton or one neutron. At this (10^{-2} sec.) stage, hundreds of types of particles were created and annihilated. The weak interactions

$$\begin{aligned} \nu_e + n &\Leftrightarrow e^- + p \\ \bar{\nu}_e + p &\Leftrightarrow e^+ + n \end{aligned} \tag{6.227}$$

created an equilibrium in the cosmic fluid and roughly there were equal number of protons and neutrons at 10^{-2} sec. At $t \approx 1$ sec., the rate of above reactions became small compared to expansion and the equilibrium could not be maintained. The neutrinos decoupled from the nucleons; electron–positron annihilation began to dominate and neutron proton ratio was frozen at about 15 per cent. At about $t = 13$ sec., the temperature of the universe dropped sufficiently. The electron–positron pairs could no longer be created. About 100 seconds after the big bang, the neutrons and protons had no longer the sufficient energy to escape the attraction of strong nuclear force and they (p, n) started to combine to produce the nuclei of deuterium. The deuterium nuclei then combined with more protons and neutrons to form helium nuclei. The nuclei of some lighter elements, lithium and beryllium were also formed. With the drop of photon energy, photo-disintegration of these nuclei could not take place. About 25 per cent of total protons and neutrons at this stage had converted into helium nuclei. The formation of the nuclei of these elements, which took place at about $10^2 - 10^3$ second, is known as *nucleosynthesis of lighter elements*. At $t \approx 35$ minutes, the temperature of the universe was still too high for the atoms to be formed, this happened only after 10^5 years.

6.9.7 From 35 min. to 10^5 years

Within only a few hours of the big bang, the production of helium and other elements would have stopped and after that for many thousand years nothing very significant occurred. The universe just expanded and cooled. When the temperature of the universe dropped to a few thousand-degree K, the electrons and the nuclei had not the enough energy to overcome the electromagnetic attraction between them and the nuclei and electrons started binding themselves into atoms and thus the first chemical elements were born. Large clouds of hydrogen and helium were formed. Before this era, the photons were coupled with protons, neutrons and electrons. Now the *photons decoupled* from the constituents of atoms and thus about 10^5 years ago the radiation and matter parted company. The *microwave background radiation* discovered by Penzias and Wilson in 1965 was the remaining trace of the radiation, which had separated from matter so long ago ($\approx 10^5$ years). Naturally, it has since then cooled to 3 K, which is the present temperature of the universe. It may however, be noted that the microwave radiation was initially light, which due to the expansion of the universe got so greatly red shifted that it appeared as microwave radiation to the observer on earth.

6.9.8 From 10^5 years to 10^7 years

The large radiating clouds of hydrogen and helium began to concentrate due to gravitational attraction. With the decrease of the radius of the cloud, its angular velocity increased so that the angular momentum is conserved. These condensing, rotating masses were the *beginning of galaxies*. Within these galaxies, due to gravitation more and more matter was compressed into the spherical objects. This was the beginning of stars. More and more matter was compressed and the increased pressure of the compressed matter produced high temperatures, which initiated the fusion process in which hydrogen was converted into helium and first stars were formed. The fusion process also caused the formation of more and more chemical elements. The higher chemical elements were formed by the absorption of neutrons. In this way all the chemical elements were formed within these stars. The first massive stars, however, did not live long and died in an explosion. The explosion, which is called a supernova, spewed the matter of all these elements into space. The fragments of these early stars became the nuclei for new stars and planets.

6.9.9 From 10^7 years to Present

The remains of dead star along with hydrogen and helium gases again formed new clouds, which were once again, compressed by gravity, until the sun, our star and its planets were formed. All the matter on the earth is the left over ashes of those early stars. By now, the universe is cooled to 3 K and its energy is only about 10^{-12} GeV.

6.10 LIMITATIONS OF SM AND GUTs

6.10.1 Limitations of the Standard Model

The Standard Model appears to be mathematically self-consistent and there is a good agreement between its predictions and experimental tests. However, the model has some limitations, the major ones are:

1. In the model there are large number of undetermined parameters, in terms of which the predictions are given. The parameters are:
   (a) Three gauge couplings of the groups $SU_c(3)$, $SU_L(2)$ and $U_Y(1)$, which describe the different strengths of the interactions.
   (b) The masses of six quark flavours, the masses of electron, muon and taon m_e, m_μ and m_τ respectively. The neutrinos are massless.
   (c) The three relative mixing angles between the quark families, which specify how W bosons couple to the quarks.
   (d) The phase, which determines the magnitude of CP violation.
   (e) The Higgs coupling constant λ, which determines its mass.
   (f) v, the vacuum expectation value of the Higgs field, which together with θ_W gives the W, Z masses.
2. The total number of free parameters is 19, which is a number, too large.
3. While constructing the Standard Model, many questions arise regarding the origin of the structure assumed. That is
   (a) Why do weak interactions violate parity, implying that why do quarks and leptons only interact with weak current in a left-handed manner?
   (b) Why neutrinos are massless and why there are no right-handed neutrino components?
   (c) Why are the charges of quarks quantized in fractions of one third that of the charged leptons?
   (d) Why are there three copies of each quark and lepton, each copy having the same coupling to the gauge bosons but greatly different mass?
4. Although, the model unifies the strong, weak and the electromagnetic interactions, the SM does not complete this unification because the strengths of these interactions are not related by the model and no explanation is given, why they should be so different, when measured in laboratory?
5. There is no explanation given in SM, why the weak interactions should be short range, mediated by massive gauge bosons while the electromagnetic and strong forces are mediated by massless gauge bosons?
6. Though the Standard Model is able to parameterize the above difference by the SSB of weak gauge symmetry through the Higgs mechanism, the postulated Higgs boson has yet to be experimentally detected. Since, the above-mentioned Higgs mechanism breaks the $SU_c(3) \otimes SU_L(2) \otimes U_Y(1)$ gauge symmetry, there appears a resultant asymmetry in the nature of fundamental forces with the weak gauge bosons acquiring the mass. Moreover, in the SM, no reason is given why it is the weak, not the strong or the electromagnetic interaction that is altered in this way?
7. Most importantly, the unification is of only three of the four fundamental interactions and there is no connection with the gravitational force.
8. If gravity is included, the gravitational constant $G_N \propto M_{\text{Planck}}^{-2}$, where $M_{\text{Planck}} \approx 10^{19}\,\text{GeV}$. At low energies, the Fermi constant for weak

interactions is $G_F \propto M_W^{-2}$, where $M_W = 10^2$ GeV. This means M_W is some 17 orders of magnitude smaller than Planck mass. The quark and charged lepton masses range from 0.5×10^{-3} GeV for electron to about 4.5 GeV for bottom quark, while the neutrinos are massless. The top quark is some 38 times heavier than the bottom quark. The Standard model does not explain the cause of such hierarchy in masses. This large difference in the masses of particles as well as between G_F and G_N is called the *hierarchy problem*. This problem remains unresolved in the Standard Model.

6.10.2 Limitations of GUTs

The grand unified theories, though able to answer some of the questions raised by the Standard Model but they too have some limitations. For example, in the SU(5) grand unified theory, the strong, weak and electromagnetic interactions are just the different facets of one SU(5) local gauge interaction. The quarks and leptons are also related in GUT because they belong to the same multiplet. The five dimensional representation of SU(5) contains three down quarks, the electron and a neutrino. The traceless condition requires that the charges be quantized in units of 1/3. That is what we require. Three (3) is the number of quark colours. In SU(5) there is no room for right-handed neutrino component. Thus we see many of the questions about the multiplet structure in the Standard Model have an answer in GUT. The major limitations of GUTs are

1. The need for three families of quarks and leptons remains unanswered in GUTs as well.
2. The original GUTs fail to reduce the overall number of parameters needed, because they require a much more complicated Higgs sector with its associated couplings to explain the pattern of symmetry breaking.
3. The grand unified theories do not include the gravitational interaction and hence do not achieve full unification.
4. The GUTs also introduce new gauge particles and their associated interactions, none of which have been seen.
5. The CKM mixing angles and the possible CP violation associated with the phase in the CKM matrix remain unexplained in GUTs as well.
6. Grand unified theories built from simple groups predict the presence of magnetic monopoles with large magnetic charges and mass $>10^{16}$ GeV. A conclusive experimental evidence of magnetic monopole is awaited.

6.11 BEYOND THE SM AND GUTs

Both the Standard Model and Grand Unified Theories have limitations to become the satisfactory theories for explaining the fundamental forces of nature. Therefore, attempts have been made to develop theories, which are complete in themselves. In this regard Supersymmetric and Supergravity theories, Technicolour theories and Superstring theories are quite significant. A brief description of these theories has been given in the following.

6.11.1 Supersymmetry and Supergravity

It is well known that particles of matter like quarks and leptons are fermions and particles (quanta) associated with interaction like photon, $W^{\pm}$, Z^0, gluons and gravitons are bosons. A theory, which unites fermions and bosons, is called the theory of supersymmetry. Therefore supersymmetry by definition is symmetry between fermions and bosons. Any quantum field theory, which exhibits the fermion-boson symmetry, is called a supersymmetric quantum field theory. Not only quantum field theory, but also the quantum mechanics can be super-symmetrized. The transformation, which changes a boson into a fermion and vice versa is called a supersymmetric transformation, which is generated by a quantum operator Q_α.

$$Q_\alpha |\text{fermion}\rangle = |\text{boson}\rangle_\alpha$$

$$Q_\alpha^\dagger |\text{boson}\rangle = |\text{fermion}\rangle_\alpha$$

In order to carry out above transformations the generator Q_α must have the fermionic character. The Q_α obeys

$$\{Q_\alpha^\dagger, Q_\beta\} = p^\mu (\sigma_\mu)_{\alpha\beta}$$

where p^μ is the energy momentum four vector and σ are the Pauli matrices. The supersymmetry can be both global and local. The local supersymmetry is called *supergravity*.

For a field theory to be supersymmetric, it is required that the theory contains the specified types and numbers of fields in interaction with each other and that the various interaction strengths and particle masses have properly related values. For example, the supersymmetrization of $SU_c(3)$ gauge theory would require the inclusion of massless colour octet of spin $-\frac{1}{2}$ particles, which are their own antiparticles. The eight gluons present in $SU_c(3)$ gauge theory are spin -1 particles (bosons) so we need corresponding eight fermions. The spin $-\frac{1}{2}$ partner of gluon is called a *gluino*. Similarly the fermionic partner of photon is photino (a massless spin $\frac{1}{2}$ particle), and that of Higgs particle (spin 0) is higgisino. The fermionic partner of graviton (spin–2) is a spin $-\frac{3}{2}$ particle called graviton. Similarly supersymmetrization of a theory, which initially contains fermions, would require including the corresponding bosonic partners. The bosonic partners of fundamental fermions are spin-zero particles named, *squarks*, *sleptons*, *selectron* etc., where *s* implies scalar. When the spin of quark or electron is taken away it becomes a scalar quark (*squark*) or scalar electron.

Various grand unified theories can also be supersymmetrized. They are referred to SUSY-GUTs. It has been observed that in supersymmetric SU (5) grand unified theory, the running coupling constants meet more exactly than in ordinary grand unified theories. The supersymmetry is an attractive theory and whether this theory exists in nature or not would be known only through the experimental observation of superpartners.

6.11.2 Technicolour Theories

We have seen in the Standard Model that no experimental evidence exists for the Higgs particle. This lack of experimental evidence for Higgs particle is the reason for the creation of the Technicolour theories. In these theories the role of Higgs scalars are played by condensates, which are composed of heavy fermions called *techniquarks*. These techniquarks experience an interaction, which is stronger than usual colour interaction. Hence it is a new interaction called technicolour *interaction*. In GWS model the symmetry breaking was due to a non-zero vacuum expectation value of Higgs field, here in technicolour theories a non-zero vacuum expectation value of a condensate of very heavy techniquarks breaks the SU(2) $\otimes$ U(1) electroweak gauge theory. The technicolour theories, however, predict the technihadrons in TeV (10^{12} eV) range.

The technicolour theories can be included in the grand unified theories where they are expected to solve the hierarchy problem in a natural way. However, the extension of technicolour theories to explain the widely different masses of fermions requires new particles, which need to couple with both colour and technicolour. These new particles make the technicolour theories less attractive.

6.11.3 Superstring Theories

The Superstring theory assumes that the ultimate building blocks of nature comprise of very small vibrating strings. Since a string is capable of vibrating in several different modes, the Superstring theory assumes that each mode of vibration of a Superstring can represent a force or a particle. There can be infinite number of possible modes of vibration; the Superstring therefore can represent an infinite number of particles. Superstrings are of two types-open strings and closed strings. The open strings have ends and closed strings are circular. The photon corresponds to the lowest mode of vibration of the open string. The gluon is thought to be a string at each end of which is connected a quark. The graviton is considered as the lowest mode of vibration of a closed string. Higher modes of vibration of open string represent particles such as quarks, protons, neutrons, etc. In this theory no particle is more fundamental than other; instead each is just a different mode of vibration of the Superstring. The superstring theory assumes that originally, the universe had ten dimensions but broke into two parts. One part had four dimensions, which is our universe. The superstrings interact with other superstrings by breaking and reforming. The superstring theories assume that there is one unifying force of the superstring and the four forces are just the different manifestations of this one unifying force.

The origin of string theory through Veneziano amplitude and the description of bosonic and various superstring theories has been given in section 8.5.6.

6.12 ASSIGNMENTS

1. Discuss the reasons that supply motive to unite electromagnetic and weak interactions. What were the initial difficulties in their unification? Develop logically, the gauge symmetry group that would provide the electromagnetic and weak embedding.
2. Show that $SU_L(2) \otimes U_Y(1)$ is the gauge symmetry group of electromagnetic and weak interactions. Pointing out the requirement of a covariant derivative, suggest the covariant derivative for the group and develop the locally gauge invariant Lagrangian density. Explain that this Lagrangian density cannot contain the mass term.
3. Explaining the reasons to unite electromagnetic and weak interactions, derive the electroweak Lagrangian density. Discuss the various interaction terms and identify the neutral weak current gauge field interaction. Why neutral interactions are difficult to observe?
4. Show that

   (1) $\bar{\psi}\psi = \bar{\psi}_R\psi_L + \bar{\psi}_L\psi_R$

   (2) $\tau_a W_\mu^a = \begin{pmatrix} W_\mu^3 & \sqrt{2}\ W_\mu^- \\ \sqrt{2}\ W_\mu^+ & -W_\mu^3 \end{pmatrix}$

   (3) $Z_\mu = (\cos\theta_W)W_\mu^3 - (\sin\theta_W)B_\mu$

   (4) $A_\mu = (\sin\theta_W)W_\mu^3 + (\cos\theta_W)B_\mu$

   (5) $g_2\cos\theta_W = e = g_1\sin\theta_W$

   (6) $\sin\theta_W = g_2/\left(\sqrt{g_1^2 + g_2^2}\right)$.

   Explain the meaning and significance of the symbols used in above equations.
5. Proceeding from a massless fermion Lagrangian density of electroweak unification, derive the expression for the neutral gauge fields Z_μ and A_μ. Explain the difference between them. Estimate θ_W and show that in this process the weak coupling constant emerges stronger than the electromagnetic constant. How this contradiction is attended to?
6. Distinguish neutral weak interactions from the charged ones and explain the reasons for their late detection. Describe, how neutral weak interactions are detected?

7. Explain the structure of neutral weak current and show that it is expressed as $J^{\mu}_{\text{neutral}} = J^{\mu}_{3} - \sin^2\theta_W \, J^{\mu}_{\text{em}}$. Comment on the handednesses of J^{μ}_{neutral} and J^{μ}_{3}. How weak neutral current is experimentally verified?
8. Show that the required symmetry for electroweak unification is $SU_L(2) \otimes U_Y(1)$ and explain the necessity to break this symmetry. Carry out the spontaneous breaking of the local electroweak gauge symmetry and show how gauge bosons acquire masses. Discuss the choice of the field required to break the symmetry.
9. Proceeding from the $SU_L(2) \otimes U_Y(1)$ Lagrangian density, apply Higgs mechanism to show that W bosons have mass $\frac{1}{2} g_1 v$ and Z boson has mass $\frac{m_W}{\cos\theta_W}$. What were the masses of W and Z bosons before above symmetry breaking?
10. Discussing the choice of Higgs field and proceeding from the Lagrangian density, apply Higgs mechanism to break the electroweak gauge symmetry and achieve the mass spectrum for gauge bosons and leptons. Explain how despite symmetry breaking, the photon and neutrino remain massless. Why Higgs particle is difficult to observe?
11. In the gauge theory of the unification of electromagnetic and weak interactions, we start with a Lagrangian density of massless particles. Why? Then how do we explain that gauge bosons, leptons and quarks have acquired mass!
12. On the basis of the following headings, give a descriptive account of GWS model. 1. Motivation. 2. Gauge group. 3. Lagrangian density. 4. Neutral current interaction. 5. Neutral gauge fields. 6. Estimation of interaction strength. 7. The propagator and strength of weak interaction. 8. Gauge symmetry breaking. 9. Choice of Higgs field. 10. Masses of gauge bosons. 11. Masses of fermions. 12. Renormalizability of electroweak theory. 13. Experimental discovery of $W^{\pm}$, Z^0. 14. Other experimental evidences of electroweak theory.
13. Give a brief account of GWS electroweak theory and show that the theory is renormalizable. Discuss the experimental status of the theory.
14. Defining quantum chromodynamics, obtain its Lagrangian density.
15. What is Standard Model? Mention its gauge group and the Lagrangian density. Explain the various terms of the Lagrangian density.
16. What is Standard Model? How can it be experimentally verified?
17. Describe $\sin^2\theta_W$ experiments to show that various $\sin\theta_W$ values agree. What inference do we draw from this agreement?

18. Explain that an important evidence for quantum chromodynamics is the evidence for colour. Provide evidences to show that quarks have three colours.
19. What are the essentials of a perturbation theory? Explain that at low energies the *QCD* becomes non-perturbative. Describe asymptotic freedom and quark confinement qualitatively.
20. Discuss the inadequacies of the Standard Model and show that SU(5) grand unification theory can explain the unification of three interactions in a better way. At what energy all the three interactions are supposed to unify.
21. Explain the behavior of coupling constants of the gauge groups U(1), $SU_L(2)$ and $SU_c(3)$ with energy. Show that $M_x = 5 \times 10^{14}$ GeV. Explaining M_x, show that an implication of Grand Unification Theories is that proton may decay.
22. What is Grand Unified interaction? Calculate the energy at which the gravitational interaction may unify with the Grand Unified interaction. What are Planck energy, Planck length and Planck time? Explain their significance.
23. Show that 1 GeV $\equiv 10^{13}$ K. Hence estimate the temperature at which
    (1) Electromagnetic and weak interactions unify.
    (2) Electromagnetic weak and strong interactions unify.
    (3) All the four interactions unify.
    Compare the energy corresponding to (1) (electroweak) above with the masses of $W^{\pm}$ and Z^0 bosons.
24. What is the Big Bang Model of universe? Discuss the evidence that provided authenticity to this Model.
25. Narrate in brief the events since Big Bang till the present time. Identify the significant events. Till/At-what time since the big bang (i) all the forces were united (ii) gravitational force parted (iii) three forces remained united (iv) strong interaction decoupled (v) electroweak force splitted.
26. The early universe provides the necessary laboratory to test the unification of all forces. Comment. Describe super heavy bosons.
27. Discuss the limitations of Grand Unified Theories and mention the theories which are the possible answers to these limitations.
28. Write notes on
    (a) Supersymmetry and supergravity
    (b) Technicolour theories
    (c) Superstring theories
29. Based on the following titles, write an essay on *Particle Physics and Thermodynamics in the Early Universe*: 1. Introduction 2. The Expanding Universe 3. The First Three Minutes of Universe 4. Matter-Antimatter Asymmetry 5. Electroweak Phase Transitions 6. Hadron Formation 7. Primeval Synthesis of the Elements. 8. Current Status of Knowledge on Early Universe.

6.13 OBJECTIVE QUESTIONS

1. The Standard Model provides
   (a) a unified theory of all the four interactions.
   (b) a unified theory of electromagnetic and weak interactions.
   (c) a unified theory of em,weak and strong interactions.
   (d) separate theories of em, weak and strong interactions.
2. The electroweak interaction is described by
   (a) Glashow–Weinberg–Salam Model.
   (b) Standard Model.
   (c) GIM Model.
   (d) Spectator Model.
3. A new quantum number is introduced in the theory of electroweak interaction. The quantum number is
   (a) isospin (b) hypercharge
   (c) weak isospin (d) G-parity
4. Which among the following do not couple to W bosons?
   (a) All fermions (b) Left-handed fermions
   (c) Right-handed fermions (d) All bosons and fermions
5. The gauge group that unifies the electromagnetic and weak interactions is
   (a) U(1) (b) SU(2)
   (c) SU(2) ⊗ U(1) (d) $SU_L(2) \otimes U_Y(1)$
6. If T denotes the weak isospin, the transformation $\psi'(x) = e^{iT.0(x)}$ denote
   (a) SU(2) global gauge transformations.
   (b) $SU_L(2)$ global gauge transformations.
   (c) SU(2) local gauge transformations.
   (d) $SU_L(2)$ local gauge transformations.
7. Local $SU_L(2) \otimes U_Y(1)$ gauge symmetry requires
   (a) only one gauge field, which should be massless isovector.
   (b) two gauge fields; one should be massless isovector and other massless vector.
   (c) two gauge fields; both should be massless vector.
   (d) two gauge fields; one should be massive isovector, other massless vector.
8. Given below are two covariant derivatives for electroweak theory:

(1) $$(D_\mu)_\alpha = \partial_\mu + ig_1 T_a W_\mu^a + ig_2 \frac{Y_\alpha}{2} B_\mu$$

(2) $$(D_\mu)_\alpha = \partial_\mu + ig_2 \frac{Y_\alpha}{2} B_\mu$$

9. What among the following is true?
   (a) (1) is for left-handed particles and (2) is for right-handed particles.
   (b) (2) is for left-handed particles and (1) is for right-handed particles.

(c) Both are for left-handed particles.
(d) Both are for right-handed particles.

9. The gauge particles of the electroweak gauge theory are
   (a) electron, $W^{\pm}$ and Z^0 bosons.
   (b) $W^{\pm}$, Z^0 and photon.
   (c) $W^{\pm}$, Z^0, photon and gluon.
   (d) neutrino, $W^{\pm}$, Z^0 and photon.
10. What is true about photon and Z^0 gauge fields?
   (a) Both are charged gauge fields because they are the linear combinations of W^+ and W^-.
   (b) Photon field is charged because it is a linear combination of W^+ and W^-. The Z^0 field, being the linear combination of gauge fields B^0 and W^0, is neutral.
   (c) Both are neutral because they are identical linear combinations of B^0 and W^0.
   (d) Both are neutral because they are the orthogonal linear combinations of B^0 and W^0.
11. Choose the correct answer.

   (a) $|\gamma\rangle = \cos\theta_W |B^0\rangle, |Z^0\rangle = \sin\theta_W |W^0\rangle$

   (b) $|\gamma\rangle = \cos\theta_W |B^0\rangle + \sin\theta_W |W^0\rangle$

   $|Z^0\rangle = \cos\theta_W |W^0\rangle - \sin\theta_W |B^0\rangle$

   (c) $|\gamma\rangle = \sin\theta_W |B^0\rangle + \cos\theta_W |W^0\rangle$

   $|Z^0\rangle = \sin\theta_W |W^0\rangle + \cos\theta_W |B^0\rangle$

   (d) $|\gamma\rangle = \cos\theta_W |W^0\rangle, |Z^0\rangle = -\sin\theta_W |B^0\rangle$
12. θ_W is called
   (a) electroweak mixing angle.
   (b) Weinberg angle.
   (c) both (a) and (b).
   (d) Cabibbo angle.
13. The angle θ_W can be determined from
   (a) $\nu - e$ scattering.
   (b) electroweak interference in $e^+ e^-$ scattering.
   (c) the ratio of the masses of $W^{\pm}$ and Z^0.
   (d) all the above three methods (a), (b) and (c).
14. Which of the following combinations is correct?

   (a) $\dfrac{M_W}{M_Z} = \cos\theta_W, \ \sin^2\theta_W = 0.213$

(b) $\dfrac{M_W}{M_Z} = \sin\,\theta_W,\ \cos^2\theta_W = 0.213$

(c) $\dfrac{M_W}{M_Z} = 1 - \sin\,\theta_W,\ \sin^2\theta_W = 0.00024$

(d) $\dfrac{M_W}{M_Z} = 1 - \cos\,\theta_W,\ \cos^2\theta_W = 0.00024$

15. The GWS model predicted the existence of
   (a) charged weak currents.
   (b) neutral weak currents.
   (c) electromagnetic currents.
   (d) parity violation in weak interactions to a maximal degree.

16. What is not true?
   (a) The charged current only couples to left-handed fermions and right-handed antifermions.
   (b) Only the charged current of weak interaction transforms one type of quark into another and one type of lepton into another.
   (c) The neutral currents are pure V–A currents.
   (d) The neutral weak current couples to left-handed and right handed fermions and antifermions with different strengths.

17. Given below are two matrix elements M_1 and M_2 for the weak interactions:

$$M_1 = \left(\frac{g}{\sqrt{2}}J_\mu\right)\left(\frac{1}{M_W^2}\right)\left(\frac{g}{\sqrt{2}}J^\mu\right)$$

$$M_2 = \left(\frac{g}{\cos\theta_W}J_\mu\right)\left(\frac{1}{M_Z^2}\right)\left(\frac{g}{\cos\theta_W}J^\mu\right)$$

Choose the correct answer from the following.
   (a) M_1 is the amplitude for a charged weak interaction process provided J_μ in M_1 is charged weak current and M_2 is the amplitude for a neutral weak interaction process provided J_μ in M_2 is neutral weak current.
   (b) M_1 is the amplitude for a neutral weak interaction process provided J_μ in M_1 is neutral weak current and M_2 is the amplitude for a charged weak interaction process provided J_μ in M_2 is charged weak current.
   (c) M_1 is the amplitude for a charged weak interaction process provided J_μ in M_1 is neutral weak current and M_2 is the amplitude for a neutral weak interaction process provided J_μ in M_2 is charged weak current.
   (d) Both M_1 and M_2 represent neutral weak interaction processes if the currents J_μ are neutral.

18. The relative strength of neutral and charged weak interactions is given by a parameter ρ expressed by

(a) $\rho = \dfrac{M_Z^2}{M_W^2 \sin^2 \theta_W}$ (b) $\rho = \dfrac{M_Z^2}{M_W^2 \cos^2 \theta_W}$

(c) $\rho = \dfrac{M_W^2 \cos^2 \theta_W}{M_W^2}$ (d) $\rho = \dfrac{M_W^2 \sin^2 \theta_W}{M_W^2}$

19. Match the quantities in Column I with their expressions in Column II and select the correct code.

	Column I		*Column II*
A	J_μ^{em}	1	$B_\mu^0 \cos\theta_W + W_\mu^0 \sin\theta_W$
B	J_μ^{NC}	2	$J_\mu^3 - \sin^2\theta_W J_\mu^{em}$
C	A_μ	3	$W_\mu^0 \cos\theta_W - B_\mu^0 \sin\theta_W$
D	Z_μ	4	$g_1 g_2 \big/ \sqrt{g_1^2 + g_2^2}$
E	e	5	$J_\mu^3 + \dfrac{1}{2} J_\mu^Y$

Codes:

	A	B	C	D	E
(a)	2	5	3	1	4
(b)	5	2	1	4	3
(c)	5	1	2	4	3
(d)	5	2	1	3	4

20. The mass term introduced in the Lagrangian density of the electroweak gauge theory has the form

(a) $m\overline{\psi}\psi$ (b) $m\overline{\psi}_R\psi_L + m\overline{\psi}_L\psi_R$

(c) $m\overline{\psi}_R\psi_R + m\overline{\psi}_L\psi_L$ (d) no mass term is introduced

21. In electroweak gauge theory
    (a) the masses of fermions and bosons are generated through Higgs mechanism.
    (b) the masses of bosons are generated through Higgs mechanism but the masses of fermions are put by hand.
    (c) the masses of fermions are generated through Higgs mechanism but the masses of bosons are put by hand.
    (d) the masses of both fermions and bosons are put by hand.

22. In electroweak gauge theory, the Higgs field $\phi(x) = \frac{1}{\sqrt{2}}\begin{pmatrix} 0 \\ v + h(x) \end{pmatrix}$ is responsible to generate
    (a) the masses of all fermions and $W^{\pm}$ and Z^0 bosons.
    (b) the masses of weak isospin $T_3 = -\frac{1}{2}$ fermions and $W^{\pm}$, Z^0 bosons.
    (c) the masses of weak isospin $T_3 = -\frac{1}{2}$ fermions and all gauge bosons.
    (d) no mass.
23. In electroweak gauge theory, the Higgs field $\phi(x) = \frac{1}{\sqrt{2}}\begin{pmatrix} v + h(x) \\ 0 \end{pmatrix}$
    (a) generates no mass in the theory.
    (b) generates masses of weak isospin $T_3 = +\frac{1}{2}$ fermions.
    (c) generates masses of weak isospin $T_3 = +\frac{1}{2}$ quarks.
    (d) generates masses of weak isospin $T_3 = -\frac{1}{2}$ fermions and Higgs particle.
24. In the GWS model
    (a) the spontaneous breaking of $[SU_L(2) \otimes U_Y(1)]$ symmetry takes place.
    (b) the spontaneous breaking of $SU_L(2)$ symmetry takes place.
    (c) the spontaneous breaking of U(1) symmetry takes place.
    (d) no spontaneous symmetry breaking takes place.
25. In the GWS model
    (a) due to the proper choice of Higgs field neutrino remains massless and due to its vector (spin-1) nature photon remains massless.
    (b) due to the proper choice of Higgs field photon remains massless and due to the absence of right handed neutrino, neutrino remains massless.
    (c) due to the proper choice of Higgs field both photon and neutrino remain massless.
    (d) due to spin-1, photon remains massless and due to non-gauge invariant term neutrino remains massless.
26. What is true among the following?
    (a) Due to the coupling strength being inversely proportional to the mass of fermion, the Higgs particle would be produced by light fermions.
    (b) Due to the coupling strength being proportional to the mass of fermion, the Higgs particle would be produced by light fermions.

(c) Due to the coupling strength being proportional to the mass of fermion, the Higgs particle would be produced by heavier fermions.
(d) Due to the coupling strength being inversely proportional to the mass of fermion, the Higgs particle would be produced by heavier fermions.

27. In a ferromagnet the symmetry which is spontaneously broken at low temperatures is restored at high temperatures. Applying this analogy to the high temperature behaviour of electroweak interaction, we should expect that at high temperatures
(a) there would be no change in the range of weak interactions.
(b) the range of weak interactions would diminish further.
(c) the weak interactions would have a long range.
(d) the weak interactions would have a small but fixed range.

28. The matrix element for the process $electron \rightarrow electron$ would be
(a) equal to 1, if the interaction of electron with its self-field takes place.
(b) not equal to 1, if the interaction of electron with its self-field takes place.
(c) equal to 1, if the electron does not interact with its self-field.
(d) as described by (b) and (c) above.

29. What is true about the self-energy of an electron?
(a) The self energy of a non-relativistic electron diverges linearly with the virtual photon momentum.
(b) The self-energy of a relativistic electron diverges logarithmically.
(c) Both (a) and (b) are correct.
(d) The self-energy of a non-relativistic electron converges linearly and that of relativistic electron diverges logarithmically.

30. While calculating matrix element for a process, we encounter with integrals of the type $\int \frac{d^4q}{(2\pi)^4} \frac{1}{q^2+m^2}$, with q denoting the internal momentum in the Feynman diagram. What is true about such integrals?
(a) . . . . $q \rightarrow \infty$. Such a convergence is called ultraviolet convergence.
(b) The integral diverges when $q \rightarrow \infty$. Such a divergence is called infrared divergence.
(c) The integral converges when $q \rightarrow \infty$. Such a convergence is called infrared convergence.
(d) The integral diverges when $q \rightarrow \infty$. Such a divergence is called ultraviolet divergence.

31. What is true about the integrals of the type?

$$I = \frac{1}{(2\pi)^4} \frac{\lambda^2}{2} \int \frac{d^4q}{(2\pi)^4} \frac{1}{(q^2+m^2)[(q-p_1-p_2)^2+m^2]}$$

(a) As $q \to \infty$, the integral I diverges as it behaves as $\int \frac{d^4q}{q^4}$ which is equal to log q.
(b) When $m^2 = 0$, $(p_1 + p_2)^2 = 0$, the integral diverges for low q.
(c) The divergence (b) above is called infrared divergence.
(d) All the above three statements are true.

32. The process by which the divergences or the infinities are removed is called
(a) normalization (b) regularization
(c) renormalization (d) vacuum polarization

33. The observable mass m of an electron is equal to
(a) m_0, where m_0 is the bare mass.
(b) $m_0 + \delta m$, where m_0 is the bare mass and δm is the contribution to m_0 due to electromagnetic interaction.
(c) δm
(d) $\frac{\delta m}{m_0}$

34. g_R, given by the expression $\left[g_R = g_e\sqrt{1 - \frac{g_e^2}{12\pi^2}\log\left(\frac{M^2}{m^2}\right)}\right]$, where M denotes the cutoff in the integral $\int_{m^2}^{\infty} \frac{dp^2}{p^2}$ with $M \to \infty$, represents
(a) the renormalized coupling constant.
(b) the bare coupling constant.
(c) the renormalized mass of electron.
(d) the bare mass of electron.

35. The parameters (mass, charge, coupling constants, etc.), which we can measure and call them physical are
(a) bare
(b) renormalized
(c) rationalized
(d) none of the above

36. The physical charge of electron is equal to the bare coupling constant g_e (question 34) only when, in the theoretical analysis, we look at
(a) one loop Feynman diagrams.
(b) two loop Feynman diagrams.
(c) three loop Feynman diagrams.
(d) zero loop (tree level) Feynman diagrams.

37. The effective charge of the electron depends on the momentum transferred in the collision. It implies that the charge of electron is not constant. This fact does not disturb the observations because

(a) due to $\alpha(q^2) = \alpha(0)\left\{1 + \frac{\alpha(0)}{3\pi} f\left(\frac{-q^2}{m^2c^2}\right)\right\}$, where α is fine structure constant, q momentum transferred; the variation being small we have $\alpha(q^2) \approx \alpha(0)$.

(b) $\alpha(q^2)$ does not play any role, it is $\alpha(0)$ that plays part.

(c) due to the reason (a) above, for most purposes $\alpha(q^2) \approx \alpha(0). = \frac{1}{137}$.

It is only in the Lamb shift that the second term, in the RHS of (a) above, makes a detectable contribution.

(d) due to the reasons cited in (a) and (c) above.

38. Given below are a Lagrangian density £ and relations R. Choose the correct answer from the given alternatives.

$$£ = \frac{1}{2}\left[(\partial_\mu \phi_0)^2 - m_0^2\phi_0^2\right] + \left[-\frac{\lambda_0}{4}\phi_0^4\right]$$

$$R \Rightarrow \begin{cases} \phi = Z_\phi^{-\frac{1}{2}}\phi_0 \\ \lambda = Z_\lambda^{-1} Z_\phi^2 \lambda_0 \\ m^2 = m_0^2 + \delta m^2 \end{cases}$$

(a) £ is the Lagrangian density of free neutral scalar field. The parameters with zero subscript represent bare quantities. The substitution of R into £ will renormalize the Lagrangian density.

(b) £ is the Lagrangian density of interacting neutral scalar field. The parameters with zero subscript represent bare quantities. The substitution of R into £ will renormalize the Lagrangian density.

(c) £ is the renormalized Lagrangian density of $\lambda\phi^4$ theory. The substitutions R would bring back £ to its unrenormalized form.

(d) £ is the unrenormalized Lagrangian density of free complex scalar field. The substitutions R would renormalize it.

39. The theories are required to be renormalizable because

(a) in non-renormalizable theories the infinities arising from higher order diagrams make the quantities of physical interest infinite.

(b) in the renormalizable theories the infinities can be cancelled consistently to all orders with a finite number of multiplicative factors and parameters and thus the quantities of physical interest acquire finite (measurable) values.

(c) in the renormalizable theories, the infinities are converted into poles through dimensional regularization and these poles are then eliminated order by order by introducing counter terms to obtain a finite result.
(d) of the reasons (a), (b) and (c) above.

40. What is true?
(a) *QED* is a renormalizable theory.
(b) *QED* and electroweak theory are renormalizable.
(c) All the gauge theories are renormalizable.
(d) All the statements (a, b, c) are true.

41. Spontaneously broken gauge symmetries can be made renormalizable
(a) by the usual method.
(b) by choosing a particular gauge.
(c) by ignoring infinities.
(d) Spontaneously broken gauge symmetries cannot be renormalized.

42. The limitations of the methods of renormalizable perturbation theory renormalization can be overcome by using
(a) the method of dispersion relations.
(b) the method of functional or path integration.
(c) the renormalization method.
(d) all the above three methods.

43. The $(g-2)$ experiments provide an experimental test of
(a) electroweak theory
(b) quantum chromodynamics
(c) quantum electrodynamics
(d) charge renormalization

44. The experimental tests of electroweak theory come from
(a) $(g-2)$ experiments
(b) neutral current verification experiments
(c) $\sin^2\theta_W$ measurements
(d) both (b) and (c) above

45. Study of cross section ratio $R = \dfrac{\sigma(e^+e^- \to \text{hadrons})}{\sigma(e^+e^- \to \mu^+\mu^-)}$ provides an evidence for the existence of
(a) three colours of quarks
(b) three generations of quarks
(c) three generations of leptons
(d) three degrees of freedom of a massive vector field

46. Studies of quarkonium systems and glueballs indicate the experimental evidence of
(a) quantum electrodynamics
(b) quantum flavourdynamics
(c) quantum chromodynamics
(d) classical electrodynamics

47. What among the following statements is not true about grand unified theories?
   (a) A grand unified theory should have three coupling constants corresponding to three interactions.
   (b) The rank of grand unification group should be at least 4 because $SU_c(3) \otimes SU_L(2) \otimes U_Y(1)$ has rank 4.
   (c) The grand unification group has to be either simple or semisimple.
   (d) SU(5) is the only semi simple group of rank 4.
48. If g_3, g_2, g_1 be the coupling constants of $SU_c(3)$, $SU_L(2)$ and $U_Y(1)$ respectively and g_G be the coupling constant of grand unified group, the correct relation for them is
   (a) $g_3 = g_2 = g_G$ and $g_1 = \sqrt{\frac{3}{5}}\, g_G$
   (b) $g_3 = g_1 = g_G$ and $g_2 = \sqrt{\frac{4}{5}}\, g_G$
   (c) $g_3 = g_2 = g_1 = g_G$
   (d) $g_3 = g_2 = g_G$ and $g_1 = \sqrt{\frac{1}{5}}\, g_G$
49. The particles of the SU(5) grand unified theory are
   (a) Higgs scalars, $W^{\pm}$, Z^0 bosons, gluons, quarks and leptons.
   (b) $W^{\pm}$, Z^0 bosons, gluons, lepto–quark gauge bosons X and Y, quarks and leptons.
   (c) Higgs scalars, photon, $W^{\pm}$, Z^0 bosons, gluons, X and Y lepto–quark gauge bosons, quarks and leptons.
   (d) Higgs scalars, $W^{\pm}$, Z^0 bosons, gluons, X and Y gauge bosons, quarks and leptons.
50. What is true about X and Y gauge bosons?
   (a) They are superheavy with mass $\approx 10^{15}$ GeV and electrical charges $\mp\frac{4}{3}e$ and $\mp\frac{1}{3}e$ respectively.
   (b) Coming in three colours, they cause quark lepton interchange.
   (c) They violate baryon and lepton number conservation.
   (d) All the above three statements are true.
51. What remains conserved in the process $[p \rightarrow \pi^0 + e^+]$?
   (a) Baryon number
   (b) Lepton number
   (c) Baryon number minus lepton number
   (d) Baryon number plus lepton number

52. The superheavy bosons X and Y decay as $X \rightarrow q + q$ and $\overline{X} \rightarrow \overline{q} + \overline{q}$. In the early Universe X and $\overline{X}$ had equal densities, but we now have a baryon asymmetry. All the quarks in the early Universe did not annihilate because
    (a) the partial decay rates of X and $\overline{X}$ differ because they violate all conservation laws.
    (b) the partial decay rates of X and $\overline{X}$ differ because conservation laws of Charge conjugation and CP are violated.
    (c) the partial decay rates of X and $\overline{X}$ differ because charge conservation law is violated.
    (d) the partial decay rates of X and $\overline{X}$ differ because they are superheavy.
53. What is true about the three coupling constants g_c of $SU_c(3)$, g_L of $SU_L(2)$ and g_Y of $U_Y(1)$?
    (a) All the coupling constants increase logarithmically with energy.
    (b) All the coupling constants decrease linearly with energy.
    (c) g_c and g_L decrease logarithmically, g_c at a rapid rate than g_L. g_Y increases slowly logarithmically with energy.
    (d) g_c and g_Y decrease logarithmically, g_c at a faster rate than g_Y. g_L increases slowly logarithmically with energy.
54. The grand unification mass scale corresponds to about
    (a) $10^{!5}$ GeV (mass of X bosons).
    (b) 10^2 GeV (mass of $W^{\pm}$, Z^0 bosons).
    (c) $10^{!9}$ GeV (Planck energy).
    (d) 10^{-12} GeV (energy of present Universe).
55. The experimental verification of grand unification is difficult because of high energy at which it takes place. Still we can rely on the grand unification because
    (a) we have four interactions as the low energy manifestations of grand unified interaction.
    (b) we have three interactions as the low energy manifestations of grand unified interaction.
    (c) the grand unified interaction agrees with experiments at $\approx 10^2$ GeV, the mass of $W^{\pm}$, Z^0 bosons.
    (d) below the energy of $\approx 10^2$ GeV, the electroweak interaction splits.
56. The Universe in which we live has to either expand or contract. This was predicted theoretically by
    (a) Albert Einstein
    (b) Alexander Freidmann
    (c) Edwin P. Hubble
    (d) Stephen W. Hawking
57. Cosmology is the science of
    (a) comets
    (b) blackholes
    (c) cold stars
    (d) entire Universe

58. Match the following two columns and select the correct code.

	Column I		*Column II*
A	M_x (GeV)	1	1.6×10^{-35}
B	*proton* (*years*)	2	1.3×10^{31}
C	*Planck energy* (GeV)	3	5×10^{-44}
D	*Planck length* (*meter*)	4	5×10^{14}
E	*Planck time* (*second*)	5	1.2×10^{19}

Codes:

(a)	A	B	C	D	E
	5	2	4	3	1
(b)	A	B	C	D	E
	4	2	5	1	3
(c)	A	B	C	D	E
	2	4	5	3	1
(d)	A	B	C	D	E
	4	2	5	3	1

59. What is not true about the Universe?
(a) The Universe has net non-zero baryon number.
(b) The electric charge of Universe is zero.
(c) The mass of Universe is estimated to be about 10^{80} m, where m is the mass of electron.
(d) All the quantum numbers of the Universe are zero, hence conserved.

60. In the Big–Bang model of Universe, the events which are imagined to have taken place when the temperature of Universe dropped to 10^{32} K, 10^{27} K and 10^{15} K are
(a) delinking of gravity from superforce; delinking of strong force from *GUT* force: and, nucleosynthesis respectively.
(b) delinking of strong force from *GUT* force; delinking of weak nuclear force from electroweak force; and, nucleosynthesis respectively.
(c) delinking of gravity from superforce; delinking of weak nuclear force from electroweak; and, delinking of radiation and matter respectively.
(d) delinking of gravity from superforce; delinking of strong force from *GUT* force; and, delinking of weak and electromagnetic forces from electroweak force respectively.

61. The cosmic microwave background radiation (CMBR) is
(a) the radiation emitted by the present day Universe.
(b) the radiation emitted by early Universe that has cooled to $3K$ temperature.
(c) the radiation emitted by stars and galaxies that has cooled to $3K$ temperature.
(d) the radiation emitted by the solar system.

62. The CMBR was discovered by
    (a) Mac Kellar in 1941
    (b) George Gamow in 1948.
    (c) Arno Penzias and Robert Wilson in 1965.
    (d) Shmaonov in 1983.
63. The discovery of CMBR tells that
    (a) initially the Universe was hot.
    (b) initially the Universe was cold.
    (c) Big–Bang is the origin of the present day Universe.
    (d) both (a) and (c) are true.
64. Supersymmetric theories unify
    (a) gravitational interaction with strong interaction
    (b) matter and interactions.
    (c) fermions and bosons.
    (d) both (b) and (c) above.
65. The transformations $Q|\text{fermion}\rangle = |\text{boson}\rangle$ and $Q^\dagger|\text{boson}\rangle = |\text{fermion}\rangle$ are
    (a) supersymmetric transformations in which Q and $Q^\dagger$ are quantum operators obeying a commutation relation.
    (b) supersymmetric transformations in which Q and $Q^\dagger$ are quantum operators obeying an anti commutation relation.
    (c) supersymmetric transformations in which Q and $Q^\dagger$ commute with the Hamiltonian operator.
    (d) supersymmetric transformations obeying (b) and (c) above..
66. If $Q = a^\dagger b\,(\hbar\omega)^{\frac{1}{2}}$, where a and b are respectively the bosonic and fermionic operators and $H = \hbar\omega\,(a^\dagger a + b^\dagger b)$, then
    (a) $\{Q, Q^\dagger\} = H,\ [Q, H] = 0$
    (b) $[Q, Q^\dagger] = H,\ \{Q, H\} = 0$
    (c) $\{Q, Q^\dagger\} = 0,\ [Q, H] = H$
    (d) $[Q, Q^\dagger] = 0,\ \{Q, H\} = H$
67. Supergravity
    (a) is a new symmetry unrelated to supersymmetry.
    (b) is local supersymmetry.
    (c) aims at the unification of gravitational with other interactions.
    (d) obeys both (b) and (c) above.

68. The mass generation in gauge theories is carried out by Higgs scalar field through the spontaneous symmetry breaking. The theories which can generate mass by considering Higgs particle as composite are

(a) supersymmetric theories (b) technicolour theories
(c) superstring theories (d) Euclidean gauge theories

69. In the question 68 above, the Higgs boson is considered to be a composite of

(a) techniquarks (b) technileptons
(c) preons (d) rishons

70. In recent composite models
(a) quarks are considered to be the composites of preons.
(b) leptons are considered to be the composites of preons.
(c) Higgs scalars are considered to be the composites of preons.
(d) quarks,leptons and Higgs scalars are considered to be the composites of preons.

ANSWERS

1. c, 2. a, 3. c, 4. c, 5. d, 6. d, 7. b, 8. a, 9. b, 10. d, 11. b, 12. c, 13. d, 14. a, 15. b, 16. c, 17. a, 18. b, 19. d, 20. d, 21. a, 22. b, 23. c, 24. a, 25. b, 26. c, 27. c, 28. d, 29. c, 30. d, 31. d, 32. c, 33. b, 34. a, 35. b, 36. d, 37. d, 38. b, 39. d, 40. d, 41. b, 42. d, 43. c, 44. d, 45. a, 46. c, 47. a, 48. a, 49. c, 50. d, 51. c, 52. b, 53. c, 54. a, 55. c, 56. b, 57. b, 58. d, 59. d, 60. d, 61. b, 62. c, 63. d, 64. d, 65. d, 66. a, 67. d, 68. b, 69. a, 70. d.

7

MANDELSTAM VARIABLES AND SYMMETRIES OF SCATTERING AMPLITUDE

7.1 INTRODUCTION

In chapter 1, while describing antiparticles and crossing symmetry, it was emphasized that if a particular reaction, say

$$a + b \rightarrow c + d \tag{7.1a}$$

is known to occur then according to the principle of crossing symmetry, the reactions like

$$a + \bar{c} \rightarrow \bar{b} + d \tag{7.1b}$$

$$a + \bar{d} \rightarrow c + \bar{b} \tag{7.1c}$$

would also hold. In the above reactions a, b, c, d represent particles and $\bar{a}, \bar{b}, \bar{c}, \bar{d}$ are their antiparticles. The four-momentum of antiparticle is opposite to that of a particle.

If a particle has a four-momentum $p^{\mu} = \left[(m^2 + \mathbf{p}^2)^{\frac{1}{2}}, \mathbf{p}\right]$, the four-momentum of its antiparticle would be

$-p^{\mu} = \left[-(m^2 + \mathbf{p}^2)^{\frac{1}{2}}, -\mathbf{p}\right]$. $(m^2 + \mathbf{p}^2)$ is the square of relativistic energy of particle of rest mass m in $\hbar = c = 1$ units, $\mu = 0, 1, 2, 3$.

Let us assume that the four momenta of particles a and b are p_1 and p_2 and those of c and d as p_3 and p_4, respectively. The first process would then be written as

$$\underset{p_1}{a} + \underset{p_2}{b} \rightarrow \underset{p_3}{c} + \underset{p_4}{d} \tag{7.2}$$

For brevity, we write it as

$$p_1 + p_2 \rightarrow p_3 + p_4 \tag{7.3a}$$

The process $a + \bar{c} \rightarrow \bar{b} + d$ would be expressed as

$$p_1 - p_3 \rightarrow -p_2 + p_4 \tag{7.3b}$$

and the process $a + \bar{d} \rightarrow c + \bar{b}$ would be

$$p_1 - p_4 \rightarrow p_3 - p_2 \tag{7.3c}$$

where we recall that antiparticles $\bar{c}$ and $\bar{d}$ have opposite (–ve) four momenta which are $-p_3$ and $-p_4$, respectively. In the process (7.3a) all four momenta are positive, whereas in (7.3b) p_2, p_3 and in (7.3c) p_2, p_4 are negative. For negative four momenta we have already mentioned the relation $-p^{\mu} = \left[-(m^2 + \boldsymbol{p}^2)^{\frac{1}{2}}, -\boldsymbol{p}\right]$, where zeroth component represents the relativistic energy and space components represent the three-momentum.

Focussing on the zeroth components, the signs (±) of four momenta imply that in the process (7.3a) $p_1^0, p_2^0, p_3^0, p_4^0$ all are positive (> 0), in (7.3b) both p_1^0 and p_4^0 are positive but both p_2^0 and p_3^0 are negative (< 0). Similarly, in the process (7.3c) both p_1^0 and p_3^0 are positive and p_2^0 and p_4^0 both are negative.

7.2 MANDELSTAM'S *s, t, u* VARIABLES

We now construct the Lorentz scalars from eqn. (7.3) as

$$s = (p_1 + p_2)^2 = (p_3 + p_4)^2 \tag{7.4a}$$

$$t = (p_1 - p_3)^2 = (p_2 - p_4)^2 \tag{7.4b}$$

$$u = (p_1 - p_4)^2 = (p_2 - p_3)^2 \tag{7.4c}$$

It can be verified that right-hand sides of these equations are Lorentz scalars. For example,

$$\begin{aligned}(p_1 - p_3)^2 &= p_1^2 - 2p_1p_3 + p_3^2 \\ &= p_1^{\mu} p_{1\mu} - 2p_{1\mu}p_3^{\mu} + p_3^{\mu} p_{3\mu}\end{aligned} \tag{7.4d}$$

Since each term on the RHS of the above equation is Lorentz invariant and hence a Lorentz scalar, making $(p_1 - p_3)^2$ a Lorentz scalar. For a two-body interaction, eqns. (7.4a-c) define the Mandelstam's *s, t, u* variables It has also to be noted that in eqns. (7.4a-c), the order of *s, t, u* cannot be interchanged. These variables obey the relation.

$$s + t + u = m_a^2 + m_b^2 + m_c^2 + m_d^2 \tag{7.5}$$

that can be proved as follows:

$$\begin{aligned}s + t + u &= (p_1 + p_2)^2 + (p_1 - p_3)^2 + (p_1 - p_4)^2 \\ &= p_1^2 + p_2^2 + p_3^2 + p_4^2 + (2p_1^2 + 2p_1p_2 - 2p_1p_3 - 2p_1p_4)\end{aligned} \tag{7.6}$$

Expressing the parenthesis of eqn. (7.6) as $2p_1\ (p_1 + p_2 - p_3 - p_4)$, we find it vanishes because of the energy momentum conservation law

$$p_1 + p_2 = p_3 + p_4 \tag{7.7}$$

Equation (7.6) then becomes

$$s + t + u = p_1^2 + p_2^2 + p_3^2 + p_4^2 \tag{7.8}$$

Using the definition that the squared length of a four vector (p_{μ} are four vectors) is the scalar product of the four vector with itself, we can show that

$$\begin{aligned}p_1^2 = p_{1\mu}p^{1\mu} &= p_{10}p^{10} + \mathrm{p}_{1i}p^{1i} \qquad (i = 1, 2, 3) \\ &= p_1^0 p_1^0 - p_1^i p_1^i \qquad \text{(metric is } + - - -) \\ &= p_1^{02} - \boldsymbol{p}_1^2 \\ &= m_a^2 + \boldsymbol{p}_1^2 - \boldsymbol{p}_1^2 = m_a^2\end{aligned} \tag{7.9}$$

where $p^{02} = (p^0)^2 = m^2c^4 + \boldsymbol{p}^2c^2$ is $m^2 + \boldsymbol{p}^2$ in natural units. Substituting eqn. (7.9) and $p_2^2 = m_b^2$, etc., in eqn. (7.8) the relation (7.5) is easily arrived at.

7.2.1 *s, t, u* for Equal Mass Case

If $m_a = m_b = m_c = m_d = m$, we have from eqn. (7.5)

$$s + t + u = 4m^2 \tag{7.10}$$

The processes described by eqns. (7.3) can now be defined in terms of Mandelstam variables *s, t, and u*. The process (7.3a) is shown in Fig. (7.1).

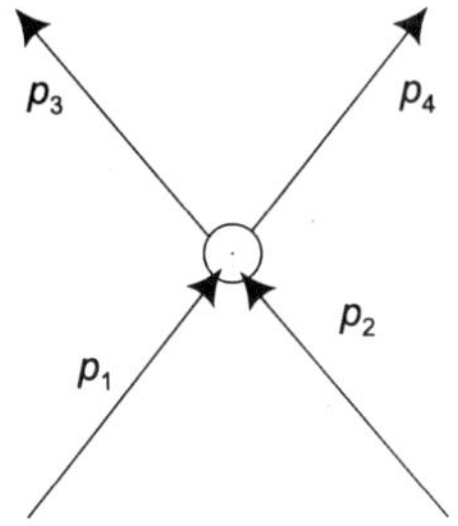

Fig. 7.1 *s*-channel process
$p_1 + p_2 \rightarrow p_3 + p_4$
$e^- + \pi^+ \rightarrow e^- + \pi^+$

The Feynman diagram for the *s*-channel process can be drawn as in Fig. (7.2).

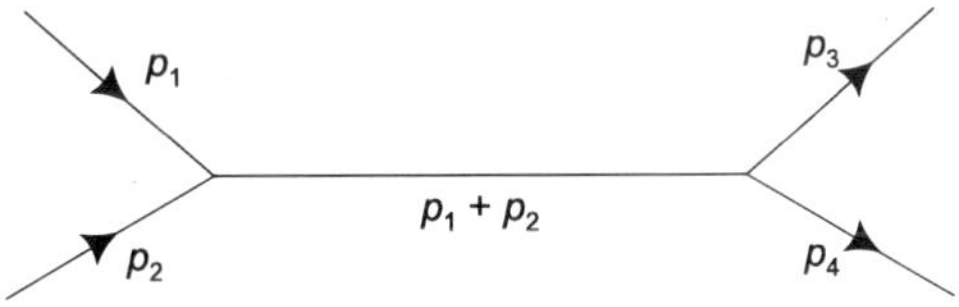

Fig 7.2 Feynman diagram for *s*-channel (space-channel) process.

In the *s*-channel process (eqn. 7.2), which is also called space-channel process, the particles *a* and *b* join together to form an intermediate particle which splits into particles *c* and *d*. The four momenta of the intermediate particle is $p_1 + p_2$. Electron positron scattering known as Bhabha scattering ($e^- + e^+ \rightarrow e^- + e^+$) is an example of *s*-channel process of equal mass particles. It may however, be observed that swapping second and third particles of the above scattering according to eqn. (7.1b), the same ($e^- + e^+ \rightarrow e^- + e^+$) scattering expression is reproduced. Since, (7.1b) represents a *t*-channel process, it is understood that Bhabha scattering would take place through both the *s*-channel and *t*-channel processes. Therefore, Bhabha scattering can be considered the example of both *s*- and *t*-channel processes. However, at the moment we are concerned with the *s*-channel.

In Fig. (7.1 or 7.2) p_1, p_2 are the four momenta of incoming particles and after the scattering process these particles go out with different four momenta p_3, p_4, respectively. Therefore, p_3, p_4 are the four momenta of outgoing particles. These four momenta are with respect to the laboratory frame of reference. Let us consider the above *s*-channel process in centre of mass (CM) system. As shown in Fig. (2.11), in the CM system.

$$\mathbf{p}_1 + \mathbf{p}_2 = 0 = \mathbf{p}_3 + \mathbf{p}_4 \tag{7.11}$$

When the mass of each of the particles a, b, c, d is m, from eqn. (7.9) and similar other equations we can write

$$p_1^{02} - \boldsymbol{p}_1^2 = p_2^{02} - \boldsymbol{p}_2^2 = p_3^{02} - \boldsymbol{p}_3^2 = p_4^{02} - \boldsymbol{p}_4^2 = m^2 \tag{7.12a}$$

From the equality of first and second terms of above equation, we can write

$$p_1^{02} = p_2^{02} + (\boldsymbol{p}_1^2 - \boldsymbol{p}_2^2) = p_2^{02} \tag{7.12b}$$

where the parenthesis has vanished due to CM equation (7.11). The equality of third and fourth terms of eqn. (7.12) would likewise lead to

$$p_3^{02} = p_4^{02} \tag{7.12c}$$

From the energy momentum conservation law (7.7), the energy conservation law is extracted as

$$p_1^0 + p_2^0 = p_3^0 + p_4^0 \tag{7.13}$$

Taking the positive roots of energy equations (7.12b, c) and combining them with eqn. (7.13), it follows that in CM system

$$p_1^0 = p_2^0 = p_3^0 = p_4^0 = \omega \tag{7.14}$$

where ω is the energy of any of the four particles in CM system.

In the light of the above, the Mandelstam variable s (eqn. 7.4) becomes

$$\begin{aligned} s = (p_1 + p_2)^2 &= (p_1 + p_2)_\mu (p_1 + p_2)^\mu = (p_{1\mu} + p_{2\mu})(p_1^\mu + p_2^\mu) \\ &= (p_{10} + p_{20})(p_1^0 + p_2^0) + (p_{1i} + p_{2i})(p_1^i + p_2^i) \\ &= (p_1^0 + p_2^0)^2 - (p_1^i + p_2^i)^2 \\ &= (2\omega)^2 - (\boldsymbol{p}_1 + \boldsymbol{p}_2)^2 \end{aligned} \tag{7.15}$$

which on using eqn. (7.11) simplifies to

$$s = (2\omega)^2 = 4\omega^2 \tag{7.16}$$

Since, the process (7.3a) has been considered in the centre of mass frame, eqn. (7.16) gives the square of total ($\omega + \omega$) energy in CM frame. Thus, the variable s is the square of the total energy in centre of mass frame. In other words $\sqrt{s}$ is the centre of mass energy and $\sqrt{s}/2$ is the energy of each equal mass particle in CM frame. The four momenta of the process (7.3a) would be

$p_\xi = \left(\frac{\sqrt{s}}{2}, \boldsymbol{p}_\xi\right), \xi = 1,2,3,4$. The three momenta $\boldsymbol{p}_\xi$ would obey the relation (7.11).

Similar description has been drawn in Fig. (2.11). In terms of the rest mass m of one of the particles whose three-momentum, i.e., $\boldsymbol{p}_\xi = \boldsymbol{k}$, with $|\boldsymbol{k}| = k$,

$\boldsymbol{k}^2 = k^2$, the relativistic energy equation in natural units gives

$$\left(\frac{\sqrt{s}}{2}\right)^2 = \boldsymbol{k}^2 + m^2 = k^2 + m^2 \tag{7.17}$$

implying that

$$2k = \sqrt{s - 4m^2} \tag{7.18a}$$

From eqn. (7.16)

$$2\omega = \sqrt{s} \tag{7.18b}$$

Since s, as defined in eqn. (7.4a) is a Lorentz invariant quantity and m the rest mass of a particle remains same in every inertial frame, dividing (7.18a) by (7.18b), a new Lorentz invariant quantity can be obtained as

$$\frac{k}{\omega} = \sqrt{\frac{s - 4m^2}{s}} \tag{7.19}$$

Since LHS of above eqn. is positive, to let remain the RHS positive, a condition that

$$s \geq 4m^2 \tag{7.20a}$$

has to be obeyed. This also means that

$$TCME \geq 2\,m \tag{7.20b}$$

where $TCME$ is the short for total centre of mass energy. The significance of s lies in the fact that it is the total centre of mass energy squared and it has same value (because s is Lorentz invariant) in every inertial frame of reference. Thus, instead of energy, the Lorentz invariant quantity s is used for energy as a kinetic variable. Utilizing s, results of many experiments can be easily combined without carrying out Lorentz transformation.

From the above, it can be understood that for the s-channel process (7.4a) to be physical, Mandelstam's variable s must obey the condition (7.20a). This means s must be positive and its threshold value is $4m^2$. What values the other two Mandelstam variables (t, u) would take in the s-channel process will be discussed below.

The variable t has been defined by eqn. (7.4b) that can now be expressed as

$$\begin{aligned} t = (p_1 - p_3)^2 &= (p_1 - p_3)_\mu (p_1 - p_3)^\mu = (p_{1\mu} - p_{3\mu})(p_1^\mu - p_3^\mu) \\ &= (p_1^0 - p_3^0)^2 - (\boldsymbol{p}_1 - \boldsymbol{p}_3)^2 \end{aligned} \tag{7.21}$$

Using eqn. (7.14), this reduces to

$$t = -(\boldsymbol{p}_1 - \boldsymbol{p}_3)^2 \tag{7.22}$$

The expression (7.22) tells that t is the negative square of the three-momentum transfer from paricle a to particle c, i.e., from particle 1 to particle 3.

Likewise, from eqn. (7.4c)

$$\begin{aligned} u = (p_1 - p_4)^2 &= (p_{1\mu} - p_{4\mu})(p_1^\mu - p_4^\mu) \\ &= (p_1^0 - p_4^0)^2 - (\boldsymbol{p}_1 - \boldsymbol{p}_4)^2 \end{aligned}$$

where too the first parenthesis vanishes due to eqn. (7.14), providing

$$u = -\,(\boldsymbol{p}_1 - \boldsymbol{p}_4)^2 \tag{7.23}$$

The expression (7.23) signifies that u is the negative square of the three-momentum transfer from particle a to particle d, i.e., from particle 1 to particle 4.

Thus for the s-channel process the t and u values are given by eqns. (7.22) and (7.23), respectively. Because of the presence of a square preceded by a minus sign, it is clear that the t and u values are negative definite (< 0). However, for the equal values of three momenta ($\boldsymbol{p}_3 = \boldsymbol{p}_1 = \boldsymbol{p}_4$) both the variables would be zero.

Therefore, it is concluded that for the s-channel process (7.4a) to be realizable, the Mandelstam variables should obey the following conditions.

$$s \geq 4m^2, \quad t \leq 0, \quad u \leq 0 \tag{7.24}$$

7.2.1.1 Range of s, t, u Variables in s-Channel Process

Equation (7.24) grossly defines the values of s, t, u, variables for the s-channel process. For the variable s, the range of its values (the physical domain for s-channel process) is obtained from the eqn. (7.18a) which on squaring gives

$$s = 4(k^2 + m^2) \tag{7.25}$$

It has to be noted that the k, the three-momentum in centre of mass frame, can have values from zero to infinity. The minimum or the threshold value corresponds to $k = 0$ and when the incident particle carries a nonzero three-momentum and so does the scattered particle, s would be higher than the threshold value. Thus for a given rest mass of one of the four particles participating in the s-channel process, the magnitude of s would be determined by the square of the magnitude of the three-momentum of the particle. s values may thus range from $4m^2$ to ∞

For the range of t and u, we consider the scattering process through the scattering angle drawn in Fig. (7.3).

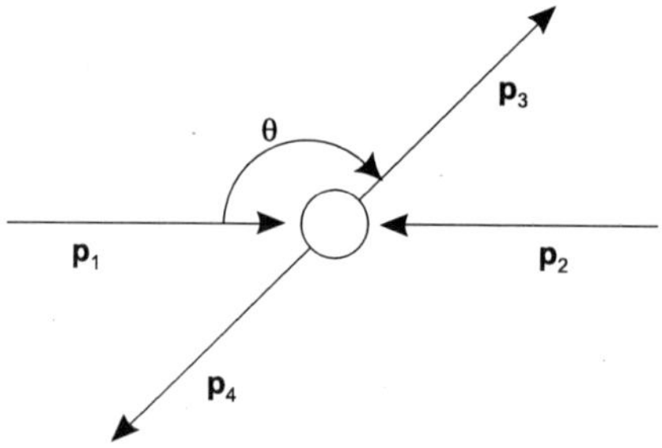

Fig. 7.3 Scattering angle in centre of mass frame.

From eqn. (7.22)

$$t = -(p_1 - p_3)^2 = -(p_1^2 + p_3^2 - 2p_1 p_3) \tag{7.26}$$

Now $$|p_1| = k = |p_3| \tag{7.27a}$$

and hence from the figure

$$p_1 \cdot p_3 = k^2 \cos\theta \tag{7.27b}$$

Substitution of eqns. (7.27) into (7.26) gives

$$t = -2k^2(1 - \cos\theta) \tag{7.28}$$

Using eqns. (7.28) and (7.25), the s-channel scattering angle (θ) can also be expressed, in terms of t, as

$$\cos\theta_s = 1 + \frac{2t}{s - 4m^2} \tag{7.28a}$$

where the scattering angle θ has been replaced by θ_s, suffix s denotes the channel (s-channel) to which the scattering angle refers.

The scattering angle θ can vary from 0° to 180° and hence $(1 - \cos\theta)$ varies from 0 to 2. This puts t between the lowest of $-4k^2$ to the highest of 0. Therefore, the range of variable t in the s-channel process is set as

$$-4k^2 \le t \le 0 \tag{7.29}$$

where, k can have values from 0 to ∞. For the Mandelstam variable u, its relation (7.23) is

$$u = -(\boldsymbol{p}_1 - \boldsymbol{p}_4)^2 = -(\boldsymbol{p}_1^2 + \boldsymbol{p}_4^2 - 2\boldsymbol{p}_1 \cdot \boldsymbol{p}_4) \tag{7.30}$$

Here

$$|\boldsymbol{p}_1| = k = |\boldsymbol{p}_4| \tag{7.31a}$$

and

$$\boldsymbol{p}_1 \cdot \boldsymbol{p}_4 = k^2 \cos(180 - \theta) = -k^2 \cos\theta \tag{7.31b}$$

These eqns. lead (7.30) to

$$u = -2k^2 (1 + \cos\theta) \tag{7.32}$$

Similar to eqn. (7.28a), in terms of u, the scattering angle of s-channel can also be expressed as

$$\cos\theta_s = -\frac{2u}{s - 4m^2} - 1 \tag{7.32a}$$

It may be mentioned here that eqn. (7.32a) for u, can also be obtained by using

$$u = 4m^2 - s - t \tag{7.32b}$$

where s and t are expressed by eqns. (7.25) and (7.28), respectively. Because of the constraint (7.10), only two of the three Mandelstam variables are independent.

For the range of scattering angle θ from 0° to 180°, $(1 + \cos\theta)$ varies from 2 to 0. Therefore, the variable u too has the lowest value of $-4k^2$ and a highest value of 0. Therefore in an s-channel process for the range of u, we have

$$-4k^2 \le u \le 0 \tag{7.32c}$$

Due to the allowed values of k, u can have values $-\infty \to 0$. Thus, eqns. (7.25), (7.29) and (7.32c) determine the range of values of s, t, u variables. Putting together, the ranges of Mandelstam variables s, t and u in the s-channel process and the scattering angle are

$$s = 4m^2 \to \infty;\ -4k^2 \le t \le 0 \text{ and } -4k^2 \le u \le 0 \tag{7.33a}$$

$$\cos\theta_s = 1 + \frac{2t}{s - 4m^2} = -\left(1 + \frac{2u}{s - 4m^2}\right) \tag{7.33b}$$

7.2.1.2 *t and u Channel Processes*

The process (7.2) in which centre of mass energy squared is given by s is called the s-channel process. The s, t, u of this process have been defined by eqn. (7.4). The s-channel process is called the basic process and its set of (s, t, u) (eqn. 7.4) is called the basic set or the defining set of Mandelstam variables. For the s-channel process the variables s, t, u obey the constraint (7.24).

Involving the same four particles (a, b, c, d) and using the principle of crossing symmetry, other processes (eqns. 7.3b & c) are observed. These processes have their own set of s, t, u variables, where s continues to express the centre of mass energy squared. To determine the channel of the process, the s, t, u set of the process is obtained and the obtained s, t, u set is compared with the defining set of s, t, u (7.4). The obtained s is either equal to t or to u of the defining set. When the obtained s is equal to t of the defining set, the process is called the t channel process, if obtained s is equal to u of the defining set, the process is u-channel process. Explicitly, the process (7.3b) can be expressed as

$$a_{p_1} + c_{-p_3} \to b_{-p_2} + d_{p_4} \tag{7.34}$$

Let us denote the s, t, u of this process as s_2, t_2, u_2, respectively where, suffix 2 indicates that (7.3b) is the second process, first is (7.2). We compare (7.34) with (7.2) which is the first or the basic or the s-channel process. We rewrite (7.2)

$$a_{p_1} + b_{p_2} \to c_{p_3} + d_{p_4} \tag{7.2}$$

The comparison shows that the four momenta p_2 & p_3 of the basic process (7.2) are replaced by $-p_3$ & $-p_2$, respectively in the process (7.34). These changes are incorporated in the basic set of s, t, u given by eqn. (7.4) to obtain the Mandelstam variables for the process (7.34) as

$$s_2 = (p_1 - p_3)^2 \tag{7.35a}$$
$$t_2 = (p_1 + p_2)^2 \tag{7.35b}$$
$$u_2 = (p_1 - p_4)^2 \tag{7.35c}$$

Comparing the set (s_2, t_2, u_2) with the basic or defining set (eqn. 7.4), we find

$$s_2 = t, \quad t_2 = s, \quad u_2 = u \tag{7.36}$$

The process (7.34) would become a physical, i.e., realizable process, if (s_2, t_2, u_2) fulfil the requirement (7.24) that means

$$s_2 \geq 4m^2, \quad t_2 \leq 0, \quad u_2 \leq 0 \tag{7.37}$$

The condition (7.37) is however, expressed in terms of (s, t, u) of the basic (s-channel) process. Therefore, using eqn. (7.36) in eqn. (7.37), the condition of making the process (7.34) physical is expressed as

$$t \geq 4m^2, \quad s \leq 0, \quad u \leq 0 \tag{7.38}$$

Since $4m^2$ represents the centre of mass energy squared, which for the process (7.34) is expressed, according to eqn. (7.38), by t, the process (7.34) is called the t-channel process.

Comparing (7.38) with (7.24), it can be observed that the expression of t-channel process (7.38) can be directly obtained from the s-channel expression (7.24) by merely swapping s and t variables. Therefore, interchanging s and t variables, the t-channel relations corresponding to (7.25), (7.28) and (7.32) are

$$t = 4(k^2 + m^2) \tag{7.38a}$$
$$s = -2k^2 (1 - \cos\theta) \tag{7.38b}$$
$$u = -2k^2 (1 + \cos\theta) \tag{7.38c}$$

where the angle θ now is the scattering angle θ_t of t-channel scattering. Either from above eqns. (7.38 a, b, c) or by just swapping s and t in (7.33b), the t-channel scattering angle (θ_t) can be expressed as

$$\cos\theta_t = 1 + \frac{2s}{t - 4m^2} = -\left(1 + \frac{2u}{t - 4m^2}\right) \tag{7.38d}$$

Similar to (7.33a), the range or the physical domain of t-channel process can be ascertained from eqns. (7.38a,b,c) or can be written directly from (7.33a) by swapping s and t. The physical domain of t-channel process is therefore

$$t = 4m^2 \to \infty;\ -4k^2 < s \le 0 \text{ and } -4k^2 \le u \le 0 \tag{7.38e}$$

The t-channel process can be shown by Fig. (7.4).

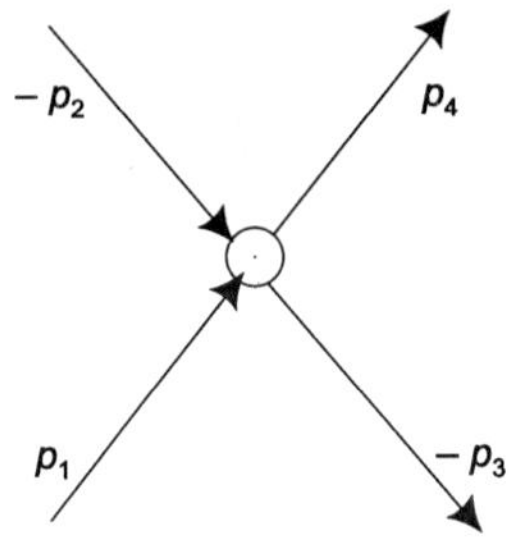

Fig. 7.4 t-channel process

$p_1 - p_3 = -p_2 + p_4$

$e^- + e^+ \to \pi^- + \pi^+$

The Feynman diagram for the t-channel process has been drawn in Fig. (7.5).

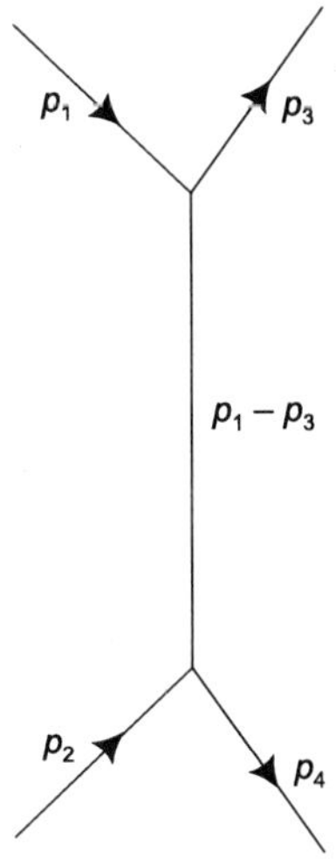

Fig. 7.5 Feynman diagram for t-channel (time-channel) process.

The Feynman diagram for t-channel represents the process in which particle a (p_1) emits the intermediate particle and becomes the final particle c (p_3), while the particle b (p_2) absorbs the intermediate particle and becomes the other final particle, d (p_4).

As described in the s-channel process, Bhabha scattering is realized through a t-channel process as well. So in the equal mass case, electron-positron scattering is an example for t-channel process. Electron-electron (Moller) scattering ($e^- + e^- \to e^- + e^-$) is another scattering process that involves the particles of same mass. If this (Moller scattering) is taken as the s-channel process its t-channel process would yield the Bhabha scattering. The u-channel process of Moller scattering too gives Bhabha scattering. Therefore, Bhabha scattering is the example of both t and u channel processes of Moller scattering.

Now, we talk about the third or the u-channel process (7.3c). Explicitly, this process is

$$a_{p_1} + d_{-p_4} \to c_{p_3} + b_{-p_2} \tag{7.39}$$

Through the comparison of (7.39) with the s-channel process (7.2)

$$a_{p_1} + b_{p_2} \to c_{p_3} + d_{p_4} \tag{7.40}$$

it can be found that replacing p_2 & p_4 by $-p_4$ & $-p_2$, respectively, in the basic s, t, u set (7.4), we would get the set of s, t, u for the process (7.39), the latter is denoted by (s_3, t_3, u_3). These replacements lead to

$$s_3 = (p_1 - p_4)^2 \tag{7.40a}$$

$$t_3 = (p_1 - p_3)^2 \tag{7.40b}$$

$$u_3 = (p_1 + p_2)^2 \tag{7.40c}$$

Comparing the variables (7.40) with the basic set (7.4), we get

$$s_3 = u, \quad t_3 = t, \quad u_3 = s \tag{7.41}$$

Keeping the condition (7.24) in view, the process (7.39) would be physically realizable, provided the variables obey (s_3, t_3, u_3) the following condition

$$s_3 \geq 4m^2, \quad t_3 \leq 0, \quad u_3 \leq 0 \tag{7.42}$$

From the comparison of eqns. (7.42) and (7.41), the condition, that the process (7.39) will be physical, emerges as

$$u \geq 4m^2, \quad t \leq 0, \quad s \leq 0 \tag{7.43}$$

Since, in the above equation the square of the centre of mass energy is expressed by the Mandelstam variable u, the process (7.39) is called the u-channel of the s-channel process (7.2).

For this channel too, the relations similar to (7.38 a, b, and c, d) can be obtained by comparing (7.43) with (7.24). This comparison notices a $s \leftrightarrow u$ swapping in the s-channel expression (7.24) and the u-channel expression (7.43). We therefore replace s by u and u by s in eqns. (7.25), (7.28) and (7.32) to obtain the corresponding u-channel expressions as

$$u = 4\,(k^2 + m^2) \tag{7.43a}$$

$$t = -\,2k^2\,(1 - \cos\theta) \tag{7.43b}$$

$$s = -\,2k^2\,(1 + \cos\theta) \tag{7.43c}$$

where the angle θ now is the scattering angle θ_u of u-channel scattering. Either from above eqns. (7.43a, b, c) or by just swapping s and u in (7.33b), the u-channel scattering angle θ_u can be expressed as

$$\cos\theta_u = 1 + \frac{2t}{u-4m^2} = -\left(1+\frac{2s}{u-4m^2}\right) \tag{7.43d}$$

The physical domain of u-channel process, like those of s and t-channel processes expressed by (7.33a) and (7.38e), respectively, can be described by

$$u = 4m^2 \to \infty; \quad -4k^2 \leq t \leq 0 \text{ and } -4k^2 \leq s \leq 0 \tag{7.43e}$$

Since, the size of physical domain is k dependent, and k can vary from $0 \to \infty$, the domain size of the process would change accordingly.

The general outline of the u-channel process can be drawn as in Fig. (7.6) while its Feynman diagram by Fig. (7.7)

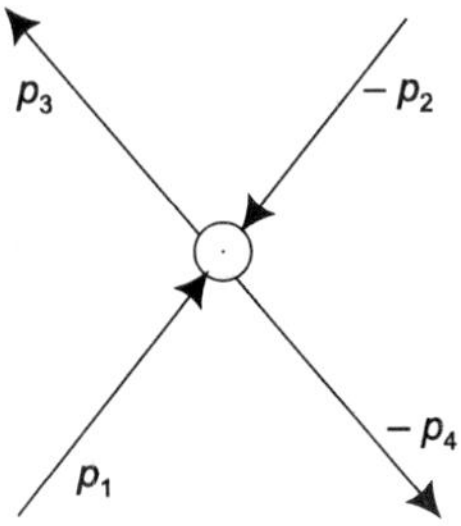

Fig. 7.6 u-channel process

$$p_1 - p_4 \to p_3 - p_2$$
$$e^- + \pi^- \to e^- + \pi^-$$

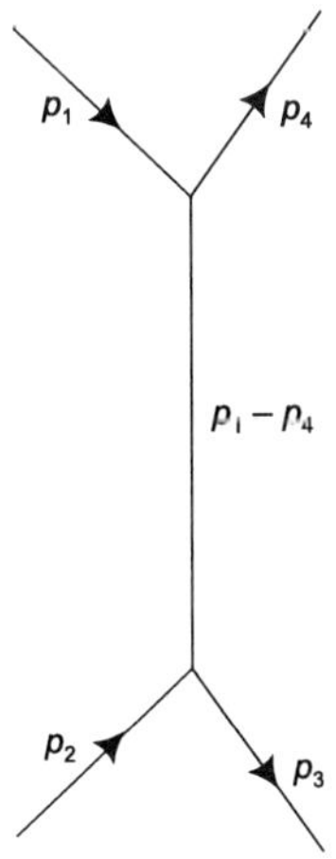

Fig. 7.7 Feynman diagram for u-channel process.

The u-channel process is basically the t-channel process in which roles of c (third particle) and d (fourth particle) with four momenta (p_3 & p_4), respectively are interchanged. The Feynman diagram for u-channel represents the process in which particle a (p_1) emits the intermediate particle and becomes the final particle d (p_4), while the particle b (p_2) absorbs the intermediate particle and becomes the other final particle c (p_3).

As an example of the u-channel process where particles of equal mass are involved is the Bhabha scattering obtained from the Moller scattering, the latter taken as s-channel process. The Moller scattering is ($e^- + e^- \rightarrow e^- + e^-$). Swapping second and fourth electrons according to the rules of crossing symmetry, we get the u-channel process ($e^- + e^+ \rightarrow e^- + e^+$), the Bhabha scattering.

7.2.2 *s, t, u* for Unequal Mass Case

In the previous section, the s, t, u variables of the processes involving particles of equal mass were discussed. However, there are many processes in which masses of participating particles are unequal. For, example, electron positron annihilation

$$e^- + e^+ \rightarrow \gamma + \gamma \quad s\text{-channel} \tag{7.44a}$$

$$e^- + \gamma \rightarrow e^- + \gamma \quad t\text{-channel} \tag{7.44b}$$

$$e^- + \gamma \rightarrow \gamma + e^- \quad u\text{-channel} \tag{7.44c}$$

and electron pion scattering

$$e^- + \pi^+ \rightarrow e^- + \pi^+ \quad s\text{-channel} \tag{7.45a}$$

$$e^- + e^+ \rightarrow \pi^- + \pi^+ \quad t\text{-channel} \tag{7.45b}$$

$$e^- + \pi^- \rightarrow e^- + \pi^- \quad u\text{-channel} \tag{7.45c}$$

are the processes in which particles of unequal masses are participating. In general, as in the above processes, out of the four, two particles have identical masses and remaining two, too have identical but different masses.

In this case, if we consider the s-channel process $a + b \rightarrow c + d$ and assume, in view of the example (7.45), $m_a = m_c = m$ and $m_b = m_d = M$, where $M > m$, we shall have

$$s + t + u = m_a^2 + m_b^2 + m_c^2 + m_d^2 = 2(m^2 + M^2) \tag{7.46}$$

When the particles had equal mass, the centre of mass energy of each particle according to eqn. (7.14) was ω. Now, instead of eqn. (7.12a), we shall have

$$p_1^{02} = \mathbf{p}_1^2 + m_a^2, \quad p_2^{02} = \mathbf{p}_2^2 + m_b^2 \tag{7.47a}$$

$$p_3^{02} = \mathbf{p}_3^2 + m_c^2, \quad p_4^{02} = \mathbf{p}_4^2 + m_d^2 \tag{7.47b}$$

and, because of unequal masses (in CM $\mathbf{p}_1^2 = \mathbf{p}_2^2$) of initial state particles a and b, their centre of mass energies would be $p_1^0 = \omega_1$ & $p_2^0 = \omega_2$, respectively. From the definition (7.4a) for s we can calculate ω_1 and ω_2 as written below

From (7.4a)

$$\sqrt{s} = (p_1 + p_2) \tag{7.48}$$

Bringing p_1 to the LHS and then squaring, we get

$$s - 2\sqrt{s}\ p_1 + p_1^2 = p_2^2 \tag{7.49}$$

Now, $p_1^2 = (m_a c)^2$, $p_2^2 = (m_b c)^2$ and $p_1 = (\omega_1/c)$, which when substituted in eqn. (7.49) in natural units $\hbar = c = 1$ yield

$$\omega_1 = \frac{s + m_a^2 - m_b^2}{2\sqrt{s}} \tag{7.50}$$

Writing eqn. (7.48) as $\left(\sqrt{s}-p_2\right)^2 = p_1^2$ and putting $p_2 = (\omega_2/c) = \omega_2$, we likewise obtain

$$\omega_2 = \frac{s + m_b^2 - m_a^2}{2\sqrt{s}} \tag{7.51}$$

It is easy to observe that for $m_a = m_b$, both the relations (7.50) and (7.51) revert to the equal mass relation (7.16).

Adding the two relations for ω_1 and ω_2, we find that

$$s = (\omega_1 + \omega_2)^2 \tag{7.52}$$

The RHS of eqn. (7.52) gives the square of the total centre of mass energy of particles a and b in CM frame. Thus in the case of unequal masses as well, s represents the square of total centre of mass energy. When $\omega_1 = \omega_2 = \omega$, eqn. (7.52) becomes, as expected, $s = 4\omega^2$, the equation (7.16) of equal mass case.

Let us now determine, for the unequal mass case, the three momenta of particles a and b in centre of mass frame. Let the modulus of three-momentum of particle a be k_1 and that of particle b be k_2. This means $|\boldsymbol{p}_1| = k_1$ & $|\boldsymbol{p}_2| = k_2$. In the centre of mass frame, $\boldsymbol{p}_1 = -\boldsymbol{p}_2$ but $\boldsymbol{p}_1 \cdot \boldsymbol{p}_1 = \boldsymbol{p}_2 \cdot \boldsymbol{p}_2$, implying that $k_1^2 = k_2^2$. Using the relativistic energy relations

$$k_1^2 = \omega_1^2 - m_a^2 \tag{7.53a}$$

$$k_2^2 = \omega_2^2 - m_b^2 \tag{7.53b}$$

we can calculate k_1 and k_2 and see that they obey the CM condition. Substituting ω_1 from eqn. (7.50) in (7.53a), we get

$$4sk_1^2 = s^2 - 2s\,(m_a^2 + m_b^2) + (m_a^2 - m_b^2)^2 \tag{7.54}$$

Adding and subtracting $2sm_am_b$ and splitting the second term as

$-2s\,(m_a^2 + m_b^2) = -s\,(m_a^2 + m_b^2) - s\,(m_a^2 + m_b^2)$, eqn. (7.54) can be written as

$4sk_1^2 = s^2 - s\,(m_a - m_b)^2 - s\,(m_a + m_b)^2 + (m_a - m_b)^2\,(m_a + m_b)^2$

Factorizing the RHS , the square of three-momentum is obtained as

$$k_1^2 = \frac{1}{4s}\left[s - (m_a - m_b)^2\right]\left[s - (m_a + m_b)^2\right] \tag{7.55}$$

For calculating k_2^2, it can be noted from eqns. (7.50, 7.51 and 7.53), that interchanging the masses m_a and m_b in (7.55), we would get k_2^2. However, because $(m_a - m_b)^2 = (m_b - m_a)^2$, the interchange does not affect the value. So,

$$k_2^2 = k_1^2 \tag{7.56}$$

as expected in the centre of mass system where $\boldsymbol{p}_2 = -\boldsymbol{p}_1$. This implies that, in the centre of mass frame, although the squares of three momenta (eqn. 7.56) are equal, their directions would be opposite. It is easy to observe that when $m_a = m_b = m$, eqn. (7.55) reverts to eqn. (7.25) of equal mass case. This also implies that eqn. (7.55), along with (7.56) is the unequal mass analogue of $k^2 = \dfrac{s}{4} - m^2$ of equal mass case.

In the above discussion of unequal mass case, we considered the s-channel process and obtained the corresponding energy(s) and three-momentum (k).

For the t and u variables of the above s-channel process, it has to noted that t involves the first and third particles whose masses have been assumed to be same, that is, $m_a = m_c = m$. The variable u involves the first and fourth particles whose masses are not same. Thus, for the variable t, the expression in this unequal mass case too would be similar to equal mass case where it has been expressed by eqns. (7.22) and (7.28). However, for the Mandelstam variable u, in the above-mentioned unequal mass s-channel case, from eqn. (7.46), we shall have

$$u = 2m^2 + 2M^2 - s - t \tag{7.57}$$

where t, as described above is

$$t = -2k^2 (1 - \cos\theta) \tag{7.58}$$

s is the positive definite root of the quadratic equation (7.57) which is written as

$$s^2 - 2s(2k^2 + m^2 + M^2) + (M^2 - m^2)^2 = 0 \tag{7.59}$$

in which $k^2 = k_1^2 = k_2^2$. Accordingly, from (7.59),

$$s = 2k^2 + m^2 + M^2 + 2\sqrt{(k^2 + m^2)(k^2 + M^2)} \tag{7.60}$$

Therefore, for the unequal mass case s-channel process like (7.45a), the Mandelstam variables s, t, u would be given by their expressions (7.60), (7.58) and (7.57), respectively where k^2 from eqn. (7.55) is

$$k^2 = \frac{1}{4s}\left[s - (M - m)^2\right]\left[s - (M + m)^2\right] \tag{7.61}$$

7.2.2.1 *Range of s, t, and u Variables in Unequal Mass s-channel Process*

Since, k, the three-momentum in centre of mass frame can have values from zero to infinity, the $k = 0$, would give the minimum or the threshold value of the Mandelstam variables (7.60), (7.58) and (7.57). This sets the following range for s, t, u in the unequal mass case when the process is through the s-channel.

$$s \geq (M + m)^2, \qquad t \leq 0, \qquad u \leq (M - m)^2 \tag{7.62}$$

where putting $k = 0$ in (7.60), the threshold value of $(M + m)^2$ is easily obtained for the variable s. For $k > 0$, s has to have a value greater than $(M + m)^2$. For the variable t, because of equal mass of the first and third particles in the process like (7.45a), the equation (7.61) is same as (7.28) and the consequent eqn. (7.29) provides $t \leq 0$. For u, putting $s = (M + m)^2$ and $t = 0$ in (7.57), we get $u = (M - m)^2$ which is the $k = 0$ value of u and is positive. u becomes zero when s increases to $2M^2 + 2m^2$ and for its values larger than this, u becomes negative. Unlike the equal mass case, where u had only negative values, in the unequal mass case, the physical region of the s-channel spreads into positive u region as well. However, most of the physical region for s-channel lies in the $u \leq 0$ region, the condition (7.62) is expressed as

$$s \geq (M + m)^2, \qquad t \leq 0, \qquad u \leq 0 \tag{7.63}$$

Now, we take the t-channel process of unequal mass case. Equation (7.45b) is an example of this process. As in equal mass case (eqns. 7.35 & 7.36), the Mandelstam variables of this process are

$$s_2 = (p_1 - p_3)^2 \qquad t_2 = (p_1 + p_2)^2, \quad u_2 = (p_1 - p_4)^2 \tag{7.64}$$

with

$$s_2 = t, \qquad t_2 = s, \qquad u_2 = u \tag{7.65}$$

The range of Mandelstam variables for this channel would be determined by (7.64), where it can be noticed that the expression for s_2 involves paricles of same mass m, so from $s \geq 4m^2$ of equal mass case, $s_2 \geq 4m^2$. However, treating the first and third particles as heavier ones, each having mass M, $s_2 \geq 4M^2$. For the (pure) t-channel the acceptable requirement becomes $s_2 \geq 4M^2$ and from eqn. (7.63), $t_2 \leq 0$, $u_2 \leq 0$, which when equated with (7.65), provide the following s, t, u for the unequal mass t-channel process.

$$s \leq 0, \qquad t \geq 4M^2, \qquad u \leq 0 \tag{7.66}$$

For the u-channel process of unequal mass case, we have from (7.40)

$$s_3 = (p_1 - p_4)^2, \qquad t_3 = (p_1 - p_3)^2, \qquad u_3 = (p_1 + p_2)^2 \tag{7.67}$$

which, in view of s-channel or the basic process (7.63) must obey

$$s_3 \geq (M + m)^2, \qquad t_3 \leq 0, \qquad u_3 \leq 0 \tag{7.68}$$

From (7.41), we observe that

$$s_3 = u, \qquad t_3 = t, \qquad u_3 = s \tag{7.69}$$

Comparision of eqns. (7.68) and (7.69) provides the following s, t, u for the u-channel process in unequal mass case.

$$s \leq 0, \qquad t \leq 0, \qquad u \geq (M + m)^2 \tag{7.70}$$

Putting together the results we have the following s, t, u for three channels in unequal mass case.

$$\begin{array}{llll} s \geq (M+m)^2, & t \leq 0, & u \leq 0 & s\text{-channel} \\ s \leq 0, & t \geq 4M^2, & u \leq 0 & t\text{-channel} \\ s \leq 0, & t \leq 0, & u \geq (M+m)^2 & u\text{-channel} \end{array} \tag{7.71}$$

where m is the mass of first and third particle and M that of second and fourth.

7.3 VISUALIZATION OF MANDELSTAM VARIABLES

7.3.1 Equal Mass Case

The visualization of s, t, u variables can best be undertaken through the s, t, u plot in a triangular coordinate system in which three axes, s-axis, t-axis and u-axis are at 120° from each other as drawn in Fig. (7.8a), where it is recalled that s-axis is a direction perpendicular to $s = 0$ line and similar definitions for t and u axes.

The visualization of s, t, u variables implies that looking at the diagram, i.e, the Mandelstam plot, the conditions for the occurrence of s-channel, t-channel and u-channel processes are easily read. In the previous sections, it has been realized that for the case in which each of the four participating particles in a scattering process has same mass m, following conditions are obeyed for the three channels.

$$s \geq 4m^2, \quad t \leq 0, \quad u \leq 0 \qquad s\text{-channel} \tag{7.72a}$$

$$s \leq 0, \quad t \geq 4m^2, \quad u \leq 0 \qquad t\text{-channel} \tag{7.72b}$$

$$s \leq 0, \quad t \leq 0, \quad u \geq 4m^2 \qquad u\text{-channel} \tag{7.72c}$$

along with the constraint

$$s + t + u = 4m^2 \tag{7.73}$$

The respective regions in the s, t, u plot where above conditions are obeyed are called the physically allowed regions for holding above processes. The shaded portions of Fig. (7.8b) display the three physical regions. The vertices of the equlilateral trangle ABC have coordinates (s, t, u) as $A \to (4m^2, 0, 0)$, $B \to (0, 4m^2, 0)$ and $C \to (0, 0, 4m^2)$. Any point inside the triangle, including the vertices satisfies the relation (7.73).

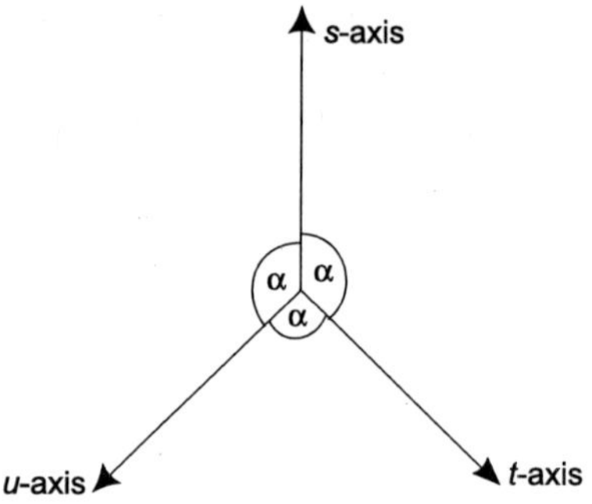

Fig. 7.8(a) Triangular coordinate frame of s, t and u axes, $\alpha = 120°$.

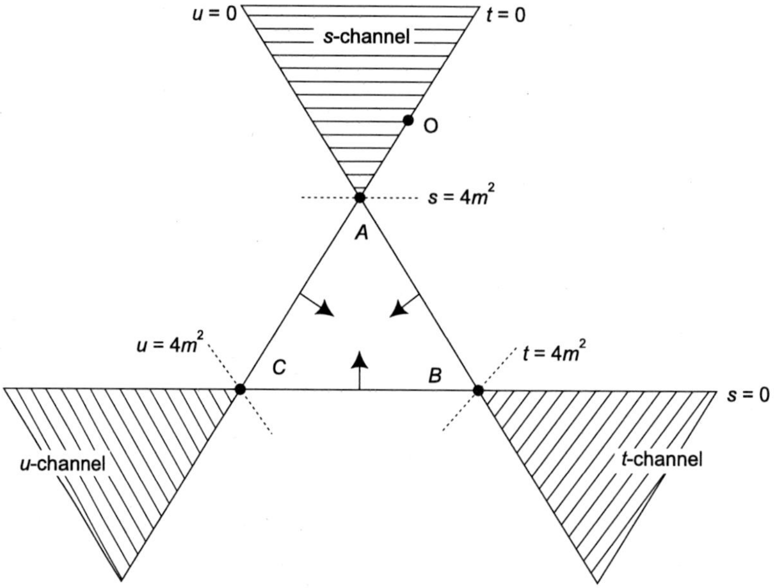

Fig. 7.8(b) Physical regions (shaded) of s, t and u channels in equal mass case. Arrows denote axes directions.

Let us discuss the s-channel region in Fig. (7.8b). The region indicates that for a reaction to take place in the s-channel according to (7.72a), minimum value of square of centre of mass energy of two participating particles must be $4m^2$. This is indicated by the point A in the s-channel region. At A, both t and u are zero. t indicates the square of three-momentum transferred from particle a to particle c ($\boldsymbol{p}_1 \rightarrow \boldsymbol{p}_3$) and u denotes the square of three-momentum transferred from particle a to particle d ($\boldsymbol{p}_1 \rightarrow \boldsymbol{p}_4$), $t = 0$, $u = 0$ obviously indicate that no momentum would be transferred to particles c and d when the centre of mass energy squared (s) of particles a and b is at its threshold of $4m^2$. The variables t and u also tell about the angles between $\boldsymbol{p}_1, \boldsymbol{p}_3$ and $\boldsymbol{p}_1, \boldsymbol{p}_4$, respectively. From eqns. (7.28) & (7.32), we can write

$$\cos\theta = 1 + \frac{t}{2k^2} \tag{7.74a}$$

as well as

$$\cos\theta = -\left(1 + \frac{u}{2k^2}\right) \tag{7.74b}$$

when $t \rightarrow 0$, $\theta \rightarrow 0°$ and for $u \rightarrow 0$, $\theta \rightarrow 180°$, implying that at the minimum of CM energy squared, the angles between $\boldsymbol{p}_1, \boldsymbol{p}_3$ would be zero and that between $\boldsymbol{p}_1, \boldsymbol{p}_4$ would be 180°. Obviously, for nonzero negative u, $\theta < 180°$.

In Fig. (7.8b), u is (–ve) on the right of $u = 0$ line and t is (–ve) towards the left of $t = 0$ line. Let us consider the point O in the s-channel region. At O, s is greater than $4m^2$ by an amount corresponding to perpendicular distance between A and O. At O, $t = 0$ but u is quite negative. Thus, O point in s channel describes that when centre of mass energy squared is greater than $4m^2$, an increased three-momentum is transferred to particle $d(\boldsymbol{p}_4)$ by particle $a(\boldsymbol{p}_1)$. If the point of consideration is along the $u = 0$ line, no momentum would be transferred to d, but significant to particle $c(\boldsymbol{p}_3)$. If the point of consideration is between the $u = 0$ and $t = 0$ lines, the momenta would be transferred to both c and d.

Similarly, in the physical region for the t-channel process at point B, $t = 4m^2$ that increases towards its right into the shaded region. This region obeys the restriction (7.72b) and electron positron scattering is an example.

In the u-channel at point C, $u = 4m^2$.This (u) increases towards its left into the shaded region of u-channel. Therefore, in this region that obeys the restrictions (7.72c), u-channel process like electron-electron scattering takes place.

It may be understood that when, e.g., conditions (7.72a) are obeyed, only a process through s-channel would take place because the values (7.72a) belong to the physical region of s-channel. Conversely, the values (7.72a) belong to the unphysical regions of t- and u-channels and hence in this case no t-channel or u-channel process would take place.

It is also evident from the figure that the physical region of s-channel is bounded by $u = 0$ and $t = 0$ lines; the physical region of t-channel is bounded by $s = 0$ and $u = 0$ lines and the physical region of u-channel is bounded by $s = 0$ and $t = 0$ lines. All these triangular physical regions are disconnected and the boundary of this disconnected region is expressed as the cubic equation

$$s\,t\,u = 0 \tag{7.75}$$

whose three obvious solutions are (1) $s \neq 0, t = 0, u = 0$ (2) $s = 0, t \neq 0, u = 0$ and (3) $s = 0, t = 0, u \neq 0$, which represent the boundaries of the physical regions of s-, t- and u-channel processes, respectively. The nonzero value of these solutions is $4m^2$.

7.3.2 Unequal Mass Case

Figure (7.8b) represents the physical regions for three channels when the mass of each particle in the two-body process $a + b \to c + d$ is same. In the case of unequal masses, however, the physical regions of three channels are not that simple. As seen in the previous sections, in general, $m_a = m_c = m$ and $m_b = m_d = M$, where $M > m$. In this case

$$s + t + u = 2(m^2 + M^2) \tag{7.76}$$

The Mendelstam variables for s, t and u channel processes for this case have been expressed by eqn. (7.71), where we can observe that the variable t depends on M only, whereas s depends on both m and M. The constraint (7.76) for $t = 0$ gives $s + u = 2(m^2 + M^2)$ from which we have $s = (M - m)^2$ if $u = (M + m)^2$ and conversely $u = (M - m)^2$ if $s = (M + m)^2$. As mentioned earlier, only two variables are independent, one of them is the variable t. For a given t, the s and u values as seen above are interralated. These values of s and u play an important role in determining the boundaries of physical regions of s, t, u channel processes in unequal mass case. The boundaries are expressed by

$$t = 0 \quad \& \quad su = (M + m)^2 (M - m)^2 = (M^2 - m^2)^2 \tag{7.77}$$

The physical regions for unequal mass case have been shown by shaded portion in Fig. (7.9)

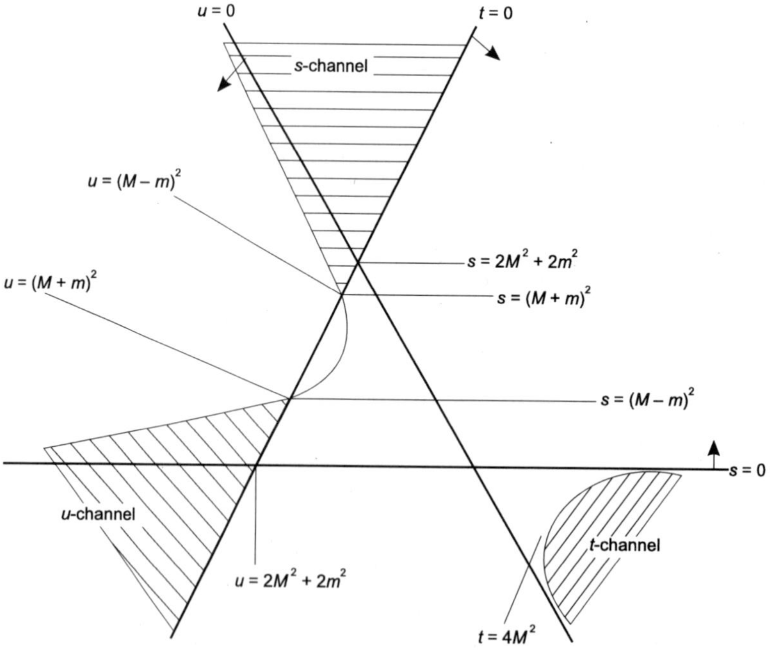

Fig. 7.9 Physical regions (shaded) of s, t and u channels in unequal mass case. Arrows denote axes directions.

In Fig. 7.9, we have $s = 0$, $t = 0$ and $u = 0$ lines perpendicular to which will be respective s, t and u axes shown by arrows. In the direction of arrows on the axes increase the value of that variable. For example, s increases perpendicular to $s = 0$ line in the direction of arrow and so on.

Contrary to equal mass case, the physical regions in unequal mass case are not confined to triangular regions formed by two straight lines. Here, as seen in Fig. 7.9, the physical regions are marked by curves. The curve that marks the physical region of s-channel also marks the physical region of u-channel. The t-channel is marked by a separate curve.

As shown in Fig. 7.9, the physical region of s-channel is marked by a curve whose branch after cutting the $t = 0$ line at a point $s = (M + m)^2$ approaches asymptotically the line $u = 0$. The point $s = (M + m)^2$ is also the point $u = (M - m)^2$. The s-channel region therefore is not sharply within $u = 0$ and $t = 0$ lines, but spreads on one side beyond $u = 0$ line into +ve u region upto $u = (M - m)^2$. Other boundary of this region continues to be the $t = 0$ line. Therefore, when the centre of mass energy s is $(M + m)^2$, the s-channel process would allow +ve values for $(p_1 - p_4)^2$. However, when s increases to $2M^2 + 2m^2$, u becomes zero and for $s > 2M^2 + 2m^2$, u, as in equal mass case becomes – ve.

The physical region of the u-channel process of unequal mass case too is not confined to $s = 0$ and $t = 0$ lines, but spreads on its s side up to $s = (M - m)^2$ which corresponds to the equality of $u \geq (M + m)^2$ of the u-channel process requirements (eqn. 7.71). The branch of the curve after cutting the $t = 0$ line at a point $u = (M + m)^2$ approaches asymptotically the line $s = 0$. Obviously, when $u \geq 2M^2 + 2m^2$; $s \leq 0$. Only for larger than $2M^2 + 2m^2$ values of u, the physical region of u-channel process for unequal mass shall have conditions similar to corresponding physical region of equal mass case.

The physical region of the t-channel process for the unequal mass case, is within the triangular region formed by $s = 0$ and $u = 0$ lines. However, the branches that asymptitocally approach these lines begin at $t = 4M^2$. Other possibility, $t = 4m^2$ does not represent a pure t channel process as this value lies between the u- and t-channels. Such regions are called double spectral regions. Contrary to s- and u-channels, the t-channel of unequal mass case lies between $s < 0$, $u < 0$ and $t > 4M^2$.

7.4 SYMMETRIES OF SCATTERING AMPLITUDE

7.4.1 Crossing Symmetry

Let us recapitulate the scattering process $a + b \rightarrow c + d$ in which particle a of mass m_1 and four-momentum p_1 is scattered by a particle b of mass m_2 and four-momentum p_2. After scattering a goes off as a particle c of mass m_3 and four-momentum p_3 and b goes off as a particle d of mass m_4 and four-momentum p_4. The s-channel process and its crossing symmetric t- and u-channel processes are expressed in terms of four-momenta as

$$p_1 + p_2 \rightarrow p_3 + p_4 \qquad (s\text{-channel}) \qquad (7.78a)$$

$$p_1 - p_3 \rightarrow -p_2 + p_4 \qquad (t\text{-channel}) \qquad (7.78b)$$

$$p_1 - p_4 \to p_3 - p_4 \qquad (u\text{-channel}) \tag{7.78c}$$

The Mandelstam variables for these channels, already mentioned earlier, are rewritten to put together.

$$\begin{aligned} s_1 &= (p_1 + p_2)^2 = s \\ t_1 &= (p_1 - p_3)^2 = t \qquad (s\text{-channel}) \\ u_1 &= (p_1 - p_4)^2 = u \end{aligned} \tag{7.79}$$

$$\begin{aligned} s_2 &= (p_1 - p_3)^2 = t \\ t_2 &= (p_1 + p_2)^2 = s \qquad (t\text{-channel}) \\ u_2 &= (p_1 - p_4)^2 = u \end{aligned} \tag{7.80}$$

$$\begin{aligned} s_3 &= (p_1 - p_4)^2 = u \\ t_3 &= (p_1 - p_3)^2 = t \qquad (u\text{-channel}) \\ u_3 &= (p_1 + p_2)^2 = s \end{aligned} \tag{7.81}$$

It can be observed from the above set of equations (7.79-7.81) that for the s-channel scattering the Mandelstam variables are (s, t, u). For the t-channel scattering these variables are (t, s, u) and for the u-channel scattering they are (u, t, s).

We, now introduce the scattering amplitude. The scattering process is undertaken by a linear operator (scattering or S-matrix) which transforms the initial $|i\rangle$ state into the final state $|f\rangle$. That is S $|i\rangle = |f\rangle$. By definition, the probability, that starting from the state $|i\rangle$ the system will be found, after scattering, in the state $|f\rangle$, is

$$P_{i\to f} = |\langle f|S|i\rangle|^2 \tag{7.82}$$

The elements of S-matrix, i.e., $\langle f|S|i\rangle = M_{fi}$, are called the scattering amplitudes as their square gives the probability.

Let us now denote the scattering amplitudes for the s-channel scattering as A, for the t-channel scattering as B and for the u-channel scattering as C. This, from eqns. (7.78-7.81) implies

$$\begin{aligned} A &\equiv A(s_1, t_1, u_1) \\ B &\equiv B(s_2, t_2, u_2) \\ C &\equiv C(s_3, t_3, u_3) \end{aligned} \tag{7.83}$$

Since, equations (7.78-7.81) relate the Mandelstam variables of t- and u-channel scatterings with the Mandelstam variables of the basic or the s-channel scattering, the corresponding amplitudes can also be expressed in terms of amplitude A as

$$\begin{aligned} A\,(s_1, t_1, u_1) &= A\,(s, t, u) \\ B\,(s_2, t_2, u_2) &= A\,(t, s, u) \\ C\,(s_3, t_3, u_3) &= A\,(u, t, s) \end{aligned} \tag{7.84}$$

Above equations imply that $A\,(s, t, u)$ is the scattering amplitude of s-channel scattering, $A\,(t, s, u)$ is the scattering amplitude of t-channel scattering and $A\,(u, t, s)$ is the scattering amplitude of u-channel scattering. Since, t- and u-channels are the crossing symmetric scatterings of the s-channel, we can say that $A\,(t, s, u)$ and $A\,(u, t, s)$

are the crossing symmetric scattering amplitudes of $A\ (s, t, u)$. The crossing reactions may therefore be regarded as different manifestations of the same fundamental process.

The scattering amplitudes, as mentioned above, are the matrix elements of scattering matrix also called the S-matrix. Feynmann graph tecnique is the standard method to calculate these matrix elements. If we analyse the Feynmann graphs and the corresponding matrix elements of these three scattering processes, it is found that the three processes are described by the same analytic expression. This means the three processes have same functional structure. This also means if we know the analytic expression for the amplitude A, the same analytic form would govern all three processes. Since, A is known to be an analytic function, the crossing symmetry relations (7.84) and the ability of one analytic function to describe three processes allow us to write

$$A\ (s, t, u) = A\ (t, s, u) = A\ (u, t, s) \tag{7.85}$$

which represents the crossing symmetry of scattering amplitude. The eqn. (7.85) implies that the amplitude of t-channel scattering process can be obtained by interchanging s and t variables in the s-channel amplitude. Therefore, $s \leftrightarrow t$ symmetry exists between the s- and t-channel amplitudes. Similarly, $s \leftrightarrow u$ symmetry exists between the s- and u-channel scattering amplitudes. Therefore, $s \leftrightarrow t$ and $s \leftrightarrow u$ symmetries represent crossing symmetry of scattering amplitude.

Since, in crossing symmetry particles are replaced by antipaticles and antiparticles have to exist for the fulfilment of CPT theorem, there is a strong connection between crossing symmetry and CPT theorem and this connection provides crossing symmetry, the status of being a fundamental property of quantum field theory.

7.4.2 Bose Symmetry

Let us imagine that in the reaction,

$$a + b \rightarrow c + d \tag{7.86}$$

the particles a and b are indistinguishable bosons. As told before, the four momenta of a, b, c, d are p_1, p_2, p_3, p_4, respectively. The scattering amplitude, i.e., the matrix element M_{fi}, must remain invariant on interchanging a and b which implies replacement of p_1 by p_2 and p_2 by p_1. As a result of this interchange, the Mandelstam variables

$$\begin{aligned} s &= (p_1 + p_2)^2 = (p_3 + p_4)^2 \\ t &= (p_1 - p_3)^2 = (p_2 - p_4)^2 \\ u &= (p_1 - p_4)^2 = (p_2 - p_3)^2 \end{aligned} \tag{7.87}$$

become

$$\begin{aligned} s &= (p_2 + p_1)^2 = s \\ t &= (p_2 - p_3)^2 = u \\ u &= (p_2 - p_4)^2 = t \end{aligned} \tag{7.88}$$

This means, interchanging the bosons a and b, the variable t becomes u and the variable u becomes t, while s remains s. Thus, $u \leftrightarrow t$ implies interchanging of indistinguishable bosons. As a result of $u \leftrightarrow t$, the scattering amplitude, M_{fi} must remain same, implying

$$M_{fi}(s, t, u) = M_{fi}(s, u, t) \tag{7.89}$$

The symmetry $u \leftrightarrow t$ is called Bose exchange symmetry or Bose symmetry of amplitude. Since, $M_{fi} \equiv A$, in terms of A, eqn. (7.89) is expressed by following Bose symmetry relation

$$A(s, t, u) = A(s, u, t) \tag{7.90a}$$

and the relation (7.85) due to Bose symmetry becomes

$$A(s, t, u) = A(t, s, u) = A(u, t, s) = A(s, u, t) = A(u, s, t) = A(t, u, s) \tag{7.90b}$$

Thus, interchanges $s \leftrightarrow t$ and $s \leftrightarrow u$ in the argument (s, t, u) represent crossing symmetry and the interchange $u \leftrightarrow t$ represents Bose symmetry. While interchanging two variables, it has to be kept in view that the third variable remains same. In the interchange $s \leftrightarrow t$, u is kept same; in $s \leftrightarrow u$, t remains t and in $u \leftrightarrow t$, s does not change. If in a process relating to a particular channel incoming and outgoing particles are indistinguishable, Bose symmetry is applicable. For the pair of distiguishable particles Bose symmetry does not hold true. For example, in the nucleon-pion (π°) scattering, the s-channel process is

$$N(p_1) + \pi^\circ(p_2) \rightarrow N(p_3) + \pi^\circ(p_4) \tag{7.91}$$

where brackets denote four momenta. The u-channel process is

$$N(p_1) + \pi^\circ(-p_4) \rightarrow N(p_3) + \pi^\circ(-p_2) \tag{7.92}$$

where antiparticle of π° is π° itself. Since the nucleon (N) and pion (π°) are distinguishable particles which constitute the in particles as well as the out particles, the Bose symmetry is not applicable in the s- and u-channels of $N - \pi^\circ$ scattering. However, the t-channel of basic (7.91) process is

$$N(p_1) + \bar{N}(-p_3) \rightarrow \pi^\circ(-p_2) + \pi^\circ(p_4) \tag{7.93}$$

where indistinguishable neutral pions are in the final (out) state. Therefore, Bose symmetry is applicable to two pions in the t-channel of $N\pi^\circ \rightarrow N\pi^\circ$ scattering process.

7.5 ANALYTICITY

In the previous sections, we have seen that $A(s, t, u)$ is the s-channel or the basic scattering amplitude. The t-channel and u-channel scattering amplitudes $A(t, s, u)$ and $A(u, t, s)$, respectively can be obtained when $A(s, t, u)$ is known. A requirement of the scattering amplitude is that it must be analytic.Thus, $A(s, t, u)$ must be analytic and its analytic continuation would lead it to the scattering amplitudes $A(t, s, u)$ and $A(u, t, s)$. A recapitulation of the basic definitions of analytic function and analytic continuation would help understand the matter better.

A function, here $A(s, t, u)$, is said to be analytic on a region say, R, if it is differentiable at every point in R. The analytic functions are both real and complex. An analytic function is infinitely differentiable function such that the Taylor series

$$T(x) = \sum_{n=0}^{\infty} \frac{f^n(x_0)}{|n}(x - x_0)^n \tag{7.94}$$

at any point x_0 in its domain converges to $f(x)$ for x in a neighbourhood of x_0 (pointwise). Due to infinite differentiability the analytic function is smooth. At the points called singularities and the lines known as branch cuts, a function ceases to be analytic. A singularity is a point where a function reaches unmeasurable or infinite value. Branch cut or branch line is a line that connects two branch points. Branch points are associated with multiple valued functions. If a function takes a small round about a point and on return, the value of function is different from its initial value, the point about which the round was taken is called a branch point and a line joining two branch points is a branch cut. For a function $f = \sqrt{z}$, $z = 0$ is the branch point and $+\sqrt{z}$ and $-\sqrt{z}$ are two branches of the function.

The exponential, hypergeometric, Bessel and gamma functions are analytic functions.

Analytic continuation provides a way of extending the domain over which a given analytic function is defined. Suppose in a region the analytic function, according to its definition, can be expanded into an infinite series that converges. In a new region, however, this expansion of the analytic function diverges.Through the technique of analytic continuation, further values of the said analytic function are defined such that its analylicity is brought to the new region as well. The analytic continuation is generally done by first establishing some functional equation on the small domain and then using the equation to extend the domain. A functional equation relates the value of a function at some point with its values at other points. For example,

$$f(z)f(1-z) = \frac{\pi}{\sin \pi z} \tag{7.95}$$

is a functional equation that is satisfied by gamma function.

$$f(z) = \Gamma(z)\ , f(1-z) = \Gamma(1-z) \tag{7.96}$$

The equality of crossed t- and u-channel scattering amplitudes with the s-channel scattering amplitude (eqn. 7.85) indicates that the s-channel scattering amplitude which gives rise to s-channel scattering in its physical region (Fig. 7.8a) analytically continues to the physical region of t-channel or to the physical region of u-channel to create the t or u-channel process. Since, the physical regions of s-, t- and u-channel scatterings are disconnected, the analytic continuation is through the unphysical region shown by unshaded triangle lying in between the three physical regions. Referring to Fig. (7.8a), this unphysical region in terms of parameters lies between $s = 0$ and $s = 4m^2$ and thus expressed as $0 \leq s \leq 4m^2$. The s-channel amplitude $A(s, t, u)$, which is analytic in the physical region of s-channel, while passing through the unphysical triangular region, if swaps the variables s and t, becomes the scattering amplitude $A(t, s, u)$ which continues to be analytic in the physical region of t-channel. Swapping of s and u would convert $A(s, t, u)$ into $A(u, t, s)$, the latter would continue to be analytic in the physical region of u-channel. The analytic continuation of a function (here scattering amplitude) from one physical region to other is smooth provided the function does not encounter discontinuities like poles or the branch cuts in the connecting unphysical region. The calculations do show the absence of singularities in the unphysical region.

While calculating the scattering amplitude or the matrix element of S-matrix through integral method, $i\varepsilon$ is added in the denominator (Appendix B) of Feynman propagator, to ensure analytic continuation.

As an example of analytic continuation of scattering amplitude, we may refer to Moller and Bhabha scatterings described in Chapter 2. Assuming Bhabha scattering $e^- + e^+ \rightarrow e^- + e^+$ as the s-channel process, its crossed u-channel process $e^- + e^- \rightarrow e^- + e^-$ is the Moller scatterimg. The respective scattering amplitudes have been derived in chapter 2.

$$M_{\text{Bhabha}} = -\frac{e^2}{(p_1 - p_3)^2}\left[\bar{v}(2)\gamma^{\mu}v(4)\right]\left[\bar{u}(3)\gamma_{\mu}u(1)\right]$$

$$+\frac{e^2}{(p_1 + p_2)^2}\left[\bar{u}(3)\gamma^{\mu}v(4)\right]\left[\bar{v}(2)\gamma_{\mu}u(1)\right] \tag{7.97}$$

and

$$M_{\text{Moller}} = -\frac{e^2}{(p_1 - p_3)^2}\left[\bar{u}(4)\gamma^{\mu}u(2)\right]\left[\bar{u}(3)\gamma_{\mu}u(1)\right]$$

$$+\frac{}{(p_1 - p_4)}\left[\bar{u}(3)\gamma\ u(2)\right]\left[\bar{u}(4)\gamma\ u(1)\right] \tag{7.98}$$

It can be observed that the replacements

$$p_2 \rightarrow -p_4,\quad p_4 \rightarrow -p_2,\quad \bar{v}(2) \rightarrow \bar{u}(4),\ v(4) \rightarrow u(2) \tag{7.99}$$

in M_{Bhabha} lead to M_{Moller}. In other words due to the changes (7.99), a s-channel ($e^- e^+$ scattering) process converts into its crossed u-channel ($e^- e^-$ scattering) process. The analytic scattering amplitude M_{Bhabha} which is analytic in s-channel continues to be analytic in u-channel region where it becomes M_{Moller}. This analytical continuance is expressed as

$$M_{e^-e^+}(p_1, p_2, p_3, p_4) = M_{e^-e^-}(p_1, -p_4, p_3, -p_2) \tag{7.100}$$

Changing the arguments on both sides in a like manner, eqn. (7.100) can be expressed as

$$M_{e^-e^+}\left[(p_1 + p_2)^2, (p_1 - p_3)^2, (p_1 - p_4)^2\right]$$

$$= M_{e^-e^-}\left[(p_1 - p_4)^2, (p_1 - p_3)^2, (p_1 + p_2)^2\right] \tag{7.101}$$

which on using the definitions of s, t, u (eqn. 7.87) provides

$$= M_{e^-e^+}(s,t,u) = M_{e^-e^-}(u,t,s) \tag{7.102}$$

representing the crossing symmetry between the s and u channels of the process and analytical continuity of their physical regions.

7.6 ASSIGNMENTS

1. Explain laboratory and centre of mass coordinate systems. In a general, binary collision $m_1 + m_2 = m_3 + m_4$, if $\sigma_0(\theta_0, \varphi_0)$ and $\sigma(\theta, \varphi)$ be the differential scattering

cross sections in laboratory and centre of mass coordinate systems respectivily, show that

$$\sigma_0(\theta_0, \varphi_0) = \frac{(1+\gamma^2+2\gamma\cos\theta)^{1/2}}{(1+\gamma\cos\theta)}\sigma(\theta,\varphi)$$

where $\gamma = +\sqrt{\frac{m_1 m_3}{m_2 m_4}\frac{E}{E+Q}}$, $E = \frac{m_1 m_2 v^2}{2(m_1+m_2)}$, Q is (+ve) for exothermic collisions and (–ve) for endothermic, v is the initial speed of m_1.

[**Hint:** Refer e.g., Schiff (qm) section 18]

2. For a two-body scattering event $a + b \to c + d$, define Mandelstam variables (s, t, u).
   (i) Show that $s + t + u = m_a^2 + m_b^2 + m_c^2 + m_d^2$. In terms of s, t, u and the masses,
   (ii) Find the centre of mass energy of particle a.
   (iii) Find the energy of a when b is at rest (lab coordinates).
   (iv) Find the total centre of mass energy.

   [**Ans.** (ii) $(s + m_a^2 - m_b^2)\, c^2/2\sqrt{s}$ (iii) $(s + m_a^2 - m_b^2)\, c^2/2m_b$ (iv) $c^2\sqrt{s}$]

3. In a two-body scattering event, the particles have same mass m and energy E. Show that in the centre of mass system the Mandelstam variables are given by

$$s = 4E^2$$

$$t = -\,2|\boldsymbol{p}|^2(1-\cos\theta) = \left|-\frac{s}{2}(1-\cos\theta)\right|_{m=0}$$

$$u = -\,2|\boldsymbol{p}|^2(1+\cos\theta) = \left|-\frac{s}{2}(1+\cos\theta)\right|_{m=0}$$

   [**Hint:** $s = 4(k^2 + m^2)$, $E_{CM} = 2E$]

4. Given that

$$\frac{1}{4}\sum_{S_i}\sum_{S_f}|M|^2 = 8e^4\left[\frac{(p_1\cdot p_2)(p_3\cdot p_4)+(p_1\cdot p_4)(p_2\cdot p_3)}{(p_1-p_3)^4}\right.$$

$$+\frac{2(p_1\cdot p_2)(p_3\cdot p_4)}{(p_1-p_3)^2(p_1-p_4)^2}$$

$$\left.+\frac{(p_1\cdot p_2)(p_3\cdot p_4)+(p_1\cdot p_3)(p_2\cdot p_4)}{(p_1-p_4)^4}\right]$$

   Show that the differential Moller scattering cross section under extreme relativistic approximations (era) is

$\frac{d\sigma}{d\Omega} = \frac{1}{4(4\pi)^2 (E_{tot})^2} \frac{|p_3|}{|p_1|} |M|^2$ is

$$\frac{d\sigma}{d\Omega} = \left[\frac{e^4}{2(4\pi)^2 (2E)^2}\right]\left[\frac{s^2+u^2}{t^2} + \frac{2s^2}{tu} + \frac{s^2+t^2}{u^2}\right]$$

where E is the energy of one of the incoming particles.

[**Hint:** Under era $s \cong 2p_1 \cdot p_2 = 2p_3 \cdot p_4 = 4E^2$, Similar approximations for t and u on equating with their scattering angle expressions can be substituted.]

5. What are s, t and u channel processes? Derive the conditions for their occurrence in identical mass case. Explain the difference when masses of the particles are not same.
6. Draw the Mandelstam plot and explain the three physical regions for equal mass case. Show that the boundaries of these physical regions obey $stu = 0$.
7. Answer question 6 for unequal mass case and find the corresponding boundary equation. Identify and explain double spectral regions.
8. Explain and distinguish the symmetries of scattering amplitude. What is analyticity and explain why it is so important that the scattering amplitude should be analytic.
9. Explain the significance of Mandelstam variables. Choosing the equal mass case, derive the expressions that indicate the physical domains of s, t and u channel processes and show their regions in the Mandelstam plane.
10. Giving physical examples distinguish s, t and u channel processes. For the s-channel process $N + \pi \to N + \pi$, prove $s + u + t = 2m_N^2 + 2m_\pi^2$ and obtain the expressions for the physical domains of s, t and u processes.
11. Derive all crossing relations for the process $p + \pi^- \to n + \pi^0$, where p refers to proton and n to neutron. Repeat the exercise for the process $n + \pi^+ \to n + \pi^+$. Discuss the difference between two cases.

    (**Hint:** Refer e.g., Paul Roman: Intro to Field Theory, sec. 10.1)
12. Prove that $s + t + u = 4m^2$
13. If $e^- + e^+ \to e^- + e^+$ is the s-channel process, prove that $s = 4(k^2 + m^2)$, $t = -2k^2(1 - \cos\theta)$ and $u = -2k^2(1 + \cos\theta)$.
14. If $e^- + \mu^- \to e^- + \mu^-$ is the s-channel process, show that the boundaries of the physical regions of this and the crossed channel reactions are given by $t = 0$,

    $su = \left[m_\mu^2 - m_\pi^2\right]^2$. Construct the Mandelstam plot.

    [**Hint:** see eqn. (7.80)]
15. Consider a scattering process of two particles with centre of mass energy $ECM = \sqrt{s}$. Show that the incident momentum $\boldsymbol{p}_1$, of any of the two particles of masses m_1 and m_2, in the centre of mass frame satisfies

$$|p_i|^2 = \frac{1}{4s}[s-(m_1+m_2)^2][s-(m_1-m_2)^2]$$

[**Hint** eqn. (7.64)]

7.7 OBJECTIVE QUESTIONS

1. The statement, that the sum of probabilities of all processes which are possible at a given energy is equal to unity, is called the
   (a) Analyticity condition
   (b) Unitarity condition
   (c) Causality condition
   (d) Condition for crossing symmetry
2. It is assumed that the scattering amplitude A is an analytic function of its arguments. This assumption is a manifestation of
   (a) Crossing symmetry
   (b) Bose symmetry
   (c) Causality principle
   (d) Unitarity principle
3. If analyticity of scattering amplitude is violated, it would mean
   (a) No waves are emitted or scattered
   (b) The scattered waves appear at their source after being emitted
   (c) The scattered waves appear at their source before being emitted
   (d) The scattering is through the s-channel only
4 If $S = 1 + iA$ is the scattering matrix and A is the scattering amplitude, then AA^+ equals
   (a) $2A$ (b) $2\, Im\, A$ (c) $2\, Re\, A$ (d) 0
5. If the four-momentum of a particle is $p^\mu = \left[(m^2+\boldsymbol{p}^2)^{\frac{1}{2}}, \boldsymbol{p}\right]$, the four-momentum of its antiparticle would be

   (a) $\left[(m^2+\boldsymbol{p}^2)^{\frac{1}{2}}, \boldsymbol{p}\right]$ (b) $\left[-(m^2+\boldsymbol{p}^2)^{\frac{1}{2}}, -\boldsymbol{p}\right]$

   (c) $\left[-(m^2+\boldsymbol{p}^2)^{\frac{1}{2}}, \boldsymbol{p}\right]$ (d) $\left[(m^2+\boldsymbol{p}^2)^{\frac{1}{2}}, -\boldsymbol{p}\right]$

6. The physical domain of a t-channel process can be expressed by
   (a) $t \geq 4m^2, \quad -4k^2 \leq s \leq 0, \quad -4k^2 \leq u \leq 0$
   (b) $t \geq 2m^2, \quad -2k^2 \leq t \leq 0, \quad -2k^2 \leq u \leq 0$
   (c) $u \geq 2m^2, \quad -2k^2 \leq t \leq 0, \quad -2k^2 \leq s \leq 0$
   (d) $s \geq 4m^2, \quad -4k^2 \leq t \leq 0, \quad -4k^2 \leq u \leq 0$
7. The Mandelstam variable s is the
   (a) Four-momentum transferred in centre of mass frame
   (b) Square of four-momentum transferred in centre of mass frame
   (c) Total energy in centre of mass frame
   (d) Square of total energy in centre of mass frame
8. The Mandelstam variable t represents
   (a) Four-momentum transferred in the s-channel process
   (b) Square of total energy in t-channel process
   (c) A negative value variable in u-channel process
   (d) All the above
9. Among the Mandelstam variables s, t, u
   (a) No variable is independent.
   (b) Only one variable is independent
   (c) Two variables are independent
   (d) All the three variables are independent
10. For a process involving four particles of equal masses, *two*, out of the three lines $s = 0$, $t = 0$ and $u = 0$ bound a channel's tringular region. Select the correct answer from the following:
    One of the lines that bounds the
    (a) s-channel region is $s = 0$ and the boundary of three regions is $stu = 0$
    (b) t-channel region is $s = 0$ and the boundary of three regions is $stu^2 = 0$
    (c) u-channel is $s = 0$ and the boundary of three regions is $stu = 0$
    (d) s-channel is $t = 0$ and the boundary of three regions is $stu^2 = 0$
11. If E be the energy of incoming particles, θ be the scattering angle, the Mandelstam variables in the centre of mass frame in the high energy limit would be
    [**Hint:** In high energy limit masses can be neglected)
    (a) $s = -4E^2, t = 2E^2(1 - \cos\theta), u = 2E^2(1 + \cos\theta)$
    (b) $s = -4E^2, t = -2E^2(1 - \cos\theta), u = -2E^2(1 + \cos\theta)$
    (c) $s = -4E^2, t = -2E^2(1 + \cos\theta), u = -2E^2(1 - \cos\theta)$
    (d) $s = -4E^2, t = -2E^2(1 - \cos\theta), u = -2E^2(1 + \cos\theta)$
12. If the s-channel amplitude of $(e^- \mu^- \rightarrow e^- \mu^-)$ scattering is denoted by $A(s, t, u)$ then the scattering amplitudes of $(e^- \mu^- \rightarrow \mu^- e^-)$, $(e^- e^+ \rightarrow \mu^+ \mu^-)$ $(e^- \mu^+ \rightarrow e^- \mu^+)$ and $(e^+ \mu^+ \rightarrow e^+ \mu^+)$ scatterings would be denoted by
    (a) $A(t, s, u)$, $A(u, s, t)$, $A(u, t, s)$ and $A(s, t, u)$, respectively.
    (b) $A(s, u, t)$, $A(t, s, u)$, $A(u, t, s)$ and $A(s, t, u)$, respectively.

(c) $A(s, u, t)$, $A(s, t, u)$, $A(u, s, t)$ and $A(s, t, u)$, respectively.

(d) $A(s, t, u)$, $A(t, s, u)$, $A(u, s, t)$ and $A(u, t, s)$, respectively.

13. The scattering amplitude for the process ($e^- \mu^- \rightarrow e^- \mu^-$) is given by $M = -e^2 (p_1 + p_3)(p_2 + p_4)\frac{1}{t}$. In terms of Mandelstam variables, it can be expressed as

(a) $M = -e^2 \dfrac{(s+u)}{t}$ (b) $M = -e^2 \dfrac{(s-u)}{t}$

(c) $M = -e^2 \dfrac{(s^2+u^2)}{t}$ (d) $M = -e^2 \dfrac{(s^2+t^2)}{t}$

[**Hint:** $p_1 + p_3 = (p_1 + p_2) + (p_3 - p_2) = (p_1 + p_2) + (p_1 - p_4)$
$p_2 + p_4 = (p_1 + p_2) - (p_1 - p_4)$]

14. In the elastic scatteng $\pi^0 \pi^0 \rightarrow \pi^0 \pi^0$, the two-pion intermediate state is formed when the threshold energy is $4m^2$, m, is the mass of pion. A four-pion intermediate state would be formed, when the threshold energy is

(a) $4m^2$ (b) $9m^2$

(c) $16m^2$ (d) $25m^2$

15. The scattering amplitude is

(a) Real (b) Complex

(c) Both real and complex

(d) Sometimes real and sometimes complex

16. If a particle is scattered into a volume V and into a momentum interval between p and $p + dp$, the number of final states available is

(a) $\dfrac{V \cdot (2\pi\hbar)^3}{4\pi p^2} dp$ (b) $\dfrac{V \cdot (2\pi\hbar)^2}{4\pi p^2} dp$

(c) $\dfrac{V \cdot 4\pi p^2}{(2\pi\hbar)^3} dp$ (d) $\dfrac{V \cdot 4\pi p^2}{2\pi\hbar} dp$

ANSWERS

1. b, 2. c, 3. c, 4. b, 5. b, 6. a, 7. d, 8. d, 9. c, 10. c,
11. d, 12. b, 13. b, 14. c, 15. b, 16. c.

8

REGGE POLES

8.1 RESONANCES AND REGGE POLES

In chapter 1 on elementary particles, a description of particles and resonances was given. Most of the particles described in the chapter had lifetimes of 10^{-10} second or more. In a detector, they travel a perceptible distance before decaying. The development of particle accelerators and the measurement of scattering cross sections revealed new particles in the form of resonances. The resonances correspond to particles with extremely small lifetimes as measured through the uncertainty relation $\Delta t\,\Delta E = \hbar$. The energy uncertainty ΔE is reflected in the width of the resonance, usually 10 to 200 MeV implying that the lifetimes are nearly $\hbar/(100 \text{ MeV}) \approx 10^{-25}$s. These short-lived hadron states get the name resonances because at certain energy the probability of their formation is large. The small lifetime of resonances is also manifested in their decay process that is through strong interactions. The states, which too have a small lifetime but decay through weak interactions, are usually called particles. Resonances are unstable bound states that stay only for a short time and can be produced in scattering experiments in which the amplitude of a wave (partial) or the scattering cross section becomes large at a particular energy. The largeness of amplitude is due to the vanishing of denominator of its expression at a particular energy and therefore resonances are called poles.

In the partial wave analysis of scattering processes, the angular momentum eigenfuctions are represented by $P_l(\cos\theta)$, where l is taken real having descrete values (l + 0. + 1. + 2, ... ∞). Resonance, therefore represents the sudden increase in the amplitude of a partial wave belonging to a real angular momentum, l. When the l values are assumed complex and continuous, the real l value scattering amplitude is found inadequate. However, the Watson-Sommerfeld transformation of the real l scattering amplitude is able to work in the complex angular momentum plane. This transformed scattering amplitude has partial waves different from the real l value partial waves. When, at various energies, the amplitudes of these partial waves corresponding to different complex angular momenta acquire large values, we have the poles called Regge poles. Thus, when angular momentum is real and descrete the poles are called resonances and when the angular momentum is complex and continuous, the poles are known as Regge poles.

8.2 SCATTERING PROCESSES

Many of the experimental observations and theoretical propositions demand a detailed understanding and analyses of resonances that can be undertaken by studying the collision of particles or a scattering process.

Let us consider the scattering process

$$a + b \rightarrow c + d \tag{8.1}$$

in which a is assumed to be the incident particle and b the target. c and d denote the product particles. If (8.1) is an elastic scattering

$$a + b \rightarrow a' + b' \tag{8.2}$$

the particles after the scattering are same as before the scattering [Figure (8.1a)]. The primes (′) indicate that the particles in the initial and final states are identical up to momenta and energy. This implies the four momenta $p_a^2 = p_b^2 = m_a^2 c^2$ and $p_b^2 = p_{b'}^2 = m_b^2 c^2$. However,

$$E_{a'} = \frac{E_a}{1 + \dfrac{E_a(1 - \cos\theta)}{m_b c^2}} \tag{8.3}$$

is observed* in elastic scattering, where $E_{a'}$ is the energy of a' which is the particle a (projectile) scattered through an angle θ and

$$\left(\frac{E_a}{c}, \boldsymbol{p}_a\right) = p_a, \quad \left(\frac{E_{a'}}{c}, \boldsymbol{p}_{a'}\right) = p_{a'} \tag{8.4a}$$

$$(m_b c, 0) = p_b, \quad \left(\frac{E_{b'}}{c}, \boldsymbol{p}_{b'}\right) = p_{b'} \tag{8.4b}$$

The difference $E_a - E_{a'}$ is the energy transferred to the target b as recoil.

In inelastic scattering, we can have [Figs. (8.1b, c, d)]

$$a + b \rightarrow c + d \tag{8.5a}$$

$$a + b \rightarrow c + d + e \tag{8.5b}$$

$$a + b \rightarrow a' + b^*, \; b^* \rightarrow c + d \tag{8.5c}$$

where new particles are produced.

The resonances can be observed in both elastic and inelastic scatterings.

When pions (π^+) are elastically scattered off the protons, a resonant state Δ^{++} is formed that lives for about 0.5×10^{-23} s and then decays back into $\pi^+ p$

$$\pi^+ + p \rightarrow \Delta^{++}(1232) \rightarrow \pi^+ + p \tag{8.6a}$$

The $\Delta^{++}(1232)$ is the baryon (fermion) resonance. Another baryon resonance formed elastically in $\pi^+ p$ scattering is N^* (1236).

$$\pi^+ + p \rightarrow N^*(1236) \rightarrow \pi^+ + p \tag{8.6b}$$

Some baryon resonances formed in inelastic scattering are

$$K^- + p \rightarrow \Sigma(1520) \rightarrow \Lambda + \pi^- + \pi^+ \tag{8.7a}$$

---

* Equation (8.3) with $E_a = h\nu$ and $E_{a'} = h\nu'$ gives the result we obtain in Compton scattering, where $m_b = (m_e)_0$ is the rest mass of electron.

$$K^- + p \rightarrow K^+ + \pi^0 + \Xi^*,\ \Xi^* \rightarrow \Xi^0 + \pi^- \tag{8.7b}$$

$$K^+ + p \rightarrow p + K^*,\ K^* \rightarrow K^0 + \pi^+ \tag{8.7c}$$

and

$$e^+ + e^- \rightarrow \rho^0(770) \rightarrow \pi^+ + \pi^- \tag{8.7d}$$

is the meson resonance. Some other meson resonances have been given in chapter 1.

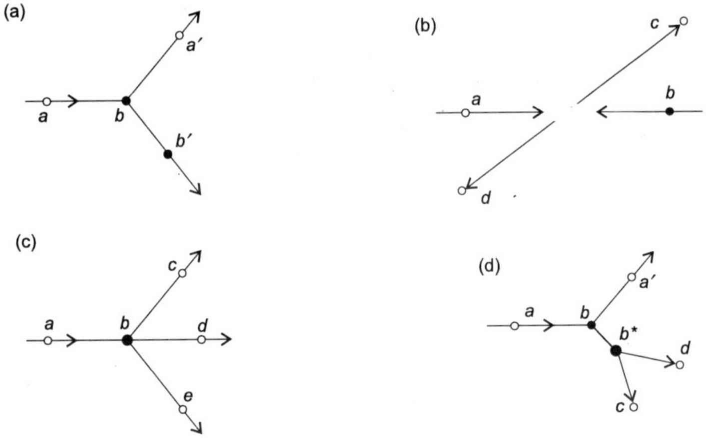

Fig. 8.1 Scattering processes : (a) elastic; (b), (c) & (d) inelastic; (b) reaction of colliding beams; (c) production of new particles; (d) production of excited state that decays into two particles.

8.2.1 Partial Waves and Scattering Amplitude

Let us begin with a situation in which two particles or two beams moving in opposite directions collide. Working in the centre of mass coordinate system in which the centre of mass of two colliding particles is at rest, the process can be thought as an elastic collision of a fictitious particle of mass μ, initial speed v and kinetic energy $E = \frac{1}{2}\mu v^2$ with a fixed scattering centre. The scattering centre is described by a potential energy $V(\boldsymbol{r})$, where $\boldsymbol{r}$ is the vector distance from the particle of (reduced) mass $\mu = m_1 m_2/(m_1 + m_2)$ to the point of origin of the scattering potential.

The wave equation for the motion of particle of mass μ in an external potential energy $V(\boldsymbol{r})$ is same as the wave equation that describes the relative motion of two (colliding) particles. This wave equation is

$$-\frac{\hbar^2}{2\mu}\nabla^2 u + Vu = Eu \tag{8.8}$$

in which the wave function, $u = u(r, \theta, \varphi)$, where r is the radial distance between the two particles, θ is the polar angle and φ is the azimuthal angle. The potential $V = V(r)$ is usually considered spherically symmetric, implying that V is a function only of the magnitude r of $\boldsymbol{r}$ measured from some origin (scattering centre). This indicates that the potential V is central, i.e., $V = V(r)$. This scattering potential has some additional characteristics that it is isotropic and localized, the latter implies a short range of $V(r)$. By short range, it is understood that the potential must decrease faster than $(1/r)$. Coulomb, the $(1/r)$ potential has infinite range which in the Coulomb scattering, however, is treated as the $\mu \to 0$ limit of finite Yukawa $(e^{-\mu r}/r)$ potential.

At large distances from the scattering centre, the effect of the potential $V(r)$ is negligible and the incident (i) particle can be represented by a plane wave

$$u_i = e^{ik.r} \quad (V(r) \to 0) \tag{8.9a}$$

where the amplitude of incoming plane wave has been assumed unity and in the exponential $i = \sqrt{-1}$. In this equation

$$\boldsymbol{k} = \frac{\boldsymbol{p}}{\hbar} = \frac{\mu v}{\hbar}, \boldsymbol{k} \cdot \boldsymbol{r} = k\,z \tag{8.9b}$$

The relation $\boldsymbol{k} \cdot \boldsymbol{r} = kz$ comes because the incident wave is assumed to be moving along the polar (z), i.e., along the $\theta = 0$ direction. Accordingly, the incident wave is expressed as

$$u_i = e^{ikz} \tag{8.10}$$

If a beam of particles is incident, the scattered particles would move along r from the scatterer, i.e., radial scattering would take place. The number of particles passing per second through an area perpendicular to any r would give the radial flux. This flux would be $(n/4\pi r^2)$, where n, is the number of particles crossing, per second, an imaginary sphere of area $4\pi r^2$. This implies that the flux would vary as $(1/r^2)$. This suggests that the amplitude of the scattered wave must be proportional to $(1/r)$ so that its square accounts for the flux $(\propto \frac{1}{r^2})$. Thus, the scattered wave that is spherically diverging can be expressed as

$$u_{\text{scatt}} = f(\theta, \varphi)\, \frac{e^{ikr}}{r} \tag{8.11}$$

where $f(\theta, \varphi)$ is called the scattering amplitude.

Thus in a scattering process a particle's passage is from a region of $V = 0$ to a region of $V \neq 0$ and then into a region of $V = 0$. $V \neq 0$ is the potential presented by the scatterer. Since, $V = 0$ implies the asymptotic $(r \to \infty)$ region, the total wave function should have the asymptotic form. This form is obtained by adding the incident plane wave (8.10) and the scattered spherical wave (8.11) to provide

$$u(r, \theta, \varphi)_{r \to \infty} \to e^{ikz} + f(\theta, \varphi)\, \frac{e^{ikr}}{r} \tag{8.12}$$

It can be easily verified, that the asymptotic ($V = 0$) form (8.12) satisfies the wave equation (8.8). However, the scattering centre presents a spherically symmetric (we have assumed it) $V \neq 0$ potential. The wave equation (8.8) for this spherically symmetric potential is solved by the method of separation of variables as $u(r, \theta, \varphi) = R(r)\,\Theta(\theta)\,\Phi(\varphi)$. Thus, we have two kinds of solutions $u(r, \theta, \varphi)_{V \neq 0}$ and $u(r, \theta, \varphi)_{V = 0}$, the former is obtained through the separation of variables method. We have to establish a connection between them. This is achieved by obtaining the asymptotic solution corresponding to $u(r, \theta, \varphi)_{V \neq 0}$ and equating it with the asmptotic solution $u(r, \theta, \varphi)_{V=0}$, which is (8.12).

The procedure that establishes the above-mentioned connection between the solutions separated in spherical polar coordinates $[u(r, \theta, \varphi)_{V \neq 0}]$ and the asymptotic form $[u(r, \theta, \varphi)_{V=0}]$ is known as the method of partial waves. The potential $V(r)$ would be considered real, implying that the scattering is elastic and for elastic scattering, the method of partial waves is best suited.

Since the incident beam is along z-axis, which is also the polar axis of (r, θ, φ) coordinates, φ becomes constant making the wave function independent of φ. Therefore,

$$u(r, \theta, \varphi) \rightarrow u(r, \theta) \rightarrow R(r)\,\Theta(\theta) \tag{8.13}$$

in which the radial wave function $R(r)$ obeys the differential equation

$$\frac{1}{r^2}\frac{d}{dr}\left(r^2\frac{dR}{dr}\right) + \left[\frac{2\mu}{\hbar^2}\{E - V(r)\} - \frac{l(l+1)}{r^2}\right]R = 0 \tag{8.14}$$

where $l(l + 1)$ is the number, acceptable in the process of separation of variables, to which both R equation and $Y(\theta, \varphi)$ [in our case $\Theta(\theta)$] equation become equal to. The $\Theta(\theta)$ obeys the differential equation

$$\frac{1}{\sin\theta}\frac{d}{d\theta}\left(\sin\theta\frac{d\Theta}{d\theta}\right) + l(l+1)\,\Theta = 0 \tag{8.15}$$

If we put

$$\Theta = P(w), \qquad w - \cos\theta, \qquad \frac{d}{d\theta} = -\sin\theta\frac{d}{dw} \tag{8.16}$$

equation (8.15) manifests itself as the differential equation for Legendre function $P = P(w)$ which is

$$\frac{d}{dw}\left[(1-w^2)\frac{dP}{dw}\right] + l(l+1)P = 0 \tag{8.17}$$

Since, l represents the orbital angular momentum quantum number which can have values 0, +1, +2, +3, +..., P represented by $P_l(w)$ are the Legendre polynomials. $P_l(w)$ are the angular momentum eigenfunctions. It may be noted that these values of angular momentum quantum number l are real.

Therefore, Legendre polynomials $P_l(\cos\theta)$ provide the solutions of the differential Θ equation (8.15).

Now, let us look at (8.14) for the solution of radial wave function, R. This equation contains the potential $V(r)$ that is presented by the scattering centre. The equation (8.14) with $V(r) \neq 0$ becomes solvable if we know the exact form of $V(r)$. Since in the present case the form of $V(r)$ is not known, eqn. (8.14) cannot be solved. Our objective, however, is to know the asymptotic value of $u(r, \theta, \varphi)$ given by (8.13) for which the asymptotic value of $R(r)$ in presence of scattering potential is required. In the absence of a direct solution of (8.14), we employ an indirect method to arrive at the asymptotic solution of $R(r)$ of eqns. (8.14) when $V(r) \neq 0$.

The method is to consider the range of $V(r)$. As already mentioned, $V(r)$ is of short range. Let us denote the range by a, then we can express that for

$$r < a, \quad V(r) = V(r)$$

and for

$$r > a, \quad V(r) \to 0 \tag{8.18}$$

Now solve the radial wave equation (8.14) in the range $r > a$ and obtain the asymptotic value of $R(r)$ as $r \to \infty$ belongs to $r > a$ region. Equation (8.18) tells that in this region $V(r) \to 0$, implying that the solution of $R(r)$ in the region $r > a$ does not carry the effect of $V(r)$. However, despite the inadequacy of not carrying the effect of $V(r)$, the solution $R(r)$ in $r > a$ region is obtained from which the corresponding $r \to \infty$ (asymptotic) solution is found. Now, to compensate for the inadequacy, the effect of scattering potential $V(r)$ is incorporated in this asymptotic solution. Thus obtained asymptotic solution would be the asymptotic value of $R(r)$ when $V(r) \neq 0$. Therefore, we first solve (8.14) for $V(r) = 0$ and obtain the corresponding asymptotic solution. We rewrite the radial wave equation (8.14) as

$$\frac{d^2R}{dr^2} + \frac{1}{r}\frac{dR}{dr} + \left[\frac{2\mu E}{\hbar^2} - \frac{2\mu V(r)}{\hbar^2} - \frac{l(l+1)}{r^2}\right]R = 0 \tag{8.19}$$

Putting

$$V(r) = 0,\ \frac{2\mu E}{\hbar^2} = k^2,\ \rho = kr,\ \frac{d}{dr} = k\frac{d}{d\rho},\ \frac{d^2}{dr^2} = k^2\frac{d^2}{d\rho^2} \tag{8.20}$$

(8.19) becomes

$$\frac{d^2R}{d\rho^2} + \frac{2}{\rho}\frac{dR}{d\rho} + \left(1 - \frac{l(l+1)}{\rho^2}\right)R = 0 \tag{8.21}$$

Since, $l(l+1)\hbar^2$ with $\hbar^2 = 1$ in natural units are the eigenvalues of L^2, where $\boldsymbol{L}$ is the orbital angular momentum operator and the orbital angular momentum quantum number l can have values $l = 0, +1, +2, +3, \ldots$, the corresponding radial wave functions are denoted by R_l and henceforth, the radial wave function R will be replaced by $R_l(\rho) = R_l(kr)$.

The second order differential equation (8.21) resembles the Bessel's differential equation whose spherical Bessel functions $j_l(\rho)$ and $\eta_l(\rho)$ are two independent solutions for $R_l(\rho)$. From these two solutions one general solution can be constructed as

$$R_l(\rho) = A_1 j_l(\rho) + B_1 \eta_l(\rho) \tag{8.22}$$

where A_1 and B_1 are constants. Since, (8.22) corresponds to $V(r) = 0$, these $R_l(\rho)$ are to be the asymptotic ($r > a \Rightarrow r \to \infty$) solutions. We, therefore use the asymptotic values of $j_l(\rho)$ and $\eta_l(\rho)$, which are

$$j_l(\rho)_{\rho \to \infty} \to \frac{1}{\rho}\sin(\rho - \frac{l\pi}{2}) \tag{8.23a}$$

and

$$\eta_l(\rho)_{\rho\to\infty} \to -\frac{1}{\rho}\cos(\rho - \frac{l\pi}{2}) \tag{8.23b}$$

Using these values in (8.22), we obtain the asymptotic value of wave functions $R_l(\rho)$ for $V(r) = 0$. Thus,

$$[R_l(\rho)]_{\rho\to\infty}^{v(r)=0} = \frac{A_1}{\rho}\sin(\rho - \frac{l\pi}{2}) - \frac{B_1}{\rho}\cos(\rho - \frac{l\pi}{2}) \tag{8.24}$$

Now, we incorporate the effect of potential $V(r)$ on the $V(r) = 0$ asymptotic solutions (8.24). Since $V(r)$ is the scattering potential and when it is ***on*** it would naturally change the amplitude and phase of scattered wave. In other words, when $V(r) \neq 0$, solutions (8.24) would be changed through the amplitude as well as phase. The obtained solutions would be the asymptotic values of $R_l(\rho) \equiv R_l(kr)$ in presence of scattering potential $V(r)$. It is pertinent to mention here that the phase difference between the scattered and unscattered waves contains the information about the shape and strength of the potential. Here, we have employed the converse property.

Let the amplitude of scattered wave with $V(r) \neq 0$, be A_l and let δ_l be the phase difference created over (8.24) by a nonzero $V(r)$. We can express

$$A_1 = A_l\cos\delta_l \quad \text{and} \quad -B_1 = A_l\sin\delta_l \tag{8.25}$$

such that

$$A_l = \sqrt{A_1^2 + B_1^2} \quad \text{and} \quad \delta_l = -\tan^{-1}\frac{B_1}{A_1} \tag{8.26}$$

Substituting (8.25) in (8.24), we get

$$[R_l(\rho)]_{\rho\to\infty}^{V(r)\neq 0} = \frac{A_l}{\rho}\left[\cos\delta_l\sin(\rho - \frac{l\pi}{2}) + \sin\delta_l\cos(\rho - \frac{l\pi}{2})\right]$$

$$= \frac{A_l}{\rho}\sin(\rho - \frac{l\pi}{2} + \delta_l) \tag{8.27}$$

which can be expressed as

$$[R_l(kr)]_{r\to\infty}^{V(r)\neq 0} = \frac{A_l}{kr}\sin(kr - \frac{l\pi}{2} + \delta_l) \tag{8.28}$$

where $l = 0, +1, +2, +3, \ldots \ldots$

Equation (8.26) tells that the scattering phase is negative. This implies that the scattered wave lags behind the unscattered wave. The negative phase also implies that the potential $V(r)$ is repulsive.

We may now return to eqn. (8.13)

$$u(r,\theta) = \sum_{l=0}^{\infty} R_l(kr)\,\Theta(\theta) \tag{8.29}$$

in which from equation (8.16) $\Theta = P_l(\cos\theta)$ and $R_l(kr)$ is given by eqn. (8.28). Substituting them in (8.29), the $r \to \infty$ (asymptotic) wave function obtained from the separation of variables ($V(r) \neq 0$) is

$$u(r, \theta) = \sum_{l=0}^{\infty} \frac{A_l}{kr} \sin\left(kr - \frac{l\pi}{2} + \delta_l\right) P_l(\cos\theta) \tag{8.30}$$

Now, we equate (8.30) with (8.12) to obtain the expression for the scattering amplitude $f(\theta, \varphi)$. The equated equation is

$$e^{ikz} + f(\theta,\varphi)\frac{e^{ikr}}{r} = \sum_{l=0}^{\infty} \frac{A_l}{kr} \sin\left(kr - \frac{l\pi}{2} + \delta_l\right) P_l(\cos\theta) \tag{8.31}$$

At this stage partial waves come into picture.

Using a fact that a plane wave can be represented as a superposition of incoming and outgoing spherical waves, the plane wave exponential e^{ikz} can be expressed in terms of spherical Bessel functions $j_l(kr)$ and Legendre polynomials $P_l(\cos\theta)$ as

$$e^{ikz} = \sum_{l=0}^{\infty} i^l (2l+1) j_l(kr) P_l(\cos\theta) \tag{8.32}$$

The expression (8.32) is also known as Bauer's formula. As discussed earlier, the plane wave e^{ikz} implied $r \to \infty$, the asymptotic form of spherical Bessel function $j_l(kr)$ would be used. From eqn. (8.23a), where $\rho = kr$, it can be expressed as

$$j_l(kr)_{r\to\infty} \to \frac{1}{kr}\sin\left(kr - \frac{l\pi}{2}\right)$$

$$= \frac{1}{2ikr}\left[e^{i\left(kr - \frac{l\pi}{2}\right)} - e^{-i\left(kr - \frac{l\pi}{2}\right)}\right] \tag{8.33}$$

Using (8.33), the expression (8.32) becomes

$$e^{ikz} = \frac{1}{2ik}\sum_{l=0}^{\infty} i^l (2l+1) P_l(\cos\theta)\left[\frac{e^{i\left(kr - \frac{l\pi}{2}\right)}}{r} - \frac{e^{-i\left(kr - \frac{l\pi}{2}\right)}}{r}\right] \tag{8.34}$$

Equation (8.34) is a summation of infinite terms obtained by giving real integer values 0, +1, +2, +3, ..., ∞ to l. The square bracket represents the superposition of an outgoing and an incoming spherical wave of equal amplitude. (e^{ikr}/r) is the outgoing spherical wave and (e^{-ikr}/r) is the incoming spherical wave. Each term corresponding to a particular l value is called a partial wave.

Coming back to eqn. (8.31), we shall now use (8.34) for e^{ikz} and in the RHS, express sine function in exponential form $\left[\sin\xi = \frac{1}{2i}(e^{i\xi} - e^{-i\xi}\right]$ and then equate on both sides the coefficients of e^{-ikr} and e^{ikr} separately.

The equated equation is

$$\frac{1}{2ikr}\sum_{l=0}^{\infty} i^l(2l+1) \quad P_l(\cos\theta)\left[e^{ikr}e^{-il\frac{\pi}{2}} - e^{-ikr}e^{il\frac{\pi}{2}}\right] + f(\theta,\varphi)\frac{e^{ikr}}{r}$$

$$= \sum_{l=0}^{\infty}\frac{A_l}{2ikr}P_l(\cos\theta)\left[e^{ikr}e^{-il\frac{\pi}{2}}e^{i\delta_l} - e^{-ikr}e^{il\frac{\pi}{2}}e^{-i\delta_l}\right] \tag{8.35}$$

Equating coefficients of e^{-ikr} on both sides of (8.35), we get

$$\frac{1}{2ikr}\sum_{l=0}^{\infty} i^l \quad (2l+1)P_l(\cos\theta)e^{il\frac{\pi}{2}}$$

$$= \sum_{l=0}^{\infty}\frac{A_l}{2ikr}P_l(\cos\theta)e^{il\frac{\pi}{2}}e^{-i\delta_l} \tag{8.36}$$

From which we get

$$A_l = i^l(2l+1)e^{i\delta_l} \tag{8.37}$$

Similarly, equating coefficients of e^{ikr} on both sides of (8.35), we get

$$\frac{1}{2ikr}\sum_{l=0}^{\infty} i^l(2l+1)P_l(\cos\theta)e^{-il\frac{\pi}{2}} + \frac{f(\theta,\varphi)}{r}$$

$$= \sum_{l=0}^{\infty}\frac{A_l}{2ikr}P_l(\cos\theta)e^{-il\frac{\pi}{2}}e^{i\delta_l}$$

from which $f(\theta) = f(\theta, \varphi)$ (as φ is constant) is obtained as

$$f(\theta) = \frac{1}{2ik}\sum_{l=0}^{\infty}P_l(\cos\theta)e^{-il\frac{\pi}{2}}\left[A_l e^{i\delta_l} - i^l(2l+1)\right]$$

which on using eqn. (8.37) for A_l gives

$$f(\theta) = \frac{1}{2ik}\sum_{l=0}^{\infty}(2l+1)i^l \quad e^{-il\frac{\pi}{2}}\left[e^{2i\delta_l} - 1\right]P_l(\cos\theta) \tag{8.38}$$

On employing the identity, $i^l = e^{il\frac{\pi}{2}}$ the expression (8.38) can be reduced to

$$f(\theta) = \frac{1}{k}\sum_{l=0}^{\infty}(2l+1)\left[\frac{e^{2i\delta_l} - 1}{2i}\right]P_l(\cos\theta) \tag{8.39}$$

Equation (8.38) or (8.39) is the principal result of the partial wave analysis. The partial wave analysis tells that for scattering due to a short range potential, partial wave decomposition is used to describe the scattering amplitude. The scattering amplitude

$f(\theta)$ and hence the scattered wave $f(\theta)\,(e^{ikr}/r)$ is expressed as the sum of partial waves for various values of l from zero to infinity.

The result (8.39) tells that the scattering amplitude is the sum of infinite partial waves. The amplitude of these partial waves depends on the phase angle δ_l, which itself depends on the value of orbital angular momentum quantum number, l. The amplitude of a partial wave corresponding to a particular l value is a complex number a_l given by

$$a_l = \frac{e^{2i\delta_l} - 1}{2i} = e^{i\delta_l} \sin\delta_l \tag{8.40}$$

where δ_l is real. k in the expansion (8.39) is

$$k = \frac{2\pi}{\lambda} = \frac{p}{\hbar} = \frac{\sqrt{2mE}}{\hbar} \tag{8.41}$$

where E is the kinetic energy of the incident particle of mass and de Broglie wavelength λ. The lth order Legendre polynomial $P_l(\cos\theta)$ in (8.39) is the angular momentum eigenfuction.

Whereas, the amplitude of a partial wave is a_l (eqn. 8.40), its phase angle is δ_l. This means similar to amplitude a_l, the phase angle δ_l too is different for different partial waves. δ_l describe the phase difference between the scattered and unscattered waves. They (δ_l) contain the information about the shape and strength of the scattering potential and the energy dependence of the scattering cross section.

The impact of δ_l can be understood from equations (8.30) and (8.37). Substituting A_l from (8.37) in (8.30), the total wave function $u(r, \theta)$ becomes

$$u(r,\theta) = \sum_{l=0}^{\infty} \frac{i^l}{kr}(2l+1)\, e^{i\delta_l} \sin\left(kr - \frac{l\pi}{2} + \delta_l\right) P_l(\cos\theta) \tag{8.42}$$

Writing sine function in terms of exponentials, it becomes $\delta_l + \delta_l = 2\delta_l$ in the first exponential and $\delta_l - \delta_l = 0$ in the second exponential. Equation (8.42) then becomes

$$u(r,\theta) = \sum_{l=0}^{\infty} \frac{i^l}{2ik}(2l+1)$$

$$\left[\frac{e^{i\left(kr - \frac{l\pi}{2} + 2\delta_l\right)}}{r} - \frac{e^{-i\left(kr - \frac{l\pi}{2}\right)}}{r}\right] P_l(\cos\theta) \tag{8.43}$$

In which the first term in the square bracket is the outgoing spherical wave and the second term is the incoming spherical wave. The outgoing wave carries an additional phase of $2\delta_l$, while the incoming has no additional phase.

The above observation signifies that, when scattering takes place, the scattering potential changes or shifts the phase of outgoing wave and leaves incoming wave undisturbed. Due to the shift in the phase, partial wave analysis is also called phase shift analysis.

In the above consideration, the scattering process is elastic in which k remains same before and after scattering. When the scattering is elastic, the partial wave amplitude has the form (8.40), where the magnitude of exponential (scattered) is obviously one. However, when the scattering is not purely elasic, implying presence of some component of inelastic scattering, the partial wave amplitude is

$$a_l = \frac{\eta_l e^{2i\delta_l} - 1}{2i} \tag{8.44}$$

where $0 \leq \eta_l \leq 1$. For elastic scattering $\eta_l = 1$.

The shift in the phase of outgoing wave can also be understood from the principle of conservation of probability (principle of unitarity). Corresponding to allowed values of l; each e^{ikr} term is an outgoing spherical wave, whereas each e^{-ikr} term is an incoming wave. Since conservation of probability (unitarity) requires that

$$\int \left|u_{\text{outgoing}}\right|^2 d\Omega = \int \left|u_{\text{incoming}}\right|^2 d\Omega \tag{8.45}$$

where $d\Omega$ is the elemental solid angle which for outgoing (scattered) particles is corresponding to r and θ. Since, $|u|^2 = u^* u$, the equality in (8.45) tells that only the phase of outgoing wave can change, not the modulus. Because of conservation of angular momentum, eqn. (8.45) is observed for each partial wave. Thus, in elastic scattering only the phase of the outgoing (scattered) wave changes whereas in an inelastic scattering both the amplitude and phase of outgoing wave change.

8.2.2 Scattering Cross Section

The most important quantity for the description and interrelations of scattering is the cross section, σ, which is a measure of the probability of an interaction between two colliding particles. The total scattering cross section σ_{total} is defined as

$$\sigma_{\text{total}} = \frac{\text{No. of reactions}}{\text{Beam particles} \times \text{Scattering centres}} \tag{8.46a}$$

where

No. of reactions imply number of reactions per unit time

Beam particles imply beam particles per unit time (8.46b)

Scattering centres imply scattering centres per unit area

For elastic and inelastic scatterings, the scattering cross section

$$\sigma_{\text{total}} = \sigma_{\text{elastic}} + \sigma_{\text{inelastic}} \tag{8.47}$$

The cross section is a physical quantity with dimensions of area. Usually differential scattering cross sections are defined because in practice only a fraction of the reaction is measured. A detector of area A is placed at a distance r and at angle θ from the beam direction. Thus, the detector covers a solid angle (A/r^2). The rate of reaction seen by this detector is then proportional to $[d\sigma(E, \theta)/d\Omega]$ which is called the differential scattering cross section.

To relate the differential scattering cross section with the scattering amplitude, let us imagine, as described above, that the detector subtends a solid angle $d\Omega$ at the scattering centre. If N is the number of particles scattered into the direction (θ, φ) per unit time per unit solid angle, then the number of particles scattered per unit time and thus entering the detector $= N\, d\Omega$.

Let Φ be the flux of particles in the incident beam. Φ is defined as the number of particles per unit time crossing a unit area placed normal to the direction of incidence.

The differential scattering cross section then is

$$\frac{d\sigma}{d\Omega} = \frac{N}{\Phi} \tag{8.48}$$

The total scattering cross section σ_{total} is then defined as the integral of differential scattering cross section over all solid angles, i.e.,

$$\sigma_{\text{total}} = \int \left[\frac{d\sigma}{d\Omega}\right] d\Omega \tag{8.49}$$

where in spherical polar coordinates

$$d\Omega = \sin\theta\, d\theta\, d\varphi \tag{8.50}$$

which when substituted in (8.49) provides

$$\sigma_{\text{total}} = \int_0^{2\pi} d\varphi \int_0^{\pi} d\theta\ \sin\theta \frac{d\sigma}{d\Omega} \tag{8.51}$$

or

$$\sigma_{\text{total}} = 2\pi \int_0^{\pi} d\theta\ \sin\theta \frac{d\sigma}{d\Omega} \tag{8.52}$$

Both $\frac{d\sigma}{d\Omega}$ and σ_{total} have the dimensions of an area

To obtain the desired relationship between the differential scattering cross section and the scattering amplitude we make use of the probability current density

$$j(x) = \frac{\hbar}{2im}(u^*\nabla u - u\nabla u^*) \tag{8.53}$$

which represents the flow of particles, and provides the flux of incident as well as of scattered particles. In view of the wave function (8.11), which we rewrite

$$u(r, \theta)_{r\to\infty} = u_{\text{in}} + u_{\text{scatt}} = e^{ikz} + f(\theta, \varphi)\ \frac{e^{ikr}}{r} \tag{8.54}$$

we substitute $u = u_{\text{in}} = e^{ikz}$ and $\nabla u = \boldsymbol{e}_z \frac{\partial u_{\text{in}}}{\partial z}$ ($\boldsymbol{e}_z$ is a unit vector) in eqn. (8.53) to obtain $\boldsymbol{j}_{\text{in}}(\boldsymbol{x})$ whose modulus gives the incoming flux. Straightforward calculation leads to

$$\text{Incoming flux} = \Phi = |\boldsymbol{j}_{\text{in}}(\boldsymbol{x})| = \left|\frac{\hbar k}{m}\boldsymbol{e}_z\right| = \frac{\hbar k}{m} \tag{8.55}$$

Similarly, substituting in eqn. (8.53) $u_{\text{scatt}} = f(\theta)\,\dfrac{e^{ikr}}{r}$ and using $\nabla u = \boldsymbol{e}_r \dfrac{\partial u_{\text{scatt}}}{\partial r}$ ($\boldsymbol{e}_r$ too is a unit vector), we get

$$|\boldsymbol{j}_{\text{scatt}}(\boldsymbol{x})| = \frac{\hbar k}{mr^2} f^*(\theta) f(\theta) = \frac{\hbar k}{mr^2} |f(\theta)|^2 \tag{8.56}$$

It may be noted that total wave function, $u_{\text{in}} + u_{\text{scatt}}$, as a whole is not used for u in eqn. (8.53) because the experimental observations do not show the effect of cross terms.

Equation (8.56) gives the number of particles scattered into the detector placed at a distance r in θ direction from the scattering centre. The area of detector is $r^2 d\Omega$. Therefore, the number of particles entering the detector per unit time is

$$N d\Omega = \frac{\hbar k}{mr^2} |f(\theta)|^2 r^2 d\Omega = \frac{\hbar k}{m} |f(\theta)|^2 d\Omega \tag{8.57}$$

Using eqns. (8.55) and (8.57) in (8.48), we obtain

$$\frac{d\sigma}{d\Omega} = \frac{N}{\Phi} = \frac{\dfrac{\hbar k}{m}|f(\theta)|^2}{\dfrac{\hbar k}{m}}$$

or

$$\frac{d\sigma}{d\Omega} = |f(\theta)|^2 \tag{8.58}$$

which is an important result of scattering theory. The result (8.58) tells that the differential scattering cross section is expressed as the modulus squared of the scattering amplitude. Equation (8.39) gives the partial wave expansion of scattering amplitude.

The total scattering cross section from eqns. (8.52) and (8.58) is

$$\sigma_{\text{total}} = -2\pi \int_{\theta=0}^{\theta=\pi} |f(\theta)|^2 \, d(\cos\theta) \tag{8.59}$$

where $d\theta \sin\theta = -d(\cos\theta)$ has been used. The $|f(\theta)|$ for (8.59) may be written, on using (8.40) in (8.39) along with $\left|e^{i\delta_l}\right| = 1$, as

$$|f(\theta)| = \frac{1}{k} \sum_{l=0}^{\infty} (2l+1) \sin\delta_l P_l(\cos\theta) \tag{8.60}$$

Substituting (8.60), (8.59) becomes

$$\sigma_{\text{total}} = +\frac{2\pi}{k^2} \sum_{l=0}^{\infty} (2l+1)^2 \sin^2\delta_l \int_{-1}^{1} P_l^2(\cos\theta)\, d(\cos\theta) \tag{8.61}$$

where limits of integral have been inverted for θ to range cosθ from –1 to +1. The minus sign has been absorbed in the inversion of limits.

Now employing an identity of Legendre polynomials

$$\int_{-1}^{1} P_l(\cos\theta)\, P_{l'}(\cos\theta)\, d(\cos\theta) = \frac{2\delta_{ll'}}{2l+1} \tag{8.62}$$

that makes $\int_{-1}^{1} P_l^2(\cos\theta)d(\cos\theta) = [2/(2l+1)]$, eqn. (8.61) becomes

$$\sigma_{\text{total}} = +\frac{4\pi}{k^2}\sum_{l=0}^{\infty}(2l+1)\sin^2\delta_l \tag{8.63}$$

Thus all the effect of potential on a given partial wave is contained in a single number, the phase shift δ_l. The partial waves are pulled in by the potential if $\delta_l > 0$ and pushed out if $\delta_l < 0$. The amount of pulling in or pushing out is determined by δ_l.

The maximum value of scattering cross section is for $\sin^2\delta_l = 1$, and the value is

$$\sigma_{\text{total}}^{\text{max}} = +\frac{4\pi}{k^2}\sum_{l=0}^{\infty}(2l+1) \tag{8.64}$$

Since the above scattering is elastic, (8.64) implies

$$\sigma_{\text{elastic}}^{\text{max}} = +\frac{4\pi}{k^2}\sum_{l=0}^{\infty}(2l+1)$$

and $$\sigma_{\text{inelastic}} = 0 \tag{8.65}$$

For very low energies, we need to consider only the first term ($l = 0$) in the partial wave expansion. This is known as *s*-wave scattering for which the differential scattering cross section can be calculated from (8.58) and (8.60) as $(d\sigma/d\Omega) = \sin^2\delta_0/k^2$ and from (8.63) the total *s*-wave scattering cross section is 4π times its differential scattering cross section. The *s*-wave scattering cross section is independent of θ and hence is isotropic. When energy is increased higher partial waves contribute and produce their angular dependence. The partial wave analysis works well if only a few *l's* have to be taken into account.

8.2.3 Optical Theorem

In the scattering amplitude $f(\theta)$, if $\theta = 0^0$, the incident beam is said to be 'scattered' in the forward direction and then $f(\theta = 0) = f(0)$ is said to be the forward scattering amplitude. $f(0)$ is the amplitude of the wave scattered to the centre of the distant screen in $\theta = 0^0$ direction.

From the expression (8.39) of $f(\theta)$ which is rewritten here on using (8.40) as

$$f(\theta) = \frac{1}{k}\sum_{l=0}^{\infty}(2l+1)e^{i\delta_l}\sin\delta_l P_l(\cos\theta) \tag{8.66}$$

it can be observed that $f(\theta)$, being complex, has real and imaginary parts. Expanding $e^{i\delta_l} = \cos\delta_l + i\sin\delta_l$, the real and imaginary parts of $f(\theta)$ are

$$Re\, f(\theta) = \frac{1}{k}\sum_{l=0}^{\infty}(2l+1)\cos\delta_l \sin\delta_l P_l(\cos\theta) \tag{8.67a}$$

$$Im\, f(\theta) = \frac{1}{k}\sum_{l=0}^{\infty}(2l+1)\sin^2\delta_l P_l(\cos\theta) \tag{8.67b}$$

The σ_{total} of (8.63) may now be expressed in terms of $Im\, f(\theta)$.

Putting $\theta = 0°$ and using the property $P_l(+1) = +1$ for any l, we can write from (8.67b)

$$Im\, f(0) = \frac{1}{k}\sum_{l=0}^{\infty}(2l+1)\sin^2\delta_l \tag{8.68}$$

which when substituted in (8.63) provides the optical theorem

$$\sigma_{\text{total}} = \frac{4\pi}{k} Im\ f(0) \tag{8.69}$$

The usefulness of optical theorem can be understood from the following.

The total cross section contains contributions from scattering into all possible angles.We may therefore have to evaluate the scattering cross section for every scattering angle and add up the indivudial cross sections. The optical theorem says we need not do this, instead, we should evaluate the imaginary part of scattering amplitude in forward direction, i.e., evaluate $Im\, f(0)$ and use (8.69) to obtain the σ_{total}.

The optical theorem was invented by Lord Rayleigh, who recognized the forward scattering amplitude in terms of index of refraction, which was used in a study of colour and polarization of sky. Being an important theorem of optics, it was called optical theorem. Later, the theorem was applied to the theory of quantum scattering.

8.3 BREIT-WIGNER RESONANCES

We have seen in the partial wave analysis, that the amplitude of a partial wave corresponding to a particular l value is a_l which for elastic scattering is

$$a_l = \frac{e^{2i\delta_l} - 1}{2i} = e^{i\delta_l}\sin\delta_l \tag{8.70}$$

The phase shifts (δ_l) depend on l and E. For a given l, δ_l varies with E, the kinetic energy of incident particle. If the beams collide, E is the centre of mass energy, that is, the total energy of both the colliding particles.

From (8.70), it can be observed that the partial wave amplitude a_l is a complex number,

$$|a_l|^2 = a_l^* a_l = \sin^2\delta_l \tag{8.71}$$

which is maximum when $\delta_l = \frac{\pi}{2}$ and δ_l becomes $\frac{\pi}{2}$ at an energy, say $E = E_R$. So

$$\delta_l(E_R) = \frac{\pi}{2} \tag{8.72}$$

and accordingly

$$|a_l|^2_{max} = 1 \tag{8.73}$$

However, in most of the particle-particle interactions, the elastic scatterings do have some component of inelastic scatterings and then a_l has the form (8.44), from which

$$|a_l|^2 = \frac{1}{4}(\eta_l^2 - 2\eta_l \cos 2\delta_l + 1) \tag{8.74}$$

which, in view of $0 \leq \eta_l \leq 1$ and allowed values of δ_l would set

$$|a_l|^2 \leq 1 \tag{8.75}$$

Equation (8.75) is called the partial wave unitarity limit; because it can also be obtained by the unitarity of S-matrix, which when operates on the initial (incoming) states transforms them into final (scattered) states. Equation (8.75) also tells that the maximum amplitude of a partial wave cannot exceed unity, incoming wave had the amplitude unity. For pure elastic scatterings (8.71) is the equation for $|a_l|^2$.

Now, let us consider the pure elastic scattering for which the scattering amplitude, $f(\theta)$, is expressed in terms of infinite partial waves, each belonging to a particular l value (eqn. 8.39).

Out of these infinite partial waves, let us consider the partial wave whose $l = l_1$. Assume that the phase shift δ_{l_1} of this partial wave becomes $(\pi/2)$ as the energy E becomes E_R, i.e., ($E \to E_R$, $\delta_{l_1} \to \pi/2$). It is assumed that the maximum contribution to phase shift of the scattered wave with respect to incoming wave is due to $l = l_1$ partial wave, other partial waves make negligible contributions to phase shift.

The amplitude of $l = l_1$ partial wave from (8.70) is

$$a_{l_1} = e^{i\delta_{l_1}} \sin \delta_{l_1} \tag{8.76}$$

For brevity, let us assume

$$a_{l_1} = a, \ \delta_{l_1} = \delta \tag{8.77}$$

Equation (8.76) for partial wave amplitude can then be expressed as

$$a = \frac{\sin \delta}{e^{-i\delta}} = \frac{\sin \delta}{\cos \delta - i \sin \delta} = \frac{1}{\cot \delta - i} \tag{8.78}$$

Now, resonance takes place when the amplitude (a) is maximum and that occurs at $E \cong E_R$, $\delta \cong \pi/2$. Therefore, near resonance, $\cot \delta \cong 0$. Since, δ is a function of E, $\cot\delta(E)$ can be expanded about E_R, according to the Taylor series

$$f(x) = f(x_0) + f'(x_0)(x - x_0) + \frac{f''(x_0)}{2!}(x - x_0)^2 + .. \tag{8.79}$$

in which substituting, $x = E$, $x_0 = E_R$, $f(x) = \cot\delta(E)$, we have the Taylor expansion

$$\cot\delta(E) = \cot\delta(E_R) + (E - E_R)\left[\frac{d}{dE}\cot\delta(E)\right] + .. \tag{8.80}$$

Using $\cot\delta(E_R) = \cot\frac{\pi}{2} = 0$ and defining

$$\left[\frac{d}{dE}\cot\delta(E)\right]_{E=E_R} = -\frac{2}{\Gamma} \tag{8.81}$$

the expansion (8.80) becomes

$$\cot\delta(E) \cong -(E - E_R)\frac{2}{\Gamma} \tag{8.82}$$

Substituting (8.82) in (8.78), the partial wave amplitude $a \equiv a(E)$ can be expressed as

$$a(E) - \frac{1}{-(E - E_R)\dfrac{2}{\Gamma} - i} = \frac{\dfrac{\Gamma}{2}}{(E_R - E) - i\dfrac{\Gamma}{2}} \tag{8.83}$$

where in the second step both numerator and denominator have been multiplied by $\frac{\Gamma}{2}$ and minus sign has swapped E and E_R.

It may be noted that, the definition (8.81) is in fact

$$-\text{cosec}^2\delta(E_R)\frac{d\delta(E_R)}{dE} = -\frac{2}{\Gamma} \tag{8.84a}$$

The phase shift $\delta(E_R)$ as described above is $\frac{\pi}{2}$, which is a resonance point. This makes $\text{cosec}^2\delta(E_R) = 1$. However, near resonance $\frac{d\delta(E_R)}{dE} \neq 0$. This nonzero value defines the width (Γ) of resonance peak.

$$\frac{2}{\Gamma} = \frac{d\delta(E_R)}{dE} \tag{8.84b}$$

In terms of energy (E_R) of resonance, the partial wave amplitude has the form (8.83). Using this form of $a(E)$, we now derive the expression for total scattering cross section. It is recalled that the scattering is elastic. Rewriting the expression (8.63), we have

$$\sigma_{\text{total}} = \frac{4\pi}{k^2}\sum_{l=0}^{\infty}(2l+1)\sin^2\delta_l \tag{8.85}$$

Substituting in it $\sin^2\delta_l = |a_l|^2$ (eqn. 8.71), (8.85) can be expressed as

$$\sigma_{\text{total}} = \frac{4\pi}{k^2}\sum_{l=0}^{\infty}(2l+1)|a_l|^2 \tag{8.86}$$

Now, the square of the modulus of amplitude of the lth partial wave $|a_l|^2$ for $l = l_1$ can be calculated by using (8.83) as

$$|a_l|^2 = a_l^* a_l = \frac{\frac{\Gamma^2}{4}}{(E_R - E)^2 + \frac{\Gamma^2}{4}} \tag{8.87}$$

Substituting (8.87) in (8.86), we get an important expression for σ_{total} as

$$\sigma_{\text{total}} = \frac{4\pi}{k^2}\sum_{l=0}^{\infty}(2l+1)\frac{\frac{\Gamma^2}{4}}{(E_R - E)^2 + \frac{\Gamma^2}{4}} \tag{8.88}$$

This relation (8.88) is known as Breit-Wigner formula. This is also expressed as

$$\sigma_{\text{total}} = \sigma_{\max}\left[\frac{\frac{\Gamma^2}{4}}{(E_R - E)^2 + \frac{\Gamma^2}{4}}\right] \tag{8.89}$$

where

$$\sigma_{\max} = \frac{4\pi}{k^2}(2l_1 + 1) \tag{8.90}$$

and because the contributions from l values other than l_1 are negligible, (8.90) does not carry the summation sign.

The relation (8.88) or (8.89) is also called the Breit-Wigner resonance with Γ as the full width at half maximum. In the elastic scattering, e.g., $a + b \to c^* \to a + b$, c^* is the resonance state, a short-lived bound state. This resonance state forms at an energy $E = E_R$ which means E_R is an energy level of c^*. It is assumed that other energy levels of the resonance (state) c^* are quite far from the energy level E_R, such that E_R is almost an isolated energy level. This situation is also described by telling that the resonance c^* has only one energy level $E = E_R$. Accordingly, the formulae (8.88) or (8.89) are also known as Breit-Wigner single level formula.

The Breit-Wigner resonance curve has been shown in Fig. (8.2).

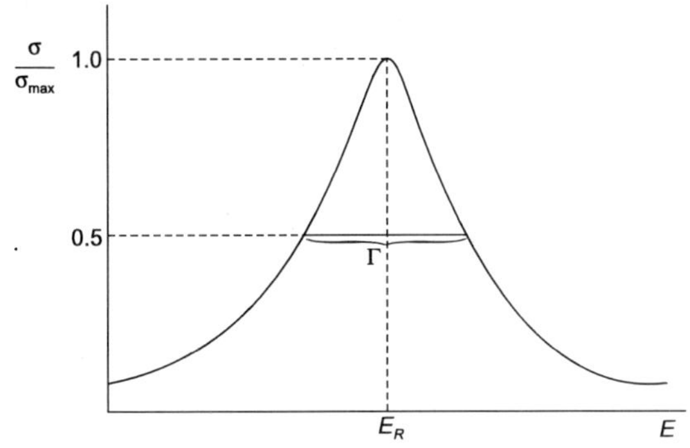

Fig. 8.2 Breit-Wigner resonance curve.

From the figure it can be inferred that Γ is the energy difference, so the energy time uncertainty relation $\Delta E \Delta t \sim \hbar$ is used to estimate the lifetime of resonance, i.e., using $\Gamma\tau = \hbar$, the lifetime τ of resonant state is

$$\tau = \frac{\hbar}{\Gamma} \tag{8.91}$$

If the resonance peak is sharp, implying small Γ, (8.91) predicts such resonance as long-lived and a resonance having a broader peak would be short-lived.

In eqn. (8.90), $(2l_1 + 1)$ appears for the spinless case, when the incident and target particles both are spinless. If a spinless particle is incident on spin (1/2) nucleon, $(2l + 1)$ is replaced by $(2J + 1)$ where J is the total angular momentum of the resonant state. If both the particles have spin say s_a and s_b, $(2J + 1)$ is divided by $(2s_a + 1)(2s_b + 1)$. In this $(a + b \rightarrow c^* \rightarrow a + b)$ case

$$\sigma_{\max} = \frac{4\pi}{k^2}\frac{2J+1}{(2s_a+1)(2s_b+1)} \tag{8.92}$$

Thus, the Breit Wigner resonance shows energy dependence of scattering cross section and the peak in the curve at a particular energy shows a strong attraction at that energy. This peak implies a resonance, which means at that energy, beam and target particles bind temporarily and then break up again. That bound state represents the resonant 'particle'. The resonances are the quasi-bound states.

If we look at eqn. (8.83), we can easily observe that the amplitude tends to infinity if the denominator vanishes. This means when

$$E = E_R - i \ — \ \ a_l(E) \rightarrow \infty \tag{8.93}$$

implying that at energy $E = E_R - i\,\frac{\Gamma}{2}$ there is a pole in the amplitude $a_l(E)$ of the partial waves. Since, eqn. (8.93) represents a resonance; we may tell that resonances are poles in the amplitudes of partial waves $a_l(E)$.

8.4 RESONANCES AS POLES OF SCATTERING MATRIX

As described in the previous sections partial wave method is an important method to describe quantum mechanical scattering, particularly at low energies. Another powerful and more general technique to describe the quantum mechanical scattering is through the concept of *S*-matrix. In a scattering process, there is an initial state $|i\rangle$, which is the state in the distant past ($t = -\infty$) when the particles are well separated and thus are non-interacting. As time proceeds, the particles approach, interact (*at* $t = 0$) and then they recede to become again noninteracting and are detected in distant future at $t = +\infty$. This state at $t = +\infty$ is the final state $|f\rangle$. *S*-matrix is a unitary operator $S = U(\infty, -\infty)$ that determines the evolution of the initial state at $t = -\infty$ to the final state at $t = +\infty$. That is,

$$|f(t = +\infty)\rangle = S|i(t = -\infty)\rangle \tag{8.94}$$

The above equation represents not a single final state but a superposition of various kinds of final states. The amplitude for finding a particular final state $|f\rangle$ is given by the matrix element $\langle f|S|i\rangle$. Thus, the element of matrix *S*

$$S_{fi} = \langle f|S|i\rangle \tag{8.95}$$

is the (scattering) amplitude for the transition $i \to f$. In chapter 2, such matrix elements have been calculated for various scattering processes. These scattering processes were described through the Feynman diagram technique in which Feynman rule no. 5 (Section 2.3.1) provided a multiplicative factor of $\frac{i(\gamma^\mu q^\mu + m)}{m^2 - q^2}$. Obviously, when its denominator vanished ($q = \pm m$), the matrix element blew up implying that the matrix element has poles at $q = +m$ and at $q = -m$. In general, due to the vanishing of denominator, a particular parameter gets a certain value, called the pole of the parameter at that value. In case of resonance, the parameter is energy, which when takes the value (8.93), the scattering amplitude goes to infinity. Since in *S*-matrix theory, the *S*-matrix element represents the scattering amplitude, we can say that *S*-matrix has poles given by eqn. (8.93) at which $S_{fi} \to \infty$. Since (8.93) is also the condition of resonance, due to unitarity of scattering matrix, the form of S_{fi} near resonance is of the kind $\left(E - E_R - i\frac{\Gamma}{2}\right)\Big/\left(E - E_R + i\frac{\Gamma}{2}\right)$. Referring to the relation (8.93), it can be noticed that near a resonance the energy *E* is complex consisting of a real part, $Re(E) = E_R$ and an imaginary part, $Im(E) = (-)\frac{\Gamma}{2}$. The axes $Re(E)$ and $Im(E)$ construct a complex energy plane. In complex analysis, one deals with Riemann surfaces. Every Riemann surface is a two-dimensional real analytic manifold. It is a surface-like configuration that covers the complex plane, in general, with infinitely many "sheets."

A manifold is a topological space that locally resembles Euclidean space near each point. The manifold of k^2 [= $(2mE/\hbar^2)$] is a two-sheeted Riemann surface whose sheets, for fixed k^2, are distinguished by the two values of k. The sheet for which the imaginary part of k is positive ($Im\ k > 0$), is called the physical sheet or the first sheet and the sheet for which the imaginary part of k is negative ($Im\ k < 0$) is the unphysical sheet or the second sheet. Usually, the analytic structure of a scattering amplitude is studied on the physical Riemann sheet. However, for the complete study of the analytic structure of the scattering amplitude, the study has to be made on the second or the unphysical sheet as well. In the physical sheet, energy is negative and hence the bound states correspond to this sheet. The bound states are the poles of S-matrix on the physical sheet. On the unphysical sheet, the energy is positive and hence as such stable bound states cannot be expected here. However, quasibound states with a small finite lifetime can correspond to this unphysical sheet. These are resonances which are poles of S-matrix on the unphysical sheet. These poles lie on the unphysical sheet in its fourth quadrant such that its energy is like (8.93).In other words, the resonances are characterized by complex pole positions. The measure of coordinate of pole along the real energy axis (a positive value) gives the mass of resonance and twice the measure along negative imaginary axis gives the full width of the resonance.

The Particle Data Group gives the mass and width of resonances in terms of their complex pole position. For example, for $\Delta(1232)\frac{3^+}{2}$ resonance, its complex pole position is $\left(1210 - j\frac{100}{2}\right)\text{MeV}$, implying that [$Re$ (pole position) $\cong 1210\,MeV$] and [$-2\,Im$ (pole position) $\cong 100\ MeV$], former denotes the mass and latter the full width of the Δ resonance. The Breit-Wigner mass and Breit-Wigner full width of this resonance are given approximately as 1232 MeV and 117 MeV respectively.

8.5 REGGE THEORY

Developed by Tullio Regge for non-relativistic quantum theory, Regge theory is the study of the analytic properties of scattering amplitude as a function of angular momentum. The angular momentum values, however, are not restricted to integer multiples of $\hbar$, but allowed to take any complex value. The poles in the complex plane of angular momentum are called Regge poles and in non-relativistic case of Schrödinger equation, composite particles correspond to Regge poles in scattering amplitudes. This observation extends to relativistic theory as well.

8.5.1 Concept of Complex Angular Momentum

The concept of complex angular momentum appears in the quantum theory of hydrogen atom where the energy of an electron in N^{th} energy level is $E_N = -\{(2m'\pi^2e^4)/(h^2N^2(4\pi\varepsilon_0)^2)\} = -(13.6\text{ eV}/N^2)$. Here $m' = (mM/(m+M))$, where m is the mass of electron and M the mass of proton. From this equation, we may write $N^2 = -(\pi^2e^4/h^2(4\pi\varepsilon_0)^2)\,(2m'/E_N)$, the square root of which gives $N = i(\pi e^2/4\pi\varepsilon_0 h)\sqrt{(2m'/E_N)}$, where $i = \sqrt{-1}$ makes N imaginary. In the Schrödinger's theory of hydrogen atom principal quantum number $N = n + l + 1$, where $n = 0, 1, 2, 3$ ... is the radial quantum number; $l = 0, 1, 2, 3, ...$ is the orbital angular momentum quantum number. From the equation for N and $N = n + l + 1$, we can easily write that $l = -\{n + 1\} + i\left\{(\pi e^2 / 4\pi\varepsilon_0 h)\sqrt{(2m' / E_N)}\right\}$, which shows that the orbital angular momentum l is a complex function of energy $E(E_N)$ and thus emerges as a complex quantity. This expression describes in the complex l-plane a path which is called a Regge trajectory.

Introducing complex angular momenta for the first time, Watson and Sommerfeld effeciently solved the diffraction of radio waves around the surface of the earth. For this, the method of expansion of wave function (scattering amplitude) into partial waves turned out to be impractical. This was because an unmanageably large number of partial waves were required to describe the angular dependence of the wave function. Providing the solution to this problem, Watson and Sommerfeld, analytically continued the series of partial waves into the complex plane of angular momentum and represented the angular dependence of the wave function by a simple analytic function. That analytic function was the Legendre function $P_l(-\cos\theta)$, where l was a complex number. Thus, $P_l(-\cos\theta)$ appear when Watson Sommerfeld transformation of real l value scattering amplitude is undertaken. Regge used this concept to prove that the partial wave amplitudes remained analytic in various potential scatterings.

8.5.2 One Particle Exchange Model

One particle exchange model (OPEM) precedes the Regge pole model, an example of former is the process of nucleon-nucleon scattering (strong interaction) via single pion exchange (e.g., Fig. 1.21). In the single pion exchange model, a virtual pion mediates the force between two interacting nucleons. Since the mass of pion is least among the mesons, its interaction range is longest and therefore the single pion exchange has the potential of being a dominant mechanism of nucleon-nucleon scattering processes. However, OPEM works well only at low energies. This fact was observed when the experimental values of total scattering cross section for proton-proton elastic scattering, obtained from the Particle Data Group, were compared with the values derived from a theoretical curve for total scattering cross section. The theoretical curve itself was drawn on the basis of one pion exchange formulation. The energy range of the incident nucleon (proton) in the above comparison was 0 to 10 GeV. The comparison clearly showed that the experimental data fitted well with the theoretical curve only in the low energy range of below 0.5 GeV. When the incident energy increases from 0.5 GeV, the theoretical curve separates (falls off) quickly from the experimental data. This means that the one particle exchange model fails at energies higher than 0.5 GeV.

If the interaction (scattering process) is represented by the Yukawa potential $V = (ge^{-\mu r})/r$, where μ is the mass of pion, the matrix element or the scattering amplitude has the form $f(q^2) = k\,g^2/(\mu^2 + q^2)$, where k is a constant, g represents pion-nucleon coupling constant and $q^2 (> 0)$ is the four-momentum transfer squared. From this expression it can be observed that $f(q^2)$ has a singularity or pole at $q^2 = -\mu^2$. The scattering amplitude and hence the scattering cross section becomes immeasurable (∞) at the pole. This singularity lies in the unphysical region because in the physical region $q^2 > 0$ and here $q^2 = -\mu^2$ is negative, mass μ of pion being positive. The appearance of poles in the scattering amplitude is a problem with OPEM as the infinity in the scattering amplitude makes the OPEM procedure unrenormalizable.

The expression for $f(q^2)$ tells that, it will have a large (measurable) finite value when both μ^2 and q^2 are small implying that the exchange particle should not have a large mass. Mass of pion which is 140 MeV fits this requirement. Rho meson has a much larger mass (μ) of 765 MeV, which makes its participation in transferring the interaction less probable. Smallness of q^2, sets the energy range as low. This, in turn, implies that in the pion exchange scattering process, only the partial waves with higher value of l (angular momentum) would participate and the more central range of partial waves would remain dormant. This feature points to the inadequacy of OPEM.

In the nonrelativistic theory, a particular process, say a two-body elastic scattering, has only one process ($a + b \rightarrow a + b$). However, in relativistic case three processes, viz., the s-channel ($a + b \rightarrow a + b$), the t-channel $(a+\bar{a} \rightarrow \bar{b}+b)$ and the u-channel process $(a+\bar{b} \rightarrow a+\bar{b})$ occur. These processes have been described in chapter 7, where for four momenta; we may refer to eqns. (7.2), (7.34) and (7.39). These are related yet different processes, one process can be reached from other by carrying out its analytic continuation in the corresponding region of Mandelstam plane. If in OPEM, the mediating particle has spin, asymtotically, the scattering amplitude in t-channel process turns out to be varying as s^j. For $s \rightarrow \infty$, this leads to a scattering cross section growing like s^{2j-2}. This shows that when the spin of the mediating particle $j > 1$, and also when the energy s is high, the scattering cross section increases indefinitely. This behaviour is unacceptable as it violates unitarity (conservation of probability). Thus, when spin of the intermediate particle is higher than one and in some cases even equal to one, OPEM becomes divergent or unrenormalizable.

Therefore, due to the above reasons, the one pion (particle) exchange model has to be replaced by another model. A suitable replacement is the Regge pole model.

8.5.3 The Regge Pole Model

As mentioned in section 8.1, the scattering amplitude is expressed as the sum of partial waves belonging to different angular momenta. When at certain energies, the amplitude of a partial wave becomes very large (∞), a pole is said to be existing corresponding that energy. If the angular momentum is taken as complex and continuous, having a nonzero positive imaginary part and the amplitude of a partial wave becomes infinite at a certain energy and angular momentum, we say there is a Regge pole corresponding to that energy. Thus, Regge poles are introduced when in the partial waves, the real and discrete angular momentum values are replaced by the complex and continuous values. Of course, here the partial waves are not the old ones, but obtained by the Watson-Sommerfeld transformation of old partial waves. As in resonances, we refer a resonance as a short-lived particle, the Regge pole is referred to a Reggeon. The model, in which the mechanism of interaction (scattering) is through the exchange of Regge poles, is called the Regge pole model. Significant features of Regge pole model are given below.

8.5.3.1 Basis for Watson and Sommerfeld Transformation

Watson and Sommerfeld introduced complex angular momenta in a method to solve the problem of diffraction of radio waves around the surface of the earth. They found that here the method of partial waves becomes inapplicable because of the requirement of unmanagably large number of partial waves.

For real angular momenta (l = 0, +1, +2, +3, ... ∞), the scattering amplitude expanded in terms of partial waves is given in eqn. (8.39) with partial wave amplitudes as (8.40). Retaining k in the partial wave amplitudes (8.40), we may express the partial wave expansion of scattering amplitude (8.39) as

$$f_l(k, \cos\theta) = \sum_{l=0}^{\infty}(2l+1)\, a_l(k)\, P_l(\cos\theta) \tag{8.96}$$

where the complex quantities

$$a_l(k) = \frac{e^{2i\delta_l}-1}{2ik} = \frac{e^{i\delta_l}}{k}\sin\delta_l \tag{8.97}$$

are the amplitudes of partial waves. Coming to channel processes, the partial wave s-channel expansion of the scattering amplitude can be expressed from (8.96) as

$$f_l(k, \cos\theta_s) = \sum_{l=0}^{\infty}(2l+1)\, a_l(k,s)\, P_l(\cos\theta_s) \tag{8.98}$$

where, from eqns. (7.28) and (7.18a) with $\theta \to \theta_s$,

$$\cos\theta_s = 1 + \frac{2t}{s-4m^2} \tag{8.99}$$

Since s values range from $4m^2 \to \infty$ and $\cos\theta_s$ range from –1 to +1, the physical s-channel domain is

$$s \geq 4m^2 \quad \text{and} \quad -1 \leq \cos\theta_s \leq 1 \tag{8.100}$$

However, for asymptotic values that are considered in scattering, the domain would be larger than the physical domain (8.100) of s-channel process. The scattering amplitude (8.98), whose square would give the scattering cross section, is therefore required to converge (have a finite value) in this larger domain. In s-channel scattering, for complex $\cos\theta$, such a convergence is possible. The choice of complex $\cos\theta$ is because we have to shift to complex angular momentum and demand analyticity of scattering amplitude, the latter requiring a complex plane. The convergence in s-channel can be understood from the following.

When $\cos\theta_s$ is complex, the Legendre functions $P_l(\cos\theta_s)$ grow exponentially as $l \to \infty$. To make the expansion (8.98) converge, the partial wave amplitudes $a_l(k, s)$ should diminish exponentially. If the scattering potential is Yukawa type, the amplitudes $a_l(k, s)$ do diminish exponentially, no such diminishing of amplitudes takes place for arbitrary potentials. Therefore, for the complex $\cos\theta_s$, when the scattering is in the s-channel and the potential is Yukawa type, the scattering amplitude (8.98) remains converged for the real angular momentum (l) values.

Now, we examine above situation (convergence) in a crossed channel. t and u are the crossed channels. As all the three channels are not independent (7.32a), we may choose t-channel scattering as the crossed channel process. The t-channel scattering amplitude corresponding to (8.98) would be

$$f_t(k, \cos\theta_t) = \sum_{l=0}^{\infty} (2l+1)\, a_l(k,t)\, P_l(\cos\theta_t) \tag{8.101}$$

where (eqn. 7.38d)

$$\cos\theta_t = 1 + \frac{2s}{t - 4m^2} \tag{8.102}$$

The scattering amplitude is required to be analytic. This requirement is, in fact, the manifestation of the causality principle. If the scattering amplitude is not analytic; it would mean that the scattered waves have appeared at their source without even having emitted. Therefore, the t-channel scattering amplitude (8.101) has to be analytic and should be analytically continued into other regions of the Mandelstam plane. When $t > 4m^2$, the imaginary part of scattering amplitude is not zero. This condition does not make (8.101) analytic in the variable, t. However, the same scattering amplitude (8.101) is analytic in the variable s and thus can be continued analytically to the region of large positive, s. This is because in the s-region $t \leq 0$ and in the t-region $t \geq 4m^2$. At large s the relative distance between $t > 4m^2$ and $t < 0$ is small $[(4m^2/s) << 1]$. Therefore, for fixed $t > 4m^2$, the scattering amplitude (8.101) is an analytic function of s and therefore can be analytically continued to the region of large positive s. The region of large positive and finite s is the asymptotic region of s. Thus in the analytic continuation we move from a physical t region to an asymptotic physical s region. However, this transition from t region to the region of $s \to \infty$ and $t < 0$, though allowed, cannot make eqn. (8.101), the scattering amplitude of t-channel, convergent. Qualitatively, this can be understood from the following.

In the t-channel scattering amplitude (8.101), $\cos\theta_t$, expressed by (8.102) is contained in $P_l(\cos\theta_t)$. Equation (8.102) can be written as

$$\cos\theta_t = \frac{2s + t - 4m^2}{t - 4m^2} \tag{8.103}$$

For $s \to \infty$ and t negative, $\cos\theta_t \to -\infty$. Now for

$$\cos\theta_t \to \infty, \quad P_l(\cos\theta_t) \to [\cos\theta_t]^l \tag{8.104}$$

Under the condition (8.104), it can be worked out that the partial wave expansion (8.101) converges quite poorly. This implies that the t-channel scattering amplitude in the form (8.101) does not fulfil the demand of convergence. This form needs to be changed. The necessary transformation of (8.101), which would provide a converging partial wave series for t-channel scattering amplitude was brought about by Watson and Sommerfeld. Hence, such transformation is known as Watson-Sommerfeld transformation.

8.5.3.2 *Watson and Sommerfeld Transformation*

Watson-Sommerfeld transformation is a technique for converting a series into contour integrals using the Cauchy's theorems. The basic characteristic of this approach is the conversion of the slowly converging series to a rapidly converging "Watson" series, which is defined as the sum of residues corresponding to an integrand's complex poles.

In the present case, Watson-Sommerfeld transformation will be used to obtain a converging partial wave series for the scattering amplitude in the t-channel. This transformation transforms the diverging scattering amplitude (8.101) of t-channel to a converging one. For this purpose, the angular momentum is assumed complex and continuous. In (8.101) the angular momenta were real positive and discrete.

It is assumed that the partial wave amplitudes $a_l(k, t)$ of (8.101) can be continued analytically from the integer values of angular momentum (l) into the complex plane of continuous variable alpha, which is a function of E and is denoted by $\alpha(E)$ where E is the kinetic energy of the incident particle. If the beams collide, E is the centre of mass energy which is the total energy of both the colliding particles. Thus,

$$\alpha(E) = Re\,\alpha(E) + i\,Im\,\alpha(E) \tag{8.105}$$

represents the complex angular momentum with its real part as $Re\,\alpha(E)$ and imaginary part as $Im\,\alpha(E)$. The real angular momentum $l = Re\,\alpha(E)$. When $E = E_n$, $Re\,\alpha(E_n) = n$, where n is an integer. $Re\,\alpha(E_n)$ therefore, provides integer value of real angular momentum. Writing a first order Taylor expansion for $E = E_n$

$$Re\,\alpha(E) = Re\,\alpha(E_n) + (E - E_n)\left[\frac{d}{dE} Re\,a(E)\right]_{E=E_n} \tag{8.106}$$

On the RHS of eqn. (8.106) putting $\left[\frac{d}{dE} Re\,a(E)\right]_{E=E_n} = \varepsilon$ and using $Re\,\alpha(E_n) = n$

and then substituting it into (8.105), we have

$$\alpha(E) = n + \varepsilon[(E - E_n) + i\,\frac{\Gamma}{2}] \tag{8.107a}$$

where
$$\Gamma = \frac{2}{\varepsilon}\,Im\,\alpha(E) \tag{8.107b}$$

Equation (8.107a) can be used in (8.83), which describes Breit-Wigner resonance. For that we express it as

$$-\varepsilon\left[(E - E_n) + i\frac{\Gamma}{2}\right] = n - \alpha(E) \tag{8.108}$$

Substituting (8.108) into (8.83), the form of partial wave amplitude near a resonance of angular momentum (n) and energy (E_n) can be expressed as

$$a(n, E) = \frac{R(E)}{n - \alpha(E)} \tag{8.109}$$

where $n = l$ and $R(E)$ is some function that we may need to know later. Equation (8.109), tells that whenever l and E have values corresponding to a resonance, the amplitude $a(n = l, E)$ shall have a singularity or pole. Therefore, partial wave amplitudes have poles, which have been denoted by $\alpha(E)$.

To transform the discrete sum (8.101) into an integral, we recall Cauchy's integral theorem, which for a function $F(z)$ reads

$$F(b) = \frac{1}{2\pi i}\int_c \frac{F(z)}{z-b}\,dz \tag{8.110}$$

To relate (8.110) with our case, we rewrite the partial wave expansion (8.101)

$$f_l(k, \cos\theta_t) = \sum_{l=0}^{\infty}(2l+1)\, a_l\,(k,t)\, P_l\,(\cos\theta_t) \tag{8.101}$$

Establishing a correspondence with (8.110), the RHS of expansion (8.101) is the $F(z)$. Since integration requires a continuous variable, that in our case is angular momentum $\alpha(E) = \alpha$, which is complex and continuous. So $z = \alpha$ and $dz = d\alpha$. In (8.110) $z = b$ represents a pole, which in our case is $\alpha = l$ so $b = l$ and $\dfrac{1}{z-b} = \dfrac{1}{\alpha - l}$.

In view of these replacements, $F(z)$ would be

$$F(z) = (2\alpha + 1)\,\alpha(\alpha, t)\, P_\alpha(\cos\theta_t) \tag{8.111}$$

and from (8.101), $F(b)$ would be ($b = l$)

$$F(l) = (2l + 1)\, a_l\,(k, t)\, P_l(\cos\theta_t) \tag{8.112}$$

where the partial wave amplitude $a_l(k, t)$ in discrete expansion (8.101), is $a(\alpha, t)$ in the continuous form (8.111). Before arriving at the integral form of (8.101), it will be shown that $\dfrac{1}{\alpha - l} = \dfrac{\pi(-1)^l}{\sin\pi\alpha}$. For that we write

$\pi\alpha = \pi\alpha - \pi l + \pi l$, where $\pi\alpha - \pi l$ is small. Therefore,

$$\begin{aligned}\sin\pi\alpha &= \sin(\pi\alpha - \pi l)\cos\pi l + \cos(\pi\alpha - \pi l)\sin\pi l \\ &\approx \pi(\alpha - l)\cos\pi l + \sin\pi l\end{aligned} \tag{8.113}$$

Now, for integer values of l, $\sin\pi l$ is always zero. Putting $l = 0, +1, +2, +3$ ... the first term would have the values

$$+\pi\alpha,\ -\pi(\alpha - 1),\ +\pi(\alpha - 2),\ -\pi(\alpha - 3)\ \dots\ \dots$$

The general term for these values is $\pi(-1)^l(\alpha - l)$. Equating this general term to $\sin\pi\alpha$, we would get

$$\frac{1}{\alpha - l} = \frac{\pi(-1)^l}{\sin\pi\alpha} \tag{8.114}$$

where in the RHS l will be replaced by its continuous counterpart, α.

The Watson-Sommerfeld transformation of discrete partial wave expansion of (8.101) into a continuous form would now be obtained by substituting the relavent eqns. {(8.112), (8.111) and (8.114)} into (8.110) to write

$$\text{L H S} = \text{R H S} \tag{8.115a}$$

where

$$\text{L H S} = \sum_{l=0}^{\infty}(2l+1)\, a_l(k,t)\, P_l(\cos\theta_t) \tag{8.115b}$$

and

$$\text{R H S} = \frac{1}{2i}\int_{c_i} \frac{(-1)^{\alpha}(2\alpha+1)\, a(\alpha,t)\, P_{\alpha}(\cos\theta_t)}{\sin\pi\alpha} d\alpha \tag{8.115c}$$

where π in RHS has cancelled. C_l, $i = 1, 2, 3$ ... denote the integration contours. When $\alpha = 0, 1, 2\ 3$ ..., the denominator $\sin\pi\alpha = 0$. Thus, at $\alpha = 0, 1, 2, 3$ ...we have poles, around which contours, C_0, C_1, C_2, ... C_i ...have to be drawn. Using the property of Legendre functions

$$(-1)^{\alpha} P_{\alpha}(\cos\theta_t) = P_{\alpha}(-\cos\theta_t) \tag{8.116}$$

the transformation (8.115) can be expressed as

$$\sum_{l=0}^{\infty}(2l+1)\, a_l\, (k,t)\, P_l\, (\cos\theta_t) = \frac{1}{2i}\int_{c_i} \frac{(2\alpha+1)a(a,t)P_a(-\cos\theta_t)}{\sin\pi\alpha} d\alpha \tag{8.117}$$

Equation (8.117) in which, employing Cauchy's integral theorem, a partial wave series in terms of real and discrete angular momentum values (LHS) is transformed into a partial wave series expressed in terms of complex and continuous angular momenta (RHS), is the desired Watson-Sommerfeld transformation. Both the partial wave series represent scattering amplitude. Conversely, RHS of (8.117) can be converted into LHS, by taking residues of ($1/\sin\pi\alpha$), at $\alpha = l = 0, 1, 2, 3$, ... and making use of Cauchy's residue theorem

$$\oint_C F(z)dz = 2\pi i \sum_n \text{Res}\, F(z_n) \tag{8.118}$$

The relation (8.118) tells that when the integral (RHS of 8.117)is taken along the contours C_i, the contribution of residues of poles of ($1/\sin\pi\alpha$) in the plane of complex α, gives rise to various terms of the partial wave series (LHS of 8.117).

8.5.3.3 Contour and Its Modification

The Watson-Sommerfeld (WS) transformation is carried out in two steps. In the first step, already taken in the above, a series is converted into an integral. In the second step, to be taken now, the contour that encircles the poles of ($1/\sin\pi\alpha$) has to be deformed to enclose the poles of $a(\alpha, t)$. As mentioned above, $a(\alpha, t)$ is the partial wave amplitude when angular momentum (α) is complex and continuous. Similar to its discrete counterpart $a_l\,(k, t)$, it [$\alpha(\alpha, t)$] too has sigularities or poles.

The scattering amplitude (8.101), in view of W-S transformation (8.117) can now be expressed as a contour integral

$$f_\alpha(t, \cos\theta_t) = \frac{1}{2i}\int_{C_i} \frac{(2\alpha+1)a(\alpha,t)P_\alpha(-\cos\theta_t)}{\sin\pi\alpha}\, d\alpha \tag{8.119}$$

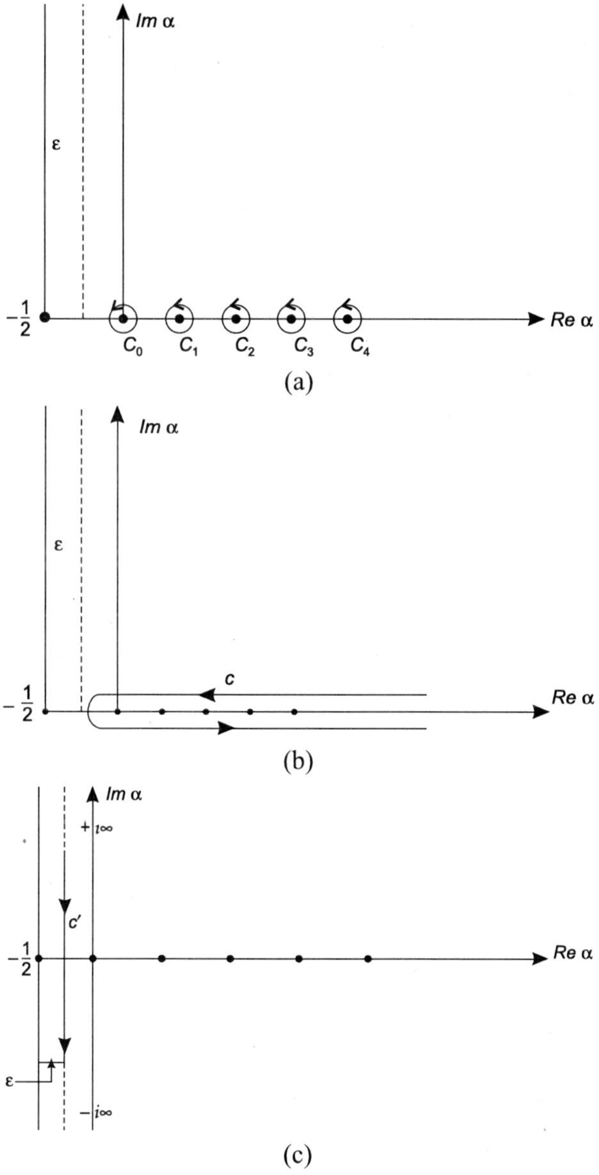

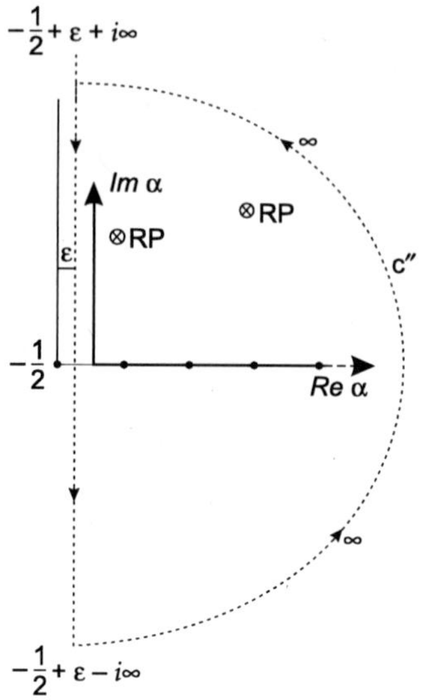

(d)

Fig. 8.3 Contour modifications: (a) Contours encircling poles (dots) along real alpha axis; (b) stretching of contours of (a) into single contour c; (c) shape c' after swinging contour (b); (d) ends of contour c' joined by infinite semicircle (dots – simple poles, crosses – Regge poles)

In Fig. 8.3a, a complex plane with $Re\ \alpha$ and $Im\ \alpha$ as real and imaginary axes has been drawn and along the positive real axis are the poles of $(1/\sin \pi\alpha)$ corresponding to $\alpha = 0, +1, +2, +3 \ldots$. The poles of $(1/\sin \pi\alpha)$ are of the first order so they are simple poles and have real l ($Re\ \alpha = 1$, $Im\ \alpha = 0$) values.

Encircling counterclockwise these poles, (Fig. 8.3a), are the contours C_0, C_1, C_2, ... C_i ... . In (Fig. 8.3b) these contours have been stretched into a single contour C to surround all these real poles. The contours C_i in eqn. (8.119) may now be replaced by the contour C. The contour C_0 and the contour C should obey $Re\ \alpha > -(1/2)$. The reason for the limit $Re\ \alpha = -(1/2)$ is found in $r \to 0$ solutions of radial wave equation, where we have two independent solutions that go like $r^{\alpha+1}$ and $r^{-\alpha}$. If we equate them we get $\alpha = -(1/2)$. Along the line $Re\ \alpha = -(1/2)$ these two solutions exchange their roles. $Re\ \alpha > -(1/2)$ implies a vertical line $Re\ \alpha = -(1/2) + \varepsilon$, where ε is small. The contour C_0 and the contour C should not cross this vertical line. A key objective of Watson Sommerfeld transformation is that the integral (8.119) should converge. The condition that $Re\ \alpha > -(1/2)$ applies for convergence as well. It is observed that, when $Re\ \alpha$ is greater than minus (1/2), the partial wave amplitude $a(\alpha, t)$ tends to zero exponentially for $|\alpha| \to \infty$, resulting in the convergence of integral (8.119).

From the discussion below eqn. (8.41), it may be recalled that the partial wave amplitudes depend on the scattering potential. For Yukawa type potentials, the counterclockwise contour (Fig. 8.3b) can be distorted such that its arms which were parallel to $Re\ \alpha$, now become parallel to $Im\ \alpha$ and extend along the positive and negative directions of *Im* α to form a vertical line (Fig. 8.3c). Since, *Re* α obeys *Re* $\alpha = -(1/2) + \varepsilon$, the vertical line parallel to *Im* α shall extend from $\alpha = -\frac{1}{2} + \varepsilon + i\infty$ to $\alpha = -\frac{1}{2} + \varepsilon - i\infty$. This vertical line then becomes the final contour of integration

for real poles [poles of $(1/\sin \pi\alpha)$]. This contour (vertical line) has been denoted by C'. The contour integral (8.119) can then be expressed as

$$f_\alpha(t, \cos\theta_t) = \frac{1}{2i}\int_{C'} \frac{(2\alpha+1)a(\alpha,t)P_\alpha(-\cos\theta_t)}{\sin \pi\alpha}\, d\alpha \tag{8.120a}$$

where as discussed above,

$$\int_{C'} \equiv \int_{-(1/2)+\varepsilon+i\infty}^{-(1/2)+\varepsilon-i\infty} \tag{8.120b}$$

in which sometimes ε is omitted.

Now, apart from these simple poles of $(1/\sin \pi\alpha)$, there are poles of partial wave amplitude $a(\alpha, t)$. These poles of $a(\alpha, t)$ are complex and have *Im* $\alpha > 0$. The number of these poles is finite and their domain is to the right of the line *Re* $\alpha = -(1/2) + \varepsilon$. This domain can be marked by joining the ends of contour by an infinite semicircle. This has been done in (Fig. 8.3d). Within the semicircle, the poles have been shown by dot and cross. The dot poles are simple poles and cross poles are the Regge poles. The α for simple poles is real and for Regge poles complex. Further discussion on real and imaginary parts of complex α has been taken in 8.5.3.7.

8.5.3.4 Separation of Simple and Regge Poles

As told above, in eqn. (8.120), the integrand has two kinds of poles, the simple and the Regge poles. The Regge poles are the poles of partial wave amplitude $a(\alpha, t)$, that appear due to the dynamical consequences of scattering potential, while simple poles arise from the zeros of $\sin \pi\alpha$. The Regge poles too could be of first order and hence they may also be simple poles. However, here by simple poles we mean the poles of $(1/\sin \pi\alpha)$. These two kinds of poles can be separated by using (8.120) and also invoking Cauchy's residue theorem (8.118). As against the contour of (8.120), the contour for integration now is the closed contour C'' of Fig. (8.3d), in which the infinite semicircle connects the infinite vertical line.

To separate the simple and Regge poles, we apply the residue theorem (8.118) to (8.120). As the poles are isolated, there is summation (Σ) of residues at the poles. For the closed contour C'', we write (8.120) as

$$\oint_{C''} F(\alpha,t)\,d\alpha = \frac{1}{2i}\oint_{C''} \frac{(2\alpha+1)a(\alpha,t)P_\alpha(-\cos\theta_t)}{\sin \pi\alpha}\, d\alpha \tag{8.121}$$

where the LHS of (8.120) has been written in the form of a closed contour integral of an integrand $F(\alpha, t)$, which from (8.121) is

$$F(\alpha, t) = \frac{1}{2i}\left[\frac{(2\alpha+1)a(\alpha,t)P_\alpha(-\cos\theta_t)}{\sin\pi\alpha}\right] \tag{8.122}$$

Applying Cauchy's residue theorem (8.118) to equation (8.121), we can write

$$\oint_{C''} F(\alpha,t)d\alpha = 2\pi i \sum_{\text{Re}\,\alpha=1}^{\infty} Res\, F(\alpha,t) + 2\pi i \sum_{RP} Res\, F(\alpha,t) \tag{8.123}$$

where *RP* is for Regge poles. In the above equation, the first term on the RHS, $2\pi i$ times the sum of residues at the simple poles of $(1/\sin\pi\alpha)$, is equal to the RHS of (8.120). The second term is also $2\pi i$ times the sum of residues at other poles called Regge poles. The difference between the simple and Regge poles is that whereas simple poles are along the $Re\ \alpha$ axis obviously having $Im\ \alpha = 0$, the Regge poles will have a complex value of α where, in general, $Re\ \alpha \neq 0$. $Im\ \alpha \neq 0$. For Regge poles $Im\ \alpha > 0$. Convention is that the Regge poles are distinguished from each other by their $Re\ \alpha$ values.

Thus, eqn. (8.123) separates the simple and Regge poles. The significance of (8.123) is that it allows the enlargement of convergence zone of the *t*-channel process. Though, the physical domain of a *t*-channel process is $t \geq 4m^2$, $-1 \leq \cos\theta_t \leq +1$, in regard to scattering angles, (8.123) extends its applicability to $\cos\theta_t \to \infty$. In view of the discussion below eqn. (8.102), the asymptotic region $\cos\theta_t \to \infty$, belongs to high-energy region, where $s \to \infty$. Therefore, Regge pole formulation (8.123), because of allowing $s \to \infty$ (high-energy) region, is relevant to a large number of elementary particle reactions that obviously are in the range of high energies (~GeV).

8.5.3.5 Regge Pole Residue

From (8.122) and (8.119), we may write

$$2\pi i \sum_{Re\,\alpha=1}^{\infty} Res\, F(\alpha,t) = f_\alpha(t,\cos\theta_t) \tag{8.124a}$$

where

$$f_\alpha(t, \cos\theta_t) = \frac{1}{2i}\int_{C'} \frac{(2\alpha+1)a(\alpha,t)P_\alpha(-\cos\theta_t)}{\sin\pi\alpha}d\alpha \tag{8.124b}$$

Therefore, from (8.123) and (8.124a), we may write

$$f_\alpha(t, \cos\theta_t) = 2\pi i \sum_{Re\,\alpha=1}^{\infty} Res\, F(\alpha,t) = -2\pi i \sum_{RP} Res\, F(\alpha,t) + \oint_{C''} F(\alpha,t)\, d\alpha \tag{8.125}$$

Using (8.121) and (8.124), above equation is written as

$$\frac{1}{2i}\oint_{C''}\frac{(2\alpha+1)a(\alpha,t)P_\alpha(-\cos\theta_t)}{\sin\pi\alpha}d\alpha -$$

$$\frac{1}{2i}\int_{c'}\frac{(2\alpha+1)a(\alpha,t)P_\alpha(-\cos\theta_t)}{\sin\pi\alpha}d\alpha = 2\pi i\,\Sigma_{RP}\,\text{Res}\,F(\alpha,t) \qquad (8.126)$$

Equation (8.126) shows that the difference of the integrals of same integrand over a closed contour C'', and an open contour, C' equates to a Regge pole term. This implies that on enlarging contour C' to a closed contour C''. Regge poles (RP) appear. This observation is natural, as contour C' covers the poles having only $Im = 0$, whereas Regge poles have $Im \neq 0$. The closed contour C'' (Fig. 8.3d) covers the poles having $Im \neq 0$ as well. Symbolically, (8.126) implies

(Regge poles + Real poles) – Real poles = Regge poles

The RHS of (8.126) denotes the sum of residues of $F(\alpha, t)$ (eqn. 8.122) at the Regge poles. As mentioned earlier, the Regge poles are the poles of $a(\alpha, t)$, the partial wave amplitudes, when angular momentum (α) is complex and continuous. Therefore, in equation (8.122) the residue of $F(\alpha, t)$ would imply the residue of $a(\alpha, t)$. Let the residue of $a(\alpha, t)$, at n^{th} Regge pole is β_n with running from 0 to N, such that the number of Regge poles is $N + 1$, a finite number. The α values, which determine their coordinates in complex α plane, change from one Regge pole to other. For n^{th} Regge pole it is α_n. The RHS of (8.126), on using (8.122) can then be expressed as

$$2\pi i\sum_{RP} Res\ F(\alpha,t) = 2\pi i\sum_{n=0}^{N}\frac{1}{2i}\left[\frac{(2\alpha_n+1)\beta_n P_{\alpha_n}(-\cos\theta_t)}{\sin\pi\alpha_n}\right] =$$

$$\pi\sum_{n=0}^{N}\left[\frac{(2\alpha_n+1)\beta_n P_{\alpha_n}(-\cos\theta_t)}{\sin\pi\alpha_n}\right] \qquad (8.127)$$

Thus, $\pi\sum_{n=0}^{N}\left[(2\alpha_n+1)\beta_n P_{\alpha_n}(-\cos\theta_t)(\sin\pi\alpha_n)^{-1}\right]$ is $2\pi i$ times the sum of $(N + 1)$

Regge pole residues. Asymptotically,

$$\cos\theta_t\to\infty,\quad P_\alpha(\cos\theta_t)\to[\cos\theta_t]^\alpha = [\cos\theta_t]^{Re\,\alpha + i\,Im\,\alpha} \qquad (8.128)$$

The summation (8.127) is therefore dominated by the Regge pole, for which $Re\ \alpha$ has the largest positive value. Such a Regge pole is situated at the farthest right point in the Regge pole domain. Thus, for large $\cos\theta_t$ the scattering amplitude (8.125) can be expressed in terms of the farthest right single Regge pole. It has, however, to be understood that $\cos\theta_t \to \infty$, acquires physical meaning only in the relativistic scattering In non-relativistic potential scattering 0 to π remain the limits of $\cos\theta_t$.

8.5.3.6 *Signatures of Regge Poles*

In Watson-Sommerfeld transformation (8.117), that we rewrite below, the partial wave

$$\sum_{l=0}^{\infty}(2l+1)a_l(k,t)P_l(\cos\theta_t) =$$

$$\frac{1}{2i}\int_{C_i}\frac{(2\alpha+1)a(\alpha,t)P_\alpha(-\cos\theta_t)}{\sin\pi\alpha}\,d\alpha \tag{8.129}$$

amplitude $a(\alpha, t)$ is expected to be analytic in the plane of complex and continuous angular momentum, α. The behaviour of partial wave amplitude depends on the nature of scattering. In the non-relativistic potential scattering, if no exchange potential (exchange force) is present, a single function, $a(\alpha, t)$, can be used to represent all the partial wave amplitudes corresponding to $Re\ \alpha = 0, +1, +2, +3, \ldots$. However, in the relativistic scattering, i.e., *s, t, u* channel scatterings or in a nonrelativistic scattering where an exchange potential is present, a single $a(\alpha, t)$ cannot represent the partial wave amplitudes corresponding to all positive integer values of real part of α. To understand this point, let us consider the *t*-channel reaction $\pi^- + \pi^0 \rightarrow \bar{p} + n$, which is the crossed process of *s*-channel reaction $\pi^- + p \rightarrow \pi^0 + n$. The pions in the *t*-channel process obey Bose-Einstein statistics and also the isospin is conserved. These two elements, viz., the BE statistics and the conservation of isospin, combine in such a way that all the partial waves corresponding to even values of $Re\ \alpha$ vanish. The Carlson's theorem now becomes applicable. This theorem states that if a function, whose growth is bound at infinity vanishes at non-negative integer values, shall be identically zero. Therefore, in the above discussed *t*-channel process, in view of Carlson's theorem, the partial wave amplitude $a(\alpha, t)$ shall vanish in the entire domain of $Re\ \alpha$ values. To remedy this situation, two separate partial wave amplitudes are employed for even and odd angular momentum ($Re\ \alpha$) values. We may use $a_+(\alpha, t)$ as the partial wave amplitude for $Re\ \alpha = 0, +2, +4, +6, \ldots$. and $a_-(\alpha, t)$ for $Re\ \alpha = 1, +3, +5, +7, \ldots$. The (+) and (–) signs in $a_+(\alpha, t)$ and $a_-(\alpha, t)$ are the signatures of partial wave amplitude $a(\alpha, t)$ for even and odd angular momentum values respectively. The signature factor. (S.F.) is expressed as

$$S.\,F. = \frac{1}{2}(1 \pm e^{-i\ \pi Re\ \alpha}) \tag{8.130}$$

and accordingly $a(\alpha, t)$ can be expressed as

$$a(\alpha, t) = \frac{1}{2}a_+(\alpha,t)(1+e^{-i\pi\ Re\ \alpha}) + \frac{1}{2}a_-(\alpha,t)(1-e^{-i\pi\ Re\ \alpha}) \tag{8.131}$$

which reduces to $a(\alpha, t) = a_+(\alpha, t)$ for even values of $Re\ \alpha$ and to $a(\alpha, t) = a_-(\alpha, t)$ for odd values of $Re\ \alpha$. In view of the appearance of the signature factor, the Regge poles, i.e., the poles of $a(\alpha, t)$, shall too have even (+) and odd (–) signatures. Regge poles having even angular momentum values shall be on one trajectory and those having odd angular momenta shall reside on a separate trajectory.

8.5.3.7 *Regge Pole Trajectories*

The path of a Regge pole on the complex α plane is the trajectory of a Regge pole. Since, our focus, mainly, is on t-channel process (scattering), let us recall (7.38) for the conditions of its occurance. In t-channel process $t \geq 4m^2$, $s \leq 0$, $u \leq 0$, where t represents the centre of mass energy squared. The square of minimum energy is $t = 4m^2$ and therefore called the threshold energy squared. When t is varied the Regge poles of the scattering amplitude trace out curves. These curves are called the Regge trajectories.

Figure (8.4) shows the behaviour of a Regge trajectory in the complex angular momentum plane having $Re\ \alpha$ and $Im\ \alpha$ as its two axes.

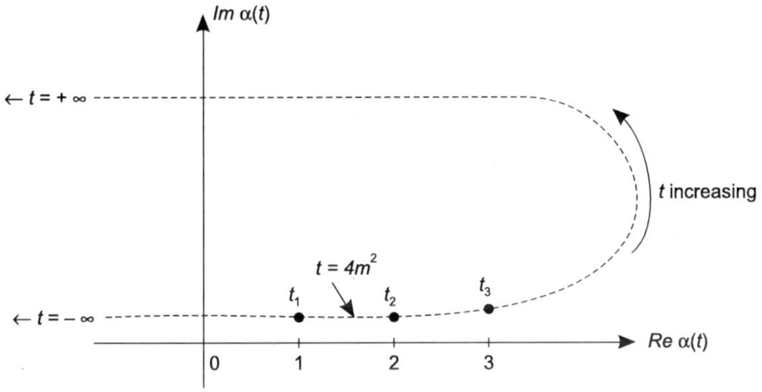

Fig. 8.4 General Regge trajectory in the $t =\ \infty$ to $t = +\infty$ rangc.
t_1 a bound state; t_2, t_3 resonances, $t = 4m^2$ threshold.

Equation (8.109) for the t-channel can be expressed as

$$a(l, t) = \frac{R(t)}{l - \alpha(t)} \tag{8.132}$$

where $a(l, t)$ is the partial wave amplitude corresponding to $Re\ \alpha = l$ and $R(E) \to R(t)$. $l = 0, +1, +2, \ldots$ The partial wave amplitude (8.132) manifests a pole at $l = \alpha(t)$. In the following (8.132) shall be analyzed for $t < 4m^2$ and $t > 4m^2$ cases.

When $t < 4m^2$, the centre of mass energy squared of the input section of t-channel is below the threshold and in such a case, the partial wave amplitude $a(l, t)$ is real and therefore $R(t)$ and $\alpha(t)$ are also real. Since l has positive integer values and if at its value, say $l = +n$, the partial wave amplitude $a(l, t)$ has a pole at $t = M^2 < 4m^2$, then the partial wave amplitude corresponds to a bound state (stable particle of mass M) with angular momentum, $+n$. This defines bound states as the poles of partial wave amplitude on the *Real* α axis when the energy is below the threshold. On the $Re\ \alpha$ axis, $Im\ \alpha = 0$. Thus, below the threshold, the pole $l = \alpha(t)$ of (8.132) behaves as a bound state.

Above the threshold, $t > 4m^2$. In this case, it turns out that the imaginary part of $\alpha(t)$ does not vanish ($Im\ \alpha > 0$). Therefore, for $t > 4m^2$, the partial wave amplitude (8.132), in view of (8.105), has the form

$$a(l, t) = \frac{R(t)}{l - Re\,\alpha(t) - iIm\ \ (t)} \tag{8.132a}$$

Let us assume now that at $t = t_1 = M^2 > 4m^2$, $Re\,\alpha(t_1) = n$, where $+n$ is an integer value of l. It may be noted here that when $l = n = Re\,\alpha(t_1)$ the denominator does not vanish due to $Im\,\alpha \neq 0$ and $a(l, t)$ though does not go to infinity, yet acquires a large value. The appearance of large amplitude shows a resonance state that can further be understood from the following.

Using (8.79) we may write upto first order $Re\,\alpha(t) = Re\,\alpha(t_1) + Re\,\alpha'(t_1)\,(t - t_1)$ where the prime denotes differentiation with respect to t, which when used in (8.132a) gives

$$a(l, t) = \frac{R(t_1)}{-Re\ \ \alpha'(t_1)(t - t_1) - iIm\,\alpha(t_1)} \tag{8.133}$$

where $l - Re\,\alpha(t_1) = n - Re\,\alpha(t_1) = 0$ has been used. Incorporating the known proportionality of $R(t_1)$ with $Im\,\alpha(t_1)$, (8.133) can be cast as

$$a(l, t) \sim \frac{\left[Im\,\alpha(t_1) / Re\,\alpha'(t_1)\right]}{(t_1 - t) - i[Im\,\alpha(t_1) / Re\,\alpha'(t_1)]} \tag{8.134}$$

which resembles the Breit-Wigner formula (8.83)

$$a(l, t) \sim \frac{\Gamma / 2}{(t_1 - t) - i\,\Gamma / 2} \tag{8.135a}$$

where

$$\Gamma \sim 2\ Im\,\alpha(t_1)/Re\,\alpha'\,(t_1) \tag{8.135b}$$

It may therefore be concluded that above the threshold of t, the pole $l = \alpha(t)$ of (8.132a) behaves as a resonance (unstable particle) of mass $M = \sqrt{t_1}$ angular momentum $a(t_1) = n$ and width $\Gamma \sim [2\,Im\,\alpha(M^2)/Re\,\alpha'\,(M^2)]$.

In (Fig. 8.4), a general outline of Regge trajectory has been drawn by undertaking the variation of t. The Regge trajectory starts along the *Real* $\alpha(t)$ axis and continues along the real axis as long as $t < 4m^2$. During this passage, at the locations, where the trajectory finds integer values of *Real* $\alpha(t)$, bound states or the stable particles are found. Along the real axis imaginary part of $\alpha(t)$ is zero. Threshold is reached when $t = 4m^2$. At this point the trajectory would begin to leave the *Real* $\alpha(t)$ axis and when $t > 4m^2$ it would acquire positive values for imaginary part of $\alpha(t)$ and increasing values of imaginary part would take the trajectory away from the real axis of $\alpha(t)$. During the passage of trajectory through the domain $t > 4m^2$, the sites at which *Real* $\alpha(t)$ has an integer value, resonances or the unstable particles would be found. The particular point of departure and details of trajectory depend on the reaction considered.

From the above description, it may be realized that a Regge trajectory may contain both the bound states and resonances, bound states refer to stable particles and resonances to unstable particles. Conversely, it may be stated that, if a stable particle lies on a Regge trajectory, it would be a bound state or non-elementary.

8.5.3.8 Chew-Frautschi Diagram

As seen in the previous subsection, Regge trajectory is the path of a Regge pole in the complex α plane in which $Re\ \alpha$ is the abscissa and $Im\ \alpha$ is the ordinate. Now, we change this ($Re\ \alpha$, $Im\ \alpha$) plane by (M^2, J) plane. Since, in the t-channel $t > 4m^2$, and α denotes the angular momentum, we may take M^2 (resonance mass squared) values forming the abscissa and angular momentum J values, the ordinate of the (M^2, J) plane. The diagram or the plot drawn in the (M^2, J) plane is called the Chew-Frautschi (C-F) diagram of Regge trajectory after the names of physicists G.F. Chew and S.C. Frautschi. The C-F plots are in the form of straight lines in the (M^2, J) plane. The angular momentum J refers to any angular momentum, whether pure spin, pure orbital or a combination.

In view of the requirements of the subsection, *signatures of Regge poles*, there will be separate Chew-Frautschi diagrams for even and odd angular momentum values. The difference ΔJ between two consecutive J values in either case would be 2. Thus, the Chew-Frautschi diagram may be defined as the plot of M^2 versus J with resonances (or particles) at steps of $\Delta J = 2$. The resonances that are spaced by two units of angular momentum have same parity, same isospin and same strangeness. C-F plot is therefore a graphic representation of Regge trajectories for given quantum numbers. It has been found that the known particles and resonances fall on straight line in the plot. Since, hadrons (mesons and baryons) are the source of resonances, the C-F plots, which segregate the resonances, can be used to group mesons and baryons in families with definite quantum numbers.

Though, we have described in the above that in the C-F plots the angular momenta would be spaced by two units and only the resonances that have either positive parity or odd parity would lie on a particular C-F plot, there are some resonances with even and odd parities that lie on the same C-F straight line. In this case, the spacing of angular momentum is one unit ($\Delta J = 1$) only. Such situation arises due to *exchange degeneracy*, due to which the even parity C-F plot coincides with the odd parity C-F plot and we see a single straight line for both even and odd parity resonances. In case of exchange degeneracy, the even parity C-F plot can be *exchanged* with the odd parity C-F plot and vice versa.

8.5.3.9 Examples of Chew-Frautschi Diagram

According to Chew-Frautschi postulate, the strongly interacting particles or resonances are expected to obey a linear relation that the angular momentum of the resonance (particle) is proportional to the square of its mass, i.e., $J \propto M^2$. Based on the available data on baryons and mesons, it would be seen in the following that these strongly interacting particles do observe the $J \propto M^2$ relation.

8.5.3.9.1 Baryon resonances

In the baryon resonances we list the N, Δ, Λ and Σ resonances. N, Λ, Σ resonances are the excited states of nucleon (N), Λ and Σ particles respectively, whereas Δ resonances comprise the Δ resonance and its higher states. The higher states may also be called the Regge recurrences of their lower states.

Baryon (N) Resonances [Isospin $I = (1/2)$, Strangeness $S = 0$]			
Resonance Name Mass M in MeV	Square of mass M^2 in $(\mathrm{GeV})^2$	Spin J, Parity P $(J)^P$	Breit-Wigner full width Γ in MeV
N (1440)	2.07	$(\frac{1}{2})^+$	350
N (1860)	3.46	$(\frac{5}{2})^+$	–
N (2220)	4.93	$(\frac{9}{2})^+$	
N (2700)	7.29	$(\frac{13}{2})^+$	400
N (1520)	2.31	$(\frac{3}{2})^-$	115
N (2190)	4.80	$(\frac{7}{2})^-$	500
N (2600)	6.76	$(\frac{11}{2})^-$	–

In the above table, **N (1440)**, **N (1860)**, **N (2220)** and **N (2700)** are the even parity resonances separated by spin $\Delta J = 2$. Similarly, **N (1520)**, **N (2190)** and **N (2600)** are the odd parity $\Delta J = 2$ nucleon (N) resonances. If separate graphs are plotted for these resonances between M^2 (column 2) and $2J$ (column 3), it is found (Figs. 8.5a and 8.5b) that the graphs are almost straight lines that proves the $J \propto M^2$ conjecture.

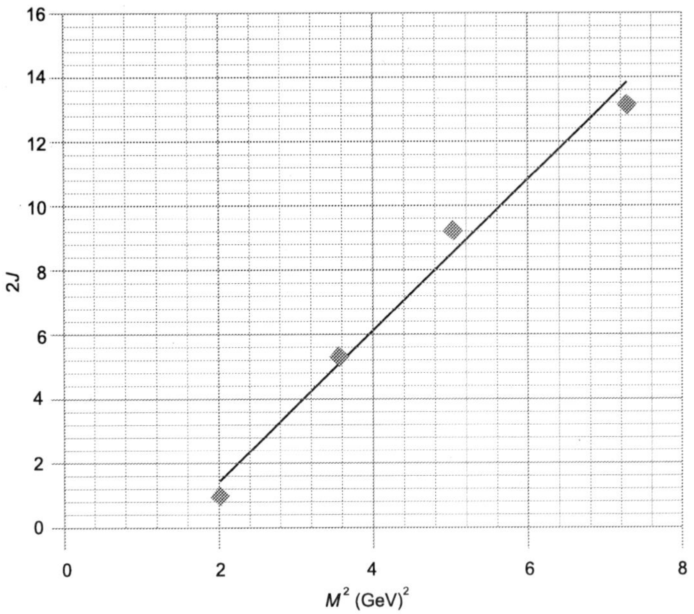

Fig. 8.5a C-F plot of even parity nucleon (N) resonances.

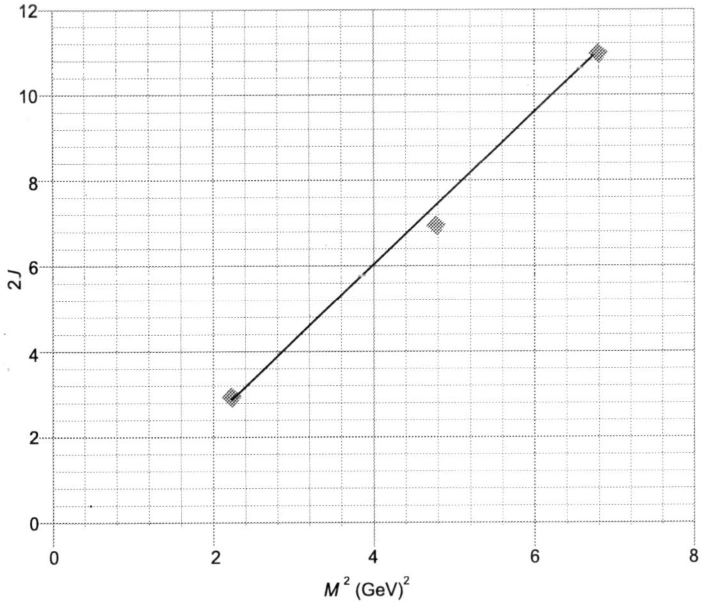

Fig. 8.5b C-F plot of odd parity nucleon (N) resonances.

The Chew-Frautschi plot of positive parity delta resonances (Fig. 8.6) listed in the following table also show the proportionality of spin (J) values to the square of mass (M^2) of these resonances.

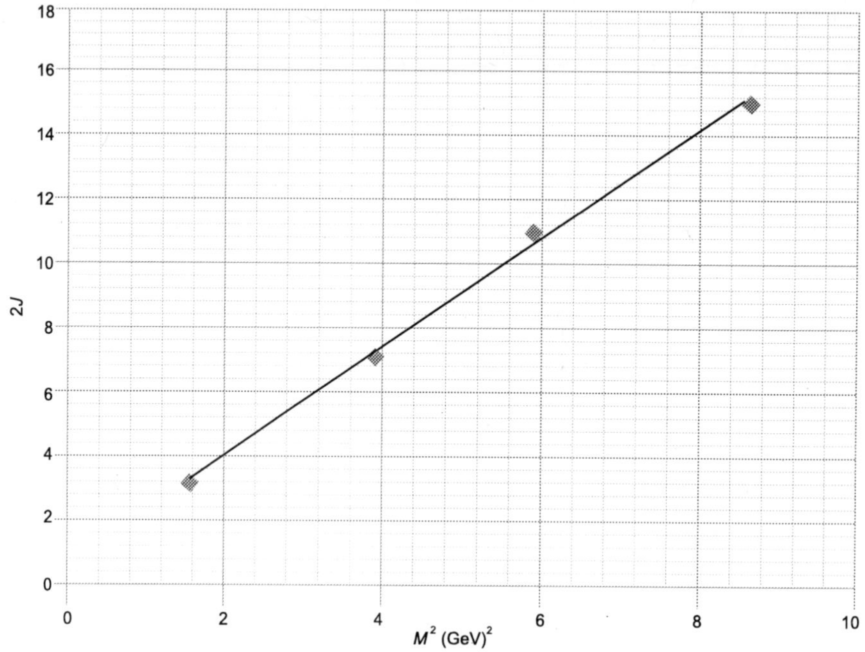

Fig. 8.6 C-F plot of even parity delta resonances.

Baryon (Δ) Resonances [Isospin $I = (3/2)$, Strangeness $S = 0$]			
Resonance Name Mass M in MeV	Square of mass M^2 in (GeV)^2	Spin J, ParityP $(J)^P$	Breit-Wigner full width Γ in MeV
Δ (1232)	1.52	$(\frac{3}{2})^+$	117
Δ (1950)	3.80	$(\frac{7}{2})^+$	285
Δ (2420)	5.86	$(\frac{11}{2})^+$	400
Δ (2950)	8.70	$(\frac{15}{2})^+$	–

8.5.3.9.2 Exchange degenerate baryon resonances

In the lambda (Λ) and sigma (Σ) baryon resonances (Figs. 8.7a &b) both the positive parity and negative parity resonances fit the same straight line. Here the resonances are separated by $\Delta J = 1$, instead of the usual $\Delta J = 2$. As described in the previous subsection, these are *exchange degenerate* baryon resonances. Following table lists them.

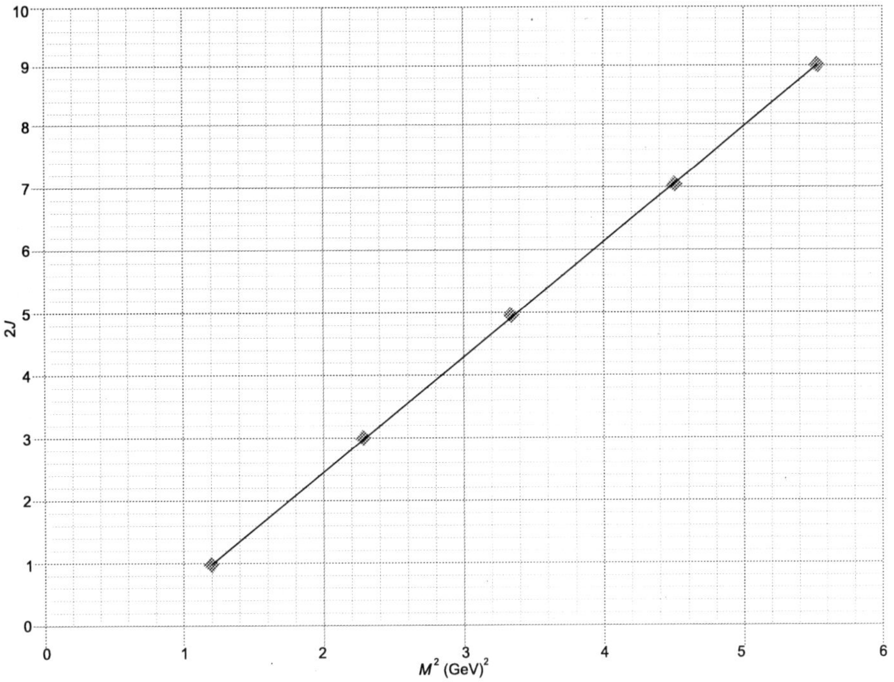

Fig. 8.7a C-F plot of exchange degenerate (both parities) lambda resonances.

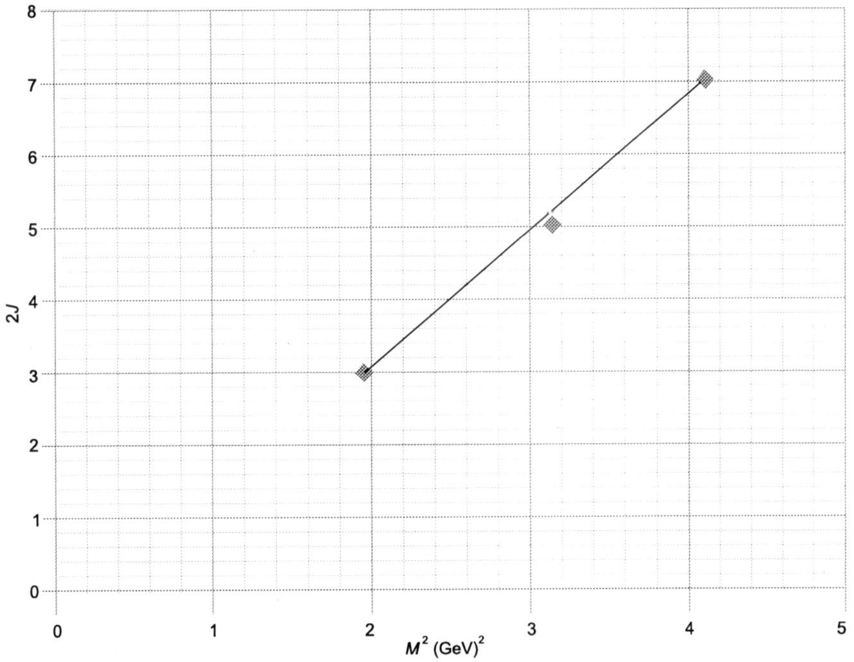

Fig. 8.7b C-F plot of exchange degenerate sigma resonances.

Baryon (Λ) Resonances [Isospin $I = (0)$, Strangeness $S = -1$]			
Resonance Name Mass M in MeV	Square of mass M^2 in (GeV)^2	Spin J, Parity P $(J)^P$	Breit-Wigner full width Γ in MeV
Λ (1115)	1.24	$(\frac{1}{2})^+$	Particle
Λ (1520)	2.31	$(\frac{3}{2})^-$	15.6
Λ (1820)	3.31	$(\frac{5}{2})^+$	80
Λ (2100)	4.41	$(\frac{7}{2})^-$	200
Λ (2350)	5.52	$(\frac{9}{2})^+$	150

Baryon (Σ) Resonances [Isospin $I = (1)$, Strangeness $S = -1$]			
Resonance Name Mass M in MeV	Square of mass M^2 in (GeV)^2	Spin J, Parity P $(J)^P$	Breit-Wigner full width Γ in MeV
Σ (1385)	1.92	$(\frac{3}{2})^+$	36
Σ (1775)	3.15	$(\frac{5}{2})^-$	120
Σ (2030)	4.12	$(\frac{7}{2})^+$	180

8.5.3.9.3 Meson resonances

Similar to baryons, meson resonances too obey the Chew-Frautschi $J \propto M^2$ law. For example, ρ(770; 1 –), ρ(1690; 3 –), ρ(2350; 5 –) fall on a Chew-Frautschi plot. This line connecting three odd parity ρ states is called the ρ trajectory. Here (1 –, 3 –, 5 –) indicate the spin parity of the meson resonance and 770, 1690, etc., are the masses of resonances in MeV. Similarly, ω(782; 1 –), ω(1670; 3 –) constitute another M^2, J straight line called ω trajectory. Likewise, even parity *f* meson resonances *f*(1270; 2 +), *f*(2050; 4+), *f*(2510; 6+) are on a straight Chew-Frautschi line (Fig. 8.8). Included with these are a meson resonances [(a_2(1320; 2 +), a_4(2040; 4 +), a_6(2450; 6+)] which lie on the a trajectory.

Though, individually these meson resonances have their own C-F plots with definite parity, they all coincide when plotted on a single chart. This implies that all these eleven resonances fall on a single Chew-Frautschi line on a J versus M^2 plot (Fig. 8.8). This presents a display of *exchange degeneracy* of these meson resonances. The exchange of Regge trajectories among ρ, ω, f and α mesons is allowed.

These observations on fermionic (baryon) and bosonic (meson) resonances show that in general they obey the Chew-Fraustchi $J \propto M^2$ law. The observed resonances that fit the straight lines have almost the same slope $\alpha' = dJ / dM^2 = 0.9\ GeV^{-2}$. These straight lines that lie in the (M^2, J) plane can be described by an equation $J = \alpha(0) + \alpha' M^2$, where the intercept $\alpha(0)$ depends on the type of resonance (Δ, Λ, Σ, etc.) but the slope α' has more or less the same value ~ 1 GeV^{-2}. The Chew-Frautschi plots thus indicate a linear relation between the spin and the square of mass of the particle or resonance. A classical derivation for this relation would be given in a later section.

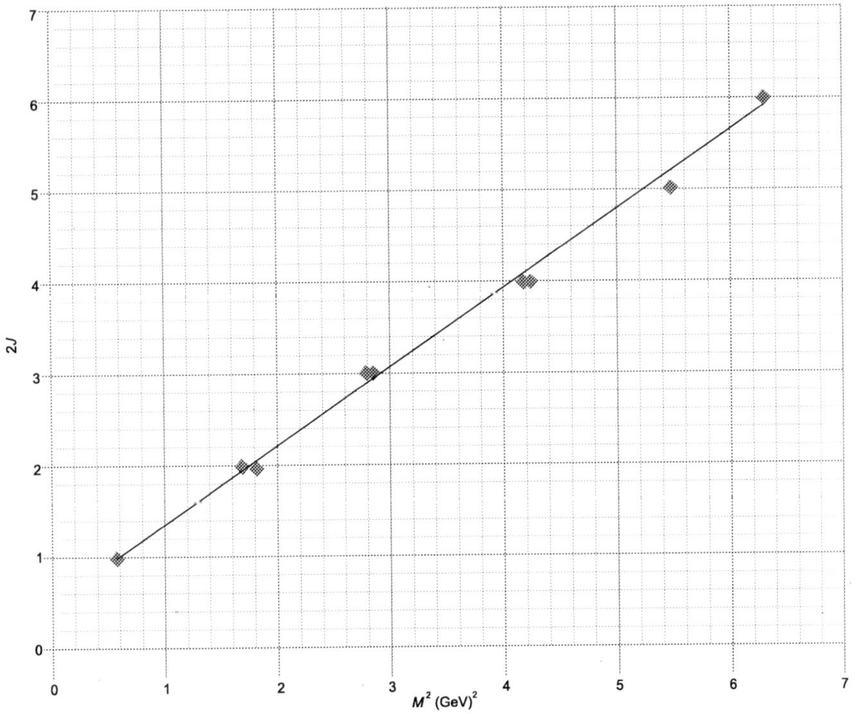

Fig. 8.8 C-F plot of exchange degenerate rho, omega, f and a meson resonances.

8.5.4 Duality

Duality may be defined as a state of combining two different things or as the equivalence of two seemingly different descriptions of a physical process. The invariance of Maxwell's equations in vacuum under the relacement of (***E***, ***B***) by (***B***, –***E***) is a well-known example of classical electromagnetic duality. In the present context, the two different phenomena are the low energy and high-energy scatterings.The mechanisms involved in these two scatterings are not same.

The low energy scattering is dominated by the formation of unstable particle states, called resonances, which subsequently decay into various channels. At resonant energies the resonances give bumps in total scattering cross sections. Due to the rapid changes in the resonant partial wave amplitudes, there is a rapid fluctuation in the differential cross sections. As the energy increases, the resonances become wider and more closely spaced resulting in their overlapping. The overlapping smooths out the previously mentioned fluctuations. *The total scattering cross section at higher energies is thus approximately independent of energy*. In higher energy scattering, the mechanism is that here a virtual particle is exchanged between the projectile and the target. This virtual particle has virtual energy-momentum vector as well as a virtual spin. Such a virtual particle is Regge pole. Thus we can say that *Regge pole mechanism causes the high-energy scattering, whereas, resonance formation causes low energy scattering*. The detailed analysis leads to the fact that two mechanisms are different and apparently disconnected. It is duality that relates these two dissimilar mechanisms.

In terms of channel processes, duality represents the relationship between the s and t channel behaviours of scattering amplitudes. The scattering amplitude at low energies denoted by $A(s, t)$ could be described by a sum over s-channel resonant partial wave amplitudes, $A(s, t) = \Sigma_{\text{res}} A_{\text{res}}(s, t)$. When s (energy) increases, a t-channel partial wave series becomes more useful and the high energy behaviour of $A(s, t)$ could be described in terms of Regge poles. At higher energies the (t-channel) scattering amplitude $A(t, s) = \Sigma_{RP} A_{RP}(t, s)$, where RP stands for Regge pole(s). The duality demands the equivalence between the s-channel resonances and the t-channel Regge poles. This means when duality is observed, $A(s, t) = A(t, s)$, implying $\Sigma_{res} A_{\text{res}}(s, t) \approx \Sigma_{RP} A_{RP}(t, s)$. When $s \to \infty$, this relation reads as $A_{\text{res}}(s, t) \approx A_{RP}(t, s)$, which is known as the local version of duality relation. The equality of s-channel scattering amplitude $A(s, t)$ with the t-channel scattering amplitude $A(t, s)$ forms the basis for dual models.

As observed above, the resonance formation guides the scattering amplitude at low energies, whereas the Regge pole mechanism determines the same at higher energies. This implies separate scattering amplitudes guide the low and high energy behaviours. A natural expectation would be that, there is a single scattering amplitude that works in the entire energy range from low to high. Such amplitude should be built up on resonances so that low energy scattering gets explained and must end up in depicting the Regge behaviour so that high energy scattering is described in terms of Regge poles. It is also expected that such a unified scattering amplitude would observe the duality relation, the latter connecting the low- and high-energy scatterings. Such unified scattering amplitude was discovered by Gabriele Veneziano and hence known as Veneziano amplitude.

8.5.5 The Veneziano Amplitude

Veneziano discovered in 1968 that physical properties like symmetry and duality of mesons can be described by a function called Euler beta function, if it is interpreted as scattering amplitude. Veneziano, in an attempt to derive the scattering amplitude for a process involving four particles (two incoming and two outgoing) stumbled at the B [beta (uppercase)] function, which was called the four-particle scattering amplitude. This scattering amplitude that became known as Veneziano amplitude is symmetric under the interchange of s and t Mandelstam variables. The Veneziano scattering amplitude is also able to establish equivalence between the low energy resonance scattering and high energy Regge pole scattering and thus observes duality. The discovery of Veneziano amplitude is regarded as the birth of string theory.

Since, the Euler beta function (B) has its roots in the Euler gamma function (Γ); the relevant description of both would be presented in the following subsection.

8.5.5.1 Euler Gamma and Beta Functions

Gamma function

The gamma function [$\Gamma(u)$] is defined by Euler's integral

$$\Gamma(u) = \int_0^{\infty} t^{u-1} e^{-t} dt \tag{8.136}$$

where u may be real or complex. For real u, we should have $u > 0$. If u is complex, the real part of u must be positive, i.e., $Re\ u > 0$. This condition for $u \to (u + n)$, where n is a positive integer, would be $Re\ u > -n$.

Replacing $u \to (u + 1)$ the Euler's integral becomes $\Gamma(u) = -\int_0^{\infty} t^u \frac{d}{dt}(e^{-t})\, dt$

which on using the formula $\int x\, dy = xy - \int y\, dx$ for integration by parts with $x = t^u$, $y = e^{-t}$, provides the identity

$$\Gamma(u + 1) = u\Gamma(u) \tag{8.137}$$

where xy has vanished under the integration limits and the integral term (8.136) has provided back $\Gamma(u)$. Identity (8.137) is a recurrence relation in which we may, for example, replace u by $(u - 1)$ and successively continue the replacement on obtained relations, to get

$$\begin{aligned}
\Gamma(u) &= (u-1)\,\Gamma(u-1) \\
\Gamma(u-1) &= (u-2)\,\Gamma(u-2) \\
\Gamma(u-2) &= (u-3)\,\Gamma(u-3) \,\ldots\ldots. \\
\Gamma\{u-(n-2)\} &= \{u-(n-1)\}\,\Gamma\{u-(n-1)\} \\
\Gamma\{u-(n-1)\} &= (u-n)\,\Gamma(u-n).
\end{aligned} \tag{8.138}$$

The above relations allow us to write

$$\Gamma(u) = (u-1)(u-2)\ldots(u-n+1)(u-n)\,\Gamma(u-n) \tag{8.139}$$

from which we may express

$$\frac{\Gamma(u)}{\Gamma(u-n)} = (u-1)(u-2)(u-3)\ldots\ldots(u-n) \tag{8.140}$$

Likewise, writing (8.137) as $\Gamma(u) = [\Gamma(u+1) / u]$ and making successive replacements of u by $(u + 1)$, we would get

$$\Gamma(u) = \frac{\Gamma(u+n)}{u(u+1)(u+2)(u+3)......(u+n-1)} \tag{8.141}$$

from which we observe that, when , $u = 0, -1, -2, -3, \ldots -n + 1$, $\Gamma(u) \to \infty$. This implies that the gamma function, $\Gamma(u)$, has simple (of first order) poles at negative integer values of u including zero. In other words, if we wish to show poles in a gamma function, u shall have to be assigned negative (integer) values. Simple poles are the only singularities that are allowed in lowest level amplitudes in relativistic quantum mechanics.

Keeping in view the relation (8.141), we wish to know the behaviour of u near $-n$, where n is a positive integer. For this purpose we calculate the residue (*Res*) of the pole at $u = -n$. Multiplying both sides of (8.141) by $(u + n)$ and writing $(u + n)\,\Gamma(u + n) = \Gamma(u + n + 1)$, we may write

$$Res\,[\Gamma(u), u = -n] = \text{Lim}_{u\to -n}\,(u+n)\,\Gamma(u)$$

$$= \text{Lim}_{u\to -n}\frac{\Gamma(u+n+1)}{u(u+1)...(u+n-1)} = \frac{(-1)^n}{n!} \tag{8.142a}$$

Therefore,

$$\text{Lim}_{u\to -n}(u+n)\,\Gamma(u) = \frac{(-1)^n}{n!}$$

implying that

$$\Gamma(u) \sim \frac{1}{u+n}\frac{(-1)^n}{n!} \tag{8.142b}$$

where, substituting $u = -n$, the numerator of (8.142a) reduced to $\Gamma(1) = 1$ and the denominator became

$-n(-n + 1) \ldots (-2)\,(-1) = (-1)^n\,[n(n-1)\,(n-2) \ldots 3.2.1] = (-1)^n\,n! = (-1)^{-n}\,n!$, resulting in the expression (8.142b). Therefore eqn. (8.142) shows the behaviour of u near $-n$.

Another important feature of gamma function is its asymptotic behaviour. When u is large, the integral (8.136) $[\Gamma(u)]$ is dominated by the region $t \approx u - 1$. In this region the gamma function can be evaluated through a technique called saddle point evaluation. The evaluation gives the asymptotic value of gamma function as

$$\Gamma(u)_{u\to\infty} \sim \left(\sqrt{2\pi}\right)u^{(u-1/2)}e^{-u} \tag{8.143}$$

This is called Sterling's formula.

Beta function

An association between the sets of input and output values is known as beta function. Beta function is the Euler integral of first kind expressed as

$$B(u, v) = \int_0^1 t^{u-1}(1-t)^{v-1}\,dt \tag{8.144}$$

where u and v can be both real and complex. When they are real, they have to be positive numbers $u > 0$, $v > 0$. In case of complex u and v, $Re\ u > 0$ and $Re\ v > 0$. For other values of u, v, the integral (8.144) does not converge so values other than the above are disallowed. The beta function (8.144) is symmetric under the interchange of u and v. This implies $B(v, u) = B(u, v)$. To prove this, the convergent property of definite integrals may be employed, which tells

$$\int_0^a f(t)dt = \int_0^a f(a-t)dt \tag{8.145}$$

Applying (8.145) to (8.144) with $a = 1$, $t \to (1-t)$ and $(1-t) \to t$ (8.144) interchanges u and v and becomes

$$B(u, v) = \int_0^1 t^{u-1}(1-t)^{v-1}\,dt \to \int_0^1 (1-t)^{u-1} t^{v-1} dt = B(v,u) \tag{8.146}$$

Another property relevant for us is the relation between the beta and gamma functions. The relation is

$$B(u, v) = \frac{\Gamma(u)\Gamma(v)}{\Gamma(u+v)} \tag{8.147}$$

To prove this, we begin with the definition of gamma function (8.136) and use two variables t and s for $\Gamma(u)$ and $\Gamma(v)$ respectively, and write

$$\Gamma(u) = \int_0^\infty t^{u-1}e^{-t}dt \text{ and } \Gamma(v) = \int_0^\infty s^{v-1}e^{-s}ds$$

$$\Gamma(u)\,\Gamma(v) = \int_0^\infty\int_0^\infty e^{-(t+s)}t^{u-1}s^{v-1}dt\,ds \tag{8.148}$$

Changing the variables t and s through $t = xy$ and $s = x(1-y)$, we may express

$$dt\,ds = \left|\frac{\partial(t,s)}{\partial(x,y)}\right| dx\,dy \tag{8.149a}$$

where,

$$\left|\frac{\partial(t,s)}{\partial(x,y)}\right| = \begin{vmatrix} \frac{\partial t}{\partial x} & \frac{\partial t}{\partial y} \\ \frac{\partial s}{\partial x} & \frac{\partial s}{\partial y} \end{vmatrix} = [\frac{\partial t}{\partial x}\frac{\partial s}{\partial y} - \frac{\partial t}{\partial y}\frac{\partial s}{\partial x}] \tag{8.149b}$$

is the two-dimensional Jacobian determinant. Substituting

$$\frac{\partial t}{\partial x} = y, \frac{\partial t}{\partial y} = x, \frac{\partial s}{\partial x} = 1-y, \frac{\partial s}{\partial y} = -x$$

obtained from the change of variable relations, the determinant turns out to be $(-x)$. Equation (8.149a) then becomes

$$dt\,ds = -x\,dx\,dy \tag{8.150}$$

The change of variable equations also provides the relations

$$t + s = x \quad \text{and} \quad \frac{1}{1+\frac{s}{t}} = y \tag{8.151}$$

Since t and s vary from $0 \to \infty$, the variable x from first of (8.151) also varies from $0 \to \infty$. However, from second equation of (8.151) when $s = 0$, $y = 1$ and when $s = \infty$, $y = 0$. Equation (8.148) then becomes

$$\Gamma(u)\,\Gamma(v) = \int_0^\infty e^{-x} x^{(u+v-1)} dx \int_0^1 y^{u-1}(1-y)^{v-1} dy \tag{8.152}$$

in which the limits of y from 1 to 0 have been inverted due to a minus sign in (8.150). Comparing the integrals of (8.152) with (8.136) and (8.144), we find they are gamma and beta functions respectively. We can then express (8.152) as

$$\Gamma(u)\,\Gamma(v) = \Gamma(u+v)\,B(u, v) \tag{8.153}$$

from which we can easily view the relation between beta and gamma fuction as in (8.147).

Another identity related to beta function is

$$B(u+1, v) + B(u, v+1) = B(u, v) \tag{8.154}$$

This identity can easily be proved by using (8.147) for both the B functions on the LHS and then using (8.137) in numerators and denominators. $(u + v)$ cancels out to leave the form (8.147) for $B(u, v)$.

8.5.5.2 Beta Function as Scattering Amplitude

We refer to beta function $B(u, v)$ (eqn. 8.147) and examine its analytic behaviour. To be analytic, $B(u, v)$ should have a derivative at each point of the complex plane and must have a single value at every point. Rewriting $B(u, v)$

$$B(u, v) = \frac{\Gamma(u)\Gamma(v)}{\Gamma(u+v)} \tag{8.155}$$

it can be noted that when u and v have negative integer values $\Gamma(u)$, $\Gamma(v)$ and hence $\Gamma(u+v)$ each shall tend to infinity. Hence at these negative integer values, each of the gamma function shall have a pole (simple pole). The presence of pole shall not render the Beta function $B(u, v)$ analytic. We may analyze it as follows.

Let there be a pole at $v = -n$ implying that when $v \to -n$

$\Gamma(v) \to \infty$. At $v = -n$, the behaviour of $\Gamma(v)$, according to (8.142), shall be given by

$$\Gamma(v) \sim \frac{1}{v+n}\frac{(-1)^n}{n!} \tag{8.156}$$

and then from (8.155), $B(u, v)$ at $v = -n$ shall be

$$B(u, v) \sim \frac{1}{v+n} \frac{(-1)^n}{n!} \frac{\Gamma(u)}{\Gamma(u-n)} \tag{8.157}$$

Substituting (8.140), the behaviour of $B(u, v)$ at $v \to -n$ (u fixed) shall be

$$B(u, v) \sim \frac{1}{v+n} \frac{(-1)^n}{n!} (u-1)(u-2) \ldots\ldots (u-n) \tag{8.158}$$

In the form (8.158), for $v \to -n$ and for large $|v|$, $B(u, v)$ diverges while the requirement of analyticity is convergence. So we need to change the form of $B(u, v)$ from (8.158) to one that converges.

It has been found that for $Re\, u > 0$, $B(u, v)$ described by the infinite sum [$n = 0$ to $n = \infty$] of the RHS of (8.158) does indeed converge. So the beta function that will be analytic in the domain of complex v should have the form

$$B(u, v) = \sum_{n=0}^{\infty} \frac{1}{v+n} \frac{(-1)^n}{n!} (u-1)(u-2) \ldots\ldots (u-n) \tag{8.159}$$

Since Veneziano amplitude represents scattering amplitude and the scattering amplitude should be analytic, the Euler beta function $B(u, v)$ of equation (8.159) has the potential to represent the Veneziano amplitude.

Now u and v have to be replaced by the relevent parameters. For that we refer to Regge trajectories. Described in the preceding subsections, the s-channel Regge trajectory is expressed as

$$\alpha(s) = \alpha(0) + \alpha' s \tag{8.160a}$$

where $\alpha(0)$ is the intercept and α' is the slope of the trajectory. Since on the Regge trajectory that we are considering, lie, particles and resonances, the slope $\alpha' > 0$. The t-channel Regge trajectory, likewise, has the equation

$$\alpha(t) - \alpha(0) + \alpha' t \tag{8.160b}$$

The left-hand sides of these equations describe the complex and continuous angular momentum, the real part of which was represented by J in the C-F plots. The angular momentum α varies with s and t in (a) and (b) equations of (8.160) respectively. The equation (8.160a) describes a linear relation between α and s, the latter being the square of energy in s channel. In t-channel, the square of energy is t and equation (8.160b) shows the linear variation between α and t. Veneziano too, had postulated the linear equations (8.160) for Regge trajectory.

We may recall the comments below eqn. (8.141), wherein, it was stated that if $\Gamma(u)$ had poles, u had negative values. Now, we have to see that $B(u, v)$ possesses poles. If u and v have negative values, $B(u, v)$ shall represent the scattering amplitude that shows poles. Therefore, for the s-channel scattering process, u in (8.159) is replaced by $-\alpha(s)$ and v by $-\alpha(t)$. For obtaining scattering amplitude for t-channel process the reverse replacements would be undertaken in (8.159), i.e., $u \to -\alpha(t)$ and $v \to$

$-\alpha(s)$. We may represent s-channel $B(u, v)$ by the scattering amplitude $A(s, t)$ and the t-channel $B(u, v)$ by the scattering amplitude $A(t, s)$. Thus, from (8.159) we get

$$A(s, t) = \sum_{n=0}^{\infty} -\frac{1}{n!} \frac{(\alpha(s)+1)(\alpha(s)+2)...(\alpha(s)+n)}{(\alpha(t)-n)} \tag{8.161}$$

where the negative sign has come from the denominator and in the numerator $(-1)^n$ has combined with the $(-1)^n$ coming from the brackets to give a value 1. Making reverse replacements ($u \to -\alpha(t)$ and $v \to -\alpha(s)$) in (8.159), the t-channel scattering amplitude $A(t, s)$ would be

$$A(t, s) = \sum_{n=0}^{\infty} -\frac{1}{n!} \frac{(\alpha(t)+1)(\alpha(t)+2)...(\alpha(t)+n)}{(\alpha(s)-n)} \tag{8.162}$$

From the above, it is observed that

$$A(s, t) = B[-\alpha(s), -\alpha(t)] \tag{8.163a}$$

and

$$A(t, s) = B[-\alpha(t), -\alpha(s)] \tag{8.163b}$$

where the right-hand sides of eqn. (8.163) are both Euler beta functions. It has been proved in eqn. (8.146) that the beta function is symmetric in its arguments. So from (8.163) and (8.146), we may write

$$B[-\alpha(s), -\alpha(t)] = B[-\alpha(t), -\alpha(s)] \tag{8.164}$$

which, in turn, proves from eqn. (8.163), the equality of s-channel and t-channel scattering amplitudes as

$$A(s, t) = A(t, s) \tag{8.165}$$

Therefore, it is concluded that

$$A(s, t) = B[-\alpha(s), -\alpha(t)] = A(t, s) = B[-\alpha(t), -\alpha(s)] \tag{8.166}$$

is the **Veneziano amplitude**, whose expressions are (8.161) and (8.162).

The relation (8.165) shows the duality of Veneziano amplitude that a scattering process can be realized through both s-channel and t-channel processes. The t-channel scattering process has been shown equivalent to the s-channel scattering process. In the s-channel process, the denominator $(\alpha(t) - n)$ of (8.161) tells that here the scattering amplitude is expressed as the sum of t-channel Regge poles determined by $\alpha(t) - n = 0$. n is the angular momentum that the Regge pole shall possess. Using (8.160b) and $t = 4M^2 \sim M^2$, the mass of Regge pole can be estimated as $M^2 \sim [n - \alpha(0)] / \alpha'$. Since, n is the largest number, M is the maximum mass of Regge pole. If J is the angular momentum (spin) of Regge pole having minimum mass, then $M^2_{\min} \sim [J - \alpha(0)] / \alpha'$. Thus, the s-channel scattering process would be undertaken by the exchange of a t-channel Regge pole of say, spin n and mass M. Likewise, a s-channel pole would undertake the t-channel scattering process. It may be added that s-channel poles are resonant states and t-channel poles are Regge poles. Therefore, the equality (8.165) is also the exhibition of resonance-Regge pole duality.

Using gamma function the Veneziano amplitude can also be expressed as

$$A(s, t) = B(-\alpha(s), -\alpha(t)) = \frac{\Gamma(-\alpha(s))\Gamma(-\alpha(t))}{\Gamma(-\alpha(s)-\alpha(t))} \tag{8.167}$$

8.5.5.3 No Ghost Theorem

From the expressions (8.161) and (8.162) of the amplitudes $A(s, t)$ and $A(t, s)$ the residues of poles are limiting values of $(\alpha(t) - n)\, A(s, t)$ and $(\alpha(s) - n)\, A(t, s)$ respectively. Their values are n^{th} order polynomial in s and n^{th} order polynomial in t respectively. However, the minus signs on the RHS do not guarantee that the residues would be positive. In a relativistic quantum theory, the residues should be positive whereas, we are encountering the possibility of negative residues. Presence of negative residues implies presence of ghosts. To get rid of this anomaly, a theorem called no ghost theorem is invoked. *This theorem tells that, to avoid ghosts or negative residues, the dimension of the space-time should be* 26 *and* $\alpha(0)$ *in the Regge trajectory should be 1.* $\alpha(0)$ is the intercept of Regge trajectory. Therefore, Veneziano amplitude admits a twenty-six dimensional space-time along with the unit intercept of Regge trajectory, so that no ghost theorem becomes applicable.

8.5.5.4 High Energy Behaviour of Veneziano Amplitude

To understand the high-energy behaviour, we refer to Regge pole formulation of equation (8.123). This equation extends its applicability to $\cos\theta_t \to \infty$. The asymptotic region $\cos\theta_t \to \infty$ belongs to high energy region where $s \to \infty$. We shall examine the scattering amplitude for $\cos\theta_t \to \infty$. From, e.g., eqn. (8.122), we may observe that the scattering amplitude, in regard to $\cos\theta_t$ dependence, depends on $P_\alpha(-\cos\theta_t) = (-1)^\alpha\, P_\alpha(\cos\theta_t)$. Now $\cos\theta_t \to \infty$, $P_\alpha(\cos\theta_t) \to (\cos\theta_t)^\alpha$ and from (7.38d) $\cos\theta_t = 1 + \dfrac{2s}{t - 4m^2}$. Therefore, for a fixed t, $\cos\theta_t \to \infty$ implies $s \to \infty$. This allows replacing $\cos\theta_t$ by s, so that asymptotically, scattering amplitude in the t-channel can be expressed as $A(t, s) \sim [\cos\theta_t]^\alpha \sim s^\alpha$, where $\alpha \Rightarrow \alpha(t)$ is the t-channel Regge trajectory. Therefore, the high-energy (asymptotic) behaviour of t-channel scattering amplitude may be expressed as

$$A(t, s) \sim s^{\alpha(t)} \tag{8.168}$$

Now, we come to Veneziano amplitude (8.168) expressed in terms of gamma functions and examine it for $s \to \infty$, which as discussed above, refers to high energy. Rewriting Veneziano amplitude

$$A(t, s) = B(-\alpha(t), -\alpha(s)) = \frac{\Gamma(-\alpha(t))\Gamma(-\alpha(s))}{\Gamma(-\alpha(t) - \alpha(s))} \tag{8.169}$$

we take the asymptotic values of gamma functions that contain $[-\alpha(s)]$. From (8.143), $\Gamma(u)_{u\to\infty} \sim (\sqrt{2\pi})u^{(u-1/2)}e^{-u}$. Putting $u = -\alpha(s)$ for the numerator and $u = -\alpha(t) - \alpha(s)$ for the denominator $A(t, s)$ on using (8.143) becomes

$$A(t, s) = \frac{\Gamma(-\alpha(t))[-\alpha(s)]^{\alpha(t)}e^{-\alpha(t)}}{\left[1 + \dfrac{\alpha(t)}{\alpha(s)}\right]^{-\alpha(t)-\alpha(s)-(\frac{1}{2})}} \tag{8.170}$$

Using $\alpha(t) = \alpha(0) + \alpha' t$ and $\alpha(s) = \alpha(0) + \alpha' s$, with fixed t and $s \to \infty$, the denominator becomes ~ 1 and using $\alpha(t) \sim \alpha' t$, the exponential is rendered a non-variable. Equation (8.170) then becomes

$$A(t, s \to \infty) \sim \Gamma(-\alpha(t))\,[-\alpha(s)]^{\alpha(t)} \tag{8.171}$$

which too on using $\alpha(s) \sim \alpha' s$, approximates to

$$A(t, s \to \infty) \sim [s]^{\alpha(t)} \tag{8.172}$$

The Veneziano amplitude (8.167), therefore, shows its high-energy behaviour through eqn. (8.172) which matches with (8.168). The expression (8.172) is the exhibition of Regge behaviour of Veneziano amplitude.

8.5.6 The Veneziano Amplitude and String Theory

Veneziano amplitude is credited for the birth of string theory. A theory based on strings can account qualitatively for various features of the strong nuclear force and the strongly interacting particles (or hadrons). Examining Veneziano amplitude in physical terms, Yoichiro Nambu,Tetsuo Goto, Holger Nielson and Leonard Susskind interpreted that the Veneziano amplitude described the behaviour of one-dimensional objects, the strings. According to this picture, the hadrons were strings rather than point particles.The interaction of hadrons is not via the exchange of other point particles but via the joining and splitting of strings. Different modes of vibrations (quantum states) of string give rise to the observed particles. Just as the particle has intrinsic mass, the string has intrinsic stiffness or tension. If a string were plucked, it would vibrate into various harmonics. The fundamental mode of vibration would correspond to one type of hadron (mass), the second harmonic, etc., would correspond to other hadrons. The tension of string determines its energy scale. If the tension is high, more energy would be required to set the string into vibration whose harmonics would correspond to more massive hadrons. These masses, which belong to different modes of the string, constitute the mass spectrum of the string. In this theory, a single fundamental object namely, the string is postulated to explain the various observed hadrons. Just as a particle is subject to the laws of special relativity, a string would also be relativistic. We can devise a quantum mechanics of strings like the one for particles.

The string theory allows two kinds of fundamental objects, the open and closed strings. Open string is like a line segment with two free ends, where entities like electric or colour charge would reside. Closed string is like a small loop with no free ends. Here charges reside in the form of a coating on the tiny loop. The endpoints of an open string can join to form a closed string; however, a closed string cannot break into open strings.

Orientability is another significant feature of strings. If moving along the string, we can tell the direction of our travel, the string is said to be oriented. If, while moving along the string, the direction of travel remains unknown, the string is unoriented.

When a particle moves through space-time, it traces out a one-dimensional world line but when a string moves through the space-time, it traces out a two-dimensional world sheet. The world sheet of open string is like the surface of a piece of paper, whereas the world sheet of a closed string is like the surface of a cylinder. An open string sweeps out a surface like an infinite ribbon and the closed string a surface like an infinite cylinder in space-time.

In space-time, a particle moves along the world line that minimises the action. The action of a particle is directly proportional to the length of its world line. In case of a string also, the string moves in such a way that the action is minimum. The action of the string is directly proportional to the area of the world sheet that the string sweeps.

The original strings whether open or closed described bosons and hence were called bosonic strings. The bosonic string theory, however, had the following difficulties.

(a) The bosonic string model did not describe fermions like quarks, etc.

(b) A higher dimensional (26-dimensional) space-time is required for bosonic strings. This is because the first excited bosonic string states have mass $M^2 = (4/\alpha')[1 - \{(D - 2)/24\}]$, where α' is the universal Regge slope and D is the dimension of the space-time. Since, to preserve the desired Lorentz symmetry, the string states have to be massless, we set $M = 0$ and obtain the dimension of space-time, $D = 26$. Additionally, the area of the world sheet, to which the action of string is proportional to, should be independent of the world sheet coordinates. The invariance of physics under the transformations of world sheet coordinates is known as the conformal symmetry. The action should exhibit conformal invariance. However, the conformal symmetry gets broken, the process causing the break is called conformal anomaly. The conformal anomaly needs to be cancelled. For the desired cancellation, a bosonic string needs to work in twenty-six dimensional space-time. Time has only one dimension rest are spatial dimensions. The string is required to vibrate in all spatial dimensions. No-ghost theorem also demands 26 dimensions. The difficulty is that the number of dimensions is too many to be realizable.

(c) Third problem with the bosonic strings is that the lowest energy states of bosonic string are tachyons, the unphysical particles whose squared mass is negative and they travel in vacuum faster than light. If a string theory that is supposed to describe the universe, permits imaginary mass, the measurements in the universe would themselves be unreal. Absence of tachyons is therefore a desirable feature of a credible string theory. In addition to tachyon, an undesired massless spin-2 particle was the constituent of the spectrum of fundamental string.

In view of the problems cited above, the bosonic string theory was an unrealistic model to describe the hadrons. A string theory that is believed to be realistic is the superstring theory. Following points describe its realistic nature.

(a) Supersymmetry is the symmetry of bosons and fermions. In this symmetry, interconversion of bosons and fermions is possible. Therefore, to incorporate fermions in the bosonic string theory it is essential to have its supersymmetrized version. Joel Scherk, a pioneer string theory physicist, suggested that the spinning string state could be fully supersymmetrized and thus paved way for the incorporation of supersymmetry in the string theory. The string theory that incorporates supersymmetry is called superstring theory. However, depending upon the amount of supersymmetry possessed by the theory, there are different superstring theories. The generators of supersymmetry (SUSY) are translations P^{μ}, the Lorentz generators $M^{\mu\nu}$ in which M^{0i} denote the Lorentz transformation and M^{ij} denote rotations and the spinor generators Q_{α}^{i} and $\bar{Q}_{\dot{\alpha}}^{i}$ where $\alpha, \dot{\alpha} = 1, 2$ and $i = 1, 2, 3, .. N$. When the superstring theory possesses $N = 1$ supersymmetry, it is called *type I* superstring theory and $N = 2$ supersymmetry makes the superstring theory as *type II* theory. Parity may be violated in weak interactions, so if superstring theory for $N = 2$ corresponds to parity violating interaction (chiral interaction) the theory is *type IIB* and when it corresponds to parity preserving interaction (non-chiral) the theory is *type IIA* superstring theory.

(b) Higher dimensional space-time is required in every string theory. The superstring theory mentioned above requires ten- and in some cases eleven-dimensional space-time. This may again appear unrealistic. However, an explanation that six or seven dimensions in the superstring theory are compactified thus rendering the extra dimensions defunct restores realism in the theory.The compactification is on an internal manifold (Calabi-Yu manifold) whose size is too small to be detected. A manifold is a collection of points forming a topologically closed surface. Therefore, though 10 or 11 dimensions are required, we actully feel the usual four-dimensional space-time, the rest of the dimensions remain undetected due to compactification. There are no tachyons in the superstring's spectrum.

(c) The problem of managing the undesired massless spin-2 particle of the bosonic string gets solved by identifying this particle as graviton, the quantum of gravitation and, in turn, making string theory suitable for the quantum theory that unifies gravitational force with other fundamental forces of nature. General relativity, is thus naturally incorporated in string theory. When the principles of supersymmetry are combined with general relativity, the combination is called supergravity.

(d) In order to be the theory to describe all interactions, the superstring theory is required to incorporate the features of weak interactions as well. Since parity violation is admitted in weak interactions, the superstring theory should also be chiral or parity violating. The superstring theory is ten dimensional, but even if we try to form a chiral superstring theory in four dimensions, we come across the Feynman diagrams that cause gauge symmetry to break that in turn leads to the violations of fundamental laws like conservation of electric charge. The situation becomes more complex in higher dimensions. This anomalous behaviour of superstring theory with respect to weak interactions is called chiral anomaly. Such an anamoly needs to be countered. The superstring theory should be chiral but free from chiral anomaly. Most physicists found that it would be impossible to construct a chiral anomaly-free chiral superstring theory.

(e) The difficulty posed by chiral anomalies (Lorentz and Yang Mills) was however, removed by the Green-Schwarz mechanism. According to this mechanism these undesired anomalies are exactly cancelled provided one chooses SO(32) or the $E_8 \times E_8$ as the gauge group. SO(32) is the special orthogonal group or the group of rotations in a Cartesian space of dimension 32. It has a rank of 16. E_8 is an exceptional group which has rank 8 and the rank of $E_8 \times E_8$ is 16, same as that of SO (32). These are anomaly-free gauge groups. However, with type I ($N = 1$) superstring theory, that admits both open and closed strings, the chiral anomalies can be cancelled only if the gauge group is SO (32). The anomaly cancellation through type I superstrings, that Green and Schwarz considered, does not take place when the gauge group is $E_8 \times E_8$. Type II theories have $N = 2$ supersymmetric closed strings. This $N = 2$ ten-dimensional superstring theory does not contain Yang-Mills theory whose quantized versions produce the QED, electroweak theory and the standard model that are desired in our scheme. In such a situation $N = 2$ supersymmetric string theories alone cannot be employed. The failure of type I superstring theory to explain the anomaly cancellation in $E_8 \times E_8$, allows the emergence of a new string theory. This is heterotic string theory.

(f) The heterotic strings are a hybrid of 26-dimensional non-supersymmetric bosonic strings and 10-dimensional type II supersymmetric strings. This combination of $N = 0$ bosonic and $N = 2$ superstring theory makes the hybrid theory as $N = 1$ superstring theory known as heterotic string theory. The vibrations in the heterotic string can move clockwise (right moving vibration) and anticlockwise (left moving vibration). The left moving vibrations are non-supersymmetric and live in 26-dimensional space-time. They are like bosonic strings. The right moving vibrations of heterotic string are supersymmetric and live in 10-dimensional space-time. They are like the closed superstrings. Now, both the left and right movers must live in the space-time of same dimension that is 10. This demands that 16 (26-10) dimensions of left movers (bosonic) must curl up or compactify. The space-time of a heterotic string is therefore a product of 10-dimensional space-time and compactified 16-dimensional space. *Only the space dimensions compactify*. To determine whether, the heterotic string is open or closed, we note that gauge groups can be introduced by attaching lebels (Chan-Patan) to the edges of *open* strings. This procedure can yield the group SO (32) not $E_8 \times E_8$. Alternatively, the gauge group can be produced by the compactification of a *closed* string theory. The compactification of 16 internal dimensions in the heterotic string produces the gauge group SO (32) or $E_8 \times E_8$. This tells that *heterotic strings are closed strings* and their gauge group is SO (32) or $E_8 \times E_8$. Being the gauge group of heterotic strings the chiral anomaly of $E_8 \times E_8$ can be cancelled through these strings. This situation provides us two types of heterotic string theories, viz., the SO (32) heterotic theory and $E_8 \times E_8$ heterotic theory.

Therefore, the interpretation that the Veneziano amplitude describes the behaviour of one-dimensional objects, the strings, led to the development of bosonic string theory and five kinds of superstring theories. Among the superstring theories, the first is *type I theory*, which is a theory of unoriented open strings. Since, the ends of open strings can join together to form a closed string, *type I theory becomes the theory of unoriented open and closed strings*. The gauge symmetry of type I superstrings is SO (32) gauge group. The theory has the desired features of chirality and $N = 1$ supersymmetry. The low energy approximation of this theory is supergravity. The spectrum of the theory contains massless graviton, and many gauge fields that behave like quanta of electroweak and strong fields of Standard Model. There are fermion particles as well. *The type I superstring theory has the potential of being a realistic unified theory.*

Type IIA theory is a theory of unoriented, closed strings with $N = 2$ supersymmetry. In this theory the anticlockwise vibrations and clockwise vibrations cannot be distinguished and therefore the theory is non-chiral (parity conserving). Non-chirality is an undesired feature of superstring theory. Though there are fermions in this type of superstring theory, the real fermions that may be either left handed or right handed cannot be its part. This theory cannot be a unified theory, the theory of real world, as it has no gauge symmetry that could give rise to Standard Model

Type IIB theory too is a theory of unoriented, closed strings with $N = 2$ supersymmetry. In this theory, the anticlockwise vibrations and clockwise vibrations can be distiguished and thus the theory has the desired element of chirality. However, this theory too has no gauge symmetry to account for the various gauge fields, though the massless graviton is present. The *type IIB* theory too is therefore not a candidate for the realistic unified theory.

Heterotic SO(32) *theory* is a hybrid string theory. It is a theory of oriented, chiral, closed strings with $N = 1$ supersymmetry. The gauge symmetry group of this theory is SO(32) which being larger than the Standard Model gauge symmetry group SU(3) × SU(2) × U(1) can predict various interactions and their carriers. The difficulty with this theory is that the chirality, a desired element, that is present in 10-dimensional theory, can be destroyed when compactification is undertaken. This would render the theory non-chiral. Therefore, this theory too is not a candidate for the theory of real world.

Heterotic $E_8 \times E_8$ *theory* is other hybrid string theory whose gauge symmetry group is $E_8 \times E_8$. Since, the group E_8 contains E_6 as its subgroup and E_6 is a strong candidate for grand unified group, heterotic $E_8 \times E_8$ string theory has the potential of unifying gravity with other three interactions. This makes the *heterotic* $E_8 \times E_8$ *theory as the strongly favoured unified theory.*

The string theory, which originated from the explanation of Veneziano amplitude, faced problems like those of anomalies and multiple dimensions. The progress on string theories further diminished, when an alternative, the quantum chromodynamics, gained recognition as the theory of strong interactions. Under such unconducive circumstances, the *three discoveries*, namely, the *anomaly cancellation, compactification* and the *heterotic strings* resurrected the superstring theory as a strong candidate for the unified theory of all interactions. These three discoveries are therefore, known as *three waves of 1984 revolution* that reestablished superstring theories.

8.5.7 The Spin and the Square of Mass Relation of a Particle

Under section 8.5.3.8 (*Chew Frautschi diagrams*) it was observed that (M^2, J) plots of various hadrons were straight lines, that indicated a direct proportionality between these two quantities, M^2 and J. In the following, we classically prove the relation that is based on the assumption that the particle is not a point but one-dimensional object, the string. Assume the string as relativistic.

Imagine an open string whose two ends are separated by a length $2R$. The centre of the string is at R distance from the ends. The string rotates (Fig. 8.9). While rotating the centre remains stationary and the ends (massless objects assumed atteched) move with velocity, c.

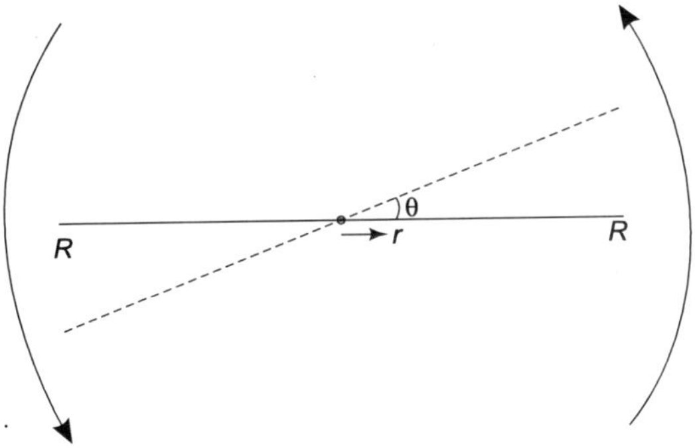

Fig. 8.9 Rotating string. $r = R\cos\theta$

The velocity of a point, situated along the string at a distance r from the centre would be

$$v = \frac{cr}{R} \tag{8.173}$$

where (c/R) is the velocity acquired in unit length. The velocity v is relativistic. Let k be the energy per unit length of the string. The energy of the string is

$$E = mc^2 = 2c^2 \int_0^R \frac{k\,dr}{\sqrt{1-\frac{v^2}{c^2}}} \tag{8.174}$$

where the *integrand* $\times\, c^2$ is the energy of the dr element of string and the limits $\left[\int_{-R}^{R} I = 2\int_0^R I\right]$. If the string rotates through an angle θ, we have

$$r = R\cos\theta,\ dr = -R\sin\theta\, d\theta.$$

Substituting this in (8.174) and carrying out the simple integration, we get

$$m = 2kR\int_0^{\frac{\pi}{2}} d\theta = \pi kR \tag{8.175}$$

On the other hand, the angular momentum, $J = mvr$ is

$$J = 2\int_0^R \frac{k\,dr}{\sqrt{1-\frac{v^2}{c^2}}}\, vr = ckR^2 \int_0^{\frac{\pi}{2}} (\cos 2\theta + 1)\, d\theta$$

or
$$J = \frac{1}{2}\pi ckR^2 \tag{8.176}$$

Eliminating R in the above two equations, we get

$$J = \left(\frac{c}{2\pi k}\right) m^2 \tag{8.177a}$$

implying

$$J \propto m^2 \tag{8.177b}$$

that shows the linearity of M^2, J plots where $m \leftrightarrow M$. Comparing (8.177a) with $J = \alpha(0) + \alpha' M^2$, we get

$$\alpha' = \frac{c}{2\pi k} \tag{8.178}$$

an expression for the slope of Regge trajectory. The experimental value of α' is ~ 0.9 GeV^{-2}.

Alternately, the string is assumed elastic in addition to be relativistic. When the string is stretched, there is a constant tension T. Tension is, in fact, the energy per unit length of the string. The quantity Tr, where r is the length of the string, is regarded as the potential energy of the stretched string. Adding to this potential energy, the relativistic kinetic energy, cp, the total relativistic energy becomes $E = cp + Tr = mc^2$, where m is the relativistic mass of the particle or resonance (now assumed a string) whose relativistic kinetic energy is cp and potential energy is Tr. The relativistic momentum p is related to orbital angular momentum through the Bohr's relation $pr = l\hbar$, which when plugged in the above relativistic energy relation converts it to $E = (\hbar cl/r) + Tr = mc^2$. Minimizing this relation with respect to r, i.e., letting $(dE/dr = 0)$, the average length r of the string turns out to be $r = \sqrt{(\hbar cl/T)}$. Substituting this in the above energy relation, we obtain $E = \sqrt{4\hbar cTl} = mc^2$, from which the proportionality of m^2 with l (implying spin J) is established as $m^2 = (4\hbar T/c^3)l$, where $(4\hbar T/c^3)$ is obviously a constant. That shows the desired proportionality.

8.6 ASSIGNMENTS

1. Explain the difference between resonances and Regge poles. Describe the resonances that are observed in elastic and inelastic scatterings.
2. Define the scattering amplitude and explain that the scattered wave has $(1/r)$ form. Describe the significance of asymptotic region in scattering.
3. Explain the scattering potential. Identify and explain its characteristics. Discuss the modification of Coulomb potential to describe Coulomb scattering.
4. What is a partial wave? Which two solutions does the method of partial waves connect? Discuss the applicability of this method.

5. Starting from

$$e^{ikz} + f(\theta, \varphi)\frac{e^{ikr}}{r} = \sum_{l=0}^{\infty} \frac{A_l}{kr} \sin\left(kr - \frac{l\pi}{2} + \delta_l\right) P_l(\cos\theta)$$

derive the principal result of partial wave analysis. Explain the significance of this result.

[**Hint:** Derive eqn. (8.39)].

6. Starting from $f(\theta) = \frac{1}{k}\sum_{l=0}^{\infty}(2l+1)\left[\frac{e^{2i\delta_l} - 1}{2i}\right]P_l(\cos\theta)$, show that in elastic scattering, only the phase of scattered wave changes not the modulus. [**Hint:** Derive eqn. (8.45)]

7. Distinguish differential and total scattering cross sections. Derive the relation $\frac{d\sigma}{d\Omega} = |f(\theta)|^2$.

8. Derive the following relations:

   (a) $\sigma_{\text{elastic}}^{\text{max}} = +\frac{4\pi}{k^2}\sum_{l=0}^{\infty}(2l+1)$

   (b) $(d\sigma / d\Omega)^{s\text{-wave}} = \sin^2\delta_0 / k^2$

9. What is an optical theorem? Why it is called so? Derive its expression and explain its significance.

10. Derive and explain the following expressions:

    (a) $|a_l|^2 \le 1$ (b) $a_l = \frac{1}{\cot\delta_l - i}$

    (c) $|a_l|^2 = \dfrac{\frac{\Gamma^2}{4}}{(E_R - E)^2 + \frac{\Gamma^2}{4}}$

11. Derive the formula that describes the energy dependence of the cross section of a reaction between two particles close to a resonance energy. What is the mechanical analogue of this formula?

    [**Hint:** Derive Breit-Wigner formula. This formula is similar to that for the resonance of a forced oscillator with large damping.]

12. Derive $\dfrac{\sigma_{\text{total}}}{\sigma_{\text{max}}} = \dfrac{\frac{\Gamma^2}{4}}{(E_R - E)^2 + \frac{\Gamma^2}{4}}$. Explain it. Why this relation is called a single level formula?

13. Explain physical and unphysical Riemann sheets. How bound states and resonances are identified on these sheets?
14. Describe the origin of the concept of complex angular momentum and its role in Regge theory. Explain the significance of Legendre function $P_l(-\cos\theta)$.
15. Describe the salient features of one particle exchange model and discuss its limitations. How these limitations are removed?
16. Explain the necessicity of introducing the Watson-Sommerfeld transformation.
17. Derive Watson-Sommerfeld transformation and explain how it introduces Regge poles.
18. What are Regge pole signatures? Explain the reasons that Regge poles have signatures.
19. Derive and explain the t-channel equation $a(l, t) = \dfrac{R(t)}{l-\alpha(t)}$. Analyze this equation for t values that are below and above its threshold.
20. Describe a general Regge trajectory and explain how a Chew-Frautschi diagram of this trajectory is obtained.
21. Describe the characteristics of Chew-Frautschi diagrams. Giving examples show some CF plots.
22. Write short notes on:
    (a) Regge pole residue (b) $M^2 - J$ relation
    (c) Exchange degeneracy in C-F plots. (d) Duality
23. Explain the background that led to the introduction of Veneziano amplitude. Derive the expression for this amplitude in terms of two Mandelstam variables s and t.
24. Derive and explain Veneziano amplitude and prove its duality. Explain the significance of Vineziano amplitude.
25. Write brief notes on:
    (a) Gamma and beta functions (b) No ghost theorem
    (c) High energy behaviour of Veneziano amplitude
26. Veneziano amplitude is credited for the birth of string theory. Elucidate.
27. What are bosonic strings? Explain their limitations.
28. Explain the necessity of introducing superstrings. Describe various superstring theories and point out, giving reasons, which superstring theories qualify for being the theory of everything.
29. Write notes on:
    (a) Quantum oscillations of strings
    (b) Heterotic strings
    (c) Compactification
    (d) String spectrum and its massless states.
30. Derive $M^2 - J$ relation and explain its significance.

8.7 OBJECTIVE QUESTIONS

1. The scattering potential is usually considered
   (a) Spherically symmetric.
   (b) Spherically symmetric and central.
   (c) Spherically symmetric, central and isotropic.
   (d) Spherically symmetric, central, isotropic and localized.
2. The amplitude of a partial wave is expressed as

$$a_l = \frac{\eta_l e^{2i\delta_l} - 1}{2i}, \text{ where}$$

   (a) $\eta_l = 1$ for elastic scattering and η_l is between 0 and 1 for inelastic scattering.
   (b) $\eta_l = 1$ for inelastic scattering and η_l is between 0 and 1 for elastic scattering.
   (c) $\eta_l = 1$ for elastic scattering and η_l is between 1 and 2 for inelastic scattering.
   (d) $\eta_l = 1$ for inelastic scattering and η_l is between 1 and 2 for elastic scattering.
3. Between elastic scattering (ES) and inelastic scattering (IES), what among the following is true for the scattered wave (SW)?
   (a) In ES, the phase of SW changes, whereas in IES, the amplitude of SW changes.
   (b) In ES, the amplitude of SW changes, whereas in IES, the phase of SW changes.
   (c) In ES, the phase of SW changes, whereas in IES, both the amplitude and phase of SW change.
   (d) In both ES and IES the amplitude as well as phase of SW changes.
4. In partial wave analysis, the scattering wave amplitude $f(\theta)$ is expressed as

   (a) $f(\theta) = \frac{1}{k}\sum_{l=0}^{n} i^l (2l+1) \sin\delta_l P_l(\cos\theta)\, P_{n-l}(\cos\theta)$

   (b) $|f(\theta)| = \frac{1}{k}\sum_{l=0}^{\infty} (2l+1)\, \delta_l P_l(\cos\theta)$

   (c) $f(\theta) = \frac{1}{2ik}\sum_{l=0}^{n} (2l+1)(e^{2i\delta_l} - 1)\, P_l(\cos\theta)\, P_{n-1}(\cos\theta)$

   (d) $|f(\theta)| = \frac{1}{k}\sum_{l=0}^{\infty} (2l+1) \sin\delta_l P_l\, (\cos\theta)$
5. For the $l = 0$ partial wave, the ratio of maximum total scattering cross section to total scattering cross section is
   (a) $\operatorname{cosec} \delta_0$ (b) $\sin \delta_0$
   (c) $\operatorname{cosec}^2 \delta_0$ (d) $\sin^2 \delta_0$

6. According to the optical theorem, the total scattering cross section is
   (a) $f(\theta)$ when $\theta = 0°$
   (b) $Re\, f(\theta)$ when $\theta = 0°$
   (c) $Im\, f(\theta)$ when $\theta = 180°$
   (d) $Im\, f(\theta)$ when $\theta = 0°$
7. Choose the correct statement regarding Riemann sheets, from the following.
   (a) Physical sheet is one for which $Im\ (k) < 0$ and bound states exist on this sheet.
   (b) Physical sheet is one for which $Im\ (k) > 0$ and bound states exist on this sheet.
   (c) Unphysical sheet is one for which $Im\ (k) < 0$ and resonant states exist on this sheet.
   (d) Both (b) and (c) are true.
8. The partial wave amplitude has a Breit-Wigner form in terms of energy E_R and full width Γ of resonance. For the resonance to take place, we should have $[i = \sqrt{-1}]$

   (a) $E = E_R - \dfrac{\Gamma}{2}$ (b) $E = E_R - i\dfrac{\Gamma}{2}$

   (c) $E = E_R + \Gamma$ (d) $E = E_R + i\Gamma$

9. The resonances are also the poles of S-matrix. What among the following would correctly describe the pole position of $N(1440)\left(\dfrac{1}{2}\right)^{+}$ resonance?

   (a) $[1365 - i\dfrac{190}{2}]$MeV (b) $[1365 + i\dfrac{190}{2}]$MeV

   (c) $[-1365]$MeV (d) $[i\dfrac{190}{2}]$MeV

10. Non-relativistic quantum theory of scattering amplitude, as a function of complex and continuous angular momentum, was developed by
    (a) Murray Gell-Mann
    (b) Tullio Regge
    (c) Geoffrey Chew
    (d) Steven C. Frautschi
11. Transformation of partial waves having real and discrete angular momenta into partial waves having complex and continuous angular momenta is called
    (a) Gribov-Froissart transformation
    (b) Carlson transformation

(c) Watson-Sommerfeld transformation

(d) Duffin-Kemmer transformation

12. In Watson-Sommerfeld transformation expressed as

$$f_\alpha(t, \cos\theta_t) = \frac{1}{2i}\int_{c_i} \frac{(2\alpha+1)a(\alpha,t)P_\alpha(-\cos\theta_t)}{\sin\pi\alpha} d\alpha$$

what among the following is true?

(a) Simple poles of $\left(\frac{1}{\sin\pi\alpha}\right)$ and Regge poles of $a(\alpha, t)$, lie along the real alpha axis.

(b) Simple poles of $a(\alpha, t)$ and Regge poles of $\left(\frac{1}{\sin\pi\alpha}\right)$, lie along the imaginary alpha axis.

(c) Simple poles of $\left(\frac{1}{\sin\pi\alpha}\right)$ lie along $Re\,\alpha$ axis and Regge poles of $a(\alpha, t)$ lie at $Re\,\alpha > 0$, $Im\,\alpha > 0$.

(d) Simple poles of $\left(\frac{1}{\sin\pi\alpha}\right)$ lie at $Re\,\alpha > 0$, $Im\,\alpha < 0$ and Regge poles of $a(\alpha, t)$ lie along $Im\,\alpha < 0$.

13. If the scattered waves have appeared at their source, without even having emitted, the scattering amplitude is

(a) Analytic (b) Not analytic

(c) Unitary (d) Not unitary

14. The partial wave amplitudes $a(\alpha, t)$ in the complex and continuous angular momentum domain are required to have (+) and (–) signatures, because in this domain

(a) $a(\alpha, t)$ acquires a very large value.

(b) $a(\alpha, t)$ acquires a finite value.

(c) $a(\alpha, t)$ acquires unreal values.

(d) $a(\alpha, t)$ vanishes.

15. Regge trajectory is the path of

(a) Regge pole along real angular momentum axis and it contains resonances.

(b) Regge pole along imaginary angular momentum axis and it contains bound states.

(c) Regge pole on the complex angular momentum plane and it contains both the bound states and resonances.

(d) Regge pole on the real angular momentum plane and it contains both the bound states and resonances.

16. Following terms characterize the Chew-Frautschi (C.F.) plot. Identify the relevant terms and select the correct code.

1. (M^2, J) 2. (M^2, J^2)

3. $(\Delta J = 2)$ 4. $(\Delta J = 3)$

5. Parabolic 6. Straight line
7. Non-degenerate 8. Exchange degenerate

Codes:

(a) (2,3,5,7) (b) (1.3.6.8)
(c) (1,4,6,8) (d) (2,3,6,7)

17. The meson resonances $\rho(770, 1-)$, $\rho(1690, 3-)$; $\omega(782, 1-)$, $\omega(1670, 3-)$, $f(2050, 4+)$, $\omega(2510, 6+)$, $a(2040, 4+)$, $a(2450, 6+)$
(a) Lie on separate C.F. plots.
(b) ρ, ω lie on one C.F. plot and rest on other C.F. plot.
(c) ρ, f, a lie on one C.F. plot and rest on other.
(d) All lie on a single C.F. plot.

18. The expression for the Chew-Frautschi law is
(a) $J = \alpha(0) + \alpha' M$, where α', the Regge slope has a value 0.9 GeV^{-2}.
(b) $J = \alpha(0) + \alpha' M^2$, where α', the Regge slope has a value 0.9 GeV^{-2}.
(c) $J = \alpha(0) + \alpha' M^2$, where α', the Regge slope has a value 1.8 GeV^{-2}.
(d) $J = \alpha(0) + \alpha' M^{-2}$, where α', the Regge slope has a value 0.9 GeV^{-2}

19. Match the columns I and II and select the correct code.

Column I	*Column II*
A. Veneziano amplitude	1. Regge pole mechanism
B. α'	2. $A(s, t) = A(t, s)$
C. Duality	3. Resonance formation
D. Low energy scattering	4. Unified amplitude
E. High energy scattering	5. (dJ/dM^2)

Codes:

	A	B	C	D	E
(a)	4	5	2	3	1
(b)	5	1	4	2	3
(c)	1	4	2	3	5
(d)	4	5	1	2	3

20. The function that Veneziano interpreted as scattering amplitude, was
(a) Bessel's function (b) Euler's gamma function
(c) Euler's beta function (d) Euler's phi function

21. Gamma function $\Gamma(u)$ has poles, when u is
(a) Zero (b) Infinite
(c) Positive (d) Negative

22. In terms of two Mandelstam variables, s and t, the Veneziano amplitude is expressed as
(a) $A(s, t) = B(-\alpha(s), -\alpha(t))$
(b) $A(s, t) = B(-\alpha(s), \alpha(t))$
(c) $A(s, t) = B(\alpha(s), -\alpha(t))$
(d) $A(s, t) = B(\alpha(s), \alpha(t))$

23. No-ghost theorem demands that
    (a) Space-time should be 10 dimensional and $\alpha(0) = (1/2)$.
    (b) Space-time should be 10 dimensional and $\alpha(0) = 1$.
    (c) Space-time should be 26 dimensional and $\alpha(0) = (1/2)$
    (d) Space-time should be 26 dimensional and $\alpha(0) = 1$.
24. The relation $A(t, s \to \infty) = [s]^{\alpha(t)}$ expresses
    (a) The low-energy behaviour of Veneziano amplitude.
    (b) The high-energy behaviour of Veneziano amplitude.
    (c) The Regge behaviour of Veneziano amplitude.
    (d) Both (b) and (c).
25. If supersymmetry is not observed, if proton decay does not take place and if higher spatial dimensions are not found, the most probable approach to develop a quantum theory of gravity would be
    (a) M-theory (b) Conformal field theory
    (c) Superstring theory (d) Loop quantum theory
26. A framework that unites the five 10-dimensional superstring theories with 11-dimensional supergravity is known as
    (a) Superstring theory (b) Conformal field theory
    (c) M-theory (d) Grand unified theory
27. Low energy limit of superstring theories is
    (a) Quantum field theory (b) Supersymmetric theory
    (c) Supergravity theory (d) Regge theory
28. The Veneziano inspired string was an extended object with size. This size was about
    (a) 10^{-4} metre (b) 10^{-8} metre
    (c) 10^{-10} metre (d) 10^{-15} metre
29. Estimating string tension from the graviton mode, it turns out to be as large as 10^{39} tonnes. This tension would contract the string to the size of
    (a) 10^{-15} metre (b) 10^{-20} metre
    (c) 10^{-35} metre (d) 10^{-60} metre
30. If the string size were as in question (29), the fundamental mode of vibration of string would give rise to a particle with a mass of the order of
    (a) Planck mass of 10^{19} GeV.
    (b) Twice the Planck mass
    (c) Half the Planck mass.
    (d) One third of Planck mass.
31. Quantum oscillations of a string (closed) correspond to particles of higher masses and higher spins. If α' be the slope of $M^2 - J$ trajectory of these particles, then to have zero mass spin-2 graviton in the spectrum of string, the slope α' must be equal to

(a) Planck constant (b) Newtonian constant
(c) Faraday constant (d) Fermi constant

32. Select the correct statement from the following. For bosonic strings
    (a) The space-time has 10 dimensions and tachyons are its lowest energy states.
    (b) The space-time has 10 dimensions and gravitons are its lowest energy states.
    (c) The space-time has 26 dimensions and tachyons are its lowest energy states.
    (d) The space-time has 26 dimensions and gravitons are its lowest energy states.
33. In relation to superstring theories, match the two columns and select the correct code.

Column I		*Column II*
A. Type I	1.	$E_8 \times E_8$, $(N = 0 + N = 2)$, 10
B. Type IIA	2.	$SO(32)$. open & closed, 10
C. Type II B	3.	closed, non-chiral,no tachyons
D. Heterotic I	4.	$SO(32)$, closed, $N = 1$
E. Heterotic II	5	closed, chiral, 10

Codes:

	A	B	C	D	E		A	B	C	D	E
(a)	3	4	1	2	5	(b)	2	3	5	4	1
(c)	1	2	4	5	3	(d)	1	3	2	5	4

34. A ten-dimensional superstring theory can reduce to real-world space-time theory, if
    (a) Six spatial dimensions are compacted on a Calabi-Yu manifold.
    (b) Nine spatial dimensions are compacted on a Calabi-Yu manifold.
    (c) Ten dimensions are compacted on a Chen-Patan manifold.
    (d) Nine dimensions are compacted on a Chen-Patan manifold.

35 In the following. Physicists have been mentioned in the order doctoral advisor and doctoral student. Choose the correct answer.
    (a) Enrico Fermi, Geoffrey F. Chew, John H. Schwarz.
    (b) Enrico Fermi, Geoffrey F. Chew, John R. Taylor.
    (c) Richard H. Dalitz, S. Mandelstam, Micho Kaku
    (d) All the above are true.

ANSWERS

1. d,	2. a,	3. c,	4. d,	5. c,
6. d,	7. d,	8. b,	9. a,	10. b,
11. c,	12. c,	13. b,	14. d,	15. c,
16. b,	17. d,	18. b,	19. a,	20. c,
21. d,	22. a,	23. d,	24. d,	25. d,
26. c,	27. c,	28. d,	29. c,	30. a,
31. b,	32. c,	33. b,	34. a,	35. d.

APPENDIX A

CONVENTIONS

NATURAL UNITS

In elementary particle physics, we come across with the quantities like mass, energy, momentum, distance, lifetime, cross sections and spin. The magnitude of these quantities for the elementary particles is so small that we use special units to write them. For example, energy is written in electron volts (eV), momentum is written in units of (eV/c), where c is the speed of light in vacuum and mass is written as (eV/c^2). Length is written in units of meter, or fermis ($1F = 10^{-15}$ m) and cross section is written in m^2 or barns ($1 barn = 10^{-28} m^2$). We employ seconds to measure time while spin, isospin and angular momentum are measured in units of $\hbar = h/2\pi$, the reduced Planck's constant.

In most of the calculations in elementary particle physics, we find the use of constants c and $\hbar\omega$. A system of units in which c is one unit of speed (L/T), and $\hbar$ is one unit of action $\left(\frac{ML^2}{T}\right)$ is known as *natural units*. The immediate consequence of natural units

$$\hbar = c = 1 \tag{A1}$$

may be understood by considering the dimensions of c, i.e.,

$$[c] = [L]\,[T]^{-1} \tag{A2}$$

When we take $c = 1$, we imply

$$[L] = [T] \tag{A3}$$

That is, in natural units the unit of length is numerically equal to the unit of time. The length and time appear as equivalent dimensions. This situation of equivalence of space and time dimensions is in agreement with the theory of relativity where space and time are different kinds of same thing and hence they can be measured in same units. The choice of $c = 1$ is then *natural*. Since,

$$[E] = \hbar\,\omega, \left(\omega = \frac{2\pi}{T}\right) \tag{A4}$$

and

$$[p] = \hbar k, \left(k = \frac{2\pi}{\lambda}\right) \tag{A5}$$

the dimensions of energy (E) and momentum (p) in units $\hbar = 1$ would be

$$[E] = [\mathrm{T}]^{-1} \tag{A6}$$

and

$$[p] = [\mathrm{L}]^{-1} \tag{A7}$$

which on using equation (A3) imply

$$[E] = [\mathrm{T}]^{-1} = [p] = [\mathrm{L}]^{-1} \tag{A8}$$

Just as the relation (A3) had its connection with special theory of relativity, the relation (A8) has its connection in the fact that energy and momentum operators are the generators of translations in space + time. The translations (movements) along spatial axes (x, y, z) create linear momentum and energy refers to translation along time axis. The fundamental commutation relation

$$[x_i, p_j] = i\,\hbar\delta_{ij} \qquad (i, j = 1, 2, 3) \tag{A9}$$

also tells that a basic connection exists between the momenta and distances.

The reduced Planck's constant $\hbar$ has dimensions of energy × time, i.e.

$$[\hbar] = [\mathrm{M}]\,[\mathrm{L}]^2\,[\mathrm{T}]^{-1} \tag{A10}$$

which on setting $\hbar = 1$, implies [using equation (A8)]

$$[\mathrm{M}] = [\mathrm{L}]^{-1} = [\mathrm{T}]^{-1} \tag{A11}$$

Combining this relation with eqn. (A8) we write

$$[E] = [\mathrm{M}] = [\mathrm{L}]^{-1} = [\mathrm{T}]^{-1} \tag{A12}$$

It is customary in particle physics to choose $[M]$ as independent dimension. Usually the unit of mass is chosen as the mass of some standard particle m_0 (e.g. the proton mass m_p) ($m_p \sim 1$ GeV). Thus we take

$$m_0 = 1 \tag{A13}$$

which is called the natural unit of mass. The natural unit of length is Compton wavelength of the particle chosen (m_0), i.e.

$$\lambda\!\!\!^{-}_0 = \frac{\hbar}{m_0 c} = 1 \tag{A14}$$

The natural unit of time is the time taken by light in travelling the length $\lambda\!\!\!^{-}_0$

$$t_0 = \left(\frac{\lambda\!\!\!^{-}}{c}\right) = 1 \tag{A15}$$

The natural unit of energy is

$$E_0 = m_0 c^2 = 1 \tag{A16}$$

Just as in SI units we define 1 m, 1 kg and 1 sec, here eqns. [(A13) to (A16)] define 1 unit of mass, 1 unit of length, etc., in natural units. The advantage of using natural units is that in these units the calculations become simple. To obtain the value of a quantity, however, we have to bring back the conventional units. Equations (A13) to (A16) also tell that in natural units we describe the mass (m_0), momentum ($m_0 c$), and energy ($m_0 c^2$), in terms of GeV and measure length ($\lambda\!\!\!^{-}_0$) and time (t_0) in units of GeV^{-1}. If we wish to express these quantities in conventional units, i.e., mass in kg, length in meter (m) and time in second (s); we have to use the following conversion factors.

$$\begin{aligned} 1\text{ kg} &= 5.61 \times 10^{26}\text{ GeV} \\ 1\text{ m} &= 5.07 \times 10^{15}\text{ GeV}^{-1} \\ 1\text{ s} &= 1.52 \times 10^{24}\text{ GeV}^{-1} \end{aligned} \tag{A17}$$

For example, the Compton wavelength of the electron $\lambdabar = (\hbar.m_e)$ in SI units is

$$(\lambdabar)_{SI} = 0.386 \times 10^{-12}\text{ m} \tag{A18}$$

and in natural units ($m_e \approx 0.51 \times 10^{-3}$ GeV)

$$(\lambdabar)_{nat.} = \frac{1}{m_e} = 1.96\times10^3\text{ GeV}^{-1} = \frac{1.96\times10^3}{5.07\times10^{15}}\text{m} = 0.39\times10^{-12}\text{ m} \tag{A19}$$

Other useful conversion factors are

$$\begin{aligned} &1\,\text{TeV} = 10^3\text{ GeV} = 10^6\text{ MeV} = 10^9\text{ KeV} = 10^{12}\text{ eV} \\ &1\,\text{fermi} = 5.07\,\text{GeV}^{-1} = 10^{-15}\text{ m} = 1\,\text{F} \\ &(1\,\text{F})^2 = 10\,\text{millibarn(mb)} = 10^4\,\mu\text{b} = 10^7\text{ nb} = 10^{10}\text{ pb} \end{aligned} \tag{A20}$$

where the abbreviations stand for

$$\begin{aligned} &\text{T}-\text{Tera }(10^{12}),\text{G}-\text{Giga }(10^9),\text{M}-\text{Mega }(10^6),\text{K}-\text{Kilo }(10^3) \\ &\mu-\text{Micro }(10^{-6}),\text{n}-\text{Nano }(10^{-9})\text{ and p}-\text{Pico }(10^{-12}) \end{aligned} \tag{A21}$$

Four-Vectors

A four-vector has four components. There are two ways to write the components of a four vector. In one, the three components are taken real and the fourth imaginary. For example, in natural units, a four-vector

$$x = (\boldsymbol{x}, it) \tag{A 22}$$

where $\boldsymbol{x}$ has three components. A different manner in which a four-vector can be expressed is to take all the four components as real; we adopt the latter convention in which all the components of a four-vector are real. Accordingly, a four- vector can be expressed as $\{x^\mu\}$ where $\mu = 0, 1, 2, 3$, denote its four components.

$$\{x^\mu = (x^0, x^1, x^2, x^3) \tag{A23}$$

where

$$\begin{aligned} &x^0 \equiv ct = t\text{ (natural units)} \\ &x^1 \equiv x, x^2 \equiv y, x^3 \equiv z \end{aligned} \tag{A24}$$

Because of the appearance of μ as the upper index on x, x^μ is called the contravariant four-vector. Again it is the convention that positive components are represented by contravariant indices, i.e.

$$+x = x^1, +y = x^2, +z = x^3 \tag{A25}$$

with

$$x^0 = t$$

We shall see in the following that the lower or covariant indices represent $\{x_\mu\}$ with

$$x_0 = t$$

and

$$x_1 = -x,\ x_2 = -y \text{ and } x_3 = -z \tag{A26}$$

A diagonal metric tensor $g_{\mu\nu}$ in Minkowski space–time, which is also called the pseudo-Euclidean metric tensor, is introduced to define the infinitesimal line element ds as

$$ds^2 = g_{\mu\nu}\, dx^\mu\, dx^\nu \tag{A27}$$

where $g_{\mu\nu}$ with both μ and ν running from zero to three (0, 1, 2, 3) has sixteen components. Explicitly, in matrix form

$$g_{\mu\nu} = \begin{bmatrix} 1 & 0 & 0 & 0 \\ 0 & -1 & 0 & 0 \\ 0 & 0 & -1 & 0 \\ 0 & 0 & 0 & -1 \end{bmatrix} \tag{A28}$$

which means

$$g_{00} = +1,\ g_{11} = g_{22} = g_{33} = -1$$

and

$$g_{\mu\nu} = 0 \quad (\mu \neq \nu) \tag{A29}$$

The infinitesimal line element [eqn. (A27)] then becomes

$$\begin{aligned} ds^2 &= g_{00}(dx^0)^2 + g_{11}(dx^1)^2 + g_{22}(dx^2)^2 + g_{33}(dx^3)^2 \\ &= (dx^0)^2 - (dx^1)^2 - (dx^2)^2 - (dx^3)^2 \end{aligned} \tag{A30}$$

The covariant four-vector x_μ can then be defined by introducing $g_{\mu\nu}$ as

$$x_\mu = g_{\mu\nu} x^\nu \tag{A31}$$

from which we get the following relations between the covariant and contravariant components

$$\begin{aligned} x_0 &= g_{00}x^0 = x^0 \equiv t \\ x_1 &= g_{11}x^1 = -x^1 \equiv -x \\ x_2 &= g_{22}x^2 = -x^2 \equiv -y \\ x_3 &= g_{33}x^3 = -x^3 \equiv -z \end{aligned}$$

i.e.,

$$x_\mu = (x_0, x_1, x_2, x_3) \equiv (\boldsymbol{t}, -\boldsymbol{x}) \tag{A32}$$

The relations (A32) imply that, when we take $g_{\mu\nu}$ as in eqn. (A28), the time part of the covariant four-vector is same as the corresponding time part of the contravariant four-vector, whereas the space components undergo the change in sign.

The contravariant counterpart of $g_{\mu\nu}$, which is $g^{\mu\nu}$, is defined as the inverse of the covariant metric tensor $g_{\mu\nu}$, where $g_{\mu\nu}$, is considered a matrix. We then have

$$g^{\mu\nu} = \frac{\text{cofactor of } g_{\mu\nu}}{\text{determinant of } g_{\mu\nu}} \tag{A33}$$

It may be seen that eqns. (A28) and (A33) give

$$g^{\mu\nu} = g_{\mu\nu} \tag{A34}$$

Consequently, following relationship exists between the contravariant and covaraint metric tensors

$$g^{\mu\nu} g_{\mu\lambda} = \delta^{\nu}_{\lambda} \tag{A35}$$

where δ^{ν}_{λ} is the Kronekar delta obeying

$$\begin{aligned} \delta^{\nu}_{\lambda} &= +1 \text{ for } \nu = \lambda \\ &= 0 \text{ for } \nu \neq \lambda \end{aligned} \tag{A36}$$

The two tensors $g^{\mu\nu}$ and $g_{\mu\nu}$ may be used to raise or lower the indices of a given four-vector. For example in

$$A^{\mu} = g^{\mu\nu} A_{\nu} \tag{A37}$$

$g_{\mu\nu}$ has raised the index ν in A_{ν} and in

$$A_{\mu} = g_{\mu\nu} A^{\nu} \tag{A38}$$

$g^{\mu\nu}$ has lowered the index ν in A^{ν}. By using the metric tensor, the scalar product of one contravariant and one covariant vector can be written as the scalar product of two covariant vectors or two contravariant vectors.

$$A_{\mu} B^{\mu} = g^{\mu\nu} A_{\mu} B_{\nu} = g_{\mu\nu} A^{\mu} B^{\nu} \tag{A39}$$

The raising and lowering of both the indices of the metric tensor can also be done by $g^{\mu\nu}$ and $g_{\mu\nu}$. For example, we have

$$g^{\mu\alpha} g^{\nu\beta} g_{\alpha\beta} = g^{\mu\nu}$$

and

$$g_{\alpha\mu} g_{\beta\nu} g^{\mu\nu} = g_{\alpha\beta} \tag{A40}$$

It may also be mentioned that the metric tensor is symmetric, i.e.

$$g^{\mu\nu} = g^{\nu\mu} = g_{\mu\nu} = g_{\nu\mu} \tag{A41}$$

THE LORENTZ TRANSFORMATIONS

On changing the frame of reference (inertial) in the Minkowski spacetime, a four-vector x^{μ} transforms to x'^{μ}, where x'^{μ} denote the four components of x^{μ} in new frame of reference. The relation between x'^{μ} and x^{μ} is

$$x^{\mu} = a^{\mu}_{\nu} x^{\nu} \tag{A42}$$

where a^{μ}_{ν} form the Lorentz transformation matrices. If the frame of reference moves with a uniform velocity ν, parallel to x^1 axis then

$$(a^{\mu}_{\nu})^1 = \begin{bmatrix} \cosh\psi & \sinh\psi & 0 & 0 \\ \sinh\psi & \cosh\psi & 0 & 0 \\ 0 & 0 & 1 & 0 \\ 0 & 0 & 0 & 1 \end{bmatrix} \quad \text{(A43)}$$

where in conventional units

$$x'^0 = ct', \; x^0 = ct \quad \text{(A44)}$$

$$\cosh\psi = \frac{1}{\sqrt{1-\frac{v^2}{c^2}}}, \quad \sinh\psi = \frac{-v/c}{\sqrt{1-\frac{v^2}{c^2}}} \quad \text{(A45)}$$

in which ψ is known as the boost parameter or rapidity. Similarly for the uniform motion parallel to spatial x^2 and x^3 axes, we have respectively,

$$(a^{\mu}_{\nu})^2 = \begin{bmatrix} \cosh\psi & 0 & \sinh\psi & 0 \\ 0 & 1 & 0 & 0 \\ \sinh\psi & 0 & \cosh\psi & 0 \\ 0 & 0 & 0 & 1 \end{bmatrix} \quad \text{(A46)}$$

and

$$(a^{\mu}_{\nu})^3 = \begin{bmatrix} \cosh\psi & 0 & 0 & \sinh\psi \\ 0 & 1 & 0 & 0 \\ 0 & 0 & 1 & 0 \\ \sinh\psi & 0 & 0 & \cosh\psi \end{bmatrix} \quad \text{(A47)}$$

More familiar forms of Lorentz transformations can be obtained by using equations (A42) to (A47). For example, for the motion along $z = x^3$ axis we shall have from eqn. (A42)

$$\begin{bmatrix} x'^0 \\ x'^1 \\ x'^2 \\ x'^3 \end{bmatrix} = (a^{\mu}_{\nu})^3 \begin{bmatrix} x^0 \\ x^1 \\ x^2 \\ x^3 \end{bmatrix} \quad \text{(A47a)}$$

Substituting $(a^{\mu}_{\nu})^3$ from eqn. (A47) and using eqn. (A44), we get

$$\begin{aligned} x'^0 &= x^0 \cosh\psi + z \sinh\psi \\ x' &= x \\ y' &= y \\ z' &= x^0 \sinh\psi + z \cosh\psi \end{aligned} \quad \text{(A48)}$$

which on using eqns. (A44) and (A45) acquire the familiar form of the Lorentz transformations

$$l' = \frac{t - \frac{vz}{c^2}}{\sqrt{1 - \frac{v^2}{c^2}}},\ x' = x,\ y' = y \tag{A49}$$

$$z' = \frac{z - \frac{vt}{c^2}}{\sqrt{1 - \frac{v^2}{c^2}}}$$

In natural units $c = 1$, is implied in these equations. The Lorentz transformations (A49) are for the frame S′ moving along the z direction with a uniform velocity v. If the frame of reference S′ is moving along an arbitrary direction then x' and y' of eqn. (A49) will also change like z'. The general Lorentz transformation can then be written as

$$(x')^0 = \gamma(x^0 - \beta.x) \tag{A50a}$$

$$(\boldsymbol{x})' = \boldsymbol{x} + \frac{\gamma - 1}{\beta^2}(\beta.x)\,\beta - \gamma\,\beta x^0 \tag{A50b}$$

where

$$\gamma = (1 - \beta^2)^{-\frac{1}{2}}$$

and

$$\beta = \frac{\boldsymbol{v}}{c},\ \beta = |\beta| < 1,\ \boldsymbol{x} \equiv (x^1, x^2, x^3) \tag{A51}$$

Assuming same velocity v along the three directions

$$\beta.\boldsymbol{x} = \nu_{(i+j+k)} \cdot {}_i x^1 + {}_j x^2 + {}_k x^3 = \nu(x^1 + x^2 + x^3) \tag{A52}$$

Along, (say) z axis x^1 and x^2 are zero. Using eqn. (A52), it can be easily checked that transformations (A49) are the particular case of eqns. (A50) belonging to

$$x = {}_k x^3 = {}_k z$$

and

$$\beta = {}_k \nu \tag{A53}$$

The Lorentz transformation matrices obey the orthogonality condition

$$a^{\nu}_{\mu} a^{\mu}_{\lambda} = \delta^{\nu}_{\lambda} \tag{A54}$$

where

$$a^{\nu}_{\mu} = (a^{-1})^{\mu}_{\nu} \tag{A55}$$

is the inverse of Lorentz transformation matrix a^{μ}_{ν}. The inverse matrices are used to define the inverse Lorentz transformations

$$x^{\mu} = (a^{-1})^{\mu}_{\nu} x'^{\nu} \quad \text{(A56a)}$$

or

$$x_{\mu} = a^{\nu}_{\mu} x'_{\nu} \quad \text{(A56b)}$$

Employing metric tensors we can also construct $a^{\mu\nu}$ and $a_{\mu\nu}$ as

$$a^{\mu\nu} = g^{\mu\lambda} a^{\nu}_{\lambda} \quad \text{(A57a)}$$

and

$$a_{\mu\nu} = g_{\mu\lambda} a^{\lambda}_{\nu} \quad \text{(A57b)}$$

The Lorentz transformations (A42) then become

$$x'^{\mu} = a^{\mu\nu} x_{\nu} \quad \text{(A58)}$$

and the orthogonality conditions (A54) become

$$a^{\nu\mu} a_{\lambda\mu} = g^{\nu\lambda} a^{\mu}_{\lambda} g_{\lambda\rho} a^{\rho}_{\mu} = \delta^{\nu}_{\rho} \delta^{\rho}_{\lambda} = \delta^{\nu}_{\lambda} \quad \text{(A59)}$$

Similarly,

$$a^{\nu\mu} a_{\nu\lambda} = \delta^{\mu}_{\lambda} \quad \text{(A60)}$$

where, equations (A57), (A54) and (A35) have been used.

INFINITESIMAL LORENTZ TRANSFORMATIONS

If the four-vector x^{μ} Lorentz transforms to x'^{μ} and if x'^{μ} and x^{μ} are infinitesimally close, then such Lorentz transformations $x^{\mu} \rightarrow x'^{\mu}$ are called *infinitesimal Lorentz transformations*. For them we may write

$$x'^{\nu} = x^{\nu} + \delta x^{\nu} \quad \text{(A61)}$$

Now express

$$\delta x^{\nu} = \varepsilon^{\nu\mu} x_{\mu} \quad \text{(A62)}$$

where the infinitesimal character comes in the RHS by demanding that the determinant of the tensor $\varepsilon^{\nu\mu}$ is infinitesimal, i.e.,

$$|\varepsilon^{\nu\mu}| << 1 \quad \text{(A63)}$$

Equation (A61) can then be written as

$$x^{\nu\mu} x_{\mu} + \varepsilon^{\nu\mu} x_{\mu} = (g^{\nu\mu} + \varepsilon^{\nu\mu})\, x_{\mu} \quad \text{(A64)}$$

Thus eqn. (A64) is the equation that provides the infinitesimal Lorentz transformations. Comparing eqn. (A64) with eqn. (A58) we write

$$a^{\nu\mu} = g^{\nu\mu} + \varepsilon^{\nu\mu} \quad \text{(A65a)}$$

and also

$$a_{\nu\lambda} = g_{\nu\lambda} + \varepsilon_{\nu\lambda} \quad \text{(A65b)}$$

Multiplying these equations and retaining the terms only upto the first order of ε, we get

$$a^{\nu\mu} a_{\nu\lambda} = g^{\nu\mu} g_{\nu\lambda} + g^{\nu\mu} \varepsilon_{\nu\lambda} + \varepsilon^{\nu\lambda} g_{\nu\lambda} \quad \text{(A66)}$$

Using the equations (A60) and (A35), and changing for the moment ν by α we may write eqn. (A66) as

$$\delta^{\mu}_{\lambda} = \delta^{\mu}_{\lambda} + g^{\alpha\mu} \varepsilon_{\alpha\lambda} + \varepsilon^{\alpha\mu} g_{\alpha\lambda} \quad \text{(A67)}$$

which on multiplying by $g^{\lambda\nu}$ becomes

$$g^{\alpha\mu} g^{\lambda\nu} \varepsilon_{\alpha\lambda} + g^{\lambda\nu} \varepsilon^{\alpha\mu} g_{\alpha\lambda} = 0$$

or

$$\varepsilon^{\mu\nu} + \delta^{\nu}_{\alpha} \varepsilon^{\alpha\mu} = 0 \tag{A68}$$

Equation (A68) describes the antisymmetry of ε in μ and ν, i.e.

$$\varepsilon^{\nu\mu} = -\,\varepsilon^{\mu\nu} \tag{A69}$$

Thus the parameter ε of infinitesimal Lorentz transformations must be antisymmetric in its indices (μ, ν) and its determinant must be small.

TIME, SPACE AND LIGHTLIKE SEPARATIONS

Let us consider the two inertial frames of reference S and S′ of special theory of relativity (Fig. A1)

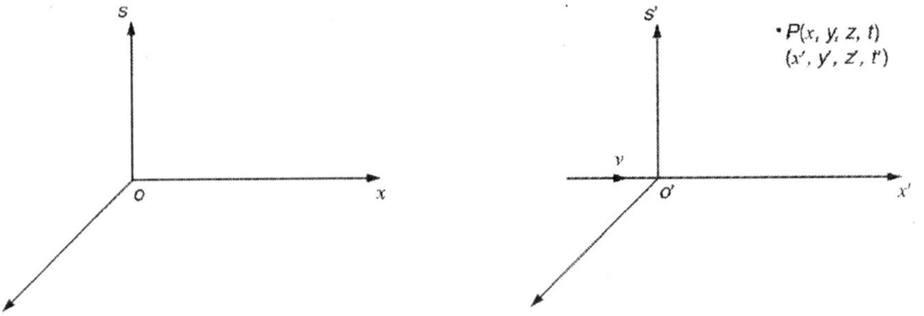

Fig. A1: Inertial frames of reference *S* and *S′*

The frame *S′* is moving with a uniform velocity *v* parallel to *x* axis with respect to S. Let P be the point whose spatial coordinates for O are (*x*, y, *z*) and for O′, (*x*′, *y*′, *z*′). The light whose speed is *c* and originates at P, reaches the observer O in time *t* sec. and the observer O′ in time
t′ second. Hence (we retain conventional units)

$$c^2 t^2 = x^2 + y^2 + z^2 \tag{A70}$$

$$c^2 t'^2 = x'^2 + y'^2 + z'^2 \tag{A71}$$

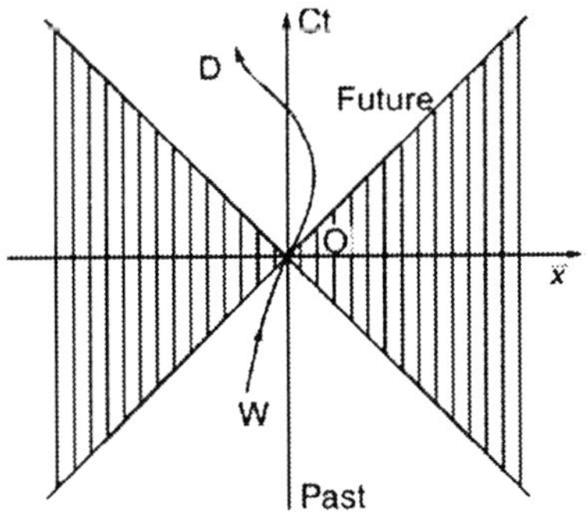

Fig. A2: The light cone and the world line of a system. The unshaded portion of the Figure represents the past and future. The region outside the cone is shaded. It is called elsewhere. A point inside the light cone is said to have a timelike separation from the origin. A point outside the light cone has spacelike separation from the origin

These are two similar equations. We describe the first one. Eqn. (A70), namely,

$$c^2t^2 - x^2 - y^2 - z^2 = 0 \tag{A72}$$

specifies a surface of light cone (FigA2). The light signals emitted from the origin at $t = 0$, would travel out along the 45° lines with a speed c. c is the upper limit of all speeds. However, a material particle has a speed less than c. Such a particle would trace out a path WOD, a wavy line. WOD is called the *world line*. The world lines, which lie on the surface of light cone, are straight.

For $t > 0$, the path of the material particle, OD, lies inside the upper half cone which is called *future*. The path OD started from O. The particle reached O through a pathline WO that lies inside the lower half cone. The lower half cone is called the *past*. The region where $t = 0$, is called the *present*. The shaded region is called the *elsewhere*. Let us consider two events which take place in space–time $(\boldsymbol{x}, t)$. The events are labeled by $P_1(t_1, \boldsymbol{x}_1)$ and $P_2(t_2, \boldsymbol{x}_2)$. As in eqn. (A52) both x_1 and x_2 have three components each. The square of the separation between P_1 and P_2 is

$$S_{12}^2 = c^2(t_1 - t_2)^2 - |(\boldsymbol{x}_1 - \boldsymbol{x}_2)|^2 \tag{A73}$$

This interval is invariant under the Lorentz transformations, implying that

$$S_{12}^2 = c^2(t_1' - t_2')^2 - |(\boldsymbol{x}_1' - \boldsymbol{x}_2')|^2 = c^2(t_1 - t_2)^2 - |(\boldsymbol{x}_1 - \boldsymbol{x}_2)|^2 \tag{A74}$$

where primed coordinates are with respect to S'. From the equation (A73) arise three possibilities

$$(a)\ S_{12}^2 > 0 \quad (b)\ S_{12}^2 < 0 \ \text{ and } (c)\ S_{12}^2 = 0 \tag{A75}$$

which we discuss in the following.

$(a)\ S_{12}^2 > 0$ (*Timelike separations*)

When $S_{12}^2 > 0$, the event $P_1(t_1, x_1)$ and the event $P_2(t_2, x_2)$ are said to have timtlike separation. To get an insight of the case, let us write the generalized Lorentz transformations (A50b) for $x_1'\ \boldsymbol{x}_1'$ and x_2'; as

$$\boldsymbol{x}_1' = \boldsymbol{x}_1 + \frac{\gamma - 1}{\beta^2}(\boldsymbol{\beta}.\boldsymbol{x}_1)\boldsymbol{\beta} - \gamma\,\boldsymbol{\beta}\,ct_1 \tag{A76}$$

$$\boldsymbol{x}_2' = \boldsymbol{x}_2 + \frac{\gamma - 1}{\beta^2}(\boldsymbol{\beta}.\boldsymbol{x}_2)\boldsymbol{\beta} - \gamma\,\boldsymbol{\beta}\,ct_2 \tag{A77}$$

where equation (A44) has been used. From these equations, we may easily obtain

$$\boldsymbol{x}_1' - \boldsymbol{x}_2' = (\boldsymbol{x}_1 - \boldsymbol{x}_2) + \frac{\gamma - 1}{\beta^2}\boldsymbol{\beta}.(\boldsymbol{x}_1 - \boldsymbol{x}_2)\boldsymbol{\beta} - \gamma\,\boldsymbol{\beta}\,c\,(t_1 - t_2)$$

from which we may write

$$\left|\boldsymbol{x}_1' - \boldsymbol{x}_2'\right| = \gamma\left(\left|\boldsymbol{x}_1 - \boldsymbol{x}_2\right| - \beta\, c\left|t_1 - t_2\right|\right) \tag{A78}$$

where $\beta = |\boldsymbol{\beta}|$ is implied. Now $\boldsymbol{x}_1'$ and $\boldsymbol{x}_2'$ are the Lorentz transformed vectors of x_1 and x_2 respectively. Let us consider a particular Lorentz transformation which makes $\boldsymbol{x}_1' = \boldsymbol{x}_2'$. For such cases eqn. (A78) reads

$$\beta = \frac{\left|\boldsymbol{x}_1 - \boldsymbol{x}_2\right|}{c\left|t_1 - t_2\right|} \tag{A79}$$

Since $\beta < 1$, eqn. (A79) implies

$$c\left|t_1 - t_2\right| > \left|\boldsymbol{x}_1 - \boldsymbol{x}_2\right| \tag{A80}$$

or

$$c^2\left|t_1 - t_2\right|^2 - \left|\boldsymbol{x}_1 - \boldsymbol{x}_2\right|^2 > 0 \tag{A81}$$

where it has to be noted that eqn. (A81) requires $\boldsymbol{x}_1' = \boldsymbol{x}_2'$. Substituting these in equation (A74), we get

$$S_{12}^2 = c^2\,(t_1' - t_2')^2 > 0 \tag{A82}$$

Thus the events, which occurred in S frame at times t_1 and t_2 at the space points x_1 and x_2, respectively, occur in frame S' at the same space point ($\boldsymbol{x}_1' = \boldsymbol{x}_2'$) but at different times t_1' and t_2'. In other words, the events which occurred in frame S were initially separated in both time and space but in S' frame they are separated only by time. Thus, for $S_{12}^2 > 0$, the events are said to have time like separations. One event occurs at the origin and other occurs in the past or future light cone.

(b) $S_{12}^2 < 0$ *(spacelike separations)*

Now use eqn. (A50a), from which we can write

$$ct_1'' = \gamma(ct_1 - \boldsymbol{\beta}.\boldsymbol{x}_1) \tag{A83}$$

and

$$ct_2'' = \gamma(ct_2 - \boldsymbol{\beta}.\boldsymbol{x}_2) \tag{A84}$$

Subtraction yields

$$\left|t_1'' = t_2''\right| = \frac{\gamma}{c}\left[c(t_1 - t_2) - \beta\left|\boldsymbol{x}_1 - \boldsymbol{x}_2\right|\right] \tag{A85}$$

From which we can observe that for

$$\beta = \frac{c\left|\, t_1 - t_2\right|}{\left|\boldsymbol{x}_1 - \boldsymbol{x}_2\right|} \tag{A86}$$

$$t_1'' = t_2'' \tag{A87}$$

The requirement of $\beta < 1$, is now achieved from equation (A86). This equation tells that for β to be less than 1,

$$|\boldsymbol{x}_1 - \boldsymbol{x}_2| \;> c|\, t_1 - t_2| \tag{A88}$$

implying that

$$c^2(t_1^2 - t_2^2) - |\boldsymbol{x}_1 - \boldsymbol{x}_2|^2 < 0 \tag{A89}$$

Hence using equations (A87) and (A89) we can express eqn. (A74) (with $x' \to x''$ and $t' \to t''$)

$$S_{12}^2 = -\left|\boldsymbol{x}_1'' - \boldsymbol{x}_2''\right| < 0 \tag{A90}$$

Thus there exists an inertial frame S″ in which two events occur at the same time ($t_1'' = t_2''$) but at different space points. One point is located at the origin O and other point lies in elsewhere region. In the inertial frame S, the two events were separated in both space and time. The events which occur in S″ frame at the same time but at different space points are said to have spacelike separation. Elsewhere is the region where points have spacelike separations.

(*c*) $S_{12}^2 = 0$ (*Lightlike separations*)

The remaining possibility of $S_{12}^2 = 0$, describes the lightlike separations. In this case the events lie on the light cone with respect to each other and can be connected by signals travelling with the speed of light.

Thus the two events occurring in space–time (frame S) can be divided into two classes. One class is of the timelike separations (frame S′) and other that of spacelike separations (frame S″). The surface of the light cone acts as a boundary.

Any point, which is inside the light cones, is timelike. The two points inside the light cone can be connected to each other by signals that travel with a speed less than the speed of light. This means the time like events are causally related. An event is said to be causal, if its effects are observed only after the event has taken place. An event at the origin can be influenced causally only by the events that occur in the past light cone.

The space–time points, which lie on the surface of the lightcone, are lightlike. These points can be connected by signals which travel exactly with the speed of light. The spacelike points lie on the region of elsewhere and such points can be connected by signals which travel with a speed greater than the speed of light. Since, no speed is greater than the speed of light (in vacuum), the spacelike points cannot be causally connected. The events, which have a spacelike separation, will not affect each other.

SCALAR PRODUCT OF FOUR-VECTORS

Equation (A1.23) defines a four vector x^μ which can also be expressed as

$$x^\mu = (x^0, x) \tag{A91}$$

where $\boldsymbol{x}$ is three component (x^1, x^2, x^3) vector or a three-vector. Let us have two four-vectors x and y where $x \Rightarrow x^\mu$ and $y \Rightarrow y^\mu$. These four-vectors transform under Lorentz transformations as

$$x^\mu \rightarrow x'^\mu = a^\mu_\nu x^\nu = a^{\mu\nu} x_\nu \tag{A92}$$

and $$y^\mu \rightarrow y'^\mu = a^\mu_\alpha y^\alpha = a^{\mu\alpha} y_\alpha$$

The scalar product of the four-vectors x and y is

$$xy = x_\mu y^\mu = x_0 y^0 + x_1 y^1 + x_2 y^2 + x_3 y^3 \tag{A93}$$

Since, $x_0 = x^0$, $x_i = -x^i$ $(i = 1, 2, 3)$ we write

$$xy = x^0 y^0 - (x^1 y^1 + x^2 y^2 + x^3 y^3) = x^0 y^0 - \boldsymbol{x.y} \tag{A94}$$

where $\boldsymbol{x \cdot y}$ is the scalar product of two, three-vectors. Notice also that the contravariant indices define the positive three-vector.

Equation (A94) can be used to define the scalar product of a four-vector with itself, i.e., the square of the four-vector. Thus

$$x^2 = x_\mu x^\mu = (x^0)^2 - \boldsymbol{x}^2 \tag{A95}$$

Equation (A95) defines the squared length of the four-vector x. Because of $g^{\mu\nu} = (+1, -1, -1, -1)$, the x^2 appears as the difference of squares of the time part x^0 and the space part $\boldsymbol{x}$.

In view of the discussion of the previous section, we may then tell for the four vector x that

(a) x is timelike if $x^2 > 0$
(b) x is spacelike if $x^2 < 0$ (A96)
(c) x is lightlike if $x^2 = 0$

Obviously in timelike x, the time part of x dominates over its space part and the opposite happens when x is spacelike. Both parts are equal in lightlike x.

LORENTZ INVARIANTS

We may write from eqn. (A92) that

$$x'^2 = x'_\mu x'^\mu = (a_{\mu\nu} x^\nu)(a^{\mu\alpha} x_\alpha) = x^\nu (a_{\mu\nu} a^{\mu\alpha}) x_\alpha = x^\nu \delta^\alpha_\nu x_\alpha = x^\nu x_\nu \tag{A97}$$

where $x^\nu x_\nu$ is the square of four-vector x wrt S and $x'_\mu x'^\mu$ is with respect to S′ and thus the square of four-vector x has same value in both S and S′ frames of reference. Since x becomes x' due to Lorentz transformation, we say that the product $x'_\mu x'^\mu$ or $x^\nu x_\nu$ is Lorentz invariant. This result can be generalized to the product of two different four-vectors each of which obeys Lorentz transformation equation like (A92). Thus

$$A'_\mu B'^\mu = (a_{\mu\nu} A^\nu)(a^{\mu\alpha} B_\alpha) = A^\nu \delta^\alpha_\nu B_\alpha = A^\nu B_\nu \tag{A98}$$

is Lorentz-invariant. Another important four-vector is four-gradient (∂ operator) which is defined in its covariant and contravariant forms as

$$\partial_\mu = \frac{\partial}{\partial x^\mu} = (\frac{\partial}{\partial t}, \text{grad}) \tag{A99}$$

and

$$\partial^{\mu} = \frac{\partial}{\partial x_{\mu}} = (\frac{\partial}{\partial t}, -\text{grad}) \tag{A100}$$

respectively. From equation (A56b)

$$x_{\nu} = a^{\mu}_{\nu} x'_{\mu} \tag{A101}$$

implying that

$$\frac{\partial x_{\nu}}{\partial x'_{\mu}} = a^{\mu}_{\nu} \tag{A102}$$

Now

$$\frac{\partial}{\partial x'_{\mu}} = \frac{\partial x_{\nu}}{\partial x'_{\mu}} \frac{\partial}{\partial x_{\nu}} \tag{A103}$$

which on using eqn. (A102) reads

$$\frac{\partial}{\partial x'_{\mu}} = a^{\mu}_{\nu} \frac{\partial}{\partial x_{\nu}} \tag{A104a}$$

or

$$\partial'^{\mu} = a^{\mu}_{\nu} \partial^{\nu} \tag{A104b}$$

Equations (A104) are the Lorentz transformation laws for the four-vector. This shows that ∂_{μ} or ∂^{μ} is a four-vector. Forming a product of four-vector ∂_{μ} with another four-vector ∂^{μ}, we get the Lorentz invariant differential operator, the d' Alembertian □,

$$□ = \partial_{\mu}\partial^{\mu} = \partial^{\mu}\partial_{\mu} = \frac{\partial^2}{\partial t^2} - \nabla^2 \tag{A105}$$

It may further be observed that if j^{μ} is a four vector, then the equation (product of ∂_{μ} and j^{μ})

$$\partial_{\mu} j^{\mu} = 0 \tag{A106}$$

is Lorentz-invariant.

In forming the Lorentz invariant products the upper indices must balance the lower indices. For an equation to be Lorentz covariant, it must be seen that all unrepeated indices (upper and lower separately) are balanced on both sides of the equation and all the repeated indices appear once as an upper and once as a lower index. The words Lorentz covariant or just covariant carry the similar meanings.

The four dimensional volume element defined by

$$dx = dx^0 dx^1 dx^2 dx^3 = dx^0 d^3\boldsymbol{x} \tag{A107}$$

is also a Lorentz invariant quantity. Incorporating time dilation and length contraction effects we can demonstrate the Lorentz invariance of eqn. (A107), i.e., by using the relations

$$\begin{aligned} dx'^0 &= \frac{dx^0}{\sqrt{1-\beta^2}}, \beta = \frac{v}{c} \\ dx'^1 &= dx^1 \ \sqrt{1-\beta^2} \\ dx'^2 &= dx^2 \\ dx'^3 &= dx^3 \end{aligned} \tag{A108}$$

we can see that

$$dx' = dx'^0 dx'^1 dx'^2 dx'^3 = dx \tag{A109}$$

Spacelike Surfaces

Let us suppose that we have a product in the form

$$(\phi, \chi)_t = i\int_t d^3\mathbf{x}[\phi^*(x)\partial_0\chi(x) - \partial_0\phi^*(x)\chi(x)] \tag{A110}$$

in which t refers to the fact that the above integral is taken over all the three dimensional space at the time $t = x^0$ (natural units). This means that the three dimensional volume element d^3x refers to a constant time $t = x^0$. To be relativistically (Lorentz) invariant, a quantity should have four components, one temporal and three spatial. The d^3x, which has only three components needs to be replaced by a quantity which has four components. For Lorentz invariance the ∂_0 in the bracket requires to be replaced by ∂_μ. However, we do not discuss the Lorentz invariance of the quantity in the bracket, we focus our attention on d^3x. We see in the following that d^3x though is a three dimensional volume element, is also an element of three-dimensional surface called the *hypersurface*. To understand hypersurface, we first take an example of two-dimensional surface.

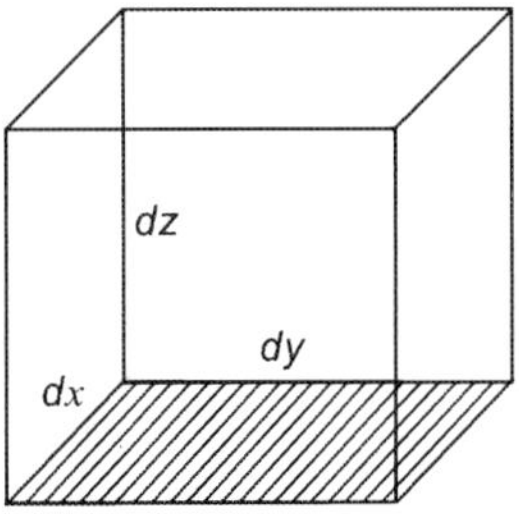

Fig. A3: The area (shaded) *dx dy* is normal to *z* axis

If we construct a three dimensional volume element (Fig.A3) $dx\,dy\,dz$, then the shaded area $dx\,dy$ is normal to z axis. Similarly, the surface element $dx\,dz$ and $dy\,dz$ are normal to y and x axes respectively. Extending this picture to four dimensions we think of a four-dimensional volume element

$$dx = d^4x = dx^0 dx^1 dx^2 dx^3 \tag{A111}$$

This four dimensional volume element will have four surface elements in which each will have three dimensions. They are called hypersurface elements. We shall denote these hypersurface elements by $d\sigma_0$, $d\sigma_2$ and $d\sigma_3$. Note that the surfaces have one dimension less than that of volume.

$$\begin{aligned} d\sigma_0 &= dx^1dx^2dx^3 = d^3\boldsymbol{x} \rightarrow \text{tdse normal to } x^0 \\ d\sigma_1 &= dx^0dx^2dx^3 = \text{tdse normal to } x^1 \\ d\sigma_2 &= dx^0dx^1dx^3 = \text{tdse normal to } x^2 \\ d\sigma_3 &= dx^0dx^1dx^2 = \text{tdse normal to } x^3 \end{aligned} \tag{A112}$$

where *tdse* is the short for the *three dimensional surface element*. Notice x^0 is the time axis. These four surface elements form a four-vector $d\sigma_\mu$ with $\mu = 0, 1, 2, 3$. We may write

$$d\sigma_\mu = \eta_\mu d\sigma \tag{A113}$$

where η_μ (x) are the unit normals to the surface σ at the space–time point x. They are constant (independent of x) timelike unit vectors obeying

$$\eta_\mu(x)\eta^\mu(x) = +1 \tag{A114}$$

for all x on σ. The surface σ, whose element $d\sigma$, according to eqn. (A113) defines $d\sigma_\mu$ (eqn.A112) is called the *spacelike surface*. For any two points x and y on the spacelike surface σ, the interval $(x - y)$ would be spacelike, i.e., it would obey

$$(x - y)^2 < 0 \tag{A115}$$

where both x and y are four-vectors. The points x and y would lie on the outside of each other's light cone and hence would not be connected by lightlike signals. It is also to be noted from eqn. (A112) that for the three dimensional surface element $d\sigma_0$; the time $x^0 = t$ is constant, for $d\sigma_1$, x^1 is constant and so on. Thus, $d\sigma_0 = d^3x$ is also called a *constant time surface*. Since the unit vector η_μ is x independent, the spacelike surface is flat and, in the special case of a constant time surface, $d\sigma_k = 0$ and the integration over such a surface reduces to the integration over a three volume at a given constant time.

Admitting the above prescription eqn. (A110) can be written in the covariant form

$$(\phi, \chi)_\sigma = i\int_\sigma d\sigma_\mu(x)[\phi^* \partial^\mu \chi - \partial^\mu \phi^* \chi(x)] \tag{A116}$$

and the eqn. (A110) becomes a particular case of eqn. (A116), i. e., when $\mu = 0$, $d\sigma_k = 0$ $(k = 1, 2, 3)$ the eqn. (A116) reduces to eqn. (A110). Thus, for bringing in the covariance $d\sigma_\mu$ would be introduced. The non-covariant version of the phrase *on the spacelike surface* **is** *at a constant time.*

The concept of spacelike surface finds its use in Gauss' divergence theorem in four dimensions. The Gauss' divergence theorem in four dimensions is

$$\int_R \partial_\mu G^\mu d^4x = \int_S G^\mu d\sigma_\mu \tag{A117}$$

Where R is the arbitrary four-volume and S the surface (three dimensional) covering the four-volume R. The four dimensional volume R may be enclosed by two arbitrary spacelike surfaces which meet at finite x as in Fig. (A4a)

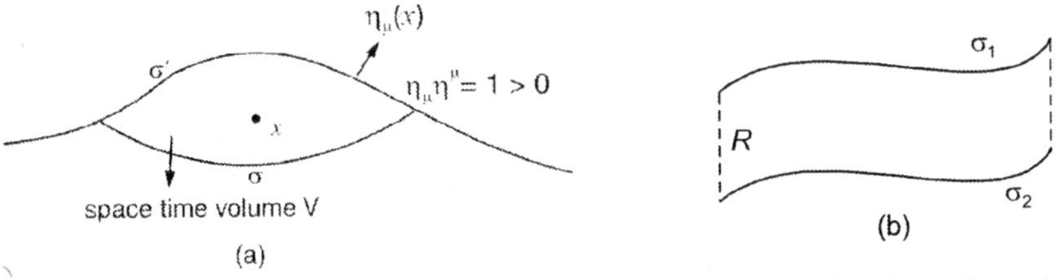

Fig. A4: The spacelike surfaces σ, σ', σ_1 and σ_2. A spacelike surface is a three dimensional urface whose normal η_μ is everywhere timelike, i.e., $\eta_\mu\eta^\mu = 1 \Rightarrow \eta_\mu\eta^\mu > 0$. In Fig. (a) the spacelike surfaces σ and σ' meet to enclose a four dimensional volume V. In **Fig. (b)**, the spacelike surfaces σ_1 and σ_2 do not meet (meet at infinity), they have been joined by nonspacelike surfaces to enclose a volume R

If the spacelike surfaces do not meet to enclose the volume R, (we adjoin) in to them at $\boldsymbol{x} = \infty$ (Fig.A4b).

If we define

$$F(\sigma) = \int_\sigma f^\mu(x)d\sigma_\mu(x) \tag{A118}$$

in which f^μ obeys

$$\partial_{\mu.} f^\mu = 0 \tag{A119}$$

then

$$\frac{\delta F(\sigma)}{\delta \sigma(x)} = 0 \tag{A120}$$

implying that $F(\sigma)$ is independent of surface σ. Eqn. (A120) denotes the functional derivative of F with respect to surface σ, the variation with respect to σ has been denoted by δ and variation wrt x (four components) by ∂ as in ∂_μ. $F(\sigma)$ is the surface functional.

To prove relation (A120), the surface σ in the neighbourhood of point x is slightly modified to enclose a small volume dV between the original surface σ and the modified surface σ' (Fig.A5). The arrows in the Fig. (A5) denote normal vectors η_μ to the surfaces.

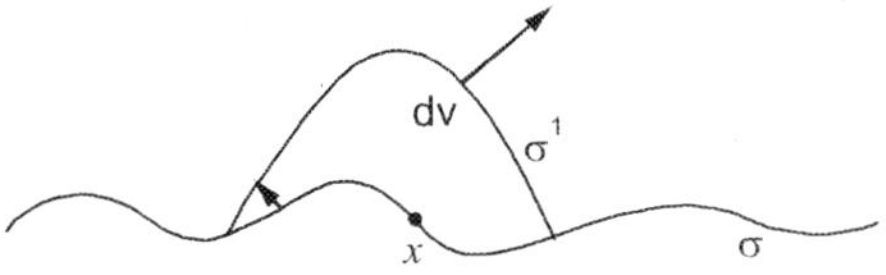

Fig. A5: A small volume dV is shown enclosed in the neighbourhood of spacelike point x by σ and σ'. Arrows denote normals to spacelike surfaces

Thus,

$$\frac{\delta F(\sigma)}{\delta\sigma(x)} = \lim_{dV \to 0} \frac{F(\sigma') - F(\sigma)}{dV} \tag{A121}$$

Using eqn. (A118)

$$\frac{\delta F(\sigma)}{\delta\sigma(x)} = \lim_{dV\to 0} \frac{\int_{\sigma}^{\sigma'} f^{\mu} d\sigma_{\mu}}{dV} \tag{A122}$$

Applying Gauss' divergence theorem (A117), in the form surface integral to volume integral, eqn. (A122) becomes

$$\frac{\delta F(\sigma)}{\delta\sigma(x)} = \lim_{dV\to 0} \frac{\int_{dV} \partial_{\mu} f^{\mu} dx}{dV} \tag{A123}$$

where $dx = d^4x$, the element of four-volume. When $\partial_{\mu} f^{\mu} = 0$,

$$\frac{\delta F(\sigma)}{\delta\sigma(x)} = 0 \tag{A124}$$

Thus for divergenceless f, F (eqn. A118) does not depend on surface.

From the above discussions on the spacelike surface, we find that σ or $d\sigma_{\mu}$ is introduced to see that expressions are covariant. For instance $\frac{\delta F(\sigma)}{\delta\sigma(x)}$ is covariant whereas $\frac{dF(t)}{dt}$ is not.

APPENDIX B

GREEN'S FUNCTIONS OF FIELD THEORY

The Lagrangian density for the free real scalar field is

$$\pounds = -\frac{1}{2}[m^2\phi^2 - \partial_\mu\phi\,\partial^\mu\phi] \tag{B1}$$

The field equations can be obtained by employing the Euler Lagrange equations

$$\frac{\partial \pounds}{\partial \psi} - \partial_\mu \frac{\partial \pounds}{\partial(\partial_\mu \psi)} = 0 \tag{B2}$$

where for the Lagrangian density (B1)

$$\frac{\partial \pounds}{\partial \phi} = -m^2\phi \tag{B3}$$

and

$$\begin{aligned}\frac{\partial \pounds}{\partial(\partial_\mu\phi)} &= \frac{\partial}{\partial(\partial_\mu\phi)}[\frac{1}{2}\partial_\mu\phi\partial^\mu\phi] + zero \\ &= \frac{1}{2}(\partial^\mu\phi)\frac{\partial(\partial_\mu\phi)}{\partial(\partial_\mu\phi)} + \frac{1}{2}(\partial_\mu\phi)\frac{\partial(\partial^\mu\phi)}{\partial(\partial_\mu\phi)} \\ &= \frac{1}{2}(\partial^\mu\phi) + \frac{1}{2}g^{\mu\nu}(\partial_\mu\phi)\frac{\partial(\partial_\nu\phi)}{\partial(\partial_\mu\phi)} \\ &= \frac{1}{2}(\partial^\mu\phi) + \frac{1}{2}g^{\mu\nu}(\partial_\mu\phi)\delta^\mu_\nu \\ &= \partial^\mu\phi\end{aligned} \tag{B4}$$

Substitution of equations (B3) and (B4) into the equation (B2) provides the homogeneous Klein–Gordon equation or the field equation for the free real scalar field as

$$\partial_\mu\,\partial^\mu\phi + m^2\phi = 0 \tag{B5}$$

or

$$(\Box + m^2)\phi = 0 \tag{B6}$$

where $\Box = \partial_\mu\partial^\mu$.

The solutions of the homogeneous Klein–Gordon equation can be obtained in terms of Green's function. It is, however, difficult to obtain the solution if we begin directly with the homogeneous K.G. equation. To circumvent this problem, we start with the inhomogeneous Klein–Gordon equation

$$(\quad + m^2)\phi(x) = \rho(x) \tag{B7}$$

[Here $\rho(x)$ is called a source function with which interacts the scalar field $\phi(x)$. This interaction is represented by the term $\rho(x)\,\phi(x)$ in the Lagrangian density

$$\pounds = \frac{-1}{2}(m^2\phi^2 - \partial_\mu\phi\,\partial^\mu\phi) + \rho(x)\phi(x) \tag{B8}$$

Using this Lagrangian density and the Euler Lagrange eqn. (B2), we can obtain eqn. (B7). We will however, not go into details of interactions, as we are dealing with free fields, i.e., fields without interaction.]

The solution of inhomogeneous Klein–Gordon equation is obtained in terms of the general Green's function $G\,(x - y)$ defined by

$$(\partial_\mu\,\delta^\mu + m^2)\;G(x - y) = \delta(x - y) \tag{B9}$$

where x, y are four-vectors, ∂_μ implies $(\partial/\partial x^\mu)$ and $\delta(x - y)$is the four dimensional Dirac delta function. $\delta(x - y)$ can be expressed as

$$\delta(x - y) = \frac{1}{(2\pi)^4}\int e^{ik(x-y)}\,dk \tag{B10}$$

where

$$dk = dk^0 dk^1 dk^2 dk^3 \tag{B11a}$$

and

$$kx = k_\mu x^\mu = k^0 x^0 - \mathbf{k.x} \tag{B11b}$$

The limits of k, the four component momentum are from $-\infty$ to $+\infty$ and four powers of (2π) indicate that here δ function is four-dimensional. We mention here the following properties of Dirac delta function $\delta\,(x - y)$, which we shall require. These are

$$\delta(x - y) = 0\; if\; x \neq y \tag{B12a}$$

$$\delta(x - y) = \infty\; if\; x = y \tag{B12b}$$

$$\delta(x - y) = \delta(y - x) \tag{B12c}$$

$$\int_{-\infty}^{\infty} \delta(x - y)\,dx = 1\; for\, x = y \tag{B12d}$$

$$\int_{-\infty}^{\infty} f(x)\,\delta(x - b)\,dx = f(b) \tag{B12e}$$

where dx is the four dimensional volume element and b is a definite value of x. The Green's function in eqn. (B9) is represented by the Fourier integral

$$G(x-y)=\frac{1}{(2\pi)^4}\int\frac{e^{ik(x-y)}}{m^2-k^2}dk \tag{B13}$$

The significance of the Green's function in the form (B13) is that it is Lorentz invariant; therefore any solution in terms of $G(x-y)$ is also covariant.

Obtaining the solution of eqn. (B7) now means that we obtain the value of $\phi(x)$ in terms of $G(x-y)$ which satisfies eqn. (B7). For this purpose we construct a solution $\phi(x)$ in the form

$$\phi(x)=\int G(x-y)\,\rho(y)\,dy \tag{B14}$$

and make two checks. First, whether eqn.(B14) satisfies eqn.(B7) and second whether $G(x-y)$ of eqn. (B13) satisfies eqn.(B9).

To check the first part,substituting eqn. (B14) into LHS of eqn.(B7),we get

$$\text{LHS} = \int(\partial_\mu\partial^\mu+m^2)\,G(x-y)\rho(y)\,dy \tag{B15}$$

where we have taken $(\partial_\mu\partial^\mu + m^2)$ inside the integral as $(\partial_\mu = \partial/\partial x^\mu)$and integral is with respect to y. Incorporating eqn. (B9) into (B15), it becomes

$$\text{LHS} = \int\delta(x-y)\rho(y)\,dy \;=\rho(x) \tag{B16}$$

where the properties (B12) (c and d) of $\delta(x-y)$ have been used. Thus $\phi(x)$ in the form (B14) satisfies the inhomogeneous K.G equation (B9).

Secondly substituting eqn. (B13) into eqn. (B9) and taking the operator $(\partial_\mu\partial^\mu + m^2)$ inside the integral and using

$$\partial_\mu\partial^\mu e^{ik(x-y)} = -k^2 e^{ik(x-y)} \tag{B17}$$

we find that (m^2-k^2) appears in the numerator and thus cancels with the denominator. The LHS of eqn. (B9) then becomes

$$=\frac{1}{(2\pi)^4}\int e^{ik(x-y)}dk \;=\; \delta(x-y) \tag{B18}$$

where eqn. (B10) has been invoked. $\delta(x-y)$ is the RHS of eqn. (B9). Thus $G(x-y)$ in the form (B13) provides the covariant solution for $\phi(x)$ in the form (B14). However, the Green's function $G(x-y)$ is undefined at $m^2=k^2$. We write this condition as

$$m^2-k^2=0 \tag{B19}$$

where

$$k^2=k_\mu k^\mu=k_0^2-\boldsymbol{k}^2 \tag{B20}$$

Eqn. (B19) then becomes

$$k_0^2-\omega_k^2=0 \tag{B21}$$

where

$$\omega_k=+\sqrt{m^2+\boldsymbol{k}^2} \tag{B22}$$

We therefore observe that the denominator (eqn. B21) of Green's function (B13) tells that there are two poles, one at $k_0 = -\omega_k$ and other at $k_0 + \omega_k$ at which the Green's function is undefined. Application of Cauchy's residue theorem to solve such an integral, demands that there should be no poles on the real k_0 axis. In order to avoid the poles from the real k_0 axis, the contour or the path of integration has to be deformed. There are four possible ways to do that. The path of integration may undertake a semicircular detour such that

(1) Both the poles, one at $k_0 = -\omega_k$ and other at $k_0 = +\omega_k$, come below the path of integration (Fig. B1a).

(2) Both the poles go above the path of integration (Fig. B1b).

(3) The pole at $k_0 = -\omega_k$ goes above and the pole at $k_0 = +\omega_k$ goes below the path of integration (Fig. B1c).

(4) The pole at $k_0 = -\omega_k$ goes below and the pole at $k_0 = +\omega_k$ goes above the path of integration (Fig. B1d).

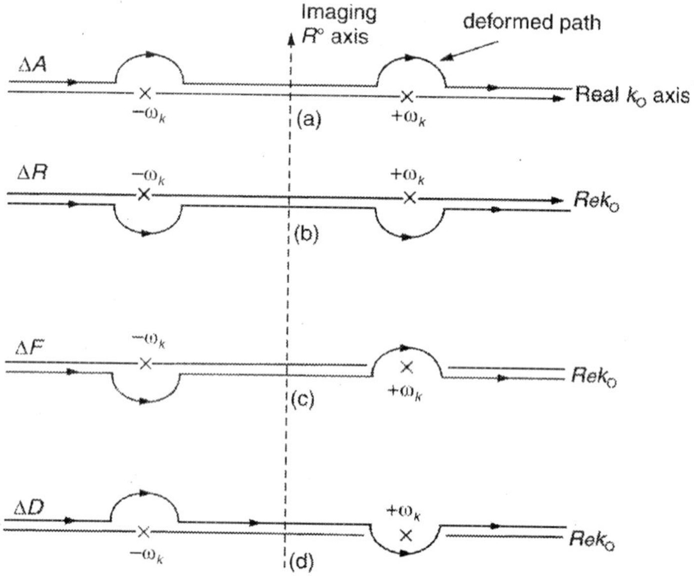

Fig. B1: The paths of integration to avoid poles on real k_0 axis at $k_0 = +\omega_k$ and $k_0 = -\omega_k$. Without these deformed paths the Green's function eqn. (B13), cannot be integrated. When the path is according to Figs. (a) the Green's function is called ΔA (Advanced Green's function). For Fig. (b) it is retarded Green's function ΔR. Paths according to Figs. (c) and (d) respectively give Feynman function (ΔF) and Dyson function (ΔD)

Depending upon the path of integration as above, the Green's function (B13) now becomes integrable everywhere through the *iε prescription*, where ε is a small positive-number. This introduces following four functions.

ADVANCED GREEN'S FUNCTION (AGF)

The, *iε prescription,* retains the path of integration along the real k_0 axis by shifting the poles along imaginary k_0 axis through the small amount ε. Thus for the path (Fig. B1a) we shall have the closed path of integration as in the following Figure (Fig. B2).

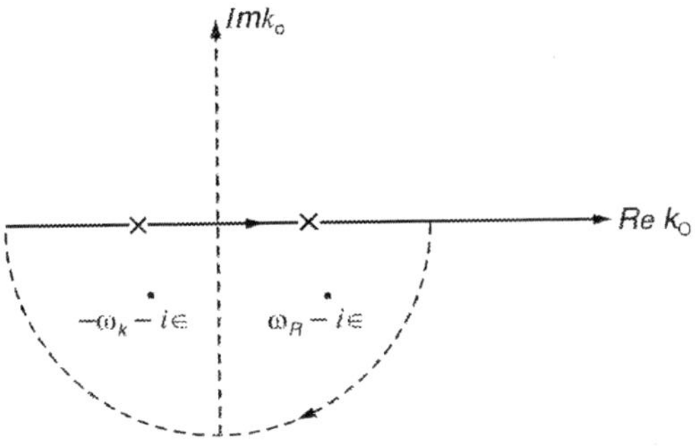

Fig. B2: The path of integration closed for $\Delta_A (x - y)$, the advanced Green's function. $i\varepsilon$ prescription has been applied to shift the poles from real k_0 axis

The poles now go below the clockwise path of integration and are shifted by $-i\varepsilon$ from $k_0 = \pm\,\omega_k$. It can be easily found that the equation quadratic in k_0 with roots $-\omega_k - i\varepsilon$ and $\omega_k - i\varepsilon$, upto the order ε, is

$$k_0^2 + i\varepsilon\, k_0 - \omega_k^2 = 0 \tag{B23}$$

This equation is equivalent to

$$m^2 - k^2 - i\varepsilon\, k_0 = 0 \tag{B24}$$

where equations (B20) and (B23) have been used. Thus when the path of integration is (Fig.B2), the denominator of $G(x-y)$ (eqn. B13) has to be replaced by eqn. (B24) such that we have the advanced Green's function as

$$\Delta_A(x-y) = \frac{1}{(2\pi)^4}\int \frac{e^{ik(x-y)}}{m^2 - k^2 - i\varepsilon\, k_0}dk \tag{B25}$$

where the integrable $G\,(x-y)$ according to the contour of (Fig B2) has been called $\Delta_A\,(x-y)$.

Properties of AGF

The advanced Green's function (B25) may be written as

$$\Delta_A(x-y) = \frac{1}{(2\pi)^4}\int d^3\boldsymbol{k} e^{-i\boldsymbol{k}.(\boldsymbol{x}-\boldsymbol{y})}\int dk_0 \frac{e^{ik_0(x_0-y_0)}}{\omega_k^2 - k_0^2 - i\varepsilon k_0} \tag{B26}$$

The k_0 integral can be solved by contour integration. Fig. (B2) shows that both the poles $-\omega_k - i\varepsilon$ and $\omega_k - i\varepsilon$ lie in the lower half of complex k_0 plane.

When $x_0 > y_0$ (future light cone) the exponent of k_0 integral is positive imaginary, the path of integration has to be closed in the upper half of complex k_0 plane. However, poles $-\omega_k - i\varepsilon$ and $\omega_k - i\varepsilon$ lie in the lower half plane and thus do not contribute for the residues. The infinite semicircle also making no contribution, the k_0 integral of eqn. (B26) vanishes for $x_0 > y_0$. The region of future light cone is expressed by $(x-y)^2 > 0$ with $x_0 > y_0$, we may express that

$$\Delta_A(x-y) = 0 \text{ for } (x-y)^2 > 0 \text{ and } x_0 > y_0 \tag{B27}$$

Now the events that lie in the future of an event as seen from frame S, lie in its future in every other frame S′, and also the events that lie in the past light cone of an event remain in past cone of that event on observing from different inertial frames. To remain within the light cones the four-vector $(x-y)$ has to be timelike, however, there is no Lorentz transformation that can change a timelike four-vector of past light cone into the timelike four-vector of future light cone. Thus, when $(x-y)$ is timelike and $x_0 > y_0$, $\Delta_A(x-y)$ will be vanishing for every frame. This means property (B27) is invariant for timelike $(x-y)$.

Now a spacelike separation in one coordinate system also remains spacelike in every other coordinate system. $\Delta_A(x-y)$ is itself Lorentz invariant function (it is slightly deviated from $G(x-y)$ which is clearly Lorentz invariant) and hence $\Delta_A(x-y)$ should have the property similar to eqn. (B27) for spacelike $(x-y)$ as well, i.e., we should have

$$\Delta_A(x-y) = 0 \text{ for } (x-y)^2 < 0 \tag{B28}$$

Now remains the past light cone which is expressed by $(x-y)^2 > 0$ and $x_0 < y_0$. Looking at the exponent of k_0 integral (eqn. B26), we find that it becomes negative imaginary due to $x_0 < y_0$, the path of integration would be the lower half of complex k_0 plane. The poles $-\omega_k - i\varepsilon$ and $\omega_k - i\varepsilon$, too lie in the lower half plane, hence residues can be calculated at these poles. In this way the integral (B26) would give a non-vanishing value in the past light cone. We express it by

$$\Delta_A(x-y) \neq 0 \text{ for } (x-y)^2 > 0,\ x_0 < y_0 \tag{B29}$$

Thus equations (B27) to (B29) describe the properties of advanced Green's function which is one solution of inhomogeneous K.G. equation. Since the support (non zero value) for AGF comes from the past light cone, and because *past* gives the sense of earlier than the present (*in advance*), the Green's function $\Delta_A(x-y)$ is called advanced (A) Green's function.

RETARDED GREEN'S FUNCTION (RGF)

When the poles at $k_0 = -\omega_k$ and $k_0 = +\omega_k$ are indented according to the path of Fig. (B1b), the poles go above the path of integration. In other words, to retain real k_0 axis as the path of integration, the poles are shifted by $+i\varepsilon$ parallel to the imaginary k_0 axis. The $i\varepsilon$ *prescription* therefore suggests the following contour (Fig. B3)

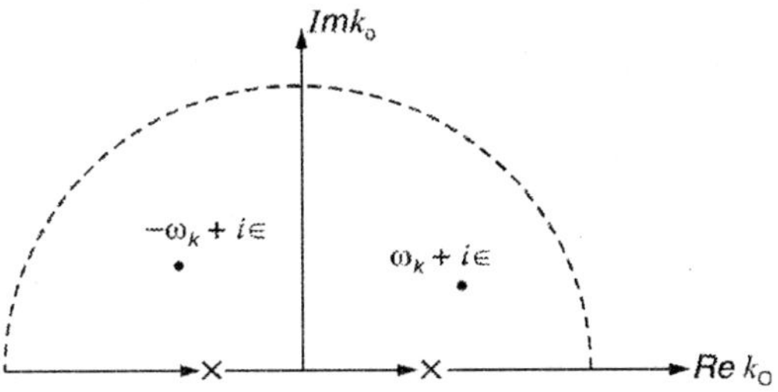

Fig. B3: The shifting of poles for the retarded Green's function Δ_R

The equation quadratic in k_0 with roots $\omega_k + i\varepsilon$ and $-\omega_k + i\varepsilon$, neglecting ε^2, is

$$k_0^2 - i\varepsilon k_0 - \omega_k^2 = 0 \tag{B30}$$

or

$$m^2 - k^2 + i\varepsilon k_0 = 0 \tag{B31}$$

Thus, for the path of integration of Figure (B3), the denominator of $G(x-y)$ (eqn.B13) is replaced by the LHS of eqn. (B31) to have the retarded Green's function as

$$\Delta_R(x-y) = \frac{1}{(2\pi)^4}\int \frac{e^{ik(x-y)}}{m^2 - k^2 + i\varepsilon k_0} dk \tag{B32}$$

PROPERTIES OF RGF

Similar to eqn. (B26) for AGF, we may split the integral (B32) as

$$\Delta_R(x-y) = \frac{1}{(2\pi)^4}\int d^3 k e^{-ik.(x-y)} \int dk_0 \frac{e^{ik_0(x_0-y_0)}}{\omega_k^2 - k_0^2 + i\varepsilon k_0} \tag{B33}$$

This integral also vanishes for spacelike arguments, i.e., when $(x-y)^2 < 0$ and can have nonzero value only for timelike $(x-y)$. However, even $(x-y)$ is timelike, $\Delta_R(x-y)$ vanishes in the past light cone. This is because the past light cone is expressed by $(x-y)^2 > 0$ and $x_0 < y_0$. For $x_0 < y_0$, the path of integration for the k_0 integral is in the lower half plane, where there are no poles. The k_0 integral therefore vanishes. Thus,

$$\Delta_R(x-y) = 0 \text{ for } (x-y)^2 > 0 \text{ and } x_0 < y_0 \tag{B34}$$

$$\Delta_R(x-y) = 0 \text{ for } (x-y)^2 < 0 \tag{B35}$$

The support cone for the retarded Green's function is therefore future light cone, i.e.,

$$\Delta_R(x-y) \neq 0 \text{ for } (x-y)^2 > 0,\ x_0 > y_0 \tag{B36}$$

Thus equations (B34) to (B36) describe the properties of retarded Green's function and because of having nonzero value in the future light cone, where *future* gives the sense of *after the proper time,* the Green's function Δ_R is called the *retarded* (R) Green's function.

FEYNMAN FUNCTION

When the path of integration is as shown in Figure (B1c), the pole at $k_0 = -\,\omega_k$ goes above, and the pole at $k_0 = +\,\omega_k$ goes below the path. Applying $i\varepsilon$ technique, the pole $-\,\omega_k$ is shifted by $+\,i\varepsilon'$ and the pole $+\,\omega_k$ is shifted by $-i\varepsilon'$. The shifts are with respect to real k_0 axis. Here we have used ε' in place of ε because we wish to retain ε in the defining equation of the function and as would become evident in the following, we shall write $2\varepsilon'\,\omega_k = \varepsilon$.

The equation quadratic in k_0 with roots $-\omega_k + i\varepsilon'$ and $+\omega_k - i\varepsilon'$, upto order ε', would be

$$k_0^2 - \omega_k^2 + 2i\varepsilon'\,\omega_k = 0 \tag{B37}$$

which is equivalent to

$$m^2 - k^2 - i\varepsilon = 0 \tag{B38}$$

where $\varepsilon = 2\varepsilon'\,\omega_k$, is still a small quantity. Then replacing the denominator of eqn. (B13) by the LHS of eqn. (B38), we get the Feynman-function $\Delta_F(x-y)$ as

$$\Delta_F(x-y) = \frac{1}{(2\pi)^4}\int \frac{e^{ik(x-y)}}{m^2 - k^2 - i\varepsilon}\,dk \tag{B39}$$

The integral (B39), too, can also be written as

$$\Delta_F(x-y) = \frac{1}{(2\pi)^4}\int d^3\boldsymbol{k}\; e^{-i\,\boldsymbol{k}.(\boldsymbol{x}-\boldsymbol{y})}\int dk_0\,\frac{e^{ik_0(x_0-y_0)}}{k_0^2 - \omega_k^2 + i\varepsilon} \tag{B40}$$

According to the $i\varepsilon$ method the path of integration for k_0 integral would be (Fig. B4),

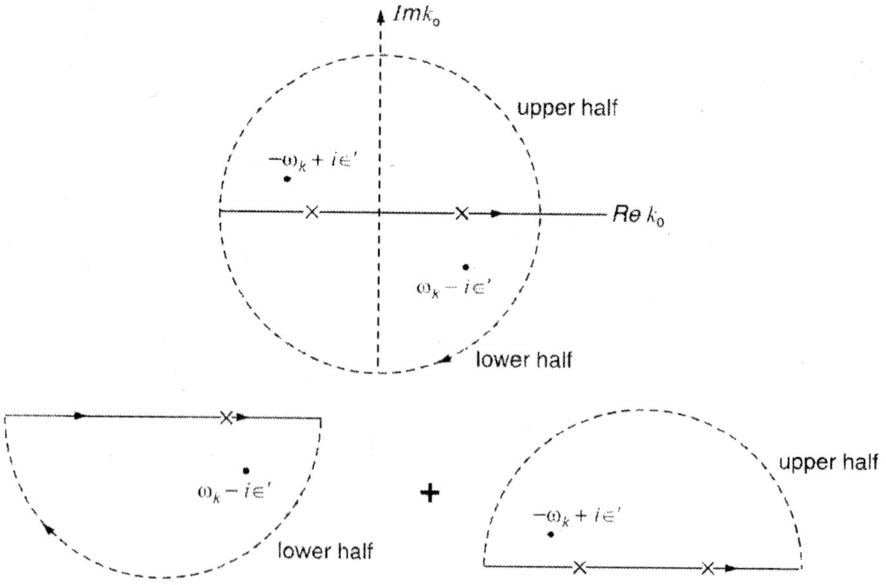

where it can be noticed that both the halves have poles, $+\omega_k - i\varepsilon'$ in the lower half and pole $-\omega_k + i\varepsilon'$, in the upper half. Though the contributions of semicircles would tend to zero as their radii tend to infinity, the residues at both the poles would give a non vanishing value of $\Delta_F(x-y)$ in both the light cones. In the past light cone $x_0 < y_0$, the residue at $+\omega_k + i\varepsilon'$, and in the future light cone where $x_0 > y_0$, the residue at $-\omega_k + i\varepsilon'$ pole will support the integral (B40). Thus

$$\Delta_F(x-y) \neq 0 \; (x-y)^2 > 0 \tag{B41a}$$

Unlike Δ_A and Δ_R, the Feynman function does not vanish outside the light cone, i.e.

$$\Delta_F(x-y) \neq 0 \quad (x-y)^2 < 0 \tag{B41b}$$

Outside the light cone the function has an exponential tail. Since, the limits of integral (B39) are from $-\infty$ to $+\infty$, implying that for every four component $+k$, there is a four-component $-k$, we can see that the Feynman function is symmetric, i.e.,

$$\Delta_F(y-x) = \Delta_F(x-y) \tag{B42}$$

DYSON FUNCTION

When the path of integration is (Fig. B1d), the pole $k_0 = -\omega_k$ goes below and $k_0 = +\omega_k$ goes above the path of integration. According to *iε technique* poles shifted from the real k_0 axis will be $-\omega_k - i\varepsilon'$ and $\omega_k + i\varepsilon'$ and for these poles the equation quadratic in k_0 is

$$k_0^2 - \omega_k^2 - 2\,i\,\varepsilon'\,\omega_k = 0 \tag{B43}$$

which is equivalent to

$$m^2 - k^2 + i\varepsilon = 0 \tag{B44}$$

where as before $\varepsilon = 2\varepsilon'\,\omega_k$. Thus from $G(x-y)$ (eqn. B13) the Dyson function $\Delta_D(x-y)$ is

$$\Delta_D(x-y) = \frac{1}{(2\pi)^4}\int \frac{e^{ik(x-y)}}{m^2 - k^2 + i\varepsilon}\,dk \tag{B45}$$

It may be noted that ε is a small real positive constant. The values of the integrals can be evaluated by contour integration. After the integration, the limit $\varepsilon \to +0$, can be performed.

SOLUTIONS OF HOMOGENEOUS KLEIN GORDON EQUATION

In the previous section, we defined four Green's functions Δ_A, Δ_R, Δ_F and Δ_D, each of which is Lorentz invariant. These are the solutions of inhomogeneous K.G. equation. Using them, we can construct solutions of the homogeneous Klein–Gordon equation. For example, $\Delta_A - \Delta_R$ provides one invariant solution of homogeneous K.G. equation and $\Delta_F - \Delta_D$ the other. The reason is that being solutions of inhomogeneous K.G. equation Δ_A and Δ_R obey

$$(\quad + m^2)\Delta_A(x-y) = \rho(x-y) \tag{B46}$$

and

$$(\quad + m^2)\Delta_R(x-y) = \rho(x-y) \tag{B47}$$

Subtracting

$$(\quad + m^2)(\Delta_A - \Delta_R) = 0 \tag{B48}$$

thus providing

$$\phi(x-y) = \Delta_A(x-y) - \Delta_R(x-y) \tag{B49}$$

as the solution of homogeneous K.G. equation (B6). The function $\Delta_A - \Delta_R$ is also called the Schwinger's delta function. We discuss its properties.

SCHWINGER'S Δ FUNCTION

By definition the Schwinger's delta function denoted by $\Delta(x-y)$ is

$$\Delta(x-y) = \Delta_A(x-y) - \Delta_R(x-y) \tag{B50}$$

Since, Δ_A vanishes for $(x-y)^2 < 0$ and also for $(x-y)^2 > 0$ and $x_0 > y_0$, we may write

$$\Delta(x-y) = -\Delta_R(x-y) \;\text{ for }\; x_0 > y_0 \tag{B51}$$

Similarly, using the properties of Δ_R, we may write

$$\Delta(x-y) = \Delta_A(x-y) \;\text{ for }\; x_0 < y_0 \tag{B52}$$

Recalling the paths of integration for Δ_A and Δ_R (Figs. B2 and B3), the path for $-\Delta_R$ would be opposite to that of Fig. (B3), so that the path of integration for $(\Delta_R - \Delta_R)$ would be a closed one as shown in Figure (B5)

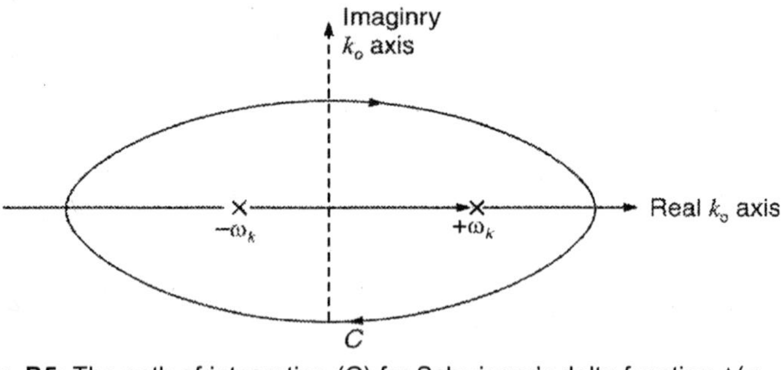

Fig. B5: The path of integration (C) for Schwinger's delta function $\Delta(x-y)$

The expression for Δ function may be written by observing the forms (B25) and (B32) for Δ_A and Δ_R respectively and the closed contour C. We are led to

$$\Delta(x-y) = \frac{1}{(2\pi)^4} \oint_c \frac{e^{ik(x-y)}}{m^2 - k^2} dk \tag{B53}$$

where

$$m^2 - k^2 = \omega_k^2 - k_0^2 \tag{B54}$$

with

$$\omega_k^2 = \boldsymbol{k}^2 + m^2 \tag{B55}$$

where ω_k is the positive root.

The Δ function can be decomposed into its positive and negative frequency components. For this purpose we write

$$\frac{1}{m^2 - k^2} = -\frac{1}{{k_0}^2 - {\omega_k}^2} = -\frac{1}{(k_0 - \omega_k)(k_0 + \omega_k)} \tag{B56}$$

Substituting, as in Δ_A and Δ_R

$$k\,(x - y) = k_0(x_0 - y_0) - \boldsymbol{k} \times (\boldsymbol{x} - \boldsymbol{y}) \tag{B57a}$$

and

$$dk = dk_0\, d^3\boldsymbol{k} \tag{B57b}$$

into eqn. (B53) and using eqn. (B56), we may write

$$\Delta(x - y) = \frac{1}{(2\pi)^4}\int d^3\boldsymbol{k} e^{-i\boldsymbol{k}.(\boldsymbol{x}-\boldsymbol{y})} \oint_c (-)\frac{e^{ik_0(x_0 - y_0)}}{(k_0 - \omega_k)(k_0 + \omega_k)} dk^0 \tag{B58}$$

in which the integral over k_0 can be solved by using Cauchy's residue theorem.

$$\text{Residue at } k_0 = \omega_k = -\frac{e^{i\omega_k(x_0 - y_0)}}{2\omega_k}$$

and

$$\text{Residue at } k_0 = -\omega_k = +\frac{e^{-i\omega_k(x_0 - y_0)}}{2\omega_k} \tag{B59}$$

and hence

$$\int (-)\frac{e^{-i\omega_k(x_0 - y_0)}}{k_0^2 - \omega_k^2}\, dk^0 = (2\pi i)\left[-\frac{e^{i\omega_k(x_0 - y_0)}}{2\omega_k} + \frac{e^{-i\omega_k(x_0 - y_0)}}{2\omega_k}\right] \tag{B60}$$

Using eqn. (B60) into the eqn. (B58), we may write

$$\Delta(x - y) = \Delta^+(x - y) + \Delta^-(x - y) \tag{B61}$$

where

$$\Delta^+(x - y) = \frac{-i}{(2\pi)^3}\int \frac{e^{i\{+\omega_k(x_0 - y_0) - \boldsymbol{k}.(\boldsymbol{x}-\boldsymbol{y})\}}}{2\omega_k} d^3\boldsymbol{k} \tag{B62}$$

and

$$\Delta^-(x - y) = \frac{+i}{(2\pi)^3}\int \frac{e^{i\{-\omega_k(x_0 - y_0) - \boldsymbol{k}.(\boldsymbol{x}-\boldsymbol{y})\}}}{2\omega_k} d^3\boldsymbol{k} \tag{B63}$$

Since, $\omega_k = \sqrt{\boldsymbol{k}^2 + m^2}$ itself is defined to be positive, eqn. (B62) in its exponential contains $+\omega_k$ and eqn. (B63) contains $-\omega_k$. These are the positive and negative frequencies in the Fourier analysis of Δ^+ and Δ^- respectively. Thus $\Delta^+(x-y)$ contains in its Fourier analysis only the positive frequencies $+\omega_k$ and the

Fourier analysis of $\Delta^-(x - y)$ allows only negative frequencies $-\omega_k$. The frequency $\omega_k = +\sqrt{\boldsymbol{k}^2 + m^2}$ is identified with the relativistic energy $(\hbar = c = 1)$ of a particle of mass m and three momentum $\boldsymbol{k}$. This relation resembles the relativistic relation $E^2 = \boldsymbol{p}^2 + m^2$ in natural units.

FOUR DIMENSIONAL REPRESENTATION OF Δ^+ AND Δ^-

Equation (B61) shows that $\Delta(x - y)$ can be decomposed into $\Delta^+(x - y)$ and $\Delta^-(x - y)$. Equations (B62) and (B63) define these functions in terms of three-dimensional integrals, while the $\Delta(x - y)$ (eqn. B53) is a four dimensional Fourier integral. We may, however, represent $\Delta^+(x - y)$ and $\Delta^-(x - y)$ in terms of four dimensional Fourier integrals, by introducing the Dirac delta function $\delta(k_0 - \omega_k)$ whose property (B22d) sets $k_0 = \omega_k$. This allows bringing back k_0, the temporal component of four momentum k and with all the four components $(k_0, \boldsymbol{k})$ within the integral, we may replace $d^3\boldsymbol{k}$ in equation (B62) by dk. Thus

$$\Delta^+(x-y) = \frac{-i}{(2\pi)^3}\int \frac{1}{2\omega_k} e^{i\{k_0(x_0-y_0) - \boldsymbol{k}.(\boldsymbol{x}-\boldsymbol{y})\}}\delta(k_0 - \omega_k)dk \qquad \text{(B64)}$$

However, $\delta(k_0 - \omega_k)$ is not Lorentz-invariant. The Dirac delta function $\delta(k^2 - m^2)$ is Lorentz invariant as it implies $k^2 = m^2$ and $k^2 = k_\mu k^\mu$ is obviously Lorentz invariant. Thus the Lorentz invariance in eqn. (B64) can be restored by introducing $\delta(k^2 - m^2)$. Other related Lorentz invariant quantities are $k\,(x - y)$, dk and $\frac{dk}{k_0}$. Using the identity,

$$\delta(\alpha^2 - \beta^2) = \frac{1}{2|\beta|}\left[\delta(\alpha - \beta) + \delta(\alpha + \beta)\right] \qquad \text{(B65)}$$

we write

$$\delta(k^2 - m^2) = \delta(k_0^2 - \omega_k^2) = \frac{1}{2\omega_k}\left[\delta(k_0 - \omega_k) + \delta(k_0 + \omega_k)\right] \qquad \text{(B66)}$$

Introducing the step functions $\theta\,(k_0)$ and $\theta\,(-k_0)$ which obey

$$\theta\,(k_0) = \begin{cases} +1 \text{ for } k_0 > 0 \\ 0 \text{ for } k_0 < 0 \end{cases} \qquad \text{(B67a)}$$

and

$$\theta\,(-k_0) = \begin{cases} 0 \text{ for } \quad k_0 > 0 \\ +1 \text{ for } \quad k_0 < 0 \end{cases} \qquad \text{(B67b)}$$

we may write $\delta(k_0 - \omega_k)$ and $\delta(k_0 + \omega_k)$ in terms of Lorentz invariant $\delta(k^2 - m^2)$ as

$$\frac{1}{2\omega_k}\delta(k_0 - \omega_k) = \theta(k_0)\delta(k^2 - m^2) \qquad \text{(B68a)}$$

$$\frac{1}{2\omega_k}\delta(k_0+\omega_k)=\theta(-k_0)\delta(k^2-m^2) \tag{B68b}$$

It may be mentioned here that $\theta(\pm k_0)$ is also Lorentz invariant. Using eqn. (B68a) in eqn. (B64), we can obtain the four dimensional form for $\Delta^+(x-y)$ as

$$\Delta^+(x-y)=\frac{-i}{(2\pi)^3}\int\theta(k_0)\delta(k^2-m^2)\,e^{ik(x-y)}dk \tag{B69}$$

where

$$k_0(x_0-y_0)-\boldsymbol{k}.\,(\boldsymbol{x}-\boldsymbol{y})=k(x-y) \tag{B70}$$

has been used. The function $\Delta^+(x-y)$ is Lorentz invariant as each of the term in the RHS of eqn. (B69) is Lorentz invariant. Similarly, $\Delta^-(x-y)$ given by eqn. (B63), which on using the Dirac delta function may be written as

$$\Delta^-(x-y)=\frac{+i}{(2\pi)^3}\int\frac{1}{2\omega_k}e^{i\{k_0(x_0-y_0)-\boldsymbol{k}.(\boldsymbol{x}-\boldsymbol{y})\}}\delta(k_0+\omega_k)\,dk \tag{B71}$$

acquires, using equations (B68b) and (B70), the following Lorentz invariant form

$$\Delta^-(x-y)=\frac{+i}{(2\pi)^3}\int\theta(-k_0)\delta(k^2-m^2)\,e^{ik(x-y)}dk \tag{B72}$$

A relationship can also be established between Δ^+ and Δ^-. From eqn. (B69), we may write

$$\begin{aligned}\Delta^+[-(x-y)]&=\frac{-i}{(2\pi)^3}\int\theta(k_0)\delta(k^2-m^2)e^{ik[-(x-y)]}dk\\&=\frac{-i}{(2\pi)^3}\int\theta(-k_0)\delta(k^2-m^2)e^{ik(x-y)}dk\\&=-\Delta^-(x-y)\end{aligned} \tag{B73}$$

where reference is made to the points above eqn. (B42). Similarly

$$\begin{aligned}\Delta^-[-(x-y)]&=\frac{+i}{(2\pi)^3}\int\theta(-k_0)\delta(k^2-m^2)e^{ik[-(x-y)]}dk\\&=\frac{+i}{(2\pi)^3}\int\theta(k_0)\delta(k^2-m^2)e^{ik(x-y)}dk\\&=-\Delta^+(x-y)\end{aligned} \tag{B74}$$

ODDNESS OF Δ FUNCTION AND ITS VANISHING FOR SPACELIKE SEPARATIONS

From eqn. (B61)

$$\begin{aligned}\Delta[-(x-y)]&=\Delta^+[-(x-y)]+\Delta^-[-(x-y)]\\&=-\Delta^-[(x-y)]-\Delta^+[(x-y)]\\&=-\Delta[(x-y)]\end{aligned} \tag{B75}$$

where equations (B73) and (B74) have been used. Eqn. (B75) shows that on changing the sign of the argument $(x - y)$, Δ function changes sign and hence Schwinger's delta function $\Delta(x - y)$ is an odd function.

When $(x - y)$ is spacelike, i.e., $(x - y)^2 < 0$, the function $\Delta(x - y)$ vanishes. As discussed in Appendix A, when $(x - y)$ is spacelike, we can choose a coordinate system in which $x_0' = y_0'$ (eqn. A87). Lorentz invariance of $\Delta(x - y)$ then demands that $\Delta(x - y) = \Delta(x' - y')$, where x' and y' are measured with respect to primed (S′) frame of reference. If $(x - y)$ is a four-vector of the form $[(x_0' - y_0' = 0), (\mathbf{x}' - \mathbf{y}')]$ we can always find a spatial rotation of 180 degrees which transforms $(\mathbf{x}' - \mathbf{y}')$ into $-(\mathbf{x}' - \mathbf{y}')$ so that

$$\begin{aligned}\Delta(x - y) &= \Delta(x' - y') = \Delta[-(x' - y')] \\ &= -\Delta(x' - y') \\ &= 0\end{aligned} \tag{B76}$$

FOUR DIMENSIONAL FOURIER REPRESENTATION OF Δ FUNCTION

From the four dimensional Fourier representation of Δ^+ and Δ^-, we can obtain the same for the Δ as well

In eqn. (B61) namely $\Delta(x - y) = \Delta^+(x - y) + \Delta^-(x - y)$

we substitute for Δ^+ and Δ^- from eqs. (B69) and (B72) respectively to write

$$\Delta(x - y) = \frac{-i}{(2\pi)^3}\int[\theta(k_0) - \theta(-k_0)]\delta(k^2 - m^2)\, e^{ik(x-y)}dk \tag{B77}$$

Now we introduce the sign function

$$\varepsilon(k_0) = \theta(k_0) - \theta(-k_0) \tag{B78}$$

which obeys

$$\varepsilon(k_0) = \begin{cases} +1 & \text{for } k_0 > 0 \\ -1 & \text{for } k_0 < 0 \end{cases} \tag{B79}$$

and substitute it in eqn. (B77) to write

$$\Delta(x - y) = \frac{-i}{(2\pi)^3}\int \varepsilon(k_0)\delta(k^2 - m^2)\, e^{ik(x-y)}dk \tag{B80}$$

Since, $\theta(\pm k_0)$ are Lorentz invariants, $\varepsilon(k_0)$ from its definition (B78) is also Lorentz invariant. Equation (B80) is then the desired Lorentz invariant four-dimensional Fourier representation of Schwinger's delta function.

However, eqn. (B80) can be changed into the three dimensional Fourier representation. From equation (B68) and eqn. (B78), we have

$$\varepsilon(k_0)\delta(k^2 - m^2) = \frac{1}{2\omega_k}[\delta(k_0 - \omega_k) - \delta(k_0 + \omega_k)] \tag{B81}$$

which when substituted in eqn. (B80) gives

$$\begin{aligned}\Delta(x - y) = \frac{-i}{(2\pi)^3}\int \frac{1}{2\omega_k}[&\delta(k_0 - \omega_k)e^{i\{k_0(x_0 - y_0) - \mathbf{k}.(\mathbf{x}-\mathbf{y})\}} \\ &-\delta(k_0 + \omega_k)e^{i\{k_0(x_0 - y_0) - \mathbf{k}.(\mathbf{x}-\mathbf{y})\}}]\,dk\end{aligned}$$

$$= \frac{-i}{(2\pi)^3} \int \frac{1}{2\omega_k} [e^{i\omega_k (x_0 - y_0)} - e^{-i\omega_k (x_0 - y_0)}] e^{-i\boldsymbol{k}.(\boldsymbol{x}-\boldsymbol{y})} d^3\boldsymbol{k}$$

implying that

$$\Delta(x - y) = \frac{1}{(2\pi)^3} \int \frac{\sin[\omega_k (x_0 - y_0)]}{\omega_k} e^{-i\boldsymbol{k}.(\boldsymbol{x}-\boldsymbol{y})} d^3\boldsymbol{k} \tag{B82}$$

From eqn. (B82), we may observe that even if $(\boldsymbol{x} - \boldsymbol{y})$; the space part of four-component $(x - y)$ changes to $-(\boldsymbol{x} - \boldsymbol{y})$, the function $\Delta(x - y)$ does not change its sign. This means the Schwinger's delta function is symmetric under space inversion. However, under time inversion $(x_0 - y_0) \to -(x_0 - y_0)$, $\Delta(x - y)$ changes its sign, hence is antisymmetric under time reflection.

From the above discussion additional properties of Schwinger's delta function can be realized. We have seen through the eqn. (B76) that $\Delta(x - y)$ vanishes for spacelike separations. Hence from eqn. (B82)

$$\frac{\partial \Delta(x - y)}{\partial x^0} = \frac{1}{(2\pi)^3} \int \frac{\cos[\omega_k (x_0 - y_0)]}{\omega_k} e^{-i\boldsymbol{k}.(\boldsymbol{x}-\boldsymbol{y})} d^3\boldsymbol{k} \tag{B83}$$

which for $x_0 = y_0$ becomes

$$\frac{\partial \Delta(x - y)}{\partial x^0} = \frac{1}{(2\pi)^3} \int e^{-i\boldsymbol{k}.(\boldsymbol{x}-\boldsymbol{y})} d^3\boldsymbol{k} = \delta\,(\boldsymbol{x} - \boldsymbol{y}) \tag{B84}$$

where $\delta\,(\boldsymbol{x} - \boldsymbol{y})$ is three dimensional Dirac delta function. We can also derive from eqn. (B82)

$$\left.\frac{\partial \Delta(x - y)}{\partial x^i}\right|_{x_0 = y_0} = 0 \tag{B85}$$

where $x^i (i = 1, 2, 3)$ indicate three space components of $\boldsymbol{x}$. Since $\Delta(x - y)$ is the solution of homogeneous Klein–Gordon equation and hence from eqn. (B48) we can express

$$(\quad + m^2)\Delta(x - y) = 0 \tag{B86}$$

EVEN SOLUTION OF HOMOGENEOUS KLEIN—GORDON EQUATION

Another Lorentz invariant solution of homogeneous K.G. equation is obtained in terms of the difference of two remaining Green's functions Δ_F and Δ_D. The functions Δ_A and Δ_R provided the Schwinger's Δ. Being the solutions of inhomogeneous K.G. equation, Δ_F and Δ_D obey

$$(\quad + m^2)\, \Delta_F\, (x - y) = \rho(x - y) \tag{B87a}$$

$$(\quad + m^2)\, \Delta_D\, (x - y) = \rho(x - y) \tag{B87b}$$

from which

$$(\quad + m^2)\, (\Delta_F - \Delta_D) = 0 \tag{B88}$$

However, this solution of homogeneous K.G. equation which is denoted as $\Delta^1(x-y)$ is defined as

$$\Delta^1(x-y) = i[\Delta_F(x-y) - \Delta_D(x-y)] \tag{B89}$$

instead of just the difference $\Delta_F - \Delta_D$. Thus, $\Delta^1(x-y)$, which also obeys

$$(\;\; + m^2)\Delta^1(x-y) = 0 \tag{B90}$$

is another solution of homogeneous K.G. equation.

Looking at the paths of integration of Δ_F and Δ_D in Figures (B1c) and (B1d), we observe that each of them is an open path. The difference, however, becomes a closed path as shown in Fig. (B6).

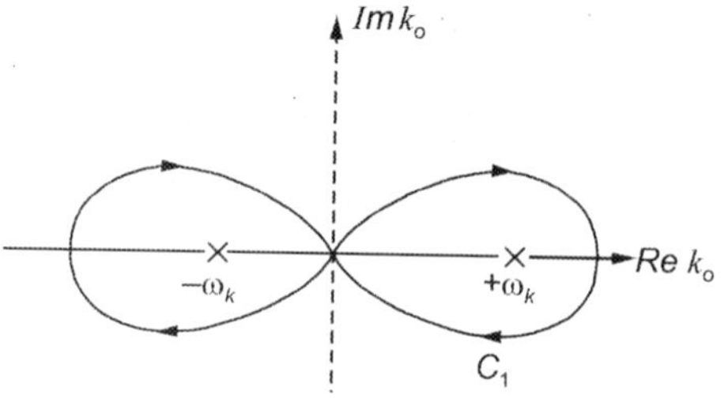

Fig. B6: The contour. (c_1) for $\Delta^1(x-y)$ function

From the equations (B39) and (B45), the contour integral representation for $\Delta^1(x-y)$ is

$$\Delta^1(x-y) = \frac{i}{(2\pi)^4}\oint_{c_1} \frac{e^{ik(x-y)}}{m^2 - k^2} dk \tag{B91}$$

Using equations (B56) and (B57), we write

$$\Delta^1(x-y) = \frac{1}{(2\pi)^4}\int d^3\boldsymbol{k} e^{-i\boldsymbol{k}.(\boldsymbol{x}-\boldsymbol{y})} \oint_{c_1} (-)\frac{e^{ik_0(x_0-y_0)}}{(k_0-\omega_k)(k_0+\omega_k)} dk^0 \tag{B92}$$

where the k_0 integral has to be evaluated for the contour of Fig. (B6). We can proceed in a manner similar to that for eqn. (B60) but here (Fig. B6) the contour closes over the pole $(-\omega_k)$ in an anticlockwise way whereas over $(+\omega_k)$ it remains clockwise. In Figure (B6), there are two closed contours. So we shall have in place of eqn. (B60)

$$\int (-)\frac{e^{ik_0(x_0-y_0)}}{(k_0^2 - \omega_0^2)} dk^0 = \frac{2\pi i}{2\omega_k}\left[-e^{i\omega_k(x_0-y_0)} - e^{-i\omega_k(x_0-y_0)}\right] \tag{B93}$$

Substitution of eqn. (B93) into eqn. (B92) leads to

$$\Delta^1(x-y) = i\,[\Delta^+(x-y) - \Delta^-(x-y)] \tag{B94}$$

where Δ^+ and Δ^- have already been given by equations (B62) and (B63) respectively.

The symmetric behaviour of $\Delta^1(x-y)$ may be observed by noting that

$$\begin{aligned}\Delta^1(y-x) &= i\,[\Delta^+(y-x) - \Delta^-(y-x)]\\ &= i\,[-\Delta^-(x-y) + \Delta^+(x-y)]\\ &= \Delta^1(x-y)\end{aligned} \tag{B95}$$

where equations (B73) and (B74) have been used. Thus on changing $(x-y)$ to $-(x-y)$, the sign of Δ^1 function does not change; therefore Δ^1 function is the even solution of homogeneous K.G. equation.

To obtain the Lorentz invariant form for $\Delta^1(x-y)$, we use the corresponding Lorentz invariant forms (B69) and (B72) for Δ^+ and Δ^-, respectively, to write,

$$\begin{aligned}\Delta^1[(x-y)] &= \frac{1}{(2\pi)^3}\int [\theta(k_0)+\theta(-k_0)]\,\delta(k^2-m^2)\,e^{ik(x-y)}\,dk\\ &= \frac{1}{(2\pi)^3}\int \delta(k^2-m^2)\,e^{ik(x-y)}\,dk\end{aligned} \tag{B96}$$

where

$$\theta\,(k_0) - \theta\,(-k_0) = 2\theta\,(k_0) - \varepsilon\,(k_0) = 1 \tag{B97}$$

has been used. Using the properties of $\varepsilon\,(k_0)$ (eqn. B79) and $\theta\,(k_0)$ (eqn. B67), it can be seen that eqn. (B97) is always 1, whether k_0 is positive or negative. Eqn. (B96) is therefore the Lorentz invariant form of $\Delta^1(x-y)$.

The three dimensional form of $\Delta^1(x-y)$ may be obtained by replacing $\delta(k^2-m^2)$ by eqn. (B66) and write eqn. (B96) as

$$\begin{aligned}\Delta^1(x-y) &= \frac{1}{(2\pi)^3}\int \frac{1}{2\omega_k}[\delta(k_0-\omega_k)+\delta(k_0+\omega_k)]\,e^{ik_0(x_0-y_0)}e^{-i\,\boldsymbol{k}.(\boldsymbol{x}-\boldsymbol{y})}\,dk\\ &= \frac{1}{(2\pi)^3}\int \frac{d^3\boldsymbol{k}}{2\omega_k}\left[e^{i\omega_k(x_0-y_0)}+e^{-i\omega_k(x_0-y_0)}\right]e^{-i\,\boldsymbol{k}.(\boldsymbol{x}-\boldsymbol{y})}\\ &= \frac{1}{(2\pi)^3}\int \frac{\cos\omega_k(x_0-y_0)}{\omega_k}\,e^{-i\,\boldsymbol{k}.(\boldsymbol{x}-\boldsymbol{y})}\,d^3\boldsymbol{k}\end{aligned} \tag{B98}$$

It can be observed from eqn. (B98) that $\Delta^1(x-y)$ does not change its sign for both $(x-y) \to (y-x)$ and $(\boldsymbol{x}_0 - y_0) \to (y_0 - x_0)$, implying the symmetry of Δ^1 function under both space and time inversions. Compared to Δ^1 function, the Schwinger's delta function was symmetric under space reflections only. We may also note from eqn. (B98) that

$$\left(\frac{\partial\Delta^1(x-y)}{\partial x^0}\right)_{x_0=y_0} \equiv \frac{1}{(2\pi)^3}\int\left[\sin\omega_k(x_0-y_0)e^{-i\,\boldsymbol{k}.(\boldsymbol{x}-\boldsymbol{y})}d^3\boldsymbol{k}\right]_{x_0=y_0} \tag{B99}$$

APPENDIX C

SCHWINGER'S ACTION PRINCIPLE

We consider a field system of operators $\psi_\alpha(x)$ for which we define the action operator as

$$A = \int_R £[\psi_\alpha(x), \partial_\mu \psi_\alpha(x)]dx \tag{C1}$$

where

$$dx = dx^0 dx^1 dx^2 dx^3 \tag{C2}$$

is the element of four-volume R. It is assumed that the four-volume R is infinite; however, we assume that this infinite volume is bounded by two spacelike surfaces σ_2 and σ_1 (FigA4b). These are the limiting spacelike surfaces from two sides. $\psi_\alpha(x)$ have the property that they fall off rapidly when spatial coordinates x^1, x^2, and x^3 tend to infinity.

Now we wish to form a new system, which is equivalent to the system described by operators $\psi_\alpha(x)$. Since ψ_α is a function of x, the four-coordinate, the new system can be formed by transforming both x and ψ_α. The change in x can be brought by going to new coordinate frame, which is shifted infinitesimally from the old one

$$x^\mu \to x'^\mu \equiv x^\mu + \delta x^\mu \tag{C3}$$

The change in $\psi_\alpha(x)$ may be brought by an infinitesimal transformation

$$\psi_\alpha(x) \to \psi'_\alpha(x) = \psi_\alpha(x) + \delta_0\psi_\alpha(x) \tag{C4}$$

$\delta_0\psi_\alpha(x)$, is called the total variation of the field operators

$$\delta_0\psi_\alpha(x) = \psi'_\alpha(x) - \psi_\alpha(x) \tag{C5}$$

$\psi'_\alpha(x)$ are the field operators *measured* with respect to S′ frame (the shifted frame) and $\psi_\alpha(x)$ are with respect to S frame.

The change in $\psi_\alpha(x)$ by $\delta_0\psi_\alpha(x)$ is due to two reasons. First the change of frame by eqn. (C3) and second the $\psi'_\alpha(x)$ may be the linear combination of the components of $\psi_\alpha(x)$.

The transformations (C3) and (C4) constitute a new field system $\psi'_\alpha(x)$ and when $\psi'_\alpha(x)$ is equivalent to the original $\psi_\alpha(x)$; the transformations (C3) and (C4) are called the equivalent transformations or the canonical transformations. By equivalence we mean that the system described by the transformed field $\psi'_\alpha(x)$ will have the same physical properties as described by $\psi_\alpha(x)$. The new field operators $\psi'_\alpha(x)$ have the same eigenvalues as $\psi_\alpha(x)$ have. These canonical transformations are generated by a self-adjoint operator $F = F(\sigma)$. The generator $F(\sigma)$, which represents infinitesimal canonical transformations, induces an infinitesimal increment $\delta_0\psi_\alpha(x)$ in the field operators $\psi_\alpha(x)$ expressed by

$$\delta_0\psi_\alpha(x) = i[F(\sigma), \psi_\alpha(x)] \tag{C6}$$

The generator $F(\sigma)$ takes the field system from one spacelike surface to other. The surfaces in which the system may proceed are shown in Fig. (C2). The generators corresponding to limiting surfaces of four-volume are $F(\sigma_2)$ and $F(\sigma_1)$ (see Fig. A4b).

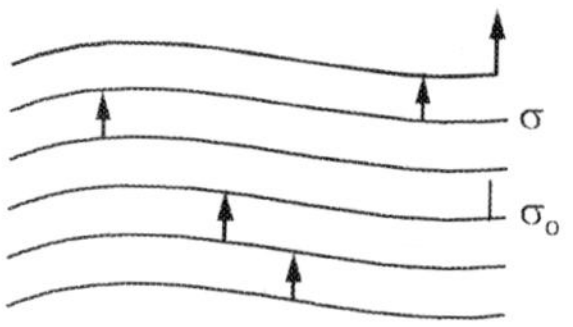

Fig. C1: Spacelike surfaces. Two spacelike surfaces have been shown by σ_0 and σ. The generator of the canonical transformations takes the system from one spacelike surface to other

Since, the action operator A, (eqn. C1) depends on $\psi\alpha(x)$ and $\partial_\mu\psi_\alpha(x)$, an increment on $\psi_\alpha(x)$ by $\delta_0\psi_\alpha(x)$ due to the canonical transformations (C3) and (C4), would also produce an increment of δA in the action operator A such that

$$A \rightarrow A + \delta A \tag{C7}$$

Thus, we have on the one hand $F(\sigma_2)$ and $F(\sigma_1)$, the generators of canonical transformations corresponding to limiting surfaces of four-volume R and on the other, the increment δA in the action operator due to canonical transformations. The Schwinger's action principle equates δA with the difference of generators $F(\sigma_2)$ and $F(\sigma_1)$. The Schwinger's action principle is thus expressed by

$$\delta A = F(\sigma_2) - F(\sigma_1) \tag{C8}$$

and may be stated as–*the change δA in the operator A due to the canonical transformations is equal to the difference in the generators of these transformations corresponding to the limiting surfaces of the enclosed four-volume.*

EXPRESSION FOR δA

Having seen that the canonical transformations produce an increment δA in A, we shall now obtain explicit form of δA.

The volume element in the action operator (C1) is dx and the whole volume enclosed between σ_2 and σ_1 is R. Due to the transformations (C3), x^μ shifts to x'^μ .The spacelike surface has the equation

$$\eta_\mu x^\mu - \tau = 0 \tag{C9}$$

from which it can be inferred that when x shifts to x', the spacelike surfaces σ_2 and σ_1 also shift, as a result, the four-volume enclosed by these surfaces changes. Let the new volume be R'. Due to the transformations (C3), x^μ shifts to x'^μ implying that σ_2 and σ_1 also shift, as a result, the four-volume enclosed by the spacelike surfaces changes. Let the new volume be R'. The transformations (C3) and (C4) have shifted the field operators $\psi_\alpha(x)$ as well. Hence the action operator corresponding to these shifted parameters would be

$$A + \delta A = \int_{R'} £[\{\psi_\alpha(x) + \delta_0\psi_\alpha(x)\}, \partial_\mu\{\psi_\alpha(x) + \delta_0\psi_\alpha(x)\}]dx'$$

$$= \int_{R'} £(\psi_\alpha + \delta_0\psi_\alpha, \partial_\mu\psi_\alpha + \delta_0\partial_\mu\psi_\alpha)dx' \qquad \text{(C10)}$$

where we have used

$$\partial_\mu\, \delta_0\psi_\alpha\,(x) = \delta_0\partial_\mu\psi\alpha(x) \qquad \text{(C11)}$$

Using the Taylor expansion we may write eqn. (C10) upto first order as

$$A + \delta A = \int_{R'} [£(\psi_\alpha, \partial_\mu\psi_\alpha) + \delta_0\psi_\alpha \frac{\partial £}{\partial\psi_\alpha} + \delta_0\partial_\mu\psi_\alpha \frac{\partial £}{\partial(\partial_\mu\psi_\alpha)}]dx' \qquad \text{(C12)}$$

Now we write the volume element dx' in terms of dx. We rewrite the transformation (C3)

$$x'^\mu = x^\mu + \delta x^\mu \qquad \text{(C3)}$$

from which we may write

$$dx' = dx + \frac{\partial(\delta x^\mu)}{\partial x^\mu}dx. \qquad \text{(C13)}$$

Substituting eqn. (C13) into eqn. (C12) and noting that for dx the enclosed volume is R, we write eqn. (C12) as

$$A + \delta A = \int_R \{£(\psi_\alpha, \partial_\mu\psi_\alpha) + £\frac{\partial(\delta x^\mu)}{\partial x^\mu}\}dx$$
$$+ \int_R \{\delta_0\psi_\alpha \frac{\partial £}{\partial\psi_\alpha} + \delta_0\partial_\mu\psi_\alpha \frac{\partial £}{\partial(\partial_\mu\psi_\alpha)}\}dx + \text{Other terms} \qquad \text{(C14)}$$

Neglecting other terms which are the products of two infinitesimals and substituting for A from eqn. (C1), i.e.

$$A = \int_R £(\psi_\alpha, \partial_\mu\psi_\alpha)dx \qquad \text{(C1)}$$

we get from eqn. (C14)

$$\delta A = \int_R [\frac{\partial £}{\partial\psi_\alpha}\delta_0\psi_\alpha + \frac{\partial £}{\partial(\partial_\mu\psi_\alpha)}\delta_0\partial_\mu\psi_\alpha + £\frac{\partial(\delta x^\mu)}{\partial x_\mu}]dx \qquad \text{(C15)}$$

Using

$$\partial_\mu(\frac{\partial £}{\partial(\partial_\mu\psi_\alpha)}\delta_0\psi_\alpha) = \frac{\partial £}{\partial(\partial_\mu\psi_\alpha)}\partial_\mu\delta_0\psi_\alpha + \delta_0\psi_\alpha\partial_\mu \frac{\partial £}{\partial(\partial_\mu\psi_\alpha)} \qquad \text{(C15a)}$$

in the middle term of eqn. (C15) and also using again eqn.(C11), we may write eqn. (C15) as

$$\delta A = \int_R [\frac{\partial £}{\partial \psi_\alpha} \delta_0 \psi_\alpha + \partial_\mu (\frac{\partial £}{\partial(\partial_\mu \psi_\alpha)} \delta_0 \psi_\alpha) - \delta_0 \psi_\alpha . \partial_\mu \frac{\partial £}{\partial(\partial_\mu \psi_\alpha)} + £ \partial_\mu \delta x^\mu] dx \quad \text{(C16)}$$

By rearranging the terms, the eqn. (C16) can be rewritten as

$$\delta A = \int_R [\frac{\partial £}{\partial \psi_\alpha} - \partial_\mu \frac{\partial £}{\partial(\partial_\mu \psi_\alpha)}] \partial_0 \psi_\alpha \, dx + \int_R \partial_\mu [\frac{\partial £}{\partial(\partial_\mu \psi_\alpha)} \delta_0 \psi_\alpha + £ \, \delta x^\mu] dx \quad \text{(C17)}$$

in which we have used

$$£ \, \partial_\mu \, \delta x^\mu = \partial_\mu \, (£ \, \delta x^\mu) \quad \text{(C18)}$$

because £ does not explicitly depend on x^μ.

Second integral of eqn. (C17) is a volume (four) integral of (four) divergence. Hence Gauss' divergence theorem can be applied. Replacing,

$$\frac{\partial £}{\partial(\partial_\mu \psi_\alpha)} \delta_0 \psi_\alpha + £ \, \delta x^\mu = \Im^\mu \quad \text{(C19)}$$

we may then write for the second integral of eqn. (C17)

$$\int_R \partial_\mu \Im^\mu dx = \int_{\sigma_1}^{\sigma_2} \Im^\mu d\sigma_\mu = \int_{\sigma_2} \Im^\mu d\sigma_\mu - \int_{\sigma_1} \Im^\mu d\sigma_\mu \quad \text{(C20)}$$

where spacelike surfaces σ_1 and σ_2 constitute the boundary of volume R (Fig. A4a, with $\sigma \to \sigma_2$ and $\sigma' \to \sigma_1$)

At this stage we refer to localizability. Classically, a field is conceived as a mechanical system that describes physical conditions at any point in the space-time continuum. The quantum mechanical counterpart of such a field is the concept of the localizable field. A localizable dynamical variable is a quantum mechanical operator that describes the physical conditions at *one particular point*, x, in space and time.

Thus considering ψ_α and ψ'_α as the operators referring to the *same geometrical point* the arguments of both, being measured from two different frames, cannot be taken as the same x. Therefore when we refer to the same geometrical point, we write $\psi_\alpha \equiv \psi_\alpha(x)$ and $\psi'_\alpha \equiv \psi'_\alpha(x')$, where x' and x are related by the transformation (C3). In this case the change from $\psi_\alpha(x)$ to $\psi'_\alpha(x')$ may be denoted by $\delta\psi_\alpha(x)$. It means

$$\delta\psi_\alpha (x) = \psi'_\alpha(x') - \psi_\alpha(x) \quad \text{(C21)}$$

As a comparison, we rewrite eqn. (C6) which refers to the *different geometrical points* and hence ψ'_α and ψ_α retain the same x,

$$\delta_0\psi_\alpha(x) = \psi'_\alpha (x) - \psi_\alpha(x) \quad \text{(C6)}$$

Equation (C21) is called the local (a particular point) variation. Now, using eqn. (C3),

$$\psi'_\alpha (x') = \psi'_\alpha (x + \delta x^\nu) = \psi'_\alpha (x) + [\frac{\partial}{\partial x^\nu} \psi'_\alpha x)] \delta x^\nu \quad \text{(C22)}$$

where we have made Taylor expansion upto the first order. Substituting eqn. (C22) into eqn. (C21), we get

$$\delta\psi_\alpha(x) = \psi'_\alpha(x) - \psi_\alpha(x) + \partial_\nu \psi'_\alpha(x)\cdot\delta x^\nu \tag{C23}$$

Using eqn. (C6) and also observing from it that $\delta_0\psi_\alpha(x)$ is infinitesimal (its $\partial_\nu = 0$), hence

$$\partial_\nu \psi'_\alpha(x) \approx \partial_\nu \psi_\alpha(x) \tag{C24}$$

we get from eqn.(C23) that

$$\delta\psi_\alpha(x) = \delta_0\psi_\alpha(x) + \partial_\nu\psi_\alpha(x).\delta x^\nu \tag{C25a}$$

or

$$\delta_0\psi_\alpha(x) = \delta\psi_\alpha(x) - \partial_\nu\psi_\alpha(x).\delta x^\nu \tag{C25b}$$

Substituting eqn. (C25b) in eqn. (C19), we get for

$$\Im^\mu = \frac{\partial £}{\partial(\partial_\mu\psi_\alpha)}\delta\psi_\alpha - \frac{\partial £}{\partial(\partial_\mu\psi_\alpha)}\partial_\nu\psi_\alpha\delta x^\nu + £\delta x^\mu \tag{C26}$$

Expressing, $£\delta x^\mu = £\delta^\mu_\nu \delta x^\nu$ and taking δx^ν as common, eqn. (C26), becomes

$$\Im^\mu = \frac{\partial £}{\partial(\partial_\mu\psi_\alpha)}\delta\psi_\alpha - \left[\frac{\partial £.\partial_\nu\psi_\alpha}{\partial(\partial_\mu\psi_\alpha)} - \delta^\mu_\nu £\right]\delta x^\nu \tag{C27}$$

Using the notation

$$\pi^\mu_\alpha = \frac{\partial £}{\partial(\partial_\mu\psi_\alpha)} \tag{C28}$$

eqn. (C27) becomes

$$\Im^\mu = \pi^\mu_\alpha\delta\psi_\alpha - (\pi^\mu_\alpha\partial_\nu\psi_\alpha - \delta^\mu_\nu £)\delta x^\nu \tag{C29}$$

Keeping the RHS of eqn. (C20) in view, we write,

$$\int_{\sigma_2} \Im^\mu d\sigma_\mu - \int_{\sigma_1} \Im^\mu d\sigma_\mu = F(\sigma_2) - F(\sigma_1) \tag{C30}$$

where

$$F(\sigma) \equiv \int_\sigma \{\pi^\mu_\alpha\delta\psi_\alpha - (\pi^\mu_\alpha\partial_\nu\psi_\alpha - \delta^\mu_\nu £)\delta x^\nu\} d\sigma_\mu \tag{C31}$$

It is to be noted that we have used, in the text $F(\sigma)$ as the generator of canonical transformations. A separate symbol could be used to replace the RHS of eqn. (C31). However, anticipating that it would emerge as the generator of canonical transformations (C3) and (C4), we have retained the symbol $F(\sigma)$.

Thus the final expression for δA (eqn. C17), on using equations (C20) and (C30), becomes

$$\delta A = F(\sigma_2) - F(\sigma_1) + \int_R \left(\frac{\partial £}{\partial \psi_\alpha} - \partial_\mu \frac{\partial £}{\partial(\partial_\mu \psi_\alpha)} \right) \delta_0 \psi_\alpha dx \tag{C32}$$

where $\psi_\alpha = \psi_\alpha(x)$ and $dx = dx^0 dx^1 dx^2 dx^3$.

THE CONJUGATE MOMENTUM

Equation (C28) introduces π_α^μ, which are N four-vectors, because $\alpha = 1, 2, \dots N$,

$$\pi_\alpha^\mu(x) = \frac{\partial £}{\partial[\partial_\mu \psi_\alpha(x)]} \quad \text{C28)}$$

Thus with every field component $\psi_\alpha(x)$, is associated a four-vector $\pi_\alpha^\mu(x)$. The time component, i.e., the $\mu = 0$ component, of π_α^μ, i.e., π_α^0 plays a significant role as it provides the momentum canonically conjugate to the field $\psi_\alpha(x)$. From eqn. (C28), putting $\mu = 0$, we get

$$\pi_\alpha^0 = \frac{\partial £}{\partial[\partial_0 \psi_\alpha(x)]} = \frac{\partial £}{\partial \dot{\psi}_\alpha} \tag{C33}$$

Usually, π_α^0 is written as π_α, thus

$$\pi_\alpha = \frac{\partial £}{\partial[\partial_0 \psi_\alpha(x)]} = \frac{\partial £}{\partial \dot{\psi}_\alpha} \tag{C34}$$

The equation (C34) is analogous to the similar expression for the momenta canonically conjugate to the generalized coordinates q_r in classical mechanics, which are defined by

$$p_r = \frac{\partial L}{\partial \dot{q}_r} \tag{C35}$$

where, L is the classical Lagrangian. Comparing equations (C35) and (C34), we find that π_α play the same role in field theory as p_r play in classical mechanics. That is why we call π_α, the canonically conjugate momenta of $\psi_\alpha(x)$. For each field component there is a corresponding conjugate momentum. Corresponding to $q_r(t)$, the generalized coordinates of classical mechanics, we have $\psi_\alpha(x)$ in field theory.

The significance of π_α also becomes manifest because of its use in obtaining the Hamiltonian of the system. Again, in analogy with the classical mechanics, where the Hamiltonian of the dynamical system is defined as

$$\mathrm{H} = \sum_r p_r \dot{q}_r - L \tag{C36}$$

the corresponding expression in field theory is given by

$$\mathrm{H}=\int d^3\boldsymbol{x}\sum_{\alpha=1}^{N}\pi_\alpha\dot{\psi}_\alpha(x)-\int d^3\boldsymbol{x}\pounds(x) \qquad \text{(C37)}$$

where £(x) is the Lagrangian density whose three dimensional volume integral gives the Lagrangian of the system. Eqn. (C38) can be written as

$$\mathrm{H}=\int d^3\boldsymbol{x}\,[\sum_{\alpha=1}^{N}\pi_\alpha\dot{\psi}_\alpha(x)-\pounds(x)]$$

or

$$\mathrm{H}=\int d^3\boldsymbol{x}H(x) \qquad \text{(C38)}$$

where

$$H(x)=\sum_{\alpha=1}^{N}\pi_\alpha(x)\dot{\psi}_\alpha-\pounds(x) \qquad \text{(C39)}$$

is called the Hamiltonian density.

Obtaining the Hamiltonian from the Lagrangian density is an important step of the canonical quantization formalism. Hence, the canonically conjugate momenta, which appear in the Hamiltonian, play an equally important part in the above formalism.

The Schwinger's action principle can be applied to

(1) show that the Lagrangian operator is self-adjoint.
(2) obtain Euler Lagrange equations
(3) determine the rules of field quantization.
(4) obtain energy momentum, angular momentum and charge conservation laws.

ENERGY MOMENTUM CONSERVATION LAW

To obtain the energy momentum conservation law, we consider the canonical transformation (C3) and demand that $\delta A=0$, under this transformation. We write eqn. (C3) as

$$x'^\nu=x^\nu+\varepsilon^\nu \qquad \text{(C40)}$$

with

$$\varepsilon^\nu=\delta x^\nu=\text{constant} \qquad \text{(C41)}$$

At the spacetime point x the field operators are $\psi_\alpha(x)$ and at the spacetime point x', they are $\psi'_\alpha(x')$. Here x and x' are related by the canonical transformations (C40). Since canonical transformations describe the equivalent systems, we shall have

$$\psi'_\alpha(x')=\psi_\alpha(x) \qquad \text{(C42)}$$

Now, keeping equation (C40) in view and describing the field system from the shifted coordinate frame, we may replace eqn. (C42) by

$$\psi'_\alpha(x)=\psi_\alpha(x-\varepsilon) \qquad \text{(C43)}$$

Now according to eqn. (C5)

$$\delta_0\, \psi_\alpha(x) = \psi'_\alpha(x) - \psi_\alpha(x) \tag{C5}$$

into which the substitution of eqn. (C43) for $\psi'_\alpha(x)$ gives

$$\delta_0\psi_\alpha(x) = \psi_\alpha(x - \varepsilon) - \psi_\alpha(x) \tag{C44}$$

Since, $\varepsilon \equiv \varepsilon^\nu$, is infinitesimal, we may use Taylor's expansion for $\psi_\alpha\,(x - \varepsilon)$ which up to the first order is

$$\psi_\alpha\,(x - \varepsilon) = \psi_\alpha(x) - \partial_\nu\psi_\alpha(x)\varepsilon^\nu \tag{C45}$$

Substituting eqn. (C45) into eqn. (C44), we have

$$\delta_0\psi_\alpha(x) = -\,\partial_\nu\psi_\alpha(x)\varepsilon^\nu \tag{C46}$$

Now use eqn. (C25a), which is

$$\delta\psi_\alpha(x) = \delta_0\psi_\alpha(x) + \partial_\nu\psi_\alpha(x)\delta x^\nu \tag{C47}$$

where $\delta x^\nu = \varepsilon^\nu$. From equations (C46) and (C47), we then get

$$\delta\psi_\alpha(x) = 0 \tag{C48}$$

Due to eqn. (C48), the generator

$$F(\sigma) \equiv \int_\sigma \{\pi^\mu_\alpha(x)\delta\psi_\alpha - (\pi^\mu_\alpha\partial_\nu\psi_\alpha - \delta^\mu_\nu\pounds)\delta x^\nu\}d\sigma_\mu \tag{C48a}$$

becomes

$$F(\sigma) = -\varepsilon^\nu \int_\sigma (\pi^\mu_\alpha\partial_\nu\psi_\alpha - \delta^\mu_\nu\pounds)d\sigma_\mu \tag{C49}$$

Now,

$$\pi^\mu_\alpha\partial_\nu\psi_\alpha - \delta^\mu_\nu\pounds \equiv \pi^\mu_\alpha\partial^\nu\psi_\alpha - g^{\mu\nu}\pounds \tag{C50}$$

We may therefore replace the bracket of eqn. (C49) by the RHS of eqn. (C50) and thus introduce a tensor $T^{\mu\nu}$

$$T^{\mu\nu}(x) = \pi^\mu_\alpha(x)\partial^\nu\psi_\alpha(x) - g^{\mu\nu}\pounds \tag{C51}$$

which is called the *canonical energy momentum tensor*. Thus

$$F(\sigma) = -\varepsilon^\nu \int_\sigma T^{\mu\nu}d\sigma_\mu \tag{C52}$$

The conservation of $T^{\mu\nu}(x)$ follows from

$$\frac{\delta F(\sigma)}{\delta\sigma(y)} = 0 \tag{C53}$$

the differential form of which in view of $F(\sigma)$ of equation (C52) is

$$\partial_\mu T^{\mu\nu} = 0 \tag{C54}$$

gives the energy momentum conservation law.

The energy momentum tensor, eqn. (C51), contains expressions for the energy of the field as well as the momentum of the field. Field is represented by operators $\psi_\alpha(x)$. π^μ_α are the four-vectors defined according to eqn. (C28) whose zeroth component is the canonical momentum. The conservation law (C54) is in fact a set of four conservation laws corresponding to $\nu = 0, 1, 2, 3$. From $T^{\mu\nu}$, we may obtain the components $T^{00}(x)$, $T^{0k}(x)$ and $T^{ik}(x)$, where $i, k = 1, 2, 3$. T^{00} is called the energy density or the Hamiltonian density of the field system at the spacetime point x. T^{0k} gives the momentum density at x. The mixed components $T^{ik}(x)$ are called the components of stress density at the spacetime point x. When these densities are integrated over a constant time surface ($d^3 x$), they provide the total field Hamiltonian or total energy, total field momentum and the total stress of the field system respectively. It is therefore concluded that the conservation of the energy and momentum of the field system follows from demanding that the system remains invariant under translations. The operators $P^{\nu} = \int_{\sigma} T^{\mu\nu} d\sigma_{\mu}$ act as translation operators.

ANGULAR MOMENTUM CONSERVATION LAW

The angular momentum conservation law is obtained by demanding that the action operator remains invariant ($\delta A = 0$) under the infinitesimal Lorentz transformations.

$$x'^{\nu} = (g^{\nu\mu} + \varepsilon^{\nu\mu})x_{\mu} \tag{C55}$$

The variation $\delta\psi_{\alpha}(x)$ in the field operators $\psi_{\alpha}(x)$, when they undergo infinitesimal Lorentz transformations is

$$\delta\psi_{\alpha}(x) = \psi'_{\alpha}(x') - \psi_{\alpha}(x) = \frac{1}{2} I_{\nu\mu}^{\alpha\beta} \varepsilon^{\nu\mu} \psi_{\beta}(x) \tag{C56}$$

where $I_{\nu\mu} = -I_{\mu\nu}$ denote six constant 4×4 matrices and indices α, β refer to their matrix elements. $\varepsilon^{\nu\mu} = -\varepsilon_{\mu\nu}$ with $|\varepsilon^{\nu\mu}| << 1$ denote the infinitesimal character of Lorentz transformations.

Substituting $\delta\psi_{\alpha}(x)$ from equation (C56) and replacing δx^{ν} by $\varepsilon^{\nu\mu} x_{\mu}$ in the expression for the generator $F(\sigma)$, that is in the expression

$$F(\sigma) \equiv \int_{\sigma} \{\pi_{\alpha}^{\rho}(x)\delta\psi_{\alpha} - (\pi_{\alpha}^{\rho}\partial_{\nu}\psi_{\alpha} - \delta_{\nu}^{\rho}\pounds)\delta x^{\nu}\} d\sigma_{\rho} \tag{C57}$$

we get

$$F(\sigma) = \int_{\sigma} \{\pi_{\alpha}^{\rho} \tfrac{1}{2} \varepsilon^{\nu\mu} I_{\nu\mu}^{\alpha\beta} \psi_{\beta}(x) - (\pi_{\alpha}^{\rho}\partial_{\nu}\psi_{\alpha} - \delta_{\nu}^{\rho}\pounds)\varepsilon^{\nu\mu} x_{\mu}\} d\sigma_{\rho} \tag{C58}$$

Now, look at the second term of eqn. (C58). We may write for it ($\mu \leftrightarrow \nu$)

$$\begin{aligned}(\pi_{\alpha}^{\rho}\partial_{\nu}\psi_{\alpha} - \delta_{\nu}^{\rho}\pounds)\varepsilon^{\nu\mu} x_{\mu} &= \tfrac{1}{2}[(\pi_{\alpha}^{\rho}\partial_{\nu}\psi_{\alpha} - \delta_{\nu}^{\rho}\pounds)\varepsilon^{\nu\mu} x_{\mu} \\ &\quad + (\pi_{\alpha}^{\rho}\partial_{\mu}\psi_{\alpha} - \delta_{\mu}^{\rho}\pounds)\varepsilon^{\mu\nu} x_{\nu}]\end{aligned} \tag{C59}$$

the RHS of which on using $\varepsilon^{\mu\nu} = -\varepsilon^{\mu\nu}$ can be expressed as

$$\text{RHS} = \tfrac{1}{2}\varepsilon^{\nu\mu}[\pi_{\alpha}^{\rho}(x_{\mu}\partial_{\nu}\psi_{\alpha} - x_{\nu}\partial_{\mu}\psi_{\alpha}) - \pounds(\delta_{\nu}^{\rho} x_{\mu} - \delta_{\mu}^{\rho} x_{\nu})] \tag{C60}$$

Substituting eqn. (C60), into eqn. (C58), we get for $F(\sigma)$

$$F(\sigma)=\tfrac{1}{2}\varepsilon^{\nu\mu}\int_{\sigma}\{\pi_{\alpha}^{\rho}I_{\nu\mu}^{\alpha\beta}\psi_{\beta}+\pi_{\alpha}^{\rho}(x_{\nu}\partial_{\mu}-x_{\mu}\partial_{\nu})\psi_{\alpha}+\pounds(\delta_{\nu}^{\rho}x_{\mu}-\delta_{\mu}^{\rho}x_{\nu})\}d\sigma_{\rho} \quad \text{(C61)}$$

Since, each of the three terms of above eqn. (C61) is anti symmetric in μ and ν, we may replace all the three terms in the curly bracket by one anti symmetric tensor

$$M_{\nu\mu}^{\rho}=-M_{\mu\nu}^{\rho} \quad \text{(C62)}$$

where

$$M_{\nu\mu}^{\rho}(x)=\pi_{\alpha}^{\rho}I_{\nu\mu}^{\alpha\beta}\psi_{\beta}(x)+\pi_{\alpha}^{\rho}(x_{\nu}\partial_{\mu}-x_{\mu}\partial_{\nu})\psi_{\alpha}(x)+(\delta_{\nu}^{\rho}x_{\mu}-\delta_{\mu}^{\rho}x_{\nu})\pounds(x) \quad \text{(C63)}$$

This leads eqn. (C61) to

$$F(\sigma)=\tfrac{1}{2}\varepsilon^{\nu\mu}\int_{\sigma}M_{\nu\mu}^{\rho}(x)d\sigma_{\rho} \quad \text{(C64)}$$

The whole integral involving antisymmetric tensor $M_{\nu\mu}^{\rho}$ can itself be replaced by another antisymmetric tensor $J_{\nu\mu}$, where

$$J_{\nu\mu}(x)=\int_{\sigma}M_{\nu\mu}^{\rho}(x)d\sigma_{\rho} \quad \text{(C65)}$$

and

$$J_{\nu\mu}=-J_{\mu\nu} \quad \text{(C66)}$$

Eqn. (C64) then becomes

$$F(\sigma)=\frac{1}{2}\varepsilon^{\nu\mu}J_{\nu\mu} \quad \text{(C67)}$$

Now using Schwinger's action principle (C8) for $\delta A = 0$, $F(\sigma)$ must be independent of surface and hence

$$\frac{\delta}{\delta\sigma(y)}(\tfrac{1}{2}\varepsilon^{\nu\mu}J_{\nu\mu})=0 \quad \text{(C68)}$$

and since $\varepsilon^{\nu\mu}$ are arbitrary, we have the following conservation law

$$\frac{\delta J_{\nu\mu}}{\delta\sigma(y)}=0 \quad \text{(C69)}$$

Applying Gauss' divergence theorem to eqn. (C65) in the order surface integral to volume integral we have,

$$J_{\nu\mu}=\int_{R}\partial_{\rho}M_{\nu\mu}^{\rho}(x)dx \quad \text{(C70)}$$

where $dx = dx^0 dx^1 dx^2 dx^3$. The demand of eqn. (C69) that $J_{\nu\mu}$ be independent of σ would be fulfilled if $M_{\nu\mu}^{\rho}$ (eqn. C70) be divergenceless, i.e.

$$\partial_{\rho}M_{\nu\mu}^{\rho}(x)=0 \quad \text{(C71)}$$

Recalling the form (C63) for $M^{\rho}_{\nu\mu}(x)$, we find that each of the three terms, provides six components, hence $M^{\rho}_{\nu\mu}$, for ν, μ = 0, 1, 2, 3, also has six components. Accordingly, eqn. (C71) are the six conservation laws in their differential forms. From eqn. (C63), we also infer that for $\nu = \mu$, $M^{\rho}_{\nu\mu}$ vanish. ν is either greater or less than μ. The components with $\nu > \mu$ are opposite in sign than those with $\nu < \mu$.

A look at eqn. (C63), particularly the middle term gives the idea that it has the $(\boldsymbol{r} \times \boldsymbol{p})$ content in it, which is, angular momentum. Thus, $M^{\rho}_{\nu\mu}(x)$ is the tensor, which describes angular momentum and is called the *covariant angular momentum density of the field system.*

If we take here also the constant time surface, then in place of eqn. (C69), we shall have

$$\frac{dJ_{\nu\mu}}{dt} = 0 \tag{C72}$$

and $J_{\nu\mu}$, obtained by putting $\rho = 0$, $d\sigma_0 = d^3\boldsymbol{x}$ in eqn. (C65), i.e.

$$J_{\nu\mu} = \int M^{0}_{\nu\mu}(x) d^3\boldsymbol{x} \tag{C73}$$

is a constant in time or a conserved quantity. $J_{\nu\mu}$ $(\nu < \mu)$ are also six components, all of which are conserved. $J_{\nu\mu}$ act as Lorentz rotation operators.

CHARGE CONSERVATION LAW

In quantum field theory, a real field is described by a hermitian field operator and a complex classical field becomes a non hermitian operator, for which, $\psi^{\dagger}_{\alpha} \neq \psi_{\alpha}$ where $\psi_{\alpha}(x)$ are the field operators. We now consider the non-hermitian field operators, $\psi_{\alpha}(x)$ and introduce the canonical transformations which produce a change only in the phase of the field operators. Such transformations produce no change in x, the argument of operators $\psi_{\alpha}(x)$. The latter condition is expressed by the equation

$$\delta x^{\nu} = 0 \tag{C74}$$

The phase transformations may be expressed by

$$\psi_{\alpha}(x) \to \exp[+iq\theta(\alpha)]\psi_{\alpha}(x) \tag{C75a}$$

and

$$\psi^{\dagger}_{\alpha}(x) \to \exp[-iq\theta(\alpha)]\psi^{\dagger}_{\alpha}(x) \tag{C75b}$$

These are called the phase transformations, because they introduce a phase of $+iq\theta$ in the field operators $\psi_{\alpha}(x)$ and a phase of $-iq\,\theta$ in $\psi^{\dagger}_{\alpha}(x)$. Here q is a finite (fixed) constant and $\theta(\alpha)$ is an angle parameter. For hermitian fields, however,

$$\psi^{\dagger}_{\alpha}(x) = \psi_{\alpha}(x) \tag{C76}$$

We may define

$$\theta(\alpha) = \varepsilon(\alpha)\theta \tag{C77}$$

where θ has now become independent of α; $\varepsilon(\alpha) = 0$ for hermitian fields, it is (+1), for non hermitian fields and (–1) for conjugates of non hermitian fields. With the notation (C77), both the transformations (C75) can be expressed by a single equation namely,

$$\psi_{\alpha}(x) \rightarrow \exp[+i\theta\, q\varepsilon(\alpha)]\psi_{\alpha}(x) \tag{C78}$$

from which $\varepsilon(\alpha) = +1$, gives eqn. (C75a) and $\varepsilon(\alpha) = -1$ the eqn. (C75b). Taking θ infinitesimal ($\theta << 1$) the corresponding infinitesimal transformations become

$$\psi_{\alpha}(x) \rightarrow [1 + i\theta q\varepsilon(\alpha)]\psi_{\alpha}(x) \tag{C79}$$

From eqn. (C25a), we have

$$\delta\psi_{\alpha}(x) = \delta_0\psi_{\alpha}(x) + \partial_{\nu}\psi_{\alpha}(x)\delta x^{\nu} \tag{C80}$$

However, due to $\delta x^{\nu} = 0$ (eqn. C74), here

$$\delta\psi_{\alpha}(x) = \delta_0\psi_{\alpha}(x) \tag{C81}$$

Since, we need to know the generator of phase transformations (C78), we as before rewrite the expression for $F(\sigma)$, from eqn. (C31)

$$F(\sigma) = \int_{\sigma} \{\pi^{\mu}_{\alpha}\delta\psi_{\alpha} - (\pi^{\mu}_{\alpha}\partial_{\nu}\psi_{\alpha} - \delta^{\mu}_{\nu}£)\delta x^{\nu}\}d\sigma_{\mu} \tag{C81a}$$

which for the present case, where $\delta x^{\nu} = 0$, becomes

$$F(\sigma) = \int_{\sigma} \pi^{\mu}_{\alpha}\delta\psi_{\alpha}(x)d\sigma_{\mu} \tag{C82}$$

Now, we have for $\delta\psi_{\alpha}$

$$\delta\psi_{\alpha}(x) = \psi'_{\alpha}(x') - \psi_{\alpha}(x) \tag{C83}$$

where ψ'_{α} is the transformed ψ_{α} and x' is the transformed x. The transformations are according to the demand, e.g., here, these are the ones, which change or introduce phase in the field operators. We repeat it here that in the phase transformations, the spacetime is left unchanged ($x' = x$), and hence $\psi'_{\alpha}(x') = \psi'_{\alpha}(x)$. Here ψ'_{α}, on the other hand is described by eqn. (C78), i.e., ψ_{α} multiplied by phase term. Therefore, for infinitesimal phase transformations (C79), we have in fact

$$\psi'_{\alpha}(x') = \{1 + i\theta q\varepsilon(\alpha)\}\psi_{\alpha}(x) \tag{C84}$$

where of course $x' = x$. Thus

$$\delta\psi_{\alpha} = \psi'_{\alpha}(x') - \psi_{\alpha}(x) = +i\theta q\varepsilon(\alpha)\psi_{\alpha}(x) = \delta_0\psi_{\alpha}(x) \tag{C85}$$

where we have called upon the eqn. (C81). Therefore, from eqn. (C82)

$$F(\sigma) = \theta\int_{\sigma} iq\varepsilon(\alpha)\pi^{\mu}_{\alpha}(x)\psi_{\alpha}(x)d\sigma_{\mu}(x) \tag{C86}$$

which can be written as

$$F(\sigma) = \theta \int_{\sigma} j^{\mu}(x) d\sigma_{\mu}(x) \tag{C87}$$

where

$$j^{\mu}(x) = iq\varepsilon(\alpha)\pi_{\alpha}^{\mu}(x)\psi_{\alpha}(x) \tag{C88}$$

in which repeated index α, implies summation.

When δA of eqn. (C8) is zero, $F(\sigma)$ of eqn. (C87) would be independent of surface, implying that (θ is arbitrary)

$$\frac{\delta Q}{\delta\sigma(y)} = 0 \tag{C89}$$

where

$$Q = \int_{\sigma} j^{\mu}(x) d\sigma_{\mu}(x) \tag{C90a}$$

$$= iq\int_{\sigma} \varepsilon(\alpha)\pi_{\alpha}^{\mu}(x)\psi_{\alpha}(x) d\sigma_{\mu}(x) \tag{C90b}$$

is called the total charge of the field system. Choosing as in the previous cases, the constant time surface $d\sigma_0$, we have for eqn. (C89)

$$\frac{dQ}{dt} = 0 \tag{C91}$$

in which Q from eqs. (C90) is

$$Q = iq\int j^0 d^3\mathbf{x} = iq\int \varepsilon(\alpha)\pi_{\alpha}(x)\psi_{\alpha}(x) d^3\mathbf{x} \tag{C92}$$

where $\pi_{\alpha}^{0} = \pi_{\alpha}$. Applying Gauss' divergence theorem in four dimensions in eqn. (C90) (*surface* $\rightarrow$ *volume*), eqn. (C89) would give the following familiar differential form for the charge conservation law

$$\partial_{\mu} j^{\mu}(x) = 0 \tag{C93}$$

It has to be noted that Q acts as charge operator.

APPENDIX D

ELECTROMAGNETIC FIELD

The electromagnetic field is a massless real vector field. A vector field consists of four components, which form a four-vector and transform under Lorentz transformations namely,

$$\phi'^{\mu}(x') = a^{\mu}_{\nu}\, \phi^{\nu}(x) \tag{D1}$$

where ϕ^{ν} and ϕ'^{μ} with $\mu = 0, 1, 2, 3$ represent the vector (four-vector) field. The electromagnetic field is described by a four component potential A^{μ}, whose time component A^0, is the scalar potential and the space part $\boldsymbol{A}$ is the magnetic vector potential. In terms of these potentials, it is well known that in natural units

$$\boldsymbol{E} = -\frac{\partial \boldsymbol{A}}{\partial t} - \nabla A^0 \tag{D2a}$$

$$\boldsymbol{B} = \nabla \times \boldsymbol{A} \tag{D2b}$$

Using $A^{\mu} = (A^0, \boldsymbol{A})$ and $\partial^{\mu} = (\partial^0, -\nabla)$ we may write the x-components of electric and magnetic fields as

$$E_x = -\frac{\partial A_x}{\partial t} - \frac{\partial A_0}{\partial x} = -(\partial^0 A^1 - \partial^1 A^0) = E^1 \tag{D3a}$$

and

$$B_x = \frac{\partial A_z}{\partial y} - \frac{\partial A_y}{\partial z} = -(\partial^2 A^3 - \partial^3 A^2) = B^1 \tag{D3b}$$

where $\partial^0 = \frac{\partial}{\partial x^0} = \frac{\partial}{\partial t}, \partial^1 = \frac{\partial}{\partial x_1} = -\frac{\partial}{\partial x}$ etc. are implied. In all, the electric and magnetic fields shall have six components which in the form of eqns. (D3) can be written as the elements of a second rank antisymmetric field strength tensor

$$F^{\mu\nu} = \partial^{\mu} A^{\nu} - \partial^{\nu} A^{\mu} \tag{D4}$$

The tensor $F^{\mu\nu}$ is antisymmetric because interchanging the indices μ and ν it changes its sign, i.e.

$$F^{\nu\mu} = -F^{\mu\nu} \tag{D5}$$

The covariant form of the Maxwell equations written as

$$\partial_{\mu} F^{\mu\nu} = J^{\nu} \tag{D6}$$

where J^{ν} is the four-component electric current density, the source, provides us the inhomogeneous Maxwell equations.

The homogeneous Maxwell equations, which assume that magnetic charge and magnetic current densities do not exist in nature; can be derived from another covariant equation

$$\partial_{\mu}\, \Im^{\mu\nu} = 0 \tag{D7}$$

where $\Im^{\mu\nu}$ is called the dual field strength tensor and is expressed by

$$\Im^{\mu\nu} = (1/2)\, \delta^{\mu\nu\rho\sigma} F_{\rho\sigma} \tag{D8}$$

where $\delta^{\mu\nu\rho\sigma}$ is a totally antisymmetric fourth rank tensor with each index running from 0 to 3. This tensor is defined by

$$\delta^{\mu\nu\rho\sigma} = \begin{cases} +1 \text{ if } \mu\nu\rho\sigma \text{ is an even permutation of } 0123 \\ -1 \text{ if } \mu\nu\rho\sigma \text{ is an odd permutation of } 0123 \\ 0 \text{ if at least any two indices are equal} \end{cases} \tag{D9}$$

The even permutation means an even number of interchanges of two indices.

LAGRANGIAN DENSITY FOR FREE ELECTROMAGNETIC FIELD

The electromagnetic field equations can be obtained from the Lagrangian density for the electromagnetic field. In the free field case there is no J^{ν} term in the Lagrangian density, which is

$$£ = -\frac{1}{4} F_{\mu\nu} F^{\mu\nu} \tag{D10}$$

where

$$F_{\mu\nu} = \partial_{\mu} A_{\nu} - \partial_{\nu} A_{\mu} \tag{D11}$$

In order to derive the field equations, we employ Euler Lagrange equations

$$\frac{\partial £}{\partial A_{\nu}} - \partial_{\mu} \frac{\partial £}{\partial(\partial_{\mu} A_{\nu})} = 0 \tag{D12}$$

To calculate each term in eqn. (D12) for the Lagrangian density (D10), we use a new set of indices α and β for $F_{\mu\nu}$ and write the Lagrangian density £ as

$$\begin{aligned} £ &= -\frac{1}{4} F_{\alpha\beta} F^{\alpha\beta} = -\frac{1}{4} (\partial_{\alpha} A_{\beta} - \partial_{\beta} A_{\alpha})(\partial^{\alpha} A^{\beta} - \partial^{\beta} A^{\alpha}) \\ &= -\frac{1}{4} [\partial_{\alpha} A_{\beta} \partial^{\alpha} A^{\beta} - \partial_{\alpha} A_{\beta} \partial^{\beta} A^{\alpha} - \partial_{\beta} A_{\alpha} \partial^{\alpha} A^{\beta} + \partial_{\beta} A_{\alpha} \partial^{\beta} A^{\alpha}] \end{aligned} \tag{D13}$$

Since there is no A_{α} or A_{β} term in £, we have only the gradient terms $\partial_{\alpha} A_{\beta}$ etc., the variation of Lagrangian density (D13) with respect to field components A_{ν} vanishes. That is

$$\frac{\partial £}{\partial A_{\nu}} = 0 \tag{D14}$$

The variation of the Lagrangian density (D13) with respect to the field gradients may be calculated as follows.

$$\frac{\partial \pounds}{\partial(\partial_\mu A_\nu)} = \frac{1}{4}\left\{-\frac{\partial}{\partial(\partial_\mu A_\nu)}[\partial_\alpha A_\beta \partial^\alpha A^\beta] + \frac{\partial}{\partial(\partial_\mu A_\nu)}[\partial_\alpha A_\beta \partial^\beta A^\alpha]\right.$$
$$\left. + \frac{\partial}{\partial(\partial_\mu A_\nu)}[\partial_\beta A_\alpha \partial^\alpha A^\beta] - \frac{\partial}{\partial(\partial_\mu A_\nu)}[\partial_\beta A_\alpha \partial^\beta A^\alpha]\right\} \quad \text{(D15)}$$

We demonstrate the calculation of the first term on the RHS of eqn. (D15) only, and on the basis of similarity write the end results for other three terms.

$$-\frac{\partial}{\partial(\partial_\mu A_\nu)}[\partial_\alpha A_\beta \partial^\alpha A^\beta] = -\left[\frac{\partial(\partial_\alpha A_\beta)}{\partial(\partial_\mu A_\nu)} \cdot \partial^\alpha A^\beta + \partial_\alpha A_\beta \cdot \frac{\partial(\partial^\alpha A^\beta)}{\partial(\partial_\mu A_\nu)}\right] \quad \text{(D16)}$$

In the first term of eqn.(D16) if $\alpha = \mu$ and $\beta = \nu$, the variation is 1, otherwise zero. This may be indicated by introducing Kronekar deltas $\delta^\alpha_\mu \delta^\beta_\nu$. In the second term the variation is w.r.t. lower indices μ,ν whereas α, β in $\partial\,(\partial^\alpha \partial^\beta)$ are upper indices. We lower them by using the metric tensor. So for the second term of eqn. (D16)

$$\partial_\alpha A_\beta \frac{\partial(\partial^\alpha A^\beta)}{\partial(\partial_\mu A_\nu)} = \partial_\alpha A_\beta \frac{\partial(g^{\alpha\beta}\partial_\beta g^{\beta\alpha} A_\alpha)}{\partial(\partial_\mu A_\nu)}$$
$$= g^{\alpha\beta}\partial_\alpha g^{\beta\alpha} A_\beta \frac{\partial(\partial_\beta A_\alpha)}{\partial(\partial_\mu A_\nu)} \quad \text{(D17)}$$
$$= \partial^\beta A^\alpha \delta^\beta_\mu \delta^\alpha_\nu$$

Using the above calculations in eqn. (D16), we may write it as

$$-\frac{\partial}{\partial(\partial_\mu A_\nu)}[\partial_\alpha A_\beta \partial^\alpha A^\beta] = -[\delta^\alpha_\mu \delta^\beta_\nu \partial^\alpha A^\beta + \delta^\beta_\mu \delta^\alpha_\nu \partial^\beta A^\alpha]$$
$$= -2\partial^\mu A^\nu \quad \text{(D18a)}$$

Similarly,

$$+\frac{\partial}{\partial(\partial_\mu A_\nu)}[\partial_\alpha A_\beta \partial^\beta A^\alpha] = \delta^\alpha_\mu \delta^\beta_\nu \partial^\beta A^\alpha + \delta^\alpha_\mu \delta^\beta_\nu \partial^\beta A^\alpha$$
$$= +2\partial^\nu A^\mu \quad \text{(D18b)}$$

$$+\frac{\partial}{\partial(\partial_\mu A_\nu)}[\partial_\beta A_\alpha \partial^\alpha A^\beta] = \delta^\beta_\mu \delta^\alpha_\nu \partial^\alpha A^\beta + \delta^\beta_\mu \delta^\alpha_\nu \partial^\alpha A^\beta$$
$$= +2\partial^\nu A^\mu \quad \text{(D18c)}$$

$$-\frac{\partial}{\partial(\partial_\mu A_\nu)}[\partial_\beta A_\alpha \partial^\beta A^\alpha] = -\delta^\beta_\mu \delta^\alpha_\nu \partial^\beta A^\alpha - \delta^\beta_\nu \delta^\alpha_\mu \partial^\alpha A^\beta$$

$$= -2\, \partial^\mu A^\nu \tag{D18d}$$

Substituting, equations (D18) into eqn. (D15) we get,

$$\frac{\partial £}{\partial(\partial_\mu A_\nu)} = -(\partial^\mu A^\nu - \partial^\nu A^\mu) = -F^{\mu\nu} \tag{D19}$$

Using equations (D14) and (D19) in eqn. (D12) we get the free electromagnetic field equations as

$$\partial_\mu F^{\mu\nu} = 0 \tag{D20}$$

It may be noted that these equations, though homogeneous are a particular case of inhomogeneous equations (D6).

If we add a term ($-J_\nu A^\nu$), called the interaction term, in the Lagrangian density (D10), we shall have, instead of eqn. (D14)

$$\frac{\partial £}{\partial A_\nu} = \frac{\partial}{\partial A_\nu}(-J_\nu A^\nu) = -J^\nu \tag{D21}$$

and in that case equations (D19), (D21) and (D12) would give the inhomogeneous Maxwell equations,

$$\partial_\mu F^{\mu\nu} = J^\nu \tag{D22}$$

THE LORENTZ CONDITION

The inhomogeneous Maxwell equations (D22) in terms of the four potential A^μ, using eqn. (D4) is

$$\partial_\mu\, \partial^\mu A^\nu - \partial^\nu\, (\partial_\mu A^\mu) = J^\nu \tag{D23}$$

where we have used $\partial_\mu \partial^\nu A^\mu = \partial^\nu \partial_\mu A^\mu$. It is significant to note that the four-potential A^μ, implying A^0 and $\boldsymbol{A}$, is not uniquely determined. It means, if we replace A^0 (temporal part) by $[A^0 + (\partial\lambda/\partial t]$ and $\boldsymbol{A}$ (the space part) by $[\boldsymbol{A} + \nabla\lambda]$, the equations (D23) are still satisfied. These two transformations are combined into one as

$$A'^\mu = A^\mu + \partial^\mu \lambda \tag{D24}$$

where λ is a parameter, which is a function of space ($\boldsymbol{x}$) and time (t). Replacing A^μ by A'^μ in eqn. (D23), we get the LHS as

$$\partial_\mu \partial^\mu (A^\nu + \partial^\nu \lambda) - \partial_\mu \partial^\nu (A^\mu + \partial^\mu \lambda)$$
$$= (\partial_\mu \partial^\mu A^\nu - \partial_\mu \partial^\nu A^\mu) + (\partial_\mu \partial^\mu \partial^\nu \lambda - \partial_\mu \partial^\nu \partial^\mu \lambda)$$
$$= \partial_\mu \partial^\mu A^\nu - \partial_\mu \partial^\nu A^\mu \tag{D25}$$

where $\partial_\mu \partial^\mu \partial^\nu = \partial_\mu \partial^\nu \partial^\mu$ has been used. Eqn. (D25) implies that even if we replace A^μ by $A^\mu + \partial^\mu \lambda$, LHS of eqn. (D23) does not change. This means that we have a freedom to use any A^μ obeying the transformations (D24). Such transformations which allow changing the potentials in such a way that the fields remain unchanged are called gauge transformations.

From the above, we see that we have a gauge freedom to choose a potential A^{μ}. We can make use of this freedom to employ only those potentials, which obey

$$\partial_{\mu} A^{\mu} = 0 \tag{D26}$$

This condition is called the Lorentz condition. Obviously, $\partial_{\mu}A^{\mu}$, a product of two four vectors, is Lorentz invariant.

COULOMB GAUGE QUANTIZATION OF ELECTROMAGNETIC FIELD

The field equations for free electromagnetic field are

$$\partial_{\mu} F^{\mu\nu} = 0 \tag{D27}$$

where

$$F^{\mu\nu} = \partial^{\mu} A^{\nu} - \partial^{\nu} A^{\mu}$$

The canonically conjugate vector associated with A_{ν} is

$$\pi^{\mu\nu} = \frac{\partial \pounds}{\partial(\partial_{\mu} A_{\nu})} = -F^{\mu\nu} \tag{D28}$$

from which the canonical conjugate of A_{ν}, i.e., $\mu = 0$ component of $\pi^{\mu\nu}$ is

$$\pi^{\nu} \equiv \pi^{0\nu} = -F^{0\nu} = \partial^{\nu} A^{0} - \partial^{0} A^{\nu} \tag{D29}$$

Then

$$\pi^{0} = \partial^{0} A^{0} - \partial^{0} A^{0} = 0 \tag{D30}$$

and

$$\pi^{i} = \partial^{i} A^{0} - \partial^{0} A^{i} = E^{i} \tag{D31}$$

where π^{i} is the canonical conjugate of $A_{i} = -A^{i}$. The canonical conjugate π^{0} of $A_{0} = A^{0}$ vanishes.

The equal time commutation relations (ETCR), that the field operators and their canonical conjugates must obey, are

$$[\pi^{i}(\boldsymbol{x}), A_{j}(\boldsymbol{x}')] = -i\delta_{ij}\delta(\boldsymbol{x} - \boldsymbol{x}') \tag{D32a}$$

$$[A_{i}(\boldsymbol{x}), A_{j}(\boldsymbol{x}')] = 0 \tag{D32b}$$

$$[\pi^{i}(\boldsymbol{x}), \pi^{j}(\boldsymbol{x}')] = 0 \tag{D32c}$$

Any commutation relation with π^{0} is obviously zero. Now, we impose the coulomb gauge condition

$$A^{0} = 0,\ \nabla \cdot A = 0 \tag{D33}$$

where $\nabla \cdot A = \partial^{j} A_{j} = 0$ ($j = 1, 2, 3$) is the transversality condition. Thus when the vector potential A_{j} obeys $\partial^{j} A_{j} = 0$, the commutation relation between the operator π^{i} and the operator $\partial^{i} A_{j}$ namely $[\pi^{i}, \partial^{j} A_{j}]$ must vanish. However, this is not the case when we use the relation. (D32a). We can see this in the following.

Taking ∂^j of eqn. (D32a) we have

$$[\partial^j \pi^i(\boldsymbol{x}), A_j(\boldsymbol{x}')] + [\pi^i(\boldsymbol{x}), \partial^j A_j(\boldsymbol{x}')] = -i\partial^j \delta_{ij}\delta(\boldsymbol{x}-\boldsymbol{x}') \qquad \text{(D34)}$$

which, on setting $i = j$ due to δ_{ij} and using $\partial^j \pi^j = \nabla \cdot E = 0$ for free field and thus making the first commutator vanish, becomes

$$[\pi^i(\boldsymbol{x}), \partial^j A_j(\boldsymbol{x}')] = -i\delta_{ij}\partial^j \delta(\boldsymbol{x}-\boldsymbol{x}') \qquad \text{(D35)}$$

where the RHS denotes the derivative of Dirac delta function. It may be noted, for example, that $\partial_1 \delta(x^1) = \dfrac{\delta(x^1)}{x^1}$. Obviously RHS is not always zero, whereas $\partial^i A_j$ on the LHS acts because of eqn. (D33) as a null operator, making the LHS vanish. The ETCR (D35) then needs to be modified so that if its LHS is zero, the RHS must also be zero. For this purpose, we replace δ_{ij} by a symmetric tensor Δ_{ij} and write the three dimensional Dirac delta function $\delta(x - x')$ in its integral form namely we use

$$\delta(\boldsymbol{x}-\boldsymbol{x}') = \frac{1}{(2\pi)^3} \int e^{i\boldsymbol{k}.(x-x')} d\boldsymbol{k} \qquad \text{(D36)}$$

where $d\boldsymbol{k} = dk^1 dk^2 dk^3$. Noting from eqn. (D31) that $\pi^i = E^i$, we may then write ETCR (D32a) as

$$[E^i(\boldsymbol{x}), A_j(\boldsymbol{x}')] = -i\Delta_{ij} \cdot \frac{1}{(2\pi)^3} \int e^{i\boldsymbol{k}.(x-x')} d\boldsymbol{k} \qquad \text{(D37)}$$

Now taking the divergence (i.e. $\partial^j = \dfrac{\partial}{\partial x_j}$) of this equation, LHS becomes

$$\text{LHS} = [E^i(\boldsymbol{x}), \partial^j A_j(\boldsymbol{x}')] \qquad \text{(D38a)}$$

and the RHS

$$\text{RHS} = -i\frac{\Delta_{ij}}{(2\pi)^3} \int [\frac{\partial}{\partial x_j} e^{i\,(k_j x_j - k_j x'_j)}] d\boldsymbol{k}$$

$$= \frac{1}{(2\pi)^3} \int \sum_j [k_j \Delta^{ij}] e^{i\,\boldsymbol{k}.(x-x')} d\boldsymbol{k} \qquad \text{(D38b)}$$

Equating these (a) and (b) equations we have

$$[E^i(\boldsymbol{x}), \partial^j A_j(\boldsymbol{x}')]$$

$$= \frac{1}{(2\pi)^3} \int \sum_j [k_j \Delta^{ij}] e^{i\,\boldsymbol{k}.(x-x')} d\boldsymbol{k} \qquad \text{(D39)}$$

Since $\partial^j A_j = 0$, LHS vanishes and RHS does not, we have to find the condition that makes RHS of eqn. (D39) vanish. We see that if we replace

$$\Delta^{ij} = \delta^{ij} - \frac{k^i k^j}{\boldsymbol{k}^2} \tag{D40}$$

then

$$\begin{aligned} \sum_j k_j \Delta^{ij} &= \sum_j k_j \delta^{ij} - \frac{k^i}{\boldsymbol{k}^2} \sum_j k_j k^j \\ &= k_i + k^i = k_i - k_i = 0 \end{aligned} \tag{D41}$$

where we have used

$$\sum_{j=1}^{3} \frac{k_j k^j}{\boldsymbol{k}^2} = -1 \tag{D42}$$

Thus if we replace Δ_{ij} in eqn. (D37) by eqn. (D40) with its indices lowered, the divergence of this new RHS of eqn. (D37) would vanish, i.e.

$$\partial^j \left[\frac{1}{(2\pi)^3} \int d\boldsymbol{k} (\delta_{ij} - \frac{k_i k_j}{\boldsymbol{k}^2}) e^{i\boldsymbol{k}.(\boldsymbol{x}-\boldsymbol{x}')} \right] = 0 \tag{D43}$$

and accordingly we call

$$\frac{1}{(2\pi)^3} \int d\boldsymbol{k} (\delta_{ij} - \frac{k_i k_j}{\boldsymbol{k}^2}) e^{i\boldsymbol{k}.(\boldsymbol{x}-\boldsymbol{x}')} = \delta_{ij}^{tr} (\boldsymbol{x} - \boldsymbol{x}') \tag{D44}$$

the divergenceless or transverse (*tr*) delta function.

Then to meet the demand of transversality instead of the commutation relation (D32a) we must have

$$[E^i(\boldsymbol{x}), A^j(\boldsymbol{x}')] = +i \int \frac{d\boldsymbol{k}}{(2\pi)^3} (\delta^{ij} - \frac{k^i k^j}{\boldsymbol{k}^2}) e^{i\boldsymbol{k}.(\boldsymbol{x}-\boldsymbol{x}')} = \delta_{tr}^{ij} (\boldsymbol{x} - \boldsymbol{x}') \tag{D45}$$

for which we used $\pi^i = E^i$ and $A^j = -A_j$ in eqn. (D32a) . Employing back the definition (D36) and using $k^j = i\partial^j$, $\boldsymbol{k}^2 = k^i k^i = -(\partial^i)(\partial^i) = -\nabla^2$ eqn. (D45) may also be written as

$$[E^i(\boldsymbol{x}), A^j(\boldsymbol{x}')] = i(\delta^{ij} - \frac{\partial^i \partial^j}{\nabla^2}) \delta(\boldsymbol{x} - \boldsymbol{x}') \tag{D46}$$

Thus the employment of the commutation relation (D46) instead of (D32a), (ETCRs (b) and (c) remaining same), is the difference which we have to notice in quantizing the electromagnetic field in Coulomb gauge.

Equations (D46), (D32b) and (D32c) with eqn. (D31) are the rules of quantization which the operator of electromagnetic field must obey to obtain particle interpretation of the electromagnetic field. The energy and momentum tensor $T^{\mu\nu}(x)$ is

$$\mathrm{T}^{\mu\nu}(x) = \pi^{\mu\lambda}(x) \partial^\nu \phi_\lambda(x) - g^{\mu\nu} \pounds \tag{D47}$$

where in the present case

$$\phi_\lambda = A_\lambda$$
$$\pi^{\mu\lambda} = -F^{\mu\lambda}$$
$$\pounds = -\frac{1}{4}F_{\lambda\xi}F^{\lambda\xi} \tag{D48}$$

in which $\mu, \lambda, \xi = 0,\ 1,\ 2,\ 3$. Then the Hamiltonian density $T^{00}(x)$ is

$$T^{00}(\boldsymbol{x}) = \pi^{0\lambda}\partial^0 A_\lambda + \frac{1}{4}F_{\lambda\xi}F^{\lambda\xi} \tag{D49}$$

Since $\lambda = 0 = \xi$ is not allowed, we may write

$$T^{00}(\boldsymbol{x}) = \pi^{0k}\partial^0 A_k + \frac{1}{4}(F_{ok}F^{ok} + F_{ko}F^{ko}) + \frac{1}{4}F_{k\gamma}F^{k\gamma} \tag{D50}$$

where each of k and γ have the space values 1, 2, 3. Now

$$\begin{aligned}\frac{1}{4}F_{k\gamma}F^{k\gamma} &= \frac{1}{2}(F_{12}F^{12} + F_{23}F^{23} + F_{31}F^{31}) \\ &= \frac{1}{2}[(\nabla\times\boldsymbol{A})^3\cdot(\nabla\times\boldsymbol{A})^3 + (\nabla\times\boldsymbol{A})^1\cdot(\nabla\times\boldsymbol{A})^1 + (\nabla\times\boldsymbol{A})^2\cdot(\nabla\times\boldsymbol{A})^2] \\ &= \frac{1}{2}(\nabla\times\boldsymbol{A})\cdot(\nabla\times\boldsymbol{A}) \\ &= \frac{1}{2}\boldsymbol{B}^2\end{aligned} \tag{D51}$$

where $\boldsymbol{B}$ is the magnetic field. In the Coulomb gauge where $A^0 = 0$, we have from equations (D4) and (D31)

$$F^{0k} = \partial^0 A^k = -E^k \tag{D52}$$

Thus, using $\pi^{0k} = \pi^k = E^k$ and equations (D51) and (D52) in eqn. (D50), we get

$$\begin{aligned}T^{00}(\boldsymbol{x}) &= -E_k E^k + \tfrac{1}{2}E_k E^k + \tfrac{1}{2}\boldsymbol{B}^2 \\ &= \tfrac{1}{2}E^k E^k + \tfrac{1}{2}\boldsymbol{B}^2 \\ &= \tfrac{1}{2}(\boldsymbol{E}^2 + \boldsymbol{B}^2)\end{aligned} \tag{D53}$$

or the total field energy of the electromagnetic field is

$$\begin{aligned}P^0 = \mathrm{H} &= \int T^{00}(\boldsymbol{x})\,d^3\boldsymbol{x} \\ &= \int \tfrac{1}{2}(\boldsymbol{E}^2 + \boldsymbol{B}^2)\,d^3\boldsymbol{x}\end{aligned} \tag{D54}$$

The density of the electromagnetic field momentum from eqn. (D47) is

$$T^{0j} = \pi^{0\lambda}\partial^j A_\lambda \tag{D55}$$

such that the total field momentum is

$$P = \sum P^j = \int T^{0j} d^3x = \int \pi^\lambda \partial^j A_\lambda d^3x \tag{D56}$$

where λ and j can have only the space (1, 2, 3) values.

The electromagnetic field is a vector (spin-1) field, the spin operators for this field are.

$$S_1 = \int (F^{20} A_3 - F^{30} A_2) d^3x \tag{D57}$$

$$S_2 = \int (F^{30} A_3 - F^{10} A_3) d^3x \tag{D58}$$

and

$$S_3 = \int (F^{10} A_3 - F^{20} A_1) d^3x \tag{D59}$$

where A_1, A_2, A_3 are the three space components of A_ν ($\nu = 0, 1, 2, 3$) and F^{10}, F^{20}, F^{30} are three space–time components of $F^{\mu\nu}$. In the radiation gauge, $A^0 = 0$ and then above expressions become

$$S_1 = \int [(\partial^0 A^3) A_2 - (\partial^0 A^2) A_3] d^3x \tag{D60}$$

$$S_2 = \int [(\partial^0 A^1) A_3 - (\partial^0 A^3) A_1] d^3x \tag{D61}$$

$$S_3 = \int [(\partial^0 A^2) A_1 - (\partial^0 A^1) A_2] d^3x \tag{D62}$$

However, because of the absence of longitudinally polarized photons S_1 and S_2 vanish.

The electromagnetic field is a real field, hence there will be no charge operator associated with it.

The electromagnetic field ($\boldsymbol{E}$, $\boldsymbol{B}$) is denoted in terms of A_μ, the four potential. The Aharonov–Bohm effect shows that the potential be treated as a physical field hence A_μ would be seen as electromagnetic field.

We have following Fourier expansion for $A^\mu(x)$

$$A^\mu(x) = \frac{1}{V^{\frac{1}{2}}} \sum_k \sum_\alpha \frac{\varepsilon^\mu(\boldsymbol{k}, \alpha)}{\sqrt{2|\boldsymbol{k}|}} \left[a_{k\alpha} e^{-ikx} + a_{k\alpha}^\dagger e^{ikx} \right] \tag{D63}$$

where $\varepsilon^\mu(\boldsymbol{k}, \alpha)$ with $\mu = 0, 1, 2, 3$ are four unit polarization vectors, $\boldsymbol{k}$ denotes the momentum and α the spin of *photon*. $a_{k\alpha}$ and $a_{k\alpha}^\dagger$ are Fourier coefficients. When A^μ is treated as an operator, these Fourier coefficients, which appear in the commutation relations, act respectively as annihilation and creation operators of photons with momentum $\boldsymbol{k}$ and spin α. As already noted, the said commutation relations provide the required rules of quantization of the electromagnetic field.

We now view the Fourier expansion (D63) for A^μ in the light of Coulomb gauge condition, namely $A^0 = 0$ and $\nabla . A \equiv \partial^i A_i = 0$. $A^0 = 0$, means $\varepsilon^0 (k, \alpha) = 0$, implying the absence of $\mu = 0$, i.e., timelike or scalar photons. $\nabla . A = 0$ means $\boldsymbol{k}.\varepsilon(\boldsymbol{k}, \alpha) = 0$. If, we assume that the momentum of photon $\boldsymbol{k}$ is directed along the third axis, i.e., z-axis, then $k^3.\varepsilon^3(\boldsymbol{k}, \alpha) = 0$ can be observed only with $\varepsilon^3(\boldsymbol{k}, \alpha) = 0$, implying the absence of longitudinal photons as well. Thus, to preserve the Coulomb gauge condition $\boldsymbol{k}.\varepsilon(\boldsymbol{k}, \alpha) = 0$, only the polarization vectors $\varepsilon^1(k, \alpha)$ and $\varepsilon^2(k, \alpha)$ exist which imply the existence of transverse photons, which are actually the physical photons.

Since, eqn. (D63) provides expansion for A^μ or for A, whereas the expression (D54) for H is in terms of $\boldsymbol{E}$ and $\boldsymbol{B}$, we shall rewrite H in terms of $\boldsymbol{A}$. From eqn. (D31), in Coulomb gauge

$$E^i = -\,\partial^0 A^i \tag{D64}$$

so that

$$\boldsymbol{E}^2 = E^i.E^i = (\partial^0 A^i)^2 \tag{D65}$$

Again from eqn. (D51)

$$\begin{aligned}\boldsymbol{B}^2 &= (\nabla \times \boldsymbol{A})^2 \\ &= \varepsilon_{ijk}\varepsilon_{ipq}\partial_j A_k \partial_p A_q \\ &= (\delta_{jp}\delta_{kq} - \delta_{jq}\delta_{kp})\partial_j A_k \partial_p A_q \\ &= \partial_j A_k.\partial_j A_k - \partial_j A_k.\partial_k A_j\end{aligned} \tag{D66}$$

where ε_{ijk} and ε_{ipq} are the Levi–Civita symbols with properties $\varepsilon_{123} = \varepsilon_{231} = \varepsilon_{312} = 1$; $\varepsilon_{132} = \varepsilon_{321} = \varepsilon_{132} = -1$ and zero if any two indices have same values. Remember k as an index. The second term of eqn. (D66) may be written in terms of a total divergence term, for which we look at

$$\begin{aligned}\partial_j(A_k\partial_k A_j) &= \partial_j A_k.\partial_k A_j + A_k.\partial_j\partial_k A_j \\ &= \partial_j A_k.\partial_k A_j + A_k.\partial_k\partial_j A_j \\ &= \partial_j A_k.\partial_k A_j\end{aligned} \tag{D67}$$

where $\partial_j\partial_k = \partial_k\partial_j$ and $\partial_j A_j = -\nabla.\boldsymbol{A} = 0$ in Coulomb gauge, have been used. Thus substituting eqn. (D67), the eqn. (D66) becomes

$$\begin{aligned}\int \boldsymbol{B}^2 d^3\boldsymbol{x} &= \int \partial_j A_k \partial_j A_k d^3\boldsymbol{x} - \int \partial_j (A_k \partial_k A_j) d^3\boldsymbol{x} \\ &= \int \partial_j A_k \partial_j A_k d^3\boldsymbol{x}\end{aligned} \tag{D68}$$

where the second term has vanished due to the application of Gauss divergence theorem. Again form

$$\partial_j(A_k\partial_j A_k) = \partial_j A_k \partial_j A_k + A_k \partial_j \partial_j A_k \tag{D69}$$

we have

$$\begin{aligned}\int \partial_j A_k \partial_j A_k d^3\boldsymbol{x} &= \int \partial_j (A_k \partial_j A_k) d^3\boldsymbol{x} - \int A_k \nabla^2 A_k d^3\boldsymbol{x} \\ &= -\int A_k \nabla^2 A_k d^3\boldsymbol{x}\end{aligned} \tag{D70}$$

where $\partial_j\,\partial_j = \partial^j\partial^j = \nabla^2$ has been used and first term on the RHS has again vanished due to volume integral changing into surface integral and surface terms vanishing on the infinitely large boundary of the surface. Thus using equations (D65) and (D70) we have

$$\begin{aligned} \mathrm{H} &= \int \frac{1}{2}(\boldsymbol{E}^2 + \boldsymbol{B}^2)d^3\boldsymbol{x} \\ &= \int \frac{1}{2}[(\partial^0 A^j)^2 - A^j\nabla^2 A^j]d^3\boldsymbol{x} \end{aligned} \tag{D71}$$

In this form, we may employ the Fourier expansion (D63) from which we observe

$$\partial^0 A^j(x) = \frac{1}{V^{\frac{1}{2}}}\sum_{\boldsymbol{k}}\sum_{\alpha=1}^{2}\frac{-i|\boldsymbol{k}|}{\sqrt{2|\boldsymbol{k}|}}\varepsilon^j(\boldsymbol{k},\alpha)\left[a_{\boldsymbol{k}\alpha}\, e^{ik.x} - a_{\boldsymbol{k}\alpha}^{\dagger}\, e^{-ik.x}\right] \tag{D72}$$

where $k^0 = |\boldsymbol{k}|$ has been used and x^0 has been set zero after differentiation of A^k with respect to time x^0. Similarly,

$$\nabla^2 A^j(x) = \frac{1}{V^{\frac{1}{2}}}\sum_{\boldsymbol{k}}\sum_{\alpha=1}^{2}\frac{-\boldsymbol{k}^2}{\sqrt{2|\boldsymbol{k}|}}\varepsilon^j(\boldsymbol{k},\alpha)\left[a_{\boldsymbol{k}\alpha}\, e^{ik.x} + a_{\boldsymbol{k}\alpha}^{\dagger}\, e^{-ik\cdot x}\right] \tag{D73}$$

Now to calculate the first and second terms of eqn. (D71) we adopt the process similar to that of previous sections to get

$$\int (\partial^0 A^j)^2 d^3\boldsymbol{x} = \sum_{\boldsymbol{k}}\sum_{\alpha}\frac{1}{2}|\boldsymbol{k}|\,(a_{\boldsymbol{k}\alpha}a_{\boldsymbol{k}\alpha}^{\dagger} + a_{\boldsymbol{k}\alpha}^{\dagger}a_{\boldsymbol{k}\alpha}) \tag{D74}$$

and

$$\int A^j\nabla^2 A^j d^3\boldsymbol{x} = \sum_{\boldsymbol{k}}\sum_{\alpha} -\frac{1}{2}|\boldsymbol{k}|(a_{\boldsymbol{k}\alpha}a_{\boldsymbol{k}\alpha}^{\dagger} + a_{\boldsymbol{k}\alpha}^{\dagger}a_{\boldsymbol{k}\alpha}) \tag{D75}$$

Substituting these into eqn. (D71) we get

$$\mathrm{H} = \sum_{\boldsymbol{k}}\sum_{\alpha}\frac{1}{2}|\boldsymbol{k}|(a_{\boldsymbol{k}\alpha}a_{\boldsymbol{k}\alpha}^{\dagger} + a_{\boldsymbol{k}\alpha}^{\dagger}a_{\boldsymbol{k}\alpha}) \tag{D76}$$

The operators in equation (D76) obey the commutation relations

$$[a_{\boldsymbol{k}\alpha}, a_{\boldsymbol{k}'\alpha'}^{\dagger}] = \delta_{\alpha\alpha'}\delta_{\boldsymbol{k}\,\boldsymbol{k}'} \tag{D77}$$

and

$$[a_{\boldsymbol{k}\alpha}, a_{\boldsymbol{k}'\alpha'}] = [a_{\boldsymbol{k}\alpha}^{\dagger}, a_{\boldsymbol{k}'\alpha'}^{\dagger}] = 0 \tag{D78}$$

the first of these when used in itself (eqn. (D76)) gives

$$\mathrm{H} = \sum_{\boldsymbol{k}}\sum_{\alpha=1}^{2}|\boldsymbol{k}|a_{\boldsymbol{k}\alpha}^{\dagger}a_{\boldsymbol{k}\alpha} + \sum_{\boldsymbol{k}}1.|\boldsymbol{k}| \tag{D79}$$

where 1 is the unit operator. Obviously when we take the eigenvalues of H operator, the term $\sum_{\boldsymbol{k}} 1.|\boldsymbol{k}|$ adds up infinite value (all $\boldsymbol{k}_s$ are possible) to E'. As already discussed this infinite contribution can be eliminated by taking the **normal ordered**

product of $a^{\dagger}_{k\alpha}$ and $a_{k\alpha}$, following which we get the operator

$$\mathrm{H} = \sum_{\boldsymbol{k}} \sum_{\alpha=1}^{2} |\boldsymbol{k}| a^{\dagger}_{k\alpha} a_{k\alpha} \tag{D80}$$

where $|\boldsymbol{k}| = k_0$ corresponds to ω_k of real massive vector field.

For field momentum we use eqn. (D56), which is

$$P^i = -\int \partial^0 A^j \partial^i A_j \tag{D81}$$

where we used eqn. (D31) in Coulomb gauge. Using eqn. (D63) for A_j we may obtain

$$\partial^i A_j(x) = \frac{1}{V^{\frac{1}{2}}} \sum_{\boldsymbol{k}} \sum_{\alpha} \frac{-ik^i}{\sqrt{2|\boldsymbol{k}|}} \varepsilon_j(\boldsymbol{k},\alpha) \left[a_{k\alpha}\, e^{ik.x} - a^{\dagger}_{k\alpha}\, e^{-ik.x} \right] \tag{D82}$$

where $\boldsymbol{k} \cdot \boldsymbol{x} = k^i x^i$ and $\partial^i = \frac{\partial}{\partial x_i} = -\frac{\partial}{\partial x^i}$ are implied. Using equations (D72) and (D82) in eqn. (D81) and employing

$$\frac{1}{V} \int_V e^{\pm i\, (\boldsymbol{k}+\boldsymbol{k}').\boldsymbol{x}} d^3\boldsymbol{x} = \delta_{\boldsymbol{k},-\boldsymbol{k}'} \tag{D83a}$$

and

$$\frac{1}{V} \int_V e^{\pm i\, (\boldsymbol{k}-\boldsymbol{k}').\boldsymbol{x}} d^3\boldsymbol{x} = \delta_{\boldsymbol{k},\boldsymbol{k}'} \tag{D83b}$$

along with the equation,

$$\sum_{\alpha,\alpha'=1}^{3} \varepsilon_\lambda(\boldsymbol{k},\alpha) \varepsilon^\lambda(\boldsymbol{k}',\alpha') = -\delta_{\boldsymbol{k},\boldsymbol{k}'} \delta_{\alpha\alpha'} \tag{D84}$$

where $\lambda = 1, 2, 3$, we get

$$P^i = \sum_{\boldsymbol{k}} \sum_{\alpha=1}^{2} \frac{k^i}{2} (a_{k\alpha} a^{\dagger}_{k\alpha} + a^{\dagger}_{k\alpha} a_{k\alpha}) \tag{D85}$$

which on using the normal ordered product of $a^{\dagger}$ and a, becomes

$$P^i = \sum_{\boldsymbol{k}} \sum_{\alpha=1}^{2} k^i a^{\dagger}_{k\alpha} a_{k\alpha} \tag{D86}$$

where $i = 1, 2, 3$. We may also write it as

$$\boldsymbol{P} = \sum_{\boldsymbol{k}} \sum_{\alpha=1}^{2} \boldsymbol{k}\, a^{\dagger}_{k\alpha} a_{k\alpha} \tag{D87}$$

Combining eqn. (D80) and eqn. (D86)), we may write for the Energy-momentum operator P^ν ($\nu = 0, 1, 2, 3$, $P^0 = H$, $P^i = \boldsymbol{P}$) for the electromagnetic field

$$P^\nu = \sum_{\boldsymbol{k}} \sum_{\alpha=1}^{2} k^\nu a_{\boldsymbol{k}\alpha}^\dagger a_{\boldsymbol{k}\alpha} \tag{D88}$$

where $k^0 = |\boldsymbol{k}|$ is implied.

To write the expression for the spin operators in terms of creation operator $a_{\boldsymbol{k}\alpha}^\dagger$ and the annihilation operator ak_α we use eqn. (D62) for S_3

$$S_3 = \int (\partial^0 A^2) A_1 d^3\boldsymbol{x} - \int (\partial^0 A^1) A_2 d^3\boldsymbol{x} \tag{D89}$$

From eqn. (D72)

$$\partial^0 A^2(x) = \frac{1}{V^{\frac{1}{2}}} \sum_{\boldsymbol{k}} \sum_{\alpha=1}^{2} \frac{-i|\boldsymbol{k}|}{\sqrt{2|\boldsymbol{k}|}} \varepsilon^2(\boldsymbol{k},\alpha)\left[a_{\boldsymbol{k}\alpha}\, e^{ik.x} - a_{\boldsymbol{k}\alpha}^\dagger e^{-ik.x}\right] \tag{D90}$$

$$\partial^0 A^1 = \text{above relation with } \varepsilon^2 \to \varepsilon^1 \tag{D91}$$

$$A_1 = \text{eqn. (D63) with } \varepsilon^\mu \to \varepsilon_1 \tag{D92}$$

$$A_2 = \text{eqn. (D63) with } \varepsilon^\mu \to \varepsilon_2 \tag{D93}$$

where it is understood that x^0 has been set zero, after differentiation in eqs., like (D90) and (D91). Substitution of equations (D90 to D93) into eqn. (D89) and making use of eqn. (D83) as usual, we get

$$\int (\partial^0 A^2) A_1 d^3\boldsymbol{x} = \sum_{\boldsymbol{k}} \frac{i}{2}(a_{\boldsymbol{k}2} a_{\boldsymbol{k}1}^\dagger - a_{\boldsymbol{k}2}^\dagger a_{\boldsymbol{k}1}) \tag{D94a}$$

and

$$\int (\partial^0 A^1) A_2\, d^3\boldsymbol{x} - \sum_{\boldsymbol{k}} \frac{i}{2}(a_{\boldsymbol{k}1} a_{\boldsymbol{k}2}^\dagger \quad a_{\boldsymbol{k}1}^\dagger a_{\boldsymbol{k}2}) \tag{D94b}$$

such that eqn. (D89) becomes

$$S_3 = \sum_{\boldsymbol{k}} i(a_{\boldsymbol{k}1}^\dagger a_{\boldsymbol{k}2} - a_{\boldsymbol{k}2}^\dagger a_{\boldsymbol{k}1}) \tag{D95}$$

where the commutation relations

$$a_{\boldsymbol{k}2} a_{\boldsymbol{k}1}^\dagger = a_{\boldsymbol{k}1}^\dagger a_{\boldsymbol{k}2} \tag{D96a}$$

and

$$a_{\boldsymbol{k}1} a_{\boldsymbol{k}2}^\dagger = a_{\boldsymbol{k}2}^\dagger a_{\boldsymbol{k}1} \tag{D96b}$$

have been used. There will be no S_1 and S_2 as they contain the polarization vector $\varepsilon^3(k, \alpha)$, which in view of the transversality condition $\boldsymbol{k}.\varepsilon(\boldsymbol{k}, \alpha) = 0$, vanishes. Introducing the operators

$$a_{k+} = \frac{a_{k1} + ia_{k2}}{\sqrt{2}} \quad \text{(D97)}$$

and

$$a_{k-} = \frac{a_{k1} - ia_{k2}}{\sqrt{2}} \quad \text{(D98)}$$

the expression for S_3 can be obtained as

$$S_3 = \sum_k (a^\dagger_{k+} a_{k+} - a^\dagger_{k-} a_{k-}) \quad \text{(D99)}$$

Above quantization of electromagnetic field in Coulomb gauge has given us the following results

$$\text{H} = \sum_k \sum_{\alpha=1}^{2} |k| a^\dagger_{k\alpha} a_{k\alpha} \quad \text{(D100)}$$

$$\boldsymbol{P} = \sum_k \sum_\alpha \boldsymbol{k} a^\dagger_{k\alpha} a_{k\alpha} \quad \text{(D101)}$$

$$S = S_3 = \sum_k (a^\dagger_{k+} a_{k+} - a^\dagger_{k-} a_{k-}) \quad \text{(D102)}$$

From these operators we can write the corresponding eigenvalues. However, we write the equations for H and P in terms of operators a_{k+} and a_{k-}. Since, α in the above equations can take only the values 1 and 2, we have

$$N_{k+} + N_{k-} = N_{k1} + N_{k2} \quad \text{(D103)}$$

where

$$N_{k1} = a^\dagger_{k1} a_{k1} \quad \text{(D104a)}$$

and

$$N_{k2} = a^\dagger_{k2} a_{k2} \quad \text{(D104b)}$$

The eigenvalues of eqn. (D103) are

$$n_{k+} + n_{k-} = n_{k1} + n_{k2} \quad \text{(D105)}$$

Using these equations, we may obtain the following eigenvalues

$$E' = \sum_k |\boldsymbol{k}| (n_{k+} + n_{k-}) \quad \text{(D106)}$$

$$\boldsymbol{P}' = \sum_k \boldsymbol{k} (n_{k+} + n_{k-}) \quad \text{(D107)}$$

and

$$S' = \sum_k (n_{k+} - n_{k-}) \quad \text{(D108)}$$

where n_{k+} and n_{k-} are the eigenvalues of number operators

$$N_{\boldsymbol{k}+} = a^{\dagger}_{\boldsymbol{k}+} a_{\boldsymbol{k}+} \tag{D109}$$

and

$$N_{\boldsymbol{k}-} = a^{\dagger}_{\boldsymbol{k}-} a_{\boldsymbol{k}-} \tag{D110}$$

and each can have values 0 ,1, 2, 3, ... ∞. Equation (D108) predicts only two kinds of photons, one whose spin angular momentum along the $\boldsymbol{k}$ direction (z-axis) is +1 and other whose spin angular momentum along $\boldsymbol{k}$ axis is –1. The former, in view of eqn. (D97) is called the right circularly polarized photon and the latter, the left circularly polarized (eqn. D98) photon. Thus the quantization of electromagnetic field, carried out as above, predicts two kinds of spin one massless particles (photons), whose energy and momentum are given by equations (D106) and (D107).

APPENDIX E

CONTINUOUS AND DISCRETE SYMMETRIES

It can be explained that P_μ, $J_{\mu\nu}$ and Q are respectively the generators of translations, Lorentz rotations and the phase transformations brought into the fields. When these transformations are infinitesimal, P_μ emerges as the generator of infinitesimal translations, $J_{\mu\nu}$ that of infinitesimal Lorentz transformations and Q the generator of infinitesimal phase transformations of field operators $\psi\alpha(x)$. The infinitesimal changes introduce continuity into the system and as such all the above infinitesimal transformations become the continuous transformations. Moreover, when we demand that the action operator should remain invariant under the above continuous transformations, the corresponding generator is found to remain constant in time and thus follows the corresponding conservation law. The invariance of the action operator [or the Lagrangian density $(A = \int \pounds dx)$] under the continuous (infinitesimal) transformations is called continuous invariance of the action operator. By remaining invariant (of a system) under certain transformation we also mean that the system is symmetric under that transformation. The continuous invariance is therefore associated with a continuous symmetry. The energy momentum, angular momentum and charge conservation laws are obviously associated with continuous symmetries. The connection between continuous symmetries and the conserved quantities was first found by Emmy Noether and accordingly this relationship that *to every continuous symmetry corresponds a conservation law* is also known as Noether's theorem. Conversely, every conservation law reveals an underlying symmetry.

However, all the symmetries are not continuous. There are discrete symmetries as well. Contrary to continuous, these are discontinues symmetries. Parity (P) or space inversion, Time reversal (T) and the Charge conjugation (C) are the examples of discrete symmetries. By parity we mean that on the operation by the parity operator P, the space of the system gets inverted, i.e., x^i become $-x^i = x_i$. If the system still is same, we call it (the system) is invariant under the operation of parity or just say that parity remains conserved. Since, the space inversion of space inversion would bring the system back to the original space, or to the original coordinate system, we may write $P^2 = I$ where I is the identity operator.

$P^2 = I$ tells that the operator P has two eigenvalues $+1$ and -1. On space inversion; the field system may remain same or pickup a minus sign. There is no intermediate state for the field system between $+1$ and -1 states. Thus discreteness gets introduced.

Similarly, the charge conjugation operator C, when operating on a particular particle state would produce its antiparticle state, i.e.,

$$C|particle\rangle = |antiparticle\rangle \tag{E1a}$$

and

$$C^2|particle\rangle = C|antiparticle\rangle = |particle\rangle$$

implying that

$$C^2 = I \tag{E1b}$$

and the operator C, too has eigenvalues $+1$ and -1. Like parity, C, therefore is also a discrete symmetry operator. More details of C operator have been given in Chapter 1,

The third discrete symmetry operator T, the time reversal operator, implies that if we make a movie of some physical process (an interaction) and after having made it, run the movie backward and if the steps which we observe in the backward motion are as probable as the steps in the forward motion, we say that the interaction is invariant under time reversal operation. Running the movie backward means, running the time backward, which implies, the future will be that which in fact is past. This is reversing the time. Like parity and charge conjugation operations, two successive time reversal operations imply no time reversal at all, i.e.,

$$T^2 = I \tag{E2}$$

The further details regarding T operator too can be found in Chapter 1.

Whereas, for every continuous symmetry, there is a conservation law, this is as such not the case with discrete symmetries. Strong and electromagnetic interactions are invariant under both P and C whereas weak interactions violate these symmetries. The time reversal symmetry is also considered to have been violated in weak interactions. Under the combined operation of C and P, i.e., CP, the weak interactions remained invariant till the decay of neutral kaons showed the violation of CP invariance (a discrete symmetry). Of course it is impossible to construct a quantum field theory in which the product CPT (in any order) is violated. Thus any interaction must remain invariant under the product of CPT (it too is obviously a discrete symmetry) operators.

BIBLIOGRAPHY

Bailin, D. and Love, A. (1993): Introduction to Gauge Field Theory, Taylor & Francis Group, New York.

Becker, Katrin; Becker, Melanie and Schwarz, John H. (2007): String Theory and M-Theory, Cambridge University Press, Cambridge.

Bjorken, J.D. and Drell, S.D. (1965): Relativistic Quantum Fields, McGraw-Hill, New York.

Bogoliubov, N.N. and Shirkov, D.V. (1983): Quantum Fields, Benjamin/ Cummings, Reading, Massachusetts.

Burchum, W.E. and Jobes, M. (1995): Nuclear and Particle Physics, Addison-Wesley, Harlow, England.

Capra, Fritjof (1975): The Tao of Physics, Flamingo, London.

Carmeli, Moshe (1982): Classical Fields, General Relativity and Gauge Theory, John-Wiley, New York.

Chang, T.P. and Li, L.F. (1984): Gauge Theory of Elementary Particle Physics, Oxford University Press, Oxford.

Cheng, D.C. and O' Neill, G.K. (1979): Elementary Particle Physics, Addison-Wesley, Reading (Mass).

Cohen, B.L. (1971): Concepts of Nuclear Physics, Tata McGraw-Hill, New Delhi.

Dass, Tulsi and Sharma, Satish K. (1998) Mathematical Methods in Classical and Quantum Physics, University Press (India) Limited.

Davies, P.C.W. (1974): The Forces of Nature, Cambridge University Press, Cambridge.

Feynman, R.P. (1961): The Theory of Fundamental Processes, Benjamin. New York.

Frautschi, S.C., Gell-Mann, M. and Zachariasen, F. (1962): Experimental consequences of hypothesis of Regge poles, *Physical Review*, 126(6): 2204-18.

Green, Michael B. (1987): Proceedings of 11th International Conference on General Relativity and Gravitation, Stockholm, Cambridge University Press, Cambridge.

Green, Michael B., Schwarz, John H., Edward Witten (1987): Superstring Theory, Cambridge University Press, Cambridge.

Gribov, V.N. (2003): The Theory of Complex Angular Momenta, Cambridge University Press, Cambridge.
Griffiths, D. (1987): Introduction to Elementary Particles, John Wiley, New York.
Halzen, F. and Martin, A.D. (1984): Quarks and Leptons, Wiley, New York.
Hammermesh, M. (1963): Group Theory, Addison-Wesley, Reading (Mass).
Harris, E.G. (1975): Introduction to Modern Theoretical Physics, John Wiley, New York.
Hawking, S.W. (1988): A Brief History of Time, Bantam, Toronto.
Huang, Kerson (1982): Quarks, Leptons and Gauge Fields, World Scientific, Singapore.
Itzykson, C. and Zuber, J.B. (1980): Quantum Field Theory, McGraw-Hill, New York.
Jakson, J.D. (1962): Classical Electrodynamics, Wiley, New York.
Jauch, J.M. and Rohrlich, F. (1976): The Theory of Photons and Electrons, Springer-Verlag, Berlin.
Joshi, A.W. (1973): Elements of Group Theory for Physicists, Wiley Eastern, New Delhi.
Joshi, D.C. (2004): Elements of Field Theory, Transmedia, Srinagar-Garhwal, Uttarakhand.
Kallen, G. (1964): Elementary Particle Physics, Addison-Wesley, Reading (Mass).
Khanna, M.P. (1999): Introduction to Particle Physics, Printice-Hall of India, New Delhi.
Leon, M. (1973): Particle Physics: An Introduction, Academic Press, New York.
Levich, B.G. (1970): Theoretical Physics, North-Holland, Amsterdam.
Lurie, D. (1968): Particles and Fields,Wiley-Interscience, New York.
Mandl, F. (1966): Introduction to Quantum Field Theory, Wiley-Interscience, New York.
Marshak, R.E. and Sudarshan, E.C.G. (1961): Introduction to Elementary Particle Physics, Interscience Publishers, New York.
Matthews, P.T. (1968): Introduction to Quantum Mechanics, Tata McGraw-Hill, New Delhi.
Moriyasu, K. (1983): An Elementry Primer to Gauge Theories, World Scientific, Singapore.
Muller-Kirstein, H.J.W. (2006): Introduction to Quantum Mechanics, World Scientific, Singapore.
Muller-Kirstein, H.J.W. and Wiedemann, A. (1987): Supersymmetry, World Scientific, Singapore.
Nishijima, K. (1969): Fields and Particles, Benjamin, New york.
Olive, K.A. et al. (2014): Review of particle physics (Particle Data Group), *Chinese Physics*, C38, 090001.
Omenes, R. (1970): Introduction to Particle Physics, Wiley, New York.

Perkins, D.H. (1972): Introduction to High Energy Physics, Addison-Wesley, Reading (Mass).

Poghosyan, Martin (2013): Introdction to Regge Field Theory, school-diff 2013. physi.uni-heldelberg.de

Polchinski, Joseph (2001): String Theory, Cambridge University Press, Cambridge.

Popov, V.S. and Dolinskii, E.I. (1964): Regge poles and nuclear resonance reactions I, *Soviet Physics JETP*, 19(3): 661-70.

Povh, B., Rith, K., Scholz, C. and Zetsche, F. (1995): Particles and Nuclei, Springer, Berlin.

Predazzi, Enrico (2007): From Basic Kinematics to Regge Poles, www.indiana.edu

Rajput, B.S. (1990): Advanced Quantum Mechanics, Pragati Prakashan, Meerut.

Ramond, P. (1990): Field Theory: A Modern Primer, Addison-Wesley, Reading (Mass).

Rolnick, W.B. (1994): The Fundamental Particles and their Interactions, Addison-Wesley, Reading (Mass).

Roman, P. (1965): Advanced Quantum Theory, Addison-Wesley, Reading (Mass).

Roman, Paul (1969): Introduction to Quantum Field Theory, John-Wiley, New York.

Ryder, L.H. (1984): Quantum Field Theory, Cambridge University Press, Cambridge.

Sakuari, J.J. (1967): Advanced Quantum Mechanics, Addison-Wesley, Reading (Mass).

Salam, A. (1988): Particle physics today, *Annales de l'Institut Heneri Poincare-Physique theorique*, 49(3): 369-85.

Scadron, M.D. (1979): Advanced Quantum Theory, Springer-Verlag, New York.

Scheck, Florian (2013): Quantum Physics: Scattering of Particles by Potentials, Springer-Verlag, Berlin.

Schellekons, A.N. (2014): Introduction to String Theory, www.worldle.net

Schiff, L.I. (1968): Quantum Mechanics, McGraw-Hill, New York.

Schweber, S.S. (1961): An Introduction to Relativistic Quantum Field Theory, Harper & Row, New York.

Segre, E. (1965): Nuclei and Particles, Benjamin, New York.

Tang, Alfred and Norbury, John W. (2000): Properties of Regge Trajectories, arxiv.org>hep-ph/0004078v3.

Taylor, John R. (2000): Scattering Theory: The Quantum Theory of Nonrelativistic Collisions, Dover Publications Inc., New York.

Urban, Paul (1966): Elementary Particle Theories, Springer-Verlag, New York.

Webb Stephen (2004): Out of This World, Springer-Verlag, Berlin.

Weinberg, Steven (1996): The Quantum Theory of Fields, Volume Two (Modern Applications), Cambridge University Press, Cambridge.
West, Peter (1990): Introduction to Supersymmetry and Supergravity, World Scientific, Singapore.
Zee, A. (1982): Unity of Forces in the Universe, World Scientific, Singapore.
Zichichi, A. (Editor) (1955): Symmetries in Elementary Particle Physics, Academic Press, New York.

INDEX

A

Abelian field theories 91
Adjoint representation 355
Advanced Green's function 598
  properties of 599
Allowed and forbidden transitions 280
Amplitude 125
Analyticity 502
Angular momentum 51
  conservation law 621
Antiparticles 6, 99
  and crossing symmetry 9
Asymmetry measurements in cross sections 440
Asymptotic freedom 93
Atomic hydrogen, energy levels of 434

B

Baryon magnetic moments 248
Baryon masses 260
Baryon number 52
Baryon resonances 77, 85, 548
Baryons 76
  quark content of 85
Basic fermions 98
Beta functions 555, 556, 558
Bhabha (positron–electron) scattering 171
Bose symmetry 501
Bosonic string theory 563
Breit-Wigner resonances 525

C

Cabibbo angle 305
Cabibbo theory 303
  consequences of 305
Cabibbo-Kobayashi-Maskawa (CKM) matrix 325
Charge conjugation 54
Charge conservation law 623
Charged and neutral weak currents 312
Charged weak quark current 320
Chew-Frautschi diagram 547
Colour quark model 255
Colour, evidence for 442
Combined covariant derivative 405
Compact group 189
Complex angular momentum, concept of 531
Compton scattering 129
Conjugate momentum 618
Conservation law 51, 339
Conserved vector current hypothesis 304
Continuous and discrete symmetries 643
Contour and its modification 538
Coupling, kinds of 373
CP invariance 64
CP violation 64
CPT theorem 72
Crossing symmetry 499

D

Deep inelastic neutrino quark scattering 439
Δ function, four dimensional Fourier representation of 608
Δ^+ and Δ^-, four dimensional representation of 606
Differential scattering cross section 123
Differential scattering section, calculation of 141
Discrete/continuous groups 188

Duality 553
Dyson function 603

E

Effective charge 91
Eightfold way 78
Einstein's general theory of relativity 30
Elastic neutrino electron scattering 437
Electric and magnetic fields 357
Electric charge 51
Electrodynamics 31
Electromagnetic coupling constant 412
Electromagnetic embedding 402
Electromagnetic field 627
  coulomb gauge quantization of 631
Electromagnetic interaction vertex 38
Electromagnetic interaction, range and strength of 39
Electromagnetic interactions 31
Electron and muon magnetic moments 434
Electron number 51
Electron–positron annihilation 145
Electron-positron scattering 35
Electro-photon scattering 33
Electroweak gauge theory, renormalizability of 428
Elementary particles 1
  classification of 74
Energy and momentum 51
Energy momentum conservation law 619
Equal mass case 495
Euler gamma 555
Exchange degenerate baryon resonances 550

F

Fermi transition 281, 282
Fermi's golden rule 126
Fermi's relativistic theory of beta decay 290
Fermi's theory of beta decay (non-relativistic) 275
Feynman diagrams 32
Feynman function 602
Feynman rules 126
Finite/infinite groups 188
Flavour changing interactions 274
Flux 123
Four vectors 579
Free electromagnetic field, Lagrangian density for 628
Fundamental interactions 29
Fundamental representation 218

G

Gamma function 555
Gamow-Teller (GT) transition 281, 282
Gauge bosons 77, 98
  masses of 418
Gauge symmetries 339
GIM mechanism and charged quarks 321
Global gauge symmetry and Noether's theorem 340
  spontaneous breaking of 369
Goldstone bosons 373
G-parity 60
  of pion state 61
Grand unification 447
  and the Big Bang 454
  implications of 451
Gravitational interaction 29, 453
Green's functions of field theory 595
Groups 187
  properties of 187
GUTs, limitations of 460
GWS model 401

H

Hadron masses 253. 255
Hadrons, quark structure of 82
Helicity and handedness 283
Helicity and Lorentz invariance 60
Higgs boson 99
Higgs field 379
  choice of 416
Higgs mechanism 373, 374
Higher order diagrams 127
Homogeneous Klein Gordon equation, even solution of 609
  solutions of 603

I

Infinitesimal Lorentz transformations 584
Infrared slavery 93
Interactions, calculation of 408
Intermediate vector boson 96, 299
Isospin 53

J

J, discovery of 94

K

Klein-Nishina formula 144
Klein-Nishina relation 141

L

Lagrangian density and mass term 406
Law of composition 195
Leptons 75, 98
    masses of 421
Lie algebra 191
Lie groups 190, 191
    generators of 191
    representations of 197
Lightlike separations 585
Local [$SU_L(2) \otimes U_Y(1)$] gauge symmetry, spontaneous breaking of 415
Local SU(2) gauge symmetry, spontaneous breaking of 380
Lorentz condition 630
Lorentz invariants 589
Lorentz transformations 581

M

Mandelstam variables 81
Mandelstam's *s*, *t*, *u* variables 482
    for equal mass case 483
    for unequal mass case 492
    visualization of 495
Mass frame, differential scattering cross section in centre of 167
Mass relation of a particle, spin and the square of 566
Mass, generation through spontaneous symmetry breaking 364
Matrix element 125, 133
Meson masses 258
Meson resonances 84, 552
Meson states, mixing of 242
Mesons 4, 75
    quark composition of 84
Microscopic reversibility 69
Moller (electron–electron) scattering 156
Mucleosynthesis of lighter elements 457
Muon number 51

N

Naïve quark model, paradoxes of 254
Natural units 577
Neutral current, structure of 414
Neutral gauge fields 410
Neutring, helicity of 59
Neutrinos 10
    sources of 14
    two-component theory of 286
No ghost theorem 561
Non-leptonic weak processes 274
Nucleon-antinucleon systems, g-parity in 63

O

O(n) group 194
One loop Feynman diagrams 36
One particle exchange model 532
Optical theorem 524
Ordinary product 120

P

Pair annihilation process, differential scattering cross section for 151
Para-positronium 155
Parity 56
Partial waves and scattering amplitude 513
Photon 3
Planck scales 453
Positronium annihilation 155
Positronum 155
Propagator 122
Purely leptonic interactions 273

Q

QCD scale arameter 446
QED, experimental tests of 434
Quantum chromodynamics 48, 431
    evidence for 442

Quantum electrodynamic processes 129
Quantum electrodynamics 32, 117
Quantum field theory 33
Quantum gravity 455
Quantum mechanical theory of gravity 31
Quantum mechanics, time reversal transformation in 69
Quantum theory of gravity 454
Quark colours 85
Quark confinement 89
Quark masses 93
Quark model 87
Quark-lepton universality 308
Quarks, masses of 424

R

Reducible and irreducible representations 199
Regge pole model 533
Regge pole residue 542
Regge pole trajectories 545
Regge poles 511
  resonances and 511
  signatures of 544
Regge theory 531
Resonances 20
  as poles of scattering matrix 530
Retarded Green's function 600
  properties of 601

S

Scalar product of four-vectors 588
Scattering amplitude 558
  symmetries of 481, 499
Scattering cross section 521
Scattering processes 512
Schwinger's action principle 613
Schwinger's Δ function 604
Semileptonic interactions 273
Simple and Regge poles, separation of 541
Simple and semi-simple groups 189
SM and GUTs, limitations of 458
Spacelike separations 587
Spacelike surfaces 591
Spontaneous symmetry 363
Standard model 401
  experimental evidences for 434
  lagrangian 433
  limitations of 458
Strange particles 15
Strangeness 52
Strangeness conserving interactions 274
Strong coupling constant with energy 445
Strong force 41
Strong interactions 41
Strong isosprin doublet 310
SU(1) local gauge symmetry and QED 343
SU(2) global gauge symmetry 347
SU(2), irreducible representations of 201
SU(2) local gauge symmetry and Yang–Mills fields 350
SU(2) product representations 204
SU(3) Global gauge invariance 359
SU(3) group, irreducible representations of 210, 213
SU(3) local gauge symmetry and QCD 361
SU(3) product representations 227
SU(3) representations, product of 220
SU(3) shift operators 211
SU(3) symmetry group,
  applications of 231
  extension of SU(4) 233
  limitations of 232
SU(3), 3 and 3* representations 217
SU(6) quark model 239
  application of 248
  hypotheses 239
SU(6), general features of 239
  baryons in 246
  mesons in 240
SU(n) group 193
Superforce 455
Supergravity 461
Superstring theories 462
Supersymmetry 461
Suppressed weak decays 320
Symmetries and groups 187
Symmetry transformation 187, 339

T

Tau number 52
t-channel process 488
Technicolour theories 462
3 ⊗ 3 representation 221, 224
3 ⊗ 6 representation 226
The Gell: Mann: Okubo mass formula 228
Time ordered product and propagators 117
Time reversal 68
Timelike separations 586
T-invariance, verification of 71
Trace, calculation of 137, 150
    evaluation of 161
Tree level diagrams 127

U

U(1) local gauge symmetry, spontaneous breaking of 374
U(n) group 193
u-channel process 488
Ultimate fundamental particles 98
Unequal mass case 498
Universality, restoration of 308
Upsilon (γ) discovery 95

V

V–A interaction 294
Vacuum polarization effects 90
Vector current minus axial vector current 57
Veneziano amplitude 554
    and string theory 562
    high energy behaviour of 561

W

W and Z bosons 441
Wastson and Sommerfeld transformation 534, 536
Weak coupling constant, universality of 297
Weak embedding 402
Weak hypercharge 316
Weak interaction vertices 37
Weak interactions 37, 48, 273
    evidence for 435
    strength of 426
    universality of 296
Weak isospin 316
Weak isospin doublet 310
Weak neutral current, verification of 436

Y

Yang–Mills fields, properties of 355
Young's tableau technique 209